ENCYCLOPEDIA OF CLIMATE AND WEATHER

ENCYCLOPEDIA OF CLIMATE AND WEATHER

Stephen H. Schneider

EDITOR IN CHIEF

VOLUME 2

New York   Oxford
OXFORD UNIVERSITY PRESS
1996

OXFORD UNIVERSITY PRESS

Oxford   New York
Athens   Auckland   Bangkok   Bombay
Calcutta   Cape Town   Dar es Salaam   Delhi
Florence   Hong Kong   Istanbul   Karachi
Kuala Lumpur   Madras   Madrid   Melbourne
Mexico City   Nairobi   Paris   Singapore
Taipei   Tokyo   Toronto

and associated companies in
Berlin   Ibadan

Published by Oxford University Press, Inc.,
198 Madison Avenue, New York, New York 10016

Oxford is a registered trademark of Oxford University Press

Library of Congress Cataloging-in-Publication Data

Encyclopedia of Climate and Weather / Stephen H. Schneider, editor in chief
p.   cm.
Includes bibliographical references and index.
1. Climatology—Encyclopedias. 2. Weather—Encyclopedias.
I. Schneider, Stephen Henry.
QC854.E523   1996
551.5'03—dc20   95-31019   CIP
ISBN 0-19-509485-9 (set)
ISBN 0-19-510440-4 (vol. 1)
ISBN 0-19-510441-2 (vol. 2)

This encyclopedia was conceived by Robert Ubell Associates, Inc.
It was edited and produced by Oxford University Press, Inc.

PROJECT STAFF

ROBERT UBELL ASSOCIATES

Project Director: Robert N. Ubell
Project Manager: Luis A. Gonzalez
Administrative Assistant: Ailene Acosta

OXFORD UNIVERSITY PRESS

Project Editor: Matthew A. Giarratano
Copyeditor: Jane McGary
Proofreader: Martha Goldstein
Indexer: Cynthia Crippen, AEIOU, Inc.
Editorial Assistant: Merilee Johnson
Production Manager: Ellen Barrie
Manufacturing Controller: Benjamin Lee
Graphic Artists: Tech-Graphics Corporation
Book Design: Joan Greenfield

Printing: 9 8 7 6 5 4 3 2 1

Printed in the United States of America
on acid-free paper

ABBREVIATIONS AND SYMBOLS

See pages 113, 114, and 115 for other abbreviations and weather symbols

A	ampere
AAOE	Airborne Antarctic Ozone Experiment
A.B.	*Artium Baccalaureus*, Bachelor of Arts
Ac	altocumulus
ACC	Antarctic Circumpolar Current
ACT^2	Advanced Customer Technology Test
ADP	adenosine diphosphate
AGL	above ground level
AIC	(H.) Akaike's information criteria
AIDS	acquired immune deficiency syndrome
A.M.	*ante meridiem*, before noon
AM	amplitude modulation
AMS	American Meteorological Society
Ar	argon
ASAP	Automated Shipboard Aerological Program
ASTER	Advanced Spaceborne Thermal Emission and Reflection Radiometer
ATP	adenosine triphosphate
AU	astronomical unit (1 AU $\simeq$ 149.6 million kilometers $\simeq$ 93 million miles)
AVHRR	Advanced Very High Resolution Radiometer
B.A.	Bachelor of Arts
BBS	(electronic) bulletin board service
B.C.E.	before common era
^{10}Be	beryllium-10 isotope
BIF	banded iron formation
BLM	(USA) Bureau of Land Management
Btu	British thermal unit
C	carbon; coulomb
°C	degrees Celsius [°C = 5/9 (°F − 32)]
^{12}C	carbon-12 isotope
^{13}C	carbon-13 isotope
^{14}C	carbon-14 isotope
ca.	circa
Ca	calcium
$CaCO_3$	calcium carbonate (limestone)
CAD	cold air damming
CAM	Crassulacean Acid Metabolism
CAPE	convective available potential energy
CAT	clear air turbulence; computerized axial (computer-assisted) tomography

Cb	cumulonimbus
Cc	cirrocumulus
CCl_4	carbon tetrachloride
CCM	Certified Consulting Meteorologist
CCN	cloud condensation nuclei
C.E.	common era
CF	Coriolis force
CFC	chlorofluorocarbon
$CFCl_3$	CFC-11
CF_2Cl_2	CFC-12
$C_2F_3Cl_3$	CFC-113
C_2F_5Cl	CFC-115
CFD	computational fluid dynamics
CFL	compact fluorescent lamp; (Richard) Courant, (Kurt) Friedrichs, and (Hans) Lewy
CH_4	methane
C_2H_2	acetylene
C_2H_6	ethane
C_5H_8	isoprene
CH_3Br	methyl bromide
CH_3CCl_3	methylchloroform
CHF_2Cl	HCFC-22
Ci	cirrus
CIN	convective inhibition
CISK	conditional instability of the second kind
Cl	atomic chlorine
Cl_2	molecular chlorine
CLIMAP	Climate: Long-range Interpretation, Mapping, and Prediction
$ClNO_3$	chlorine nitrate
$ClONO_2$	chlorine nitrate ($ClNO_3$)
cm	centimeter (10^{-2} meters)
CNG	compressed natural gas
CO	carbon monoxide
CO_2	carbon dioxide
COS	carbonyl sulfide
CRF	cloud radiative forcing
Cs	cirrostratus
Cu	cumulus
CZCS	Coastal Zone Color Scanner
DDT	dichlorodiphenyltrichloroethane

DEW	Distant Early Warning
DMS	dimethyl sulfide
DNA	deoxyribonucleic acid
DNS	direct numerical simulation
D.Sc.	Doctor of Science
DSDP	Deep Sea Drilling Project
DU	Dobson unit
e	naturally occurring constant (approximately 2.71828) used as the base of the system of natural logarithms.
°E	degrees east longitude
ECMWF	European Centre for Medium-Range Weather Forecasting
ed.	edition
EERE	everything else remains equal
e.g.	*exempli gratia*, for example
EM	electromagnetic
ENIAC	Electronic Numerical Integrator And Calculator
ENSO	El Niño-Southern Oscillation
EOS	(NASA) Earth Observing System
ERB	Earth radiation budget
ERBE	Earth Radiation Budget Experiment
ERS	Earth resources satellite
ETA	estimated time of arrival
Ex.	(Bible) Exodus
°F	degrees Fahrenheit (°F = 9/5°C + 32)
FAA	(USA) Federal Aviation Administration
FAO	(UNESCO) Food and Agricultural Organization
FDE	finite difference equation
Fe_2O_3	iron oxide
Fe_2S_2	iron sulfide
FFT	fast Fourier transform
FIR	far infrared
FIRE	First ISCCP Research Experiment
F-RIS	Filchner-Ronne Ice Shelf
FUV	far ultraviolet
GARP	Global Atmospheric Research Program
GCM	general circulation model
GDP	gross domestic product
GFD	geophysical fluid dynamics
GIS	geographical information system
GISP2	(North American) Greenland Ice-Sheet Project 2
GMT	Greenwich mean time
Gn.	(Bible) Genesis
GOES	(USA) Geostationary Operational Environmental Satellite
GPS	Global Positioning System
GRIP	(European) GReenland Ice-core Project
GTS	(WWW) Global Telecommunications System
H or ^{1}H	atomic hydrogen
^{2}H	deuterium (hydrogen-2 isotope)
^{3}H	tritium (hydrogen-3 isotope)

H_2	molecular hydrogen
HCFC	hydrochlorofluorocarbon
HCl	hydrochloric acid
HCN	hydrogen cyanide
HCO_3^-	hydrogencarbonate (bicarbonate) ion
H_2CO_3	carbonic acid
He	helium
^{3}He	helium-3 isotope
Hg.	(Bible) Haggai
HIRS	High-resolution Infrared Radiation Sounder
HIV	human immunodeficiency virus
HNO_3	nitric acid
H_2O	water
H_2O_2	hydrogen peroxide
hPa	hectopascal (1 hPa = 100 pascals = 1 millibar)
H_2S	hydrogen sulfide
H_2SO_4	sulfuric acid
Hz	hertz
IBM	International Business Machines Corporation
ICRF	infrared cloud radiative forcing
i.e.	*id est*, that is
IGBP	International Geosphere-Biosphere Program
IGY	International Geophysical Year
IN	ice nuclei
INMARSAT or Inmarsat	INternational MARitime SATellite Organization
IPCC	Intergovernmental Panel on Climate Change
IR	infrared
ISCCP	International Satellite Cloud Climatology Project
ITCZ	Intertropical Convergence Zone
J	joule
Jer.	(Bible) Jeremiah
JNWP	joint numerical weather prediction
K	K Index (stability index); kelvin (K = °C + 273.15); potassium
^{40}K	potassium-40 isotope
km	kilometer (10^3 meters)
kph	kilometers per hour
Kr	krypton
L☉	solar bolometric luminosity (3.82×10^{33} ergs per second)
LCL	lifting condensation level
LES	large-eddy simulation
LGM	last glacial maximum
LI	Lifted Index (stability index)
Lidar	(laser) light radar
Lk.	(Bible) Luke
LL.D.	*Legum Doctor*, Doctor of Laws
LLJ	low-level jet
LLWM	low-level wind maximum
LORAN	long-range navigation
LPG	liquid petroleum gas
m	meter (about 39.37 inches)

M☉	solar mass (1.989×10^{33} grams)
M.A.	Master of Arts
mb	millibar (10^{-3} bars)
MBL	microscale boundary layer
Mg	magnesium
MHD	magnetohydrodynamic
MIR	middle infrared
MIT	Massachusetts Institute of Technology
MJO	Madden-Julian Oscillation
Mk.	(Bible) Mark
MLI	Modified Lifted Index (stability index)
mm	millimeter (10^{-3} meters)
MODIS	MODerate resolution Imaging Spectroradiometer
MOS	Model Output Statistics
mP	maritime polar
MST	mesosphere, stratosphere, troposphere
MSW	municipal solid waste
mT	maritime tropical
Mt.	(Bible) Matthew
MTBE	methyl tertiary-butyl ether
N	newton; atomic nitrogen
°N	degrees north latitude
N_2	molecular nitrogen
Na	sodium
NAAQS	(USA) National Ambient Air Quality Standards
NaCl	sodium chloride
NADW	North Atlantic Deep Water
NASA	(USA) National Aeronautics and Space Administration
NCAR	(USA) National Center for Atmospheric Research
Ne	neon
NEXRAD	NEXt generation weather RADar
NGM	Nested Grid Model
NH_3	ammonia
NH_4	ammonium
NHC	(USA) National Hurricane Center
NH_4OH	ammonium hydroxide
NHRE	National Hail Research Experiment
NIR	near infrared
nm	nanometer (10^{-9} meters)
NMHC	nonmethane hydrocarbon
NNMI	nonlinear normal mode initialization
NO	nitric oxide
NO_2	nitrogen dioxide
NO_x	nitrogen oxides
N_2O	nitrous oxide
NOAA	(USA) National Oceanic and Atmospheric Administration
NPP	net primary productivity
Ns	nimbostratus
NSSL	(USA) National Severe Storms Laboratory
NUV	near ultraviolet
NW	northwest
NWP	numerical weather prediction
NWS	(USA) National Weather Service
O	atomic oxygen
O_2	molecular oxygen
O_3	ozone
^{16}O	oxygen-16 isotope
^{18}O	oxygen-18 isotope
ODE	ordinary differential equations
ODP	Ocean Drilling Project
OECD	Organization of Economic Cooperation and Development
OH	hydroxyl molecule
OPEC	Organization of Petroleum Exporting Countries
Pa	pascal
PAH	polyaromatic hydrocarbon
PAL	present atmospheric level
PAN	peroxyacetyl nitrate
PBL	planetary boundary layer
PCB	polychlorinated biphenyl
PDE	partial differential equation
pdf	probability density function
PDSI	Palmer Drought Severity Index
PET	potential evapotranspiration
PG&E	Pacific Gas and Electric Company
PGF	pressure gradient force
pH	potential of hydrogen ions (measure of acidity and alkalinity)
PH_3	phosphine
Ph.D.	*Philosophiae Doctor*, Doctor of Philosophy
PIC	products of incomplete combustion
P.M.	*post meridiem*, after noon
PNA	Pacific-North American
ppbv	parts per billion by volume
ppmv	parts per million by volume
pptv	parts per trillion by volume
PRF	pulse repetition frequency
Ps. or Pss.	(Bible) Psalms
PSC	polar stratospheric cloud
PSME	polar summer mesospheric echoe
PV	potential vorticity
QBO	Quasi-Biennial Oscillation
QG	quasi-geostrophy
R	thermal resistance
R☉	solar radius (6.96×10^{10} centimeters)
Radar	radio detecting and ranging
RANS	Reynolds Averaged Navier-Stokes
RH	relative humidity
RIS	Ross Ice Shelf
RNA	ribonucleic acid
RWIS	road weather information system
°S	degrees south latitude
SAM	source area mathematical model

SAMS	Stratospheric And Mesospheric Sounder
SAO	semiannual oscillation
SAR	Synthetic Aperture Radar
Sc	stratocumulus
Sc.D.	*Scientiae Doctor*, Doctor of Science
SCOPE	Scientific Committee On Problems of the Environment
SCRF	solar cloud radiative forcing
SELS	SEvere Local Storms
SI	Showalter Index (stability index); *Systéme International d'Unités*, International System of Units
SIC	sensitivity to initial conditions
SiO_2	silica
SMMR	Scanning Multichannel Microwave Radiometer
SO_2	sulfur dioxide
SO_x	sulfur oxides
SOM	soil organic matter
Sr	strontium
^{86}Sr	strontium-86 isotope
^{87}Sr	strontium-87 isotope
SST	sea-surface temperature
St	stratus
SWEAT	Severe WEAther Threat (stability index)
TDRSS	(NASA) Tracking and Delay Relay Satellite System
TFR	total fertility rate
^{230}Th	thorium-230 isotope
THI	temperature-humidity index
TIROS	television and infrared observation satellite
TM	Thematic Mapper
TNT	2,4,6-trinitrotoluene
TOMS	Total Ozone Mapping Spectrometer
TOTO	TOtable Tornado Observatory
trowal	trough of warm air aloft
TT	Total Totals (stability index)
TTO	Ten-to-Twelve-year Oscillation
^{234}U	uranium-234 isotope
^{235}U	uranium-235 isotope
^{238}U	uranium-238 isotope
UCAR	(USA) University Corporation for Atmospheric Research
U.K. or UK	United Kingdom
UNEP	United Nations Environment Program
UNESCO	United Nations Educational, Scientific, and Cultural Organization
UO_2	detrital uraninite
U.S.	United States
U.S.A. or USA	United States of America
UTC	*universel temps coordonné*, coordinated universal time
UV	ultraviolet

UVB	ultraviolet-B
V	volt
VOC	volatile organic carbon
vol.	volume
VOS	Voluntary Observing Ship
VUV	vacuum ultraviolet
W	watt
°W	degrees west longitude
WBF	Wegener-Bergeron-Findeisen
WCIP	World Climate Impacts Program
WCIRP	World Climate Impact assessment and Response strategies Program
WMO	World Meteorological Organization
WSI	weather stress index
WSR-88D	Weather Surveillance Radar, 1988, Doppler
WWW	(WMO) World Weather Watch
XUV	extreme ultraviolet
ybp	years before present
yr	year
Zec.	(Bible) Zechariah
μm	micron (10^{-6} meters)
Ω	ohm

MEASURES AND CONVERSIONS

centimeter	0.39 inches
foot	12 inches (0.3048 meters)
gram	0.0353 ounces
inch	2.54 centimeters
kilogram	1,000 grams (2.2046 pounds)
kilometer	1,000 meters (3,280.8 feet or 0.62 miles)
meter	39.37 inches
metric ton	1,000 kilograms (2,204.62 pounds)
mile	5,280 feet (1.6093 kilometers)
ounce	28.3495 grams
pound	16 ounces (0.4536 kilograms)
ton	(USA) 2,000 pounds (907.18 kilograms); (UK) 2,240 pounds (1,016.06 kilograms)

PREFIXES FOR INTERNATIONAL SYSTEM OF UNITS

yotta (Y)	10^{24}	deci (d)	10^{-1}
zetta (Z)	10^{21}	centi (c)	10^{-2}
exa (E)	10^{18}	milli (m)	10^{-3}
peta (P)	10^{15}	micro (μ)	10^{-6}
tera (T)	10^{12}	nano (n)	10^{-9}
giga (G)	10^{9}	pico (p)	10^{-12}
mega (M)	10^{6}	femto (f)	10^{-15}
kilo (k)	10^{3}	atto (a)	10^{-18}
hecto (h)	10^{2}	zepto (z)	10^{-21}
deca (da)	10^{1}	yocto (y)	10^{-24}

ENCYCLOPEDIA OF CLIMATE AND WEATHER

L

LAKE EFFECT STORMS. A type of snowstorm produced when very cold air flows across relatively warm waters of lakes is called a *lake effect storm*. Common in winter around the Great Lakes of North America, these storms are confined to regions over the lakes and inland 16 to 80 kilometers (10 to 50 miles) along their downwind shores (Figure 1). The main source of energy for these storms is the temperature difference between the air and the lake surface. Therefore, they are most frequent and tend to be most severe during the late fall and early winter months, when the lakes are still fairly warm and incursions of very cold air masses are rather common. In the most severe storms the snow is very heavy and accompanied by bitterly cold temperatures, strong gusty winds, and near-zero visibility. Two feet (50 centimeters) of snow accumulation in 24 hours is not uncommon. Severe lake effect storms can occur almost anywhere around the Great Lakes, but they are most common near the eastern ends of lakes Erie and Ontario.

A winter's accumulation of snow from many lake effect storms gives rise to areas along downwind lake shores where annual snowfall amounts are 2 to 3 times those in areas at the same latitudes but away from the lakes. These areas are known as *snowbelts*. Major snowbelts occur along the southern and eastern shores of each of the Great Lakes.

In the formation of a lake effect snowstorm, cold dry air flowing across a warm lake picks up heat and moisture from the lake surface. Warm moist parcels of air rise like bubbles from near the lake surface and mix into the colder air. This results in vigorous mixing, warming, and moistening of the overflowing air. The mixed layer deepens as the air continues to flow across the lake.

Some of the moisture evaporated from the lake surface condenses into streamers of steam fog near the lake surface. Several fog streamers may come together into nearly vertical cloud columns called *steam devils*. (In many ways, these are similar to dust devils.) Some of the rising mois-ture condenses to form a stratocumulus cloud in the upper half of the mixed layer. Cloud drops may remain unfrozen for long periods even though their temperatures are below 0°C (32°F); this condition is known as *super-cooling*. When air temperatures are below about −15°C (5°F) some of the supercooled drops freeze and become small ice crystals. These grow into snow through vapor deposition and the freezing of supercooled cloud drops with which they collide. As the snow grows, it consumes the supercooled cloud.

Typically half or more of the water evaporated from the lake goes to moisten the warm air downstream of the lake. If, subsequently, this air encounters higher terrain, further lifting results in more clouds and snow. The very heavy snowfalls that occur in some of the snowbelts of the Great Lakes, especially in southwestern New York state and northwestern Pennsylvania, result from a combination of lake effect heating and moistening of the air followed by orographic lifting of the air (over rising or mountainous terrain).

Lake effect storms over the Great Lakes have been extensively studied using instrumented satellites, aircraft, radar, and numerical models, to determine their interior structure and dynamics. It has been shown that the warm moist air parcels rising from the lake surface usually become organized into patterns of updrafts and downdrafts. The most common pattern is one of bands of clouds spaced at intervals of 5 to 20 kilometers (3 to 12 miles) and oriented nearly parallel to the low-level winds. These are known as *wind-parallel bands* or *cloud streets*; they are readily observed on satellite cloud pictures. When the wind direction is roughly parallel to the major axis of the lake and the air is very cold, lake effect storms often take the form of a single band of clouds and snow located along the major axis of the lake. These are known as *mid-lake bands*. They are considerably wider than individual wind-parallel bands and produce much heavier snow. Prolonged periods of heavy snow can occur where a mid-lake band intercepts the downwind shore. The

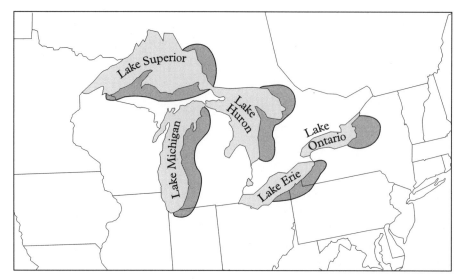

LAKE EFFECT SORMS. Figure 1.   *Snowbelts (shaded area) around the Great Lakes.* (From Eichenlaub, 1979.)

heavy lake effect snows that often occur in the vicinity of Buffalo, New York, are of the mid-lake variety.

The socioeconomic impact of lake effect storms can be considerable. Heavy snow and blizzard conditions disrupt public utilities, transportation, schools, business, and industry. Cleanup and restoration of normal activities can be expensive. At the same time, however, lake effect snow provides many benefits. It is an important source of water for the lakes and surrounding areas. Large amounts of snow benefit winter sports activities. The snow-covered surfaces, in combination with a nearby cold lake that warms slowly, delay the spring growth of plants until after the normal time of late-winter cold spells. This combination makes snowbelt areas favorable places for fruit and grape culture.

Lake effect storms differ in several important ways from the large migratory weather systems that produce most of the world's snow. Lake effect storms are fixed in location by the lakes from which they derive their energy and moisture. They are rather shallow; whereas clouds associated with large migratory winter storms may reach heights of 12 kilometers, lake effect storm clouds rarely extend to 3 kilometers and are often less than 1 kilometer in depth. Closely related to lake effect storms, in terms of causative physics and internal structure, are the snowstorms that occur on the Japanese islands of Honshu and Hokkaido when very cold air from Asia crosses the warm Japan Current before encounteringthe islands.

[*See also* Snowstorms; *and* Storms.]

BIBLIOGRAPHY

Braham, R. R., Jr., and R. D. Kelly. "Lake-effect Snow Storms on Lake Michigan, USA." In *Cloud Dynamics*, edited by E. M. Agee and T. Asai. New York, 1982, pp. 87–101.

Chang, S. S., and R. R. Braham, Jr. "Observational Study of a Convective Internal Boundary Layer over Lake Michigan." *Journal of Atmospheric Sciences* 48 (1991): 2265–2279.

Court, Arnold. "The Climate of the Conterminous United States." In *Climates of North America*, edited by R. A. Bryson and F. K. Hare. Amsterdam, 1973, p. 219.

Eichenlaub, V. L. *Weather and Climate of the Great Lakes Region.* Notre Dame, Ind., 1979.

Passarelli. R. E., and R. R. Braham, Jr. "The Role of the Winter Land Breeze in the Formation of Great Lakes Snowstorms." *Bulletin of the American Meteorological Society* 62 (1981): 482–491.

ROSCOE R. BRAHAM, JR.

LAKES. As part of the hydrological cycle, lakes are a component in the movement of water among the land, the atmosphere, and the oceans. Lakes both supply water to the atmosphere and store a region's moisture supply. They are a vital element of the Earth's weather and climate. Evaporation of water from lakes affects the regional climate; lakes also may affect weather by exerting a cooling effect on the surrounding area in summer and warming it in winter. Lake size is related to the moisture supply of a geographical region or river watershed, so changes in lake size may occur as a region becomes more arid or moist, revealing long-term climate changes.

Local weather is affected by evaporation of water from lakes and by the temperature contrast between the lake water and land. More water evaporates as the lake water temperature rises, as wind speed over the lake increases, and as the temperature of the overlying air increases. Therefore, as an air mass moves over a lake, it may pick up moisture. The difference in temperature between the lake surface and the land can cause a lake breeze to develop, blowing from the water to the land. The breeze is strongest when the temperature contrast is greatest—usually in the afternoon, after heat and energy have been absorbed from the Sun. The contrast in temperature is greatest at the shoreline, and it is there that lake breezes are strongest. [*See* Small-Scale or Local Winds.]

The Great Lakes of North America are large enough to have a significant impact on weather. Lake effect snows are a striking example. In winter, cold dry air, often originating from the polar regions of Canada, blows over the Great Lakes, picking up moisture. The lakes are relatively warmer than the polar air mass, so air blowing over the lakes warms slightly. The air's increased moisture and warmth makes it lighter in weight, or more buoyant, than the surrounding air mass. The buoyant air rises and forms clouds, which then produce snow showers. Lake effect snows are most common in early winter when the temperature contrast between land and water is greatest. The snowstorms occur on the downwind side of the lake; since polar masses typically move from the north and west, the southeastern shores of the Great Lakes are most vulnerable to lake effect snows. These snowstorms typically occur up to 100 kilometers from the lakes. [*See* Lake Effect Storms.]

The size and extent of a lake may be a result of long-term climatic conditions. The amount of water in a lake is determined in part by the balance between precipitation and evaporation in the lake's *catchment area* (the surrounding areas that contribute runoff to the lake). The balance between precipitation and evaporation changes as the climate becomes more arid or more moist. For example, studies of dry lakebeds in the Sahara and Sahel of Africa reveal that they were filled with water in the early Holocene, from about 9,000 to 7,000 years ago. The deep lakebed levels indicate past humid climatic conditions.

Determining that climate shifts have caused lake levels to change is not simple; other factors, such as tectonic activity and groundwater movement, may also affect a lake's water supply.

[*See also* Water Resources.]

BIBLIOGRAPHY

Benson, L., D. Currey, Y. Lao, and S. Hostetler. "Lake-size Variations in the Lahontan and Bonneville Basins between 13,000 and 9,000 ^{14}C yr B.P." *Palaeogeography, Palaeoclimatology, Palaeoecology* 95 (1992): 19–32.

Carpenter, D. M. "Utah's Great Salt Lake, a Classic Lake Effect Snowstorm." *Weatherwise* 38 (1985): 309–311.

Gasse, F., V. Lédée, M. Massault, and J. -C. Fontes. "Water-level Fluctuations of Lake Tanganyika in Phase with Oceanic Changes during the Last Glaciation and Deglaciation." *Nature* 342 (1989): 57–59.

Reinking, R. F., et al. "The Lake Ontario Winter Storms (LOWS) Project." *Bulletin of the American Meteorological Society* 74 (1993): 1828–1849.

Sousounis, P. J., and J. M. Fritsch. "Lake-aggregate Mesoscale Disturbances. Part II: A Case Study of the Effects on Regional and Synoptic-scale Weather Systems." *Bulletin of the American Meteorological Society* 75 (1994): 1793–1811.

MARGARET KNELLER

LAMB, HUBERT HORACE (b. 1913), English climatologist. Arguably the greatest climatologist of the twentieth century, H. H. Lamb was born in Bedford, England, in 1913. He attended the Oundle School and read natural sciences and geography at Trinity College, Cambridge, receiving his B.A. degree in 1935 and his M.A. in 1947. In 1982, Cambridge University awarded him the Sc.D. He has received two honorary degrees—a D.Sc. from the University of East Anglia and an LL.D. from the University of Dundee, both in 1981.

In 1936 Lamb commenced his service with the U.K. Meteorological Office. His broad interests in the field of climatology began during his early years as a meteorologist and received further impetus during spells abroad. In 1940 he transferred to the Irish Meteorological Office, where from 1941 he was in charge of forecasting for transatlantic flying, then in its infancy. His great skill at forecasting is reflected in the fact that crews crossing the Atlantic from the Irish flying base during World War II had a perfect safety record. In 1946–47, Lamb served as a meteorologist aboard a Norwegian whaling ship bound for Antarctica. Experiences and study during this voyage led him to realize that climate was by no means constant and unchanging. From 1951 to 1952 Lamb was the senior British forecaster in Germany, and from 1952 to 1954 he was in charge of the British weather forecasting office in Malta.

From 1945 to 1971, Lamb was based in England with the Meteorological Office, with responsibilities in the fields of long-range weather forecasting, global climatology, and climate change. His achievements during this period included a seminal paper on weather types and spells and the development of a classification system for British daily weather types (now referred to as *Lamb weather types*).

While working in the Climatology Division, Lamb also had access to the Meteorological Office archives, where he discovered a wealth of data on past climate that had never been seriously examined. From these data he produced monthly mean sea-level pressure maps from 1750 on for a large area of the North Atlantic and Europe. Data from ship's logs in these and other archives led Lamb to realize that variations in sea-surface temperature must be related to climatic variations. Another interest that developed at this time was a possible connection between volcanism and climate change. This led to a classic study in which Lamb compiled a list of all known volcanic eruptions since 1500, together with an estimate of the dust each eruption had ejected into the atmosphere—now known as *Lamb's Dust Veil Index*.

In 1972 Lamb left the Meteorological Office to establish and direct the Climatic Research Unit at the University of East Anglia in Norwich, England. At this time, the Climatic Research Unit was unique—a center focusing entirely on research into climatology, with a strong emphasis on climatic change. This was also the first setting in which climatologists and historians could work together in the analysis of data on past climate. The current resurgence of interest in historical climatology clearly draws its inspiration from Lamb's pioneering work in the use of documentary historical evidence to reconstruct the details of past climates.

Lamb retired as director of the Climatic Research Unit in 1978 but has continued as emeritus professor. His magnum opus, the second volume of *Climate: Present, Past, and Future,* was published at about this time. In addition to his books, he is the author of more than 100 scientific articles.

With his interest, enthusiasm, and knowledge, Lamb has been the inspiration for much of the climatological research being undertaken today. His services to climatology are reflected in the many awards given to him. These include the L. G. Groves Memorial Prize (1960 and 1970); the Darton Prize of the Royal Meteorological Society (1963); the Murchison Award of the Royal Geographical Society (1974); the Vega medal of the Royal Swedish Geographical Society (1984); and the Symons Memorial Medal of the Royal Meteorological Society (1987). He has also been honored by the Danish Natural History Society (1978) and the Royal Academy of Sciences and Arts, Barcelona (1983).

Apart from his research, Hubert Lamb's favorite activity has been to spend time in hilly or mountainous country, in particular in Scotland or Scandinavia. It was in Scotland that he met his wife, Moira Milligan. Their long and happy marriage has resulted in three children and six grandchildren. Lamb is still actively involved in pursuing his research.

BIBLIOGRAPHY

Lamb, H. H. "Types and Spells of Weather around the Year in the British Isles: Annual Trends, Seasonal Structure of the Year, Singularities." *Quaternary Journal of the Royal Meteorological Society* 76 (1950): 393–438.
Lamb, H. H. *The English Climate.* London, 1964.
Lamb, H. H. *The Changing Climate.* London, 1966.
Lamb, H. H. "Volcanic Dust in the Atmosphere: With a Chronology and Assessment of its Meteorological Significance." *Philosophical Transactions of the Royal Society,* Series A, 266 (1970): 425–533.
Lamb, H. H. *Climate: Present, Past and Future.* Vol. 1. *Fundamentals and Climate Now.* London, 1972.
Lamb, H. H. *Climate: Present, Past and Future.* Vol. 2. *Climate History and the Future.* London, 1977.
Lamb, H. H. *Climate, History and the Modern World.* London, 1982.
Lamb, H. H. *Weather, Climate and Human Affairs.* London, 1988.
Lamb, H. H. *Historic Storms of the North Sea, British Isles, and Northwest Europe.* Cambridge, 1991.
Lamb, H. H., and A. I. Johnson. "Climatic Variation and Observed Changes in the General Circulation." Parts I, II, and III. *Geograpfiska Annaler* 41 (1959): 94–134; 43 (1961): 363–400.

ASTRID E. J. OGILVIE

LAND BREEZES. *See* Small-Scale or Local Winds.

LAND ICE. *See* Glaciers.

LA NIÑA. *See* El Niño, La Niña, and the Southern Oscillation.

LAPSE RATE. *See* Convection; *and* Static Stability.

LARGE SCALE. *See* Scales.

LARGE-SCALE METEOROLOGY. *See* Meteorology.

LATENT HEAT. Energy is required to melt ice or to evaporate liquid water. This energy, however, need not change the temperature of the water: it only causes the molecules to change state, from a solid to a liquid, or from a liquid to a gas. Because this is stored energy that is released through subsequent freezing or condensation, it is called *latent heat*. Although latent heat is a form of internal energy, it is usually considered separately from sensible heat. [*See* Sensible Heat.]

The energy required to melt ice is called the *latent heat of fusion* and equals 334 joules per gram of water at 0°C. When ice melts, this energy is absorbed by the water molecules; it is released when the liquid water subsequently freezes. Similarly, the energy required to evaporate liquid water is called the *latent heat of vaporization* and equals 2,501 joules per gram at 0°C. This energy is released when the water vapor later condenses. By comparison, only 4.19 joules are required to raise the temperature of one gram of water by 1°C. Thus nearly 6 times as much energy is required to evaporate water at 0°C than to raise its temperature to 100°C.

Under certain conditions, it is possible for processes called *deposition* and *sublimation* to occur. In deposition, water vapor changes directly to ice, as when frost forms; in sublimation, ice changes directly to a vapor without passing through the liquid state, as occurs with dry ice (frozen carbon dioxide). Since energy must be conserved, the latent heat of sublimation equals 2,835 joules per gram at 0°C—the sum of the latent heat of fusion and the latent heat of vaporization at that temperature.

These latent heats are not a constant, however, but vary as a function of temperature. At warmer temperatures, water molecules in the liquid phase contain more internal energy, and consequently less energy is required for them to evaporate. The latent heat of vaporization thus decreases as temperature increases. For example, the latent heat of vaporization equals 2,454 joules per gram at 20°C, and 2,261 joules per gram at 100°C. Similarly, these molecules contain less internal energy at colder temperatures, so less energy is released when they freeze, the latent heat of fusion decreases to 289 joules per gram at −20°C.

The transfer of energy through latent heat plays a significant role in the global energy balance. For example, more than 3 times as much energy is transferred from the surface of the Earth to the atmosphere by the evaporation of water than through the combined processes of conduction and convection. Globally, the removal of energy by evaporating water, the transport of water vapor to other locations, and the subsequent release of the latent heat through condensation and precipitation help to drive the Hadley circulation as well as to provide the necessary poleward transport of heat.

[*See also* General Circulation.]

BIBLIOGRAPHY

Henderson-Sellers, A., and P. J. Robinson. *Contemporary Climatology.* New York, 1986.
Peixoto, J. P., and A. H. Oort. *Physics of Climate.* New York, 1992.
Sellers, W. D. *Physical Climatology.* Chicago, 1965.

DAVID R. LEGATES

LIFTING CONDENSATION LEVEL. *See* Conditional and Convective Instability; *and* Thermodynamics.

LIGHTNING. If you shuffle across a rug on a dry winter day, you acquire an excess of electrical charge; reach for a light switch or a person's hand, and a spark is discharged. In some ways, lightning is similar. Cloud regions acquire an excess of electrical charge, and lightning may be discharged within the cloud or, more dangerously, from the cloud to the ground. Lightning is a long spark that travels several kilometers through the air to relieve the excess of electrical charge within the cloud. It occurs often; it is estimated that 100 lightning flashes happen each second somewhere on Earth, or more than 8 million per day. [*See* Global Electric Circuit.]

Lightning is usually associated with cumulonimbus clouds, but it may also occur in nimbostratus clouds, snowstorms, dust storms, and even in the erupting gas of an active volcano. The flash occurs because separate regions of net positive and net negative electric charge are formed in a cloud. This separation produces an electrical potential, and when the potential exceeds the dielectric strength (insulator strength) of air, a lightning flash occurs. Within a thunderstorm, the flash can pass between clouds (called *cloud-to-cloud discharge*), within a cloud *(intracloud discharge),* between a cloud and air *(air discharge),* or from a cloud to ground *(cloud-to-ground discharge).*

Lightning usually occurs within a cloud or from a cloud to the ground. It is a high electrical current of bright luminosity flowing through the atmosphere. Currents in the lightning channel are measured in thousands of amperes, and temperatures within the channel have been measured

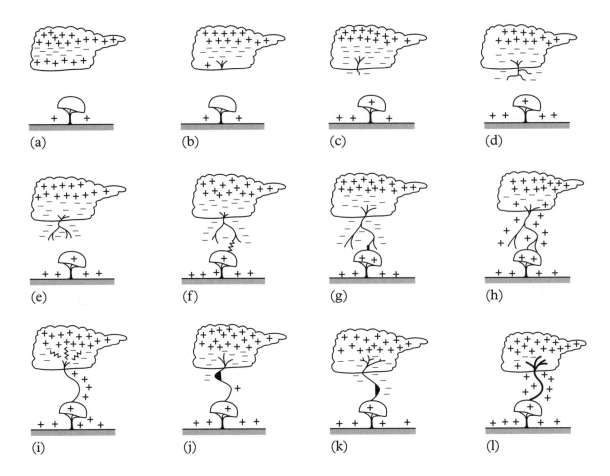

LIGHTNING. Figure 1. *Schematic of the lightning flash (diagrams are not drawn to scale).* The formation of the stepped leader from cloud to ground (a)-(f), is followed by the return stroke from ground to cloud (g) and (h), brief K and J streamers (i), the dart leader from cloud to ground (j), followed by the second return stroke (l). (Adapted from Uman, 1986, and Wallace and Hobbs, 1977.)

at around 50,000°F (28,000°C), Earth's highest naturally occurring temperature, 5 times hotter than the surface of the Sun. The most common type of lightning is intracloud lightning.

The first lightning flash within a thundercloud is typically an intracloud flash. When an intracloud flash occurs, the cloud becomes luminous for approximately 0.2 seconds. A discharge is initiated by a breakdown process that appears as a luminous leader, which travels between charge centers in the cloud. The 0.2-second luminosity is continuous and has several pulses of higher luminosity of about 1 millisecond duration superimposed on the initial luminosity. We believe that the amount of charge involved is usually about 20 coulombs, but it may range from 0.3 to 100 coulombs. The mean velocity of propagation of the intracloud flash luminosity ranges from 10,000 to 20,000 meters per second, or about 5 to 10 miles per second. Lightning strikes to instrumented

aircraft indicate that the currents in the intracloud flashes are around a few thousand amperes. This is approximately an order of magnitude less than the current in cloud-to-ground lightning.

Cloud-to-ground lightning flashes are composed of one or more strokes, which the eye often perceives as a flicker. Although these discharges represent only about 20 percent of all lightning flashes, they cause more than 100 deaths and $40 million in property damage, and they set 10,000 fires destroying $50 million worth of marketable timber each year in the United States alone.

The flash to ground is initiated by electrical breakdown between regions of electrical charge in the thundercloud (Figure 1a), typically between the small positive charge region and the negative charge region (Figure 1b). On a time scale measured in milliseconds, high-speed cameras record a luminous leader, called the *stepped leader,* emerging from the thundercloud (Figure 1c) and propagating toward the Earth (Figure 1d-e). This faint luminous process, too fast for the eye to see, descends in regular, distinct steps in a downward branching pattern toward the ground. The stepped leader propagates at an average speed of 1 one-thousandth the speed of light, or 100,000 meters per second, carrying currents on the order of hundreds of amperes. As the branching process nears the ground (Figure 1f), about 5 coulombs of charge have been deposited in the channel, inducing an opposite charge on the ground and increasing the electric field at the ground between the leader and the point about to be struck. An upward discharge occurs to complete the path to ground, and the return stroke phase begins (Figure 1g). The cloud is short-circuited to ground and a brilliant, luminous return stroke of high current occurs. The luminosity returns to the cloud (Figure 1h) at one-third the speed of light, producing average currents of 30,000 amperes, but occasionally exceeding 200,000 amperes. Peak currents at the ground are reached in less than a few millionths of second. The rapid rise in current produces temperatures in the channel exceeding 50,000°F.

The resulting shock wave of air is eventually heard as thunder.

After the first return stroke, breakdown processes, called K and J streamers, often occur in the cloud, perhaps as a precursor to the next stroke in the flash (Figure 1i). A luminous dart of light, the dart leader, is observed to emerge from the cloud. It follows the previous channel to ground without branching (Figure 1j-k) and a second return stroke occurs (Figure 1l). Thus, only the first stroke in a flash exhibits branching. There may be even more strokes in the flash; cloud-to-ground flashes typically have 3 to 4 strokes, but one unusual flash was recorded with 26 strokes.

Most cloud-to-ground lightning flashes lower the negative charge (and are termed negative flashes) from the cloud to the ground, but a small percentage lower the positive charge (positive flashes). The negative or positive identity of a flash is known as its polarity. We know from recent measurements that in the United States, on an annual basis, 96 percent of the cloud-to-ground flashes lower negative charge, and 4 percent lower positive

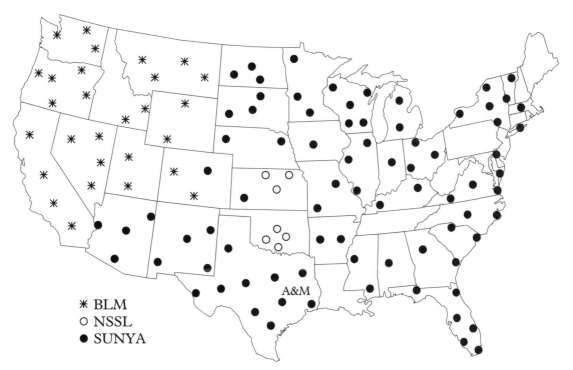

LIGHTNING. Figure 2. *The locations of over 100 magnetic direction finders comprising the National Lightning Detection Network are plotted with symbols indicating the organization that installed the detection finders.* These organizations are the Bureau of Land Management (BLM), the National Severe Storms Laboratory (NSSL), and the State University of New York at Albany (SUNYA). GeoMet Data Services Inc., Tucson, Arizona, operates the network. The location of Texas A&M University, where much of the research is performed on the meteorological aspects of the lightning flash, is approximately at the "A" in the A&M label. Data are as of 1992.

charge. On a storm-by-storm basis, the percentages can vary widely. A storm in the Midwest in May can begin with over 70 percent positive flashes and then change to 100 percent negative in less than an hour. The reverse is also true, in that a storm can begin with negative flashes and end with positive ones. There is increasing evidence that in many storms, an increasing percentage of positive flashes presages the end of the storm.

A small percentage of discharges between the cloud and ground begin at the ground, with a stepped leader propagating upward to a charged region in the cloud. These discharges are often initiated (or *triggered*) by tall structures, towers, or—in rare cases—rockets launched at the Kennedy Space Center in Florida. These triggered lightning flashes have branching in the upward direction, giving the impression of an upside down lightning flash. Triggered lightning studies at the NASA Kennedy Space Center, using rockets to trigger lightning intentionally in storms, have revealed that this type of lightning is composed of many tens of strokes, and that each stroke has small current amplitude, usually less than 15,000 amperes. Lightning is a hazard to rockets and airplanes passing through electrified clouds because the presence of the rocket or airplane may itself trigger the lightning.

Before the components of lightning were known, studied, or given scientific definitions, people defined the types of lightning according to their appearance. Names such as *forked, streak, sheet, heat, hot and cold, ribbon,* and *bead lightning* can be explained by reference to cloud-to-ground or intracloud lightning.

Forked lightning is just the cloud-to-ground lightning discussed above, with its many branches directed toward the ground. Sheet lightning describes the illumination of a cloud by an intracloud flash whose branches remain inside the cloud so that no channels are visible. Heat lightning, which appears red or orange, is cloud-to-ground lightning seen from a distance on a warm summer evening. All colors of the spectrum are emitted by lightning; blue light, however, is scattered more efficiently than red, so that relatively more red light is seen by the distant observer, given the impression of red lightning. (For the same reason, the setting sun appears red.) Thunder is not heard with this type of lightning only because the lightning is so far from the observer.

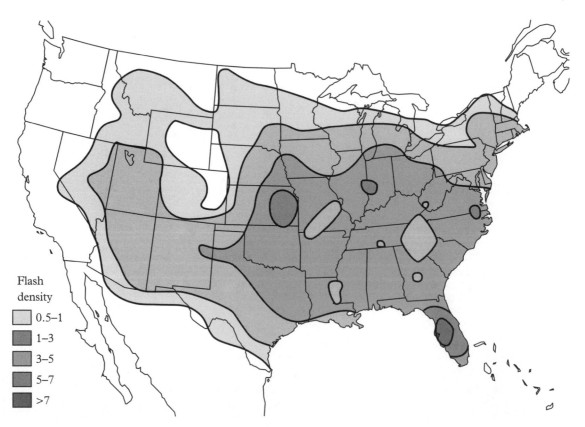

Flash density
- 0.5–1
- 1–3
- 3–5
- 5–7
- >7

LIGHTNING. Figure 3. *Annual lightning-flash density contours are drawn for 1989.* The units are in cloud-to-ground lightning flashes per square kilometer. Maximum flash densities occur in Florida and in Kansas. Minimum flash densities are in the north and along the west coast.

The term *hot and cold lightning* refers to the observation that some cloud-to-ground flashes cause forest fires and others do not. To understand why, we must note that although the nature of the material struck is important, the return stroke current is more important. Some flashes have strokes with continuing currents; that is, the current continues to flow at the rate of a few hundred amperes for many tens of milliseconds. This sustained current is known to cause ignition and set fire to trees. We call this *hot lightning*. Cold lightning, in contrast, does not have continuing currents; the heat from the high current beneath the bark usually causes only explosive damage.

Ribbon and bead lightning are rare forms of cloud-to-ground lightning. If a flash is composed of several strokes, and a strong wind is blowing perpendicular to the line of sight, the channel is blown sideways and subsequent strokes are displaced, appearing as a ribbon of several strokes. Bead lightning is just the breaking up of the decaying lightning channel into luminous parts as some sections of the channel remain luminous longer than others. The time-extended luminosity of these sections can be an observational phenomenon as areas of the channel are viewed end on, or some sections can intrinsically have a larger diameter and therefore take longer to cool.

One of the exciting developments in the past decade in lightning research in the United States has been the creation of the National Lightning Detection Network. Started as a research project, the network has passed into the management of a commercial company where its benefits are available to the public. The network consists of magnetic direction finders for locating cloud-to-ground flashes. These are installed throughout the United States and provide information from which the location, time, and number of strokes can be calculated for each recorded flash. In addition, positive or negative polarity and peak current are determined for the first stroke in each detected flash.

The development of the National Lightning Detection Network began in 1982, when the National Science Foundation funded research with a small lightning net-

LIGHTNING. Figure 4. *Cloud-to-ground lightning near Parshall, North Dakota.* The nighttime exposure was about 30 seconds, but all the lightning in the photo was virtually simultaneous. (Photo copyright Steve Albers.)

work in eastern New York. This network expanded in the next 6 years with support from the Electric Power Research Institute to cover the contiguous United States. Figure 2 shows the present configuration of the network. Research findings show that the United States receives approximately 30 million flashes to ground each year, a rate of 1 flash per second for every second in a year. In reality, however, 75 percent of the lightning in the United States occurs in the summer months of June, July, and August.

One way of graphing the measurements of the number of flashes to ground is to plot a contour of the flash density—the number of flashes to ground per square kilometer. The first result from the National Lightning Detection Network was obtained in 1989 and is reproduced in Figure 3. Maximum flash densities for the United States occur in Florida, where the heating and moisture from the surrounding water provide the right conditions for thunderstorm development. Flash densities decrease to the north and to the west, with the exception of a relatively high density in Kansas. Few lightning flashes are recorded along the west coast. Not shown in this figure is the high flash density discovered over the relatively warm waters of the Gulf Stream off the Carolina coast.

[See also Atmospheric Electricity; and Religion and Weather.]

BIBLIOGRAPHY

Kramer, S. P. *Lightning.* Minneapolis, Minn., 1991, p. 48.
Salanave, L. E. *Lightning and Its Spectrum.* Tucson, Ariz., 1980.
Uman, M. A. *All about Lightning.* New York, 1986, p. 167.
Viemeister, P. E. *The Lightning Book.* Garden City, N.Y., 1961.
Wallace, J. M., and P. V. Hobbs. *Atmospheric Science: An Introductory Survey.* New York, 1977, Chapter 4.

RICHARD E. ORVILLE

LITERATURE AND WEATHER. The moment the Sun rises, there is the illumination of a story. From the creation myths of the Sumerians, Assyrians, and Babylonians to the Hellenic tales of Odysseus and Jason to the biblical account of the birth of the seasons and beyond, the Sun has been the source of creative energy. Casting its light upon the Earth and its elements, it influences not only the movement of the seasons, the temperature of the air, and the distribution of water but also the writing of literature. Its presence or absence can determine the very temperature of literature—the heat of a passionate tale or the chill of an indifferent narrator, the trials of a Chaucerian pilgrimage or the struggles of an Arthurian hero. Unable to be looked at directly, it

requires the indirection of metaphor. To say that the Sun is related intimately to the possibility of literature, however, is only to state a more general truth about the relation of literature to the weather at large. Climatic and meteorological imagery has always been more than a mere motif or theme in literature. It has been a means for writers to express various conceptions of language, the imagination, religion, history, and nature. Its winds, its variable temperatures, and its storms have often been the source and measure of a character's temperament or emotional state. In some instances, as when a James Thomson claims that poetry is seasonal or when a Samuel Coleridge or a Charles Baudelaire identifies mist or frost with figurative language, it is even what the word *literature* seems to mean. At the same time, literature is not only influenced by the weather—by a "mind of winter" or the lightning flash of an insight—but it is often what makes the weather possible. We need only think here of Shakespeare's *The Tempest,* in which Prospero fabricates the tropical storm that in turn initiates the play's action, or of John Ruskin's *The Storm-Cloud of the Nineteenth Century,* in which Ruskin describes the ways in which we create the weather that then determines what is possible for the literature of any given era.

Still, if the weather would seem to be everywhere in literature, we may wonder why the weather and literature are often thought to have nothing to do with one another. Not only is the weather something that we generally assume takes place outdoors while the reading and writing of literature take place indoors, but the weather's reputation as perhaps the most mundane and trivial topic of conversation would seem to make it unsuitable as a topic for great works of literature. As Leo Tolstoy would have it, writers speak of the weather only when they have nothing else to say. The assumption that the weather is not an important literary subject, however, is historically a relatively recent development and can perhaps be understood as part of the legacy of both the Enlightenment and the progress of modern science. If the weather had always been a major concern of literature, the science of the Enlightenment attempted to clear away the clouds obscuring the objects of its understanding. Nevertheless, the skies of the Enlightenment revealed that they could not exist without the weather. The French writer most closely associated with the ideals of the Enlightenment, Denis Diderot, would explain that it was only out of the chaotic dispersion—or "cloud"—of human knowledge that understanding could begin, an understanding that itself could only be articulated in "a language of clouds."

The history of the relation between literature and weather must therefore also begin in this cloud and in this language. Not only because the weather's variable and unpredictable nature makes it difficult to circumscribe, but also because in order to give an account of the pervasiveness of the weather in literature we must also read literature itself as a kind of weather. Literature too is a medium that moves and occurs according to the rhythms and crises of its own atmospheres, storms, or pressure zones. It too is read and interpreted. If literature has never been able to avoid its meteorological figures, it is perhaps because this has been its way of telling us what it is. Within the realm of literature, the weather helps us understand the role that language and metaphor play within the constitution of literary history.

Weather as Language. When Mark Twain states that "everyone talks about the weather, but nobody does anything about it," he points not only to our incapacity to govern or control the weather but also to the essential relation between the weather and language. The quickness with which the weather becomes a subject of conversation suggests its linguistic character. A nebulous space of communication, literary weather constantly produces language, conveys messages, and serves as the condition for narrative. From messages carried by the wind and blown astray in storms to the rainbow that signals the covenant between God and his chosen people, from the tempests that indicate the turbulent emotions of a hero to the "dark and stormy night" without which there could be no gothic novel, the weather has been a medium of communication. In Kalidasa's long, fourth-century poem *Meghaduta (The Cloud Messenger),* a cloud is sent by an exiled lover to carry a message to his beloved. In biblical literature and commentary as well as in Puritan writing of the seventeenth century, thunder and lightning are constantly identified with the voice of God. In Coleridge's *Rime of the Ancient Mariner,* the poem's title joins language to the climate by evoking the pun between rime as verse with like endings and rime as hoarfrost or frozen mist.

This link between language and the weather is also evident in our understanding of the weather as a kind of text, a system of signs to be read—even if these signs are the nebulous texts of wind velocity, cloud formations, and vapors. The tradition of reading the weather—from the ancient weather prophets on through J. W. Goethe, Victor Hugo, Benjamin Franklin, and beyond—names a tendency to read providential omens and hints of the future in the signs of nature. That we read the weather is no doubt due to its shifting forms and modes of unpredictability—all of which require some manner of interpretation. Moreover, like literary language, meteorological events can bear a number of different meanings. The

Sun can evoke the divine as well as the demonic, the rain can signal birth or death, the dew can refer to God's doctrines or tears of mourning, the snow can suggest the power of annihilation and a means of purification, a force of transformation and even the descending voice of the divine. This is why the indeterminacy of the weather—our inability to determine in advance what the weather might be—remains an apt figure for the richness of a literary text.

Weather as Topos. The weather has generally been understood or read in two ways, neither of which is entirely separate from the other. It is either continuous and regular or discontinuous and unpredictable. The former reading emphasizes the weather's cyclical and repetitive character—the annual passing of its seasons, for example—while the latter focuses on its mobility and constant alteration. While a William Hazlitt praises the similarity of the Sun that "we see and remember year after year" or an Alexander Pope claims that "all discord" in nature is "harmony not understood," an Edmund Spenser in his "Mutability Cantos" or a Percy Shelley in his "Ode to the West Wind" accent the weather's changing movement. If the weather passes through annual cycles, however, it does so with infinite variation. William Wordsworth records the interplay in the weather between the permanent and the transitory in Book II of *The Prelude* when he writes: "The seasons came, and every season, wheresoe'er I moved, / Unfolded transitory qualities / Which . . . left a register / Of permanent relations, else unknown"—an interplay he perhaps captures more economically in his *Preface to the Lyrical Ballads* when he refers to the "revolution of the seasons," that is, to what is both cyclical and radically new within seasonal movement.

If the weather oscillates between these two perspectives, it has nevertheless more often been associated with uncertainty, disorder, and change than with stability or order. The weather's tendency to incline or drift away from the understanding can even be read in the word *climate.* Derived from the ancient Greek word *klîma,* it refers not only to a latitudinal zone of the Earth but also to "an inclination or slope." Climate therefore figures not only what falls from the sky but also what falls away from understanding. It names whatever is incalculable and uncontrollable. It is at times another word for "chance" and "time."

Weather as Metaphor. Although literature has recourse to meteorological metaphors for a variety of reasons, we can perhaps delineate its major fronts and pressure zones by considering the most common areas within which such metaphors arise. These categories should not, however, be considered comprehensive. Moreover, like the weather events to which they refer, they often cross the lines that would keep them distinct.

Time. Within the meteorology of literature, climatic phenomena are often conceptualized in relation to time. Within the Judeo-Christian tradition, for example, the Fall names a fall into both time and the weather. Adam and Eve's sin makes them mortal and exiles them to a world of seasonal change. This is why Shakespeare can refer in *As You Like It* to "The season's difference" as "the penalty of Adam" and John Milton, in the tenth book of *Paradise Lost,* can recall how, once sin and death enter the world, God's angels rearrange the universe and the temperate climate of Paradise at once gives way to extremes of heat and cold, and furious winds begin to blow across the ruined Earth. The passing of time wrought by the introduction of the seasons also has made the weather a name for the transitory and fugitive character of existence. Wallace Stevens makes this association in his long weather poem, *The Auroras of Autumn,* when he states that existence "is of cloud transformed / To cloud transformed again . . . the way / A season changes color to no end." Immediately different from itself, always taking another form, the weather speaks to us of our own finitude. As Shelley suggests, "We are as clouds" that are soon lost forever. It is perhaps no accident that in French the same word—*le temps*—names both time and weather.

Epistemology. References to meteorological phenomena such as mist, clouds, and fog are conventionally employed as metaphors for the indeterminacy or instability of knowledge. These phenomena interrupt the dream of knowing and seeing by introducing a certain obscurity or darkness into the system of perception. Diderot's obsession with clouds can be read within this register: from the blinding cloud of his *Letter on the Blind* to his discussion of the figure of the cataract that clouds the eye and prevents the passage of light, Diderot links the weather to the possibility of perception. Friedrich Hölderlin will use the blinding flash of lightning to signal a similar disruption of sight. Thoman Mann in *The Magic Mountain* and Paul Celan in *Language Mesh* will speak of the blurring force of a blizzard. On the other hand, the Sun that gives way to clarity, reflection, speculation, and lucidity—that is, to knowledge in general—can also cause blindness and death if looked at directly. Evidence of the Sun's duplicity is given in the opening pages of Joseph Conrad's *Heart of Darkness,* a novella that begins during the moment of twilight, between light and darkness. In the long run, the groundless and aleatory medium of the weather perhaps prevents knowledge and perception from being established on firm ground.

Psychology. Climatic metaphors are often used to express psychological and emotional states. In Shakespeare's *King Lear,* "the tempest" in Lear's mind takes all feeling away from him except that of "filial ingratitude." In Mary Shelley's *Frankenstein,* for example, the fluctuations of the weather represent the unpredictability of her characters' moods. In Herman Melville's *Moby Dick,* Father Mapple takes on the character of a storm as he gives a sermon on Jonah: "while he was speaking these words, the howling of the shrieking, slanting storm without seemed to add new power to the preacher, who, when describing Jonah's sea-storm, seemed tossed by a storm himself." Moreover, the internal climate of a narrator can influence his or her subjective state and thereby determine the course of a story. As Jonathan Swift suggests in *Tale of a Tub,* meteorological disturbances can even refer to a kind of creative madness: "as the face of nature never produces rain but when it is overcast and disturbed, so human understanding, seated in the brain, must be troubled and overspread by vapors, ascending from the lower faculties to water the invention and render it fruitful." In Albert Camus's *The Stranger,* the narrator is drawn into a senseless murder because of the disorientation and confusion brought on by the burning Sun on an Algerian beach and, in John Steinbeck's *Grapes of Wrath,* the drought comes to figure the sterile and empty lives of poor farmers. That writers so often draw their metaphors from the weather in order to describe inner states not only suggests that such states require some metaphorical interpretation but also that writers, when expressing interior experience, are obliged to draw their vocabulary from a dynamic, shifting element of the world.

Imagination. The power and workings of the imagination are often figured in terms of the power and workings of the weather. In Coleridge, Baudelaire, and Edgar Allan Poe, the imagination is frequently understood as a kind of mist or vapor. In Nathaniel Hawthorne's "The Snow-Image" and Ralph Waldo Emerson's "Snow-Storm," the transformative power of the imagination is pictured in the capacity of snow to transform the appearance of a landscape. In Wordsworth's *The Prelude,* an autobiographical project that links the self to nature, "a gentle breeze" marks the imagination's first appearance. Elsewhere in the poem, Wordsworth refers to the imagination as an "unfathered vapor" and situates the poem's apocalyptic recognition of the imagination's power in the face of the natural world in a weather scene at the top of Mount Snowden.

Inspiration. That the weather is connected to the production of literature can be noted by the various breezes and showers that name the force of poetic inspiration. From the "fresh and fruitful showers" that Sir Philip Sidney wishes would flow upon his "sunburnt brain" in *Astrophel and Stella* to the breeze of inspiration that initiates the "gentle agitations" of the Romantic mind to the ever-changing wind that moves through the Aeolian lyre of a Homer, a Coleridge, or a Shelley, the weather figures the external and internal impressions that move a poet to write. The Aeolian harp, named for Aeolus, the god of the winds, is in fact one of the most recurrent images for the mind during the moment of poetic inspiration.

Politics. When Emerson writes of "the influences of climate and soil in political history," he invokes various eighteenth-century Augustan and Enlightenment debates over the effects of climate on political systems. These debates have their beginnings in writings by Herodotus, Plato, and Aristotle, which discussed the role of climate in determining forms of government. Such theories are reinforced by the Bible as well as by sixteenth- and seventeenth-century writers such as Shakespeare and Jean Bodin, but they blossomed most fully in eighteenth-century writers such as Arbuthnot, Clarendon, and Temple in England and Buffon, Fontanelle, Rousseau, and, perhaps most importantly, Montesquieu in France. The force of the analogy between the weather and politics comes in part from the incalculability that structures each of the two terms. One reason that an investigation into the weather often provides such a good entry into the realm of the political may be that in both forecasting the weather and speculating on the future in politics we are dealing with what Emerson calls "a fluxional protean incalculable element." Walt Whitman makes this analogy in *Specimen Days* when he asks whether or not the weather "sympathizes" with the events of the Civil War: "since this war, and the wide and deep national agitation, strange analogies, different combinations, a different sunlight, or absence of it; different products even out of the ground. After every great battle, a great storm. Even civic events the same. . . . Indeed, the heavens, the elements, all the meteorological influences, have run riot for weeks past." The recourse to weather imagery to figure the events of a war is generally linked to an effort to account for the violence, the bloodshed, and the defeats of the war in terms of a language that ties these with the volatile and sometimes violent storms of nature itself. Rendering war in natural or climatic terms helps to naturalize the war. It helps us to explain events or crises that would otherwise appear traumatic in their contingency.

Religion. As what happens between the heavens and Earth, the weather is often a medium of communication between the divine and the human. From the lightning bolts of a Zeus to the sunlight of a Helios, from the winds of an Aeolos to the sea-storms of a Neptune, the weather has often been a means for the gods to express themselves. [*See* Religion and Weather.] We have already mentioned the ways in which clouds and lightning may be the bearers of messages or the visible signs of God's voice. We have also suggested the weather's relation to the Judeo-Christian notion of the Fall. In addition, the breeze or wind may represent God's divine breath. The snow can suggest the movement of God's hosts as they carry His will downward. The effects of the weather may also suggest a character or narrator's place within a destiny that is both natural and providential. This is particularly true in medieval texts such as Chaucer's *Troilus and Criseyde* and Dante's *Divine Comedy* or Renaissance texts such as Ben Jonson's *The Alchemist* and Shakespeare's *Macbeth*, but it is also evident in the Puritan writing of a Jonathan Edwards or a Michael Wigglesworth. In Edwards's "Sinners in the Hands of an Angry God," for example, the wrath of God comes in the form of whirlwinds, black clouds, and dreadful storms. Within biblical literature, God's anger and retribution are in fact often figured in terms of the weather. From the forty days and nights of rain that lead to the Flood, to the thunder and hail that comes down upon the Egyptians, to the rain of fire and sulfur that falls upon Sodom and Gomorrah, the weather has been the sign of providential power. This power is evident in the lightning that strikes through passages from Exodus to Zechariah, from the Psalms to Jeremiah and Ezekiel, in the rain that brings us God's doctrines in Deuteronomy and the frost that in Job comes with the breath of God. Biblical weather can of course also carry positive connotations, as when the rainbow appears to seal the covenant between God and all living creatures or when the Sun offers its holy rays to His chosen people. In the long run, if the weather that so often traverses religious literature signifies divine power, it is no doubt linked to the fact that, like the words of the gods, it too comes down from above.

A Brief Meteorology of Literature. To assess the weather's significance in the history of literature would require much more space than that afforded by an encyclopedia article, especially since the weather is implicated in the entirety of this history. Nevertheless, we can sketch the contours of a meteorology of literature by mentioning some of the writers for whom the weather has played an important and singular role.

In *Beowulf,* the oldest of the great long poems written in English, the weather constitutes part of the trial that Beowulf must undergo. In the account of his swimming match, he speaks of the flood, the sea-storm, and the north wind as elements of an unknowable destiny that he must confront. In Chaucer's *Canterbury Tales,* the spring marks the beginning of the pilgrimages that will supply his stories. Literary almanacs and calenders such as Spenser's *The Shepheardes Calender* often organize themselves according to the months and, like Spenser's "Mutability Cantos" and *The Faerie Queene,* emphasize the season's mutability. Petrarchan sonnets figure the changing moods of love in terms of the weather's alterations. In Shakespeare, we find the weather song at the end of *Love's Labour's Lost,* the opening storm of *The Tempest,* the winter wind of *As You Like It* and *The Winter's Tale,* the storms in *King Lear* and *Macbeth,* and the various weather events of the sonnets. If the weather is everywhere in Shakespeare, it is perhaps because, as he suggests in the fourth act of *Cymbeline,* only death frees one from the weather, only then can one "fear no more the heat of the sun, the lightning flash, the thunder stone." In John Donne's "The Sun Rising," the eternity of love is said to be unknown to the seasons or the climate, "which are the rags of time." Milton speculated about the climate's influence throughout his life, urging everyone to study the clouds, the snow and rain, lightning and thunder. Novels such as Daniel Defoe's *Robinson Crusoe* and *Roxana* and Swift's *Gulliver's Travels* take their points of departure from the storms that shipwreck their heroes. Thomson's *The Seasons* figures poetry itself in terms of the instability and violence of the weather. William Blake writes "To Autumn" and "To Spring" and his *Book of Thel* is full of speaking clouds. The Romantics consistently figured poems as plants and thereby suggested the dependence of poetry on the weather. In addition to Coleridge, Wordsworth, and Shelley, Keats too wrote poetry to the weather. In France Rousseau was writing a meteorology of the self and reading the "barometer" of his soul, while in Germany Goethe was inventing the weather forecast. Indeed a book could be written on Goethe's relation to the weather, from his essays "The Causes of Barometric Fluctuations" and "Cloud Formations" to his essay "The Theory of Weather," from the rainbow that figures Faust's renewed understanding to the thunderstorms that bring Werther and Lotte together in *The Sorrows of Young Werther.*

In America, in addition to the Puritan obsession with the weather, St. Jean de Crèvecoeur was praising the weather's capacity to encourage industry and community in *Letters from an American Farmer,* and Washington Irving was linking the changes in the weather to the changes in "the magical hues and shapes" of the mountains in "Rip Van Winkle." Hawthorne's weather stories—"The Snow-Image," "Snow-flakes," and "A Visit to the Clerk of the Weather"—are all meditations on the power of the imagination to create the weather and John Whittier's *Snow-Bound* is perhaps one of the most extraordinary reflections on what it means to be bound by snow or whiteness. Henry Thoreau's *Walden* not only is organized around the seasons but uses the seasons to figure Thoreau's relation to both nature and language. His famous description of the thawing of a sandbank in spring, for example, enables him to link his effort to understand the multiplicity of natural forms with his effort to penetrate the language of convention and discover within it an original language of nature. Emerson's famous transparent eyeball passage in *Nature* occurs while he is standing in a snow puddle. In Harriet Beecher Stowe's *Uncle Tom's Cabin,* the runaway slave Eliza escapes from her pursuers by running across blocks of ice. T. S. Eliot's *The Waste Land* evokes the cruelty of the spring, the surprise of rain, the brown fogs of winter noons, and is even in part spoken by the thunder. In Eudora Welty's *The Golden Apples,* the October rain falls on the whole South, blessing it and cleansing it, and, in Toni Morrison's *Sula,* the weather's excesses—too much heat, too much cold, too little rain, too much rain—figure the hardships that the black community of Medallion must endure. Finally, Wallace Stevens's poetry would be nothing without the weather. From *The Snow Man* to *The Auroras of Autumn* to *Transport to Summer,* Stevens traces the relations between literature, existence, and the weather. His poetry is full of winters, frosts, snow, wind, storms, mist, and what he calls "the theatre of clouds." It more than once refers to "the inconceivable idea of the sun," to clouds as "pedagogues" that teach us who we are, and to the wind as a figure for our "will to change." His writings themselves constitute a brief meteorology of literature, a way of talking about the irreducible relation between literature and weather.

[*See also* Drama, Dance, and Weather.]

BIBLIOGRAPHY

Bloom, H. *Wallace Stevens: The Poems of Our Climate.* Ithaca, N.Y. 1976.

Delius, F. C. *Der Held und sein Wetter: Ein Kunstmittel und sein ideologischer Gebrauch im Roman des bürgerlichen Realismus.* Munich, 1971.

Johnson, J. W. "Of Differing Ages and Climes." *Journal of the History of Ideas* 21.4 (1960): 465–480.

Lodge, D. "Weather." In *The Art of Fiction*, by David Lodge. New York, 1992. 84–88.

Mergen, B. "Winter Landscape in the Early Republic: Survival and Sentimentality." In *Visions of American Landscapes*, edited by M. Gidley and R. Lawson-Peebles. Cambridge, 1989.

Reed, A. *Romantic Weather: The Climates of Coleridge and Baudelaire.* Hanover, 1983.

Ross, A. *Strange Weather: Culture, Science and Technology in the Age of Limits.* New York, 1991.

Serres, M. *La naissance de la physique dans le texte de Lucrèce: Fleuves et turbulences.* Paris, 1977.

EDUARDO CADAVA

LITTLE ICE AGES. Since the last glacial maximum there have been periods during which glaciers enlarged and their fronts oscillated around forward positions. The most recent such period, generally occurring between about the mid-sixteenth century and the mid-nineteenth, was worldwide, affecting glaciers on every continent. This is frequently referred to as the *Little Ice Age*. It ended with fluctuating recession of the ice, also worldwide. In Europe and many other parts of the world, the advances of the past 500 years were preceded by another enlargement of equal magnitude around 1300 C.E. Because conditions between the two cold intervals were not as benign as they had been before 1300, some authorities view the whole period between the end of the thirteenth century and the start of recession in the nineteenth century as a single Little Ice Age.

Glaciers enlarge when accumulation exceeds loss of volume (ablation) of ice and snow. Although the Little Ice Age was global in extent, the dates at which advances began and ended were not exactly the same everywhere. The European Alps were affected about a century before the glaciers in western Norway, while in Iceland maximum extent was not reached until the late nineteenth century. Large valley glaciers typically advanced about 1 or 2 kilometers; mean temperatures varied by about 1°C or less. Temperatures were not lower throughout the Little Ice Age; rather, they varied decade by decade, with a series of warmer and colder, wetter and drier decades following each other. The balance between accumulation and ablation continued to be tilted so that glaciers remained enlarged, although they were not constant in volume, thickness, and frontal position. Figure 1 shows examples of these fluctuations.

Little Ice Age history is revealed by historical documents, by the dating of moraines, by studies of ice cores, and by dendroclimatic investigations. The timing of fluctuations in the European Alps is best known because of the wealth of historical records there. The sequence of positions of the fronts of the Grosser Aletsch has been traced in the most detail from written descriptions, pictures, photographs, radiocarbon dating, tree rings, and deliberate measurements, which have been made in Switzerland for more than a century (see Figure 1a).

In North America, dating of moraines has been of greater importance than in Europe (see Figure 2). This is achieved using radiocarbon dating of organic materials underlying, overlying, or within moraines; by counting the number of rings in trees growing on moraines to give minimum dates for their formation; and by measuring the size of lichens growing on the rocks forming the moraines. Radiocarbon dating, unfortunately, cannot provide dates young enough within the Little Ice Age period, but it is the most important dating method for the cold period prior to the fourteenth century. Lichenometry, based on the assumption that the largest lichen

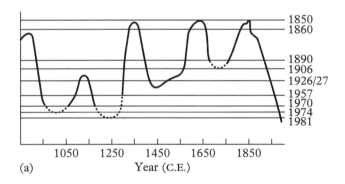

(a)   Year (C.E.)

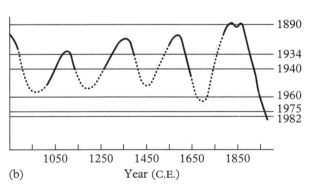

(b)   Year (C.E.)

LITTLE ICE AGES. Figure 1. *Glacier fluctuations during the past 1,000 years.* (a) Aletsch Glacier, Switzerland; dates on the left indicate the positions of the front since 1850, allowing comparison with earlier positions. (b) Tasman Glacier, New Zealand; fluctuations of the Tasman Glacier seem to have been very similar to those of the Aletsch, but this diagram is less firmly based. (Sources: (a) Adapted from an original diagram by H. Holzhauser; (b) from an original diagram by A. F. Gellatly, 1985.)

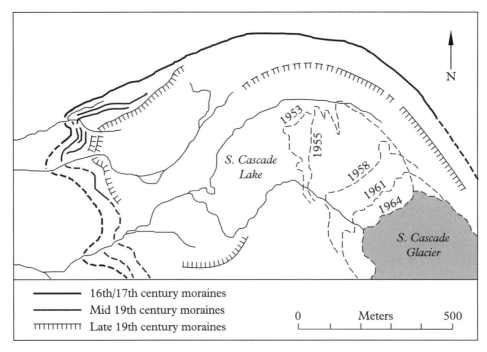

LITTLE ICE AGES. Figure 2. *South Cascade Glacier.* The history of the South Cascade Glacier, northwestern United States, has been established by moraine dating and twentieth-century observations. (Adapted from an original illustration by C. D. Miller.)

growing on a surface is the oldest, has to be applied with great care because different lichen species do not grow at the same rate, and because growth rates are affected by age, the nature of the underlying rock, and the microclimate of the site. A growth curve must be set up for each region using surfaces exposed at known dates. Dendrodating also has limitations, because a tree growing now may not be first-generation, and because there is an inevitable interval between moraine deposition and stabilization and the growth of vegetation. Evidence from North America and other continents shows conclusively that glaciers elsewhere advanced in general accord with those in Europe.

The nature of the climatic fluctuations involved can be determined by using tree-ring sequences, by investigating ice cores, and by using phenological (relating to the growth stages of plants) and other historic data to reconstruct former meteorological conditions. Trees do not flourish at times when glaciers are enlarging. Cold summers tend to produce narrow rings. Information about past climates can also be obtained by investigating the physical and chemical characteristics of cores taken through glaciers, especially those in arctic and subarctic regions. Ice cores from North America, Peru, Tibet,

Greenland, and Antarctica provide independent evidence of fluctuations of past climate on a decadal as well as a century scale. The quantity of paleoclimatic evidence from all these sources has greatly increased recently, to reveal distinct patterns of temperature anomaly during the past thousand years, rather than a simple pattern. The overall rise in temperature over the past century has involved a complex pattern of local variations—even temperature fall in some places at certain times. A reconstruction of global temperature over the past thousand years can only be a generalization concealing important regional anomalies, but it can demonstrate the general trend, as does the graph of global mean temperature during the past century (Figure 3).

The Little Ice Age was not unique in the Holocene. Much evidence suggests that earlier events on similar scales occurred during the Holocene's 10,000 years. Information about these is sparser, and it is not certain that they were all global in extent. They appear to have been associated with low-index circulation patterns and with atmosphere–ocean interactions leading to or led by lower sea-surface temperatures, spread of sea ice, and extension of mean snow cover.

There is no single generally accepted explanation for

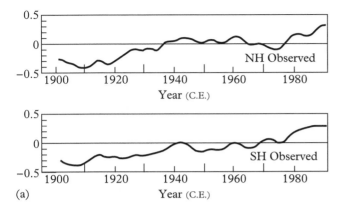

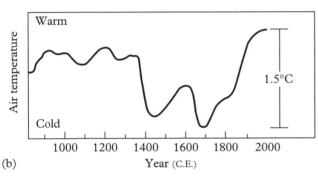

LITTLE ICE AGES. Figure 3. *Fluctuations in global mean temperature.* (a) Observations during the past century reveal that details of fluctuations in mean temperature have not been exactly the same in the two hemispheres. These two graphs are generalized, so local anomalies are concealed. (b) The generalized trends in global temperature during the past 1,000 years, shown here, do not take into account local anomalies of great importance. Present knowledge of conditions, especially in the period preceding 1500 C.E., is insufficient to give them weight. (Sources: (a) Adapted from an original figure by P. D. Jones; (b) from an original diagram by Webb.)

the Little Ice Age and its forerunners. Some attribute the ending of the last one to the increase of greenhouse gases in the atmosphere in the past 100 years, but this overlooks the comparable degree of warming that took place around 950 C.E. Some suggest that the atmospheric system may be subject to random variation from one mode to another. Many have noticed a possible correlation between periods of low solar activity, when there are few sunspots, and the incidence of lower temperatures. The coincidence of timing is striking, but not perfect. The importance of volcanic eruptions, which inject gases and dust into the stratosphere and so reduce incident radiation at the Earth's surface, cannot be ignored. It seems that a full explanation must involve several factors,

including both variations in solar activity and the influence of volcanic eruptions.

In highland regions, the direct effects of glacial advances include the covering of farms and farmland, the blocking of rivers to form ice-dammed lakes that are liable to flood and devastate downstream settlements, and the disruption of long-established routes. They are often associated with increased danger from flooding, rockfalls, landslides, and avalanches in surrounding regions. But the influence of Little Ice Ages is felt over regions well beyond the vicinity of ice.

Minor climatic fluctuations on decadal and century-long time scales can be very important for human populations, especially in regions marginal for agriculture, either latitudinally or altitudinally. Temperature lowering has more important ecological and agricultural consequences in maritime areas than in continental areas because the growing season is shortened more in the maritime than in the continental. Harvest failures increase —especially if one cool, short summer follows another —and altitudinal limits to agriculture decrease. In preindustrial societies, sequences of such failures cause famine. The year 1816 examplifies the effects of an extreme year in a period of severe conditions, which caused a subsistence crisis in both North America and northwestern Europe. In New England most of the vegetables were frozen, the corn crop was almost totally lost, and the hay crop was so small that fewer animals could be kept over the following winter. The 1530s and the 1690s were among the coldest periods ever recorded in Europe. Crop failures in the 1590s affected even the Mediterranean. The years around 1740 and 1816–1819 were also years of famine and dearth.

[*See also* Ice Ages; Paleoclimates; *and* Younger Dryas.]

BIBLIOGRAPHY

Bradley, R. S., ed. *Global Change of the Past: Papers from the 1989 OIDES Global Change Institute.* Boulder, Colo., 1991.
Bradley, R. S., and Jones, P. D., eds. *Climate since* A.D. *1500.* London and New York, 1992.
Gellatly, A. F., F. Röthisberger, and M. A. Geyh. "Holocene Glacier Variations in New Zealand (South Island)." *Zeitschrift für Gletscherkunde und Glazialgeologie* 21 (1985): 265–273.
Grove, J. M. *The Little Ice Age.* London and New York, 1988.
Jones, P. D. "The Climate of the Past Thousand Years." *Endeavour* n.s. 14 (1990): 129–136.
Lamb, H. H. *Climate, History and the Modern World.* London, 1982.
Parry, M. L., and T. R. Carter. "The Effects of Climatic Change on Agriculture." *Climatic Change* 7 (1985): 95–110.
Post, J. D. *The Last Great Subsistence Crisis.* Baltimore, 1977.

Stommel, H., and E. Stommel. "The Year without a Summer." *Scientific American* 240 (1979): 134–140.

Tandong, Y., and L. G. Thompson. "Trends and Features of Climatic Changes in the Past 5,000 Years Recorded by the Dunde Ice Core." *Annals of Glaciology* 16 (1992): 21–24.

Wigley, T. M. L., M. J. Ingram, and G. Farmer, eds. *Climate and History: Studies of Past Climates and Their Impact on Man.* Cambridge, 1981.

JEAN M. GROVE

LOESS. *See* Ice Ages.

LONG-RANGE WEATHER FORECASTING.
See Weather Forecasting.

LONGWAVE RADIATION. The portion of radiation known as *longwave, thermal infrared,* or *terrestrial infrared* radiation consists of electromagnetic waves propagating at the velocity of light (3×10^8 meters per second), with wavelengths greater than about 4×10^{-6} meters (4 microns). Another term commonly used in describing this radiation is *wavenumber,* which is inversely proportional to wavelength and is expressed in inverse centimeter units. In the theory of quantum mechanics, the propagation of electromagnetic radiation at any wavelength can also be thought of as occurring in the form of discrete units—elementary particles called *photons.* The spectrum of Earth's longwave radiation lies in the infrared (heat) portion of the electromagnetic spectrum and is almost completely separated from that of solar radiation. The longwave spectrum extends from 4 to 100 microns in wavelength space, with most of the energy contained between 5 and 50 microns. Longwave radiation is absorbed and emitted by the planetary surface and by atmospheric constituents including gases, aerosols, and clouds. On an annual basis, the longwave flux (energy per unit area per unit time, also termed *irradiance*) emitted by the Earth-atmosphere system very nearly balances the net input of solar radiation. [*See* Radiation.]

The fundamental laws governing longwave radiation are formulated with respect to a black body—that is, an object that is a perfect absorber of radiation of all wavelengths incident on it and that emits the maximum energy possible at a given temperature. Planck's law states that the intensity (energy per unit time per unit area per unit solid angle per unit wavelength) of radiation emitted by a black body at any wavelength is uniquely determined

by its temperature and is isotropic—that is, the intensity is independent of direction. The total flux emitted by a black body integrated over all wavelengths, given by the Stefan-Boltzmann law, is proportional to the fourth power of the absolute temperature. According to Wien's law, the wavelength of maximum radiation by a black body is inversely proportional to its absolute temperature. The peak emission corresponding to the surface temperatures on Earth occurs at a wavelength of about 10 microns. The emission and absorption of radiation by non-black-body materials are described with reference to a black body. A measure of how strongly a body radiates at any wavelength is given by the *emissivity* (a fraction between 0 and 1), which is the ratio of the actual emission to a black body emission at the same wavelength and temperature. Likewise, the *absorptivity* of a material is defined as the ratio of the radiation absorbed by that material to that absorbed by a black body. According to Kirchhoff's law, in thermodynamic equilibrium (valid below an altitude of about 70 kilometers), any selective absorber of radiation at a given wavelength is also a selective and an equally effective emitter of radiation at the same wavelength. At any wavelength, the intensity of the radiation emitted by a non-black-body is the product of its emissivity and the black body emission at that temperature.

The absorption and emission of longwave radiation by atmospheric gases occurs at specific wavelengths depending on their atomic and molecular structure. The atmospheric gaseous constituents can be viewed as isolated entities possessing energies associated with the vibrations of the atoms about their mean positions in a molecule and the rotation of the molecules about their center of mass. Further, according to quantum mechanics, electrons can occupy only certain orbital configurations within atoms, and only certain vibrational and rotational states are permitted for a molecule. Specific combinations of the electron orbits and the vibrational and rotational states determine the energy level of a molecule. A molecule may make a transition to a higher energy level (an *excited* state) by absorbing photons, or through collisions with other molecules. Likewise, it may proceed from a higher to a lower energy level by emitting photons either spontaneously or via collisions. The interaction of a molecule with photons is quantized, in that only certain discrete changes in energy levels are permitted, and these remain the same for both absorption and emission. Both the wavelengths (or frequencies) at which such transitions are possible and the energies associated with them are characteristic for each molecule, giving rise to line-like features of distinct strength (termed *absorption lines*) in the observed spectra of the molecule. The individual spectral lines of the molecules are not sharp, but rather they have finite widths in wavelength space, according to the Heisenberg uncertainty principle. Within the atmosphere, the width of the spectral lines is further broadened owing to the random motions of the molecules (an effect called *Doppler broadening*) and to the frequent collisions between molecules *(collisional broadening)*. Below about 30 kilometers altitude, the widths are determined largely by collisional broadening. For any specific molecule, a large series of closely spaced spectral lines in a narrow interval of the spectrum is referred to as its *absorption band.*

If radiation at a particular wavelength incident on a gas cannot excite its atoms and molecules, energy is neither absorbed nor emitted at that wavelength. The major constituents of the atmosphere by volume—nitrogen (N_2) and oxygen (O_2)—have no absorption bands in the longwave spectrum and hence are termed *infrared-inactive.* Thus only gases present in trace amounts in the atmosphere are *infrared-active* and responsible for the longwave radiative process. Molecules that have absorption bands in the 5-to-50 micron wavelength region include water vapor (H_2O), carbon dioxide (CO_2), ozone (O_3), methane (CH_4), nitrous oxide (N_2O), and the halocarbons (such as chlorofluorocarbons, CFCs); these constitute the so-called *greenhouse gases.* Water vapor is the strongest longwave gaseous absorber, absorbing virtually throughout the longwave spectrum. It has strong absorption bands extending from 14 microns toward longer wavelengths, and also in the 5-to-8 micron region. The region between 8 and 18 microns displays a continuous absorption feature that is not entirely understood. The strongest carbon dioxide absorption bands are centered in the 15-micron region, with a minor band at 10 microns. Among the gaseous absorbers, water vapor and carbon dioxide together dominate the longwave radiative process in the troposphere. Ozone has an important band centered at 9.6 microns, which becomes important for the longwave radiative process in the stratosphere. Among the species present in lesser concentrations, methane has a band centered at 7.8 microns, nitrous oxide has bands centered at 4.5, 7.8, and 17 microns, and the halocarbons absorb in the 7-to-14 micron region. The 8-to-12 micron spectral region exhibits the weakest gaseous absorption. At any wavelength, the absorptivity of a gas depends on its concentration and on the strength of its absorption band. The radiative intensity emitted depends on the absorptivity and the temperature of the gas.

Particles in the atmosphere also absorb and emit longwave radiation. Particles active in this process include aerosols (such as sulfate, sea salt, dust, and soot) and the water drops and ice crystals in clouds. [*See* Aerosols.] The absorption and emission of longwave radiation by these particles depends on their size, shape, and composition. Aerosols, with the exception of dust and soot, consist of spherical particles; water clouds also consist of spherical drops, but ice clouds contain crystals of nonspherical shapes. The longwave absorptivity of the particles varies with wavelength, although this is less pronounced than with the gaseous absorbers. Within liquids and solids, the interactions between the electromagnetic properties of individual molecules are so strong that absorption and emission takes place throughout a continuous spectrum of wavelengths, in contrast to molecular absorption and emission. The region of most significance for longwave radiative interactions with particles is the 8-to-12 micron portion of the spectrum, because this is where gaseous absorption is weakest, while aerosols and clouds have moderate to strong absorption features here. As in the case of gases, the absorptivity of particles at any wavelength is governed by the absorption strength and concentration of the species. Clouds composed of water drops in the lower portions of the atmosphere tend to act as black bodies; that is, they are almost completely absorbing at all wavelengths in the longwave spectrum. In general, clouds composed of ice crystals and aerosol layers tend not to behave as black bodies and have absorptivities less than unity. Again as for gases, the radiative intensity emitted by a particle at any wavelength depends on absorptivity and temperature. For most practical applications, the various types of surfaces on Earth can be considered to have uniform properties and to absorb and emit longwave radiation like black bodies. This assumption holds true for water, snow, and vegetated surfaces; the exceptions are desert surfaces, whose longwave properties depart from that of a black body.

In addition to the selective characteristics of spectral absorption and emission by atmospheric species and the surface, the amount of longwave radiation absorbed, transmitted, and emitted in the atmosphere is governed by the vertical variation of the atmospheric mass, the species' concentrations, and the temperature. If the total absorptivity at any wavelength due to all the species is less than unity, a portion of the radiation at that wavelength is transmitted. If there is no absorption at that wavelength, the radiation passes unimpeded. Thus, radiation emitted under cloudless skies from a black body surface is strongly absorbed in the major bands of the various gases, notably water vapor and carbon dioxide; a certain amount is emitted by these gases in the same bands. The net effect at the top of the atmosphere is a decrease of the radiation in these bands compared with that emitted by the surface. In contrast, a large part of the 8-to-12 micron radiation emitted from the surface escapes to space. The portion of the longwave radiation that does not escape to space is "trapped" by the infrared-active gases, leading to a warming of the surface and the troposphere (the so-called *greenhouse effect*). This trapping is important in forming the Earth's climate. In the absence of water vapor and carbon dioxide, the Earth's surface would be colder by 33°C. Paleoclimatic evidence from ice-core records indicates that temperature variations in the past have paralleled changes in carbon dioxide and methane, suggesting that the presence or absence of these greenhouse gases may have led to warmer or colder conditions, respectively. When clouds are present in the atmosphere—particularly, thick clouds that act as near-black-bodies—even more longwave radiation is trapped than in clear conditions. Thus, water in all its phases is a dominant regulator of the amount of longwave radiative energy trapped in the planet's atmosphere. [*See* Greenhouse Effect.]

At different altitudes, the absorption and emission processes, when integrated over all wavelengths, result in either a net loss of longwave radiative energy *(radiative cooling)* or a net gain *(radiative warming)*. This is in contrast to solar radiation, which always acts to warm the atmosphere. If more radiation is emitted than is absorbed in any region, radiative cooling occurs; the converse leads to radiative warming. In the clear-sky troposphere (below about 12 kilometers), longwave cooling is essentially due to water vapor and carbon dioxide. The cooling decreases with height, becoming small near the tropopause. Layers containing clouds in the lower troposphere experience strong radiative cooling; the thicker the cloud is, the more it acts like a black body and the stronger the cooling. In layers with cirrus clouds, however, in the upper troposphere (around 12 to 17 kilometers), there tends to be radiative warming. In the lower stratosphere (around 20 kilometers), ozone absorption causes warming; in the rest of the stratosphere (above about 25 kilometers), there is cooling owing to water vapor, carbon dioxide, and ozone. On an annual basis, the global strat-

ospheric longwave cooling very nearly balances the solar heating owing to absorption of solar radiation by ozone, resulting in a state of radiative equilibrium for that region. This is in contrast to the troposphere, where longwave cooling and solar heating do not balance each other; the residual energy is manifest in nonradiative processes such as dynamical motions and latent and sensible heat fluxes. Above about 60 kilometers (the mesosphere), the longwave cooling is due to carbon dioxide.

Earth's surface is also affected by atmospheric longwave radiation. Most of the radiation received at the surface is due to emission by tropospheric water vapor and carbon dioxide. In the 8-to-12 micron region, there is a strong contribution resulting from clouds, particularly from low-lying clouds that radiate virtually as black bodies.

Perturbations in amounts of longwave radiation occur in response to changes in the concentrations of the various absorbers—gases, aerosols, and clouds. As a result of human activities, the concentrations of certain greenhouse gases (CO_2, CH_4, N_2O, and the halogenated species) have been increasing over the past century or so. These have increased the trapping of longwave energy in the atmosphere, giving rise to concerns about possible global climatic warming. If increases in water vapor accompany the increases in trace gases, this would exacerbate surface warming. Changes in the concentrations of the longwave absorbers, including aerosols and clouds, also perturb the radiative cooling of the atmosphere. The destruction of ozone in the stratosphere by the CFCs (the *ozone hole* phenomenon), increases in greenhouse gases, and growing concentrations of stratospheric aerosols following volcanic eruptions perturb the longwave radiative cooling of the stratosphere, causing changes in temperatures there as well. [*See* Greenhouse Gases; Ozone Hole.]

Instruments used to measure longwave radiation fall into two broad categories: some measure radiant energy at different wavelengths; others measure the total longwave flux. In recent years, satellites have permitted the precise global monitoring of the longwave flux emitted by the Earth. In particular, the Earth Radiation Budget Experiment (ERBE) has confirmed that, in clear skies, water vapor is the dominant infrared absorber. Longwave trapping is most effective in the moist tropics and least effective in the drier subtropical regions. Clouds are even more effective in trapping longwave radiation, especially those with higher tops. This effect is pronounced in convectively active tropical areas. The present global effect of clouds in reducing the longwave flux is equivalent to about 7 times the effect that would result from a doubling of carbon dioxide, and about 14 times more than the effect of increases in greenhouse gases over the past century. The effect of clouds on longwave radiation is the opposite of their effect on solar radiation. As a consequence of the absorption and the emission processes, the longwave radiation spectrum at the different altitudes and at the top of the atmosphere is a function of the vertical distribution of temperature and the absorptivity of the various species. For this reason, remote sensing by satellites at different wavelengths in the longwave spectrum may be useful in extracting information about the atmosphere. This principle has been exploited to infer temperature and species concentrations, especially above the tropopause. It is also being used to deduce physical properties of clouds—for example, cloud cover and particle sizes—as part of the International Satellite Cloud Climatology Project (ISCCP).

[See also Black Body and Grey Body Radiation; Clouds and Radiation; and Shortwave Radiation.]

BIBLIOGRAPHY

Peixoto, J. P., and A. H. Oort. *Physics of Climate.* New York, 1992.
Ramanathan, V., B. R. Barkstrom, and E. F. Harrison. "Climate and the Earth's Radiation Budget." *Physics Today* 42 (1989): 22–32.
Wallace, J. M., and P. V. Hobbs. *Atmospheric Science: An Introductory Survey.* New York, 1977.

V. RAMASWAMY

LORENZ, EDWARD (b. 1917), American meteorologist. A pioneer in the development of chaos theory, Edward Lorenz helped to revolutionize the study of physical sciences in the process of examining unexpected results that had appeared in his work on weather systems.

As a weather forecaster for the U.S. Army Air Corps during World War II, Lorenz worked with bomber pilots, whose lives often depended on his calculations. When he became a research meteorologist at the Massachusetts Institute of Technology, he concentrated on developing a computerized model of the weather that used differential equations to express relationships among temperature, pressure, wind speed, and other factors. Although researchers knew that measurements could never be perfect, Lorenz gradually came to realize that his computer models showed weather patterns repeating themselves —but never in precisely the same way. In fact, there seemed to be a pattern of disturbances.

In 1960 Lorenz took a mathematical shortcut in a calculation and discovered that the effect was far-reaching; an omission of one part in a thousand resulted in the generation of greatly diverging weather patterns. Continuing work with simpler computer models, he began to see a kind of geometrical structure emerging that a decade later would give birth to the study of chaos, or patterns of irregularity and unpredictability. [See Chaos.]

Lorenz's discovery of the so-called *butterfly effect* had far-ranging consequences for mathematicians, physicists, and physical scientists. Formally known as *sensitive dependence on initial conditioning*, the butterfly effect described a system, such as the weather, that was both unpredictable and also extremely sensitive to tiny changes or perturbations. In theory, a butterfly stirring in China could affect weather patterns over the continental United States several weeks later—not in a predictable way, of course, but only in the sense that a microdisturbance in the atmosphere would grow dramatically with time.

Lorenz theorized, for example, that a complex Earth system can have preferred modes of behavior that most scientists would describe as a *forced change*—for example, an ice age or an interglacial period. He suggested, however, that these behaviors might not be forced at all. Rather, they could be two preferred states of an extremely complicated nonlinear system. (In a nonlinear system, a response is not a simple multiple of the strength of the forcing: for example, if you push it with 1 pound of force, it responds 1 inch, but if you push it with 2 pounds of force, it may respond with 6 inches, or only a half inch.)

Lorenz argued that what we perceive as changes in a weather system may not be forced by anything, but may rather be part of the internal structure of a complex system. That was one of his contributions to the continuing debate in the scientific community between what is external and what is internal, and whether a system is *deterministic* (responds continually to forcing), *stochastic* (obeys statistical laws), or *chaotic* (may respond in a nonrandom, nondeterministic way).

Lorenz asked the kinds of big questions that motivated other physicists and mathematicians to pursue new avenues of science. He wondered, for example, if there were such a thing as a long-term average in weather. In the end, he decided that long-term weather forecasting was, in fact, impossible, even if long-term climate prediction (for example, the effects of greenhouse gas increases) may be predictable in an average sense.

BIBLIOGRAPHY

Gleick, J. *Chaos: Making a New Science.* New York, 1987

DIANE MANUEL

LOWER ATMOSPHERE. *See* Atmosphere, *overview article.*

LOW PRESSURE. Atmospheric pressure, also called barometric pressure, was one of the first meteorological variables to be measured. On any given surface, air pressure is the result of the weight of the atmospheric air column above it; it is expressed in units of force per unit area. By international agreement, the standard unit of pressure measurement is the *bar*, which is equivalent to 10^6 dynes per square centimeter. (A dyne is the force that will accelerate a one-gram mass by one centimeter per second per second.) In atmospheric pressure observations, the usual unit of measurement is the *millibar* (abbreviated mb), or one-thousandth of a bar. Sea-level pressure measurements around the world during the past two centuries have yielded an absolute range of extreme sea-level pressure values from about 850 mb to 1,084 mb. A standard reference for atmospheric variables, called the *standard atmosphere*, gives a sea-level pressure value of 1,013.25 mb, which can be taken as an average for the globe (Oliver and Fairbridge, 1987).

Atmospheric pressure is measured by a barometer. In 1660, seventeen years after the invention of the barometer by Italian scientist Evangelista Torricelli, measurements of atmospheric pressure were first noted by the English scientist Robert Boyle. Because air density decreases with elevation, barometric pressure steadily decreases upward. The rate of change of atmospheric pressure with height can be derived from the hydrostatic equation and the equation of state for dry air (Holton, 1992). Atmospheric pressure decreases exponentially with height, reaching one-tenth of its surface value at an altitude of around 16 kilometers; at approximately 6 kilometers, barometric pressure is half its value at sea level —that is, half the mass of the total atmosphere exists between sea level and 6 kilometers altitude. [*See* Barometric Pressure.]

Atmospheric pressure at the surface exhibits periodic variations associated with solar and lunar tidal forces and with the diurnal (daily) and seasonal heating and cooling cycles. Climatologically speaking, there are preferred regions around the globe where low or high sea-level pressure is predominant. The subtropical latitudes in each of the major ocean basins of the Northern and Southern hemispheres are regions of mean high sea-level pressure. These subtropical highs, a response to the mean meridional (north-south) Hadley circulation, are typically situated beneath areas of mean upper-level convergence and associated subsidence in the atmosphere. [*See* Meridional Circulations.]

There are four main regions of relatively low sea-level pressure. The convergence of extratropical storm tracks in subpolar regions of the North Atlantic and Pacific oceans helps to form two broad troughs of low pressure; that in the Atlantic is known as the Icelandic Low, and that in the Pacific as the Aleutian Low. Although these are sometimes referred to as semipermanent centers of action, the configuration of the sea-level pressure field on any given day does not closely resemble that of the corresponding mean field. Nevertheless, these two areas are well known for their storminess, high winds, and extensive cloud cover.

In the strong belt of westerly winds between the Antarctic continent and the subtropical high presssure belt lies a nearly continuous zone of low pressure. Cyclogenesis—the processes that generate lows or cyclones—is relatively frequent in this region because of the strong meridional temperature gradients found throughout most of the year. Maximum cyclone frequency occurs near 60° south latitude in both winter and summer. In contrast, the latitude of mean maximum extratropical cyclone frequency in the Northern Hemisphere is displaced toward the equator (closer to 50° north latitude) during the northern winter. [*See* Cyclones.]

Along the equatorial regions of the oceans, areas of low pressure generally develop in response to strong surface heating by the Sun and warming by the release of latent heat in the middle and upper troposphere during the process of deep convection (intense thunderstorm activity). These oceanic low pressure areas are associated with mean upward motion and upper-level divergence—the spreading out of air lifted by the intense thunderstorm activity—and with convergent winds near the surface. Two other tropical low pressure systems develop in response to the seasonal cycle. During the summer season, *heat lows* develop in the subtropical desert regions of large continents owing to intense solar heating. However, these heat lows decay with elevation and tend to dissipate above about 2,800 meters.

One of the more important climatic features associated with the formation of seasonal large-scale low pressure areas is the summer monsoons of both hemispheres. The most vigorous of these is the South Asian monsoon, which extends from the Indian subcontinent eastward through Indochina. Intense surface heating during summer induces a broad trough of low pressure to form there, and the convergence of moist low-level winds into this monsoonal trough leads to the development of deep thunderstorms and very high amounts of rainfall. These processes in turn help to maintain and intensify the monsoon low pressure system. Similar monsoonal-type troughs are experienced in Africa, the Amazon River basin in South America, and over northern Australia (Fein and Stephens, 1987; Hastenrath, 1985).

Types of Storms. In the extratropical latitudes, the energy for generating the low pressure systems commonly seen on ordinary weather maps is derived from interactions between air masses with differing thermal characteristics. The generation of potential energy, which is ultimately tapped during cyclogenesis, arises from the processes of differential heating and cooling of the earth-atmosphere system between high and low latitudes and between the ocean and adjacent land masses. Because the earth is a rotating sphere, conservation of absolute angular momentum causes the air trajectories to curve counterclockwise around low pressure centers in the Northern Hemisphere, and clockwise in the Southern Hemisphere. This deflective force is called the *Coriolis effect*. [*See* Coriolis Force.]

The most intense extratropical low pressure systems generally occur during the winter months (from about November to March in the Northern Hemisphere, and April to October in the Southern Hemisphere), when the temperature differences between air masses originating in high and low latitude are greatest. Extremes of low pressure in extratropical cyclones in the Northern Hemisphere are around 920 mb. Fewer data are available for the Southern Hemisphere; observed low barometric readings in its middle latitudes are also around 920 mb.

Tropical cyclones derive their energy from the release of latent heat of condensation inside the tall cumulonimbus clouds that form spiral rain bands around the center of low pressure. The principal tropical cyclogenetic regions are naturally the areas with the highest mean sea surface temperatures. Most tropical cyclone formation occurs in regions with water temperatures exceeding 27°C. In the Northern Hemisphere these include the tropical North Atlantic—including the Caribbean Sea and the Gulf of Mexico—the eastern tropical North Pacific east of about 120° west longitude, and the western

tropical North Pacific, as well as the Bay of Bengal in the Indian Ocean. In the Southern Hemisphere, tropical storm formation is active in the region of the tropical South Pacific west of about 160° west longitude, along northern Australia and adjacent tropical areas to the west, and in the general vicinity of Madagascar eastward to about 70° east longitude.

Most tropical disturbances or depressions do not intensify into tropical storms or hurricanes unless certain environmental conditions are present; these factors include low values of vertical wind shear, weak upper-level winds, and enhanced low-level wind convergence. The strongest tropical cyclones occur in the western tropical Pacific, where they are called *typhoons*. The lowest sea-level pressures ever recorded have been associated with so-called super typhoons, with central pressures nearing 850 mb and wind speeds estimated near 200 knots (approximately 100 meters per second). In the North Atlantic, a few twentieth-century hurricanes approached central pressures of 890 mb. In 1988, central pressure in Hurricane Gilbert was measured at 888 mb while located in the western Caribbean Sea, establishing a record low-pressure reading for North Atlantic hurricanes. The largest of these storms can attain diameters of more than 1,000 kilometers and can hurl storm-surge tides more than 6 meters above the high-tide line in some coastal areas. Because of their high winds, heavy rainfall, and powerful tidal surges, these storms have been associated with great loss of life and property throughout history. [*See* Hurricanes.]

Intense thunderstorms often accompany cold frontal passages in mid-latitude continental regions. In some cases, thunderstorms are associated with the development of intense vortices known as *tornadoes*. Very steep horizontal pressure gradients are a charateristic of these intense funnel-shaped storms, and the approach of a tornado may induce pressure changes at a given site on the order of 50 to 100 mb per second. Surface wind speeds in the most intense tornadoes can exceed 100 meters per second (200 miles per hour), and the largest tornadoes may be a few kilometers in diameter (Kessler, 1985). The central regions of the United States are prone to tornado activity, particularly during the spring months. These storms are caused by a combination of forces, including strong vertical instabilities due to the presence of moist low-level air and a relatively steep vertical temperature lapse rate aloft, together with triggering mechanisms such as an approaching upper-level trough or the presence of a dryline, which acts as a temporary lid; when this lid is bridged by the incipient thunderstorm, explosive storm development occurs. [*See* Tornadoes.]

There are a number of other local storms that represent special cases of cyclonic low pressure systems. The severe blizzards in middle- and high-latitude regions of North America, the intense tropical American squalls known as *temporales*, monsoon depressions, tropical squall lines, and easterly waves all owe their development to the same kinds of driving mechanisms discussed above for extratropical and tropical cyclones.

[*See also* High Pressure.]

BIBLIOGRAPHY

Fein, J. S., and P. L. Stephens, eds. *Monsoons.* New York, 1987.

Hastenrath, S. *Climate and Circulation in the Tropics.* Dordrecht, Netherlands, 1985.

Holton, J. R. *An Introduction to Dynamic Meteorology.* San Diego, Calif., 1992.

Kessler, E. "Severe Weather." In *Handbook of Applied Meteorology,* edited by D. D. Houghton. New York, 1985, pp. 133–204.

Oliver, J. E., and R. W. Fairbridge, eds. *Encyclopedia of Climatology.* New York, 1987.

HENRY F. DIAZ

M

MADDEN-JULIAN CIRCULATION. *See* Tropical Circulations.

MAGNETOSPHERE. The Earth's magnetosphere is the region of space near the Earth in which the motion of charged particles is determined primarily by the Earth's magnetic field. The magnetosphere is formed by the interaction of this geomagnetic field with the solar wind, a flow of ionized gas streaming away from the Sun. The solar wind travels at 400 to 1,000 kilometers per second. [*See* Solar Wind.]

The magnetosphere is in the domain of solar-terrestrial science, the study of physical systems from the Sun and solar wind to the upper atmosphere and ionosphere down to and including Earth. Geospace is the largest primary Earth-system domain (which includes solid Earth, ocean, atmosphere, and near-Earth space). The middle atmosphere and upper atmosphere link the low atmosphere with the magnetosphere and the remainder of geospace by providing a transition region from neutral gas to the partially or fully ionized gas (plasma) that characterizes most of geospace.

Plasmas in Geospace. Most of geospace, including all of the magnetosphere, is a *plasma;* therefore, understanding the geospace environment requires an understanding of the electromagnetic processes and behavior of plasmas. Plasmas are ionized gases and constitute one of the four primary states of matter; solid, liquid, and gas are the other three. In the plasma state, a particle's energy of motion (or *kinetic energy*) is greater than the energy of attraction between the positively and negatively charged particles closest to each other (the *potential energy*). Electrons have very low mass and can move quickly to reduce any net charge; this, in turn, keeps the electric forces as low as possible. In this process, a charged particle within a plasma collects a "screening" cloud of oppositely charged particles, and this cloud operates to eliminate the net charge. Each particle is simultaneously an element in many different screening clouds; this overlap in screening leads to the simultaneous coupling of each particle with many other particles. This is a collective process: the basic interaction is one-to-many rather than one-to-one. Thus, interactions of particles in the magnetosphere are almost never by direct collision. Because the densities of magnetospheric plasmas are so low and their temperatures so high, they are effectively collisionless.

Ordinary gases and fluids in the near-Earth system are typically neutral, collisional, isotropic (that is, particle motion has no preferred direction), and in equilibrium, with one temperature and one sound speed at any given location. In contrast, magnetospheric plasmas are ionized, collisionless, anisotropic (particle motion has a preferred direction, often related to the orientation of the magnetic field), and not in equilibrium, having multiple temperatures, particle and wave speeds, and characteristic frequencies. Gravity exerts negligible force on magnetospheric plasmas, compared to the electromagnetic forces. In spite of these profound differences, collective interactions often allow us to treat the plasma as an electrically conducting, fluidlike medium.

Near-Earth Geospace Environment. The concept of the magnetosphere was first formulated by Sydney Chapman and V. C. A. Ferraro in 1931, but it was not until the 1970s that the magnetosphere began to be explored extensively. Until the late 1950s, it was generally believed that space was empty, except for intermittent particle bursts from the Sun and higher-energy cosmic rays from beyond the solar system. In January 1958, James A. Van Allen discovered the Earth's radiation belts (later named after him), using an energetic particle detector, originally designed for small rockets and balloons, which had been mounted on the first artificial satellite to carry scientific instruments. When he saw Van Allen's early data, researcher Ernie Ray exclaimed, "My God, space is radioactive!" Since then, direct spacecraft observations have measured plasmas, electromagnetic fields, and energetic particles throughout the solar system. There is strong evidence for the existence of plasmas in other star systems and in most regions between the stars (the interstellar medium), suggesting that plasmas occur throughout the universe.

The Earth's magnetosphere is continuously created by the onslaught of solar-wind plasma. Unlike a supersonic aircraft, information about the upcoming obstacle in the flow cannot be provided through collisions, because the plasma is collisionless. Nonetheless, a planetary bow shock is formed though plasma and electromagnetic field interactions, which slows and heats the plasma and also deflects it around the magnetosphere (see Figure 1). Although the Earth's bow shock is without collisions, its shape is surprisingly similar to that predicted by a simple gas dynamic model which neglects magnetic fields and plasmas. How is this possible? As noted above, plasmas behave collectively, so they can often be treated as electrically conducting "fluids" through the *magnetohydrodynamic* (MHD) *approximation.* Solar wind particles at the Earth's bow shock are deflected by electromagnetic forces, and the overall plasma-field interaction can be closely modeled as a fluidlike MHD system.

Ionized particles move easily along the direction of the magnetic field—like beads on a wire—while gyrating around the field, but they resist motion across the field. When plasma is forced to flow across a magnetic field, a force arises which deflects particles in a direction perpendicular to both the particle velocity and the local magnetic field. This sets up a boundary current, called the *magnetopause,* at the outer magnetospheric boundary, as a result of partial penetration of the shocked solar wind, here termed the *magnetosheath plasma.* About 1 to 3 percent of the magnetosheath further penetrates the magnetopause to produce a magnetospheric boundary layer. These structural elements are depicted in Figure 1.

Along the direction from the Earth to the Sun, the magnetopause is typically situated at about 8 to 13 Earth radii toward the Sun; the bow shock is at 10 to 15 Earth radii along the same direction. At the dawn–dusk meridian (the location where dusk or dawn is experienced at a given time), the magnetopause forms a roughly circular cross section in the solar wind flow, about 24 to 34 Earth radii in diameter. The magnetotail extends downstream, in the direction away from the Sun, for 200 to more than

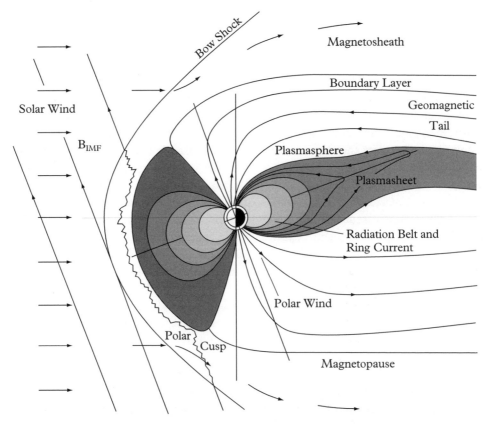

MAGNETOSPHERE. Figure 1. *A schematic diagram of Earth's magnetosphere in the noon-midnight plane.* The basic particle and magnetic field features are representative of other planetary magnetospheres although the details can be different. (From Parks, 1991, p. 8.)

500 Earth radii; its cross section is about 50 to 60 Earth radii in diameter. Thus, the entire magnetosphere is 10^5 to 10^6 times the size of the Earth. Although its average density is very low, both volume and average energy per particle are high.

A Cosmic Ping-Pong Game. Magnetospheric particles come from both the solar wind and the ionosphere, and most magnetospheric particles have been energized relative to their sources. These particles have energies ranging from about one electron volt to thousands or even millions of electron volts. The most energetic particles are the radioactive decay products of cosmic rays. Other energetic particles arise from previous episodes of energization and transport to populate the ring current and the Van Allen radiation belts. Particles in these radiation belts are trapped in the equivalent of a large magnetic bottle, in that they travel back and forth between the north and south magnetic poles while gyrating about the field. A fairly typical 40-kiloelectron-volt electron makes the journey from pole to pole in less than one second! Closer to the Earth, high magnetic field strengths act as magnetic "mirrors" that reflect particles and sustain this cosmic ping-pong game.

Relatively quiescent plasmas of high density and low temperature populate a relatively stable plasma region in the inner magnetosphere; this is called the *plasmasphere* (see Figure 1). Particles in this region rotate with the Earth and are closely linked with their ionospheric source. Most of the magnetosphere at higher latitudes and larger radial distances includes plasmas from both the ionosphere and the solar wind. These plasmas become energized through successive storm episodes (discussed below) and are stored in the *plasma sheet,* a region near the center of the magnetotail. This region is magnetically connected to the auroral regions and is consequently a major source for auroral particle precipitation, which creates the aurora (northern and southern lights). A large portion of the geomagnetic tail, called the *magnetotail lobe region,* extends from the polar cap regions. Both polar caps are encircled by auroral regions and are roughly centered on the corresponding magnetic poles, which are displaced 12 degrees from the rotational poles. A *polar wind* accelerated out of the ionosphere provides a plasma of very low density and high variability in the lobe regions of the magnetotail.

Northern and Southern Lights. The aurora borealis of the far north and the aurora australis of the far south are among the Earth's most spectacular and beautiful phenomena. Much as phosphors of a television screen display a coherent pattern in response to successive sweeps of an electron gun, the ions and atoms of the ionosphere display coherent patterns when struck by precipitating particles (mostly electrons) from the magnetosphere. The aurora is closely coupled via fields and particles to whichever portion of the magnetosphere constitutes the active source—usually the magnetotail. This active region is immense, sometimes extending well beyond the orbit of the Moon. [*See* Aurora Borealis.]

Geomagnetic Storms. Dramatic changes can occur in the geospace environment when coronal mass ejections or solar flare effluents arrive at the Earth's orbit. The precipitating particle fluxes that produce the aurora can change rapidly in response to quick variations in the magnetotail plasma sheet and boundary-layer source regions. Such changes are most dramatic when the external, interplanetary magnetic field is directed southward and thus opposes the northward-directed geomagnetic field. The magnetic coupling of the external and internal fields can then change as the fields become disconnected and then reconnected through a process called *magnetic reconnection*. This reconnection occurs both near the frontside magnetopause and in the midplane of the magnetotail.

The combination of magnetic reconnection and MHD current generation sets up a gigantic magnetospheric "engine." This engine consists of a single convection system which is both global and coherent. The collective effects of the plasma communicated through field-aligned currents, and the rapid motion of particles and waves (propagating at speeds from about 100 kilometers per second to more than 1,000 kilometers per second), make possible a very high level of coherence on a scale far larger than the Earth. This coherence is evident in an auroral display.

When the connection between the external and internal fields is strong, the solar wind-magnetosphere generator creates voltages greater than 200 kilovolts across the polar caps, with currents of 1 to 2 million amperes linking the magnetopause to the ionosphere. This can result in a *geomagnetic storm* that disrupts the entire magnetospheric system, producing strong auroral displays and induced currents on the ground which affects ground-based systems located at both middle and high latitudes. When the field connection and voltage generation are less strong, disruption to the magnetosphere occurs as a *substorm,* generally limited to high-latitude field lines where the aurora most commonly occurs—the so-called *auroral zone*.

Space Weather. Disruptions in the geospace environment can have significant impact on human technological systems, including communications satellites, power distribution grids, and pipelines. Some of the effects that are monitored by the National Oceanic and Atmospheric Administration (NOAA) Space Environment Laboratory in Boulder, Colorado, include the following:

- Failure of electric power distribution grids.
- Sharp increases in drag and unwanted torque on orbital satellites, and radiation damage to their electronic instruments.
- Errors in radio navigation by ships and aircraft.
- Radiation hazards to astronauts.
- Anomalous conditions affecting experiments being conducted on space flights or satellites.
- Enhanced corrosion of pipelines.
- Factors affecting frequency selection for radio communication.
- Adverse effects on military surveillance satellites.

Because Earth-orbiting satellites operate primarily within the magnetosphere, an understanding of this highly variable environment is essential to their safe and efficient operation, just as understanding weather in the troposphere is essential to the operation of airplanes. As systems operating in space—and even some ground systems—become more technologically advanced, their vulnerability to space weather increases. Predictive models for space weather are quite crude at the present time; routine and reliable prediction is possible, but much research and development are needed.

BIBLIOGRAPHY

Chen, F. F. *Introduction to Plasma Physics and Controlled Fusion.* 2d ed. New York, 1984.

Dooling, D. "Stormy Weather in Space." *IEEE Spectrum* 32.6 (June 1995): 64–72.

Eather, R. H. *Majestic Lights: The Aurora in Science, History, and the Arts.* Washington, D.C., 1980.

Eliezer, Y., and S. Eliezer. *The Fourth State of Matter: An Introduction to the Physics of Plasma.* Bristol, 1989.

Herman, J. R., and R. A. Goldberg. *Sun, Weather, and Climate.* New York, 1985.

Hill, T. W., and A. J. Dessler. "Plasma Motions in Planetary Magnetospheres." *Science* 252 (1991): 410–415.

Jursa, A. S., ed. *Handbook of Geophysics and the Space Environment.* Air Force Geophysical Laboratory, ADA 167000, National Technical Information Service. Springfield, Va., 1985.

Lanzerotti, L. J. "Geospace: The Earth's Plasma Environment." *Astronautics and Aeronautics* 19.4 (April 1981): 57–63.

Lerner, E. J. "Space Weather." *Discover* (August 1995): 54–61.

McCormack, Billy M., ed. *Weather and Climate Response to Solar Variation.* Boulder, Colo., 1983.

Parks, G. K. *Physics of Space Plasmas.* Redwood City, Calif., 1991.

Peratt, A. L. *Physics of the Plasma Universe.* New York, 1992.

Siscoe, G., et al. "Developing Service Promises Accurate Space Weather Forecasts." *Earth and Space* 7.2 (October 1994): 10–14.

Van Allen, J. A. *Origins of Magnetospheric Physics.* Washington, D.C., 1983.

Wynn-Williams, G. *The Fullness of Space: Nebulae, Stardust, and the Interstellar Medium.* Cambridge, 1992.

TIM EASTMAN

MAPS. *See* Charts, Maps, and Symbols.

MARINE WEATHER OBSERVATIONS AND PREDICTIONS.

Day-to-day weather forecasting uses computer models, which must simulate the atmospheric circulation over the entire globe. Marine weather observations are needed to ensure that the initial state of these atmospheric forecast models correctly reflects the weather presently occurring over the vast ocean areas. The observations are also used in applications such as marine accident investigations and marine insurance claims, as well as to define the marine climate to aid, for example, in the design of oil rigs and coastal defenses. Collections of weather observations spanning several years or more are used for the study and prediction of climate change

Marine weather predictions are important for the safety of shipping and other ocean users. Forecasts of sea conditions allow ship captains to choose the most cost-effective routes; in coastal areas, they help in predicting possible flooding resulting from storm surges.

Collecting Observations. Over the centuries, ships' officers recorded various weather observations in the ships' logbooks. Matthew F. Maury, an American naval officer, initiated the systematic collection of wind and current information in the mid-nineteenth century, however, it was not until the advent of radio communication that observations from ships could be used to assemble a weather map useful for weather prediction. The present World Weather Watch (WWW) system of the World Meteorological Organization (WMO) was established in 1963. Observations obtained from ships, fixed platforms such as oil rigs, meteorological buoys, and weather satellites are distributed to weather centers around the globe, using the WWW Global Telecommunications System (GTS).

Observations from ships. Port meteorological officers of the national meteorological agencies recruit and equip merchant ships as Voluntary Observing Ships

(VOSs). The officers on these VOSs take observations (normally at the main *synoptic hours*, 0000, 0600, 1200, and 1800 Greenwich mean time) of the air and sea temperature, wind velocity, and air pressure, and describe the present and recent past weather. Many reports also include humidity and information on cloud cover and sea state. These data are compressed into a weather message using a standard WMO *SHIP code*, recorded in the ship's meteorological logbook, and transmitted via a shore radio station to a meteorological agency and thence to the GTS. To be useful for weather forecasting, the data must be available at meteorological centers within a few hours of the time of observation. Difficulties in radio communication limit the number of observations received; some observations that have been corrupted during transmission are rejected by quality-control procedures at the forecasting centers. For determining the marine climate, the priority is to assemble the maximum number of observations. The meteorological logbooks are collected by the port meteorological officers, and the recorded data are used to supplement the GTS observations. This collection process takes several months, but it can double the number of observations available.

In 1992 there were more than 7,000 VOSs recruited by some 50 countries, with about 4,000 observations obtained each day. With the trend toward fewer, larger merchant ships, with fewer officers on each, it is increasingly difficult to recruit new VOSs. A further problem is that the VOSs are concentrated in the major shipping routes and do not provide even coverage over the whole ocean.

With the advent of transoceanic air travel, an array of about a dozen weather ships was established in the 1950s in the North Atlantic and North Pacific oceans to collect weather observations, assist in aircraft navigation, and assist in air-sea rescue. The weather ships reported the SHIP code observations hourly and obtained data on upper-air atmospheric structure at the four synoptic hours by launching balloons. These were tracked by radar to determine wind velocity; those launched at 0000 and 1200 Greenwich mean time carried radiosondes to measure air temperature and humidity. By the 1970s, improvements in aircraft navigation and reliability had lessened the importance of the weather ships, and by the 1990s only two remained. Radiosondes designed to receive and retransmit radio navigation signals now allow wind profiles to be determined without using expensive radar systems, and merchant ships participating in the Automated Shipboard Aerological Program (ASAP) have replaced weather ships for collecting upper-air data.

Buoys. Drifting buoys, deployed from aircraft or from ships, are typically used to measure sea temperature and air pressure in areas away from the normal shipping routes. In 1990 there were about 2,000 observations each day placed on the GTS from about 200 drifting buoys. A typical buoy has a lifetime of less than 18 months, 50 new buoys constantly have to be deployed. The meteorological agencies of some countries have deployed moored buoys around their coasts. The United States has about 40 buoys deployed and maintained by the National Data Buoy Center that measure most surface meteorological variables, including waves. In addition, an array of buoys deployed across the equatorial Pacific Ocean, as part of the Tropical Ocean Global Atmosphere research program, may be maintained in the future for the purpose of forecasting El Niño events.

Satellites. A system of weather observation satellites was first established in the 1960s during the Global Atmospheric Research Program. Five geostationary satellites (GOES, GMS, INSAT, and Meteosat, operated by the United States, Japan, India, and Europe respectively), situated about 36,000 kilometers above the equator, provide continuous global coverage, although with poor spatial resolution in regions poleward of 60° latitude. Multichannel radiometers estimate the surface temperature and the atmospheric temperature and humidity profiles. Consecutive cloud images are compared to provide an estimate of the cloud-level winds. [*See* Satellite Instrumentation and Imagery; Satellite Observation.]

Pairs of satellites (for example, the U.S. TIROS-N/ NOAA series) placed in orbits passing nearly over the poles at 800 kilometers altitude, provide higher-resolution measurements, but they view a particular area only a few times each day. Radiometers measure visible and infrared radiation to provide temperature and humidity profiles and, in cloud-free areas, estimates of sea-surface temperature (SST). These data may also be analyzed to determine the surface radiation budget, to assess sea-ice conditions, and to estimate rainfall based on cloud-top temperature values. Satellites such as the DMSP series (U.S. Department of Defense) also carry microwave radiometers, which provide a more direct measurement of atmospheric water vapor and rainfall and can give estimates of sea ice, sea-surface winds, and SST even in cloudy areas. Unfortunately, the accuracy and spatial resolution of microwave radiometers are worse than those of the visible and infrared instruments.

Both geostationary and polar-orbiting satellites provide data relay facilities, receiving and forwarding meteorological data from buoys and other platforms. Systems such as the French ARGOS on the NOAA satellites also provide location information. Most oceangoing ships use satellite navigation systems (such as GPS) to fix their position, and an increasing proportion of ships' meteorological observations are transmitted to shore using INMARSAT communication satellites.

Satellites allow the whole Earth to be scanned by a single instrument; however, to attain the highest accuracy, the data must be calibrated and checked against *in situ* measurements. For example, increased amounts of aerosol particles in the upper atmosphere resulting from a volcanic eruption will cause a satellite radiometer to underestimate the SST, and the error must be detected, and if possible removed, by using SST data from ships and buoys.

Future weather satellites will carry radar-based instruments tested on both research and preoperational satellites such as the U.S. SeaSat, European ERS-1, and Japanese ADEOS. Radar altimeters measure surface wind speed and wave height directly beneath the satellite; microwave scatterometers use inclined beams to determine the surface wind velocity over a swath 500 kilometers wide, and synthetic aperture radars can determine the sea-wave spectrum. These radar instruments can work in cloud-covered regions and can also be used to determine sea-ice conditions.

Studies of Climate. In order to describe climate and to detect climatic change, researchers need observations spanning several decades. Ship observations provide the longest time series, with some data going back to about 1860 but with most data since 1900. Unfortunately, over the years there have been major changes in the types of ships, the routes followed, and the meteorological instruments used. All these factors result in systematic biases in the measurements which mask genuine climatic changes.

Early measurements of sea-surface temperature were obtained by lowering a canvas bucket over the side of the ship. Full of seawater, the bucket was hauled back to the deck, and the water temperature was measured using a thermometer. During this process the wind caused evaporation from the wet bucket, so the observations were biased low. To avoid this problem, beginning in about 1940 most meteorological agencies provided rubber-insulated buckets; however, it is difficult to lower a bucket 30 meters down to the sea from the bridge of a large modern ship traveling at 20 knots or more. Instead, the temperature of the seawater used to cool the ship's engines is recorded as the SST value, and now a warm

bias may result from the heating in the pipework that leads to the engine room. Nowadays, an increasing number of ships are being equipped with hull temperature sensors, which do not have this bias. The apparent variations in SST caused by changes in measurement techniques are each a few tenths of a degree Celsius, which is similar to the probable real change in SST over the past 100 years. Similar biases have probably also occurred in air temperature and humidity measurements, with unscreened thermometers being replaced by instruments shielded by varnished, and later white-painted, louvered screens, and with an increased use of hand-held psychrometers. [*See* Instrumentation.]

Methods of wind determination have also changed. Early anemometers (such as the Hooke swinging plate anemometer), which measured the pressure of the wind, were difficult or impossible to use on a ship. In the 1830s, Admiral Francis Beaufort defined a scale of wind estimation based on the sails that could be carried by a "well conditioned man-of-war." Now defined in terms of the appearance of the sea, the Beaufort scale is still used as the method of wind determination by VOSs recruited by the meteorological agencies of the United Kingdom and Germany. However, increasing numbers of ships now carry anemometers both for meteorological observations and to aid in docking procedures. As ships have become larger, the anemometers have been mounted higher. Because wind strength normally increases with height, this higher position results in measured winds being biased high when compared to past observations.

Marine Weather Predictions. In return for the time they spend in obtaining weather observations, the officers on merchant ships obtain weather forecasts to improve the safety and cost-effectiveness of their operations. Forecasts covering specified marine areas are prepared by nominated meteorological agencies and transmitted by radio. Weather charts showing the analyzed synoptic situation and the predicted developments over the next 1 to 3 days are available via radio facsimile. Normally these forecasts and charts are prepared 4 or 8 times daily, using the observations obtained at the synoptic hours. Factors affecting safety include wind conditions, visibility, and, in high latitudes, the possibility of ice covering the superstructure and rigging. Economic considerations include the routing of ships to avoid delays caused by rough seas, and the preservation of perishable cargo in hot weather.

Marine weather predictions are also used to warn coastal communities of potential hazards caused by storms and hurricanes. Strong winds and low atmospheric pressure can create a *storm surge,* an increase in sea level that, with rough seas and a high tide, may overflow or breach sea walls. Accurate prediction of storm surges requires ocean models that are driven by the winds forecast by the atmospheric models and that simulate the interacting effects of the waves, currents, and tides. Tropical storms (called *hurricanes* or *typhoons*) originate far from land over the warm tropical oceans. These are associated with storm surges and huge waves, which can cause more damage and loss of life than the high winds themselves. Tracking these storms is an important application of marine weather prediction.

[*See also* Weather Forecasting.]

BIBLIOGRAPHY

Dobson, F., L. Hasse, and R. Davis. *Air-Sea Interaction: Instruments and Methods.* New York, 1980.

Kent, E. C., and P. K. Taylor. *Ships Observing Marine Climate: A Catalogue of the Voluntary Observing Ships Participating in the VSOP-NA.* Marine Meteorology and Related Oceanographic Activities Report 25. World Meteorological Organization. Geneva, 1991.

World Meteorological Organization. *International List of Selected, Supplementary, and Auxiliary Ships.* WMO-47. Geneva, annually.

World Meteorological Organization. *Manual on Marine Meteorological Services,* vol. 1, *Global Aspects.* WMO No. 558. Geneva, 1981.

World Meteorological Organization. *Guide to Marine Meteorological Services.* 2d ed. WMO No. 471. Geneva, 1982.

PETER K. TAYLOR

MARITIME CLIMATE. A maritime climate occurs in regions where climatic characteristics are conditioned by their proximity to a sea or an ocean. Such regions, also known as *oceanic climates* or *marine climates,* are considered the converse of continental climates.

To explain the nature and location of maritime climates, it is first necessary to determine why climates over ocean and land differ. Thereafter, a closer look at the circulation of both the oceans and the atmosphere permits a clearer understanding of the differences between maritime and continental climates.

Land and Sea Differences. About 72 percent of the Earth is covered by water, with the Pacific, Indian, Atlantic, and Arctic Oceans containing much of that amount. The basins of the continents and oceans are unevenly distributed over the Earth's surface, with the Northern Hemisphere containing some 65 percent of the total continental surface. The vast area covered by water and its relative distribution play a very important role in the climates that occur, not only at sea but also, to varying degrees, on land.

Basic properties of water and land. The basic physical parameters that give rise to the differences in the thermal properties of water, air, and soil, are given in Table 1. Water has a much greater capacity for absorbing and storing heat energy than does other Earth materials. The basis of this difference is the high *specific heat* of water. Specific heat is defined as the amount of heat required to heat a unit mass by a unit of temperature. The temperature of water would be raised only one fifth that of the same weight of a given soil upon applying the same unit of heating.

The reflectivity of a surface, its *albedo,* varies over both land and water. Generally, the lighter, smoother, or less transparent the surface, the higher the albedo (a perfect reflector has an albedo of 100 percent.) Albedo values of land surfaces vary because of the great range of surface types; a desert surface may have an albedo of 25 percent while an asphalt parking lot may be only 5 percent. The major difference in albedo over water is the angle of inclination of the Sun. If the Sun is high in the sky very little reflection occurs, but as the angle decreases the albedo becomes greater. When the Sun reaches an inclination angle of 90 degrees (overhead), water albedo is less than 2 percent; at 10 degrees, it becomes 35 percent. [*See* Albedo.]

The amount of heat that can pass through a given thickness of material in a given time (using standardized units) is called *thermal conductivity.* It is roughly proportional to density of the material, causing the surfaces of dense material to heat slowly. If transparency is considered along with conductivity, then a marked difference is seen between land and water. *Transparency* is a measure of how much radiant energy can penetrate a surface. Since opaque materials like granite rocks have no transparency, radiant energy is transformed to heat energy at their surfaces. The transparency of water permits energy to penetrate to various depths, with most infrared energy absorbed close to the surface; with increasing depth other portions of the electromagnetic radiation energy are absorbed until, at about 10 meters, the unabsorbed blue portion of the spectrum is reflected back to give the sea its blue color.

The heat budget. To the identified physical parameters must be added the dynamic elements that make up the heat budget of a surface. In the upper part of

MARITIME CLIMATE. Table 1. *Selected thermal properties of air, soil, and water*

Material	Specific Heat	Albedo (percent)	Thermal Conductivity	Transparency to Sunlight
Air	0.17	6	very low	high
Soil	0.2	5–20	variable	low
Water	1.0	5–10	medium	medium

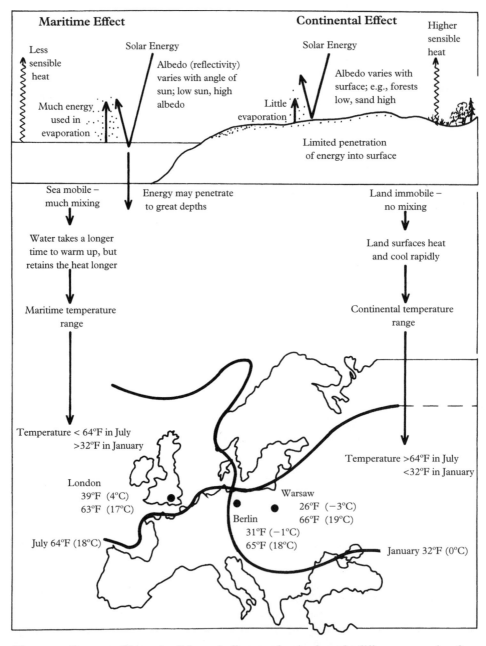

MARITIME CLIMATE. Figure 1. *Schematic diagram showing how the different properties of sea and land result in maritime and continental climates.*

Figure 1, the pathways of arriving solar energy over an ocean and continental surface are compared. The diagram illustrates the following two important concepts:

1. The amount of energy used to evaporate water is much greater over the ocean than the continent. This means that a large part of available energy goes to change water to water vapor. Because so much is used for evaporation, less is available for sensible heat (direct temperature rise). This contributes to air over the oceans being cooler than that over the continents. [*See* Sensible Heat.]
2. Oceans absorb radiation to a substantial depth; over land the penetration of radiant energy is often limited to a thin surface layer. The oceans are mobile and the absorbed energy may be advected (transported horizontally) or convected (transported vertically) to other areas by ocean currents and drifts.

The combined result of the physical and dynamic differences cause temperature variations in oceans to be much less than those of a land surface. The ocean is an equable environment that absorbs, transports, and releases stored heat to the atmosphere.

Oceans warm up more slowly than land masses, but because they have high heat capacity, they cool down more slowly. Temperature changes in the oceans are constrained while those over land masses, which warm and cool quickly, are extreme. This maritime effect will be most clearly seen in extratropical regions where seasonal extremes are expected. This effect is illustrated in the lower part of Figure 1. In tropical latitudes, seasonality is expressed more in terms of precipitation changes than in temperature cycles.

The Oceanic Influence. While maritime climates owe their characteristics primarily to the special properties of the oceans outlined above, there are other factors that play a very important role in the spatial variations that occur.

Surface ocean currents. Most surface ocean currents are wind driven, and like the wind are subject to considerable variation. The actual pattern of world ocean currents can be related to the global circulation of winds and the shape of land masses. Just as the subtropical high pressure zones are a dominant feature of the atmospheric circulation, so too are the oceanic gyres (oceanic circulation cells) associated with them. On viewing generalized maps of ocean currents of the world, the dominance of the subtropical gyres are easily identified. [*See* Oceans.]

The direction in which currents flow is important. Those flowing poleward, the warm currents, transport heat energy from low to high latitudes. Cold currents, flowing from high to low latitudes, bring cool water to warmer areas. The relative impact of this movement upon temperature is illustrated in Figure 2 which shows temperature distribution over the northern Atlantic Ocean. Note how the 10°F (−12°C) isotherm is located at about 50° north latitude on the east coast of North America but extends to almost 70° north latitude in western Europe. This is largely the result of the influence of the warm water current known as the *North Atlantic Drift.*

Upwelling. The upper layers of the ocean (the mixed layer), typically down to 100 meters, are mixed by the sinking of dense, cool, salty waters or by overturning motions from the wind and waves. Below this mixing layer there is a general decline in temperature (the thermocline) until the cold (near 0°C) bottom waters are reached. The strong temperature contrast between top and bottom waters creates a stable stratification that inhibits vertical mixing unless cold water is forced up to the surface. This upwelling occurs in a number of locations, often where surface water is swept away from the coast by winds.

The influence of upwelling, together with cold ocean currents, will effect the rainfall regime of some coastal regions. By inhibiting convection in the atmosphere, cold offshore waters help create desert climates along some coastlines. Although the dominant air is maritime, it is extremely stable and results in minimal precipitation. The Atacama and the desert of Baja California provide examples. [*See* Arid Climates.]

Maritime air masses. Air derives most of its thermal and moisture properties from the surface over which it originates. Air originating over the oceans, maritime tropical (mT) or maritime polar (mP) air, is generally characterized by (1) relatively small changes in the annual range of temperature in the source region and (2) a high moisture content derived from its origin over the oceans. These characteristics reflect the ocean properties already outlined. The relative frequency of maritime air received by a location will play a significant role in determining its characteristic climate.

Precipitation. The influences of maritime air massed provides coastal areas with abundant moisture in the atmosphere. Precipitation occurs so long as winds/storm systems remain onshore and an appropriate lifting mechanism occurs. Often, the extent of the marine influence is limited by mountains, which themselves often receive

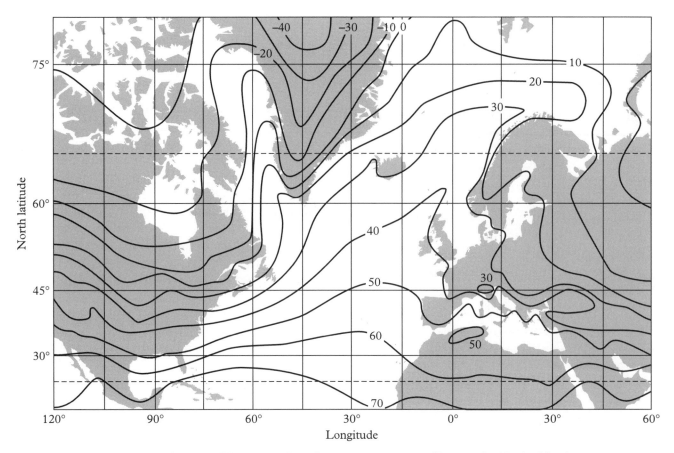

MARITIME CLIMATE. Figure 2. *Mean January temperatures (°F) over the North Atlantic Ocean.* Note the great difference between temperatures on the east and west coasts.

copious precipitation. It is possible, however, for regions adjacent to the oceans to be very dry. As noted above, dry coastal deserts occur where very stable air exists.

Climates near the Oceans. The discussion of the factors that influence climates over the oceans can now be related to the climates that occur in regions close to the oceans. It has been emphasized that temperature changes in oceans are constrained; such is also the characteristic of maritime climates.

The equable maritime climate. Small, isolated oceanic islands are most representative of the influence of maritime air masses and oceanic climate controls. Table 2 shows that a small range of annual temperature is characteristic of the places listed. Such, of course, is anticipated in the low latitude islands, but in stations as far poleward as the Falkland Islands, 51°42′ south latitude, the annual range of temperature is only 7.1°C. Precipitation can occur throughout the year at the island stations (especially in mid latitudes), the amount varying with location in relation to global pressure belts, winds, and local topographic features.

East and west coast climates. On continental areas, the influence of the oceans on climate is experienced in the west coast locations. This is a result of the general circulation of the atmosphere which, outside of the tropics, is dominated by a westerly air flow. Thus the most frequent air masses that influence the west coast of middle latitudes are from maritime source regions. Those on the east coasts experience air that moves from a continental source.

Ocean currents, which themselves reflect atmospheric circulation, influence both temperature and rainfall in some west coast locations. The cool air of the offshore water lowers annual temperatures while the resulting stable air provides little rainfall.

The west coast/east coast differences are demonstrated in Table 3, which provides comparative data for selected coastal and continental locations in the Northern Hemisphere. Of note is the much smaller annual range of temperature at west coast locations and the warmer conditions when compared by latitude. Note that Bergen, Norway (at about 60° north latitude), is colder in sum-

MARITIME CLIMATE. Table 2. *Climatic data for selected maritime stations*

Island*	Warmest Month	Temperature (°C)		Annual Precipitation (millimeters)
		Coolest Month	Annual Range	
Fiji	26.6	22.9	2.7	3026
Midway	25.9	18.4	7.5	1176
Waitangi	14.9	7.8	7.1	851
Falklands	9.1	7.8	2.0	651
Dutch Harbor	11.7	−0.3	12.0	1443

Suva, Fiji Islands, 18°08′ south latitude, 178°26′ east longitude, 5 meters elevation; *Midway Island*, 28°13′ north latitude, 177°22′ west longtiude, 13 meters elevation; *Waitangi, Chatham Isalnd*, 43°57′ south latitude, 176°34′ west longitude, 44 meters elevation; *Falkland Islands (Port Stanley)*, 51°42′ south latitude, 57°52′ west longitude, 52 meters elevation; *Dutch Harbor, Alaska*, 53°53′ north latitude, 166°32′ west longitude, 4 meters elevation.

mer and warmer in winter than Beijing, China, located some 25 degrees of latitude closer to the equator.

Continentality and oceanicity. To measure the effect of a continental land mass on climate (that is, a minimum impact of oceans), climatologists use the concept of continentality. As early as 1888, it was suggested that by measuring the average annual range of climate and adjusting for latitude, continentality could be quantified; the original approach was to cite it in percent (%) from 0 to 100 (minimum to maximum continentality) by

MARITIME CLIMATE. Table 3. *Temperature comparison of east and west coasts*

Location	Temperature (°C)		
	January Mean	July Mean	Range
	West Coast		
Tatoosh Island, Wash. 48°23′ north latitude	5.6	13.1	7.5
Eureka, Calif. 40°39′ north latitude	8.6	13.5	4.9
Bergen, Norway 60°24′ north latitude	1.5	15.0	13.5
Brest, France 48°24′ north latitude	6.1	15.6	9.5
	East Coast		
Portland, Maine 43°39′ north latitude	−5.7	20.1	25.8
Atlantic City, N.J. 39°22′ north latitude	1.6	23.9	22.3
Beijing, China 35°57′ north latitude	−4.7	26.1	30.8
Tokyo, Japan 35°41′ north latitude	3.7	25.1	21.4

use of constants. Later, workers used this idea as illustrated by the formula used by W. Gorzynsky:

$$K = 1.7 \left(\frac{A}{\sin \phi} \right) - 14,$$

where K is continentality, A is annual thermal amplitude (average annual temperature range), and ϕ is latitude. Many modifications of this formula have been suggested. Using the formula where continentality is given by

$$1.7 \left[\frac{A}{\sin (\phi + 10)} \right] - 14,$$

Thorshaven in the Faeroe Islands has a continentality index of 0 (Conrad, 1946).

While maritime climates may be expressed by low values of continental indices, oceanicity indices have also been proposed. The first, introduced by F. Kerner in 1905, is given by

$$O = 100 \left(\frac{T_o - T_a}{A} \right)$$

where O is oceanicity, T_o and T_a are the mean monthly temperatures for October and April, and A is mean annual range of temperature. Another, developed for ecological research in Norway, also includes precipitation:

$$K = \frac{N \, d_t}{100 \Delta}$$

where N is precipitation in millimeters per year, d_t is the number of days with mean temperature between 0° and 10°C, and Δ is the difference (in degrees Celsius) between the warmest and coldest month.

MARITIME CLIMATE. Table 4. *Oceanicity values for selected locations* *

Location	Average Annual Temperature Range (°C)	Oceanicity (percent)
Verkhoyanski, Russia	62.7	−4
Bismarck, N.Dak	33.8	2
Winnipeg, Manitoba, Canada	39.0	4
Tokyo, Japan	21.4	15
New York, N.Y.	23.9	17
San Diego, Calif.	8.0	34
Honolulu, Hawaii	4.3	49

*Using Kerner's index where the higher the value the more maritime the climate.

Although the scientific applicability of these empirical measures has been questioned (Driscoll and Fong, 1992), the measures do provide a general comparison of the oceanic and continental influences. This is perhaps best summed in Table 4, which provides oceanicity values for selected stations. The low values of maritime stations, reflecting the highly equable climate, present a significant contrast with stations within a continental area.

[*See also* Dynamics; *and* General Circulation.]

BIBLIOGRAPHY

Conrad, V. "Usual Formulas for Continentality and Their Limits of Validity." *American Geophysical Union Transactions* 27 (1946): 663–664.
Driscoll, D. M., and J. Fong. "Continentality: A Basic Climatic Parameter Examined." *International Journal of Climatology* 12 (1992): 185–192.
Kirkpatrick, A. H., and B. S. Rushton. "The Oceanicity/Continentality of the Climate of the North of Ireland." *Weather* 45 (1990): 322–326.
Minetti, J. L. "Continentality Indices: Methodological Revision and Proposition." *Erdkunde* 43 (1989): 51–58.
Van Loon, H., ed. *Climates of the Oceans.* New York, 1984.

JOHN E. OLIVER

MEDICINE. *See* Health.

MEDITERRANEAN CLIMATE. *See* Subtropical Climate.

MEDIUM-RANGE WEATHER FORECASTING. *See* Weather Forecasting.

MERIDIONAL CIRCULATIONS. As integral components of the atmospheric general circulation, meridional circulations are north–south overturnings that arise from the differential heating between the poles and the equator. The concept of a meridional circulation was first advanced by the eighteenth-century English scientist George Hadley. To explain the presence of prevailing trade winds in the tropics, he proposed a global-scale meridional circulation with rising motion in the equator and sinking motion in the poles, and a simple system of surface easterlies and westerlies like that shown in Figure 1. Hadley's concept of the meridional circulation was influential until new observations in the mid-nineteenth century began to cast doubt on it. One of the main discrepancies was that the prevailing westerlies in the Northern Hemisphere were observed to blow from the southwest, rather than from the northeast as Hadley had envisioned. In 1856 William Ferrel suggested, based on new observations, that there are three basic meridional circulations. He argued that Hadley's global equator-to-pole meridional circulation must be disrupted by a mid-latitude reversed meridional circulation with rising motion in the mid latitudes and sinking motion in the subtropics. This is known as the *Ferrel cell*.

Much subsequent work on the general circulation accepted the premise of the Ferrel cell, with minor modifications. At first the meridional circulations were regarded as zonally symmetric overturnings, independent of longitude, and it was thought that the zonally

MERIDIONAL CIRCULATIONS. Figure 1. *A schematic representation of the meridional circulation and accompanying system of horizontal winds.* (After Hadley, 1735.)

symmetric cells were real physical entities. By the turn of the twentieth century, there was increasing realization that the general circulation involves more than the zonally symmetric circulations, and that the meridional circulations were at best an approximate description of an averaged state of the general circulation. Many studies began to investigate the influences of cyclones and other disturbances on the meridional circulation. These studies, including the work of Jeffrey, Tor Bergeron, and Vilhelm Berjknes, was the beginning of the modern view of the meridional circulation. It is now recognized that the meridional circulations arise as a result of the conversion of potential energy derived from the equator-to-pole heating or temperature differences, which give rise to pressure differences; these in turn create kinetic energy in the form of a field of atmospheric motions. The meridional circulations play an essential role in the transport of energy, momentum, and water in the Earth's climate system. The term *meridional circulation* refers strictly to the zonally symmetric part of the general circulation in the meridional-height plane, but the asymmetric components of the motion field—such as eddy motions—contribute significantly to its distribution.

The modern picture of the mean meridional circulation is encapsulated in Figure 2. Here the term *Hadley circulation* refers to the part of the circulation with rising motion near the equator and sinking motion near the subtropics (30° north latitude). The rising motions are driven primarily by latent heat released by condensation of water via convection in the warm, moist equatorial regions. The strongest upward motions are concentrated in the heavy-rainfall region of Indonesia, the Amazon, and the Congo basin. Over the eastern Pacific and the Atlantic, the upward motions are confined to narrow zonal bands in the Intertropical Convergence Zone (ITCZ), usually between 5° and 10° north latitude.

Between the areas of rising motion are vast regions of compensatory subsiding motion. However, because the upward motions are much stronger, when averaged zonally, there is a net rising motion along the equator, as shown in Figure 2. [*See* Convection; Intertropical Convergence; Latent Heat.]

The relative humidity of the air throughout the tropics is maintained approximately constant by a near balance of heavy rainfall, surface evaporation, and moisture transported into converging regions. The rising hot air in the ITCZ also pushes the tropopause (the boundary separating the troposphere and stratosphere) higher in the tropics than elsewhere. Because the air loses much of its moisture and heat as it ascends, precipitates, and moves poleward, the descending air in the subtropics is relatively cold and dry. The Hadley circulation transports the higher angular rotational speed of tropical air upward from the Earth's surface in the tropics and poleward to between 30° and 60° latitude, where this component of the Earth's rotational speed is transferred from atmosphere to Earth in the sinking branch of the Hadley cell. As a result, the total *angular momentum* of the atmosphere is conserved over a sufficiently long period.

The buildup of westerly winds aloft is represented by the presence of a subtropical westerly jet stream. In the mid latitudes, the eddy motions arising from the instability of the mean flow provide further northward transport of heat toward the pole. The heat transport by eddies is concentrated in the lower troposphere in the region below 500 millibars pressure, or about 5 kilometers above the surface. In transporting heat and momentum poleward, the eddies also induce a reverse or thermally indirect circulation with rising motion in the mid latitudes and subsiding motion coinciding with the descending branch of the Hadley circulation in the subtropics. This indirect circulation, known as the *Ferrel cell,* is much weaker than the Hadley circulation.

The descending branch of the Hadley cell is also related to the polar front that separates the tropical air masses from the mid-latitude air masses. This region features a very strong zone of weather variability characterized by very strong surface temperature and moisture gradients. The contrast between the tropical warm, moist air mass and the mid-latitude drier, colder air masses is also reflected in a "break" in the tropopause, which occurs at a lower altitude for the mid-latitude air masses. The subtropical jet stream, which is found near the tropopause breaks, is a measure of the intensity of the interaction between the two kinds of air masses. Similarly, the

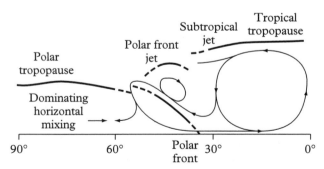

MERIDIONAL CIRCULATIONS. Figure 2. *The three-cell meridional circulation.* (After Palmen and Newton, 1967.)

contrast between the polar air mass and the mid-latitude air mass gives rise to a secondary break in the tropopause which coincides with the polar edge of the Ferrel cell, where a polar jetstream is found. The discontinuities in the tropopause that mark the equatorward and poleward limits of the Ferrel cell are also regions of strong synoptic-scale exchange of air masses between the troposphere and the stratosphere.

[See also Atmospheric Structure; General Circulation; Jet Stream; and Stratosphere.]

BIBLIOGRAPHY

Hadley, G. "Concerning the Cause of the General Trade Winds." Philosophical Transactions 29 (1735): 58–62.
Hoton, J. R. An Introduction to Dynamic Meteorology. 2d ed. New York, 1972.
Gill, A. E. Atmospheric-Ocean Dynamics. New York, 1982.
Palmen, E., and Newton, C. W. Atmospheric Circulation Systems. New York, 1967.

WILLIAM K.-M. LAU

MESOSCALE METEOROLOGY. See Meteorology.

MESOSCALE MODELING. See Models and Modeling.

MESOSCALE SYSTEMS. See Scales.

MESOSPHERE. The atmosphere of the Earth is divided into different regions, according to temperature as it varies with height. At Earth's surface is the region called the troposphere (to about 10 kilometers in height). As the distance increases away from the surface, the temperature decreases to a minimum (220 K) near 10 kilometers, at a region called the tropopause. The stratosphere starts above the tropopause with rising temperatures, reaching a maximum (300 K) in the stratopause near 50 kilometers. The mesosphere starts above the stratopause, with temperature decreasing to a minimum (140 K) at the mesopause between 90 and 100 kilometers. The thermosphere begins with increasing temperatures above the mesopause. The mesosphere is part of the middle atmosphere. [See Aeronomy.]

Over the upward extent of the mesosphere, atmospheric density decreases by a factor of more than 1,000, while pressure decreases from 1 millibar to 0.001 milli-

bars. These state variables at any given location vary with the seasons and with changes in incoming solar energy and energy from perturbations originating lower in the atmosphere. In general, there is a constant relative abundance of molecular oxygen and molecular nitrogen in the mesosphere. There are also a number of minor constituents here that influence the radiation balance in the atmosphere and have significant roles in the heating and cooling of the middle atmosphere. Many of these constituents—including water vapor, ozone, carbon dioxide, and nitrogen dioxide) are high-altitude greenhouse gases. [See Greenhouse Gases.] The major trace gas throughout the mesosphere is carbon dioxide, with a mixing ratio on the order of 100 parts per million. Water vapor is present in an amount exceeding the total of all other trace gases, with a mixing ratio of a few parts per million. Near the base of the mesopause, however, both water vapor and carbon dioxide become dominated by atomic oxygen, with a mixing ratio exceeding 100 parts per million.

Sunlight is the primary energy source for the mesosphere. Sunlight at wavelengths below 200 nanometers modifies ionization and dissociates molecules into atoms as low as the middle of the mesosphere. Therefore, variability in sunlight at these wavelengths has a major effect on the compositional state of the mesosphere.

Energetic particles entering the Earth's atmosphere and atmospheric electricity must also be considered. There is some evidence that very high-energy particles and X-rays enhance ionization and molecular dissociation, affecting not only ion and electron densities but also the chemistry involved with trace species such as ozone and nitric oxide as well. An electric circuit is set up between the ground and the region in and above the mesosphere where electron conductivity is appreciable. This highly-conductive region is called the ionosphere. The D region of the ionosphere is within the mesosphere; enhanced electron density in this region disrupts radio communications at HF frequencies. The ionosphere-to-ground electric circuit carries a current with an instantaneous value between 1,500 and 1,800 amperes. Any change in the conductivity of the mesosphere in response to changes in the Sun's ionizing radiation, particle energy ionization, or perturbations in the global magnetic field will affect the current flow through the middle atmosphere and consequently modify the physics, chemistry, and dynamics of the region. [See Ionosphere.]

The dayglow light stimulated by sunlight—a small contribution to visible daylight—is dominated by light emitted by oxygen molecules in the mesosphere. This light becomes weaker and chemical reactions among atmospheric constituents become more dominant as night progresses. Nightglow light becomes prominent in the mesopause in the 80 to 100 kilometer region. Although usually below the visible threshold, yellow-green light at 555.7 nanometers (due to atomic oxygen) and yellow light at 589 nanometers (due to sodium) appear in the nightglow spectrum. The molecular bands of the hydroxyl molecules (OH) and molecular oxygen collectively contribute most of the total intensity of the nightglow spectrum. The entire nightglow spectrum is, however, generally outside the visible region.

Understanding the mesosphere also requires the study of the physical motion of the atmosphere. Its particular composition and chemistry at any given time or location depends on how its various constituents are transported. Thus global circulation models for both horizontal and vertical motions are needed to specify the chemical state of the atmosphere. Tidal motions, internal gravity waves, and acoustic gravity waves play significant roles at different latitudes, longitudes, altitudes, and seasons The general circulation is composed mainly of a large-scale global cell caused by heating over the summer hemisphere and cooling over the winter hemisphere. The main features are an easterly zonal flow in the summer hemisphere and a westerly zonal flow in the winter hemisphere. Maximum velocities up to 100 meters per second occur near 60 kilometers. In general, the structure of the circulation in the mesosphere is much less chaotic than we experience below the tropopause. One example of coupling from below is seen in the events called *stratospheric warmings*. These events are observed only in the Northern Hemisphere. The 40 K warming of the stratosphere and cooling of the mesosphere results from sudden enhancements in the propagation of large planetary waves traveling from the troposhere to higher altitudes in the atmosphere. The temperature of the mesopause is highly variable. For example, in the summer polar regions, the mesopause temperature becomes very low (less than 140 K) coincident with the formation of very high altitude (84 kilometers) clouds. Called *noctilucent clouds* because of their appearance to a ground observer, they have been more recently viewed from satellites and renamed *polar mesospheric clouds*. One theory suggests that these clouds are the signature of enhanced production of water vapor in the middle atmosphere through the oxidation of methane, which began increasing in the middle atmosphere after the start of the Industrial Revolution.

[*See also* Atmospheric Structure.]

BIBLIOGRAPHY

Banks, P. M., and G. Kockarts. *Aeronomy*. 2 vols. New York, 1973.
Brasseur, G., and S. Solomon. *Aeronomy of the Middle Atmosphere: Chemistry and Physics of the Stratosphere and Mesosphere*. Dordrecht, Netherlands, 1984.
Herman, J. R., and R. A. Goldberg. *Sun, Weather and Climate*. New York, 1985.
Holton, J. R. *An Introduction to Dynamic Meteorology*. San Diego, 1992.
Kelley, M. C. *The Earth's Ionosphere, Plasma Physics and Electrodynamics*. San Diego 1989.
Meier, R. R. "Ultraviolet Spectroscopy and Remote Sensing of the Upper Atmosphere." *Space Science Reviews* 58 (1991): 1–185.
Shimazaki, T. *Minor Constituents in the Middle Atmosphere*. Tokyo, 1985.

GERALD J. ROMICK

MESOZOIC PERIOD. *See* Paleoclimates.

METEOROLOGICAL SYMBOLS. *See* Charts, Maps, and Symbols.

METEOROLOGY. In modern usage meteorology denotes the science of weather and includes only the study of atmospheric phenomena. In Classical Greece, however, the study of all heavenly phenomena was called *meteorologia*. The Greeks referred to any projectile falling earthward as *meteoron* ("thing in the air"); this class of celestial objects included raindrops, hail, and snowflakes as well as meteorites.

The meteorological thinking of the early Greeks was quite accurate at times. As early as 640 B.C.E. Thales ascribed the four seasons to variations in the position of the Sun. Hippocrates (fifth century B.C.E.) practiced a type of medicine that was a forerunner of modern biometeorology, the study of how weather and climate affect human health and psychology. Hippocrates advised physicians to ascertain prevailing winds and relative amounts of sunshine and cloudiness—sound counsel in caring for weather-sensitive patients.

Aristotle was the preeminent ancient meteorologist. [*See the biography of Aristotle.*] His account of weather and climate in *Meteorologica*, written around 340 B.C.E., was accepted as meteorological gospel until the Renaissance. Some of his beliefs, however, demonstrate his fallibility: for example, he believed that wind was the dry and moist exhalations of a breathing Earth. Yet his explanation of the process of rain was remarkably accurate and would essentially hold up to scrutiny even today:

The Earth is at rest, and the moisture about it is evaporated by the Sun's rays and the other heat from above and rises upwards; but when the heat which causes it to rise leaves it, . . . the vapor cools and condenses again as a result of the loss of heat and . . . turns from air into water. The exhalation from water is vapor. The formation of water from air produces clouds.

Aristotle's reference to rising air that had been heated from below and cooled into clouds aloft could well be interpreted as a forerunner of modern microscale meteorology, the study of chaotic atmospheric motions on a spatial scale ranging from a centimeter to a few hundred meters. Thermals—rising columns of air on which hawks sometimes soar—are an example of the scale of motion associated with microscale meteorology. There are millions of small-scale turbulent motions forming and dissipating each minute over the Earth. Some are caused by heating or cooling; others are produced when air is forced to flow over rough terrain like mountains. Though small compared to systems such as hurricanes, they are meteorologically significant because they can interact with and influence larger-scale weather processes. To predict their impact on larger-scale phenomena, these helter-skelter motions have to be expressed mathematically, but their chaotic nature makes this task very difficult. Therefore, approximations are made, assumptions that turn chaos into an idealized physical system that can be expressed in relatively simple mathematical terms. Such approximations are called *parameterizations,* a word unknown to Aristotle. [*See* Chaos.]

Archimedes (born 287 B.C.E.) is said to have shouted "Eureka!" as he immersed himself in a bath, suddenly realizing that a submerging solid body displaces an equal volume of water. Archimedes' principle is applied in modern research involving the buoyancy of warm, moist air within growing cumulus clouds.

Classical theories of weather and climate predominated until the Enlightenment, when philosophers such as René Descartes (1596–1650) cast doubts on received scientific wisdom. For example, the ancient Greeks believed that air was weightless, incapable of exerting pressure; in the early seventeenth century, however, Evangelista Torricelli, a friend of Galileo, demonstrated the pressure exerted by air. Torricelli followed this uncloaking of the atmosphere's invisible force with the invention of a mercury barometer, a key breakthrough in the quantification of atmospheric pressure.

Galileo made a water thermometer, and the Dutchman Cornelis Drebbel a wine thermometer, although both devices would have frozen at low temperatures. By 1670, mercury thermometers were introduced as a more reliable tool in climates having large annual temperature swings. The marriage of thermometers and barometers was a meteorological milestone for probing the invisible processes of the atmosphere.

Advances in chemistry, physics, and mathematics in the seventeenth century complemented the invention of the instruments of observation; these included Robert Boyle's 1661 discovery of a gas law relating pressure and density at a constant temperature, and Isaac Newton's 1666 development of *fluxions,* a form of differential calculus that would later prove indispensable in the complex mathematical solutions of meteorological problems. Newton's calculus became the cornerstone of the modern branch of science known as *dynamic meteorology*—the study of the motions of the atmosphere, ranging from the development and life cycles of low-pressure systems to the global circulation of winds. Dynamic meteorology utilizes the physical variables of pressure, density, temperature, and velocity to predict future states of the atmosphere. In this branch of meteorology, the fundamental laws of fluid mechanics and thermodynamics that govern atmospheric motions are expressed in terms of complex partial differential equations involving the physical variables, in a continually evolving mathematical discourse that serves as a lasting monument to Newton.

In the eighteenth century, weatherwise people like Benjamin Franklin made important contributions to the growing body of meteorological knowledge. In 1743, before his harrowing kite-flying encounter with a thunderstorm, Franklin was intrigued by a coming eclipse of the Moon over Philadelphia. His desire to view the eclipse was frustrated by the untimely arrival of thick clouds associated with a storm system. He later discussed the lost opportunity in a letter to a friend in Boston. When the reply came, Franklin learned that clouds did not arrive over Boston until after the eclipse, leading him to surmise that the storm system must have moved—a novel idea in an era when storms were thought to form and spend their lifetime over one location. [*See the biography of Franklin.*]

Another piece in the jigsaw puzzle of weather and climate fell into place around 1800, when Sir William Herschel discovered invisible infrared radiation, an important factor in weather and climate. Herschel allowed visible light to pass through a prism, a geometric glass apparatus that separates light into the primary colors of the visible spectrum. He bathed a thermometer in turn with violet, blue, green, yellow, orange, and red light. With each shift to the next longer wavelength of light, the temperature rose as Herschel had anticipated; but, after he moved the thermometer out of red light into the darkness surrounding his experiment, the temperature continued to rise, leading Herschel to postulate the existence of a new form of invisible energy. Herschel's discovery of infrared ("beyond red") radiation was critical for understanding the heat balance of the Earth–atmosphere system and the importance of the greenhouse effect.

Without greenhouse gases such as carbon dioxide, water vapor, and methane, which radiate infrared energy to Earth, the average global temperature would hover near $-20°C$ ($-4°F$) far from the twentieth-century average near $15°C$ ($59°F$). As concentrations of greenhouse gases increase owing to the burning of fossil fuels, sophisticated mathematical computer models of climate predict dramatic planetary warming in the twenty-first century. Though imperfect, these models employ budgets of infrared energy to calculate temperatures critical for assessing future trends of our planet's climate. Herschel could never have imagined the far-reaching impact of his discovery. [See Models and Modeling.]

In the nineteenth century, early weathermen did not know which way the wind blew around synoptic low-pressure systems. (Synoptic meteorology is a modern term that denotes the study of weather systems with a horizontal spatial scale of 800 to 8,000 kilometers and a lifetime of 1 day to 1 week; it is the most important branch of atmospheric science in issuing short-range weather forecasts). There were then two schools of thought. One theory, proposed by William Redfield and William Reid, held that winds blew in a circular, counterclockwise direction around low-pressure systems. The other camp, led by James Espy, believed that winds blew directly toward the centers of low-pressure systems, as if these were atmospheric "black holes" which drew air inward from all directions. Had the two camps combined their ideas, the problem would have been solved. The flow of air around a low-pressure system in the Northern Hemisphere spirals counterclockwise, inward toward the center of the low. This important advance in the study of storms did not come until 1859, when Elias Loomis unraveled the wind circulation of a storm by carefully examining a storm that had occurred in 1836.

Though Espy was only partly correct in his theory of the sense of wind circulation around lows, he later made a significant contribution to the growing body of meteorological truths. Espy identified the importance of *latent heat* in the formation of tall cumulus clouds. When invisible water vapor condenses into trillions of tiny cloud drops, a "hidden" heat is released. This latent heat, like the heat from a burner on a hot-air balloon, keeps rising parcels of moist air buoyant to great altitudes in the atmosphere, forming tall cumulus clouds and thunderstorms. This understanding led to other meteorological breakthroughs. In *The Philosophy of Storms,* Espy listed the atmospheric conditions conducive to the formation of tornadoes and waterspouts and accurately explained the process by which their ominous funnels become visible.

While the body of knowledge of how the atmosphere works was growing, the idea of a weather forecast—the bread and butter of modern meteorology—had not spread in many scientific circles. There were a few weather enthusiasts, such as Thomas Jefferson, who kept a diary of daily weather at Monticello, Virginia, in the late 1700s, but weather observations were not taken systematically in the United States until 1812, when the Surgeon General handed down an edict requiring army doctors to keep detailed records. By 1853, 97 army posts were keeping detailed records of daily weather.

Without knowing the present condition of the atmosphere, it is difficult if not impossible to predict the future state of the wind with any degree of accuracy. The growing network of weather observations led to the first weather maps, which were attempts to organize the great number of singular observations on one chart in order to glean some large-scale perspective of what organized regional data suggested about the prevailing weather pattern and its evolution. Compared to the modern computer graphics used on television weather presentations, the first weather maps appear primitive; but though devoid of essentials like warm and cold fronts, they were a beginning. Elias Loomis eventually improved the usefulness of these charts limiting the data to observations taken simultaneously at various sites, leading to the first weather analyses that included contours of variations in pressure and temperature (the prototypes of isobars and isotherms). By examining a sequence of such weather maps, early modern meteorologists soon devised a model of the life cycle of low-pressure systems. But actual weather forecasts were still not feasible because the distribution of weather observations took much too long. By the time data were collected and weather maps of storms drawn, the storm was long gone. To issue a meaningful forecast, weather data had to be processed faster. [See Charts, Maps, and Symbols.]

Forecasting moved closer in 1849, when observations were first transmitted by telegraph. By 1860, data from 500 localities became available over the wire, setting the stage for America's first weather forecasts. In 1870, the U.S. Congress created a national meteorological service under the auspices of the Army Signal Corps. The agency's duties included the collection of weather observations and the issuance of storm warnings as an aid to navigation along the nation's coasts and across the Great Lakes. The challenging art of operational forecasting had begun. The new weather service was also charged with the maintenance of the nation's climate records. In 1891 this agency was called the U.S. Weather Bureau and was administratively transferred to the Department of Agriculture, where its purview was expanded to include a wide range of forecasting for agricultural interests. Government-issued forecasts for aviation began in 1918 to serve the new industry of airmail service. [See Weather Services.]

One of the most important concepts in issuing an accurate forecast—the existence of warm and cold fronts—was undiscovered until the 1930s, although some aspiring weather forecasters had unwittingly alluded to these zones of abrupt temperature change as early as the 1800s.

Military strategy in World War I was largely restricted to the realm of trench warfare, and weather played a significant role in the movement and deployment of ground troops. In the immediate aftermath of the war, the Norwegian Jacob Bjerknes conceptualized a battle of cold and warm air masses near a boundary called a *front.* If cold air marched southward, forcing warm air to retreat, the front was termed *cold,* if warm air outflanked cold air and forced it to retreat, the front was *warm.* Stationary fronts marked the boundary where neither of the battling air masses held a distinct advantage. This conceptual model of fronts and their movement marked a major breakthrough in weather forecasting. [See the biography of Bjerknes.]

Modern military operations led to other meteorological discoveries. During World War II, radar was first used to detect enemy aircraft and ships by emitting radio waves (electromagnetic energy) and pinpointing targets when the energy was reflected back to the radar receiver. Much to the chagrin of the military, other objects—including raindrops from showers and thunderstorms—sometimes blocked the radar as it scanned for enemy aircraft. A byproduct of this wartime nuisance was the realization that radar could be used to pinpoint and scrutinize weather features. After the war, radar technology and surplus military equipment became available for civilian studies of the atmosphere. In 1959 the National Weather Bureau established a nationwide network of radars, and the fruit of military technology was finally harvested for operational forecasting on a large scale.

Although observations of wispy cirrus clouds streaking across the sky gave early meteorologists a hint of the existence of fast, high-altitude winds, they did not discover the jet stream—a narrow, serpentine river of air flowing fast from west to east between 7.7 and 10.8 kilometers altitude—until World War II. Proof came in the form of failed bombing missions undertaken by American aircraft flying westward over the Pacific toward Japanese targets. These airplanes had a maximum air speed of 175 knots, which should have overcome any head wind the atmosphere might generate. On one mission, however, the pilots encountered high-level winds of 175 knots, and found themselves literally standing still over the ocean.

The discovery of the jet stream was a milepost for dynamic and synoptic meteorology. The missing links in the development and movement of storms were on the verge of being found. For example, careful observations of the movement of storms over the years revealed that they moved at a speed approximately equal to one-half the speed of the jet stream. With the advent of high-speed computers in the 1950s, complicated, time-dependent equations from dynamic meteorology that govern the motions of the jet stream were solved, paving the way for a new branch of meteorology called Numerical Weather Prediction. Today, the 30- and 90-day outlooks issued by the National Meteorological Center located outside Washington, D.C., are based on computer predictions of large-scale features of the the Northern Hemisphere's jet stream (a series of ridges and troughs that girdle the globe, each pair of which has a spatial scale of 1,000 miles or more). [See Jet Stream.]

Following the development of the conceptual model of fronts, a number of major mileposts moved meteorology farther into the realm of exact sciences. Meteorologists had been limited to the lower atmosphere, but they needed to reach greater heights. Observations taken only at ground level often miss the high-altitude moving waves of air that play an important role in determining weather.

In the late 1930s, high-flying balloons carried aloft electronic instruments that were capable of remotely sensing temperature, pressure, and humidity at various altitudes and sending the results back to Earth via radio. Today a worldwide network of weather balloons called *radiosondes* are released each day at midnight and noon Greenwich mean time. These upper-air measurements are indispensable for operational and research meteorology. [See Instrumentation.]

The advent of high-speed computers in the 1950s revolutionized weather forecasting. The overwhelming flood of atmospheric data, coupled with the need for solutions to complex mathematical equations that governed motions of the atmosphere, defeated human calculators. Many millions of calculations were needed to obtain a reasonable short-term forecast. Where slide rules failed, computers excelled, and the age of numerical weather prediction began. [*See* Numerical Weather Prediction.]

On 1 April 1960, meteorologists finally got a glimpse of clouds from space when the first weather satellite, TIROS-1, was launched. This cosmic eye in the sky (the satellite was equipped with a television camera) orbited the planet at an altitude of 725 kilometers, giving eager meteorologists unique insight into the development, movement, and interaction of synoptic weather systems. Satellite meteorology grew rapidly. Soon polar-orbiting satellites were joined by geostationary satellites, vehicles in synchronous orbit with the Earth that were "perched" 31,220 kilometers above a fixed point on the surface, allowing continuous monitoring of a given region. Radiometers measured infrared radiation emitted from cloud

tops, allowing meteorologists to detect clouds at night. Moreover, colder tops meant higher clouds, giving forecasters a means to pinpoint tall and possibly severe thunderstorms. Today weather satellites can measure ground and ocean temperatures as well as trace atmospheric water vapor. They have the capacity to gauge rainfall rates in areas where observations are sparse, sometimes providing precious lead time for flood warnings. [*See* Satellite Observation.]

In addition to its immediate impact on operational forecasting, satellite imagery led to a new branch of science called *mesoscale meteorology,* with a spatial scale on the order of 1 kilometer to 100 kilometers. The study of gust fronts is an example of the scope of mesocale meteorology. From its vantage point in space a satellite can detect a thin ring of clouds around a large thunderstorm (Figure 1). The ring propagates away from the storm and is often the focus for later thunderstorms. This ring is the

METEOROLOGY. Figure 1. *Satellite photograph of thunderstorms and their gust front.* [*See* Gust Front; *and* Thunderstorms.]

thunderstorm's gust front, the leading edge of rain-cooled air that spreads out radially from the center of the storm.

More recently, Doppler radar has given meteorologists what X-rays, CAT scans, and magnetic resonance imaging have afforded the medical profession—the ability to peer into the inner workings of their subject. The principle by which these radars operate, known as the Doppler effect, was discovered by Christian J. Doppler in 1853. The classic example of this effect is that of a train moving toward a stationary observer beside the tracks with its whistle blowing. As the train approaches and passes, the observer will notice that the frequency of the sound of the whistle becomes lower. Near the center of a potentially tornadic thunderstorm, circulating winds (generally counterclockwise) carry raindrops first toward and then away from a nearby radar site. As the Doppler radar beam bounces off raindrops, the frequency of electromagnetic energy reflected back to the radar receiver varies, depending on whether the raindrops are moving away from or toward the radar. Thus the radar can detect rotation within the storm, giving meteorologists precious lead time to issue warnings even though a tornado may not yet have formed. [*See* Radar.]

During the 1990s, the National Weather Service is working toward the installation of Doppler radars across the United States. The agency's goal in the twenty-first century will be to concentrate on issuing detailed forecasts of severe weather for specific communities threatened by approaching storms, focusing on saving lives and preserving property.

[*See also* Climatology.]

BIBLIOGRAPHY

Ahrens, C. D. *Meteorology Today.* 4th ed. St. Paul, Minn., 1991.
Aristotle. *Meteorologica.* Translated by H. D. P. Lee, Cambridge, Mass., 1952.
Forrester, F. H. *1001 Questions Answered about the Weather,* New York, 1981.
Hughes, P. *A Century of Weather Service, 1870–1970.* New York, 1970.
Rinehart, R. E. *Radar for Meteorologists.* 2d ed. Fargo, N. D., 1991.

LEE MICHAEL GRENCI

MICROCLIMATE. Microclimatology has traditionally been the study of climate near the ground and of small areas (Geiger, 1965; Yoshino, 1975). The focus of microclimate is at the lower end of the space and time scales. [*See* Scales.] The emphasis in microclimatology is on phenomena occurring in a shallow layer of frictional influence near the Earth's surface, known as the *atmospheric boundary layer* (Oke, 1987). It is in this layer that humans, animals, and plants live, and here that meteorological conditions change most rapidly with height above the Earth's surface. A standard weather shelter housing temperature and humidity equipment used by weather networks worldwide usually consists of sensors placed approximately 4 to 5 feet above the surface. Under certain circumstances air temperatures on the surface and those 4 or 5 feet above it may differ by as much as 40°F. Thus the conditions actually experienced by plants and animals can vary significantly from conditions represented by standard weather reports. The role of microclimatologists is to understand these differences, since they affect a myriad of processes and applications in sectors such as industry, agriculture, the military, ecology and biology, chemistry, water resources, and weather hazard abatement and policy areas. Some examples of microclimates are listed in Table 1.

Micrometeorologists are primarily interested in fluctuations and short-term averages of meteorological properties in the boundary layer (Arya, 1988). Microclimatologists, by contrast, study long-term climatological averages, such as characteristic diurnal, seasonal, and long-term patterns in the near-surface layer of air and the subsurface environment. They may sample short-term conditions in the field, but they must couch these findings in terms of typical conditions of a local regio. Both subdisciplines rely on fundamental theories concerning heat, moisture, and mass exchanges between the Earth and

MICROCLIMATE. Table 1. *Examples of Earth surface conditions promoting various aspects of microclimates*

Hills, basins, valleys, bases of mountains
Sloping, rugged terrain features
Edges of adjacent surfaces (e.g., desert/cultivation)
Fog-prone areas
Cultivated or abandoned fields
Natural fields
Trees, forest, deforested areas
Oases
Rivers
Orchards
City surfaces/structures
Green, shelter, or thermal belts
Lakes, reservoirs, saline deposits
Sea or lake shores, ocean wave conditions
Forest on slopes
Grasslands on slopes
Meadows on slopes
Snow patches (flat or slopes)

atmosphere, and on the theory of atmospheric motion in the boundary layer of air near the Earth's surface. For example, both must study the process of turbulence, in which fluid flow takes on irregular and apparent random fluctuations.

Key Principles. At the core of microclimatological studies lies the understanding of exchanges of heat, moisture, and mass in the air–ground layer close to the surface, which may comprise snow, water, vegetation, bare soil, agricultural crops, rock, or surfaces affected by human use (anthropogenic effects). The field of microclimatology involves a systematic, structured view of these exchanges and related climate properties that represent typical patterns within the mesoclimates and macroclimates of the Earth.

The exchanges analyzed relate to conservation of energy—mass and momentum principles—and can be generally represented as:

$$Q^\star + Q_F = Q_E + Q_H + \Delta Q_S + \Delta Q_A \qquad (1)$$

where

$Q^\star$ = net all-wave radiation
Q_F = anthropogenic heat
Q_E = latent heat
Q_H = sensible heat
ΔQ_S = the net sensible heat storage change
ΔQ_A = the net heat advection.

Each term is a flux density (for example, in watts per square meter). Generally, for the typical mid-latitude surface and daytime conditions in summer, radiative energy surpluses accumulate on a surface and are dissipated by the other processes. Assuming that the surface is moist, a significant portion of this energy is dissipated as evapotranspiration into the lower atmosphere along with sensible heat, if the surface is warmer than the overlying air. Usually heat is gained below the immediate surface during daytime. At night, especially if it is clear, radiative losses occur from the surface upward toward space, and atmospheric fluxes of latent and sensible heat move downward to make up that deficit. Subsurface heat flow propagates toward the surface as well. Variations from this scenario commonly occur with alterations in weather and changing surface types and conditions. [*See* Evapotranspiration; Latent Heat; Radiation; Sensible Heat.]

Understanding the process of heat flow (ΔQ_S) into and out of the near-surface zones of the Earth can be an enormously difficult challenge, especially for surface conditions that vary rapidly both temporally and spatially;

examples include effects after a rain or snowstorm, or across complex surfaces such as cities, heterogeneous vegetation topography, or snow, ice, or water conditions. Understanding horizontal advective heat gains and loses (ΔQ_A) across the Earth's surface and developing a theory of how the boundary zones of the air are formed are among the remaining fundamental problems in the field. In addition, increasing focus on global environmental change has meant sharpened emphasis on studying the anthropogenic factors (Q_F) in analyses of microclimate. Processes such as desertification, drought, dust storms, deforestation, irrigation, salinization, and urbanization all have been significantly linked to human activities. These effects are increasingly being addressed with modern technological methods, including Geographic Information Systems, satellite systems and remote sensing, innovative instrumentation and special station arrays, airborne platform systems, and sophisticated cross-section analyses of the lower boundary zones of the atmosphere. Many of these analyses of anthropogenic processes are conducted at the microscale, at least initially, in terms of sampling and measurement.

Tables 2 and 3 list some controlling factors that affect the fluxes indicated in Equation 1. The problem for the microclimatologist is to measure or obtain estimates of these factors to produce model constructs of energy and moisture exchanges, in order to achieve a greater understanding of what determines the microclimatic conditions of an area.

MICROCLIMATE. Table 2. *Albedos and emissivities**

Material	Albedo	Emissivity
fresh snow	0.70–0.95	0.99
old snow	0.55	0.82
dry sand	0.25–0.45	0.91–0.95
wet sand	0.10	0.94–0.98
dry peat	0.05–0.15	0.97
asphalt	0.05–0.20	0.95
concrete	0.10–0.35	0.71–0.90
grass	0.20	0.90–0.95
water	var 0.03 ≤ but ≤ 1.0	0.99
deciduous forests	0.20 (leaves)	0.95–0.98
coniferous forests	0.10–0.15	0.97–0.98

Albedo is the ratio of reflected shortwave radiation to incoming shortwave radiation. *Emissivity* is the ratio of total energy emitted from a surface for a wavelength and temperature to that of a black body under the same conditions.
Source: Pielke and Avissar, 1990.

MICROCLIMATE. Table 3. *Thermal properties of selected materials*

Material	Thermal Conductivity watts/m²/K	Heat Capacity joules/kg/K	Density kg/m³	Thermal Diffusivity m²/sec
concrete	4.60	879	2.3×10^3	2.30×10^{-6}
rock	2.93	753	2.7×10^3	1.40×10^{-6}
ice	2.51	2100	0.9×10^3	1.20×10^{-6}
new snow	0.11	2092	0.14×10^3	0.20×10^{-6}
old snow	0.42–1.67	2092	$\sim0.6 \times 10^3$	$\sim0.70 \times 10^{-6}$
still air	0.025	1005	0.0012×10^3	21.00×10^{-6}
dry clay	0.25	890	1.6×10^3	0.18×10^{-6}
dry sand	0.30	800	1.6×10^3	0.24×10^{-6}
dry peat	0.06	1920	0.3×10^3	0.10×10^{-6}
water	0.57–0.63	4186	1×10^3	0.15×10^{-6}

Source: Modified after Pielke and Avissar, 1990.

Albedo and emissivity play a significant role in determining shortwave and longwave radiation in a microclimatic environment. The Leaf Area Index is the area of leaves per unit area of ground, taking one side of each leaf into account (Monteith and Unsworth, 1990; Rosenberg et al., 1983). The values of this index relate to radiative exchanges to, within, and from plant canopies; these values vary significantly among the Earth's biomes and agricultural crop surfaces. Soil properties determine flows of heat, gases, and moisture in the subsurface environment. For example, thermal conductivity, heat capacity, density, and thermal diffusivity drive the heat exchange up and down in the subsurface. Aerodynamic roughness length (the height above a surface datum at which wind speed is zero) plays a role in the turbulent exchanges of latent and sensible heat. It is proportional to the physical height of roughness elements on the surface, but it also depends on the shape and density distribution of those surface elements. For a given wind speed and temperature gradient, the magnitude and depth of convection from a surface would be increased over increasingly rougher surfaces. The total convective process also depends heavily on the temperature structure in the lower layers of the air.

For complex surfaces such as vegetation and crops, or urban environments, roughness lengths should increase with increasing height of canopies, but this ignores the effect of the geometry of the vegetation or structures. The shape and spacing of the elements determines the roughness as well. For example, research in plant microclimatology has illustrated that beyond some intermediate level of plant canopy density, the aerodynamic surface roughness becomes smoother and actually tends to decrease. Changes in initial wind speed over these surfaces complicate the understanding of the role of roughness length as well. Where vegetation is flexible to wind, a streamlining effect may be experienced and the surface roughness reduced.

Recent Development. Microclimatological measurements are often made from towers or aircraft (Dabberdt et al., 1993). Airplanes can provide data that are true spatial averages over surfaces such as forests and oceans. However, data have to be extrapolated to the surface, and problems in temporal representativeness may be encountered. Tower-based profile measurements can be made continuously and are more representative of surface fluxes. The use of instrumented vertical towers to observe flux changes with height involves a conflict between requirements for sensor height and site homogeneity. The source area of observed fluxes employing tower-mounted equipment must be homogeneous on the large scale to avoid horizontal advection and to permit identification of differences between true surface fluxes of interest and the measured fluxes from the tower (Dabberdt et al., 1993). Some microclimatologists use enclosures or chambers to isolate variables and study complicated flux relationships. Others have utilized source area mathematical models (SAMs) to identify the upwind areas contributing to fluxes measured at given heights on a tower in the boundary zone.

Whatever approach is utilized, previous assumptions or rules of thumb for placement of sensors with height in the boundary microclimate zone are being challenged with new methods and ideas. Extrapolating findings from microscale work to larger regions is still a challenge, and developing models that link to larger scales is a goal for the future. New weather service technologies, remote sensing, and use of geographical information system (GIS) techniques, in addition to theory-building in the discipline of micrometeorology, will assist in these endeavors.

[*See also* Instrumentation.]

BIBLIOGRAPHY

Arya, S. P. *Introduction to Micrometeorology.* San Diego, 1988.
Dabberdt, W. F., et al. "Atmosphere-Surface Exchange Measurements." *Science* 260 (1993): 1472–1481.
Geiger, R. *The Climate near the Ground.* 3d ed. Cambridge, Mass., 1965.
Monteith, J. L., and M. H. Unsworth. *Principles of Environmental Physics.* 2d ed. London, 1990.
Oke, T. R. *Boundary Layer Climates.* 2d ed. London, 1987.
Pielke, R. A., and R. Avissar. "Influence of Landscape Structure on Local and Regional Climate." *Landscape Ecology* 4 (1990): 133–155.
Rosenberg, N. J., B. L. Blad, and S. B. Verma. *Microclimate: The Biological Environment.* 2d ed. New York, 1983.
Schmid, H. P., and T. R. Oke. "A Model to Estimate the Source Area Contributing to Turbulent Exchange in the Surface Layer over Patchy Terrain." *Quarterly Journal of the Royal Meteorological Society* 116 (1990): 965–988.
Yoshino, M. M. *Climate in a Small Area.* Tokyo, 1975.

ANTHONY J. BRAZEL

MICROSCALE. *See* Meteorology; *and* Scales.

MIGRATION. *See* Animals and Climate; *and* Human Life and Climate.

MILANKOVITCH, MELUTIN (1879–1958), Serbian astrophysicist. During a career devoted to developing a mathematical theory of climate, Milankovitch calculated the seasonal and latitudinal variation of solar radiation (insolation) received by the Earth over hundreds of thousands of years. In 1941 he published his findings, that changes in insolation were responsible for the expansion and retreat of the Pleistocene ice sheets.

Born in the rural Slovenian village of Dalj in 1879, Milankovitch graduated from Vienna's Institute of Technology in 1904 with a doctorate in technical sciences. After serving as chief engineer for a large construction company, he returned to his native Serbia in 1909 to take a position as professor of applied mathematics at the University of Belgrade. He remained there for the rest of his life.

Milankovitch's theory rests on the following precepts: if we know the distance of a planet from the Sun, the shape of its orbit, and the angle between its rotational axis and the plane of its orbit, then we can calculate the seasonal insolation received by the planet at any latitude. From these calculations, the planet's climate and its seasonal variations can be predicted. Milankovitch published such calculations for the Earth, Venus, and Mars in 1920, under the title *Mathematical Theory of Heat Phenomenon Produced by Solar Radiation.* [*See* Orbital Variations.]

Milankovitch made his greatest contribution, however, by estimating past climates. For this he needed to know how the geometry of the orbit changes with time. This geometry varies in three quasi-periodic ways: change of the angle of axial tilt; change in the shape of the orbit from nearly circular to slightly elliptical; and the precession of the equinoxes. The last, caused by the gravitational tug primarily of the Sun and Moon on the tilted, oblate Earth, causes the Earth to wobble (much like a top), so that the season when the Earth comes closest to the Sun (perihelion) changes. The date of perihelion cycles through all four seasons in about 26,000 years. The other two cycles—with periods of approximately 40,000 and 100,000 years, respectively—are caused by the gravitational pull of the other planets, primarily Venus and Jupiter. [*See* Orbital Parameters and Equations.]

Previous authors had speculated that orbital variations might cause ice sheets to grow and decay. It was Milankovitch, however, who made the laborious calculations necessary to determine accurately how these perturbations change the latitudinal distribution of solar radiation through the year. Changes in precession and tilt alter the difference in temperature between seasons (such as warm winter/cool summer or cold winter/hot summer), but they do not change the total radiation the Earth receives from the Sun over a year; therefore, their control of ice age cycles was controversial during Milankovitch's lifetime. He assumed that radiation changes in some latitudes and seasons are more important to ice-sheet growth and decay than those in others. At the suggestion of the German climatologist Vladimir Köppen, he chose summer insolation at 65° north latitude as the most important latitude and season, reasoning that great ice sheets grew near this latitude and that cooler summers might reduce summer snowmelt, leading to a positive annual snow budget and ice-sheet growth.

We now know that the record of Pleistocene glacial cycles contains the periods of orbital variations, and that the timing of these rhythms is in harmony with Northern Hemisphere insolation changes. Consequently, orbital variations are now widely accepted as controlling the timing of these cycles. What is still not understood is how they do it.

Milankovitch never doubted the importance of insolation changes to glacial cycles. This shy scientist died in Belgrade in 1958, content in the knowledge that he had changed the course of climatology and established a new field—astronomical climatology.

BIBLIOGRAPHY

Hays, J. D., J. Imbrie, and N. J. Shackleton. "Variations in the Earth's Orbit: Pacemaker of the Ice Ages." *Science* 194 (1976): 1121–1132.

Milankovitch, M. "Kanon der Erdbestrahlung und seine Andwendung auf das Eiszeitenproblem." *Royal Serbian Academy Special Publications* 133. Belgrade, 1941; English translation published by Israel Program for Scientific Translations, published by U.S. Department of Commerce, 1969.

Zeuner, F. E. *The Pleistocene Period.* London, 1959.

JAMES D. HAYS

MILANKOVITCH THEORY. *See* Ice Ages; *and* Orbital Variations.

MIXED LAYER. *See* Maritime Climate; Oceans; *and* Planetary Boundary Layer.

MIXING RATIO. *See* Water Vapor.

MODELS AND MODELING. Models of weather and climate range from rather simple to very complex. The simplest are conceptual or qualitative models that merely try to sort out the mechanisms involved in an event or process. At the next level are models that employ just a simple conservation principle, usually at very large space and time scales. At the highest level are formulations based on detailed equations expressing the conservation laws of physics. [*See* Conservation Laws.]

A Simple Climate Model. We can illustrate how a model is cast into mathematical form by setting up a very simple model for the global average temperature. The model is based on the idea that the Earth is in a state of radiation balance. The rate of solar radiant energy inter-cepted and absorbed by the Earth should equal the rate at which energy is in turn radiated to space, on an annual average and averaged over many years. To a first approximation, the amount of heat absorbed per unit of time is the solar constant σ (1,370 watts per square meter) times the cross-sectional area of Earth (πr^2) times the fraction absorbed, called the *coalbedo*, a (about 0.70). The rate at which radiation flows to space from the Earth–atmosphere system is given by $4\pi r^2 (A + BT)$, where A (211.2 watts per square meter) and B (1.90 watts per square meter per degree Celsius) are empirical coefficients taken from satellite measurements, and T is the temperature in degrees Celsius. The factor $4\pi r^2$ is the surface area of the Earth. After cancelling the πr^2 factors we have

$$A + BT = \frac{\sigma a}{4}.$$

Solving for T, we obtain 15°C, a value close to the observed global average temperature. The success of this elementary calculation shows that the concept of radiative balance is a useful one, and that it might be expanded to apply in more detail to particular regions or seasons. At this stage, the modeling of weather and climate becomes much more complicated. When we ask about the balance of energy flows in a particular fixed volume in the atmosphere, we must specify not only the flows into and out of the volume owing to radiation, but many other flows as well. Specifying the detailed flows of energy into and out of a small test volume leads us to a very complicated system to model. Ordinarily, such a system cannot be solved by analytical procedures; instead, numerical algorithms must be implemented on high-speed computers.

Before proceeding, we can make one other simple calculation with the model of the global average radiation balance. By taking differentials, we can find that a 1 percent change in the solar constant leads to an increase of the global average temperature of 1.26°C, assuming that a and the other parameters are independent of temperature. It is worth noting that this value is approximately twice the value that would obtain if the Earth were a black body with no atmosphere. The enhancement of sensitivity to changes in the solar constant is due to a positive feedback in the system, embodied in the coefficient B. The feedback is thought to be due mainly to the increase in water vapor in a typical atmospheric column as the system warms. Hence, the atmosphere has internal mechanisms that operate to amplify the system's response to externally imposed perturbations affecting the radiation balance.

Time Dependence and Forecasting. Before passing to the more complicated case of smaller regions, let us consider temporal adjustment of the system when a perturbation is applied. If the rate at which heat energy is absorbed temporarily exceeds the rate at which heat is radiated to space, there will be a temperature increase with time. The rate of change of the temperature will be the difference between the absorbed and radiated rates, all divided by the effective heat capacity of the system C. This leads to an elementary differential equation to be solved for the evolution of the temperature $T(t)$, where t is the time

$$\frac{dT}{dt} = f(T)$$

where $f(T)$ is a function of the temperature. Although this equation with the form of $f(T)$ as given earlier is easily solved analytically, it can exemplify the way one would proceed if the equation were very complicated and perhaps too difficult to solve by hand.

We pose the problem as a finite difference equation that can be solved step-by-step using a computer. We imagine that the time is broken into a series of equal steps $t_0, t_1, \ldots, t_j, \ldots$. To each time corresponds a temperature $T_0, T_1, \ldots, T_j, \ldots$. The differential equation may be written approximately as

$$T_{n+1} = T_n + f(T_n)\Delta t$$

This finite difference approximation may be considered an algorithm for computing T_{n+1} if T_n is known. If we know the initial value of $T(t)$, then we can proceed to calculate an approximation for its value at the incremented time, and after that compute the next step, and so on. In the same way, we can make a forecast of the temperature at a later time based on knowledge of it at some initial time. This scheme is the root of numerical weather prediction and climate simulation. [*See* Numerical Weather Prediction.]

Laws of Momentum and Heat Change. In order to stimulate the temperature at the regional scale, one must consider a small test volume of air and take into account the flows of heat energy into and out of the volume. There will be many forms of energy flow into the test volume besides radiation energy. For example, if some moisture condenses into water droplets to form a cloud, there will be a conversion of latent heat of condensation into sensible heat, and this will be available to cause a temperature change. An important process bringing heat

into the test volume will be the advection (transport by wind) of air at a different temperature. In order to simulate such a process, we must simulate the wind field in some detail over a much larger volume than our test volume. It follows that we cannot faithfully simulate the temperature of a test volume of air without simultaneously simulating the field of winds. Besides winds, we must know about the moisture field of the surrounding air, because there may be condensation and evaporation processes influenced by the importation of air of a different composition and temperature. The occurrence or change of cloudiness in the volume will also affect the amount of radiation absorbed or reflected in the volume. Finally, the passage of wind and moisture through the volume must be such that the law of conservation of matter is strictly observed. Hence, to make a numerical simulation of the evolution of atmospheric temperature in a small test volume, we must simultaneously simulate the evolution of a large number of other fields. The equation governing the evolution of one field will depend on the current local values of all the other fields. The essential variables that must be retained are the temperature, pressure, and all three components of velocity, density, and moisture. In some numerical experiments, other fields might be carried along interactively in the calculation, such as dust concentration or chemical composition.

The thermodynamic heat equation is a generalization of the one just considered for the evolution of the temperature. The law of physics invoked here is the first law of thermodynamics. Because the internal energy for an ideal gas is proportional to the temperature, this law can be converted to an equation for the rate of change of the local temperature. Changes in temperature in a test volume can be brought about by various forms of heating, such as net radiational heating, advection, and condensation heating; in addition, the gas can be warmed by compression as it moves from one altitude to another.

The rates of change of the components of the wind field are prescribed by Newton's second law, which states that the rate of change of momentum of an object—for example, a parcel of air—is given by the vector sum of forces acting on the object. The forces that act on a parcel of moving air are gravity, viscous stress, pressure gradients, and the Coriolis force. Gravity pulls the parcel toward the Earth's surface. If the parcel has less density than the surrounding air, and therefore less mass to be affected by gravity, it will be displaced upward by denser adjacent parcels that tend to fill in below it. This so-called *buoyant effect* plays a major role in sending light air upward, expanding as it goes because of the lower pressure of the environment at the higher elevation. As rising air expands and therefore cools, it may reach its dew point, and water droplets will begin to form. The process of droplet formation leads to the release of sensible heat, and therefore additional buoyancy. The release of latent heat during cloud formation is an important source of atmospheric kinetic energy. Thus the gravitational force, through buoyancy, can lead to moisture conversion processes that form not only clouds but also precipitation and kinetic energy.

Viscous stress is the force of friction between neighboring volumes of air as they slip past one another. At the scales of resolution of numerical models, this is accomplished not by exchanges of individual molecules but rather by eddies of air passing from a neighboring parcel of one speed into one of another speed. This exchange of momentum results in a net force acting oppositely on each parcel separately. The faster air will tend to decelerate, while the slower neighboring air will tend to accelerate.

The existence of a pressure gradient causes a net force on a parcel of air because the external air pressure (force = pressure × perpendicular area) forcing inward on one end of the parcel is different from the pressure forcing inward on the opposite end of the parcel, so that there is a net force on the mass enclosed by the volume. When all 6 sides of an imaginary cube surrounding the parcel are considered, the pressure gradient force results in a net tendency for the parcel to accelerate toward the environmental air at lowest pressure. The simplest example of this is the vertical pressure gradient force. The pressure of environmental air falls nearly exponentially as one ascends into the atmosphere. According to the preceding argument, a thin slab of air will therefore experience an upward force. This force would lead to upward acceleration, except that it is nearly always balanced in the atmospheric column by the gravitational force. In fact, the exponential form of the pressure's decrease (assuming that the temperature is vertically uniform) can be calculated from such an equation, sometimes known as the *barometric law*. [*See* Barometric Pressure.]

The Coriolis force is the force exerted on a moving parcel by the rotation of the Earth. Newton's second law holds only for an inertial coordinate system. When we insist on expressing it in a rotating coordinate system—such as on the surface of the rotating Earth—we encounter some additional terms, including the centrifugal force and the Coriolis force. The centrifugal force from this source is small and can be included effectively in the gravitational force term. The Coriolis force applies to

moving air and is directed perpendicular to the motion of the air, to the right in the Northern Hemisphere and to the left in the Southern Hemisphere. For most purposes, the Coriolis force is small for spatial scales less than 1,000 kilometers and for times less than 24 hours. For longer space and time scales, however, it is important in controlling the shape of the wind field and, therefore, the advection of air properties. [*See* Coriolis Force.]

In addition to the thermodynamic heat equation and the momentum conservation equations (Newton's second law), the equations of motion must include conservation laws for the individual constituent masses. Finally, the equation of state must be included to relate the pressure, density, and temperature of air. [*See* Primitive Equations.]

Atmospheric General Circulation Models. If the partial differential equations just described are cast into finite-difference form along the lines of the global model mentioned earlier, we can begin to see how the atmospheric motions might be simulated numerically. First, consider a grid covering the surface of the Earth and having a number of layers in the vertical. Typical climate models of the 1980s had a horizontal resolution of about 6° longitude by 6° latitude (perhaps less dense near the poles), with 10 levels in the vertical. This leads to roughly 18,000 grid points where the fields are to be calculated. If we retain 8 variables at each grid point, we will need to make 144,000 evaluations at each time step. To make a climate simulation lasting 25 years with time steps of a half hour will require 438,000 time steps. If 100 arithmetic operations are required to produce one of the fields at a point for each time step, we can expect that in the climate simulation, we will require 6×10^{12} calculations. The world's fastest computers can perform calculations of this type at a rate of about 10^9 per second. A climate simulation then takes about 6,000 seconds, or 100 hours of computer time. The amount of computer time should increase as the fourth power of the resolution in the north–south direction, because increases in resolution in one direction must be accompanied by corresponding increases in the other directions and also in time. One hundred hours of supercomputer time is a typical requirement for a climate simulation employing an atmospheric general circulation model. [*See* General Circulation Models.]

Atmospheric general circulation models are used in the simulation of past and future climates. Many experiments have been conducted with these models to help us understand past climates, such as that of an ice age. Simulation of past climates and comparison with paleoclimatic data have increased our confidence in the ability of these models to simulate real climates. Currently, many other research groups are considering the problem of increasing concentrations of greenhouse gases on future climate, with the help of atmospheric general circulation models. [*See* Greenhouse Effect; Paleoclimates.]

A special type of general circulation model is the *stratospheric model,* which focuses on large horizontal-scale motions above about 10 kilometers. In these models, only a crude representation of the lower atmosphere is included. Of special interest for the stratosphere are the concentrations of ozone and other chemical constituents. Hence, in these simulations there is often an explicit accounting for the populations of various chemical and aerosol species that interact with ozone.

Use of a general circulation model for weather forecasting requires a different set of priorities. The desired forecast covers a few days and is to be broadcast within about 4 hours of the model's initialization. This allows us to employ a much finer grid and time step. Weather-forecasting models of the early 1990s are at a resolution of about 100 kilometers and time steps of a few seconds. In producing a forecast, we encounter problems different from those in climate simulation. For example, the observed values of the fields must be entered into the initialization procedure. This is by no means a trivial task; the observations come in from a global network and are collected at odd times and from a variety of sensor types, with measurements derived from radiosondes, aircraft instruments, ground station instrumentation, and satellites. These data must be smoothed and adjusted for internal consistency before the forecast simulation can begin. Often the initialization and data preparation may occupy as much computer time as the simulation itself.

Smaller-scale Atmospheric Models. Considerable research effort is currently being expended to develop models at space-time scales below those of the general circulation models. These include mesoscale models, convection models, and boundary-layer models.

The mesoscale includes processes larger than those governing the growth of a cloud but smaller than those governing a typical mid-latitude cyclone system. This is the scale in which the Coriolis force becomes an important factor. Systems of interest include squall lines, fronts, tropical storms, tropical cloud clusters, and severe weather outbreaks. The mesoscale model is usually solved on a rectangular numerical grid of a few kilometers in spacing and a few minutes in time step. The model must be embedded as a patch in the larger-scale flow of air. The boundary conditions at the edges of the patch must either be prescribed by typical data or perhaps given by the output of a general circulation model at coarser resolution. A major problem in mesoscale modeling is just how these boundary conditions are specified. The mesoscale model is a research tool at present, but it is envisioned that this class of models will someday play a role in operational forecasting, if researchers can figure out how to initialize the model through the use of radar or other remote observing systems capable of delivering information at the proper space and time scale, at the required speed to make the forecast useful.

Cloud models follow the evolution of an individual cloud or a small cluster of clouds. The spatial resolution of these models must be at or below 1 kilometer, and the temporal resolution must be in the range of a few seconds. Cloud models operate at scales where the Coriolis force can usually be neglected. These models are research tools used to further our understanding of the cloud convection process and the production of precipitation. The interiors of individual clouds are complicated by the production of many different phases of water, including a variety of ice forms. [*See* Clouds.]

Closely linked with cloud models, and at about the same scale, are boundary-layer models. These models are concerned with the interchange of energy, momentum, and mass between the atmosphere and the underlying planetary surface. The boundary layer is typically a turbulent zone in which material and properties can be stirred rapidly from the surface to the layer's poorly defined top. The boundary layer is typically 1 to a few kilometers thick, depending on latitude and the local temperature profile (called *vertical dependence*). Although useful as a research tool to improve our understanding of the general circulation of the atmosphere, boundary-layer models also have an important operational application in forecasting and simulating air quality and air pollution. [*See* Planetary Boundary Layer.]

Linking Outside Systems. As the time scale of our desired simulation lengthens, we must consider the influence of neighboring systems that are usually not of interest in weather forecasting. For example, in decadal-scale climate simulations, one cannot ignore the influence of the oceans. When conditions in the atmosphere change, so do those in the ocean, leaving an altered boundary for the atmospheric simulation. Therefore, current simulations of climate for the greenhouse problem emply simultaneous simulations of the evolving ocean. Among the other systems that have to be included are the evolution of sea ice and, on longer time scales, the changes in land-based ice. Finally, the biosphere becomes a factor, because its distribution of constituents can affect the reflectivity of the surface and the rates at which properties like moisture can be exchanged with the air above.

Future of Atmospheric Modeling. All the types of models mentioned here are in a state of evolution. As we learn more about the processes governing the motions and transformations of the atmosphere, both from observations and from increasingly detailed numerical experiments, we will continue to improve the models. In parallel with the improvement of our understanding of the physical processes, computer technology is rapidly changing. Tomorrow's computer architecture will be quite different from today's, with an added emphasis on the parallel processing of different parts of a numerical simulation. Hence, the future of modeling is tightly interwoven with the future of computer architecture.

BIBLIOGRAPHY

Schlesinger, M., ed. *Physically-Based Modelling and Simulation of Climate and Climate Change.* 2 vols. Boston, 1988.
Wallace, J. M., and P. V. Hobbs. *Atmospheric Science: An Introductory Survey.* New York, 1977.
Washington, W., and C. Parkinson. *An Introduction to Three Dimensional Climate Modeling.* Mill Valley, Calif., 1986.

GERALD R. NORTH

MONSOONS. The word *monsoon* comes from the Arabic word *mausam*, meaning "season." A monsoon is defined as a seasonal shift in direction of the surface winds between summer and winter. This seasonal shift was very important for sailing vessels on the Arabian Sea, part of the Indian Ocean. Arab traders used the summer monsoon winds to steer along the west coast of India, and the winter monsoon to move along the east coast of Africa.

Meteorologically, the monsoon influences most of the Old World. The domain of the monsoon is generally delimited by the reversal of the surface winds between the winter and the summer seasons. Figure 1 shows the outline of this domain, within which the winds alternate between southwesterly during the northern summer and northeasterly during the northern winter.

The annual cycle of the monsoon contains two strong components. One is the summer monsoon, which lasts roughly from June through September. The second component, the winter monsoon, lasts roughly from December through February. During the summer, the major circulation features of the lower troposphere include the following. The *Southern Hemisphere trade winds* along roughly 15° south latitude blow from east to west at a speed of roughly 10 meters per second. The *Mascarene High* is a high-pressure area that resides over the Mascarene Islands (includes Mauritius and Madagascar). *Cross-equatorial flow* along the Kenya highlands is a southerly stream of the monsoon current. The *Somali jet* is a continuation of this southerly stream as it exits eastward into the Arabian Sea. Here the maximum winds are of the order of 30 meters per second, at a height of roughly 1 kilometer above sea level. This stream enters the Indian subcontinent, bringing in moisture from the Arabian Sea and across the equator from the Southern Hemisphere trade winds. To the north of this westerly current, over the land area near 25° north latitude, resides the *monsoon trough*, which is characterized by heavy rain. The counterclockwise flows around this monsoon trough

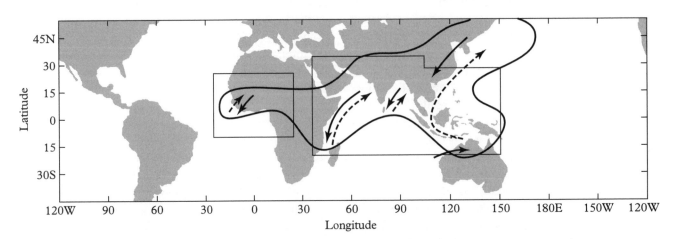

MONSOONS. Figure 1. *Domain of monsoons.* The map shows the reversal of surface wind direction in northern summer (broken arrows) and northern winter (solid arrows).

interact with the mountains of eastern Asia (the Himalayas, the Arakans, and the Khasi Hills) resulting in ascent of moist monsoon air and heavy orographic (mountainous) rains. [*See* Orographic Effects.] Above and to the north of the monsoon trough the troposphere is warm as a result of the heat fluxes from the elevated Tibetan Plateau and from the latent heat released by the monsoon rains. In the upper troposphere between roughly 50° to 130° east longitude and 15° to 40° north latitude, an anticyclonic circulation called the *Tibetan High* prevails; the clockwise flows around this anticyclone contain an easterly jet stream in their southern flank, called the *tropical easterly jet*.

During the winter monsoon, the roles of the land and ocean are reversed. The Mascarene High is replaced by the *Siberian High*. The cross-equatorial flow and the Somali jet are replaced by strong northeasterly cold surges of wind along the Pacific coast of Asia and the South China Sea. The monsoon trough and its heavy rains reside over the so-called *maritime continent*, comprised of Indonesia, Borneo, Malaysia, the South China Sea, and the Java Sea. Over the upper troposphere, a high pressure area over the western Pacific replaces the Tibetan High; the tropical easterly jet is replaced by a westerly jet stream in the northern periphery of the high near 25° north latitude.

A spectacular onset of monsoon rains occurs during the commencement of both the summer and the winter monsoons. Factors related to the sudden change are the arrival of a wind convergence line from the south (or north) during the summer (or winter) monsoon. This brings with it a deep moist layer that accompanies the onset. Other features include a strengthening of the westerly low-level flow off the Somali coast and a strengthening of lower tropospheric winds, which are southwesterly during the summer and northeasterly during the winter monsoon. The date of onset varies from place to place over the monsoon belt: the earliest onset of summer rains is experienced over Myanmar (formerly Burma). The onset over India, southern China, and southern Japan usually occurs about a month after the onset over Myanmar.

The annual cycle of the monsoon is characterized by the migration of a rainfall belt from Indonesia to the eastern foothills of the Himalayas, followed by a reverse migration. This seesaw from January to August and to the following January defines a principal axis of the monsoon. This is the axis of the major monsoon heat source, where latent heat is released over regions that experience well over 250 centimeters of rainfall during the rainy season. In addition, there are other components of monsoon rainfall that follow the migration of a broad-scale monsoon trough, which extends quasi-zonally from the Arabian Sea to the South China Sea. Southwesterly flow to the south of the monsoon trough is very moist; this flow interacts with local mountains, including the Western and Eastern Ghats of India, the Assam Hills, the Arakan Range of Myanmar, the Indochina highlands, the Sumatra Hills, and the mountain ranges of Borneo, New Guinea, and Malaysia. The seasonal summer rainfall over the Indian subcontinent is very large over mountainous regions.

A plethora of weather disturbances move through the monsoon domain. *Monsoon lows* (on scales of the order of 3,000 kilometers, with frequency of roughly every 5 days), and *monsoon depressions* (on scales of the order of 3,000 kilometers, with frequency of roughly every 15 days) are among the more common rain-producing systems. The monsoon lows are weaker in amplitude, with winds of the order of 10 meters per second and rainfall amounts around 5 centimeters per day. The monsoon depressions that generally form over the monsoon oceans move westward at a speed of roughly 5° longitude per day; the maximum winds can be as high as 15 to 20 meters per second, and rainfall amounts as high as 15 centimeters during a disturbance passage. During the onset of the monsoon, a hurricane-like disturbance called the *onset vortex* has been known to form over the Arabian Sea and over the Bay of Bengal. This storm, with hurricane-force winds of around 40 meters per second, carries a deep moist layer in its southern flank. The northward motion of this storm brings the deep moist layer northward, ushering in an onset of the monsoon rains over the Indian subcontinent. The winter monsoon contains most of the same family of disturbances. Monsoon lows enter the South China Sea from the western Pacific Ocean with similar scales and frequencies. Monsoon depressions form along the monsoon trough, producing heavy rain events. The onset of the winter monsoon, especially over northern Australia, is also preceded by the formation of hurricane-like disturbances along the monsoon trough.

A time scale important for the monsoon is around 40 days, the typical length of its wet and dry spells. Both the summer and the winter monsoon exhibit such alternation. Wave motions on this time scale exhibit an eastward propagation on the planetary scale. On the regional scales of the summer and winter monsoons, slow meridionally propagating waves—on scales of the order of 3,000 kilometers and speed of propagation around 1° latitude per day—appear to originate over the ocean and propagate meridionally over the monsoon land masses. They seem to be driven by cumulus convection and ocean–atmosphere coupling via fluxes of moisture.

The single most important time scale of interannual variability of the monsoon is that of the El Niño–Southern Oscillation, which occurs irregularly in cycles of 2 to 6 years. El Niño is characterized by the occurrence of warm sea-surface temperatures over the central and eastern equatorial Pacific Ocean. Cumulus convection and rainfall over the warm El Niño water is a characteristic feature. The convective areas exhibit a general area of rising motion over the warm waters. A branch of the compensating sinking motion occurs over the monsoon region to its east. Figure 2 shows a history of roughly 100 years of monsoon rainfall over all of India, expressed as normalized percent departures from normal shown by the vertical bars; the darkened bars identify El Niño years. This illustration shows that most El Niño years were years of negative departures, or years of extremely deficient rainfall. This aspect of interannual variability is also a pronounced feature of the winter monsoon rainfall over regions such as northern Australia, Indonesia, Malaysia, and southern India. [See El Niño, La Niña, and the Southern Oscillation.]

Currently, statistical multiple regression methods are being used by several research groups to determine parameters that can be used to predict the seasonal rainfall. Generally the parameters are selected from an antecedent period so that the regression will have predictive utility. The summer monsoon rainfall over the entire continental area of India has been subjected to detailed studies by the Indian Weather Service using these regression methods. Their studies suggest the importance of

MONSOONS. Table 1. *Monsoon rainfall predicted by the Indian Weather Service compared with observed rainfall*

Year	Percent of Normal Rainfall Predicted	Percent Observed
1989	102	101
1990	101	106
1991	94	91
1992	98	98

several parameters, including pressure, temperature, snow cover, winds, and sea-surface temperature anomalies over different locations worldwide. Using this method, they offered seasonal forecasts for 1989 to 1992 with a great measure of success, as shown in Table 1. The following parameters were considered most important for seasonal predictions: Argentina pressure (April); Northern Hemispheric pressure (January and February); Himalayan snow cover (January to March); Darwin pressure (Spring); and Indian Ocean equatorial pressure (January to May).

Given the sea-surface temperatures (based on observations from satellites and shipboard measurements), global climate models have explored the possibility of monsoon forecasts. The global models include dynamical and physical effects including the role of mountains and sea-surface temperatures. The Pacific Ocean's surface temperatures appear to have a large impact on monsoon evolution; the role of the Indian Ocean appears to be somewhat less important. Climate models are beginning to show a reasonably slow response of the atmosphere to oceanic forcing, and the dry and wet monsoon seasons are being predicted with some accuracy; how-

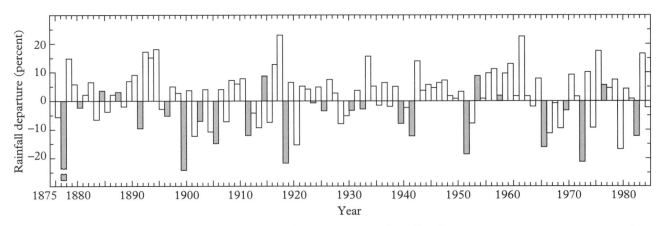

MONSOONS. Figure 2. *Year-to-year variability of monsoon rainfall.* Variability is shown as percentage departure from average. Darkened bars coincide with occurrences of El Niño.

ever, these are still not considered true predictions because the sea-surface temperatures are prescribed. Coupled ocean–atmosphere models would be needed to handle complete predictive capability, and this is an area of active research at the present time.

[*See also* Coastal Climate; *and* Maritime Climate.]

BIBLIOGRAPHY

Chang, C.-P., and T. N. Krishnamurti, eds. *Monsoon Meteorology*. New York, 1987.
Fein, J. S., and P. L. Stephens, eds. *Monsoons*. New York, 1987.
Krishnamurti, T. N., ed. *Monsoon Meteorology*. Special issue of *Pure and Applied Geophysics* 115 (1977): 1087–1529.
Lighthill, J., and R. Pearce, eds. *Monsoon Dynamics*. Cambridge, 1981.

T. N. KRISHNAMURTI

MOON. The Earth and its sole natural satellite, the Moon, were formed about 4.6 billion years ago. The orbit of the Moon around the Earth is elliptical, varying slightly from that of a circle with a radius of 239,000 miles. At its point of closest approach to the Earth (smallest perigee), the Moon is 221,500 miles away, and at its greatest distance (largest apogee) it is 252,700 miles away. There are various periodic motions associated with the Moon's orbit due to its slant with respect to the Earth's path around the Sun and also the wobble of the plane of its orbit. [*See* Orbital Variations.]

The joint gravitational attraction of the Moon and Sun on the waters of Earth's oceans produces the ocean tides; they in turn influence climate. This joint attraction also produces weak tides in the atmosphere that are insignificant in their influence on climate. Oceanic tides are well understood and have been accurately calculated since the nineteenth century.

One measure of tidal activity is the annual number of extremely low and high tides at a given location. This activity changes greatly from year to year, as can be seen in Figure 1, which shows the tidal activity series in Saint John Harbor in the Bay of Fundy in southeastern Canada. The dotted line indicates the activity, based on daily tide tables from 1899 to 1992. The solid curve is a least squares fit to the activity data using sine curves with periods that are related to the Moon's orbit. This analytical function is used to extend the data both backward and forward in time so that it may be compared with temperature records.

Loder and Garrett (1978) showed that when tides are high, tidal currents produced by the motion of ocean waters over bottom irregularities produce extremely strong turbulent columns of water.

This action is strong enough to bring considerable amounts of bottom water to the surface despite any effects of thermal and salinity gradients. Examples of these turbulent columns reaching the sea surface have been photographed. Where the ocean water is stratified, with layers of differing temperature and salinity at the surface and the bottom, this tidal mixing will produce a noticeable change in surface water conditions, with consequent effects in the climate.

Climate effects of tidal mixing are strikingly exemplified in the region of the Arctic Ocean, which is unique in

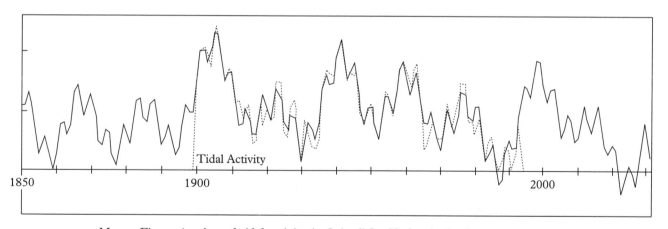

MOON. Figure 1. *Annual tidal activity in Saint John Harbor in the Bay of Fundy.* The vertical axis shows the number of extremely low tides below 1.1 feet above datum (a fixed reference marker in the bay) occurring each year. Marks on the vertical scale are at intervals of 50 counts. The dotted curve is the count as read from the daily tide tables. The solid curve is a fit with lunar tidal periods.

that it always has a cold surface layer of low salinity with a warmer, more saline layer below. It is also shallow, so that high tidal activity and consequent mixing increase surface temperature and salinity and lower the temperature and salinity below. The decadal temperature maps of Jones and Kelly (1981 and 1982) show high-temperature areas near the Arctic Circle. These are prominent only in the 1930s and 1940s (also times of high tidal activity) in all seasons.

The histograms in Figure 2 display surface air temperature departures, in degrees Celsius, from the reference period (1951 to 1970) mean for the Northern Hemisphere (a) from 1851 to 1990 and for the Southern Hemisphere (b) from 1858 to 1990. Notice from Figures 1 and 2 that changes of Northern Hemisphere temperature accompany times of great tidal activity. The high

peak in air temperature near 1940 changes with latitude; it is small at low northern latitude but rises to about 2.5 degrees in the Arctic. This peak is absent in the Southern Hemisphere.

Following the method of Broecker (1975) these temperature records have been fitted with four cyclic curves (lower smooth curves in Figure 2) and then with the four cyclic curves plus the rising exponential that represents the heating of added atmospheric carbon dioxide. The latter effect was negligible before 1900 but is now rising quite steeply. Most of the periods used for the cyclic curves are found in various climate records and have explanations in astronomical constants associated with the Moon's orbit. The same amount of greenhouse heating is required for a good fit in both hemispheres, but the amplitude of the cyclic periods required is much larger

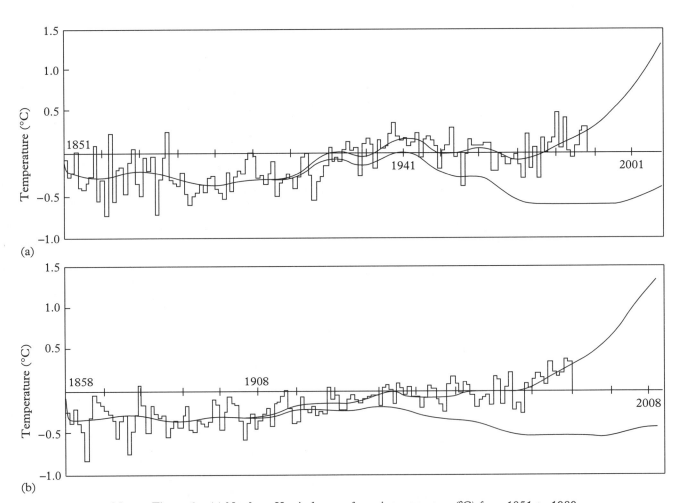

MOON. Figure 2. *(a) Northern Hemisphere surface air temperature (°C) from 1851 to 1989 fitted with carbon dioxide heating and lunar tidal periods. The lower curve is without carbon dioxide heating. (b) Southern Hemisphere surface air temperature (°C) from 1858 to 1989.* (From Jones et al., 1986a and 1986b.)

in the north than in the south. This difference suggests that tidal mixing in the Arctic Ocean is an important factor in natural climate fluctuations.

Some have suggested that the natural fluctuations may be caused by changes in solar luminosity. There is continuing study and controversy on this subject. However, this hypothesis does not account for the differences observed between the two hemispheres.

[*See also* Tides.]

BIBLIOGRAPHY

Bleakney, J. S. "Ecological Implications of Annual Variation in Tidal Extremes." *Ecology* 53.5 (1972): 933–938.
Broecker, W. S. "Climate Change: Are We on the Brink of a Pronounced Global Warming?" *Science* 189 (1975): 460–463.
EOS (Transactions of the American Geophysical Union) 69.27 (5 July 1988): cover.
Friis-Christensen, E., and K. Lassen. "Length of the Solar Cycle: An Indication of Solar Activity Closely Associated with Climate." *Science* 254 (1991): 698–704.
Jones, P. D., and P. M. Kelly. "Decadal Surface Temperature Maps for the Twentieth Century, Part I." *Climate Monitor* 10.5 (1981).
Jones, P. D., and P. M. Kelly. "Decadal Surface Temperature Maps for the Twentieth Century." *Climate Monitor* 11.1: 12–17; 11.2: 40–45; 11.3: 72–77; 11.4: 100–105 (1982).
Jones, P. D., et al. "Northern Hemisphere Surface Air Temperature Variations: 1851–1984." *Journal of Climate and Applied Meteorology* 25 (1986a): 161–179.
Jones, P. D., et al. "Southern Hemisphere Surface Air Temperature Variations: 1851–1984." *Journal of Climate and Applied Meteorology* 25 (1986b): 1213–1230.
Loder, J. W., and C. Garrett. "The 18.6 Year Cycle of Sea Surface Temperature in Shallow Seas Due to Variation in Tidal Mixing." *Journal of Geophysical Research* 83–C4 (1978): 1967–1970.
Thompson, D. J. "The Seasons, Global Temperature, and Precession." *Science* 268 (1995): 59–68.

P. R. BELL

MOUNTAIN CLIMATE. *See* Small-Scale or Local Winds.

MOUNTAIN GLACIERS. *See* Glaciers.

MOUNTAIN-VALLEY CIRCULATIONS. *See* Small-Scale or Local Winds.

MUSIC AND WEATHER. The sounds of weather and the moods and feelings associated with weather conditions are deeply embedded in the music of many cultures. From classical symphonies and chamber works to folk ballads and pop tunes, instrumental and vocal techniques have been developed to imitate the patter of rain, the crash of thunder, the howling of the wind, and other meaningful meteorological sounds; lyrics use weather imagery to represent aspects of the human emotional climate.

In many folk traditions, music both celebrates favorable weather conditions and helps people cope with frightening or damaging meteorological events. In the sea chanteys of many maritime cultures and the traditional songs of Native American tribes, weather is a societal concern dealt with by literally addressing music to the heavens through chants, drumming, and song. The Navajo *Tóó Hatáál* is a water or rain chant, while the Pima tribe has *Huhwuhli Nieh*, ("Wind Song"). Weather disasters revealing human helplessness and heroism are a staple of folk music, often crossing over into the popular realm. Mexican balladeer Ramon Luna's "San Marcial" commemorates a 1929 flood in New Mexico. A disastrous 1900 Texas hurricane is recalled in "Galveston Flood," a traditional song recorded in 1970 by U.S. folk singer Tom Rush. Canadian singer-songwriter Gordon Lightfoot memorialized a deadly Great Lakes storm of the 1970s in his "Wreck of the Edmund Fitzgerald" (1976).

The most successful music inspired by weather does not simply mimic or reproduce the sounds made by various kinds of weather. Rather, it uses the full capacities of voice and instrumentation, sometimes aided by wind-machines and other sound devices, to produce tonal colorations and rhythms that evoke emotional states.

Eighteenth-century Venetian composer Antonio Vivaldi's four violin concerti *The Four Seasons* (1725) may be the best-known climatological work of European music's Baroque period. The work portrays the progress of the months and seasons by modulations in tempo and intensity of volume. Composers from Franz Joseph Haydn in the late eighteenth century to John Cage in the twentieth have found musical inspiration in seasonal change. Haydn's oratorio *Die Jahreszeiten* ("The Seasons," 1801), based on a text by English poet James Thomson, renders the seasonal activities of country life. Peter Ilyich Tchaikovsky's piano suite of twelve "Seasons" (Opus 37; 1876) celebrates changes in weather month by month. Alexander Glazunov's ballet *The Seasons* premiered in 1899 and became the Russian composer's most popular work. Cage's 1947 ballet score, *The Seasons*, was his first piece for orchestra. Cage, a student of Hindu aesthetics and philosophy, wrote that it was "an attempt to express the traditional Indian view of the seasons as quiescence (winter), creation (spring), preservation (summer), and destruction (fall)."

The metaphoric relationship between stormy weather and stormy human feelings inspired the German-born eighteenth-century composer George Frideric Handel's early cantata *Cuopre tal volta il cielo* ("Thunder, Lightning, and Whistling Wind," composition date unknown). This 12-minute work begins with a recitative depicting black clouds and rising wind. An aria describes the storm's terrifying power. In the second part, the storm in the heavens is compared to the storm of unrequited love within a human heart. An example of the way Romantic music employed weather imagery is Ludwig van Beethoven's well-known *Symphony No. 6 in F Major* (the *Pastoral*, 1808). In its allegro movement "The Storm," physical disruption and spiritual turmoil are represented by an approaching thunderstorm, foretold by low rumblings of cello and bass, a change to the key of F-minor, and use of the thunderous timpani for the first time. The chromatic contrast of growling trombone and shrieking piccolo enhances the stormy effect.

Franz Schubert's song cycle *Die Winterreise* ("The Winter's Journey"), based on 24 poems by Wilhelm Müller, was published in 1828, the year of the Austrian composer's death. The songs, 16 of them written in a minor key, express the profound gloom of a man who simultaneously endures his lover's indifference and a harsh winter, only occasionally relieved by melodic hints of approaching spring. Polish composer Frederic Chopin's *Prelude #15 in D-flat Major* is known as the "Raindrop Prelude" because of the piano's incessant drip-like repetition of A-flat. It was composed in sunny Majorca in 1838, where Chopin hoped fine weather would alleviate the tuberculosis that killed him 11 years later at age 39.

American modernist John Cage presented weather themes in some later musical projects, finding in the capriciousness of weather nature's endorsement of his theory of musical randomness. In *Winter Music* (1957) as few as one and as many as 20 pianists have a choice of at least eight ways of playing Cage's randomly generated notes. Cage's *Lecture on the Weather* (1975) is a darkly pessimistic performance piece in which 12 men read excerpts from the writings of Henry David Thoreau while tape-recordings of wind, rain, and thunder play and lightning flashes are projected on a black backdrop.

Modern popular music, from blues to country to rock and roll, has also found weather imagery and sounds irre-

sistible. As in classical music, many pieces describe actual weather, while in others weather's variety and changeability provide metaphors for human emotions and actions. The boundary is rarely clear-cut, and examples of popular music that allude to weather are virtually inexhaustible. For example, pianist and jazz balladeer Nat King Cole included "On the Sunny Side of the Street" and "Thunder" on recordings he made during the late 1930s. More recently, the Burt Bacharach-Hal David song "Raindrops Keep Falling on My Head" made a hit of the soundtrack of the 1969 film *Butch Cassidy and the Sundance Kid*. When an Austrian, a Czech, a Brazilian, and two American musicians got together in 1970 to form a jazz fusion ensemble, they named themselves Weather Report; their 1977 album *Heavy Weather* was their best-seller.

In their 1966 hit, "California Dreamin'," the Mamas and Papas use images of gray and chilly fall weather to express their longing for California's warmth. "Here Comes the Sun," written and performed by the Beatles, was also recorded by Richie Havens (1971). Stevie Wonder recorded "You Are the Sunshine of My Life" in 1973. Influential bluesman Elmore James wrote and sang "The Sky Is Crying," issued on an album of the same name in 1971, eight years after his death. It has been recorded by rock bands Fleetwood Mac, John Mayall, and George Thorogood, among others, and was a hit for country-rock guitarist Stevie Ray Vaughan. Pioneering country performers the Carter Family had a major hit in 1964 with their radio program theme song "Keep on the Sunny Side." In 1991, country star Garth Brooks's "The Thunder Rolls" was the year's top country single and music video.

Some bands use storm imagery to proclaim their own self-assurance and disdain for nature's power. "Born to Laugh at Tornadoes," a 1983 album by Was (Not Was), and the heavy-metal "Rockin' Like a Hurricane" (1990) by Germany's Scorpions are examples. Creedence Clearwater's "Who'll Stop the Rain?" (1970) uses rain as a metaphor for a dangerous political climate. For the Eurythmics, a British group, rain symbolizes the inevitability of failure in "Here Comes the Rain Again" (1983), while in Eric Clapton's "Let it Rain" (1970) a downpour means emotional renewal after a long dry spell.

The singing group Martha and the Vandellas had a big hit in 1963 with "Heat Wave," in which heat represents youthful passion. Pat Benatar's 1979 "In the Heat of the Night" equates summer heat with the loosening of moral restraints. "Summer in the City" is the Lovin' Spoonful's

1967 anthem to the pleasures and excitements of hot urban streets.

A somewhat different relationship between weather and music has emerged in New Age music. Here, recorded or electronically synthesized natural sounds are used therapeutically to affect the mental and emotional state of listeners and mobilize the potential healing properties of natural forces. For example, the *Atmosphere Collection* includes "Midnight Rainshower" and the *Nature Recordings* album features "Thunderstorm in Big Sur." A direct and personal link between weather and music can be seen in the popularity of wind chimes, metal or ceramic mobiles that produce music-like sounds in response to every change in wind speed and direction.

BIBLIOGRAPHY

Arnold, D. *The New Oxford Companion to Music.* 2 vols. Oxford and New York, 1983.

Birosik, P. J. *The New Age Music Guide.* New York, 1989.
Burlin, N. C. *The Indians' Book.* New York, 1968. (Reprint of 1923 2d ed.)
Kivy, P. *Sound and Semblance: Reflections on Musical Representation.* Princeton, 1984.
Pareles, J., and P. Romanowski, eds. *The Rolling Stone Encyclopedia of Rock & Roll.* New York, 1983.
Plantinga, L. *Romantic Music.* New York, 1984.
Pritchett, J. *The Music of John Cage.* Cambridge and New York, 1993.
Robb, J. D. *Hispanic Folk Music of New Mexico and the Southwest.* Norman, Okla., 1980.
Schmid, R. "Rockin' Thunder." *Weatherwise* 42 (August, 1989): 192–196.
Wagner, J. "Music to Watch the Weather By." *Weatherwise* 42 (October, 1989): 249–255.

MARSHA ACKERMANN

MYTHOLOGY. *See* Religion and Weather.

N

NATIONAL WEATHER SERVICE. *See* Weather Services.

NATURAL SELECTION. *See* Ecology.

NAVIGATION. *See* Marine Weather Observations and Predictions.

NEAR INFRARED RADIATION. *See* Shortwave Radiation.

NEWS MEDIA. With the growth of information and communication technologies, weather information has become more readily available than ever before. Meteorology is perhaps the foremost science-based public service in the mass media, and its technical, scientific, and public evolution have all been shaped by developments in those media. Weather and climate services and studies have been among the first and most profitable applications of each new technology: from the installation of the first commercial telegraph (1845), wireless telegraphy (1898), telephone subscription services (1905), first radio broadcasts (1906), first commercial radio (1920), to the invention of the automatic calculator (1939), the computer (operational computer forecasts began in 1955), and the launching of the first artificial satellite (1957; first weather satellite, 1960). Each new medium has shaped the way we see and understand the weather.

Today one can obtain a weather report for a city or region simply by turning on a radio or television or looking at the weather page of the local newspaper, or by employing other private media from the home or office. In the United States, 1 billion telephone calls are made annually to check the weather. A wide choice of weather software and other services are available by subscription and direct purchase, delivered by means of computer modem or facsimile machine. Some enable users to construct their own forecasts from satellite-feed graphic and numerical data, while others transmit prepared continental weather listings or permit interactive discussions of weather- and climate-related topics.

Weather reports began to appear daily in local newspapers in the 1850s. Each subsequent medium has created new forms and markets for weather forecasts. Radio broadcasting, for instance, made regular weather updates accessible to home listeners, while scheduled weather reports benefited broadcasters by providing low-cost programing and attracting a larger audience. The commercial dimension has affected the nature of weather coverage in various ways. Today creating and disseminating weather information relies on complex technological systems that are more sophisticated and interconnected than technologies used for observing and reporting in the past. For example, satellites and radar help track hurricanes and storm systems. The many new technologies, however, are mainly effective in cutting personnel costs through automation, extending the speed and geographic scope of transmitting weather information, and improving image quality. Few meteorologists claim that these technologies ensure more reliable everyday predictions.

The process of bringing weather information to the public involves several phases. Essentially, these are creating and transmitting weather data, producing weather forecasts, and disseminating the forecasts to the public.

Creating and Transmitting Weather Data. The first phase and to some extent the second are administered by the federal government in the United States through the National Weather Service (NWS), part of the National Oceanic and Atmospheric Administration (NOAA), which reports to the U.S. Department of Commerce. In Canada this work is the responsibility of Environment Canada, which reports to the Ministry of the Environment.

To produce daily forecasts, the NWS collects and transmits weather data from hundreds of weather stations, meteorological satellites, radar sites, ships, ocean data buoys, and international forecasts. These data are generated, transmitted, and incorporated electronically into atmospheric models with the aid of the NWS central computer in Suitland, Maryland. The assembly and interpretation of large quantities of weather information now rely more on the statistical capabilities of electronic data processing than on the skills of meteorologists. Forecasts are mainly produced through numerical prediction, often by analogue with previous similar conditions, rather than by analysis of the present weather.

Since the 1970s, some meteorologists have voiced the concern that meteorology is becoming an electronic rather than an analytic science. Others maintain that the volume of weather information has become too large for existing computer programs to process accurately beyond the traditional 48-hour to 4-day forecast. Some researchers also suggest that there is an imbalance between the acute information-gathering capabilities of satellite-imaging systems, and other observational technologies, and the information-processing capabilities of computers. Finally, the technological infrastructures necessary for processing this information are seen to be creating new types of technical dependencies in less developed countries. These issues continue to confront meteorologists in the wake of rapid technological change.

Weather Forecasts and the Public. The NWS's atmospheric models are converted to forecasts that are distributed nationally to commercial weather-forecasting companies. These companies serve newspapers, radio and television stations, film and television producers, farmers, utilities, aviation, and resource and environmental firms, among others. To ensure professional standards, meteorologists are certified by the American Meteorological Society (AMS), which produces a number of publications, including *Bulletin of the American Meteorological Society, Monthly Weather Review*, and *Weather and Forecasting*. The AMS began awarding a seal of approval to radio and television forecasters in 1959, in part to counteract the antics of many popular weather personalities. Then as now, weather was often defined in entertainment terms as light relief after the bad news. This may have encouraged some forecasters to assume that there was public disapproval toward "bad weather"; however, recent anxieties about increasing ultraviolet radiation are apparently not strong enough to inspire an opposing cultural trend in the form of complaints about sunny forecasts, or "good weather."

Not all on-air forecasters are professional meteorologists. Many network and local news services, and both of the cable weather stations (the Weather Channel in the United States and Canada's Metro-Media/Weather Network) recruit most of their weather reporters from broadcasting and entertainment rather than from meteorology. Meteorologists work behind the scenes to combine forecasts from incoming data with local conditions, or they may work for independent companies selling forecasts to broadcasters. The on-air forecaster often contributes a local touch to forecasts that are generated entirely or partly from distant locations.

A substantial portion of the media-oriented weather industry is devoted to designing and distributing new computer software, particularly graphics, which together with digitally processed satellite views provide the visual dimension in television and newspaper weather forecasts. The competitive nature of commercial broadcasting is evident here in the continuous escalation of technical standards for visual information. The now conventional machine-made symbolic images derived from satellite pictures and computer-generated color graphics have become the dominant visual grammar of weather in the mass media.

Media Coverage of Climate and the Environment. Aside from scheduled forecasts, weather sometimes figures in public-safety and public-interest stories. Both tend to focus on the most dramatic events—hurricanes, tornadoes, blizzards, and floods. Many more Americans lose their lives each year because of pollution and environmental stress than because of hurricanes and blizzards, but such long-term problems are harder to depict dramatically; they do not fit into a daily assignment schedule, and their coverage involves potential controversies with government and industry.

In the 1990s the environment began to appear as a regular feature in weather reports and services. The most common reports are on air pollution indexes and measurement of the ozone layer. These have immediate local health consequences and are cheaply and easily transmitted. At present, however, global environmental news and related issues receive better coverage in electronic bulletin board services (BBSs) than in radio, television, or the daily press. For instance, news and views about the environmental effects of the Kuwait oil fires during the 1991 Persian Gulf War were widely circulated by BBSs at a time when such information did not appear in the regular press. Anyone with access to a computer modem can join a BBS; they are usually nonprofit and/or university-based. Information about commercial weather services via modem or facsimile can be found in *Weatherwise* and in *The Air and Space Catalog*.

[*See also* Charts, Maps, and Symbols; *and* Weather Forecasting, *article on* Models and Methods.]

BIBLIOGRAPHY

American Meteorological Society. *Conference on Weather Forecasting And Analysis and Aviation Meteorology.* Silver Spring, Md., 1978.

Corbett, J. B. "Atmospheric Ozone: A Global or Local Issue? Coverage in Canadian and U.S. Newspapers." *Canadian Journal of Communication* 18 (1993): 81–87.

Henson, R. *Television Weathercasting: A History.* Jefferson, N.C., 1990.

Hughes, P. *A Century of Weather Service: A History of the Birth and Growth of the National Weather Service 1870–1970.* New York, 1970.

Mack, P. E. *Viewing the Earth: The Social Construction of the Landsat Satellite System.* Cambridge, Mass., 1990.

Makower, J., ed. *The Air and Space Catalog.* New York, 1989.

Nelkin, D. *Selling Science: How the Press Covers Science and Technology.* New York, 1987.

Ross, A. *Strange Weather: Culture, Science and Technology in an Age of Limits.* New York, 1991.

Tuchman, G. *Making News: A Study in the Construction of Reality.* New York, 1978.

Weatherwise: The Magazine about the Weather. Washington, D.C.

JODY BERLAND

NORTH AMERICA. Every part of North America experiences a distinctive and fairly regular pattern of weather from year to year. Although there is much variation, the range of temperatures, the amount of precipitation, the number of hours of sunshine, the number of storms, and other climatic elements tend to be similar over long periods of time.

Climate—an area's average weather and its variability—is controlled by four relatively permanent features of the Earth: the amount of solar radiation received (especially as it varies with latitude), the distribution of land and water masses, elevation and topography, and ocean currents. North America extends over latitudes from the tropics to the Arctic, and it ranges in elevation from sea level to several kilometers. There are large inland water bodies, such as the Great Lakes and Hudson Bay, and the continent is surrounded by the Atlantic, Pacific and Arctic oceans and associated seas. The climates of North America are thus quite diverse.

There are several ways of classifying climate. *Empirical* classifications are based on observable features of climate, usually treated in combination; temperature and moisture dominate empirical classifications. *Genetic* classifications attempt to organize climates according to their causes, but often these causes are difficult to quantify. *Applied* classifications help to solve specialized problems that relate one or more climatic elements; an example is a system that relates climatic factors to agriculture.

Because there are numerous classification systems, and no one system is best, the climates of the four major political regions of North America (Canada, the United States, Mexico, and the Caribbean) are described below in terms of patterns of averages and variability of the major elements affecting human activity—temperature (Figure 1 and 2), moisture (Figure 3), and storms.

Canada. A Canadian winter is often thought of as unrelentingly frigid. The amount of bitterly cold weather (minimum temperatures below −20°C), however, varies considerably throughout the country. On average, the number of bitterly cold days varies from near zero along the Pacific Coast, around Lake Erie, and on the outer Atlantic Coast, to 210 days at Eureka in the Northwest Territories. In a typical year there are 5 very cold days in the interior of British Columbia, 60 across the eastern prairies, 20 in central Ontario, 30 over the Gaspé Peninsula, and more than 100 in the Yukon.

Temperature. Inland temperatures in winter are usually significantly colder than coastal temperatures. The west coast enjoys the shortest and mildest winters because of the moderating effect of warm Pacific air. Along the Atlantic coast the moderating maritime effect is less pronounced because the prevailing air flow is off the land. Away from the coasts and mountains, monthly temperatures vary fairly directly with latitude. In mountainous country, atmospheric inversions often make the upper slopes warmer than the valleys below; some of the coldest temperatures in Canada occur in the mountain valleys of the Yukon and Ellesmere Island.

In midsummer, afternoon temperatures are usually above 25°C over much of Canada's interior. In coastal areas, however, sea breezes keep maximum temperatures 5° to 8°C cooler than at inland localities. Summers are warmest in the dry interior valleys of British Columbia and in extreme southwestern Ontario, where the average daily afternoon temperature exceeds 27°C. The coolest areas, apart from the Arctic, are the western mountains and the Labrador coast, where daily maxima in summer average between 15° and 20°C, and daily minima between 5° and 10°C. In the Arctic, there is noticeable cooling as latitude increases, with daily maxima declining from 10°C along the Arctic Ocean coast to 5°C at the northern end of Ellesmere Island. Significant local cooling often occurs because of the cold waters of the large inland lakes and Hudson Bay, but shoreline localities can

MEAN AIR TEMPERATURE °C, YEAR

NORTH AMERICA. Figure 1. *Mean annual temperatures (°C) in North America.*

523

AVERAGE ANNUAL RANGE OF AIR TEMPERATURE °C

NORTH AMERICA. Figure 2. *Average annual range of mean monthly temperature (°C) for North America.*

524

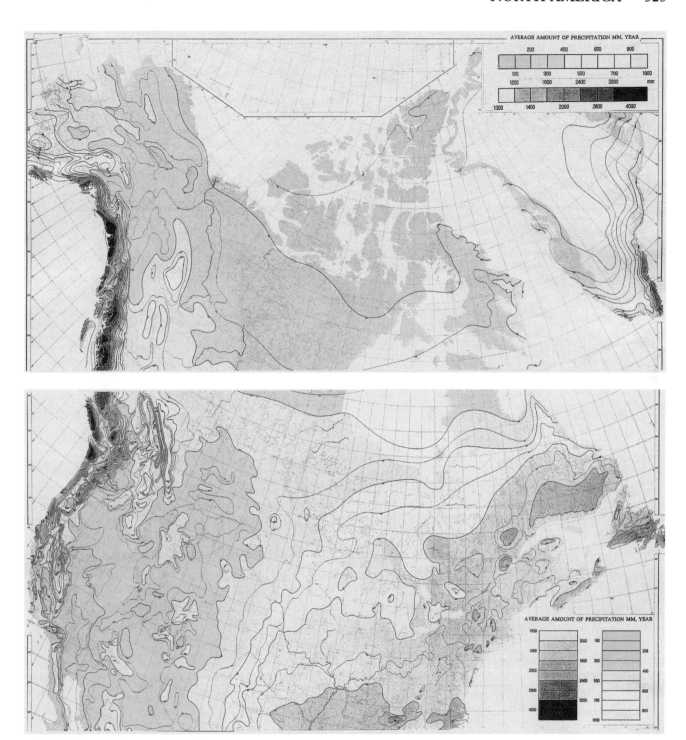

NORTH AMERICA. Figure 3. *(a), (b), (c) Average amounts of annual precipitation in North America in millimeters.* [See page 524 for (c).]

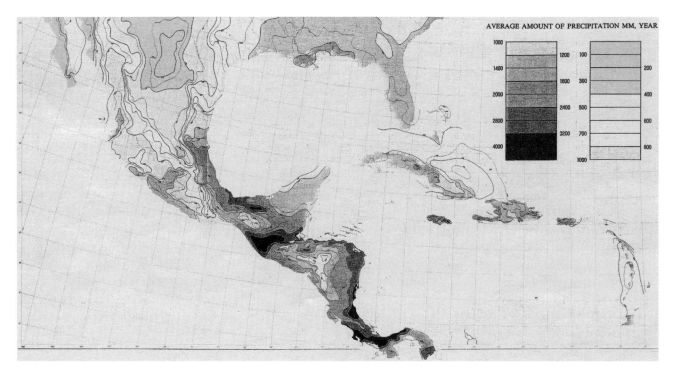

NORTH AMERICA. Figure 3. *(c) Average annual precipitation in millimeters.*

experience sudden changes of temperature with shifts of wind direction.

The daily temperature range (the difference between the highest and lowest temperatures of a day) is greatest in summer, ranging from about 10° to 14°C inland to less than 9°C in coastal areas. In winter the range diminishes everywhere, averaging between 5° and 10°C. The average annual range varies from 40°C in the central Northwest Territories and northern Manitoba to 10°C along the Pacific coast. Typical ranges are 35°C for the prairie provinces, 28°C for southern Ontario, 32°C for southern Quebec, 25°C for the Maritimes, and 20°C for Newfoundland.

Day-to-day weather change is a characteristic of Canadian climate, but one type of weather sometimes persists for a long time. Every region has a few periods each year with unusually wet, dry, cold, or hot weather. Cold spells are common on the prairies and can last for weeks or even months; they are seldom as long or as severe in the west. Heat waves occur commonly in southern Ontario and the interior of southern British Columbia. The rest of the country, however, rarely experiences more than 3 consecutive days with temperatures of at least 32°C at one location. The heat waves are much shorter in duration than winter cold waves.

Frost is of vital concern to agriculture. Local topography, including differences in elevation, aspect, and slope, as well as latitude, surface cover (vegetation, soil, or snow), and proximity to water bodies and human settlements affect the occurrence of frost. The frost-free season—the period normally free of subfreezing temperatures—varies from more than 240 days in parts of the Queen Charlotte Islands and Vancouver Island to less than 50 days over the entire Arctic and the high mountains of western Canada. In the northern forest zone the frost-free season is between 60 and 110 days, and across the prairies there are from 80 to 120 frost-free days. Seasons of 120 to 140 days occur in southwestern Canada, and around the Great Lakes, the frost-free season exceeds 160 days.

Precipitation. Rainfall and snowfall are distributed very unevenly across Canada in both space and time. The Pacific coast is the wettest area (3,200 millimeters of precipitation annually), with most of its rain and snow falling in winter. Less than 100 kilometers to the east, in the rain shadow of the Coast Mountains, lies a sagebrush-dotted semidesert, one of the driest areas of the country. The rain shadow continues across the prairies. Annual precipitation here is about 350 to 500 millimeters, with some localities receiving less than 250 millimeters. June is the wettest month in this region, but summer droughts are frequent and severe. Across northern Canada extends the Arctic desert, with meager rainfall and even scantier snowfall. Annual precipitation averages about 100 to 200 millimeters. Eastern Canada usually has ample and reliable quantities of precipitation. Moving eastward from Manitoba, annual precipitation increases about 40 millimeters for every 100 kilometers, from 500 millimeters at Winnipeg to 1,500 millimeters at Halifax. More than 1,000 millimeters falls in the elevated areas in the lee of the Great Lakes, over Quebec's Laurentian uplands, and over the Atlantic provinces. The Appalachian highlands and Atlantic coast receive 1,400 to 1,600 millimeters a year. Ontario and Quebec have no identifiable wet or dry season; October through December is the wettest time along the Atlantic coast.

In the wettest locations in British Columbia and in Newfoundland, more than 200 days a year experience precipitation. In the dry areas of western Canada where precipitation is under 400 millimeters, there are fewer than 100 days a year with precipitation. In the east, the number of wet days varies between 125 and 150. The eastern side of Baffin Island (annual precipitation, 500 millimeters) has about the same number, but over most of the Arctic, only 75 to 100 days have measurable precipitation.

Average annual snowfall shows a general pattern of lower amounts in the north and in the interior, and higher amounts in the east and in all elevated areas. The pattern is most complex in British Columbia, where Canada's highest and lowest snowfall amounts both occur. Vancouver Island rarely receives snow, but more than 1,000 centimeters a year falls in the Coast and Rocky mountains. Large amounts also occur over the St. Elias Mountains of the Yukon, Quebec's Laurentian slopes, over the Appalachian highlands, and over the Cumberland and Hall peninsulas of Baffin Island. Newfoundland is one of the snowiest provinces, with several areas receiving more than 400 centimeters a year. There are also snowbelts in the lee of large open lakes and Hudson Bay.

Snow days in British Columbia range from 10 to 20 along the Pacific coast, 20 to 50 in the southern valleys, and 75 or more in the northern portion and elevated areas. The number of days with snow in the Yukon range from 50 along the Arctic coast and over most of the territory to 100 in Kluane National Park. The prairies have between 35 and 50 snow days, and the Great Lakes region around 50. Across southern Quebec and Newfoundland, there are 50 to 75 snow days, but over most of the Maritimes, there are only about 50.

Storms. Thunderstorms are one of the most notable features of summer weather across southern Canada. The highest average annual number of thunderstorm days is 34 at locations in extreme southwestern Ontario, with secondary peaks of at least 25 occurring between Calgary and Edmonton, Alberta, in southwestern Quebec, and in southeastern Manitoba. In the boreal forest zone between the prairies and the arctic tundra there are fewer than 10 thunderstorm days, and the arctic islands and the Pacific coast are largely free of thunderstorms. Elsewhere, the count of thunderstorm days is 5 over Newfoundland, 10 to 15 over the Maritimes, 15 to 20 over central Quebec and northern Ontario, and 15 to 20 over the prairies. April through October is the normal thunderstorm season; most occur in July.

About 70 to 80 tornadoes hit Canada's populated areas each year. They are rare west of the Alberta foothills and in the north, and most frequent in extreme southwestern Ontario, southern Saskatchewan, southwestern Manitoba, and northwestern Ontario. Tornadoes occur most often during afternoons in late June and early July.

Usually only one or two Atlantic tropical storms or hurricanes enter Canadian waters each year. Although these storms have weakened considerably by the time they reach southeastern Canada, they can bring heavy rains and strong winds, usually during the period from late August to the end of October. Typhoons in the Pacific rarely move into Canadian waters.

Intense winter storms, frequently accompanied by numbing cold, ice, and heavy snow, occur throughout Canada. Days with blowing and drifting snow are common in the eastern Arctic and over open, windy areas of southern Canada east of the Rockies. The valleys of British Columbia, southern Yukon, and the northern prairie provinces average fewer than 10 such days each year. As many as 20 blowing-snow days occur over southern Ontario and southern Quebec, and more than 90 days over the arctic islands and over the central mainland of the Northwest Territories. The area around Hudson Bay experiences more than 60 such days; the area north of the tree line from the Beaufort Sea southeast to James Bay and central Quebec has 30 or more days; Newfoundland has 30 days; and southern portions of Manitoba and Saskatchewan have 20 to 30 days.

United States. Climatic patterns over the contiguous 48 United States are similar to those in Canada, but because of its more southerly latitude, the United States is warmer. The major topographic influences are the Sierra Nevada–Cascade ranges and the Rocky Mountains in the west from the Canadian to the Mexican border, and the Appalachian mountains in the east. Important water bodies include the Gulf of Mexico, the Great Lakes, and the Atlantic and Pacific oceans.

Temperature. In winter, coastal areas are usually warmer than inland regions. The Pacific and Gulf coasts and Florida experience the shortest and mildest winters, with average temperatures ranging from about 3°C in Washington state to 10°C in southern California, and 15° to 20°C along the Gulf coast and in Florida. Along the Atlantic coast north of the Carolinas, the moderating effect of the Atlantic Ocean is less pronounced. Away from the eastern and western coasts and the mountains, monthly temperatures vary directly with latitude. Average winter temperatures range from about −15°C in the northern plains to about 10° to 15°C near the Gulf coast. In mountainous regions, the slopes are often warmer than the valleys below because of atmospheric inversions.

Midsummer temperatures are highest over the southwestern desert and inland from the Gulf Coast, where afternoon readings of more than 38°C are not uncommon. In the midsection of the country, afternoon temperatures range from about 35°C north of the Gulf of Mexico to about 27°C near the Canadian border. Sea breezes and oceanic effects result in more moderate temperatures along the Pacific and Atlantic coasts. Mountain peaks in the west are generally the coolest locations in summer; on the east coast, mountain locations are also cooler than the surrounding piedmont areas.

Daily temperature ranges in winter are normally between 7° and 12°C throughout most of the eastern half of the country. Exceptions are along the Gulf and southeastern Atlantic coasts, where the range is usually less than 7°C. In the western United States, ranges are often less than 7°C along the Pacific coast, and more than 15°C in desert regions. The pattern is similar in summer, except that the daily temperature range may approach 20°C in the deserts and 10°C along the Pacific coast.

In the middle third of the country, the average frost-free period decreases northward, from about 330 days along the Gulf of Mexico to less than 90 days at the Canadian border, and less than 60 days in the northern part of the Great Lakes region. In the eastern third, the frost-free period ranges from more than 330 days in southern Florida to less than 90 days in Maine. Mountainous areas generally experience a shorter period than surrounding piedmont areas. In the west, the southwestern desert and the Pacific coast have frost-free periods of more than 300 days at low elevations, but in mountainous regions, the period here can be less than 30 days.

Precipitation. Average annual precipitation varies from less than 20 millimeters in the desert Southwest to more than 2,000 millimeters in the Pacific northwest and along the ridges of the Cascade and Sierra Nevada mountains in the west and in the Blue Ridge Mountains in the east. The windward side of the western mountains is generally wet, while the leeward side is dry. Along the

Pacific coast there are two precipitation seasons, dry in summer and wet in winter. In the central part of the country precipitation increases eastward, from about 300 millimeters just east of the Rocky Mountains to about 1,250 millimeters just west of the Appalachians. The east coast generally receives from about 1,250 to 1,750 millimeters a year.

The average annual snowfall over the ridges of the western mountains is more than 1,000 centimeters, and over the ridges of the eastern mountains, more than 250 centimeters. Over flatter land, snowfall is rare in the most southerly areas, increasing northward to more than 250 centimeters over the Great Lakes. Away from mountains, blizzards occur most often over the northern plains.

Storms. During the warmer seasons, thunderstorms are quite common in the east and along the Gulf Coast, and tornadoes occur most often in the central portions of the country from the Gulf of Mexico to the Canadian border. The number of hurricanes striking land along the Atlantic and Gulf coasts has varied from one or two a year to more than 20. Storms as intense as hurricanes, but not of tropical origin, sometime form off the North Carolina coast. Called "northeasters," they travel northward along the coast into the Canadian Maritimes. The heavy precipitation and strong winds associated with these storms often cause local flooding and coastal erosion.

Alaska. The climate of Alaska is varied. The southeast, around Juneau, is an area of high precipitation (about 2,400 millimeters annually, mostly rain). The climate of this region is similar to that of the coast of Washington. The Alaska Peninsula and Aleutian Islands are damp and foggy all year. Annual precipitation here varies from about 750 to 1,500 millimeters annually; snowfall is usually about 200 centimeters. West central Alaska is low-lying and averages about 450 millimeters of precipitation, with monthly average temperatures from about −12° to 12°C. Central Alaska is generally dry (about 350 millimeters of precipitation annually) and can experience the widest temperature ranges in Alaska with a daily minimum of −10°C and a daily maximum of 35°C. The climate of arctic Alaska is similar to that of arctic Canada.

Mexico. Although Mexico lies in the tropics and subtropics, its climate is controlled mainly by elevation; most of the country is mountainous. Mexico can be divided into three climatic zones based on elevation: hot, temperate, and cold. The hot zone runs from sea level to about 1,000 meters and includes the coastal plains, the Yucatan Peninsula, Baja California, the Isthmus of Tehuantepec, and the north-central portion of the Mesa

del Norte. The temperate zone includes the highland valleys, which range from about 1,000 to 2,000 meters in
elevation. The cold zone includes elevations above 2,000
meters. In the hot zone average annual temperatures are
above 22°C and are usually about 24° to 27°C. The
Pacific coast is slightly warmer than the Gulf coast. The
temperate zone experiences average annual temperatures
between 15° and 22°C, and the cold zone temperatures
below 15°C.

There is a slight temperature decrease from south to
north. Weak temperature gradients along the coasts are
accounted for by the low latitude of Mexico and the uniform temperatures of the coastal waters. In the interior,
the expected northward decrease along the plateau is balanced by a northward decrease in elevation. The aridity
in the northern part leads to higher temperatures in summer, but extensive cloudiness moderates winter temperatures. The northern desert experiences the greatest
range—about 20°C—between average daily maximum
and average daily minimum temperatures. Corresponding ranges are small in the south at about 5°C. Frost is
likely on about 25 days in the northern and central plateaus, about 10 to 15 days in the northwestern areas, and
less than 5 days in most of the remainder of the country.

Precipitation ranges from about 55 millimeters in the
northwestern desert to more than 4,000 millimeters on
the windward side of mountains. Throughout most of the
country there are two seasons: a rainy season from May
to October, and a dry season in winter. An exception is
in the northern parts of Baja California, where most of
the precipitation falls in winter and summers are dry.
Days with thunderstorms occur most often (about 10 per
summer month) in the southwestern region. Tropical
storms commonly enter Mexico from both Pacific and
Gulf waters in summer and fall.

A unique feature of the Mexican climate is "El Norte"
(the norther). This is a strong, cold, humid wind which
blows in winter mainly along the Gulf coast. Accompanied by low temperatures and rainfall, it is most often
experienced in Yucatan, Tabasco, and Vera Cruz. It also
affects Texas in the United States.

Bahamas and Caribbean. The climate of the Caribbean Sea and its islands is influenced mainly by elevation
and the northeast trade winds. Except for occasional
intrusions of cold continental air, warm maritime air covers the region throughout the year. Many of the islands
are mountainous, and local climates are affected by
elevation.

Monthly temperatures near sea level generally average
between 21° and 26°C. The annual range of average tem-

peratures is small—about 2° to 7°C. Daily temperatures
rarely fall below 18°C or rise above 37°C, and day-to-
day variations are minimal. Although temperatures vary
little from place to place, localities exposed to the trade
winds are generally cooler than those on leeward coasts
or surrounded by higher terrain. Temperatures are coolest at the highest elevations.

There are two precipitation seasons: wet (May to
December) and dry (January to April). It rains throughout the year, and the seasons are contrasted by rainfall
amounts. Monthly rainfall averages more than 250 millimeters in the wettest months and less than 100 millimeters in the driest. The lee slopes and the interior portions of the mountainous islands have relatively dry
winters; the bulk of the rains occur in summer, when
convection is most active. Windward slopes generally
receive more precipitation. Most of the precipitation
occurs during the afternoon hours. Thunderstorms are
most frequent in summer and rare in winter; tropical
storms are common from June through October.

The most destructive climatic feature is the hurricane.
The hurricane season lasts from late June until about January. Storm frequency and intensity varies greatly from
year to year, with an average of one or two storms per
year that cause significant damage or disruption. [See
Hurricanes.]

Effects of Climate. The climate of North America
varies greatly with seasons, latitude, and topography
(Figures 1, 2, and 3), influencing many aspects of human
life. The amount of energy used, the crops grown, the
livestock raised, and the type of housing built in an area
depend on climate. Indices have been developed which
in simple terms relate complex climatic features to various applications. For example, the Palmer Drought
Index describes conditions of anomalous dryness or wetness. The number of Normalized Difference Days
derived from satellite images, reflects the seasonal duration of green vegetation in an area; the higher the number
of days, the longer the growing season.

Changes in climate may also affect human history. In
North America, annual precipitation has increased over
the past century. The increase in the contiguous United
States is most apparent after the 1960s and is in large
part due to an increase during fall (September through
November). Records of precipitation over the twentieth
century are very difficult to evaluate in areas of significant
snowfall because of problems in measurement. Nonetheless, there is evidence of positive trends through much of
Canada and Alaska during the past 40 years, with a 5 to
15 percent increase over northern Canada (above 55°

north latitude). Data over southern Canada (below 55° north latitude) and the northern United States (above 45° north latitude) indicate an increase in precipitation of 10 to 15 percent for Canada and about 5 percent for the United States. Increases are more prevalent in the eastern two-thirds of North America than in the western third.

Except for a small region in the lower Mississippi River basin, temperatures during the twentieth century have increased across all of North America. Increases have been as large as 2°C in the Yukon and Northwest Territories of Canada. Since 1950, temperatures have cooled across the eastern third of Canada and over southwest Greenland. The temperature increases, however, have more than compensated for the areas of cooling; as a result, snow cover during the 1980s declined over North America during spring.

BIBLIOGRAPHY

National Climatic Data Center. *Climatic Atlas of the United States.* Asheville, N.C., 1983.
Page, J. L. "Climate of Mexico." *Monthly Weather Review* Supplement 33 (1930).
Phillips, D. *The Climates of Canada.* Ottawa, 1990.
Powers, E., and J. Witt. *Traveling Weatherwise in the U.S.A.* New York, 1972.
Soto Mora, C., and E. O. Jauregui. *Cartografía de Elementos Bioclimáticos en la República Mexicana.* Mexico City, 1968.
World Meteorological Organization. *Climate Atlas of North and Central America,* technical supervisor F. Steinhauser. Geneva, 1979.

NATHANIEL B. GUTTMAN

NORTHERN HEMISPHERE. *See* General Circulation.

NORTHERN LIGHTS. *See* Aurora Borealis.

NORTH POLE. The North Pole is essentially the center of the Arctic, the southern limits of which have been defined in several differing ways; of these the most practicable, from the meteorological aspect, is the +10°C isotherm (a line indicating equal average temperature) in July. Although an artificial concept, this boundary corresponds well with the ecological one of the treeline. Where these boundaries diverge, it is usually because of inadequate soils or similar constraints affecting vegetation. Although the North and South poles are geodetically comparable, there are very substantial contrasts between them. The North Pole is at sea level with 4,220 meters of water below it, and the South Pole is at 2,835 meters altitude on an ice cap almost that thick, based on bedrock which is near sea level. The concept of the Pole of Inaccessibility (the point in the Arctic Ocean most remote from land) has also been used to determine the Arctic's center; this lies between the North Pole and Wrangel Island. From the meteorological point of view, these poles are largely similar. [*See* Arctic.]

The weather of the central Arctic (like that of the Antarctic) is characterized by very low temperature, so that precipitation is nearly always in the solid state as snow, and by the exceptional annual distribution of daylight. These factors cause winters to be some 9 months long, dark, and cold. In general polar climates vary greatly; the weather systems are comparatively unstable, with sudden changes possible during any day, especially in the 3 summer months when the systems are most active. No long-term observations have been made at the North Pole, and the few sporadic ones recorded are nearly all for summer. In contrast, a meteorological station has operated at the South Pole since 1956. More data are available from other parts of the central Arctic, and these are useful in analyzing north polar conditions. To give one recent example of the weather at the North Pole, on 13 July 1992 the air temperature was −0.6 °C at 12:00 (noon), wind velocity was less than 1 meter per second; a very light rain fell for an hour before becoming flake snow by 14:00 (2 P.M.); and the cloud cover was 100 percent low stratus. This was near the height of summer.

Temperature. The North Pole and the central Arctic Ocean are not the coldest northern region; they have an annual average air temperature estimated at −23°C. The temperature is moderated by the low altitude and the effects of the sea. Much lower temperatures are encountered inland in Siberia and Greenland (and in Antarctica because of its altitude). During winter, radiative cooling at the ice surface is associated with the most extreme cold, but at elevations of about 1,000 meters, temperatures as much as 10°C to 16°C warmer may exist; such temperature inversions occur more than 90 percent of the time during winter, extending from the Siberian coast over much of the Arctic Ocean toward the pole. From the biological point of view, the rate of loss of heat (wind chill) is more serious than the absolute temperature; cooling rates in the polar regions may be perilous owing to the exceptional strength of the winds, with gusts exceeding 40 meters per second quite possible.

Light. The total duration of daylight received in a year at any point in the Arctic (or anywhere else on Earth) is equal, but the intensity and distribution of light during the year vary greatly with latitude. The thin, pure atmosphere (especially as upper-atmosphere ozone concentrations are being depleted) transmits high intensities of ultraviolet radiation, which have significant adverse biological effects, often enhanced by reflection from the substrate. Although the solar energy received during summer may be considerable, it is not effective in raising temperatures at very high latitudes; and there is rarely more than 10°C in diurnal variation. Arctic lighting, air density, and several other conditions frequently cause mirages, looming, and other uncommon optical phenomena. Although not strictly weather, a spectacular and famous optical phenomenon seen around the poles during winter is the aurora. [See Aurora Borealis.]

Ice and Precipitation. Sea ice, with the occasional polynya or lead (types of opening), is the only surface feature of the North Pole. The Arctic Ocean sea ice has an area of 15 million square kilometers at its maximum extent. This has a much higher reflectivity than any other surfaces on Earth, often radiating 80 to 90 percent of received sunlight. [See Sea Ice.]

Precipitation is scarce, and virtually desertlike conditions prevail in most of the Arctic. The greatest proportion falls during summer, mostly in the form of powdery snow. The estimated quantity received rarely exceeds 20 centimeters of water equivalent a year because the frigid atmosphere has a low capacity to retain moisture. Where blizzards blow over ridged ice, much deeper snow accumulations may be common, but this represents redistribution of previously fallen snow by the wind, rather than new precipitation.

Owing to the low temperatures there is little evaporation, and the absolute humidity is minimal. These temperatures ensure that the saturation point is low and thus frequently exceeded; this results in frequent, often dense, fogs at almost any time of year. The density of the very cold air often stratifies these fogs and confines them to barely 25 meters above the surface. Sea fogs are another special feature of the Arctic Ocean where very cold air meets a patch of open water in the pack ice (sea water begins to freeze at −1.8°C). Sometimes frost smoke may be seen ascending from water in still air owing to the lower density of water vapor. This is sometimes called *arctic sea smoke*. All these phenomena contribute to frequent periods of poor visibility, although extreme opposite conditions may rarely occur. [See Fog.]

The many varieties of precipitation in high Arctic regions (for example snow, sleet, and graupel) are best demonstrated by the vocabulary of far northern peoples.

Some dialects of Eskimo have more than a dozen words for snow forms, and more for snow and ice surfaces.

Many detailed studies of sea ice have been made in response to the practical needs of general navigation as well as for theoretical meteorology. Only very powerful icebreakers or nuclear-fueled submarines can reach the North Pole. The entire surface of the Arctic Ocean is frozen from October to June. Many discontinuous and opportunistic observations have been correlated to obtain long-term information on sea-ice distribution, but recent data are now obtained largely from instrument-bearing satellites, with confirmatory surface observations from ships.

Pressure and Winds. The winter pressure field is dominated by the vast continental high-pressure system which develops as the continental land mass cools from September onward. In summer the general pressure distribution is reversed; from July pressure falls over the continental masses and rises over the central portions of the Arctic Ocean.

A consequence of these conditions, and the exceptional arctic geography, is the wind; high velocities are common owing to cyclone effects. The poles are centers of westerly vortices, which turn counterclockwise over the North Pole. A primary mechanism for arctic wind circulation is provided by a continuous cap of low pressure in the middle and upper troposphere at about 3 kilometers altitude. In winter the westerly wind belt flows around this vortex in response to the temperature gradient toward the equator. In summer the Arctic vortex contracts and weakens, and cyclonic activity is widespread. Lows on the arctic front affect the arctic coasts of North America and Siberia. Gales are recorded in most regions of the Arctic—sometimes to hurricane force—on an annual average of 2 to 4 days a month; the worst time is around the equinoxes. [See Polar Lows.]

The very low turbidity of the air, and the consequent exceptionally blue skies when clear, indicates that the central arctic atmosphere is among the purest on the planet—despite a seasonal appearance of Arctic Haze—and visibility may be enormous. In contrast, however, is the phenomenon of whiteout, which arises when a uniformly white surface lies under low cloud that admits diffuse illumination so no shadows are formed. The effect on humans is a loss of depth perception, a dangerous state, recorded in the diaries of many who have traversed the pack ice.

Clouds. The central Arctic is generally a cloudy region, but the frigid atmosphere greatly reduces the variety of cloud formations. Stratus forms with a low base predominate, and cumuliform clouds are virtually unknown. Summer and autumn are the cloudiest periods, and March is generally the clearest. A rare specialty of high polar regions are the stratospheric nacreous clouds, where ice replaces water vapor as the refractive medium. At the appropriate lighting angle these can be most spectacular formations. Noctilucent clouds, another rare phenomenon of polar latitudes, are very high stratospheric clouds illuminated by sunlight across the central polar regions. They appear as white clouds at night when the sky is dark. [*See* Clouds.]

Historical Records. The Arctic is second only to the Antarctic in the comparative dearth of meteorological observatories and scarcity of long runs of observations. This is unfortunate, because the polar regions are unusually sensitive to climatic changes, which are exhibited much more apparently there than elsewhere on the Earth.

Among the older historical records there exist various short-term sets of observations, mainly for summers, around the periphery of the Arctic Ocean, but few are comprehensive. A most interesting and detailed series of winter meteorological observations was made in the North American Arctic during mid-nineteenth-century explorations endeavoring to discover the Northwest Passage; more than 30 vessels wintered in these waters, and virtually all collected weather data regularly.

Although the Arctic Ocean coasts of Russia were charted well before those of the other circumarctic lands, it was not until late in the nineteenth century that many observations were made. The voyage of the *Tegetthoff* (led by Julius Payer from Austria), which discovered the archipelago Franz Josef Land in 1873, recorded data over two winters. These data formed the earliest comprehensive series from such a high latitude, exceeding 80° north latitude, and were long some of the closest to the North Pole.

The International Polar Year of August 1882 to August 1883 was the first coordinated international investigation of meteorology and many other aspects in the Arctic. Twelve stations were established by different countries in the far northern latitudes; six of these were distributed around the high Arctic. The Polar Year concentrated attention on the North Pole, and subsequently scientific investigation and exploration—particularly attempts to reach the North Pole—became more frequent. Several expeditions had winter bases, and virtually all maintained comprehensive meteorological observatories. The longest record from a high-latitude station during this period was from Cape Flora, Franz Josef Land, where accurate data were recorded for 1,000 days from 1895 to 1897.

Although several northern penetrations had been made from land, extensive information was not gained over the Arctic sea ice until the voyage of the *Fram* and the journey of Fridtjof Nansen toward the North Pole in 1895. Several other northern forays were made on the surface of the ice, mainly from Franz Josef Land, Svalbard, and Greenland. Although sea ice and other features were noted, such observations were not the primary purpose for many north polar attempts. From 1895 until 1913 at least 12 attempts to reach the pole were made, but it was not indubitably reached by surface until Ralph Plaisted did so in 1968. Comprehensive knowledge of the region of the North Pole came only after exploration by air, when airships and planes flew poleward beginning in 1925. The first airship crossed the pole in 1925 and the first plane in 1926. Since the late 1950s, North Pole observations have been made by a regular series of meteorological flights; remote sensing from satellites is useful for certain data.

Observatories. No observatories have ever operated at the North Pole. A few have existed around the periphery of the Arctic Ocean from the time of World War I. The Second International Polar Year (1932–1933) was a time when many more far northern observatories were established, especially around the Arctic Ocean. The Soviet Union, in particular, set up a series of Polar Stations along the Northeast Passage, where navigation was being developed. Many are on remote, uninhabited islands and have now produced long series of data. Stations in Greenland, arctic Canada, and Alaska were more frequently associated with other outposts—settlements, Royal Canadian Mounted Police posts, missions, and other research stations—but several also have amassed long meteorological records. Many present observatories were founded for the International Geophysical Year of 1957–1958, and some are operated by military outposts such as the DEW (Distant Early Warning) line.

A specialized opportunity for arctic observation is provided by the ice itself. Since 1937 a series of floating ice stations have been deployed. The first, established by Ivan Papanin of the Soviet Union, started about 50 kilometers from the North Pole and drifted south with the East Greenland Current for 18 months. In 1946 the United States established a similar series, and in most subsequent years both countries have had ice stations operating. They are highly efficient platforms for detailed bathymetry (depth measurement) and hydrography studies. Ice stations are established by aircraft landing on an ice floe and drift slowly with the ice circulation. Over the years their courses have covered virtually all regions of the Arctic Ocean and some associated seas; several have passed near the North Pole, although they had been established well south. All carry comprehensive meteorological observatories and report data by radio. More recently, a number of automatic meteorological stations have been established on the ice, and weather buoys have been dropped. All ice stations have problems associated with their constantly changing positions, but they are valuable observatories in a most inaccessible region.

Some other arctic observations have been made from ships and submarines. The North Pole was first reached by sea in 1957 by the submarine USS *Nautilus,* which made the voyage submerged. Beginning in 1959 with the USS *Skate,* approximately 25 submarine surfacings occurred at the pole by U.S., British, and Soviet/Russian vessels; a few of these have been there during winter. In 1977 the *Arktica,* from Murmansk, was the first surface ship to sail to the pole through the arctic pack ice. There were no more such voyages until 1987, but another 16 voyages by surface vessels have been made as of March 1995, most by Russian nuclear-powered icebreakers. The submarines obtained sporadic meteorological reports which can confirm aircraft observations. The icebreakers maintain records throughout the voyages and have the opportunity to remain on station for longer periods. Both make observations on sea ice—the submarines by upward sonar techniques.

The current data from these sources are comparatively sparse. Arctic weather forecasts are available for ships and aircraft as well as for the few outposts, but their reliability corresponds closely with the amount of data available and is affected by the enormous variability.

Changes. The weather of arctic regions in general is of much theoretical interest because both polar regions exhibit greater variation in climate than elsewhere on Earth and are sensitive indicators for what might happen elsewhere. In addition to large annual differences (making average figures only approximate predictive guides), long-term fluctuations are large. A recent example is the thickness of sea ice at 90° north latitude: In 1991 it was barely 1.2 meters thick and two diesel-powered icebreakers were able to reach the Pole (the first ones to do so), while in 1992 the depth exceeded 3.0 meters and it took much work by a nuclear-powered icebreaker to reach the same position. There is much conjecture as to whether global warming may affect sea-ice thickness and permit easier navigational access through the central Arctic Ocean, but more data are required to predict this.

Only 1,000 years ago the weather permitted farming in southern Greenland in areas which are now glaciated. If present trends in global warming persist, this may again become practicable. An analysis of mean temperatures from a number of far northern and arctic stations since 1880 shows a marked warming that reached its peak in the decade 1930–1940. After that the trend reversed and the arctic climate was cooling; this too reversed during the 1960s and 1970s. Indications for the 1990s are of a warming period. The cooling of the Arctic was most marked in the same regions where previous warming was most intense. In Franz Josef Land, where there are four observatories, the 10-year temperature average fell by 3° to 4°C, and average winter temperatures were 6° to 10°C lower in the 1960s than in preceding decades. Other changes include increased flow of the cold East Greenland Current taking polar water south. In several recent years this has moved more sea ice to the coasts of Iceland than in any of the past 50 years. Various paleoclimatic investigations are being applied in the Arctic, but this is more of a global program than a regional one.

Biological Effects. The region of the North Pole is exceedingly hostile to life, more because of its sparse resources and geography than from its weather. The small number of organisms recorded there are thus of special interest; they include birds, mammals, fish, and algae. Undoubtedly records are incomplete and, as more vessels reach the region, more will be seen. Single instances of a snow bunting (*Plectrophenax nivalis*) sighted from a submarine, a black-legged kittiwake (*Rissa tridactyla*), and ringed seal (*Phoca hispida*) sighted from an icebreaker, are known from or very close to 90° north latitude. Norther fulmars (*Fulmarus glacialis*), and the tracks of polar bears (*Thalarctos maritimus*) and arctic foxes (*Alopex lagopus*) have been seen north of 88° north latitude. Diatomaceous and other algal growth occurs on the underside of the pack ice, and fish have been seen associated with it virtually to the Pole. Much of the Arctic Ocean supports organisms of these kinds.

BIBLIOGRAPHY

Armstrong, T. E., G. Rogers, and G. Rowley. *The Circumpolar North.* London, 1978.
Central Intelligence Agency. *Polar Regions Atlas.* Washington, D.C., 1978.
Hydrographer, Royal Navy. *The Arctic Pilot.* Taunton, England, 1985.
Sage, B. *The Arctic and Its Wildlife.* London, 1986.
Sugden, D. *Arctic and Antarctic.* Oxford, 1986.
Sutcliffe, R. C., ed. *Polar Atmosphere Symposium: Meteorology Section.* London, 1958.
Treshnikov, A. F. *Atlas of the Arctic.* Moscow, 1985.

R. K. HEADLAND

NUCLEAR WINTER. The term *nuclear winter,* coined by Richard Turco in 1983, refers to the effects on climate of a large-scale nuclear war, presumedly in the Northern Hemisphere. Smoke from resulting fires, especially the black, sooty smoke from cities and industrial facilities, would block out sunlight for weeks or even months over most of the hemisphere—and, if this holocaust occurred in Northern Hemisphere spring or summer, over much of the Southern Hemisphere as well. The resulting cool, dark conditions at the Earth's surface would prevent crop growth for at least one growing season, resulting in mass starvation over most of the world. More people could die in the noncombatant countries than in those where the bombs were dropped because of these indirect effects. The horrible recent famines in Somalia and Ethiopia become a model of the fate of most of the world's population after a nuclear war.

Nuclear winter is a theoretical concept, but by using numerical models of the climate system and by looking at analogs (occurrences in nature similar to certain presumptions of the theory), we can have confidence in its validity. Of course, the theory can never be completely validated without an actual experiment, but an experiment that could kill more than half the people on the planet would never be attempted deliberately. In fact, belief in this theory may have contributed to the lessening of tension between superpowers and the reversal of the nuclear arms race during the 1980s.

The Nuclear Arsenal. There were more than 50,000 nuclear weapons in the world in 1990. The total explosive power of these weapons is equal to about 15,000 megatons (15 billion tons) of TNT—the equivalent of 3 tons of TNT for each human on Earth. The atomic bomb dropped on Hiroshima, Japan, in World War II had an explosive power of 13 kilotons (= 0.013 megatons), or 0.00009 percent of the current global arsenal. If one Hiroshima-sized bomb had been dropped every hour of the 45 years from the end of World War II to 1990, more than two-thirds of the arsenal would still not have been used up! The number of weapons has slowly been falling recently as the superpowers' nuclear arms race has ended. However, the current number of weapons is still 10 or 100 times more than is needed to produce a nuclear winter.

About one-third of the energy of a nuclear explosion is in the form of light or heat. It is like bringing the Sun to the Earth's surface for a brief moment. Anything close to the explosion will burst into flames. Following the flash of light comes the blast wave (like thunder following lightning), which will break apart many structures and blow out the flames, but crumpled structures burn more easily and fires will be reignited by burning embers and electrical sparks. Fires will be fueled by broken gas lines or gasoline pumps. In fact, there are many sources of fuel for fires in cities—buildings and their contents, trees, and even asphalt. Modern materials, such as plastics, not only burn with a sooty smoke but also produce high levels of toxic chemicals. Studies of fuel inventories in cities and industrial regions show that less than 1 percent of current global nuclear arsenals would be enough to produce nuclear winter, if targeted at oil refineries. Most climate-model calculations of nuclear winter have assumed that about one-third of the current arsenal would be used.

Smoke in the Atmosphere. The black smoke cloud from devastated Northern Hemisphere sites would rise into the atmosphere and rapidly envelop the entire hemisphere. Absorption of sunlight would heat the cloud, causing it to rise and inducing winds that would blow some of it into the Southern Hemisphere. Although about two-thirds of the smoke would be fairly rapidly cleansed from the atmosphere by rain and snow, about one-third would remain aloft for many months. Initial studies suggested that the Earth's surface would become dark and as cold in the summer as in a normal winter, hence the term *nuclear winter.* More recent calculations show that temperatures in the summer would fall to levels closer to those of late fall. Nonetheless, the majority of the world's population would still be vulnerable to the agricultural disruptions of catastrophic proportions. In addition, the Asian summer monsoon would fail, and precipitation over land would be reduced throughout the Northern Hemisphere. Because of lofting of smoke into the stratosphere above the region where it can be washed out by rain, and because of snow and sea ice feedbacks, nuclear-winter effects could last for several years. (The snow and ice feedbacks involve not only albedo feedbacks but also thermal inertial feedbacks, where increased sea ice makes the ocean more land-like and allows colder winter temperatures, thus prolonging and enhancing the cooling.) Ozone depletion on a hemispheric scale of 30 to 50 percent, lasting for several years, has been predicted in the latest simulations. [*See* Albedo.]

Analogs. Events used as analogs for a nuclear winter include the firestorms of World War II, the diurnal cycle, the seasonal cycle, the asteroid impact that may have caused the extinction of dinosaurs, Saharan dust storms, Martian dust storms, forest fires, and volcanic eruptions. During World War II, firestorms were produced by conventional bombing in Dresden, Hamburg, and Tokyo, and by nuclear bombing in Hiroshima and perhaps in

Nagasaki. From these we know that burning cities can produce firestorms—super fires that spread far beyond the initial area of ignition and pump smoke high into the atmosphere.

The diurnal cycle (day and night) provides another good analog. Imagine that the Sun did not rise tomorrow, and tonight was followed by another night, followed by yet another. This would be the situation under thick clouds of smoke in plumes downwind of major fires. It is easy to see that even in summer, temperatures could plummet rapidly to below freezing. The seasonal-cycle analog, as already mentioned, has given us the term *nuclear winter.*

Sixty-five million years ago an asteroid or comet smashed into the Earth, producing the greatest mass extinction of species the planet has ever known. The best-known group of organisms that disappeared was the dinosaurs, but in addition, every land animal species larger than a cat died out, and many plant species disappeared as well. It was once thought that a global dust cloud blocked out the sunlight, producing cold and dark conditions; however, recent evidence suggests that there were also continental-scale forest fires emitting a thick smoke layer into the atmosphere—exactly what has been proposed for a nuclear winter. Smoke is much more efficient than dust at blocking sunlight, so it is now thought that conditions at the surface were even colder and darker than originally proposed; the resulting climatic changes could easily have caused these extinctions.

Periodically, clouds of dust are blown up from the Sahara Desert and transported all the way across the Atlantic Ocean. From this we learn that dust particles in the troposphere (the lowest atmospheric layer, from the surface up to 10–12 kilometers) can spread great distances around the world. It has been observed that under Saharan dust clouds it is colder, and there are fewer water clouds and less rain. This suggests a similar reaction to nuclear soot and dust.

There have been large forest fires in recent history, too, although not as large as those possibly implicated in the extinction of the dinosaurs. Research has found surface cooling of 2° to 4°C (4° to 7°F) in the midwestern United States under smoke clouds generated by forest fires in British Columbia in 1982. Similar results were found for other forest fires in British Columbia (1981), in China, Siberia, and California (1987), and Yellowstone Park (1988). In a nuclear winter, the smoke from burning cities and industrial facilities is expected to be thicker and blacker than forest-fire smoke, producing even more cooling.

When the U.S. *Mariner 9* spacecraft first flew by Mars to take high-resolution pictures of the surface, the northern hemisphere of Mars was covered by a thick cloud of dust. (Mars has an atmosphere, but it is much thinner than Earth's.) A few weeks later, the entire Martian globe was covered by this dust cloud. The heating of the atmosphere caused by the dust cloud in one hemisphere induced a circulation that transported the dust into the other hemisphere. This same effect is part of the nuclear winter scenario and implies that regions far removed from the conflict will experience climatic changes.

Volcanic eruptions provide several examples that can teach us about nuclear winter. Clouds of volcanic dust injected into the stratosphere (the layer above the troposphere, where there is no rain to wash out the dust) have been observed to spread completely around the globe in 3 weeks and to remain aloft for several years. This is the same fate postulated for nuclear smoke either initially injected into the stratosphere or lofted there by solar heating. Large volcanic eruptions can produce dust clouds in the troposphere immediately. Large surface temperature changes were observed under such tropospheric volcanic dust clouds following the Krakatau eruption of 1883 and the Mount St. Helens eruption of 1980. The long-lasting stratospheric dust clouds also have been found to produce global climatic changes for several years following large volcanic eruptions.

Studies done by the United Nations, the U.S. National Academy of Sciences, the Soviet Academy of Sciences, the Los Alamos and Lawrence Livermore national laboratories, the U.S. National Center for Atmospheric Research, the General Accounting Office of the U.S. Congress, the U.S. Office of Science and Technology Policy, the Royal Society of Canada, the United Kingdom Meteorological Office, and the U.S. Department of Defense have all supported the nuclear-winter theory. A 3-year study involving more than 300 scientists from more than 30 countries, conducted by the Scientific Committee on Problems of the Environment (SCOPE) of the International Council of Scientific Unions, has detailed the climatic, environmental, and agricultural effects of nuclear winter.

Uncertainties of the Theory. Because the theory of nuclear winter cannot be tested in the real world, there are uncertainties in it. One concerns human actions. It can never be known ahead of time the location, number, altitude, yield, season, or duration of nuclear detonations. Therefore, the approach has been to construct different scenarios of possible combinations of nuclear attacks and to study their effects. The other type of uncertainty lies in the physical theories and is associated with the amount of smoke and dust input from given explosions, and with the climatic response.

The main remaining uncertainties concern the amount of smoke that would enter and remain in the atmosphere for a given scenario. However, the combined uncertainties in our knowledge of the amount of flammable material in cities, the fraction of this material that would end up as smoke, and the effects of these smoke particles on sunlight have been considerably reduced by recent research. There is more than enough flammable material in the world, even in military target areas alone, to produce a nuclear winter as the result of a full-scale attack.

It was earlier suggested that when some of the uncertainties had been resolved, the predicted climatic effects would become less. In fact, consideration of some details made the effects less, and that of others made the effects greater. The basic results of the original theory have stood up. It seems that it would be very difficult to design a nuclear war strategy that would not produce a nuclear winter.

Policy Implications. The import of nuclear-winter theory for policy are clear. The most overwhelming implication is that nuclear weapons must not be used as an instrument of war or policy, and that their use would be suicide for the peoples of this planet. A first strike would kill the aggressors, even if there were no response. The threat of nuclear retaliation against aggression using conventional weapons is an empty one if it will kill the retaliators.

Star Wars (the Strategic Defense Initiative) would not solve the problem, even if it worked as originally envisioned. Nuclear winter has been used to argue both for and against the Strategic Defense Initiative of President Reagan. If it would work perfectly and were available to all parties, then it could prevent nuclear winter. But since even the staunchest supporters do not claim that it will stop all incoming ballistic missiles, and it will also not prevent low-trajectory submarine-launched missiles, cruise missiles, bombs dropped from airplanes, or bombs smuggled into the country, and since the computer software and hardware are too complex to work perfectly the first time, it is clear that Star Wars is not a solution to the problem.

Even a "limited" nuclear war would produce these effects. This implies that continued production and plans for the use of nuclear weapons decrease, rather than increase, a nation's security. Therefore, if the people of this planet are to survive any use of nuclear weapons, whether by accident or by intent, it is necessary to drastically reduce the number of nuclear weapons on the planet.

BIBLIOGRAPHY

Aleksandrov, V. V., and G. L. Stenchikov. "On the Modelling of the Climatic Consequences of the Nuclear War." In *The Proceeding on Applied Mathematics.* Computing Centre, USSR Academy of Sciences. Moscow, 1983.
Crutzen, P. J., and J. W. Birks. "The Atmosphere after a Nuclear War: Twilight at Noon." *Ambio* 11 (1982): 115–125.
Harwell, Mark, and Thomas C. Hutchinson. *Scientific Committee on Problems of the Environment 28, Environmental Consequences of Nuclear War.* Vol. 2. *Ecological and Agricultural Effects.* New York, 1985.
Malone, R. C., et al. "Nuclear Winter: Three-dimensional Simulations Including Interactive Transport, Scavenging and Solar Heating of Smoke." *Journal of Geophysical Research* 91 (1986): 1039–1053.
National Research Council, Committee on the Atmospheric Effects of Nuclear Explosions. *The Effects on the Atmosphere of a Major Nuclear Exchange.* Washington, D.C., 1985.
Pittock, A. B., et al. *Scientific Committee on Problems of the Environment 28, Environmental Consequences of Nuclear War.* Vol. 1. *Physical and Atmospheric Effects.* New York, 1986.
Robock, Alan. "Policy Implications of Nuclear Winter and Ideas for Solutions." *Ambio* 18 (1989): 360–366.
Robock, Alan. "Surface Cooling Due to Forest Fire Smoke." *Journal of Geophysical Research* 96 (1991): 20,869–20,878.
Turco, R. P., et al. "Nuclear Winter: Global Consequences of Multiple Nuclear Explosions." *Science* 222 (1983): 1283–1292.
Turco, R. P., et al. "Climate and Smoke: An Appraisal of Nuclear Winter." *Science* 247 (1990): 166–176.

ALAN ROBOCK

NUMERICAL METHODS. The term *numerical methods* encompasses a plethora of techniques in which equations are used to solve a diverse range of physical problems in science and engineering. Researchers in a subdiscipline of numerical methods called *computational fluid dynamics* (CFD) apply these numerical techniques specifically to solve the equations that govern the physics of fluid flow. Because atmospheric scientists and oceanographers are interested in the flow of air (air is considered to be a fluid) and water over the Earth, our interest lies primarily in this branch of numerical methods. In order to concretize some of what follows, we will apply the discussion to *geophysical fluid dynamics* (GFD), which deals with the fundamental dynamical concepts essential to the understanding of the atmosphere, the oceans, and the mechanisms by which they interact.

There are three fundamental approaches to solving a physical problem: experimental, theoretical, and numerical. Although experimental approaches have the capability of being the most physically realistic, they usually involve extremely localized investigations, with drawbacks including measurement difficulties, availability of equipment, and operating costs. The theoretical approach allows concise general information to be obtained; however, one is usually limited to simplified geometries and is often forced to simplify the physics of the problem—by linearizing an inherently nonlinear problem, for example. There is no restriction to linearity when one uses the numerical approach. Complicated physics can be accommodated, and the evolution of the flow field can easily be followed. However, these advantages may be offset by the introduction of numerical errors and computer costs.

Today's researchers have an arsenal of numerical methods with which to attack the many challenging problems arising in CFD. The three types most prevalently used are the *finite element*, *spectral*, and *finite difference* methods; less frequently used methods include *similarity solution* procedures and *integral* methods. Finite element methods are often employed to solve elliptic boundary value problems, where information is specified on the boundary of the domain under consideration. These methods usually attempt to approximate the solution using a family of what are referred to as *first-degree spline functions*. They are more convenient for approximating complicated geometric domains, or in cases where the functions involved have varying behavior in different regions of the domain. Spectral methods are widely used to analyze data sets (where desired information is often obscured), and they also are used in general circulation models (GCMs) that simulate climate. Spectral methods are most appropriate when variables admit a wavelike structure—that is, they are made up of a superposition of trigonometric functions. Here information is transformed to the spectral domain where the computations are performed; then fast Fourier transforms (FFTs) may be employed to transform information back to physical space. Although this approach is generally more efficient, it relies on the assumption that the variables or data are wavelike in nature, and this is not always the case. Application of new results from wavelet theory often allows one to circumvent this restriction. Probably the most widely employed methods are the finite difference *methods* that we will now discuss in detail.

The solution of ordinary differential equations (ODEs) and partial differential equations (PDEs) using finite difference methods began in the early 1900s. These simple methods required an overwhelming number of hand calculations to obtain a single solution. As one might imagine, the history of numerical methods is linked closely to the development of the computer, invented by John V. Atanasoff from 1937 to 1942. It was not until high-speed computers became generally available about 30 years later, however, that the explosion of computational activity took place. The origination of methods crucial to CFD is usually attributed to Lewis Fry Richardson, who in 1910 solved the Laplace and biharmonic equations using innovative point iterative schemes. A celebrated paper published in 1928 by Richard Courant, Kurt Friedrichs, and Hans Lewy addressed the crucial questions of existence and uniqueness of numerical solutions and the stability of numerical methods used to obtain them. To this day, the acronym CFL (from their initials) is used in reference to the stability requirement (a condition that guarantees that a numerical scheme will converge to a solution) for certain classes of numerical schemes. Twenty years later, John von Neumann introduced his now well-known techniques for analyzing the stability of numerical methods. Since that time vast progress has been made, and an ever increasing number of researchers have become involved in CFD. New and innovative approaches, combinations of existing schemes, and even hybrid problem-dependent methods, as well as contributions in areas such as stability and convergence theory, have armed numerical modelers in the atmospheric sciences with a vast weaponry with which to attack the numerous problems arising in numerical weather prediction (NWP) and climate prediction.

The important physical processes in oceanography and atmospheric science are governed by PDEs. It is imperative that the behavior of the physical problem modeled by the PDE is understood; consequently, information is needed concerning the mathematical properties of the governing equations as well as their solution. General second-order PDEs are classified into three categories— *elliptic*, *parabolic*, and *hyperbolic* (analogous to the quadratic function in algebra). Elliptic PDEs govern what are often called *equilibrium problems*, while parabolic and hyperbolic equations usually describe *marching problems* (in which one is usually concerned with the behavior of some quantity as time is "marched" forward). Each class of PDE requires the use of a different type of numerical method. Model equations—those for which an analytical solution is already known—are often introduced to test numerical methods applied to PDEs from each category. Typically used are the Laplace (or steady-state heat) equation for elliptic problems, the one-dimensional heat equation for parabolic problems (associated with diffusion processes), and the wave equation (termed the *advection equation* in meteorology) for hyperbolic problems.

In any numerical method, the continuous problem (the PDE) is *discretized*. For a finite difference formulation this means that dependent variables are envisioned to exist only at specified distinct points. This involves the construction of a grid over the original continuous physical domain. Dependent variables along with their derivatives at a specific point are updated using values from previous iterations in addition to values from neighboring points on the grid. This results in an algebraic representation of the original PDE. Hence, an algebraic problem has been substituted for a differential one. The solution techniques for these algebraic problems fall into two categories, *direct methods* and *iterative methods*. A direct method gives the exact solution (up to machine precision) in a finite number of operations of an algorithm. An iterative method applies a simpler algorithm repeatedly until a specified level of accuracy (called a *tolerance level*) has been reached. Iterative methods are further subdivided according to whether they are *explicit* or *implicit*. An explicit scheme is one for which only one unknown appears in the difference representation; it can therefore be updated using known quantities. An implicit scheme has more than one unknown in the finite difference equation (FDE), resulting in a more complicated algebraic formulation that requires the simultaneous solution of several equations containing these unknowns.

Several considerations determine whether the numerical solution will be an acceptably good approximation to the exact solution to the original governing equation. The *truncation error* is defined as the difference between the PDE and the associated finite difference approximation (algebraic representation) to the PDE. Truncation errors associated with all derivatives in the PDE may be obtained by expansion of the *Taylor series* about a specified point. The *order* of the finite difference approximation is the lowest power of the grid spacing or time step term in the truncation error. Because we solve the FDE and not the PDE, it is beneficial to have a small truncation error. In general, higher-order schemes are more accurate but more complicated and therefore more time-consuming. The term *consistency* describes how closely the FDE represents the PDE. FDEs are said to be *consistent* if the truncation error disappears as the res-

olution is increased—that is, as spacing between grid points is decreased. If this consistency condition is not satisfied, the FDE solution actually corresponds to the solution of a PDE completely different from that which the FDE supposedly approximated. *Round-off error* is caused by computer rounding to a finite number of digits in arithmetic operations; it can become significantly large when an increasing number of repetitive operations are performed in obtaining machine solutions. *Discretization error* is the error caused by substituting a discrete (algebraic) problem for a continuous (differential) one; it is defined as the difference between the exact solution of the PDE and the exact solution of the FDE (assumed to be free of round-off error). In short, the difference between the exact solution to the PDE and the computed solution to the FDE is the sum of the discretization error and the round-off error. The discretization error is comprised of the truncation error and any error introduced in the treatment of boundary conditions (specifications of data on the boundary of the domain).

Perhaps the most important property of a numerical scheme is *stability*. Conceptually, this is most easily thought of for a marching problem, but stability is important for any iterative method. A numerical scheme is said to be *stable* if errors introduced by truncation, round-off, and so forth do not grow in marching from one time step to the next. Thus, as the computed solution evolves, any initially introduced errors are bounded and do not cause the solution to "blow up," or *diverge*. The numerical scheme is said to *converge* if the solution to the FDE tends to the true solution of the PDE as the grid is refined.

The pertinence of the preceding discussion is made clear if we note *Lax's equivalence theorem*. This states that a consistent finite difference scheme applied to a properly posed initial value problem will converge if and only if the scheme is stable. Although this theorem holds only for linear problems, it is also used in the formulation of numerical methods for the more general class of nonlinear problems.

FDEs require the construction of (preferably uniform) rectangular grids. As the resolution of the grid increases, the higher-order numerical methods become more accurate. Owing to the irregular geometry of the Earth's coastlines, topography, and bathymetry, it is challenging to model these boundaries accurately. When uniform rectangular grids are employed, saw-tooth boundaries are introduced which create *boundary condition errors*. Boundary-conforming grids can be constructed using numerical grid generation techniques that alleviate these problems; these present a means by which grid points can be manipulated to produce varying resolution over the domain. It should be noted that the ability to adapt the density of points to the behavior of a function is also one of the main advantages of the finite element approach.

Unfortunately, as resolution increases, so does computer time. GCMs tend to use low resolution and large time steps because they simulate the evolution of climate, which implies long periods of time (years, decades, or centuries). With a resolution of 8° latitude by 10° longitude, 1 day of simulation requires about 1 minute of computer time on an IBM RISC6000 320 workstation. Increasing resolution by a factor of 2 in each direction, to 4° × 5°, the time increases to 6 minutes, while 2° × 2.5° requires 35 minutes, and 1° × 1.25° takes approximately 3.5 hours of computer time. Although the number of grid points increases only by a factor of 4 (since the vertical resolution remains unchanged at 9 layers), the increase in time by a factor of approximately 6 is due to the aforementioned CFL stability requirement, which necessitates the use of a shorter time step. In finite difference, limited-area, NWP models (which can predict up to a few days), higher-resolution grids are favored over higher-order (more accurate) numerical schemes. The National Meteorological Center has been using the limited area fine-mesh model (LFM) since 1971; it uses a 7-layer 190.5-kilometer resolution, which loosely corresponds to a 1.8° × 1.8° model. A two-day forecast using the LFM requires just over an hour of Cray computer time. The nested grid model uses 0.8° × 0.8° resolution for the United States and its immediate surroundings; this is then embedded within the 1.8° × 1.8° LFM grid. The NGM model has been used for model output statistics (MOS, which the general public knows as *weather forecasts*) since 1992, but was replaced in 1994 by the more accurate 50-layer 30-kilometer resolution Eta model. With the dedication of supercomputers such as the Cray Y-MP to GCMs and NWP models, simulations can now incorporate both high-resolution and high-order numerical methods as well as accurate physics in an attempt to generate more reliable climate and weather predictions.

[See also Numerical Weather Prediction.]

BIBLIOGRAPHY

Hansen, J., et al. "Efficient Three-dimensional Global Models for Climate Studies: Models I and II." *Monthly Weather Review* 111 (1983): 609–662.
Johnson, L. W., and R. D. Riess. *Numerical Analysis.* 2d ed. Boston, 1987.
Levine, R. D. "Supercomputers." *Scientific American* 246 (January 1982): 118–135.
Ortega, J., and W. Rheinbolt. *Iterative Solution of Nonlinear Equations in Several Variables.* New York, 1970.
Roache, P. J. *Computational Fluid Dynamics.* Albuquerque, 1982.
Washington, W., and C. Parkinson. *An Introduction to Three-dimensional Climate Modeling.* Mill Valley, Calif., 1986.
Young, D. M. *Iterative Solution of Large Linear Systems,* New York, 1971.

WILLIAM S. RUSSELL

NUMERICAL WEATHER PREDICTION.

Future changes in the state of the atmosphere are calculated by solving the equations that express the fundamental laws of fluid dynamics. These equations, called the *hydrodynamical equations,* include:

- *The Newtonian equations of motion,* which relate the change of velocity of a unit mass of fluid to the forces acting on it.
- *The thermodynamical energy equation,* which relates the change of temperature of a fluid element to the heat energy added to it and the work done by it in expanding against pressure forces.
- *The equation of mass conservation,* which relates change in air density to the net transport of mass.
- *The Boyle-Charles equation of state,* which relates the air pressure, density, and temperature.
- *The equations of continuity,* which relate the changes in the concentrations of various atmospheric constituents (such as water vapor, liquid water drops, or carbon dioxide) to the net transport of those constituents and the strength of their sources and sinks.)

Taken together with the current spatial distributions of wind velocity, air temperature, pressure, density, and the concentrations of variable constituents, the hydrodynamical equations determine the state of the atmosphere at future times.

The methods of numerical weather prediction (NWP) thus have a rational, quantitative basis in well-known and long-established physical theory. In this respect, they are in contrast with earlier, more empirical methods of prediction, which are based primarily on the premise that what happens in the future will resemble what has happened before in similar circumstances. The practical limitation of this analogical method is that one can rarely find a past state of the atmosphere that is similar to the current state in all respects; furthermore, two very similar states are often followed by very different sequences of events. As a result, empirical methods rarely yield accurate forecasts for longer than a day or two in advance.

Early History. The problem of predicting weather by solving the hydrodynamical equations was first stated explicitly by Vilhelm Bjerknes in 1904. He was aware that the essential nonlinearity of the equations excluded the possibility of constructing a linear combination of special solutions that agreed with the current state of the atmosphere, and also that one could not expect to find any simple and exact mathematical dependence of the future state on the current state. Accordingly, one must resort to approximate methods of solving the hydrodynamical equations. Bjerknes proposed a graphical method of calculation, but he did not apply it systematically to the general problem of weather prediction, in part because dense observations of the current state of the atmosphere were not available to him. [*See the biography of Bjerknes.*]

The first genuine attempt at numerical weather prediction was made by Lewis Fry Richardson in the closing months of World War I, between stints of driving an ambulance to the front. His conception of the problem was essentially the same as Bjerknes's, but his formulation was quite different. It was based on the finite difference method, in which derivatives of a variable are approximated as differences between its values at two consecutive instants in time or at two adjacent points in a regular grid covering the surface of the Earth. In this way, the hydrodynamical equations (which are *differential* equations) can be reduced to a set of *algebraic* equations, involving only the numerical operations of addition, subtraction, multiplication, and division. Each equation in this set expresses the value of one of the state variables (e.g., temperature, pressure, or wind speed) at an instant slightly later than current time in terms of the current values of the state variables at the grid points. One then regards this very short-range forecast as a new set of current conditions, and repeats the process over and over to predict conditions at a succession of times in the future. [*See* Numerical Methods.]

Richardson's results, reported in his famous book *Weather Prediction by Numerical Process* (1922), were disappointing, because they predicted changes of pressure that are larger than ever observed. Interest in numerical weather prediction—or, for that matter, any mathematical method of prediction—withered rapidly and lay dormant for 25 years. Even more discouraging was Richardson's estimate that it would take about 26,000 human calculators just to predict weather as fast as it actually happens. [*See the biography of Richardson.*]

Renaissance of Numerical Weather Prediction. The rebirth of interest in numerical weather prediction came with the development of the electronic computing machine by John Eckert and John Mauchly in the mid-1940s. Rather than the cumbersome cogged wheels and electromechanical relays of earlier calculating machines, the switching elements of the first electronic machines were vacuum tubes. Because the latter have virtually no mechanical inertia, they have very short response times. The electronic computer of that day had a typical operating speed of about 1,000 operations per second, about 10,000 times the speed of a human-operated mechanical calculator.

At the same time, John von Neumann and his associates were designing a "stored-program" machine that was to a large extent self-programming. In the common case of a series of calculations to be repeated on different sets of numbers, self-programming is achieved by storing the operands (or numbers to be operated on) at addressable locations in the machine's memory, together with the instructions, which include the addresses of the operands. The basic execution cycle (list of instructions) may then be reused, merely by concluding it with a set of instructions to change the addresses of the operands to the addresses of a new set of operands, where they enter into the operating instructions. Thus it is unnecessary to write out the basic execution cycle more than once. This feature of the stored-program machine makes it especially suitable for large hydrodynamical calculations and, more particularly, for numerical weather prediction, which typically requires that a small number of simple operations be repeated on many different sets of numbers.

Recognizing its computational suitability, practical importance, and intrinsic scientific interest, von Neumann chose the problem of numerical weather prediction for special study and established the Meteorology Project at the Institute for Advanced Study in 1946. These developments did not go unnoticed by theoretical meteorologists, a few of whom turned their attention to the cause of Richardson's errors. Jule Charney, an early member and de facto leader of von Neumann's meteorology group, devised a scale analysis to show that certain quantities that enter into the "primitive" hydrodynamical equations (the original, unmodified form of the equations) are characteristically small differences between large terms of the same sign, with each term subject to considerable observational error. This alone accounted for Richardson's poor results. It was also realized, however, that errors in the solution of the primitive equations would be amplified unless one chose an increment of time between successive steps of the solution that was smaller than the time required for either sound or gravity waves to traverse the distance between grid points. This fact made the solution of the primitive equations a practical impossibility, even with electronic computers of the speed and capacity then envisioned.

The most significant event in the early development of numerical weather prediction was Charney's discovery in 1948 that the *large-scale* motions of the atmosphere are approximately governed by the derived equations of conservation of angular momentum and entropy, taken together with the conditions of mechanical balance—that is, the conditions that the horizontal pressure force be balanced by the Coriolis force (which depends on the Earth's rotation) and that the vertical pressure force be balanced by the gravitational force. From these principles, Charney derived a single equation involving only the distribution of atmospheric pressure. This, the so-called *quasi-geostrophic equation,* is demonstrably free of the first difficulty described above, and has no solutions corresponding to sound and gravity waves. Thus, it does not require the use of extremely short time steps to prevent the amplification of error. This brought numerical weather prediction within the realm of computational feasibility.

The first successful numerical prediction was made by von Neumann's meteorology group in April 1950, using the ENIAC Computer at the Aberdeen Proving Ground in Maryland. It was based on the simplest model imaginable, a stripped-down version of the quasi-geostrophic equation that applied specifically to motions at a single altitude around 20,000 feet. Their results, reported in a paper by Charney, Fjörtoft, and von Neumann (1950), showed that methods of numerical weather prediction based on the simplest model were already comparable in accuracy to the subjective methods of a skilled forecaster, and that the whole problem of numerical prediction might not be as formidable as many had feared.

Well before the appearance of this paper, news of the Charney group's formulation and methods of calculation had spread throughout the meteorological community. Several other groups immediately set out to extend those methods to the prediction of conditions at two or more elevations. By 1953, no fewer than five major research groups—at the Air Force Cambridge Research Center, the British Meteorological Office, the International Institute of Meteorology at the University of Stockholm, the

Deutscher Wetterdienst, and the Japanese Meteorological Agency—had developed and partially tested two-level models based on Charney's quasi-geostrophic equations. [*See the Biography of Charney.*]

Routine Application. In the summer of 1953, the heads of the U.S. civil, naval, and Air Force weather services concluded that development of numerical prediction had advanced far enough to justify putting it into routine practice. The Joint Numerical Weather Prediction (JNWP) Unit was established, staffed, and supported by these three weather services. Soon after its inauguration in July of 1954, the JNWP Unit acquired an IBM 701, a stored-program machine with severely limited memory capacity, and began production of twice-daily numerical forecasts in May 1955. These were based on the finite difference formulation of the 3-level quasi-geostrophic equation; they covered the area north of about 10° north latitude, with about 350 kilometers between grid points. The JNWP Unit's numerical forecasts were significantly more accurate than conventional forecasts and were soon adopted as background forecasts for the National Weather Service's regional and local weather forecasts.

When transistors were introduced as switching elements, the speed of electronic computers suddenly increased by more than an order of magnitude. It thus became feasible to solve the primitive equations, the original form of the hydrodynamical equations. It was desirable to do so for two reasons: first, the primitive equations are free of the approximations of the quasi-geostrophic equations; and second, the numerical methods for solving them are logically simpler than those for solving the quasi-geostrophic equations.

The use of the primitive equations, however, introduced a new difficulty, in that the inaccuracy and incompleteness of the observations of the current state were manifested in the prediction of fast-moving gravity waves of spuriously large amplitude. Methods of slightly modifying (or *initializing*) the current state input to reduce the amplitude of gravity waves were proposed by Charney (1955), Hinkelmann (1959), Thompson (1956), and Bolin (1956) and were introduced into the operating routine of the JNWP Unit. Less restrictive methods of initialization, developed by Machenhauer (1977) and Baer and Tribbia (1977), had the same effect.

The rapid increase in the speed and capacity of electronic computers (about an order of magnitude from one generation of computers to the next) also permitted increases in the horizontal and vertical resolution of numerical forecasts. The JNWP Unit reduced the distance between grid points to about 175 kilometers and reduced the separation between levels to about 1 kilometer, with a corresponding decrease in the interval between time steps. In addition to providing desirable detail, the increased resolution resulted in a general decrease in the errors inherent in the finite difference method.

Recent Developments. Several European weather services soon followed the JNWP Unit's lead in producing numerical forecasts on a routine basis. By 1978, however, they realized that numerical prediction beyond a few days would severely strain the resources of a single weather service. To avoid duplication of effort, the major European weather services created the European Center for Medium-Range Weather Forecasting (ECMWF), jointly staffed and supported by the member institutions. With sole access to a supercomputer and with no preconceptions or investments in earlier computer programs, the ECMWF was able to choose the best features of the various methods available at that time. For example, it chose to use the primitive equations from the outset, together with the Machenhauer-Baer initialization of the current state. To avoid the errors of the finite difference method, however, the spatial distribution of variables was not represented by their values at an array of grid points, but rather by sums of several hundred wavelike components with variable amplitudes, analogous to the sine and cosine components of Fourier analysis. A semi-implicit method for proceeding from one time stage to the next allowed the use of longer time steps, without amplification of error.

The great speed and capacity of the Cray XMP computer was also utilized to improve the calculation of the rate of atmospheric heating. The rate of precipitation was first calculated from the Clausius-Clapeyron equation and the equations of continuity for water vapor, liquid water, and ice. A byproduct of this calculation is the rate of heating by release of the latent heat of condensation or heat of fusion, together with the distribution of clouds. The rate of absorption of solar radiation and the rate of emission and reabsorption of infrared radiation for the Earth's surface, atmospheric carbon dioxide, water vapor, and clouds were computed from an improved formulation of the laws of Kirchhoff and Stefan-Boltzmann. Finally, a more sophisticated treatment of the atmosphere's boundary layer yielded the rates of turbulent transport of heat, momentum, and water vapor from the underlying land or sea surface.

With all these improvements, the ECMWF model represents the present stage of development of numerical prediction methods.

Limitations. We have discussed numerical weather prediction as if the future state of the atmosphere were completely determined by its current state. This is true, but with an important caveat: owing to the inaccuracy and incompleteness of observations, there is always some uncertainty in our specification of the current state. This is first manifested in random errors in predicting the small-scale components of the atmosphere's flow patterns. As a result of nonlinear interactions among different scales of motion, however, this initial uncertainty is rapidly transferred to successively larger scales, so that predictions for more than about a week in advance are no better than sheer guesses. This process of degeneration would operate more slowly if observations of the current state were more accurate and denser; however, it appears that there is a theoretical limit of about 2 weeks on the ultimate range of marginally accurate predictions. For practical reasons, it would be desirable to predict the probable error of predictions and its increase as the range of forecasts increases. Several methods for doing so have been proposed by Epstein (1969), Pitcher (1977), Fleming (1971), Thompson (1988) and Leith (1971), and it is probable that some such method will be incorporated into the routine of numerical weather prediction within a few years.

By now, it is undoubtedly clear that numerical weather prediction did not arise from one or a few spectacular breakthroughs; rather, it evolved through a succession of technological and scientific advances in computing, mathematics, and meteorology. To a considerable extent this is due to the complexity and sheer magnitude of the problem, whose methods of solution are necessarily approximate and subject to continual improvement. For this reason, we can expect that new and more exact methods will emerge as soon as faster computers allow us to apply them.

[*See also* Weather Forecasting.]

BIBLIOGRAPHY

Baer, F., and J. Tribbia, "On the Complete Filtering of Gravity Modes through Nonlinear Initialization." *Monthly Weather Review* 105 (1977): 1536–1539.

Bolin, B. "An Improved Barotropic Model and Some Aspects of Using the Balance Equation for Three-dimensional Flow." *Tellus* 8 (1956): 61–81.

Bjerknes, V. "Das Problem der Wettervorhersage, betrachet von Standpunkte der Mechanik und der Physik." *Meteorologische Zeitschrift* 21 (1904): 1–7.

Charney, J. "On the Scale of Atmospheric Motions." *Geofisiske Publikasjoner* 17 (1948): 1–17.

Charney, J., R. Fjörtoft, and J. von Neumann. "Numerical Integration of the Barotropic Vorticity Equation." *Tellus* 2 (1950): 237–254.

Charney, J., R. Fjörtoft, and J. von Neumann. "The Use of the Primitive Equations of Motion in Numerical Weather Prediction." *Tellus* 7 (1955): 22–26.

Epstein, E. "The Role of Initial Uncertainties in Prediction." *Journal of Applied Meteorology* 8 (1969): 190–198.

Fleming, R. "On Stochastic-Dynamic Prediction." *Monthly Weather Review* 99 (1971): 851–892.

Hinkelmann, K. "Ein numerisches Experiment mit den primitiven Gleichungen." In *The Atmosphere and Sea in Motion,* New York, 1959, pp. 486–500.

Leith, C. E. and R. H. Kraichnan, "Predictability of Turbulent Flows." *Journal of the Atmospheric Sciences* 29 (1971): 1041–1058.

Machenhauer, B. "On the Dynamics of Gravity Oscillations in a Shallow Water Model with Applications to Normal Mode Initialization." *Beitrage der Physik der Atmosphäre* 10 (1977): 253–271.

Pitcher, E. "Application of Stochastic-Dynamic Prediction to Real Data." *Journal of the Atmospheric Sciences* 34 (1977): 2–21.

Richardson, L. F. *Weather Prediction by Numerical Process.* London, 1922.

Thompson, P. D. "A Theory of Large-scale Disturbances in Non-geostrophic Flow." *Journal of Meteorology* 13 (1956): 251–261.

Thompson, P. D. "Stochastic-Dynamic Prediction of Three-dimensional Quasi-geostrophic Flow." *Journal of the Atmospheric Sciences* 45 (1988): 2669–2679.

PHILIP D. THOMPSON

O

OCCLUDED FRONTS. An occluded front, also called an *occlusion*, is generally believed to result from the merger of a cold front and a warm front late in the life of a mid-latitude cyclone. The occluded front can sometimes be identified by a tongue of warm air connecting the low-pressure center to the pool of much warmer tropical air in the warm sector of the cyclone. At the surface, a large wind shift and pressure minimum usually coincide with the front. [*See* Cyclones, *article on* Mid-latitude Cyclones.] Controversy has surrounded the exact nature of the occlusion process ever since its discovery. Recently, as the global variability of cyclone structures and evolutions has been investigated, the occlusion process has been found to occur in several different ways. The theories of its formation are discussed below.

History of Research. The occluded front was first discovered by Tor Bergeron while he was investigating a cyclone of 18 November 1919 off the Norwegian coast. He noticed that the structure of this storm changed over time, implying that Jacob Bjerknes's model of mid-latitude cyclone structure did not always apply. Instead, Bergeron noted that the warm sector shrank and separated from the low center as the cyclone aged. He noticed that the faster-moving cold front approached and eventually appeared to ride up over a slower-moving warm front (in this case, impeded from coming inland by the coastal mountains of the Scandinavian peninsula). With the meeting of the two fronts, the warm air between them was forced aloft, leaving the occluded front as a boundary between two polar air masses. A later interpretation held that the occluded front formed when the cold front caught up to the warm front, because the cold front was rotating around the cyclone center faster than the warm front did. Bergeron's occlusion concept was incorporated by Jacob Bjerknes and Halvor Solberg into their landmark paper of 1922, "Life Cycle of Cyclones and the Polar Front Theory of Atmospheric Circulation." This was the first theory to describe mid-latitude cyclones as undergoing an evolutionary life cycle rather than being

static structural features. These Norwegian scientists called their theory the *ideal cyclone model,* suggesting the potential for variability in any given cyclone.

After this theory was proposed, certain scientists leveled serious criticisms against the Norwegians' work. Some raised the objection that the structure of the ideal occluded front could be duplicated without the cold front ever joining the warm front. It was therefore asked how representative the frontal catch-up model was for most mid-latitude cyclones, and whether the mature and decaying stage of a mid-latitude cyclone could not be arrived at by other means. More recent work with cyclones occurring in the lee of the Rocky Mountains suggests that the structure of an occluded front can be formed without undergoing the classical occlusion process.

Satellite investigations have also shown that classical occlusion-like features can form in nonclassical ways. In a process called *instant occlusion,* a polar cloud vortex approaches and merges with a polar front cloud band, forming a single comma-shaped cloud mass out of two distinct cloud masses. The final product has similarities with the classical occluded cyclone without ever having undergone the classical occlusion process.

Additional work by Canadian meteorologists in the 1950s suggested that most of the clouds and precipitation was found not with the surface occluded front, but with the elevated cold front. They therefore began to analyze the trough of warm air aloft (abbreviated *trowal*) on their surface maps. They usually found a rapid clearing of clouds, a change from steady prefrontal precipitation to airmass showers, and a minimum in pressure with the passage of the trowal.

More evidence implying that cold fronts do not merge with warm fronts in some cases of occlusion comes from recent observations of explosive marine cyclones, which are structurally different in certain details from the classical Norwegian model. In these oceanic cyclones, the cold front separates from and then moves perpendicularly to the warm front, and hence never merges with it. The strong winds ahead of the warm front carry the warm front around the low center, forming a so-called *back-bent warm front*. This back-bent warm front bears many similarities to the back-bent occlusion proposed by Bjkernes and Bergeron in the 1930s. [*See* Cyclones, *article on* Explosive Cyclones.]

Finally, some scientists have claimed that in an occlusion, the cold front never catches up with the warm front at all. Instead, the occluded front is viewed as a new front that forms as the low center deepens back into the cold air and separates from the warm air, leaving a pressure minimum or trough. At present, the validity of this hypothesis is uncertain.

Mesoscale investigations of the structure of occlusions in the 1960s and 1970s led to the discovery of many previously unknown features. Carl Kreitzberg was the first to show the detailed mesoscale structure of occlusions. He identified a prefrontal surge of dry air above the warm front, warm tongues of air associated with locally heavy precipitation, and numerous secondary cold fronts, often behind the primary surface occluded front. Many of these features were verified by later investigators. Kreitzberg also found that younger occlusions often extended vertically throughout the lower troposphere (atmospheric region closest to Earth) and that they split to form features such as squall lines, secondary cold fronts, and rain bands. In contrast, the older occlusions often dissipated, never reaching the ground.

Structure. The Bergen meteorologists believed that if, during the formation of an occluded front, the polar air behind the cold front was colder (and thus denser) than the polar air ahead of the warm front, then this less dense air and the associated warm front would ride up off the ground over the cold front as the occluded front formed. This type of structure is called a *cold-type occlusion* (Figure 1a). In contrast, if the polar air ahead of the warm front was colder than the polar air behind the cold front, then the cold front would be lifted off the ground by the warm front, thus forming a *warm-type occlusion* (Figure 1b). Recent research suggests that these subtle temperature and density differences across the developing occluded front do not determine whether the resultant structure is a cold or warm occluded front. In fact, not one well-documented occluded front possesses the structure of a cold-type occlusion—all cases have the structures of warm-type occlusions—despite the fact that many have the temperature differences across the occluded front necessary to form a cold-type occlusion.

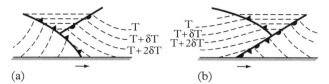

OCCLUDED FRONTS. Figure 1. *Idealized models of occluded fronts: (a) cold type and (b) warm type.* The sketches represent frontal surfaces (—) and isotherms (---) in vertical cross sections normal to occluded fronts, which are moving from left to right. (From Wallace and Hobbs, 1977: 128. Copyright 1977 by Academic Press.)

Thus it is now believed that during the evolution of a mid-latitude cyclone, the warm-type occlusion is favored, not because of the temperature differences across the occluded front, but because the general nature of an occluding mid-latitude cyclone results in a frontal structure that tilts forward with height.

Consequently, the modern view of the formation of a warm-type occlusion can be described as follows. As the upper-level shortwave trough and its associated upper-level frontal zone approach the low-level frontal zone in a developing mid-latitude cyclone, the upper-level frontal zone may descend into and ahead of the region where the cold front and warm front are meeting. A structure reminiscent of a warm-type occlusion would thus be formed, but without the cold front riding up over the warm front. The formation of an occluded front can best be summarized as comprising the meeting of the surface warm and cold fronts, their union to form a single entity, and the ascent of this whole structure off the ground. Depending on the strength of the upper-level front, this feature may or may not play a role in the occlusion process.

In some cases, a dry airflow from the upper troposphere (called the *dry slot*) may descend over the occluded front. The placement of this dry air over lower-level warm moist air may result in an unstable situation, producing cumulonimbus clouds and thunderstorms associated with the leading edge of this dry air. If this air is present over the occluded front ahead of the upper-level front, then the leading edge of the dry air is called a *pre-frontal surge*.

Weather Events. An occluded front can be described as a "back-to-back front," with the weather ahead of the occluded front being similar to that in advance of a warm front, and the weather behind it being similar to that behind a cold front. A few hundred kilometers ahead of the occluded front, the highest cirrus clouds are first visible. As the occluded front approaches, the height of the cloud base lowers and the clouds thicken and become darker. Persistent rain, freezing rain, sleet, and/or snow may fall from nimbostratus clouds, and heavier rain and thunderstorms may be embedded along the elevated warm front and at the pre-frontal surge. After the passage of the pre-frontal surge or the surface occluded front, rapid clearing usually ensues as the dry air aloft appears. Temperatures at the surface usually remain nearly constant or reach a slight maximum during the passage of the occluded front. Winds at the surface are usually from an easterly or southerly direction ahead of the occluded front and shift to blowing from the west or north after its passage. The dew-point temperature (a direct measure of the amount of moisture in the air) usually decreases with the passage of an occluded front.

On the western coast of the United States, the polar air behind an occluded front comes from the ocean and is warmer and moister than the continental air ahead of the system, and so the temperature and dew point will often rise behind an occluded front coming in off the Pacific Ocean. This is one example of the ways local topography and geography can affect the kind of weather associated with occluded fronts.

Much remains to be learned about the occlusion process in mid-latitude cyclones. Few detailed case studies have been done to illustrate the formation and structure of occluded fronts in the atmosphere. It is hoped that recent research on the occlusion process will spark additional investigation to examine the variability in the nature of the occlusion process.

[See also Fronts; and Occlusion.]

BIBLIOGRAPHY

Carlson, T. N. *Midlatitude Weather Systems.* New York, 1991.
Friedman, R. M. *Appropriating the Weather: Vilhelm Bjerknes and the Construction of a Modern Meteorology.* Ithaca, 1989.
Jewell, R. "Tor Bergeron's First Year in the Bergen School: Towards an Historical Appreciation." In *Weather and Weather Maps: A Volume Dedicated to the Memory of Tor Bergeron,* edited by G. H. Liljequist. Boston, 1981, 474–490.
Kuo, Y.-H., R. J. Reed, and S. Low-Nam. "Thermal Structure and Airflow in a Model Simulation of an Occluded Marine Cyclone." *Monthly Weather Review* 120 (1992): 2280–2297.
Newton, C. W., and E. O. Holopainen, eds. *Extratropical Cyclones: The Erik Palmén Memorial Volume.* Boston, 1990.
Schultz, D. M., and C. F. Mass. "The Occlusion Process in a Mid-latitude Cyclone over Land." *Monthly Weather Review* 121 (1993): 918–940.
Wallace, J. M., and P. V. Hobbs. *Atmospheric Science: An Introductory Survey.* New York, 1977.

DAVID M. SCHULTZ

OCCLUSION. The Norwegian cyclone model was the first to describe the basic features of mid-latitude storms. In this theoretical model, storms are disturbances that develop along fronts. In a developing storm there are two fronts: a warm front preceded by warm-air advection, and a cold front followed by cold-air advection. The warm front is ahead of the cold front; however, owing to stronger steering winds associated with the cold front, the cold front travels faster than the warm front. In mature storms, the part of the cold front nearest the storm catches up to the warm front, forming what is known as an *occluded front;* the process by which an occluded front forms is referred to as an *occlusion* (see Figure 1). The intersection of the occluded front, cold front, and warm front is known as the *triple point.*

An occlusion usually signals that the storm has finished developing and is beginning to decay. Sometimes a secondary storm will form at the triple point. Occluded storms also tend to be vertically stacked or aligned (that is, the surface cyclone lies directly beneath the upper-troposphere storm), reducing the storm's forward speed. When an occlusion occurs, the warm air near the surface

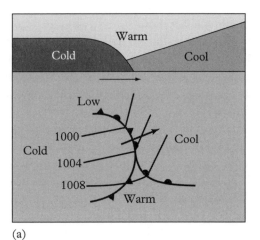

(a)

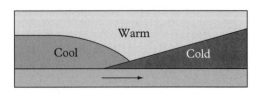

(b)

OCCLUSION. Figure 1. *(a) A cross-sectional illustration of an occluded front. First-order discontinuity in the pressure field is represented by the isobars in this cold-type occlusion. (b) A cross-sectional illustration of a warm-front type occlusion.* (Adapted from Cole, 1980.)

storm center is forced aloft and is replaced by colder air. When the warm air is forced aloft this stabilizes the atmosphere, and steady precipitation is cut off and replaced by drizzle. Unlike the case around cold fronts and warm fronts, no strong thermal gradients exist adjacent to an occluded front. The air behind an occluded front can be either colder (referred to as a *cold occlusion*) or warmer (a *warm occlusion*) than the air ahead. [*See* Cyclogenesis; Fronts; Occluded Fronts.]

Better observations and understanding of mid-latitude storms have led some to revise the classic model of occlusions. In a developing storm, warm advection is associated with rising motion and adiabatic cooling to the east of the storm, while cold advection is associated with sinking motion and adiabatic heating to the west of the storm. Therefore, the horizontal advection and vertical motions oppose the warming or cooling due to thermal advection. This process also converts potential energy to kinetic energy, which is essential for storm growth. In the Norwegian model, the cold front is viewed as eventually catching up to the warm front. According to new theory, the warm-cored storm is viewed as moving poleward and even westward into the increasingly colder air, eventually becoming totally separated or "pinched off" from the warm air to the east. This juxtaposes cold-air advection with rising motion, producing the coldest vertically integrated temperatures directly above the storm and causing the upper-level storm to become vertically aligned with the surface storm. This also converts kinetic energy to potential energy, so the storm begins to decay.

BIBLIOGRAPHY

Carlson, T. N. *Mid-Latitude Weather Systems*. London, 1991.
Cole, F. W. *Introduction to Meteorology*. 3d ed. New York, 1980.
Gedzelman, S. D. *The Science and Wonders of the Atmosphere*. New York, 1980.

JUDAH COHEN

OCEAN–ATMOSPHERE INTERACTIONS. *See* El Niño, La Niña, and the Southern Oscillation; Maritime Climate; *and* Oceans.

OCEANIC CLIMATE. *See* Maritime Climate.

OCEANICITY. *See* Maritime Climate.

OCEANS. Oceanography is a broad field of science focusing on the understanding of the oceans. It includes

four subdisciplines. *Geological oceanography* addresses the historical formation of the ocean basins and the interface of land and water. *Chemical oceanography* studies the composition of the water and its processes and interactions. *Biological oceanography* concerns marine organisms and the relationships between the organisms and the ocean environment. Finally, *physical oceanography* covers the processes of air–sea interactions, heat transfer, water cycles, and water movement. Physical oceanography is of primary concern to students of weather and climate.

Physical Dimensions. The Earth is commonly called the "Blue Planet." From space, our world, with nearly 71 percent of its surface covered by oceans, does indeed appear blue. The oceans are not equally distributed over the Earth, as nearly 70 percent of the world's oceans are in the Southern Hemisphere.

The world ocean is continuous but is divided into four principal parts—the Pacific, Atlantic, Indian and Arctic oceans. The Arctic is sometimes considered an extension of the northern Atlantic Ocean. The broad band of ocean surrounding Antarctica is sometimes termed the Southern Ocean. Numerous smaller bodies of water, including seas, gulfs, and bays, lie between the oceans and land masses.

The Pacific Ocean is by far the largest of the four oceans in terms of area, volume, and depth (see Table 1). The Pacific is twice as large by volume as any other body of water and occupies more area than all the world's land surface combined—about one-third of the total area of the Earth. It also contains the greatest depth of any ocean. The bottom of the Challenger Deep in the Marianas Trench of the Pacific is about 11,033 meters (36,200 feet) deep; for comparison, Mount Everest, the highest point above sea level, is about 8,870 meters or 29,100 feet tall.

Chemical Composition. Ocean water is a dilute solution containing probably every element that occurs on the Earth naturally (atomic numbers 1 through 96). The most abundant compound is water (H_2O), accounting for 96.5 percent of the total. Only eight other elements

OCEANS. Table 1. *Physical properties of the major oceans*

Ocean	Area (square kilometers)	Volume (cubic kilometers)	Depth (meters)
Atlantic	106,400,000	354,700,000	3,332
Indian	74,900,000	291,900,000	3,897
Pacific	179,700,000	723,700,000	4,028
All Oceans	361,000,000	1,370,300,000	3,795

in this solution exceed 10 parts per million (Table 2). The most important of these are chlorine (Cl) and sodium (Na), which combine to form the compound sodium chloride (NaCl). This gives ocean water its salty taste and its *salinity*.

The salinity of ocean water is a measure of the concentration of dissolved salts, primarily sodium chloride. Salinity values average 35 parts per thousand, or 3.5 percent of total weight. Salinity values are not constant but rather vary from place to place because of the differing amount of evaporation and the addition of fresh water into the ocean. The lowest salinity rates are typically found where rainfall is heavy, near the mouths of major rivers, and in most equatorial and arctic regions. Salinity rates are highest in partly landlocked seas in dry, hot regions, and in subtropical regions of the oceans in general.

Ocean Currents. The atmospheric and oceanic circulations directly affect one another. As air flows over the ocean surface, it causes the top layers of the water to drift, usually at a right angle to the wind (see the discussion of the Ekman Spiral below). The moving water gradually piles up, creating pressure differences within the water. This causes the deeper water to begin to move and creates the oceanic circulation.

These ocean currents range in speed from several kilometers per day to several kilometers per hour. The semiclosed circular patterns created by the surface water flow are called *gyres*. In the North Atlantic Ocean, the gyre is controlled by the subtropical (Bermuda) high pressure. This clockwise-spinning pressure system creates ocean currents that move in a clockwise, circular fashion. The moving water is deflected to the right in the Northern Hemisphere by the Coriolis force, causing the surface water to move at an angle to the wind (see the discussion of the Ekman Spiral below). [*See* Coriolis Force.]

OCEANS. Table 2. *Major constituents of ocean water*

Element	Concentration (parts per million)	Element	Concentration (parts per million)
Oxygen	857,000	Bromine	65
Hydrogen	108,000	Carbon	28
Chlorine	19,000	Strontium	8.1
Sodium	10,500	Boron	4.6
Magnesium	1,350	Silicon	3.0
Sulfur	885	Fluorine	1.3
Calcium	400	All others	<1.0
Potassium	380		

In the North Atlantic, there are four primary ocean currents. Starting off the east coast of Florida and flowing northward along the west side of the Atlantic Ocean is the Gulf Stream, which carries large quantities of warm water into the higher latitudes. The warm waters of the Gulf Stream follow the westerly winds away from North America and cross the Atlantic Ocean as the broad, slow North Atlantic Drift. When the current approaches Europe, it splits into two currents. One, the Canary Current, flows south along the east side of the Atlantic Ocean, bringing cooler water equatorward. In the tropics the Canary Current merges with the North Equatorial Current, flowing westward with the Northeast Trade Winds to complete the gyre. (See Figure 1.)

Other Types of Ocean Movement. In addition to the surface currents, which can be defined as *horizontal* movements of water, water also moves vertically in the oceans. To understand the vertical transport of relatively shallow water, we may examine the western coast of North America (the eastern side of the Pacific Ocean) in the summer. The currents are controlled by the Pacific High, which at this time of year flows in a southerly direction parallel to the coastline. The cool California Current keeps the waters from Washington southward to Mexico relatively cool all year. The coolest temperatures are not found in the northern areas, however, but normally occur along the coast of northern California. This is due to two important features of ocean circulation: the *Ekman Spiral* and *upwelling*.

The surface water driven by the wind sets the water immediately below in motion. The lower layer will move more slowly than the surface layer because of low friction coupling and will be deflected to the right (in the Northern Hemisphere) of the surface layer. At a depth of 100 to 150 meters, the water is moving very slowly relative to the surface movement and in a direction opposite to the flow of water at the surface. This turning of water with depth is known as the *Ekman Spiral*.

Because of the Ekman Spiral, the net transport of the surface water is at a 90 degree angle to the wind (to the right in the Northern Hemisphere); therefore, along the California coast, it flows out to sea. The water at the surface along the coast must be replaced, and this is accomplished by *upwelling*, a process in which the colder, deeper water rises to the surface to replace the departed surface water. Along the northern California coast, surface water is coldest because of the upwelling of the cold water that results from the wind flow parallel to the coast.

One important anomaly in the usual pattern of cold water upwelling along the eastern sides of the oceans

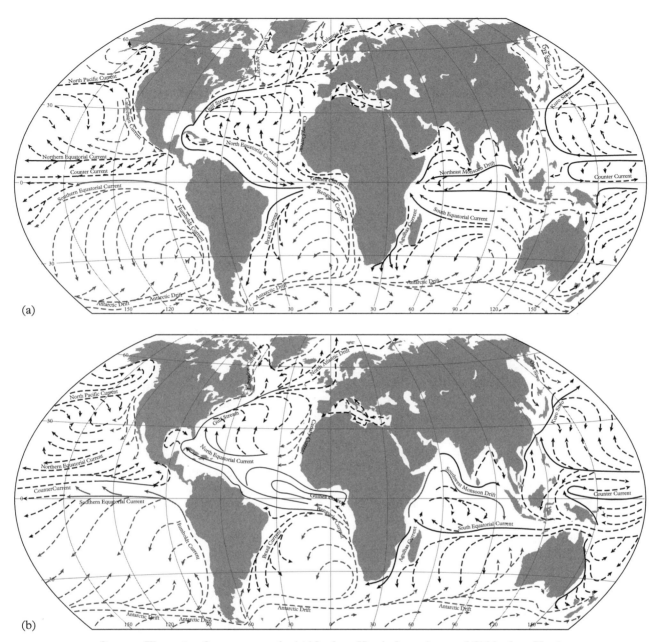

OCEANS. Figure 1. *Ocean currents for (a) Northern Hemisphere winter and (b) Northern Hemisphere summer.* Warm currents are indicated by black lines, and cold currents are represented by gray lines. (Adapted from the *Oxford Atlas of the World*, New York, 1992, p. 11.)

occurs in the South Pacific. Along the western coast of South America, the cool Peru Current moves equatorward (to the north in the Southern Hemisphere), causing the same type of upwelling along the Peru coast as we have described for the California coast. Every few years this cool current is replaced by a warmer current moving from the equator. This phenomenon, known as El Niño (Spanish for "the child," as it frequently begins around Christmas), can have devastating effects on the economy of Peru; it has also been correlated with weather deviations across the globe. Researchers are only now beginning to understand the causes of El Niño. [*See* El Niño, La Niña, and the Southern Oscillation.]

Interaction between Oceans and Atmosphere. Oceans play two important roles in relation to atmospheric circulation. First, the oceans contribute the majority of the water vapor needed for precipitation. About 396,000 cubic kilometers (94,644 cubic miles) of water move through the atmosphere each year. Because the atmosphere can only hold the equivalent of 15,300 cubic kilometers (3,580 cubic miles) at any one time, the atmospheric water must be completely replaced about 26 times a year. Most of this water comes from the oceans: about 334,000 cubic kilometers (79,826 cubic miles) of water are evaporated from the oceans each year. Without this tremendous source of water vapor, the Earth would be a desert.

The second important role of the oceans involves the relationship of sea surface temperatures to atmospheric circulation. Between the ocean surface and the atmosphere there are exchanges of heat, moisture, and momentum, depending in part on the temperature differences between water and air. In comparison with land masses, the oceans have a much greater capacity for storing heat. The solar energy absorbed near the sea surface is distributed by wind stirring and nighttime convection through a *mixed layer* approximately 100 meters in depth.

The ocean currents and winds transport the excess heat from the tropics to higher latitudes. Approximately 40 percent of the total heat transport in the Northern Hemisphere is due to surface ocean currents. The Gulf Stream, for example, adds additional heat to the waters along the western side of the Atlantic Ocean and across to Europe, providing areas such as the Grand Banks off Newfoundland with heavy fog.

In winter, when the temperature contrasts between air and water are greatest, there is a substantial transfer of sensible and latent heat from the ocean surface to the atmosphere. This energy transfer to the atmosphere helps to maintain the global air flow.

Oceanic Effects on the Atmosphere. We have discussed how the winds guide the ocean currents; to complete the cycle, we must examine how the ocean circulation affects the atmosphere. This is most easily done by examining the climate of coastal locations. Four generalizations may be made concerning the climatic characteristics of coasts that are paralleled by warm or cold currents.

- Western coasts in tropical and subtropical latitudes that are bordered by cool-water currents (southern California, for example) have low average temperatures and small annual and diurnal ranges of temperature, and they are normally foggy but arid.
- Western coasts in the middle and higher latitudes that are bordered by warm-water currents (for example, Britain) have mild winters and small annual temperature ranges.
- Eastern coasts in the lower latitudes that are bordered by warm-water currents (Virginia, for instance) have relatively cold winters and warm to hot summers.
- Eastern coasts in higher latitudes that are bordered by cool-water currents (such as Newfoundland) have cool summers and cold winters.

Precipitation patterns are greatly affected by oceans, primarily because of the large amount of evaporation from the oceans and the strong temperature contrasts that can occur between water and land. Marine air masses contain more water vapor than continental air masses do, so more precipitation tends to occur along coastal areas than inland. Extremely wet areas are normally found on the leeward sides of oceans (the windward sides of continents). Areas such as the Pacific Northwest of the United States, Britain, and the southwestern coast of South America receive large amounts of annual precipitation. This is due primarily to the flow of the wind over the water and to the temperature contrast between land and water. Ocean currents can also cause tremendous droughts in areas such as the coasts of Peru and northern Chile. Arica, Chile, averages only 1 millimeter of precipitation each year because of the highly stable atmospheric conditions caused in part by strong upwelling of cold water along the coast.

Atmospheric circulations are also affected by the oceans in terms of synoptic-scale phenomena (on the order of 2,000 kilometers in size) such as cyclogenesis, and mesoscale phenomena (around 20 kilometers in size) such as sea-breeze fronts. Cyclones (low-pressure areas) tend to form in winter off the eastern coasts of North America and Asia. In both areas, an abundance of tropical water is transported into the middle latitudes and supplies heat to the atmosphere. Steep temperature gradients are normally present, creating very strong jet streams in these areas. On the local scale, intense heating of the land in contrast to the relatively cool ocean water during summer afternoons creates miniature low-pressure regions over the land. The winds flow inland into the low pressure; the warmer air over land rises; and thunderstorms occur, as can be seen over the peninsula of Florida on most summer days. [*See* Cyclogenesis.]

Exchange of Energy between Ocean and Atmosphere. The solar energy absorbed near the ocean surface is distributed by wind stirring and nighttime convection through the mixed layer, which normally extends down 100 meters. The mixed layer is separated from the colder, deeper waters by a region called the *thermocline* (distinct from the main thermocline at 800 to 1,500 meters depth). The thermocline is a layer characterized by strong stability, analogous to an inversion in the atmosphere. The oceans can transport large amounts of heat horizontally, so that there is no requirement for a balance between the incoming and the outgoing fluxes of heat at a given location on the ocean surface.

Wind-driven ocean circulations can create and sustain large local imbalances in the energy exchanges at the ocean surface. Areas of strong upwelling (on the eastern sides of oceans) produce low ocean-surface temperatures and cause the ocean to absorb much more heat than they release to the atmosphere; thus the oceanic heat budget shows a surplus in such areas. This surplus of energy is returned to the atmosphere on the western sides of the oceans by warm-water currents, such as the Gulf Stream, flowing into higher latitudes. For this reason, an energy deficit normally occurs over western ocean regions.

Over land masses, the latent and sensible heat fluxes are roughly comparable in importance. The latent heat flux (the amount of energy released or absorbed during a change in state of a substance) dominates over areas covered by vegetation, while the sensible heat flux (the amount of energy released or absorbed by a substance during a change of temperature not accompanied by a change of state) dominates over cities and deserts. Over the oceans, the flux of latent heat is much greater than that of sensible heat, primarily because of evaporation. [See Latent Heat; Sensible Heat.]

Exchange of Gases between Ocean and Atmosphere. Gases move between the ocean and the atmosphere at the ocean surface. Atmospheric gases dissolve in seawater and are distributed to all depths by mixing processes and currents. The most abundant gases in both the atmosphere and the oceans are molecular nitrogen (N_2), molecular oxygen (O_2), and carbon dioxide (CO_2) (see Table 3). Oxygen and carbon dioxide are important for biological activity and are modified at various ocean depths. Nitrogen is not used directly by living organisms except for certain bacteria. Other gases (such as hydrogen and the inert elements argon, helium, and neon) are present in small amounts.

The oceans also play a role in regulating the oxygen balance of the planet. Photosynthesis, occurring mostly in algae, produces oxygen in the oceans as it does on land, but a smaller proportion of this oxygen is used in the oceans because decomposition and other oxygen-consuming processes occur to a lesser degree in water, especially at deep levels. About 300 million metric tons of excess oxygen are released annually into the atmosphere from the oceans. After eons, however, the undecomposed marine sediments are transformed into rock and may be uplifted above the sea level; atmospheric oxygen is then used as these deposits undergo chemical changes.

Even more important is the part of the oceans and their great mass of plant life in the carbon dioxide cycle. At present the annual oceanic uptake of carbon in the form of CO_2 is calculated at 2.5 billion metric tons per year. The rate at which the oceans absorb carbon dioxide is affected by water temperataure, pH, salinity, the chemistry of the ions, various biological processes, and patterns of water mixing and circulation. The most important way in which CO_2 is broken down into carbon and oxygen is the process of photosynthesis, which requires sunlight and in the oceans occurs at shallower levels. As a result, deeper waters have higher levels of CO_2. As deep waters well up, this gas may be returned to the atmosphere.

Climatic Change. The oceans are critical in understanding and projecting climatic change for two important reasons. First, because the oceans cover more than 70 percent of the surface of the Earth, if the atmosphere is sensitive to conditions at its lower boundary, then it must reflect changes at the ocean surface. Second, the oceans, because of their great heat storage capacity, can act as a stabilizing influence on an otherwise turbulent atmosphere. Water must absorb five times as much heat energy to raise its temperature by the same amount as a comparable mass of dry soil. The high heat capacity of water prevents extreme variations in the temperature of the oceans and explains why coastal localities generally have less extreme temperature variations than inland sites.

Atmospheric and oceanic events are intimately coupled at different climatic scales and at different locations. Major advances in understanding the mechanisms of climatic change will not occur without consideration of the oceans. Sea-surface temperatures and near-surface heat content are two of the more direct variables related to external air-sea interactions. Internal interactions such as

OCEANS. Table 3. *Primary gases in the air and sea*

Gas	Percentage by Volume in the: Atmosphere	Surface Water	Water-Air Ratio
Nitrogen	78.03	47.50	0.6
Oxygen	20.99	36.00	1.7
Carbon dioxide	0.03	15.10	503.3
Argon, Hydrogen, Neon, and Helium	0.95	1.40	1.5

salinity, current systems, and sea level can also affect the atmosphere.

[*See also* Hydrological Cycle; *and* Tides.]

BIBLIOGRAPHY

Ahrens, C. D. *Meteorology Today.* 3d ed. New York, 1988.

Binkley, M. S., and J. E. Oliver. "Ocean-Atmosphere Interaction." In *The Encyclopedia of Climatology,* edited by J. E. Oliver. New York, 1987.

Duxbury, A. C., and A. B. Duxbury. *Oceans.* 3d ed. Dubuque, Iowa, 1989.

Graedel, T. E., and P. J. Crutzen. *Atmospheric Change.* New York, 1993.

Gribbin, J. *Climatic Change.* Cambridge, 1978.

Hsu, S. A. *Coastal Meterology.* San Diego, Calif., 1988.

McKnight, T. L. *Physical Geography.* Englewood Cliffs, N.J., 1990.

Pinet, P. R. *Oceanography.* New York, 1992.

Thurman, H. V. *Introductory Oceanography.* New York, 1991.

Wallace, J. M., and P. V. Hobbs. *Atmospheric Science.* New York, 1977.

MARK S. BINKLEY

OPTICAL DEPTH. *See* Radiation.

OPTICAL MEASUREMENTS. *See* Instrumentation.

ORBITAL ECCENTRICITY. *See* Orbital Variations.

ORBITAL PARAMETERS AND EQUATIONS. *[This article treats a technical aspect of climate and weather studies; some of it is intended for readers at an advanced level.]*

The Earth's elliptical orbit around the Sun is called the *ecliptic.* The Sun is not located at the center of the ellipse, but rather at one of the foci, a fact that was demonstrated in the seventeenth century by the German astronomer Johannes Kepler (1571–1630).

- *Kepler's first law.* Each planet moves in an ellipse with the Sun at one focus.
- *Kepler's second law.* The radius vector from the Sun to any one planet describes equal areas in equal times.
- *Kepler's third law.* For any two planets, the squares of the periods are proportional to the cubes of the semi-major axes of their orbits.

As the Earth travels counterclockwise around its orbit each year, it is sometimes nearer to, sometimes farther away from the Sun. Today, the Earth reaches the point on its orbital path known as *perihelion,* the point at which it is closest to the Sun, on or about 3 January. On or about 4 July, it reaches *aphelion,* the point farthest from the Sun. At aphelion the distance between the Earth and the Sun is 5 million kilometers greater than it is at perihelion.

Figure 1 displays the various elements of the Earth's orbit. The orbit of the Earth around the Sun is represented by the ellipse drawn through the perihelion, the Earth, and the aphelion. The Earth's eccentricity, which is a measure of its elliptical shape, is given by the formula $(a^2 - b^2)^{1/2}/a$, where a is the semi-major axis and b the semi-minor axis. The vernal point, γ, the position of the Sun in the sky at the time of the vernal (spring) equinox when it crosses the celestial equator from the Southern

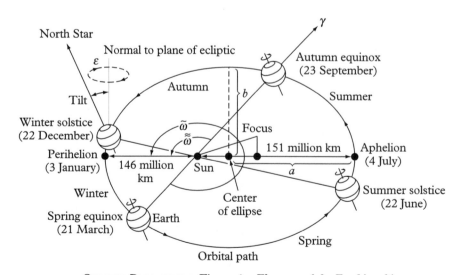

ORBITAL PARAMETERS. Figure 1. *Elements of the Earth's orbit.*

to the Northern Hemisphere, is also called the first point of Aries (it must not be confused with γ_0 in Figure 2, which is a particular γ of reference). The winter and summer solstices and γ are shown at their present-day positions. The obliquity ϵ is the inclination of the equator upon the ecliptic and is equal to the angle between the Earth's axis of rotation and the perpendicular to the ecliptic. Parameter $\tilde{\tilde{\omega}}$ is the longitude of the perihelion relative to the moving vernal point and is equal to $\pi + \psi$. The annual general precession in longitude, ψ, describes the absolute motion of γ along the Earth's orbit relative to the fixed stars. The longitude of the perihelion, π, is measured from the reference vernal point of 1950 C.E. and describes the absolute motion of the perihelion relative to the fixed stars (see Figure 2). In practice, we consider the longitudes measured from the spring equinox. Observations are made from the Earth, and the Sun is considered as if it were revolving around the Earth. In that case, the angle $\tilde{\omega}$ is defined in the geocentric coordinate system; its proper name is *longitude of the perigee*, but scientists traditionally call it *longitude of the perihelion*. In any case, the numerical values of $\tilde{\omega}$ are obtained by adding 180° to $\tilde{\tilde{\omega}}$, which is the value obtained from the trigonometrical series of Berger (1978) or Berger and Loutre (1991).

The principal cause of the seasons is not the nonuniformity of the apparent annual motion of the Sun, but rather the tilting of the Earth's axis at an angle, ϵ, presently 23°27' away from a vertical drawn to the plane of the orbit. Seasons occur because the orientation of that axis remains approximately fixed in space as the Earth revolves about the Sun in one year. Each season begins at a particular point in the Earth's orbit. Today these points are reached roughly on 22 December, 21 March, 22 June, and 23 September. In the Northern Hemi-

sphere, 22 December marks the beginning of winter because the North Pole is tipped farthest away from the Sun on that day, making it the shortest day of the year over the entire Northern Hemisphere. In this hemisphere, this point is called the winter solstice; In the Southern Hemisphere, it is called the summer solstice, the longest day of the year, marking the beginning of summer. Six months later, on 22 June, the Earth reaches the point at which summer begins in the Northern Hemisphere and winter in the Southern. At this point, the North Pole is tipped toward the Sun, making 22 June the longest day of the year in the Northern Hemisphere. [*See Seasons.*]

On 21 March and 23 September the two poles are equidistant from the Sun. On these dates, the number of daylight hours equals the number of hours of darkness at every point on the globe. These two points on the Earth's orbit are therefore known as the *equinoxes* (equal nights). In the Northern Hemisphere, the vernal equinox (21 March) marks the beginning of spring, and the autumnal equinox (23 September) marks the beginning of fall; in the Southern Hemisphere, the seasons are reversed.

Two lines, one drawn through the equinoxes and the other through the solstices, intersect at right angles to form a cross whose center is the Sun. The short arm of the cross divides the orbit into two unequal parts. The distance traveled by the Earth from 23 September to 21 March is shorter than the distance traveled from 21 March to 23 September. Therefore, in the Northern Hemisphere, spring and summer together contain 7 more days than fall and winter. In the Southern Hemisphere the situation is reversed, and the warm seasons are 7 days shorter than the cold seasons. Moreover, during the winter solstice in the Northern Hemisphere, the Earth is presently close to the perihelion and thus receives approximately 7 percent more solar radiation than at the aphelion (as measured at the outer edge of the Earth's atmosphere).

In astronomy, it is usual to define an orbit and the position of the body describing that orbit by six quantities called *elements*. As shown in Figure 2, three of the elements define the orientation of the orbit with respect to a set of axes (Ω, ω, and i), two of them define the size and the shape of the orbit (a and e respectively; see Figure 1), and the sixth (with time) defines the position of the body within the orbit at that time. In the case of a planet moving in an elliptic orbit about the Sun, it is convenient to take a set of rectangular axes in and perpendicular to the plane of reference, with the origin at the center of the Sun. The x-axis may be taken toward the ascending node

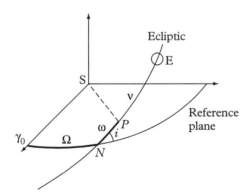

ORBITAL PARAMETERS. Figure 2. *Position of the Earth (E) around the Sun (S).*

N, the y-axis being in the plane of reference and 90° from x, while the z-axis is taken to be perpendicular to this reference plane, so that the three axes form a right-handed coordinate system. The reference point from which the angles are measured is γ_0. As the reference plane is usually chosen to be the ecliptic at a particular fixed date of reference (called the *epoch of reference* in celestial mechanics; Woolard and Clemence, 1966), γ_0 is, in such a case, the vernal point at that fixed date. P is the perihelion; Ω is the longitude of the ascending node; ω is the argument of the perihelion; π is the longitude of the perihelion and is equal to the sum of Ω and ω; i is the inclination; ν is the true anomaly; and λ, the sum of π and ν, is the longitude of the Earth in its orbit measured from γ_0 (λ, the sum of $\tilde{\omega}$ and ν, is the longitude of the Earth in its orbit measured from the spring equinox—actually, of the Sun in the geocentric coordinates system; see Figure 1). Note that ν and λ fix the Earth on its orbit respectively from the perihelion or from the vernal point of reference.

In the orbital plane, the distance, r, from the Earth to the Sun, normalized to a, is given by the equation of the ellipse:

$$\rho = \frac{r}{a} = \frac{1 - e^2}{1 + e \cos \nu} \qquad (1)$$

where $\nu = \lambda - \tilde{\omega}$. The perihelion distance is given by $a\,(1 - e)$.

Solar Energy at the Outer Edge of the Earth's Atmosphere. The Earth with its envelope of atmosphere has a diameter of about 12,800 kilometers and is approximately 150×10^6 kilometers from the Sun. The Earth intercepts approximately 0.5×10^{-12} of the Sun's total radiation. This amounts to about 1,370 watts per square meter on a surface normal to the incident radiation and situated at the mean distance, r_m, of the Earth from the Sun (defining the so-called *solar constant*, S_0). It must be noted that S_0 is not constant with time, because it is a function of e through r_m:

$$r_m^2 = a^2\,(1 - e^2)^{1/2}.$$

Figure 3 shows how to locate a point S on the celestial sphere. The astronomical horizon is the great circle in which the celestial sphere is intersected by the plane that passes through the observer, O, and is perpendicular to the direction of the local gravity. The zenith, Z, is the point vertically upward or overhead. The nadir, N, is the point diametrically opposite the zenith. The vertical cir-

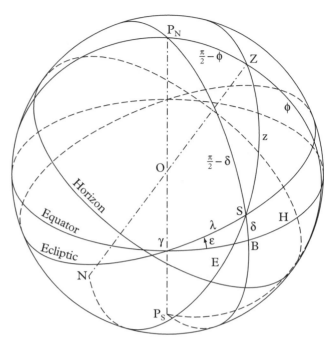

ORBITAL PARAMETERS. Figure 3. *Position of a point S on the celestial sphere.*

cles are the great circles through the zenith and nadir, and therefore necessarily perpendicular to the horizon. The two diametrically opposite points on the celestial sphere, which mark the ends of the axis of the apparent diurnal rotation of the sphere, are called celestial poles (P_N, P_S). The great circle of the celestial sphere midway between the poles is the equator. Great circles through the poles are called secondary circles. The zenith distance, z, of any point in the celestial sphere is the angular distance from the zenith measured along the vertical circle through the given point; it varies from 0 to π and is complementary to the altitude (elevation angle) E. The position of any point on the celestial sphere can also be measured by the angle between the meridian (great circle through the poles, the zenith, and the nadir) and the secondary through the point. This angle is called hour angle, H. The declination, δ, is the angular distance of the point from the equator measured on the secondary. The latitude of the observer, ϕ, is the angular distance from the equator to the zenith measured on the meridian.

Let us call W the instantaneous insolation on a horizontal surface corresponding to a zenith distance $z = \pi/2 - E$, or to a particular time in the day given by the hour angle, H. This instantaneous insolation, or irradiance, is given by the equation:

$$W = S \left(\frac{a}{r}\right)^2 \cos z. \qquad (2)$$

S is the solar constant adjusted to the distance a from the Sun:

$$S = S_0 \left(\frac{r^m}{a}\right)^2 = S_0(1 - e^2)^{1/2}.$$

Spherical trigonometry applied to the astronomical coordinates on the celestial sphere for the Earth orbital motion provides $\cos z$

$$\cos z = \sin \phi \sin \delta + \cos \phi \cos \delta \cos H \qquad (3)$$

ϕ being the latitude and δ the declination. Moreover, as

$$\sin \delta = \sin \lambda \sin \epsilon \qquad (4)$$

z, and therefore E, appears to be only a function of ϵ for a given latitude ϕ and a given hour angle H.

Using Equations 1 and 3, Equation 2 becomes

$$W = S \frac{[1 + e \cos (\lambda - \tilde{\omega})]^2}{(1 - e^2)^2} (\sin \phi \sin \delta \qquad (5)$$
$$+ \cos \phi \cos \delta \cos H).$$

Equation 5 has the advantage of showing that the precession (and eccentricity) and obliquity signals originate from two distinct factors with a clear physical meaning: the precession (and eccentricity) signal originates from the distance factor (ρ^{-2}), and the obliquity signal comes from the inclination factor ($\cos z$). Therefore, for any fixed distance from the Earth to the Sun, there is only an obliquity signal in the insolation through time, and for a fixed zenith (or elevation) angle there can be only a precessional (and eccentricity) signal.

The daily irradiation is obtained simply by integrating Equation 5 over 24 hours of true solar time, t_s. However, t_s is not regular because of the elliptical shape of the orbit and Kepler's second law. This is why the march of diurnal insolation is better described by using a regular evolving time—mean solar time, t. The relation between t and t_s is given by the so-called *equation of time*, ET, provided in the Astronomical Ephemeris for each day. ET is a measure of the difference between the time, t_s, defined from the real Sun and the time, t, related to a Sun whose apparent motion is regular, describing a circle located in the equatorial plane. ET is independent of latitude and varies presently over the year between a minimum of -14 minutes on 11 February and a maximum of $+16$ minutes on 3 November. Therefore,

$$t_s = t + ET$$

$$\frac{dt_s}{dt} = 1 + \frac{dET}{dt}.$$

Neglecting the small variations of ET compared to 1, we have $dt \sim dt_s$. The solar hour angle, H, in time units, is called t_s, and so we have:

$$H = \frac{2\pi}{\tau} t_s$$

where τ is the interval of 24 hours. Therefore:

$$dH = \frac{2\pi}{\tau} dt_s \sim \frac{2\pi}{\tau} dt$$

and the integration of Equation 5 leads to the daily irradiation W_d measured in joules per square meter:

$$W_d = \int_{24h} W \, dt = \frac{S}{\rho^2} \frac{\tau}{2\pi} \left(\int_{-12}^{-H_0} \cos z \, dH \right.$$
$$\left. + \int_{-H_0}^{H_0} \cos z \, dH + \int_{H_0}^{12} \cos z \, dH \right)$$

where the first and third terms are equal to zero because they correspond to insolation received during the night, and the second term represents insolation received between sunrise ($-H_0$) and sunset (H_0), H_0 being given by $\cos H_0 = - \tan \phi \tan \delta$ obtained from Equation 3, in which $z = \pi/2$.

Averaging W_d over a constant time period, τ, of 24 hours leads to the 24-hour mean irradiance $\overline{W}_d$ measured in watts per square meter.

$$\overline{W}_d = \frac{1}{\tau} W_d.$$

More precisely for the latitudes $|\phi| < \pi/2 - |\delta|$, where there is a sunrise and a sunset every day, we have:

$$\overline{W}_d = \frac{S}{\pi} \left(\frac{a}{r}\right)^2 (H_0 \sin \phi \sin \delta$$
$$+ \cos \phi \cos \delta \sin H_0) \qquad (6)$$

and the length of the day (in hours) is given by $\tau/\pi \cos^{-1}$ $(-\tan \phi \tan \delta)$. For the latitudes $|\phi| \geq \pi/2 - |\delta|$, where there is the long polar night, the length of the day is equal to 0 and we have:

$$\phi \cdot \delta \leq 0 \quad \text{and} \quad \overline{W}_d = 0. \qquad (7)$$

Where there is the long polar day, the length of the day is equal to 24 hours and we have:

$$\phi \cdot \delta > 0 \quad \text{and} \quad \overline{W}_d = S \left(\frac{a}{r}\right)^2 \sin \phi \sin \delta. \quad (8)$$

The poles belong to this second case (either long polar night or long polar day) all year. At the equinoxes ($\lambda = 0$ and $\lambda = \pi$), we are at the transition between the two.

Therefore, from Equations 2, 7, and 8, the seasonal and latitudinal distributions of solar radiation at the upper limit of the atmosphere can be represented with high accuracy as a mathematical function of latitude, time, and Earth's orbital and rotational elements.

We can see that the energy received on a horizontal surface at any latitude ϕ on the Earth, on the assumption of a perfectly transparent atmosphere and of a constant solar output, depends on the so-called solar constant, S_0, and on the Earth's orbital and rotational parameters, which reflect the gravitational effect of the Sun, the Moon, and the planets. These include the semi-major axis (a) of the Earth's orbit around the Sun, its eccentricity (e), the obliquity (ϵ), and the longitude of the perihelion ($\tilde{\omega}$) measured from the moving vernal equinox,

which plays a role through the climatic precession parameter ($e \sin \tilde{\omega}$).

Two of these elements play a dominant role in calculating the insolation: the precession of the equinoxes, which alters the distance between Earth and Sun at a fixed given time in the year, and the obliquity, which affects the seasonal contrast and the latitudinal gradient of insolation. The eccentricity determines primarily the amplitude of the precession cycle and is also the only parameter that can change the total energy received by the Earth each year, through altering the mean distance from the Earth to the Sun, although the latter impact is quite small quantitatively, of the order of a few tenths of 1 percent at the most for the largest changes in eccentricity.

Present-day values of these astronomical elements are $e = 0.017$, $\epsilon = 23°27'$, and $\tilde{\omega} = 281°$. The daily solar energy received at 60° north latitude at spring equinox is 218 watts per square meter, 476 at the summer solstice, 215 at fall equinox, and 24 at the winter solstice. Corresponding watts per square meter at 60° south latitude

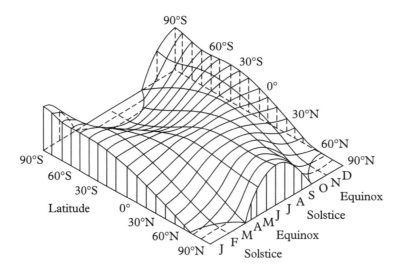

ϕ	90	70	50	30	0	−30	−50	−70	−90
3 January	0	0	90	231	413	505	509	514	547
21 March	0	149	280	378	436	378	280	149	0
22 June	524	492	482	474	384	213	80	0	0
4 July	513	482	477	472	386	216	84	0	0
23 September	0	147	276	373	430	372	276	147	0
21 December	0	0	86	227	410	507	514	526	559
Positive (negative) latitudes for the Northern (Southern) Hemisphere									

ORBITAL PARAMETERS. Figure 4. *Solar irradiance (watts per square meter) at the top of the atmosphere as a function of the latitude and seasons through the year.*

are 215 on 23 September, 515 on 22 December, 218 on 21 March, and 23 on 22 June. The asymmetry reflects the present-day situation of the seasons relative to the aphelion. At the equator, the Earth receives 436 watts per square meter on 21 March, 384 on 22 June, 430 on 23 September, and 411 on 22 December, with a double maximum and minimum reflecting the passage of the Sun twice at the zenith of any inner tropical latitude during a year. Figure 4 displays these seasonal variations diagrammatically.

[See also Orbital Variations.]

BIBLIOGRAPHY

Berger, A. "Long-term Variations of Daily Insolation and Quaternary Climatic Changes." *Journal of Atmospheric Science* 35 (1978): 2362–2367.

Berger, A. and M. F. Loutre. "Insolation Values for the Climate of the Last 10 Million Years." *Quaternary Science Reviews* 10 (1991): 297–317.

Wollard, E. W., and G. M. Clemence. *Spherical Astronomy.* New York, 1966.

ANDRÉ BERGER

ORBITAL VARIATIONS. *[This article treats a technical aspect of climate and weather studies; some of it is intended for readers at an advanced level.]*

The amount of incoming solar radiation received at any point on the Earth has an annual periodic variation caused by the Earth's elliptic translation motion around the Sun. [See Orbital Parameters and Equations.] In addition, the seasonal and latitudinal distributions of this solar radiation have a long-period variation caused by the so-called secular variations in the orbital elements. The total solar energy received by the Earth over a year also varies in relation with one of these orbital parameters—the eccentricity of the Earth's orbit. Other long-period variations affecting the solar constant and attributable to changes in the physical condition of the Sun or in the transparency of the interplanetary medium have been postulated from time to time, but their effect remains difficult to prove.

Analysis of the incoming solar radiation and of its geographic distribution shows that there exist three main parameters for calculating long-term variations (Berger et al., 1993): the eccentricity, e, a measure of the shape of the Earth's orbit around the Sun; the obliquity, ϵ, the tilt of the equator on the Earth's orbit; and the climatic precession, $e \sin \tilde{\omega}$, a measure of the distance from Earth to Sun at the summer solstice ($\tilde{\omega}$ being the longitude of the perihelion measured from the moving equinox). As

determined from celestial mechanics, the secular variations of these elements of the Earth's orbit and rotation are due to the gravitational perturbations which the Sun, the other planets, and the Moon exert on the Earth's orbit and on its axis of rotation. Two systems therefore have to be considered: one deals with the motion of the nine planetary point masses around the Sun; the other considers the rotation of the Earth as a result of the lunisolar attraction.

The first problem gives rise to the Lagrange differential equations (Brouwer and Clemence, 1961) relating all the orbital elements of the planets together and describing their motion around the Sun. The second problem is solved through the Poisson equations for the Earth-Moon-Sun system, leading to the determination of the precession in longitude (ψ), involved in the calculation of the longitude of the perihelion ($\tilde{\omega} = \pi + \psi$), and of the obliquity of the ecliptic.

The astronomical computations have allowed us to reconstruct past variations with very good accuracy. The short-term variations of the astro-climate elements (e, ϵ, $e \sin \tilde{\omega}$) have been computed for the past few millennia (Loutre et al., 1992) using one of the classical secular planetary theories built for the ephemerides and used in astronomical observatories (another well-known expression was calculated by Simon Newcomb in 1898 for the eccentricity as a function of time). Spectral analyses of the astronomical parameters for the past 5,000 years show periodicities ranging from 2.67 years to 250 years. Particularly significant for the eccentricity and climatic precession are the periods of 2.67, 3.98, 5.93 and 11.9 years. Obliquity shows, in addition, periods of 29 and 18.6 years, the last one arising from the nutations related to the Moon and well known in the theory of the Earth's tides. Accordingly, orbitally induced short-term variations in insolation have periods of 18.6, 11.9, 5.9, 4.0, and 2.7 years. [See Orbital Parameters and Equations.]

For the long-term variations over the last few million years, the equations for the motion of the planets around the Sun, for the precession, and for the obliquity cannot be integrated analytically, but their numerical solutions can be expressed in trigonometrical form as quasi-periodic functions of time:

$$e = e^{\star} + \sum E_i \cos (\lambda_i t + \phi_i) \qquad (1)$$

$$\epsilon = \epsilon^{\star} + \sum A_i \cos (\gamma_i t + \zeta_i) \qquad (2)$$

$$e \sin \tilde{\omega} = \sum P_i \sin (\alpha_i t + \eta_i) \qquad (3)$$

where the amplitudes P_i, A_i, E_i, the frequencies α_i, γ_i, λ_i and phases η_i, ζ_i, ϕ_i, are given in Berger and Loutre

ORBITAL VARIATIONS. Table 1. *Amplitudes, mean rates, phases and periods of the five largest amplitude terms in the trigonometrical expansions of climatic precession, obliquity, and eccentricity*★

Climatic Precession			
Amplitude	Mean Rate (''/year)	Phase (degrees)	Period (years)
0.018970	54.66624	32.2	23,708
0.016318	57.87275	201.3	22,394
0.012989	68.33975	153.4	18,964
0.008136	67.79501	311.4	19,116
0.003870	55.98574	78.6	23,149

Obliquity			
Amplitude ('')	Mean Rate (''/year)	Phase (degrees)	Period (years)
−1969.00	31.54068	247.14	41,090
−903.50	32.62947	288.79	39,719
−631.67	32.08588	265.33	40,392
−602.81	24.06077	129.70	53,864
−352.88	30.99683	43.20	41,811

Eccentricity			
Amplitude	Mean Rate (''/year)	Phase (degrees)	Period (years)
0.011268	3.20651	169.2	404,178
0.008819	13.67352	121.2	94,782
0.007419	10.46700	312.0	123,818
0.005600	13.12877	279.2	98,715
0.004759	9.92226	110.1	130,615

★This very limited number of terms does not allow computation of these astronomical parameters with an acceptable accuracy.
Source: Berger and Loutre, 1991.

(1991) for a new solution, for which the five largest amplitude terms are listed in Table 1.

Actually, in the term $e \sin \tilde{\omega}$, the amplitude of $\sin \tilde{\omega}$ is modulated by eccentricity, and the envelope of $e \sin \tilde{\omega}$ is given exactly by e because the frequencies of e originate strictly from combinations of the frequencies of $\tilde{\omega}$. Figure 1 shows the long term variations of these three astronomical parameters over the past 200,000 years and the next 100,000 years.

Over the past 3×10^6 years, the eccentricity of the orbit varies between near circularity ($e = 0$) and slight ellipticity ($e = 0.07$) at a period with a mean of about 100,000 years. The most important terms in the series expansion occur, however, at 413,000, 95,000, 124,000, 99,000, and 131,000 years (in decreasing order of amplitude). The tilt of the Earth's axis varies between about 22° and 25° at a period of nearly 41,000 years. Although this period corresponds in Equation 2 to the amplitude that is by far the largest, there are two other important terms, one with a period of 54,000 and one with 29,000 years.

As far as precession is concerned, two components must be considered. The first is the axial precession, in which the torque of the Sun, the Moon, and the planets on the Earth's equatorial bulge causes the axis of rotation to wobble like that of a spinning top. The net effect is that the North Pole describes a clockwise circle in space (provided the nutations—the terms with small amplitude and much smaller periods—are neglected), with a period

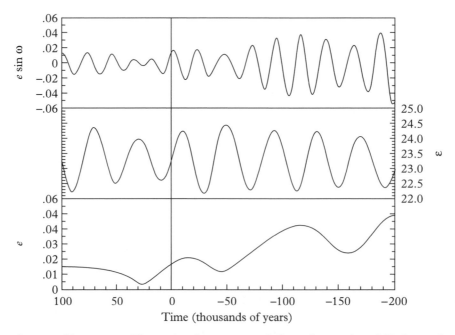

ORBITAL VARIATIONS. Figure 1. *Long-term variations of precession, obliquity, and eccentricity from 200,000 years ago to 100,000 years into the future.* (From Berger and Loutre, 1991.)

of about 25,800 years corresponding to the period of the vernal equinox against a fixed reference point (Figure 2). The second is related to the fact that the elliptical figure of the Earth's orbit is itself rotating counterclockwise in the same plane, leading to an absolute motion of the perihelion with a period, measured relative to the fixed stars, of about 100,000 years (the same as that of the eccentricity). The two effects together result in what is known as the *climatic precession of the equinoxes,* a motion mathematically described by $\tilde{\omega}$, and in which the equinoxes and solstices shift slowly around the Earth's orbit relative to the perihelion, with a mean period of 21,000 years (this period actually results from the existence of two closely similar periods of 23,000 and 19,000 years; see Table 1).

Therefore, while today the winter solstice occurs near perihelion, some 11,000 years ago it occurred near aphelion (Figure 3). Moreover, because the lengths of the seasons vary in time according to Kepler's second law, the solstices and equinoxes have occurred at different calendar dates during the geological past. Presently in the Northern Hemisphere, the longest seasons are spring (92 days, 19 hours) and summer (93 days, 15 hours); fall (89 days, 20 hours) and winter (89 days) are definitively the shortest. In about 1250 C.E. spring and summer had exactly the same length as fall and winter because the winter solstice was occurring at the perihelion. About 4,500 years from now, the Northern Hemisphere spring and winter will have the same shorter length and, consequently, summer and fall will be equally long.

Impact on Solar Radiation. The combined influence of changes in e, ϵ, and $e \sin \tilde{\omega}$ produces a complex pattern of insolation variations. A detailed analysis of the solar radiation (Berger et al., 1993) shows that it is principally affected by variations in precession, although the obliquity plays a more important role as latitudes increase,

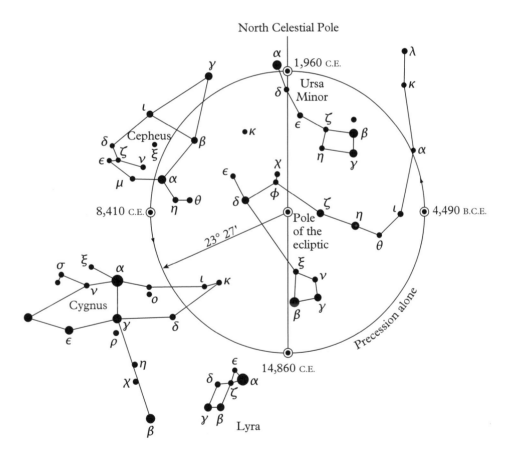

ORBITAL VARIATIONS. Figure 2. *The precession of the north celestial pole about the North Pole of the ecliptic; one cycle takes about 25,800 years.* Today, the point around which the stars seem to rotate when viewed from the Northern Hemisphere falls near the star Polaris, which forms the end of the handle of the Little Dipper; but in 2000 B.C.E. the North Pole pointed to a spot midway between the Little and the Big Dippers, and in 4000 B.C.E. it pointed to the tip of the handle of the Big Dipper.

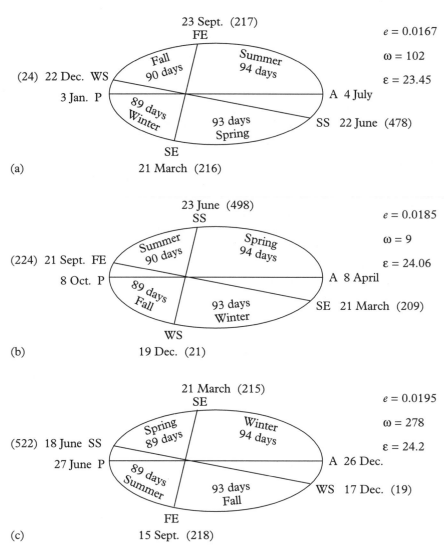

ORBITAL VARIATIONS. Figure 3. *Shift of the position of the equinoxes and the solstices around the Earth's elliptical orbit for (a) today, (b) 5,500 years ago, and (c) 11,000 years ago.* SE: spring equinox, FE: fall equinox, SS: summer solstice, WS: winter solstice, P: perihelion, A: aphelion. The orientation of this orbit has been kept arbitrarily the same and the beginning of the spring is fixed on calendar date 21 March. The beginning of the other seasons is changing in time because the length of the seasons is changing. The numerical values of the eccentricity, e, of the obliquity, ϵ, of the longitude of the perihelion, $\tilde{\omega}$ (measured from FE), and of the 60° north insolation at the solstices and equinoxes (values in parentheses) are taken from Berger and Loutre (1991). (In this figure, the shape and angles are exaggerated for clarity.)

mainly in the winter hemisphere. At the equinoxes, insolation for each latitude is only a function of precession. At the solstices, both precession and obliquity are present, although precession dominates for most latitudes (Figure 4).

Changes in incoming solar radiation caused by changes in tilt are the same in both hemispheres during the same local season: An increase of ϵ leads to an increase in insolation in the summer hemisphere and to a decrease in the winter hemisphere. Because the strength of the effect is small in the tropics and maximal at the pole, an increase in obliquity tends to amplify the seasonal cycle in the high latitudes of both hemispheres simultaneously.

The precession effect can cause warm winters and cool summers in one hemisphere while doing the opposite in the other. For example, portions of the present winter

Daily insolation (watts per square meter)

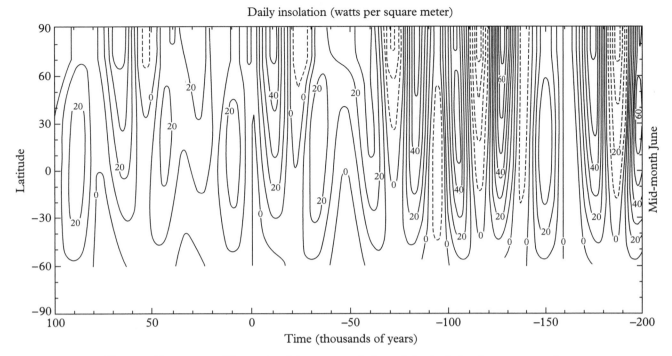

ORBITAL VARIATIONS. Figure 4. *Long-term variations of the deviation from present values of the irradiation at the Northern Hemisphere summer solstice from 200,000 years ago to 100,000 years in the future.* (From Berger and Loutre, 1991.)

Northern Hemisphere receive as much as 10 percent more insolation than they did 11,000 years ago, when the perihelion occurred in the Northern Hemisphere summer.

Milankovitch Theory. The orbital hypothesis was first quantitatively formulated by the Serbian astronomer Milutin Milankovitch in the 1920s and 1930s. His early calculations (1941) provided information on the variations in incident solar radiation as a function of latitude for the last million years in winter and summer. [*See the biography of Milankovitch.*] He argued that insolation changes in the high northern latitudes during the summer season were critical to the formation of continental ice sheets. During periods when insolation in the summer was reduced, the snow of the previous winter would tend to be preserved—a tendency that would be enhanced by the high albedo of the snow and ice fields. Eventually, the effect of this positive feedback would lead to the formation of ice sheets.

Looking at the mathematics of insolation, a minimum in the Northern Hemisphere summer insolation at high latitudes requires a Northern Hemisphere summer occurring at the aphelion, a maximum eccentricity which leads to a large distance between the Earth and the Sun at the aphelion, and a minimum obliquity implying a weak seasonal contrast and an increased horizontal gradient between the equator and the poles. Following these requirements, it may be suggested not only that the northern high latitudes would be cool enough in summer to prevent snow and ice from melting, but also that a mild winter would allow substantial evaporation in the intertropical zone and abundant snowfalls in temperate and polar latitudes, the humidity being supplied there by an intensified general circulation caused by a maximum latitudinal gradient.

A simple linear version of the Milankovitch model would therefore predict that the total ice volume and climate over the Earth would vary with the same regular pattern as the insolation. This means that the proxy record of climate variations would reflect those frequencies of the astronomical parameters that are responsible for changing the seasonal and latitudinal distribution of incoming solar radiation. Investigations during the past 20 years have indeed demonstrated that the 19,000, 23,000, and 41,000 year periodicities actually occur in long records of the Quaternary climate (Hays et al., 1976), that the climatic variations observed in these frequency bands are linearly related to the orbital forcing functions, and that there is a fairly consistent phase relationship among insolation, sea surface temperature, and ice volume (Imbrie et al., 1992). The geological observation of the bipartition of the precessional peak, confirmed to be real in astronomical computations (Berger, 1977), was one of the first and most impressive of all tests for the Milankovitch theory. The same investigations, however, identified the largest climatic cycle as 100,000 years. Because the eccentricity signal is very weak in the insolation, it cannot be related to orbital forcing by any simple linear mechanism. Actually, the variance components centered near this 100,000-year cycle seem to be in phase with the eccentricity cycle, but its exceptional strength requires some nonlinear amplification—by the glacial ice sheets themselves, involving mechanisms such as ice albedo feedback and isostatic rebound of the lithosphere; by the carbon dioxide atmospheric concentration; or by the oceanic circulation.

The interpretation of the geological record is not always so clear. The 100,000-year cycle, so dominant a feature of the ice volume record for the late Pleistocene, does not exhibit a constant amplitude over the past 2 to 3 million years; it disappears before 10^6 years ago, at a time when the ice sheets were much less developed over the Earth (Shackleton et al., 1988). The shape of the spectra and is far from being the same from one climatic parameter to another (Imbrie et al., 1992).

Since the publications of Hays et al. (1976), a number of modeling studies have attempted to explain the relation between astronomical forcing and climatic change. Both general circulation models—which provide snapshot views of the climate in equilibrium with some boundary conditions—and statistical-dynamical models, which take into account all the different parts of the climate system in an interactive and time-dependent way, have been used. For example, simulations of monsoon variability have provided insight into possible controls of precession in low latitudes (Prell and Kutzbach, 1987). Changing the orbital configurations to that of 9,000 years ago, when insolation seasonality was 14 percent higher than today, leads to an intensified southwest monsoon; under glacial conditions, however, the simulated monsoon is weakened in southern Asia, but precipitation is increased in the equatorial western Indian Ocean and in equatorial North Africa.

Most modeling studies have focused on the origin of the 100,000-year cycle in ice volume with two main classes of possible explanation: In one, ice volume fluctuations are primarily driven by orbital forcing, with nonlinear interactions in the air-sea-ice system modifying the signal; in the other, ice volume fluctuations result from inherently nonlinear interactions in the system, with

orbital variations serving only to phase-lock the variations at the preferred time scales. Although specific details in models of the first kind vary considerably, their success is related to the observation that the 100,000-year power can be generated by 19,000- and 23,000-year periods, producing, through a nonlinear system, substantial power in both harmonics (10,000 years) and subharmonics (100,000 years) (Wigley, 1976). In an example of the second kind of model, Le Treut and associates (1988) modeled internally generated glacial–interglacial fluctuations which, only when orbitally forced, produce characteristic Milankovitch periods, harmonics, and subharmonics. Saltzman and Sutera (1987), using a different approach of incorporating a small number of variables (carbon dioxide, surface and deep water temperatures, and ice volume) in a model, were also able to reproduce the 100,000-year cycle and, in addition, the transition from dominant 41,000-year fluctuations prior to 900,000 years ago to the dominant 100,000-year ice volume fluctuations after that time—a result which reflects the possible existence of multiple equilibrium states in the system.

Another set of experiments using time-dependent physical climate models is also providing significant results. Such a 2.5-dimension model, taking into account some of the most important feedbacks among the atmosphere, the upper ocean, the sea ice, the ice sheets, and the lithosphere, has been used to study the transient response of the climate system to orbital and carbon dioxide (CO_2) forcings (Gallée et al., 1992). Over the last glacial–interglacial cycles, it has simulated with some success the gross characteristics of the low-frequency part of the climatic variations reconstructed from proxy data. In particular, oxygen isotopic stages 5 to 1 are reproduced (Figure 5), and the individual ice volumes of the North American, Eurasian, and Greenland ice sheets are in agreement with reconstructions independently obtained from field data. At the last glacial maximum (LGM), the simulated temperature of the Northern Hemisphere is about 4°C colder than now, and the simulated total ice volume is 50×10^6 cubic kilometers larger than today. Moreover, owing to the ice albedo and water vapor feedbacks, 50 percent of the temperature change and 30 percent of the ice volume change in the Northern Hemisphere at the LGM can be attributed to the long-term CO_2 change measured in the air bubbles trapped in the Vostok ice core.

The Astronomical Periods over Pre-Quaternary Times. The orbital and axial parameters are probably now well defined for the last several million years, but

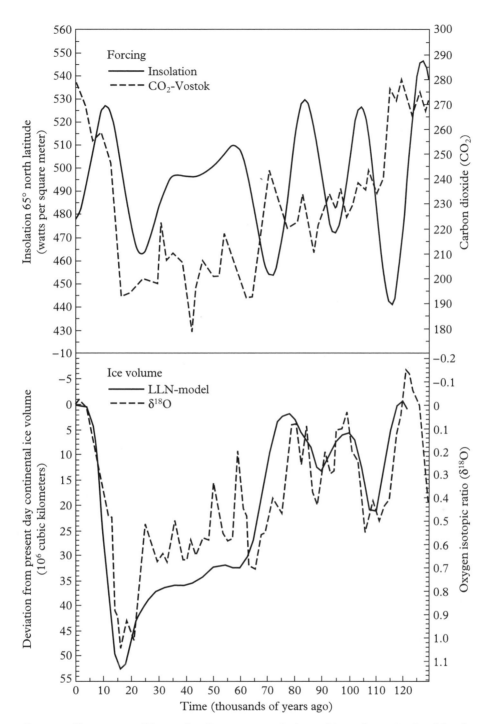

ORBITAL VARIATIONS. Figure 5.   *Long-term variations of ice volume simulated by the LLN-model (from Gallée et al., 1992) in response to insolation and CO₂ forcings over the last glacial-interglacial cycle. The dotted line of the lower panel represents the δ¹⁸O values* taken from Labeyrie et al. (1987) as a proxy for total ice volume over the Earth.

much remains to be learned about these variations early in the Earth's history. Sensitivity analyses (Berger et al., 1992) of the astronomical frequencies have suggested that the shortening of the Earth–Moon distance and of day length in past time would induce a shortening of the fundamental periods for the obliquity and climatic precession. Over the last half-billion years, these periods would change respectively from 54 to 35, 41 to 29, 23 to 19, and 19 to 16 thousand years. On the other hand, Laskar (1989) showed that the behavior of the solar system has been chaotic and that the obliquity of the Earth's orbit might have reached 90 degrees.

BIBLIOGRAPHY

Barnola, J. M., D. Raynaud, Y. S. Korotkevitch, and C. Lorius. "Vostok Ice Core: A 160,000 Year Record of Atmospheric CO_2." *Nature* 329 (1987): 408–414.

Berger, A. "Support for the Astronomical Theory of Climatic Change." *Nature* 268 (1977): 44–45.

Berger, A., and M. F. Loutre. "Insolation Values for the Climate of the Last 10 Million Years." *Quarternary Science Reviews* 10 (1991): 297–317.

Berger, A., M. F. Loutre, and J. Laskar. "Stability of the Astronomical Frequencies over the Earth's History for Paleoclimate Studies." *Science* 255 (1992): 560–566.

Berger, A., M. F. Loutre, and C. Tricot. "Insolation and Earth's Orbital Periods." *Journal of Geophysical Research* (1993).

Brouwer, D., and G. M. Clemence. *Methods of Celestial Mechanics.* New York, 1961.

Gallée, H., et al. "Simulation of the Last Glacial Cycle by a Coupled Sectorially Averaged Climate-Ice Sheet Model. II. Response to Insolation and CO_2 Variation." *Journal of Geophysical Research* 97 (1992): 15,713–15,740.

Hays, J. D., J. Imbrie, and N. J. Shackleton. "Variations in the Earth's Orbit: Pacemaker of the Ice Ages." *Science* 194 (1976): 1121–1132.

Imbrie, J., et al. "On the Structure of Major Glaciation Cycles." *Paleoceanography* 7 (1992).

Labeyrie, L. D., J. C. Duplessy, and P. L. Blanc. "Deep Water Formation and Temperature Variation over the Last 125,000 Years." *Nature* 327 (1987): 477–482.

Laskar, J. "A Numerical Experiment on the Chaotic Behavior of the Solar System." *Nature* 338 (1989): 237–238.

Le Treut, H., J. Portes, J. Jouzel, and M. Ghil. "Isotopic Modeling of Climate Oscillations: Implications for a Comparative Study of Marine and Ice Core Records." *Journal of Geophysical Research* 93 (1988): 9365–9384.

Loutre, M. F., A. Berger, P. Bretagnon, and P. L. Blanc. "Astronomical Frequencies for Climate Research at the Decadal to Century Time Scale." *Climate Dynamics* 7 (1992): 181–194.

Milankovitch, M. M. *Kanon der Erdbestrahlung und seine Anwendung auf des Eizeitenproblem.* Belgrade, 1941.

Prell, W. L., and J. E. Kutzbach. "Monsoon Variability over the Past 150,000 Years." *Journal of Geophysical Research* 92 (1987): 8411–8425.

Saltzman, B., and A. Sutera. "The Mid-Quaternary Climatic Transition as the Free Response of a Three-Variable Dynamical Model." *Journal of Atmospheric Sciences* 44 (1987): 236–241.

Shackleton, N. J., J. Imbrie, and N. G. Pisias. "The Evolution of Oceanic-Isotope Variability in the North Atlantic over the Past Three Million Years." *Philosophical Transactions of the Royal Society of London* B 318 (1988): 679–688.

Wigley, T. M. L. "Spectral Analysis and Astronomical Theory of Climatic Change." *Nature* 264 (1976): 629–631.

ANDRÉ BERGER

OROGRAPHIC EFFECTS. The Earth's mountains affect weather in many ways, on spatial scales ranging from 1 kilometer or less up to the global circulation, and on time scales ranging from a few minutes to days, weeks, seasons, years. The cumulative effects of all these processes constitute climate. The importance to weather of orographic (mountain-formed) effects is comparable to that of the difference between the Earth's two surfaces—oceans and continents. Mountains modify the large-scale atmospheric flow, change the course and evolution of storm systems, generate their own air-circulation patterns, and alter the distribution of cloudiness and precipitation. These processes are described below under the classes of global-scale effects, regional and local winds, and orographic clouds and precipitation.

Global-Scale Effects. The average general circulation of the Earth's atmosphere shows prominent features in the airflow, called *stationary planetary-scale waves*. In the Northern Hemisphere, large-scale troughs (areas of reduced pressure and cyclonic curvature of the airflow) are often found along the eastern coasts of the continents. These troughs form in response to the continuous influence on airflow of the major surface features of the planet—oceans, continents, and large mountain ranges.

In the mid latitudes (from around 30° to 60°), the persistent airflow in the troposphere (lowest atmospheric layer) is from the west. This air must rise over the major north–south mountain chains, and, in order to conserve angular momentum, it turns toward the equator. As it descends the lee side of the mountains, it turns eastward again; the result is a trough with cyclonic curvature in the lee of the mountain range. This tendency to form low-pressure areas in the lee of mountains depends on the strength and direction of the airflow. It is strongest in mid-latitude winter conditions when the westerly winds are strongest. In equatorial latitudes, however, the average airflow is weak and from the east, so the influence of mountains on large-scale atmospheric dynamics is weak. In spite of this, mountains affect tropical precipitation quite strongly, as described below. The lee-side low-pressure areas are breeding grounds for cyclones (low-pressure traveling storm systems). This birth and development of new storms is termed *cyclogenesis.* Preferred regions for cyclogenesis include the areas south of the Alps, east of the Rocky Mountains, east of the Altai Mountains in central China, and east of the Tibetan Plateau. [*See* Cyclogenesis.]

Mountains also impose drag on the large-scale airflow, causing it to decelerate and change direction. Pressure drag results from the average lower pressure in the lee and higher pressure on the windward side. There is also surface drag caused by the greater roughness of mountains and resulting friction. [*See* Friction.]

Extensive mountain ranges may also contribute to blocking the movement of traveling cyclones. The Northern Hemisphere's circulation alternates between periods of mostly west-to-east flow, called *zonal,* and periods with substantial north and south components, called *meridional.* The meridional flow is characterized by waves of large amplitude, with traveling storms or cyclones in the troughs and anticyclones (high-pressure centers) along the ridges. Often the meridional flow is strong, persistent, and relatively stationary. Under such conditions traveling storms are generated in the troughs and move downstream, but they are deflected poleward around the areas of high pressure. In this situation, termed *blocking,* storms are prevented from following their normal west-to-east path. Although the blocking phenomenon is recognized, its causes and dynamics are not well understood. The fact that it occurs preferentially in regions that are generally aligned with mountain ranges suggests that the orographic effect on airflow is a contributing factor. Another important factor is the temperature contrast between oceans and continents. [*See* Blocking.]

Monsoons are airflow circulation patterns that form in response to the difference in the heating rates of land and ocean, along with interactions between the large-scale airflow and mountains. Monsoons change seasonally, accompanied by large variations in precipitation. The southern Asian monsoon is probably the best-known example. The Himalaya Mountains are a significant block to tropospheric air flow, although they are not a long north–south mountain range. During winter, the mid-latitude atmospheric jet stream (a narrow region of strong winds) flows south of the Himalayas. In late May,

however, the jet begins moving north, and this marks the onset of the monsoon. The jet stream's transition from the southern side of the mountains to the northern side does not occur gradually; it is sudden, taking only a few days. As the jet stream moves north, the Intertropical Convergence Zone (ITCZ) and its easterly jet move over India, essentially reversing the previous circulation pattern. The ITCZ brings a considerable low-level flow of moist, warm air; as this rises against the mountains, deep convective clouds form, with heavy precipitation and release of latent heat. This latent heat contributes to driving the monsoon circulation. In summer, the Tibetan Plateau is an elevated heat source, providing sensible heat near the middle of the atmosphere rather than near sea level. This also helps to drive the monsoon circulation. The monsoon persists for several months, until seasonal changes in solar heating terminate the circulation. [*See* Monsoons; Tropical Circulations.]

Regional and Local Winds. Mountain systems create regional and local wind systems. The mountain-induced winds are very significant within the mountain valleys and ridges and can even dominate the daily patterns of airflow over extensive areas away from the higher terrain. The most familiar example of this is the mountain-valley circulation. [*See* Small-Scale or Local Winds.]

Mountain-valley circulations. This type of airflow is caused by the differences in heating and cooling rates between mountain slopes and valleys. It is a fair-weather circulation, and the forces that drive it are similar to those that drive the sea-breeze circulation. Mountain slopes that face the Sun during the early part of the day receive more incoming radiation per unit area than does surrounding flat land. Consequently, the boundary-layer air along such slopes becomes warmer and therefore more buoyant than surrounding air at comparable elevations. The air moves upvalley and upslope, and rises above the ridges. To compensate for this flow, air also flows along the surface over nearby lower-elevation regions toward the mountains. This pattern shifts during the day as the angle of the Sun changes; shortly after sunset, the pattern reverses. At night, the higher-elevation slopes cool by radiating energy to space, and shallow layers of air along the slopes become colder than the free atmosphere at comparable elevations. This cool, dense air flows downslope and downvalley; it may also pool in basins.

When sufficient moisture is available, clouds form in response to the mountain-valley circulation. During the daytime, clouds develop over the ridges in rising air currents created along the heated mountain slopes. Latent heat is released in these clouds, providing additional buoyancy to the upward-moving air currents; large convective clouds and thunderstorms can develop. This cloudiness can subsequently shade the ground from further solar heating, leading to the onset of flow downslope and away from the mountains. This downslope flow can combine with the outflow from precipitating clouds. When this downward flow meets the flow coming in from the plains, convergence regions form, fostering the development and intensification of storms. The storms generally continue to move away from the mountains, steered by the upper-level winds. They can develop into larger cloud systems later in the day.

Drainage flows. Mountain-induced cloudiness and temperature differences are particularly significant in producing daytime cloudiness and precipitation associated with upslope airflows; cooling in valleys and other low-elevation locations, however, is associated with cool, dense downslope air drainage. In winter, many valleys experience extremely cold temperatures—even colder than nearby mountains peaks. This effect is accentuated when mountain slopes are snow-covered. Pools of cold, dense air in mountain valleys and closed canyons can become trapped under temperature inversions created by radiational cooling and drainage flows. They can persist for several days or longer and be subject to further radiational cooling. In larger mountain basins, this cool dense air accumulates and after several weeks, it can lead to the formation of broad high-pressure regions. This is a fairly common occurrence in the Great Basin of the western United States, bounded by the Sierra Nevada range on the west and the Rocky Mountains on the east. In populated or industrial areas, this stagnant air can become dangerously polluted. The regional high pressure contributes to the blocking of cyclonic storms, although the cold air pool occasionally gets flushed out by the passage of strong storm systems.

Barrier jets. The precipitation in traveling storm systems releases a significant amount of latent heat. When such storms encounter extensive mountain barriers, they slow down, and the effect of latent heat release can change the local thermodynamic and dynamic balance. A so-called *barrier jet* can develop in this situation, with the jet blowing parallel to the barrier ridge line at about the same elevation. For example, winter storms over the Sierra Nevada in California often develop a barrier jet which blows from south to north along the western side of the mountain range.

Channeled air flow and strong downslope winds. As cold, stable air masses encounter extended mountain barriers, their movement can be channeled into the valleys, resulting in strong winds that blow parallel to the valleys for long periods. Deep air masses flow over mountain barriers, producing warming on the lee side. This heating is caused by two effects: compressional heating occurs along the lee slope where descending air is forced to lower levels (higher pressures); and latent heating occurs on the windward side of the mountain barrier in rising air, where moisture condenses in clouds and then falls out as precipitation. When lee troughs, with their lower pressure, are present, large pressure gradients develop, the downslope flow accelerates, and hurricane-force winds are often produced. These warm, dry, high-speed, downslope winds are called *chinooks* in the Rocky Mountains, *foehns* in the Alps, and *Santa Ana winds* in southern California. If, however, the air mass is especially cold, then its descent will not warm it appreciably; a dry, cold wind resulting from such conditions is called a *bora* or *mistral* in Europe.

Mountain waves. If the air is stable and flows at high speed over a mountain barrier, with a direction approximately perpendicular to the ridge line, waves often form in the airflow. The first wave is over the ridge line, and additional waves form downstream at regular intervals of about 10 kilometers. The waves are stationary, fixed in position with respect to the ground. The air rises into the upstream side of the waves and descends from the downstream side. This phenomenon is similar to one often seen in rivers: waves form downstream of a boulder as water flows over it. If there is sufficient moisture in the air, these waves are visible as *lenticular clouds,* smooth, lens-shaped clouds having a long axis that aligns with the ridgeline. In such a situation, *rotor clouds* sometimes form near the base of the mountain; these clouds are extremely turbulent. The so-called *hydraulic jump,* which forms downstream of shallow dams, also has an atmospheric counterpart. If the high-speed, shallow downslope flow exceeds a critical value, it becomes unstable and transforms into a deeper region of low-speed flow—a hydraulic jump. Hydraulic jump regions are also very turbulent. Whenever air flows through and over rugged mountainous terrain, turbulence—irregular and random fluctuations of airflow—is generated. The *mechanical turbulence* generated over mountains is generally distinguished from *thermal turbulence,* which is associated with cumulus clouds. Turbulence breaks down inversion layers and enhances the mixing of air. [*See* Turbulence.]

Orographic Clouds and Precipitation. Clouds that form during the forced ascent of air over mountain barriers are called *orographic clouds.* They occur throughout the year, worldwide; in some locations, they are responsible for large increases in precipitation, compared to nearby nonmountainous areas. Mountainous regions often have complicated terrain, with intersecting ridges of different elevations, interconnected valleys, and isolated peaks. Although the general airflow pattern over such a region may be relatively simple, the detailed airflow is quite complex. Satellite images show, for example, that a separate cloud can form over each ridge. These clouds, taken individually or collectively, are orographic clouds.

The large-scale, or *synoptic,* flow pattern affects the formation of orographic clouds. They tend to form when the large-scale flow is generally perpendicular to a mountain barrier. Often there are short synoptic-scale wave disturbances in this flow, with the stronger waves typically producing surface cyclones and widespread altostratus or altocumulus clouds. As this large region of cloudiness moves over a mountainous region, orographic clouds often develop within or superimposed on the other clouds. Two distinctly different types of clouds develop over orographic barriers. When the air mass is thermodynamically stable, a *blanket cloud* forms. These generally form, move, and dissipate directly along the streamlines followed by air moving over the barrier. When the air mass is unstable, however, *convective orographic clouds* develop; these have cumuliform elements which develop as convective instability is released during the ascent of air over the mountain slopes.

Stable orographic clouds. In winter, the atmosphere in mountainous areas is usually stable, with stable layers at about the height of mountain ridges; above these layers, the atmosphere is often dry and cloudless. Because of this stability, the air is negatively buoyant throughout its forced ascent over a mountain; that is, it tends to sink. When orographic clouds form in this situation, they have a smooth, well-defined top. Furthermore, temperatures in the cloud are colder than in the free atmosphere at the same elevation. As this layered air is lifted over mountain barriers, multiple layers of cloud can form, sometimes resembling a stack of pancakes.

The simplest case of an orographic cloud involves a single mountain ridge. As humid air flows up the barrier, its pressure decreases and it cools. As it cools, the relative humidity increases until it reaches 100 percent. At this point, the air can no longer hold moisture in the form of vapor, and the water begins to condense on condensation nuclei in the air, growing into cloud drops. The cloud drops may pass over the barrier and evaporate during descent on the lee side, or they may grow large enough

to fall as precipitation, especially if the air mass is of tropical and maritime origin. Furthermore, in cold clouds, these drops may support the growth of ice crystals. These ice crystals may fall as snow, or they may fall into warmer air where they melt and reach the surface as rain. Even in the absence of direct precipitation, some condensate can reach the ground. For example, the orographic clouds may consist only of upslope fog blowing over the ridges. Trees collect some of the fog, and the water trickles down the branches to the ground. At temperatures below 0°C, this fog freezes on objects such as trees or rocks, forming rime ice. In extremely arid regions, attempts have been made to harvest the water from upslope fog by erecting nets and other collection devices. Orographic precipitation in stable air masses tends to occur at small rates, but it can last for a long time—hours or even days. This is a primary, though not exclusive, characteristic of winter orographic clouds. [*See* Condensation; Fog.]

Orographically induced convective clouds. Convective (cumulus-type) clouds can form over orographic barriers as a result of mountain surface heating, forced lifting that releases conditional instability, or a combination of these processes. In the case of surface heating, the surface layer of the mountain slope is heated by the Sun and becomes warmer than the surrounding air at the same elevation. The rising warmer air can carry moisture to a height where it begins to condense into cloud, ultimately producing precipitation. The precipitation often occurs as mountain showers, but severe thunderstorms can also develop.

Upward-moving airflow *(convection)* can also be triggered by lifting as a conditionally unstable air mass moves over a mountain barrier. If the air rises to an elevation where it is warmer than the surrounding air at the same level, then the air is buoyant and will continue to rise. With sufficient lifting, a cloud forms, and under suitable conditions, convective-type showers and precipitation result.

Both stable and convective orographic clouds can form in all seasons. Stable orographic clouds are most frequent during winter, and orographically induced convective clouds are most frequent during summer. [*See* Convection.]

Mountains and Climate. Since mountains influence airflow, stability, cloud formation, and precipitation, the cumulative effects of such influences, at fixed locations and over long time periods, produce *orographic climate regimes.* Thus, there are regions of enhanced precipitation on the windward side of mountains, and reduced precipitation on the leeward side. The windward side often supports a rain forest, while the leeward side is in the arid *rain shadow*. Consequently, mountains serve as natural reservoirs of water. Winter snowpack accumulates at high altitudes; it melts and feeds the rivers throughout the rest of the year. Vegetative communities evolve in response to the local climate; with different rates of evapotranspiration feeding water vapor back into the atmosphere. The type and amount of ground cover also determines the how much solar radiation is reflected back to space (the *albedo*) and how much is absorbed. All these factors act and interact to determine climate.

BIBLIOGRAPHY

Cole, F. W. *Introduction to Meteorology*. New York, 1980.

Cotton, W. R., and R. A. Anthes. *Storm and Cloud Dynamics*. San Diego, Calif., 1989.

Grant, L. O., and A. M. Kahan. "Weather Modification for Augmenting Orographic Precipitation." In *Weather and Climate Modification*, edited by W. N. Hess. New York, 1974, pp. 282–317.

Kasahara, A. "Influence of Orography on the Atmospheric General Circulation." In *Orographic Effects in Planetary Flows*, edited by R. Hide and P. W. White. Global Atmospheric Research Programme Publ. No. 23. World Meteorological Organization, Gevena, 1980, pp. 4–54.

Kuettner, J. P., and T. H. R. O'Neill. "ALPEX, the GARP Mountain Subprogram." *Bulletin of the American Meteorological Society* 62 (1981): 793–805.

Ludlam, F. H. "Artificial Snowfall from Mountain Clouds." *Tellus* 7 (1955): 277–290.

DAVID C. ROGERS and LEWIS O. GRANT

OSCILLATION OF EARTH'S AXIS. *See* Orbital Variations.

OZONE. The trace gas ozone consists of molecules having three oxygen atoms, represented by the symbol O_3. The gas, most of which resides in the stratosphere, plays a vital role in the atmosphere's radiation budget. This stratospheric ozone layer is responsible for the nearly complete absorption of solar radiation in wavelengths from 240 to 290 nanometers, thereby protecting life forms on the Earth's surface from damaging ultraviolet (UV) radiation. The atmospheric heating resulting from this absorption drives the circulation of the middle atmosphere. Although only 10 percent of atmospheric ozone resides in the troposphere (the region closest to Earth's surface, about 10 kilometers in height), it plays a critical role in the chemistry of the region. [*See* Stratosphere.] Ozone production in the atmosphere begins with the photodissociation (breaking apart by light) of oxygen molecules (O_2):

$$O_2 + h\nu \rightarrow O + O.$$

Here $h\nu$ is a quantum of solar UV radiation of a wavelength shorter than 240 nanometers. The monatomic oxygen and the ozone cycle rapidly from one form to the other:

$$O + O_2 + M \rightarrow O_3 + M$$
$$O_3 + h\nu \rightarrow O + O_2,$$

where M is any third body and $h\nu$ is a UV quantum in the range 200 to 300 nanometers. The balance between the availability of oxygen (which decreases with height) and solar UV irradiance (which increases with height) results in peak ozone formation in the tropics above 25 kilometers altitude. The distribution of ozone also depends on atmospheric transport; a typical peak amount is 6×10^{12} molecules per cubic centimeter at 25 kilometers altitude in the tropics, or 18 kilometers near the poles. The ozone distribution is variable, showing diurnal, seasonal, and interannual fluctuations.

Ozone amounts are often expressed as a sum in a vertical column of air. One Dobson unit (DU) equals 10^{-5} meters of ozone if the column were brought to standard temperature and pressure. A typical column amount is 300 DU.

The relatively small amount of ozone in the atmosphere (peak mixing ratios are only 10 parts per million) belies the great importance of this trace gas. No other gas absorbs UV radiation in the 240 to 290 nanometer band, but ozone absorbs virtually all such incident radiation. This radiation would otherwise reach the surface, damaging genetic material and inhibiting photosynthesis. In addition, the energetically excited monatomic oxygen produced by the photodissociation of ozone is necessary for the formation of radicals such as nitric oxide (NO) and hydroxide ions (OH), which in turn are involved in chemical reactions of great importance. Moreover, ultraviolet absorption by ozone is the main agent of solar heating in the middle atmosphere. Ozone heating peaks at between 40 and 50 kilometers altitude and is responsible for the increase in air temperature with height above the tropopause (the transition layer between the troposphere and stratosphere). This temperature structure results in a stable region called the *stratosphere*. At the height of production, ozone abundance is maximum in the tropics where ultraviolet radiation is most intense. Below this, in the lower stratosphere, tropical ozone amounts are lower than mid latitudes due to the tropical upwelling of ozone-poor tropospheric air. In general, the distribution of ozone and the circulation of the stratosphere interact in a complicated way. Finally, in addition to its UV-absorbing properties, ozone is a greenhouse gas, absorbing outgoing infrared radiation of wavelength 9,600 nanometers and thus contributing to global warming.

Production of ozone in the stratosphere is balanced primarily by catalytic loss owing to oxides of nitrogen, chlorine, hydrogen, and bromine. The chlorine and bromine catalysts largely come from anthropogenic (human-generated) gases: chlorofluorocarbons and halons. Thus, increasing emissions of these compounds constitute a threat to global ozone. Satellite intruments observed a global downward trend in the column amount of ozone, of 0.27 percent per year over the decade of the 1980s. The most remarkable manifestation of ozone loss is the rapid, near-total depletion that occurs in the lower stratosphere over the Antarctic, and to a lesser extent the Arctic, each spring. These dramatic seasonal losses, discovered in 1985, are due to complicated chlorine reactions occurring on polar stratospheric cloud particles. [See Ozone Hole.]

Although representing only about 10 percent of the total column, ozone in the troposphere is involved in important chemical processes. In addition to being an oxidizing agent itself, photolysis of ozone in the presence of water (H_2O) is the primary source of the hydroxyl radical (OH). Thus the abundance of tropospheric ozone plays a critical role in establishing the oxidizing capacity of the atmosphere. This capacity, for example, partly determines the removal rates of such climatically important trace gases as methane and the chlorofluorocarbons.

Sources of tropospheric ozone are those transported toward the surface from the stratosphere and those produced by the oxidation of hydrocarbons and carbon monoxide in the presence of oxides of nitrogen. An important example of this chemical production is urban photochemical smog, in which the precursor ingredients come largely from automobile exhaust. At levels that can exist in and downwind of polluted urban regions, ozone causes respiratory ailments and is detrimental to plant growth. Ozone is also produced by biomass burnings, such as forest and savanna fires, which release the same precursors as fossil fuel combustion.

While total ozone has been decreasing due to loss in the stratosphere, there is evidence that ozone in the troposphere has been increasing, at least in the northern hemisphere. Reported trends range from zero to several percent per year, depending on location. Recent analysis suggests that the rate of increase was larger in the 1970s, and slowed to near zero in the 1980s. The tropospheric trend at the South Pole has actually been downward by about 0.5 percent per year since the mid-1970s.

Global losses of ozone will certainly be accompanied by increased UV radiation reaching the Earth's surface. However, determing the influence on the greenhouse effect of changes in ozone is more complex, since ozone absorbs both incoming UV and outgoing infrared radiation. Recent modeling work suggests that the effect of ozone loss in the stratosphere depends on the region of primary depletion; ozone loss above the lower stratosphere will contribute to global warming, while loss in the lower stratosphere will lead to cooling. The studies conclude that the stratospheric loss during the 1980s would have tended to produce a negative radiative forcing, or a cooling effect on climate. However, increases in tropospheric ozone will have the opposite effect. The net effect is quite sensitive to ozone changes near the tropopause, where uncertainty in measurement is significant.

[See also Atmospheric Chemistry and Composition; Global Warming; and Greenhouse Gases.]

TIMOTHY M. HALL

OZONE HOLE. A significant depletion of ozone (O_3) occurs in the Southern Hemisphere spring in the lower stratosphere above Antarctica. The phenomenon became detectable in the mid-1970s and since then has increased in severity; in October 1994 ozone over Antarctica was depleted by nearly 80 percent compared to October values before the ozone hole. At the most affected altitude, from 14 to 19 kilometers above the surface, more than 99 percent of the ozone is lost. Present understanding connects the loss to catalytic destruction of ozone by reactive chlorine compounds derived from anthropogenic (human generated) chlorofluorocarbons (CFCs). In fact, the seasonal timing and location of the hole must be understood in terms of the unique antarctic polar meteorology and the resulting isolation and low temperatures of the lower stratospheric air mass there.

Atmospheric chemists had suspected since the early 1970s that anthropogenic chlorine compounds could destroy ozone. The destruction rate was projected to yield a 5 percent loss in total ozone by the middle of the twenty-first century. Based on this research, in 1978 the

U.S. government banned the use of the implicated chlorofluorocarbons in commercial aerosols, and public attention subsequently waned. In 1985, however, ozone depletion was suddenly brought back into the headlines by the dramatic report of a British Antarctic Survey team: over the previous decade total springtime ozone levels above the Halley Bay station in Antarctica had fallen by as much as 50 percent. This was an unexpected discovery both on theoretical grounds and because the National Aeronautics and Space Administration had been monitoring ozone levels globally since 1979 with the Total Ozone Mapping Spectrometer (TOMS) instrument aboard the *Nimbus 7* satellite. The standard TOMS data-processing procedure automatically neglected as unreliable ozone levels below a fixed value of 180 Dobson units (a Dobson unit, or DU, equals 10^{-5} meters of ozone in a standard vertical column of the atmosphere), and hence the antarctic springtime depletion had been discarded. Only after the British survey team's report was the TOMS data reprocessed; the ozone depletion was verified and the geographic extent of the hole determined. (See Figures 1 and 2.)

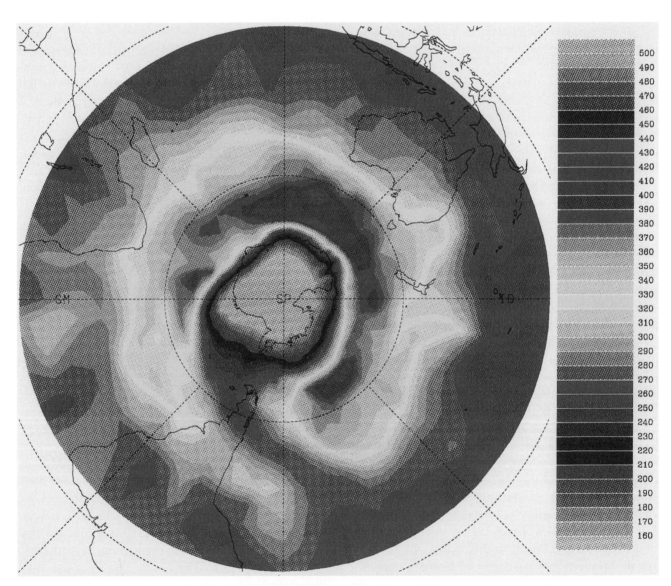

OZONE HOLE. Figure 1. *Southern Hemisphere polar stereographic projection of total column ozone from the Total Ozone Mapping Spectrometer (TOMS)* Nimbus 7 *satellite instrument for 6 October 1993.* The gray scale indicates Dobson unit (DU) contours. The area in the center over Antarctica is below 160 DU.

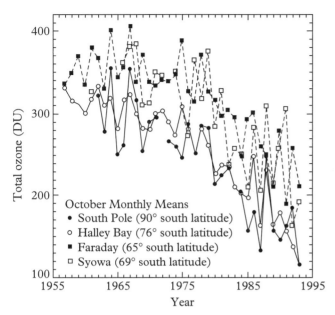

OZONE HOLE. Figure 2. *Historical records of October monthly mean column ozone (Dobson units) at four Antarctic stations.* (From World Meteorological Organization, 1995.)

These dramatic and unexpected antarctic ozone observations led to an immediate mobilization of the atmospheric science community. The Airborne Antarctic Ozone Experiment (AAOE) mission of 1987 carried out 25 airplane flights in the lower antarctic stratosphere during August and September. While researchers were planning and launching the mission, two competing theories to explain the rapid ozone depletion came into focus. One was entirely dynamical in nature, while the other saw CFC-derived chlorine as the primary agent of rapid ozone destruction. The AAOE mission greatly added to the ground-based and satellite observations and played a decisive role in implicating anthropogenic CFCs as a root cause for the ozone hole.

Antarctic Meteorology. The winter Antarctic is a unique environment, and its meteorological characteristics must be incorporated in any theory for predicting the seasonal and localized nature of the ozone hole. The lower stratosphere in winter exhibits a latitudinal peak in temperature at mid latitudes. Poleward of this peak, the decreasing temperature and the Coriolis force combine to produce a strong westerly geostrophic wind which defines the edge of a vortex centered approximately on the pole (called the *polar night vortex*). In the Southern Hemisphere the scarcity of topographic features from the mid latitudes to the antarctic continent results in relatively little planetary wave activity. Wave-eddy transport

of heat is weak and the average latitudinal temperature gradient strong in relation to the case in the Northern Hemisphere. Consequently, the southern polar vortex is more robust than the northern. This vortex inhibits transport of equatorial air to polar latitudes and surrounds the cold polar stratospheric air, which undergoes distinct chemical and dynamical processes. The polar air remains isolated until late southern spring, when it has been significantly eroded by wave-induced mixing and warmed by direct solar heating. [*See* Dynamics.]

During the winter isolation, the antarctic stratospheric air may come close to reaching thermal equilibrium, a fact that motivated a purely dynamical theory of the ozone hole. The theory pointed to evidence suggesting a substantial reduction in Southern Hemisphere planetary wave activity since 1979. Consequently, according to the theory, the lifetime of the polar night vortex was prolonged and a greater opportunity provided for the air within the vortex to reach thermal equilibrium. If thermal equilibrium is achieved, the tendency for cold air to sink is eliminated. Then, with the arrival of the first sunlight of the antarctic spring, a portion of the energy available from solar heating may be used to induce upward motion. Ozone concentrations are much lower in the troposphere than in the stratosphere, so this upward motion would mix ozone-poor air into the lower stratosphere, causing a local depletion.

The AAOE mission provided evidence contradicting the dynamical theory. The upward motion of polar air in the antarctic spring would transport not only ozone but also such long-lived trace constituents as nitrous oxide (N_2O). Nitrous oxide decreases with height throughout the atmosphere, so rising motions should be accompanied by rapidly increasing N_2O; however, polar N_2O levels were observed to be relatively constant during the period of high ozone loss. Subsequent aircraft and balloon campaigns have confirmed AAOE results and have shed additional light on chemical and dynamic processes of the polar statosphere.

Chemical Theory of the Ozone Hole. Although most chlorine in the troposphere (the atmospheric layer closest to Earth's surface, below the stratosphere) is contained in sodium chloride from sea spray and hydrogen chloride from volcanoes, these forms dissolve in water and are rained out before reaching the stratosphere. Another theory of the ozone hole draws from earlier gas-phase research that pointed to catalytic ozone destruction by chlorine. The theory also incorporates the formation in the isolated winter antarctic stratosphere of polar stratospheric clouds (PSCs). Non-gas-phase processes on

PSC particle surfaces play a critical role by releasing the destructive forms of chlorine.

Most chlorine entering the stratosphere is carried by chlorofluorocarbons (CFCs), compounds that are used as refrigerants, solvents, and aerosol propellants. The most common forms are $CFCl_3$ (CFC-11) and CF_2Cl_2 (CFC-12). The CFCs are valuable industrially because of their chemical inertness. When they are released from scrapped machinery or sprayed into the air, their atmospheric lifetime (total atmospheric content divided by mean loss rate) is decades to hundreds of years, allowing the compounds to reach the upper stratosphere without suffering significant loss. At that altitude solar UV radiation breaks down the CFCs, releasing monatomic chlorine (Cl) gas. Monatomic chlorine can catalytically destroy ozone efficiently; increasing levels of this gas in the atmosphere constitute a long-term threat to global ozone levels. However, this process is slow because most of the atmosphere's ozone resides in the lower stratosphere.

Ultimately, reactions with other trace gases bind the Cl atom to form less reactive "reservoir" species such as hydrochloric acid (HCl) and chlorine nitrate ($ClONO_2$, or $ClNO_3$). The atmosphere mixes them back into the ozone-rich lower stratosphere. Gas-phase chemistry is not fast enough to release significant amounts of reactive chlorine from HCl and $ClONO_2$; however, in winter the sunless antarctic lower stratosphere provides PSC particle surfaces on which rapid solid-phase reactions can occur. For example, in a reaction that critically perturbs the gas-phase photochemistry, surface-adsorbed HCl combines with gaseous $ClONO_2$ to produce molecular chlorine (Cl_2) gas and surface HNO_3 (nitric acid). Other surface reactions release compounds such as $ClNO_2$ and HOCl, which undergo further transformation into reactive chlorine species. These processes share two important characteristics: inactive chlorine is converted to active species, such as Cl_2, that break down rapidly under the first springtime sunlight to produce monatomic chlorine; and nitrogen species, that would otherwise be available to reconvert active chlorine to reservoir forms, are lost to the relatively inert HNO_3 in the PSC particles. In fact, there is evidence that these acidified cloud particles can coalesce into sizes large enough for sedimentation, completely removing their nitrogen and water from the lower stratosphere.

When the solar rays first strike the antarctic lower stratosphere in early spring, Cl_2 is photodissociated to two separated chlorine atoms. The presence of sunlight and Cl allows the following catalytic cycle to destroy ozone efficiently:

$$2 \times (Cl + O_3 \rightarrow ClO + O_2)$$
$$ClO + ClO + M \rightarrow Cl_2O_2 + M$$
$$Cl_2O_2 + hv \rightarrow Cl + ClO_2$$
$$ClO_2 + M \rightarrow Cl + O_2 + M$$

where M indicates the presence of any third body and hv is a quantum of solar radiation of the appropriate wavelength (near UV). The net result is the destruction of two ozone molecules, $2O_3 + hv \rightarrow 3O_2$; the chlorine atoms are free to repeat the process. A second cyclic process employs a synergism between chlorine and bromine (Br), a trace gas which occurs both naturally and in the anthropogenic halon family:

$$Cl + O_3 \rightarrow ClO + O_2$$
$$Br + O_3 \rightarrow BrO + O_2$$
$$ClO + BrO \rightarrow Br + ClO_2$$
$$ClO_2 + M \rightarrow Cl + O_2 + M$$

with the net result $2O_3 \rightarrow 3O_2$. Again, chlorine and bromine are free to continue the cycle.

During the antarctic spring of 1987 the AAOE flights recorded levels of a number of trace constituents, including important active and reservoir chlorine species, bromine, active nitrogen, nitric acid, cloud condensation nuclei, ozone, and meteorological variables. These data strongly confirm the chemical theory of the ozone hole. For example, measurements through the vortex and into the depleted region exhibit a striking spatial anticorrelation between ClO and O_3. As the spring progresses within the vortex, the reactive chlorine levels remain high while ozone drops at a rate near 2 percent per day. In further support of the chemical theory, reactive nitrogen, which would otherwise moderate the ozone-destroying potential of the chlorine, is reduced to a small fraction of its normal climatological abundance. Nitric acid is similarly reduced from the gas phase and perhaps entirely removed from the region by sedimentation.

Consequences and Future Prospects. Decreased antarctic ozone results in increased penetration of solar UV radiation to the surface. Because radiation reaching the surface is strongly dependent on the solar zenith angle, peak levels occur in the summer. Antarctic spring observations show that the ozone hole advances the onset of summertime UV levels, in effect lengthening and intensifying the summer season. Ultraviolet radiation studies from New Zealand and Australia have also shown dependence on ozone levels.

Increased UV levels at the Earth's surface will have biological repercussions. Ultraviolet radiation in the wavelength band 200 to 320 nanometers (known as

UVB) is damaging to the genetic material DNA (deoxyribonucleic acid). The antarctic ozone depletion may therefore cause an increased risk of skin cancer in southern populations experiencing episodic ozone reductions following the breakup of the hole in November and December. Laboratory research predicts an approximate 2 percent increase in melanoma risk for every 1 percent increase in UVB. In addition, excess UVB inhibits photosynthesis in phytoplankton, seriously affecting antarctic marine ecology. These ecological repercussions have not yet been quantified because of the complexity of the processes involved. For example, different species of phytoplankton tend to have differing reactions to enhanced UVB levels, and each adapts to different extents. As a result, the relative populations of the species may vary, with effects on the food chain more complex than those of a simple reduction in total biomass. Because of a relative lack of research on phytoplankton photobiology prior to the discovery of the ozone hole, comparisons to climatological norms are difficult.

The far larger human and animal populations in the Northern Hemisphere than in the Southern make the potential for rapid ozone depletion in the Arctic even more troubling. Enhanced UV levels might also be detrimental to high-latitude land ecosystems and agricultural production.

The springtime Arctic is a natural place and time to expect a phenomenon similar to the antarctic ozone hole. Recent arctic airborne observations indicate that the necessary chemical ingredients are in place. However, different weather conditions may result in a more readily disturbed polar night vortex and consequently, a less isolated stratospheric air mass. Mixing of warm low-latitude air into the northern polar region keeps the temperature 30° to 50°C warmer than in the Antarctic—too warm for much PSC formation. (In fact, during events called *sudden stratospheric warmings* that occur every second or third winter, the arctic night vortex completely breaks down and the polar air warms dramatically.) Still, some arctic ozone depletion does occur. Airborne measurements in January and February 1989 deduced an average 0.4 percent per day loss of ozone above 20 kilometers. Regions within the arctic vortex displayed depletions as large as 17 percent by early February. Future arctic depletions may deepen. Increasing levels of greenhouse gases such as CO_2 act to cool the stratosphere, and a cooler arctic stratosphere may allow more PSCs to form. Even under present climate conditions, it seems likely there will eventually be a northern winter with a stable enough vortex to allow significant ozone depletion. In fact, early 1995 has produced measurements of record low ozone levels in the Arctic, such as a depletion of 35 percent from climate averages over Siberia.

On a global scale, determination of rates of ozone depletion is complicated by the strong modulation in total atmospheric ozone caused by the quasi-biennial oscillation and the 11-year solar cycle; however, recent statistical analysis of more than a decade of TOMS ozone observations suggests a global loss rate of about 0.27 percent per year. The polar regions are losing ozone most rapidly; in equatorial regions the loss is not yet detectable, suggesting that the global loss is a result of accumulated processing by the isolated winter polar air masses.

The dramatic destruction of antarctic ozone and the conclusive scientific evidence pointing to man-made causes motivated precedent-setting international agreements to address the problem. The Montreal Protocol of 1987 called for reducing CFC emissions to half their 1986 levels by the year 2000. In 1990 this timetable was amended to eliminate all CFC emissions of industrialized countries by 2000, as well as to limit the emissions of the bromine-containing halons. It was further amended in 1992 to eliminate the most damaging gases by 1996. Recent observations show a marked decrease in the growth rates of certain CFCs and halons, and, in fact, trends in total tropospheric chlorine suggest emissions below the maximum allowed under the international agreements. Consistent with this, the abundance of substitutes (the hydrofluorocarbons and hydrochlorofluorocarbons) is increasing. Nonetheless, modeling studies predict that even if the amended protocol is completely implemented, chlorine will not return to pre-ozone-hole levels until the second half of the twenty-first century. Ozone depletion will get worse before it gets better.

[See also Health.]

BIBLIOGRAPHY

Anderson, J. G., D. W. Toohey, and W. H. Brune. "Free Radicals within the Antarctic Vortex: The Role of CFCs in Antarctic Ozone Loss." *Science* 251 (1991): 39–45.
Brune, W. H., et al. "The Potential for Ozone Depletion in the Arctic Polar Stratosphere." *Science* 252 (1991): 1260–1265.
Solomon, S. "Progress towards a Quantitative Understanding of Antarctic Ozone Depletion." *Nature* 347 (1990): 347–354.
Stolarski, R. S., P. Bloomfield, and R. D. McPeters. "Total Ozone Trends Deduced from Nimbus 7 TOMS Data." *Geophysical Research Letter* 18 (1991): 1015–1018.
World Meteorological Organization. "Scientific Assessment of Ozone Depletion: 1994." World Meteorological Organization Global Ozone Research and Monitoring Project, Report number 37. Geneva, 1995.

TIMOTHY M. HALL

P

PAINTING, SCULPTURE, AND WEATHER. From earliest times, visual artists of many cultures have found inspiration for enduring works in weather and the cycle of the seasons. Sun and clouds, violent storms, gentle rains, and heavy snowfalls have been used by painters and sculptors to evoke an array of human moods, to mirror human actions, and to give tangible shape to the struggle between good and evil. Representations of weather in all its variety have also been used to express ideals of beauty and ideas about the moral and physical relationships between humans and the natural world.

Painting. In the sixth century, Byzantine painters and mosaicists decorated religious and secular public buildings with personified representations of weather and climate. The cosmos depicted on the vaulted ceiling of a now-vanished Gaza bathhouse was said to have included images of the four seasons and the four winds, thunder, lightning, and cloud, each endowed with human features and traits. The Plains Indians of North America adorned their war shields with the thunderbird, a sacred creature armed with the power of thunder and lightning; for example, the five thunderbirds depicted on a Cheyenne leather shield (ca. 1860) were meant to imbue the warrior carrying it with the spiritual energy of nature's mightiest manifestations.

In the emerging European tradition of landscape painting, artists since the Romans used nature's own forms, colors, and movements to depict moral states and human emotions. Critics regard Bavarian painter Albrecht Altdorfer (ca. 1480–1538) as the first to make landscape his main subject. In his art, cosmic struggles between clouds and Sun reinforce the mood and message of the human activities taking place below. In Renaissance religious art, exemplified by sixteenth-century Italian Paolo Veronese's (1528–1588) *The Adoration of Kings*, shafts of sunlight breaking through dark, massed clouds are used to highlight and ennoble the humans on whom the beams fall and to emphasize such virtues as heroism and piety.

Atmospheric effects are central to works by the English master of landscape and seascape, J. M. W. Turner (1775–1851). Exploring the Thames river valley during the cold, rainy summer of 1805, Turner prepared a special watercolor sketchbook to capture the changeable moods of Sun, rain and clouds. His observations found their way into such later oil paintings as *Sun Rising through Vapour* (1807) and *Snowstorm: Hannibal and His Army Crossing the Alps* (1812). In *Rain, Steam, and Speed—The Great Western Railway* (1844), a late work that influenced the French Impressionist movement, Turner's swirls of color made his locomotive appear to be yet another natural force in a painting suffused with atmospheric motion.

Depictions of carefully selected weather conditions were central to the moral lessons embodied in the mid-nineteenth-century landscapes of the Hudson River School in the United States. Frederic Edwin Church (1826–1900) made use of weather phenomena to emphasize the heroic features of his American landscapes in such paintings as *Storm in the Mountains* (1847) and the Turner-influenced *Niagara Falls from Goat Island in the Snow* (1856). In his South American paintings, such as the 1854 *Tamaca Palms*, Church emphasized the climatic contrast of humid lowlands shadowed by snow-capped peaks.

Church's contemporary, Albert Bierstadt (1830–1902), a trailblazing painter of the American west, frequently used the contrast of brilliant sunshine and inky storm clouds to dramatize the visual excitement and moral power he found in the landscapes of the Colorado and Wyoming territories. His 1859 *Thunderstorm in the Rocky Mountains* and 1869 *The Buffalo Trail: The Impending Storm* are two examples. The cataclysmic summer thunderstorms depicted by Martin Johnson Heade (1819–1904) in four major landscapes painted between 1859 and 1870 reveal a nature that is often actively hostile to human undertakings; these works represent a darker side of the generally optimistic Hudson River viewpoint.

The Hudson River style demanded close attention to tiny details that gave finished canvases a polished perfection. In the later nineteenth century, the use of weather as a storytelling device and means of moral commentary began to give way to an emphasis on the inherent beauty and force of meteorological phenomena. Painters of the French Barbizon School of landscape painting, and Americans influenced by them, sought to depict the transitory aspects of weather rather than its picturesque effects. Edward M. Bannister (1828–1901), a one-time sailor who became America's first recognized African-American artist, used contrasts of light and dark and suggestions of shapes rather than sharp detail in such paintings as *Approaching Storm* (1886). In it, a lone figure scurries across a deliberately vague landscape, dwarfed by nature's power and beauty.

Asian artists had pioneered the use of weather images to express an aesthetic vision. The works of Kao K'o-kung of China (1248–1310) and Japanese master printmakers Katsushika Hokusai (1760–1849) and Ando Hiroshige (1797–1858) show wind and waves, snow, mist, and slashing rain as formal design elements and as key actors in a visual drama in which shivering or umbrella-clutching humans play only a bit part.

Among Western painters deeply influenced by the Japanese were the French Impressionist Claude Monet (1840–1926) and the art patron, gardener, and painter Gustave Caillebotte (1848–1894). Monet's "Grainstack" (or "Haystack") series, painted at Giverny, France, between 1888 and 1891, employed a palette of weather conditions to depict farmers' fields under every possible state of light and atmosphere. Because these 25 paintings were made *en plein air* (outdoors), weather conditions directly affected Monet's project. He later attributed attacks of rheumatism to "my time in the rain and snow." Caillebotte's famous painting *The Place de l'Europe on a Rainy Day* (1877) applies the *plein air* tradition to an urban setting. In it, the umbrellas wielded by Parisian pedestrians become a major compositional element.

The American realist Winslow Homer (1836–1910) made the changeability and unpredictability of the weather, especially at sea, his central subject. His 1898 watercolor *Hurricane, Bahamas* and 1904 *A Summer Squall* are characteristic. Weather, reduced to its abstract essence of power and motion, can be found in paintings by the American Expressionists, of whom Arthur Dove (1880–1946) is an exemplar. Dove kept an artistic almanac, daily recording temperatures, wind direction, and barometric pressure and making note of interesting cloud formations. In Dove's *Thunder Shower* (1938), a bright jagged streak rends massed dark shapes that suggest wooded mountains.

If landscapes, both traditional and abstract, have been preoccupied with nature's power and splendor, painters of urban life have used weather to explore the boundaries between the natural and the constructed in cities. John Sloan (1871–1951), in his etching *Roof, Summer Night, 1906*, shows sweltering tenement families seeking a breeze on a roof that is dirty and uncomfortable, but oddly exciting and companionable. In Sloan's oil *The White Way, 1926–7*, a winter storm on Broadway confounds New York's vaunted technology, smudging its sharp contours and turning the marvels of the neon-lit "Great White Way" into a snowy white mess.

In many works by Edward Hopper (1882–1967), the Sun beats down pitilessly on shabby or monotonous urban roofs, windows, and streets, unflatteringly exposing the features of the occasional human caught in its hot glare. Two examples are the 1944 oil *Morning in a City* and *Second-Story Sunlight*, a 1960 oil in which a man and woman on a sun-struck balcony seem at once scorched and frozen by the strong rays.

Hale A. Woodruff (1900–1980), an African-American painter and art professor, focuses on how weather affects the human order in *Tornado* (ca. 1933), which shows neither a funnel cloud nor even the sky. Rather, the tornado's results are shown as a frantic vortex of tree limbs, fences, farm animals, laundry, even an outhouse, propelled by an unseen force of tremendous power.

Sculpture. Weather has received less extensive but nonetheless important treatment in sculpture. Indeed, the humble weathervane became a New England folk-art genre in the eighteenth and nineteenth centuries. Craftsmen in metal and wood gave this uncomplicated weather instrument an array of fantastic and decorative shapes, making art out of its function of supplying wind direction to a lone farmhouse or an entire town. Two weathervanes made by tinsmith Shem Drowne (1683–1774) adorn the steeples of Boston's Christ Church and Faneuil Hall to this day. Drowne's copper *swallow-tailed banner*, fabricated in 1740 for the church, is traditional in form, mimicking a flag flapping in the breeze. His 1742 Faneuil Hall weathervane, by comparison, is a naturalistic copper Grasshopper with glass eyes. Later examples of Yankee weathervane sculpture include birds, fish, whaling ships, dogs, sheep, snakes, horses, fire engines, and the Angel Gabriel.

Practitioners of Land Art (or Earthworking) a U.S.-based movement that began in the 1970s, have created massive sculptural forms that cannot be relocated; their weather-hastened decomposition is central to their artistic meaning. Walter de Maria's *The Lightning Field*, installed in New Mexico, consisted of four hundred 23-foot-tall steel poles intentionally arranged to attract electrical storms. Dennis Oppenheim's *Circles*, traced in the New England snow, melted with the change in season. Both encouraged viewers to understand weather's manifestations as an art form shaped but not controlled by the sculptor's hand. The well-known mobiles of U.S. sculptor Alexander Calder (1898–1976) respond to every movement of the air, forming and reforming like clouds in a continuous process of creation and renewal.

BIBLIOGRAPHY

Bearden, R., and H. Henderson. *A History of African-American Artists: From 1792 to the Present.* New York, 1993.

Cohn, S. *Arthur Dove: Nature as Symbol.* Ann Arbor, Mich., 1985.

Elzea, R. *John Sloan: Spectator of Life.* Wilmington, Del., 1988.

Ferrier, J-L. *Art of Our Century: The Chronicle of Western Art, 1900 to the Present.* New York, 1988.

Goodrich, L. *Edward Hopper.* New York, 1983.

Hendricks, G. *The Life and Work of Winslow Homer.* New York, 1979.

Hill, D. *Turner on the Thames.* New Haven, 1993.

Kaye, M. *Yankee Weathervanes.* New York, 1975.

Maguire, H. *Earth and Ocean: The Terrestrial World in Early Byzantine Art.* University Park, Penn., 1987.

Maurer, E. M. *Visions of the People: A Pictorial History of Plains Indian Life.* Minneapolis, 1992.

Metropolitan Museum of Art. *American Paradise: The World of the Hudson River School.* New York, 1987.

Narazaki, M. *Hiroshige: The 53 Stations of the Tokaido.* Tokyo, 1969.

Tucker, P. H. *Monet in the '90s: The Series Paintings.* New Haven, 1989.

MARSHA ACKERMANN

PALEOCLIMATES. Paleoclimatology is the study of climates or climatic conditions during any part of the geological past (*paleo* is derived from the Greek word for "old"). The reconstructed climates of specific times in the past are called *paleoclimates*. The documentation and study of paleoclimates provides knowledge of such areas of inquiry as: the range over which climate can vary and has actually varied; how stable or unstable climate is in response to documented or suspected changes in the Earth's atmospheric composition or surface conditions; how fast and to what degree climate can vary, given such changes; what changes in plant and animal populations, such as extinctions or geographical migrations, accompany climate change; and where we can prospect for mineral resources—are all formed under specific climatic conditions.

The study of paleoclimates is also important for assessing future climatic change. Establishing the relationship between particular paleoclimates and contemporaneous environmental conditions (atmospheric composition, surface characteristics, and solar radiation intensity) provides an analogue for future climates under similar conditions. Such analogue predictions must be made with caution, however, because no period in time is exactly similar to another: continents shift slowly over time, mountains are built and eroded, the atmospheric composition changes, and other conditions change at different rates. Nonetheless, in combination with numerical (computer) modeling studies, the documentation of past climates is invaluable. The models help to establish which environmental conditions must be similar, and how similar they must be, before an analogue is possible. Paleoclimates also provide a means for testing, improving, and ultimately validating models and their ability to simulate climate under conditions different than those of today. This is important for assessing model forecasts of future climate change, because most models are developed and calibrated for modern conditions, and it is otherwise unclear how valid they are for different conditions.

Paleoclimate studies also allow the assessment of a wide variety of climatic-change impacts that are difficult to assess through models. For example, can vegetation belts migrate fast enough to keep pace with shifting climate zones? Will there be widespread extinctions of plants and animals? Is there a direct correlation between glacial ice volume and some climate indicator (for example, average global temperature) that allows prediction of sea-level change for a given climate change? These are just a few of the questions currently being addressed by paleoclimate studies.

Reconstructing Paleoclimates: Deciphering the Data. Clues revealing past climatic conditions are obtained from *proxy indicators*, types of evidence that can be used to infer climate; for example, certain kinds of pollen may indicate specific temperature ranges. Proxy indicators are found both on land and in the sea, and they may represent biological, chemical, or geological evidence. The most commonly used indicators include pollen, faunal and floral remains, chemical isotope composition in fossils or certain geological deposits, sediment types or composition (e.g., wind-blown dust or glacial deposits), and geomorphological features indicating physical conditions. In the ocean, indicators such as microplankton, pollen, and sediments settle to the seafloor, where they accumulate to provide a nearly continuous record of climate for millions of years. This record is readily accessed by the use of sediment coring or drilling devices on ships. On land, irregular deposition of sediments, erosion, weathering, land development, and other processes make it more difficult to obtain continuous climate records from any one location. Ancient lakes, marshes, bogs, or inland seas often contain the best climate records from inland regions. [*See* Cores.]

Some of the most useful information regarding paleoclimates comes by examining groups, called *assemblages,* of fauna or flora. The most common approach involves the use of sophisticated mathematical procedures called *factor analysis* and *regression analysis.* The approach requires a modern-day database consisting of floral or faunal assemblages representing a variety of climatic conditions. The factor analysis divides the assemblages into mathematically distinct subgroups (often called *factors*), and the regression analysis establishes the relationships between the major subgroups and specific climatic conditions. After these relationships are established for present conditions, assemblages from past times can be analyzed and their composition used to infer the climatic conditions prevailing at that time (see Figure 1).

This method assumes that the flora and fauna inhabited regions in the past, with the same environmental conditions as they do in the present. Until this assumption can be confirmed by corroboration through independent evidence, these results will be subject to some uncertainty.

Another fundamental technique in paleoclimatological studies involves the isotopic composition of oxygen contained in the calcium carbonate ($CaCO_3$) of fossil shells and corals. Oxygen is composed of 8 protons, and its most common form has 8 neutrons, giving it an atomic weight of 16 (^{16}O); however, a small fraction of oxygen atoms have 2 extra neutrons and an atomic weight of 18 (^{18}O). In modern seawater (H_2O), this fraction is about 1 in 500 atoms, but, as recorded in the shells of fossilized microplankton, the ratio of the two isotopes has changed in the past. This change primarily reflects the amount of water trapped in continental ice sheets. During evaporation, water molecules with the lighter ^{16}O oxygen atoms evaporate more readily than the heavier ^{18}O atoms, resulting in a slightly higher ratio of heavier ^{18}O in the surface liquid water. Today the evaporated moisture is transported in the atmosphere and deposited over continents as rain or snow, ultimately returning to the ocean through rivers and aquifers. During glacial periods, however, much of the snow remained trapped in glaciers instead of returning relatively rapidly to the ocean. As a consequence, ^{16}O was systematically removed from the oceans, leaving a larger fraction of ^{18}O in the ocean than today. How much larger depends mainly on how much water was trapped in glaciers (a small percent of the change can be attributed to changes in water temperature

and other factors). Therefore, by examining how the oxygen isotope fraction changed with time, one can estimate how much ice was present on the continents (see Figure 2). In a similar manner, information regarding the distribution of carbon between the oceans and biomass (with additional complications) can be obtained by examining the ratio of carbon isotopes in fossil shells.

On land, in addition to information from sediments, pollens, and faunal and floral assemblages, geomorphological features reveal the location and history of ice sheets. For example, glacial *striae* (surface scratches in rocks caused by ice movement), *moraines* (mounds of debris carried and deposited by ice), and other features allow researchers to reconstruct the flow direction of an ice sheet and to delineate its basic shape. Features such as ancient dunes, soils, annual layering of lake sediments, or weathering features, provide additional information regarding local climatic conditions. For the more recent past (thousands to hundreds of thousands of years), high-resolution climatic information, often on a yearly basis, is provided by tree rings, coral reefs, thermal profiles in permafrost regions, and ice cores from existing ice caps in Greenland, Antarctica, and mountain glaciers. Tree rings have recently provided a detailed record of how temperature has varied over several thousand years. Coral records reveal El Niño events for hundreds of years in the tropical Pacific Ocean. Ice cores have provided detailed records about more than 100,000 years of temperature, atmospheric gas content (using air trapped in tiny bubbles in the ice), precipitation, atmospheric dust, and other climatic indicators. [*See* Tree-Ring Analysis.]

Limitations and Problems in Reconstructing Paleoclimates. One of the most serious limitations on paleoclimate reconstruction (apart from uncertainties in interpreting the climatic implications of the proxy indicators) arises from the uncertainties associated with dating the proxy indicators or other paleoclimatic evidence. Dating is important in two respects. First, reconstructing the paleoclimate for a specific geological time requires accurate *absolute dating*—that is, the ability to identify the actual geological time represented by the evidence. Second, in order to combine data from different geographic locations, we must be certain that the evidence at each location represents the same moment in geological time—whether or not we may know it. This requires accurate *relative dating* or correlation.

Absolute dating methods are limited and rely predominantly on evaluating the amount of decay of naturally occurring radioactive isotopes. The different isotopes permit dating over a broad range of time scales. Uncer-

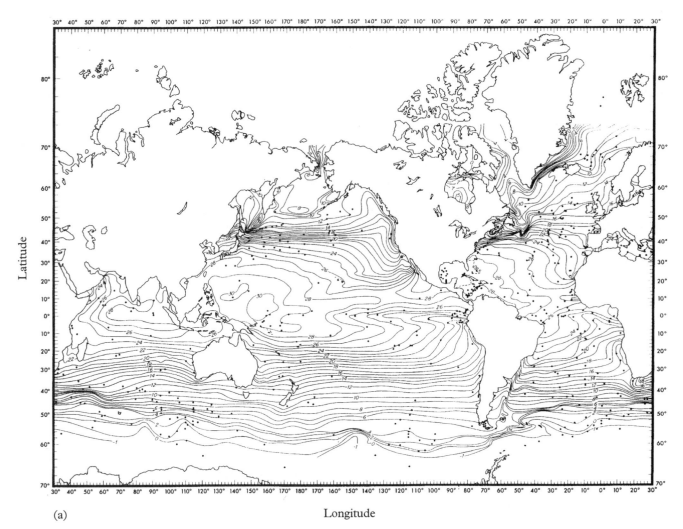

(a)   Longitude

PALEOCLIMATES. Figure 1. *An estimate of sea-surface temperature for (a) modern summer conditions, and (b) summer conditions during the last ice age.* The latter was reconstructed by researchers using factor analysis and regression analysis (aided by additional information) on microfossil assemblages collected throughout the ocean basins. There is some controversy regarding the ice age temperatures, particularly in the tropical regions where different techniques are suggesting somewhat colder conditions (by about 6°C), though still other techniques yield results consistent with those shown here. (CLIMAP Project members, 1981.)

tainties in absolute dates increase steadily with age, so the older the evidence, the more uncertain is the dating. In the absence of absolute dates, equating evidence in one location to evidence of the same age in another location is done using stratigraphy and correlation. *Stratigraphy* involves establishing a relative sequence of events or characteristics (geological, geochemical, or biological) within which the paleoclimatic evidence lies. If this same sequence can be identified in each location containing the paleoclimatic evidence, it can be used to establish the relationship between locations and the relative timing of the indicators. Depending on the nature of the stratig-

raphy, one may be able to relate evidence from different locations around the world (as is possible using oxygen isotope variations as the basis of the stratigraphy), or only from a single region (if using local features such as volcanic ash layers to establish the stratigraphy). The uncertainties vary according to the quality and resolution of the stratigraphy used. With some stratigraphies, the absolute dates of the events used have been established already; therefore, in establishing the stratigraphy, the absolute dates can also be directly inferred. A *paleomagnetic stratigraphy,* based on the recurrent reversal of the Earth's magnetic poles as recorded in certain minerals

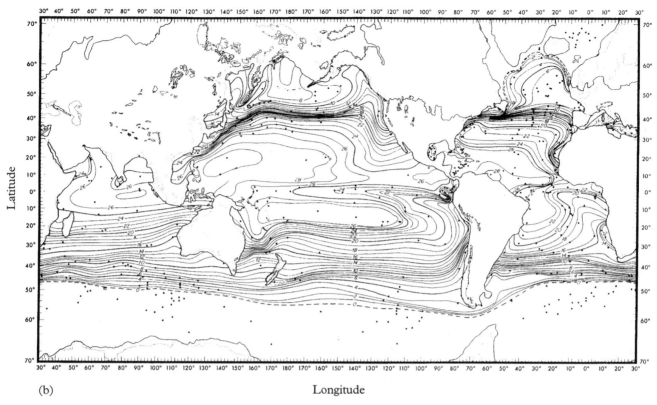

(b)   Longitude

PALEOCLIMATES. Figure 1. *(Continued)*

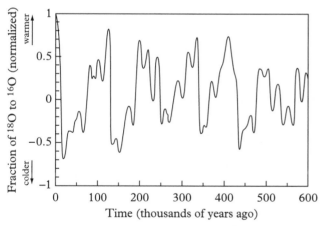

PALEOCLIMATES. Figure 2. *A representative oxygen isotope record for the past 300,000 years.* The isotope values were obtained by analyzing microfossil shells, about the size of a grain of sand, which accumulated by sinking to the sea floor after the microplankton that grew the shells died. Periods of high $\delta^{18}O$ (the fraction of ^{18}O to ^{16}O normalized to modern-day value) indicate glacial conditions, while the periods of low $\delta^{18}O$ represent interglacial conditions. Notice that periods as warm as today occur only infrequently—typically around every 100,000 years, and then only for a brief period in geological time.

contained within the sediments, is an example of such a stratigraphy.

In addition to dating problems, paleoclimate reconstructions are limited by other factors. The farther we go back in time, the larger are the uncertainties in the paleoclimatic evidence, the greater the degradation of the evidence, and the smaller the body of evidence available. For example, because life forms have evolved over millions of years, the climatic conditions preferred by certain plant or animal species today may not be representative of the conditions similar species preferred in the distant past. Therefore, our understanding of past climates degrades steadily with time. We have a very good understanding of climate over the past several thousand years, based on tree-ring records, ice cores, lake sediments, coral records, and historical records. Over the past several hundred thousand years, we have a fairly good understanding from deep-sea sediment records, land evidence, and ice cores. Over the past several million years, we are in the process of establishing a better record, because we can still obtain (albeit with more effort and expense) continuous high-resolution deep-sea sediment records. Prior to that, the records and our documentation of paleoclimates are more erratic and geographically isolated, based predominantly on land records.

Recent Findings. Some recent paleoclimatic findings of particular interest include the following. The ice ages have waxed and waned over the past few million years in a fairly consistent manner (see Figure 2). Specifically, ice-age cycles vary with periods of around 21,000, 40,000, and 100,000 years (the last cycle occurs only during the past 700,000 years). These variations coincide with variations in the precession of the Earth's equinoxes, the tilt of the Earth's axis, and the shape of the Earth's orbit. They are consistent with theories suggesting an astronomical control on climate, advanced in the 1860s and quantified by Milutin Milankovitch in the 1930s. [*See* Orbital Variations.]

Greenland ice core records indicate that during the last glacial period, the frigidly cold temperatures were often abruptly interrupted by tremendous warming events, in which temperatures rose in a few decades or less to values nearly as warm as those felt today. These events lasted hundreds to a few thousand years before abruptly dropping back to the frigid glacial values. These dramatic episodes, known as Dansgaard-Oeschger events, show an intriguing relationship with occurrences of huge iceberg armadas that surged across the North Atlantic (known as Heinrich events), and with rapid advances of glaciers in other parts of the world, suggesting that the temperature swings may have been felt world-wide. In addition, the deepest levels in some Greenland ice cores, as well as fossil-pollen evidence in Europe, suggest that such dramatic temperature swings also occurred during the last interglacial period (a nonglacial period with climate similar to today's). While the validity and implications of these observations are still being debated and studied, it is clear that during the past 10,000 years, as civilizations developed, our climate has been remarkably stable. This now appears to be highly unusual, and tremendous climate change seems likely, if not inevitable, in the future.

In addition to these exciting findings, Antarctic ice cores reveal that the atmospheric carbon dioxide (CO_2) concentration was around 30 percent lower during the last glacial maximum than just before the Industrial Revolution. The increase in atmospheric CO_2 over the past century owing to the burning of fossil fuels is of a similar order.

Annual deposits laid down around 200 million years ago show variations in the thickness of the annual layers, presumably reflecting changes in climate, with frequencies similar to those suspected for the Earth's orbit. This implies that the astronomical control on climate is not just limited to the most recent geological period. Also,

since the end of age of dinosaurs (ca. 65 million years ago), the climate appears to have been gradually deteriorating—that is, getting significantly colder.

The most recent evidence suggests that significant glaciation of Antarctica began about 35 million years ago. Additional continental glaciation may have occurred about 14 million years ago. Modeling studies suggest that much of this deterioration can be explained by the gradual uplift of mountains over this period, although plate motions, changes in atmospheric CO_2 concentration, and changes in the extent of shallow seas have also contributed.

[*See also* Ice Ages; Little Ice Ages; *and* Younger Dryas.]

BIBLIOGRAPHY

CLIMAP Project Members. *Seasonal Reconstruction of the Earth's Surface at the Last Glacial Maximum.* Map and Chart Series 36. Geological Society of America, Boulder, Colo., 1981.

Imbrie, J., and K. P. Imbrie. *Ice Ages: Solving the Mystery.* Short Hills, N.J., 1979.

Imbrie, J., et al. "The Orbital Theory of Pleistocene Climate: Support from a Revised Chronology of the Marine $\delta^{18}O$ Records." In *Milankovitch and Climate,* edited by A. Berger, et al. Boston, 1987, pp. 269–305.

Monastersky, R. "Staggering through the Ice Ages: What Made the Planet Careen between Climate Extremes?" *Science News* 146 (1994): 74–76.

Ruddiman, W. F., and J. E. Kutzbach. "Plateau Uplift and Climatic Change." *Scientific American* 264 (1991): 66–75.

DOUGLAS G. MARTINSON

PALMER DROUGHT SEVERITY INDEX.

W. C. Palmer formulated the Palmer Drought Severity Index (PDSI) in 1965. He defined *drought* (1965) as "an interval of time, generally months or years in duration, during which the actual moisture supply at a given place rather consistently falls short of the climatically expected or climatically appropriate moisture supply." The purpose of the index is to compare quantitatively the actual amount of precipitation received in an area during a

PDSI. Table 1. *Conditions represented by values of the Palmer Drought Severity Index*

PDSI	Description
≤−4.00	Extreme drought
−3.00 to −3.99	Severe drought
−2.00 to −2.99	Moderate drought
−1.99 to +1.99	Near normal
+2.00 to +2.99	Unusual moist spell
+3.00 to +3.99	Very moist spell
≥+4.00	Extremely moist

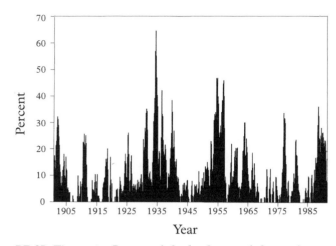

PDSI. Figure 1. *Percent of the land area of the contiguous United States experiencing severe-to-extreme drought according to the Palmer Drought Severity Index, 1895–1991.* (Data from the National Climatic Data Center.)

specified time period with the normal or average amount expected during that same period. Conceptually, the PDSI is based on a procedure of hydrologic or water-balance accounting, by which excesses or deficiencies in moisture are determined in relation to average climatic values. Variables taken into account in the calculation of the index include precipitation, potential and actual evapotranspiration (the combination of evaporation and transpiration), infiltration of water into the soil zone, and runoff. [*See* Drought.]

The PDSI is the most widely used index of drought in the United States, employed biweekly by federal and state government meteorologists to track moisture conditions nationwide. The PDSI represents excessive dryness or wetness on a scale ranging from −4.00 to +4.00, although values of −8.00 or below have been calculated. PDSI values are interpreted according to the scale shown in Table 1.

The PDSI is a good tool for examining patterns of regional moisture conditions at a particular time. It is also useful in comparing historical drought periods.

The percent of the land area of the contiguous United States affected by severe to extreme drought (PDSI ≤ −3.00) from 1895 to 1991 is shown in Figure 1. This graph includes many periods of widespread drought during the period for which index values have been calculated. The most significant drought in United States history occurred during the 1930s, commonly referred to as the *Dust Bowl*, when severe-to-extreme drought conditions persisted for nearly a decade, affecting more than 65 percent of the country in 1934. The period of maximum intensity and spatial extent occurred during July 1934 (Figure 2). More recent droughts in the 1950s, 1970s, and late 1980s to early 1990s, were of comparable intensity but shorter duration. During the 1988 drought, nearly 40 percent of the nation experienced severe to extreme drought conditions.

The PDSI has come under considerable criticism in recent years, in part because the index is often applied to situations for which it was not intended. For example, although the PDSI is best characterized as a meteorolog-

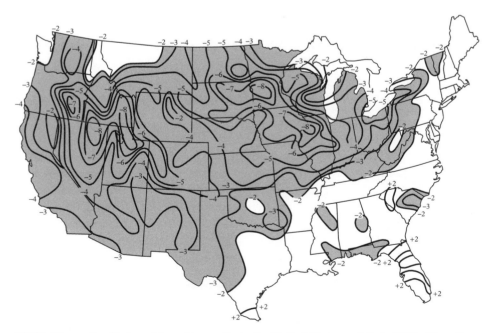

PDSI. Figure 2. *Status of drought severity in the United States on 1 July 1934.* (Data from the U.S. Department of Agriculture.)

ical or hydrological drought index, it is often used to assess the severity and impact of agricultural drought. When used in this way, the index often does not perform reliably because plant response was not considered in the derivation of the index. The index also uses a simplified method to calculate evapotranspiration, a key variable in determining the severity of agricultural drought. Although other indices of drought, such as soil-moisture and water-supply indices, have been developed in recent years, the widespread use of the PDSI will probably continue because it is the only index routinely calculated that allows regional comparisons of moisture excesses and deficiencies.

BIBLIOGRAPHY

Alley, W. M. "The Palmer Drought Severity Index: Limitations and Assumptions." *Journal of Climate and Applied Meteorology* 23 (1984): 1100–1109.

Karl, T. R., and R. W. Knight. *Atlas of Monthly Palmer Hydrological Drought Indices (1895–1930) for the Contiguous United States.* Asheville, N.C., 1985.

Karl, T. R., and R. W. Knight. *Atlas of Monthly Palmer Hydrological Drought Indices (1931–1983) for the Contiguous United States.* Asheville, N.C., 1985.

Palmer, W. C. *Meteorological Drought.* Research Paper 45, U.S. Weather Bureau. Washington, D.C., 1965.

Palmer, W. C. "Keeping Track of Crop Moisture Conditions Nationwide: The New Crop Moisture Index." *Weatherwise* 21 (1968): 156–161.

Riebsame, W. E., S. A. Changnon, Jr., and T. R. Karl. *Drought and Natural Resources Management in the United States: Impacts and Implications of the 1987–89 Drought.* Boulder, Colo., 1991.

Wilhite, D. A. "Drought." In *Encyclopedia of Earth System Science.* San Diego, Calif., 1991, vol. 2, pp. 81–92.

Wilhite, D. A., ed. *Drought Assessment, Management, and Planning: Theory and Case Studies.* Boston, 1993.

Wilhite, D. A., and M. H. Glantz. "Understanding the Drought Phenomenon: The Role of Definitions." *Water International* 10 (1985): 111–120.

Wilhite, D. A., and W. E. Easterling, eds. *Planning for Drought: Toward a Reduction of Societal Vulnerability.* Boulder, Colo., 1987.

DONALD A. WILHITE

PARTICULATES. *See* Aerosols.

PHASE CHANGES. *See* Water Vapor.

PLANETARY ATMOSPHERES. Gaseous, fluid envelopes are wrapped about some planets and satellites and are retained by the gravity of these bodies. A *planet* can be defined as a star-orbiting body whose mass is insufficient to cause energy-releasing nuclear (fusion) reactions within its interior. A *satellite*, in turn, is a smaller celestial body in orbit about a planet. Table 1 lists basic properties of the known planetary and satellite atmospheres: temperatures at the surface, at 1 bar; surface pressures, at 1 bar; surface gravities (both polar and equatorial if planetary equatorial bulging is great from rapid rotation or its oblateness); escape velocities; and primary compositions.

Our solar system has an extreme range in atmospheres, from ephemeral and dynamical ones to stable and mas-

PLANETARY ATMOSPHERES. Table 1. *Properties of planetary and satellite atmospheres*

	Temperature (degrees kelvin)	Pressure (bars)	Gravity (meters per square second)	Escape Velocity (kilometers per second)	Composition
Mercury	395	$\leq 10^{-12}$	3.95	4.3	He, H, O, Na, K
Venus	730	90	8.88	10.3	96% CO_2, 3.5% N_2, SO_2, H_2O
Earth	288	1	9.78	11.2	77% N_2, 21% O_2, 1% H_2O, 0.9% Ar
Mars	218	7×10^{-3}	2.18	5.0	95% CO_2, 2.7% N_2, 1.3% O_2, CO, H_2O
Jupiter	165	1	27.0 p, 23.1 e★	59.5	90% H_2, 10% He, NH_3, CH_4, PH_3, H_2O
Io	130	10^{-7}	1.8	2.6	SO_2
Saturn	134	1	12.1 p, 9.0 e	35.6	94% H_2, $\leq 6\%$ He, NH_3, CH_4
Titan	93	1	1.4	2.6	65–99% N_2, 0.5-3.4% CH_4, H_2
Uranus	76	1	9.2 p, 8.7 e	21.2	85% H, 13% He, CH_4
Neptune	68	1	11.5 p, 11.2 e	23.8	75% H, $\leq 25\%$ He, CH_4
Triton	38	16×10^{-6}	0.8	1.0	N_2, CH_4
Pluto	50	$1-10 \times 10^{-6}$	0.7	0.8	CH_4

★p = pole; e = equator

sive ones. The planets are grouped into the inner, or *terrestrial*, planets (Mercury, Venus, Earth, and Mars) and the outer planets (Jupiter, Saturn, Uranus, Neptune, and Pluto). The silicate-rich surfaces of the inner solar system give way to the water-ice surfaces of Jupiter's satellites and finally to very distant methane-ice surfaces like that of Pluto. The atmospheres of the inner planets are oxidized or oxidizing, while the outer planets have reduced or reducing atmospheres consisting of hydrogen-rich molecules and higher-order hydrocarbons. Several of the many satellites in the solar system also have noteworthy atmospheres: Jupiter's Io, Saturn's Titan, and Neptune's Triton. Only the Earth, however, appears to have developed an atmosphere congenial to life. (Recently discovered planets around other stars have yet to be studied because of their remoteness.)

Origin and Evolution. The planets of our solar system formed approximately 4.6 billion years ago within the condensing solar nebula, a giant disk of dust and gas. Gravitational attraction and collisions aggregated the dust grains into increasingly larger bodies. The atmospheres associated with these large bodies were created through some combination of four processes, described by the accretion, solar nebula, solar wind, and comet/asteroid theories.

In the *accretion theory,* the planetesimals and dust building the planets held small amounts of volatiles, elements and compounds, such as water, carbon dioxide, and molecular nitrogen, easily driven out of minerals with moderate heating. Gases were released from minerals as these planets heated up and vented into the young, primitive atmosphere. According to the other three theories, the planets were formed out of volatile-free material, and their atmospheres were derived from external sources. In the *solar nebula theory,* atmospheres were gravitationally caught from the gases of the primordial solar nebula during or after a planet's formation. In the *solar wind theory,* atmospheres were caught out of the solar wind through the eons. The solar wind is the flow of plasma (primarily protons and elections, with ionized gas of many elements) from the Sun, moving from 300 kilometers per second during solar quiet times to 1,000 kilometers per second during active solar flare periods (200–600 miles per second). Finally, in the *comet/asteroid theory,* atmospheres were derived from volatile-rich comets and asteroids striking the planets.

The primordial planetary atmospheres evolved into secondary atmospheres through a combination of mechanisms: outgassing from the interior, crustal interactions, photochemistry, photosynthesis, and atmospheric loss.

Outgassing of volatiles from the solid interior occurs through volcanic eruptions and venting, which supply *juvenile,* or *new, gases.* Crustal interactions include the weathering of surface rocks, the incorporation of carbon dioxide (CO_2) into carbonate rocks, and the condensation of water vapor (H_2O) or other volatiles into crustal ice reservoirs. Atmospheric oxygen (O_2) increases through photodissociation of water vapor molecules by solar ultraviolet (UV) radiation. Photosynthesis on the Earth occurred with the emergence of plants and contributed most of Earth's atmospheric oxygen.

Atmospheric loss refers to gases lost to space by several mechanisms: Jean's escape (thermal), hydrodynamic escape, atmospheric cratering, or interaction with the solar wind. To be lost, atmospheric gases must reach their escape velocities. The *escape velocity* is the speed necessary for a given particle (in this case, a gas molecule) to break completely free of a primary body's (planet's) gravitational field.

In *Jean's escape,* thermal evaporation of the atmosphere occurs when a gas follows an upward ballistic course and does not suffer a collision. This works best for the lightest gas species in the high speed side of the Maxwellian thermal speed distribution. This distribution is the most probable distribution of speeds for gas molecules at a certain temperature. As the temperature of a gas rises, the speeds of its molecules increase. Simultaneously, the distribution curve flattens and predicts more molecules having speeds higher and lower than the most probable speed, the peak of the distribution curve. The gravitational fields of the outer planets are much stronger than those of the terrestrial planets, so escape velocities for the outer planets are much higher. The giant planets have consequently retained more of their primordial atmospheres, with nearly solar levels of hydrogen and helium.

In *hydrodynamic escape,* an upper atmosphere is heated by absorption of intense extreme-ultraviolet (XUV) radiation. Primordial hydrogen gas (the lightest gas species) expands at supersonic speeds, exerting forces on heavier atmospheric constituents and lifting them upward through the atmosphere. Lighter constituents could be entrained in the hydrodynamic escape flow of hydrogen and also lost to space. Hydrodynamic escape is active in models of Pluto's and Triton's thin, remote atmospheres.

In the case of *atmospheric cratering,* escape velocities are reached when comets, stray planetesimals, or meteoroids impact an atmosphere. Finally, loss by *interaction with solar wind* occurs through ionization of heavier gases such as carbon monoxide (CO) and molecular nitrogen (N_2)

by solar ultraviolet light. The subsequent convection of these ionized gases through the planet's magnetosphere allows them to become entrained into the solar wind. Mercury's early heavier gases were, for example, mostly lost through this mechanism.

Thermal Structure and Photochemistry. Planets receive solar energy mainly in the visible radiation wavelengths near 0.5 microns (10^{-6} meters) and then reradiate energy out to space in the form of infrared (IR) radiation (in the tens of microns) wavelengths. The radiation from a body depends on its temperature. Planck's law is an equation specifying the intensity of emitted radiation at every wavelength. The resultant Planck distribution curve resembles an asymmetric hill with a peak in intensity that shifts to shorter wavelengths and increases in magnitude as the body's temperature heats up. The Sun has a photosphere temperature of roughly 6,000 K, so its photons (elementary particles, the quanta of electromagnetic radiation) have energies corresponding to short visible wavelengths; planets are cooler and radiate lower-energy photons corresponding to longer infrared wavelengths. Most of the solar insolation (solar input received) is absorbed at a planet's surface, because atmospheric gases are poor absorbers of 0.5-micron radiation, the peak solar input. Planetary infrared radiation originates mostly from the heated surface and then interacts with atmospheric molecules, which couple easily with wavelengths around 10 microns.

Vertical regions in atmospheres are labeled based on their temperature structure. Near the warm surface, convection (overturning) occurs with surface heating by sunlight. Gases are well mixed, and with increasing height air temperatures fall. Clouds occur at the higher cooler temperatures with condensation of volatile gases such as H_2O and CO_2. This lowest region is called the *troposphere*. On the Earth, the troposphere (the lowest 10 kilometers) contains all the air motions familiarly called *weather*. The temperature minima in other planets' upper tropospheres act as *cold traps*, forcing condensible gases into cloud liquid or ice particles. [*See* Atmospheric Structure.]

Above the troposphere lies the *stratosphere*, with constant or warming temperatures with altitude. Air masses are stable with very little vertical mixing or buoyancy. Above the stratosphere, there may be a *mesosphere* in which temperatures decrease with height. Vertical overturning occurs, and thin clouds condense out volatiles. Above the mesosphere lies the *thermosphere*. Solar extreme-ultraviolet radiation (wavelengths ≤ 0.1 microns) gets absorbed in this region, and consequently, its maximum temperatures depend on solar activity. The

tops of these layers are called the *tropopause, stratopause, mesopause,* and *thermopause,* respectively.

One vertical portion of the atmosphere is characterized by its composition rather than its thermal structure. The *ionosphere,* the uppermost portion of an atmosphere, begins in the thermosphere in the Earth's case. It contains significant populations of free electrons and ions resulting from photoionization.

A basic rule of atmospheres is that absorption of higher-energy wavelengths, such as solar ultraviolet radiation or cosmic rays, occurs in the upper atmosphere. Interaction with high-energy radiation leads to both thermal warming of the atmosphere and to photochemistry or chemical reactions, depending on the nature of the interaction. Both outcomes can influence the temperature structure and the composition of the atmosphere. Atmospheric atoms or molecules can undergo various processes: excitation into hot electronic or vibrational states; dissociation, in the case of molecules; or ionization, when they absorb high-energy photons. The resulting radicals, ions, or simpler gas species are often more reactive than the original atoms or molecules.

Greenhouse Effects. *Opacity* is the reduction of radiation's intensity as it passes through an atmosphere or some other medium. The greenhouse effect is a phenomenon that occurs in an atmosphere when its infrared (IR) opacity is much greater than its visible opacity. Sunlight penetrates deeply into the atmosphere and is reemitted as heat (IR photons); the reemitted IR photons, however, are caught by atmospheric gases that interact readily with these wavelengths. The photons go through repeated absorptions and emissions by the gases before escaping to space. This delayed escape sets up substantial temperature differences between the ground and the radiating atmosphere.

Mars, Earth, and Venus have atmospheres representing different levels of greenhouse warming. Mars's atmosphere is composed mainly of CO_2. This molecule absorbs infrared wavelengths though a bending mode at 15 microns, where in the oxygen atoms bend about the carbon atom and through a vibration mode at 15 microns. The surface of Mars is roughly 7° warmer owing to this atmospheric infrared opacity. The thinness of the Martian atmosphere (reflected in its low surface pressure), and its lower solar irradiance relative to that of Earth and Venus, curtail the greenhouse mechanism despite Mars's CO_2 atmosphere.

In the case of the Earth, water vapor and water clouds, along with CO_2 gas, are the dominant infrared atmospheric absorbers. The atmosphere's main constituents, N_2 and O_2, absorb infrared radiation poorly. The infrared opacity increases surface temperatures by 35 K above what would exist without atmospheric infrared absorption. Water vapor, in particular, has a rich rotation spectrum superimposed on its vibrational modes, along with a permanent electric dipole, and couples well to IR photons. The modest greenhouse warming is held in check by the fact that the vapor pressure of water vapor is reached as temperatures rise, and the volatile gas condenses out. The Earth, however, also has anthropogenic (human generated) sources of gases—infrared absorbers that may enhance the modest natural greenhouse effect and cause global warming. [See Greenhouse Effect.]

Venus exhibits an extreme of runaway greenhouse warming, with surface temperatures of 730 K, able to melt lead. This atmosphere, like that of Mars, is made up mainly of CO_2, along with smaller amounts of water vapor, sulfur dioxide (SO_2), and sulfuric acid (H_2SO_4) cloud droplets. Unlike Mars, Venus has an extremely dense atmosphere and its gaseous water vapor and SO_2 absorb infrared in those portions of the spectrum where CO_2 does not absorb. With its infrared opacity, Venus's surface is roughly 500 K warmer than its effective temperature! (A planet's effective temperature is its expected surface temperature, based on the sunlight received at its distance from the Sun combined with its size.) All of its water exists as a gas in the atmosphere, because temperatures are too hot and pressures too high to allow liquid water to form and curb this strong "runaway" greenhouse warming.

The atmospheres of Mars, Earth, and Venus exhibit different amounts of greenhouse warming. This greenhouse warming is a mechanism included in models of their past climates and the evolution of their past climates. Given a denser early atmosphere with various greenhouse gases, Mars could have had increased greenhouse warming that elevated surface temperatures above the freezing point of water. Such paleoclimatic greenhouse models, however, need to assume early atmospheric compositions, solar inputs, cloud feedbacks, dust feedbacks, ice cap feedbacks through sublimation, and heat transports—all of which have substantial impact on the early atmospheres being modeled.

Atmospheres of the Inner Planets. The inner planets have evolved secondary atmospheres and are deficient in the so-called noble solar gases helium, neon, argon, krypton, and xenon. Mercury is the exception and has an extremely tenuous atmosphere discussed later. The other inner planets' atmospheres are relatively rich in water, carbon dioxide, and nitrogen, and interact with their planets' hydrospheres (the total water substance at and above the surface) and lithospheres (the planetary

crust). The Earth's atmosphere also interacts with its unique biosphere, the total of living organisms.

Venus. The atmosphere of Venus, with a surface pressure of 90 bars, is enshrouded by clouds and is composed mainly of CO_2, along with a few percent of molecular nitrogen. Smaller amounts of SO_2 and H_2O exist in the troposphere, and trace amounts of CO exist above the clouds, derived from photodissociation of CO_2. The thick cloud deck, composed of sulfuric acid droplets, covers nearly 100 percent of the planet; it has a high albedo (reflectance) over visible wavelengths.

Near the planet surface below this dense atmosphere, meridional (equator-to-pole) temperature differences are less than a few degrees K owing to the efficient transport of heat by meridional circulation in both hemispheres. Temperature differences do exist between the equator and the pole in the upper troposphere, with constant pressure levels in the atmosphere bulging at the warm equator. A pressure gradient force on the atmosphere is directed away from the bulge toward the poles. This pressure gradient balances with centrifugal forces from retrograde zonal winds of 100 meters per second (hurricane winds on the Earth begin at 55 meters per second). Retrograde motion is motion opposite to the rotation of the solid planet, so these strong winds are very unusual. This balance of forces in Venus's troposphere is called *cyclostrophic balance.* Usually such meridional pressure differences would be balanced by Coriolis forces in an atmosphere, given a sufficiently large planet rotation—a situation called *geostrophic balance.* Venus, however, has a very slow rotation rate of 243 Earth days. These retrograde winds are called *four day winds* because they move the bulk of Venus's atmosphere around the solid planet once every 4 days. It is still uncertain how these retrograde, extremely rapid winds are maintained.

Venus's thermosphere experiences large diurnal variations (day-to-night differences) owing to its low density. The dayside thermosphere has noon temperatures of 300 K, while the nightside thermosphere has midnight temperatures of 100 K. The frigid nightside thermosphere is therefore termed the *cryosphere.* These thermospheric temperatures are much lower than those of the Earth (600–1,000 K) because of very effective infrared radiative cooling by CO_2. The diurnal difference in temperature sets up strong subsolar to antisolar (noon to midnight) winds, reaching 150 to 200 meters per second in hydrodynamic atmospheric models.

Earth. The atmosphere of Earth, with a surface pressure of 1 bar, has accumulated most of its oxygen through photosynthesis by plants. The main constituent is not CO_2, as in the other terrestrial atmospheres;

instead, nitrogen composes 77 percent of Earth's atmosphere. Most of the Earth's carbon dioxide lies bound up in carbonates such as limestone, and in the oceans. [*See* Atmosphere, *article on* Formation and Composition.]

Despite its greater orbital distance, the surface of the Earth receives more total solar energy than the surface of Venus, because of Venus's high cloud albedo. Our atmosphere receives and absorbs most of its solar energy between the equator and latitudes of 40° and radiates away more energy than it receives at latitudes poleward of 40°. Meridional circulation, in the form of three Hadley cells in each hemisphere, transports heat from the equator to the pole. A Hadley cell is a form of circulation created through the difference in solar heating between equatorial and polar regions. Air heated at the equator rises and moves poleward until it cools, sinks, and moves equatorward again. The equatorial Hadley cell on the Earth is disrupted near 30° north latitude by Coriolis forces (an inertial force from the solid planet's rotation). Coriolis forces act to deflect air motion perpendicularly from their direction of motion, producing zonal (parallel to latitude lines) components to the Hadley transport. The Earth's atmosphere consequently has zonal jets—the trade winds moving westward in the low latitudes, and the jet streams moving eastward in the high latitudes. Above the convection-controlled troposphere and the stratosphere photochemically warmed by ozone, Earth's thermosphere resides, with high night and day temperatures of 800 and 1,000 K respectively. With only slight amounts of CO_2 present, infrared cooling occurs less effectively and more through emission of nitric oxide (NO) and atomic oxygen (O).

Astronomical variations in orbital eccentricity and obliquity have affected solar insolation and thus the Earth's climate. Successive ice ages have occurred during which global surface air temperatures fell by several degrees K, and continental ice sheets spread to cover much of North America and Eurasia. [*See* Orbital Variations.]

Mars. The planet Mars resembles a windblown desert with a thin, cold atmosphere. Its atmospheric CO_2 can condense out, unlike that of the Earth and Venus. Its thin surface pressure of 7 millibars leads to large diurnal temperature swings of around 100 K at the ground. Both CO_2 and H_2O can exist as either solids or vapor on Mars. Even where temperatures rise above 273 K on Mars, surfaces are dry. The amount of water available is too slight for condensation (the triple point of water is at 6 millibars and 273 K). There may be a large H_2O reservoir bound up in permafrost as well as at the poles on Mars. There may also be a large reservoir of CO_2 bound up in car-

bonates, such as limestone, and adsorbed in regolith (uncompacted surface layers).

The Martian surface topography is asymmetric, with the southern hemisphere called *highlands* because its surface is 4 kilometers higher than the global average (the highest mountain on the Earth, Everest, rises almost 9 kilometers above sea level). Differences in topography and insolation make the global circulation tend to be asymmetric about the equator. Remarkable diurnal surface temperature changes drive thermal tides or thermal winds. The thinness of the CO_2 atmosphere also transports heat inefficiently from the equator to the poles. Mars has a single Hadley cell spanning the equator from the summer hemisphere to the winter subtropics. The poles, therefore, become extremely cold, so that CO_2 condenses out (the triple point of CO_2 is at 217 K and 5 bars). Carbon dioxide acts as an effective upper-atmospheric infrared coolant (as in the case of Venus), keeping thermospheric temperatures at roughly 310 K regardless of the time of day.

The constant presence of dust in the atmosphere is a unique characteristic of Mars. Fine dust (0.1 to 1 micron in size) is ever-present in the atmosphere because of *saltation,* the skipping of 100-micron sand grains along the surface due to strong surface winds. These colliding large sand grains kick up smaller dust grains into suspension in the atmosphere. Many local dust storms can occur within a Martian year. The dust changes the amount of solar isolation received, the polar ice albedos, and dynamics in this thin atmosphere. Mars's high orbital eccentricity (0.093) causes its solar insolation to maximize during perihelion over its southern hemisphere. Southern-hemisphere summers are shorter but hotter than northern-hemisphere summers, with maximum surface temperatures of 290 K. Consequently, dramatic dust storms occur annually, beginning locally during the extreme spring-to-summer heating of the martian southern hemisphere. These storms grow to wrap the planet in an optically thick shroud.

Mars also has permanent and seasonal polar caps where condensation and sublimation of its CO_2 atmosphere take place. The permanent or residual cap in the south is dry CO_2 ice, and that in the north is H_2O ice. These polar caps follow a seasonal cycle reflected in the total pressure of Mars's CO_2 atmosphere. Total atmospheric pressure swings to a minimum when the polar caps are largest, and to a maximum when the caps are small. This unusual condensation flow, with transport of CO_2 from low to high latitudes and formation of the polar ice caps, is part of the global circulation.

Periodic changes in Mars's axial and orbital parameters

have created climate fluctuations. Geological evidence, including dry channels apparently cut by liquid water and layered deposits in the polar regions, points to past atmospheric environments quite different from the current cold and dry climate.

Outer or Giant Planetary Atmospheres. The atmospheres of the giant outer planets are noted for colorful latitudinal bands of zonal winds, oval storm systems, high amounts of primordial gases, and internal heat sources. Two of their satellites, Titan and Triton, also have definite atmospheres.

Jupiter and Saturn comprise 92 percent of the extrasolar mass in the solar system and have no solid planet surfaces. Instead, they have metallic fluid interiors around centers composed of rock and ice. The metallic fluid zone is made up of molecules stripped of their outer electrons because of the high interior pressures. If these giant planets had been roughly 80 times more massive, the resulting internal temperatures could have triggered nuclear fusion of hydrogen. Our solar system would then have had multiple stars and very different planetary atmospheres.

Jupiter. Bright in the Earth's nighttime sky, Jupiter reflects sunlight from high cirrus ice clouds of ammonia (NH_3). The mainly hydrogen and helium atmosphere has cold traps holding NH_3, H_2O, and ammonium hydroxide (NH_4OH) clouds in its troposphere. Solar abundances of hydrogen, carbon, and nitrogen are present, indicative of Jupiter's limited atmospheric evolution. Jupiter rotates rapidly, with a day of roughly 10 hours, causing oblateness in its shape. Clouds occur in latitudinal belts (dark brown bands) and zones (bright bands). These colored latitudinal bands correspond to counterflowing zonal jets. Jupiter has roughly 5 zonal jets per hemisphere. Tracking of cloud features in images taken by the *Voyager* satellite determined that the equatorial winds—the strongest—move with the planet's rotation at top speeds of 100 meters per second. Jets at higher latitudes move with lower speeds and reverse direction in each successive band.

Within these zonal jets Jupiter's atmosphere experiences a great deal of eddy activity (huge circular storms). These eddies get their energy from buoyancy and are created by solar and interior heat. The larger storms can be very long-lived, while the smaller eddies are pulled apart by the zonal winds. The zonal jets then absorb the kinetic energy of the smaller storms, so these smaller eddies help to maintain the zonal jets. The most famous of these circular storms, The Great Red Spot, was first observed 300 years ago and covers 30° in longitude—wide enough to swallow the Earth. Its flow

moves counterclockwise at 100 meters per second, consistent with a high-pressure region in the southern hemisphere.

Jupiter emits 1.5 to 2.0 times the radiative energy it absorbs from the Sun, implying the existence of a substantial internal heat source. The emitted radiation is uniform over latitude, unlike the received solar energy, pointing to heat transport across latitudes at some depth in the atmosphere. The internal heat source is primordial, generated by the gravitational contraction of the portion of the nebula from which it condensed, and, it is still slowly cooling.

Saturn. Also a rapid rotator with a 10-hour day, Saturn has an atmosphere composed mostly of molecular hydrogen followed by helium, but with fewer observed compounds than Jupiter. The more distant giant planet has a lower gravity and cooler temperatures, as expected from its smaller solar input. Saturn consequently has a more extended atmosphere, deeper cloud structures, and cloud features more subdued than those of Jupiter. Saturn's winds are stronger than Jupiter's jets, reaching top eastward speeds of 500 meters per second at the equator and dropping to direction-reversing bands of lower speeds toward the poles. This planet has fewer zonal jets per hemisphere than Jupiter. Cold traps again limit the abundances of volatiles in the upper atmosphere, condensing out the NH_3, NH_4OH, and H_2O from the deeper atmosphere into clouds. Overall, the vertical temperature structure mimics that of Jupiter, with upper-atmosphere temperatures of roughly 140 K falling to a minimum of 80 K at 0.1 bars, and then warming to higher temperatures deeper in the atmosphere.

Saturn's atmosphere, however, has an additional internal heat source and emits 1.5 to 2.5 times the planet's solar energy input. Its internal temperatures are too cold for helium to be uniformly mixed with hydrogen in the metallic zone. Therefore, helium condenses out at the top of the metallic zone and forms helium raindrops, which fall into the center of the planet. The helium raindrops' gravitational energy is converted into heat and added to the internal heat from the primordial gravitational contraction. Internal heat is believed to be transported up to higher atmospheric altitudes through convection.

Titan. The atmosphere of this Saturnian satellite strongly resembles the Earth's, with a surface pressure of 1.5 bars and a composition mostly of molecular nitrogen. Several percent of methane are present and are chemically evolving into trace amounts of complex hydrocarbons such as ethane (C_2H_6). The surface is hidden by photochemically produced organic material or smog that extends up to 200 kilometers altitude. Stratospheric temperatures reach 170 K. Below this, a cool cloud deck of methane droplets at 70 K caps a troposphere that has modest greenhouse warming and a surface temperature of 95 K. The slight greenhouse warming of 10 K occurs because of broad infrared absorption bands for nitrogen, hydrogen, and methane in the ambient high pressures. The satellite's atmosphere is highly reducing and contains exotic molecules such as hydrogen cyanide (HCN) in addition to the organic polymers of the opaque smog layer. Dynamically, the atmosphere may be in cyclostrophic balance, similar to that of Venus, and methane moist convection may occur. Meridional wind transport of heat from equator to pole should be efficient, given the atmosphere's thickness; thus, surface temperatures should be uniform over latitude and time of day.

Titan has a substantial atmosphere because of a combination of events. This satellite formed from material that contained methane and ammonia ice. Given its size, it may have undergone internal differentiation to concentrate ices near the surface. Surface temperatures became high enough to allow methane and ammonia gas to sublimate. Nitrogen gas developed from ammonia reactions with solar UV radiation. Lighter gases such as hydrogen escaped owing to Titan's small escape velocity. Even though Titan usually resides within Saturn's magnetosphere and occasionally in the planet's magnetosheath or in the solar winds, Saturn's magnetosphere is insufficiently vigorous to cause losses from Titan's atmosphere of the kind induced on Io by its proximity to Jupiter.

Uranus. The 98° obliquity of Uranus means that the planet lies nearly on its side, so that it receives more sunlight at the poles of its rotational axis than at its equator. Uranus also has a surprisingly weak internal heat source when compared with the other giant planets. Its internal output is actually 12 percent less than the solar heat input. As it receives one-sixteenth the solar input that Jupiter receives and has no compensating internal heat, the mainly hydrogen atmosphere has little energy available to it. Uranus looks very bland and featureless in *Voyager* satellite images.

Helium and methane are also present in this atmosphere. The methane photochemically develops into acetylene (C_2H_2) and other hydrocarbons in the stratosphere. Uranus's stratosphere has temperatures up to 90 K, decreasing to tropopause temperatures of 53 K above a convective troposphere. A methane cloud may be present at the 1-bar level. The stratospheric methane chemistry develops hydrocarbons that polymerize and then precipitate into haze aerosols. Such aerosols act to warm the stratosphere through absorption of solar radiation.

This differs from the warming of Jupiter and Saturn's stratospheres by vertically propagating planetary waves.

Voyager tracking of limited cloud features found maximum zonal winds at 45° south latitude that co-rotate with the solid planet at a speed of 110 meters per second. Uranus is another oblate planet, with a rotation period of 17.24 hours. The equatorial atmosphere actually rotates more slowly than the planetary interior, at speeds below the higher-latitude maximum. Little variation over latitude could be found in tropospheric temperatures, except for cooler regions at 25° south latitude and 40° north. Meridional circulation may cause these cool latitudinal bands, perhaps created from frictional drag on the dominant eastward zonal winds. Models suggest that the atmosphere at low latitudes subrotates from the equatorial plane and superrotates at higher latitudes. Cold air would then rise at the mid latitudes to form meridional cells.

Neptune. The second icy planet appears to be extremely windy, with zonal wind patterns reminiscent of Uranus, overlain with storm systems reminiscent of Jupiter. Its hydrogen atmosphere mixes with up to 25 percent helium and larger amounts of methane than found on Uranus. The methane gas forms bright, optically thin ice clouds at 1.5 bars, high above optically thick clouds of hydrogen sulfide (H_2S) or ammonia ice under 3 bars. The methane gas absorbs red light so that Neptune looks like a blue orb in *Voyager* images. A photochemically produced haze or smog composed of hydrocarbons extends from the lower stratosphere into the troposphere. The Neptunian atmosphere has horizontal temperatures remarkably similar to those of Uranus, with cold regions in the mid latitudes.

Neptune's atmosphere is very active, with one overwhelming anticyclonic storm system called The Great Dark Spot, an oval storm spreading across 38° longitude in the southern hemisphere. The dark storm contracts and stretches as it rolls in the ambient zonal winds and may be a hole or clear area in the methene cloud tops. Bright, ephemeral streaming clouds edge The Great Dark Spot. Other smaller oval storms exist at more southern latitudes, along with a bright, sometimes square or triangular cloud called "the Scooter." The Great Dark Spot and smaller storms may or may not be stable features like The Great Red Spot on Jupiter.

Neptune's equatorial atmosphere subrotates like that of Uranus and the Earth, with zonal winds having periods shorter than the interior's rotation. Its higher latitudes superrotate like that of Uranus; again, a meridional cell appears to rise at the mid latitudes. This atmosphere actually emits 2.5 times the energy it receives from the Sun. The larger output of internal heat found on Neptune than found on Uranus may drive this atmosphere's very active nature.

Triton. This satellite of Neptune has the lowest observed surface temperature in the solar system at 38 ± 4 K. Its molecular nitrogen atmosphere of 16 microbars pressure at the surface also contains methane gas at a ratio of 10^{-4} relative to the nitrogen abundance. Both gases are supplied and buffered by nitrogen and methane ices on the surface through sublimation during Triton's summer. Surface frosts extend nearly to the equator. N_2 is the more volatile ice and acts in a manner similar to that of CO_2 in the Martian atmosphere. Nitrogen ice clouds form over warm areas of active sublimation, and sublimation winds transport N_2 from the summer to the winter hemisphere.

Several geyser-like plumes have been discovered in *Voyager* images of Triton. They occur over the summer high latitudes in narrow channels of N_2 gas carrying particulates up from the icy surface to heights of 8 kilometers. Different mechanisms exist for producing such plumes, but data are too sparse to select one.

Pluto. Based on limited ground-based observations of occultations, Pluto's atmosphere is believed to consist of methane at surface pressures of 1 to 10 microbars over methane ice. (An *occultation* is when one celestial body blocks the view of another body—in this case a star passes behind Pluto's atmosphere during its passage and is measured by terrestrial observers.) Pluto's high orbital eccentricity causes large changes in solar insolation during its orbit, and thus large changes in surface temperature and pressure. Estimates of Pluto's temperatures range about 50 K, depending on individual occultations and amounts of atmospheric gases assumed in the occultation analysis. Given these relatively warm temperatures, which are due to the planet's near-perihelion orbital position (near the Sun), the atmosphere should be escaping rapidly. The methane (CH_4) released from the warmed ice surface undergoes photodissociation by solar XUV emission, and the lighter hydrogen escapes to form an extended cloud about the planet. This hydrodynamic escape of hydrogen pulls other gases along with it and extends the atmosphere; hydrodynamic escape of CH_4 may even occur.

Atmospheric models require some surface reservoir to replenish the atmosphere's volatiles over repeated perihelia and, in some models, the presence of heavier gases (CO or N_2) to slow the diffusion of the CH_4.

Tenuous Planetary Atmospheres. Observations have discovered extremely tenuous atmospheres around Mercury and Io. The innermost planet in our solar system has an atmosphere that is more an exosphere as its gas comes from the solar wind flowing past Mercury. Io, one of Jupiter's moons, derives its uneven atmosphere from volcanic eruptions. Their insubstantial nature makes the label of "atmosphere" debatable at times.

Mercury. The innermost planet's original atmosphere was swept away by the solar wind in the form of ions. Mercury is simply too small and too hot to have a traditional atmosphere, given its extremely high eccentricity of 0.206 and its closeness to the Sun. Nightside surface temperatures can drop to 90 K while dayside temperatures can reach 700 K. Its present atmosphere is an exosphere with extremely low densities, ballistic trajectories, and rare collisions of its atoms. Composed of neutral helium atoms and smaller densities of hydrogen and oxygen, as measured by *Mariner 10* in the 1970s, this exosphere extends to several thousands of kilometers above the surface with ballistic paths (free-fall paths) directed only by gravity and possible only in low-density conditions. Ground-based observations also detect very slight amounts of sodium and potassium atoms.

Outgassing from Mercury's interior supplies helium, and both helium and hydrogen are added by neutralization of solar wind ions. The sodium and potassium atoms are contributed by surface materials and meteoritic material. The surface is vaporized by sputtering (a process wherein deposited energy ejects material from a surface) by solar winds or meteoritic impacts, and the meteoritic material vaporizes with the kinetic energy in its impact. Atmospheric loss occurs through photoionization and—given Mercury's extreme eccentricity—ensuing magnetospheric convection.

The Earth's Moon has an ephemeral exosphere that is similar to the exosphere of Mercury.

Io. This satellite of Jupiter has a very thin, volcanic-derived atmosphere of sulfur dioxide. It has active volcanoes ejecting SO_2 gas; one eruption was photographed by *Voyager 1.* Evidence also exists for H_2S at the surface of this satellite and in its tenuous atmosphere. This atmosphere is in rough equilibrium with the surface sulfur dioxide frost, and atmospheric pressures are driven by the frost temperature. When surface temperatures rise to noontime highs of 130 K, the atmosphere has a surface pressure of 0.6 microbars. The nighttime surface temperatures fall to roughly 90 K and the thin atmospheric pressures follow, dropping to a surface pressure of 2×10^{-8} microbars. Io has more gas over warm areas and near active vents, giving it lumpy constant-pressure surfaces.

The SO_2 atmosphere dissociates owing to solar UV radiation and electrons from Jupiter's magnetosphere, forming Io's thick ionosphere. Because Io exists within Jupiter's strong active magnetosphere, it has a dense ionosphere with electron densities of 10 to 100×10^3 per cubic centimeter (compared to the Earth's ionospheric densities of 10^6 electrons per cubic centimeter). The giant planet's magnetosphere causes atmospheric loss by sputtering off SO_2 gas from Io's tenuous atmosphere with bombarding magnetospheric particles in a manner similar to the solar wind's sputtering action on Mercury's surface.

Io may have a sublimation flow, similar to the Martian CO_2 flow, with transport of SO_2 gas from warm, high-pressure areas to the cold, low-pressure of the nightside. Io has no polar caps, possibly because its volcanic atmosphere cannot extend to cover the poles. Instead, SO_2 may condense out at the edge of its sublimation flow in Io's mid latitudes.

[See also Aeronomy.]

BIBLIOGRAPHY

Beatty, J. K, B. O'Leary, and A. Chaikin, eds. *The New Solar System.* 2d ed. New York, 1982.

Hartmann, W. K. *Moons and Planets.* 2d ed. Belmont, Calif., 1983.

Ingersoll, A. P., et al. "Supersonic Meteorology of Io: Sublimation-Driven Flow of SO_2." *Icarus* 64 (1985): 375–390.

Lindal, G. F., et al. "The Atmosphere of Uranus: Results of Radio Occultation Measurements with Voyager 2." *Journal of Geophysical Research* 92 (1987): 14987–15002.

Lunine, J. I. "Voyager at Triton." *Science* 250 (1990): 386.

Pepin, R. O. "On the Origin and Early Evolution of Terrestrial Planet Atmosphere and Meteoritic Volatiles." *Icarus* 92 (1991): 2–79.

Smythe, W. H. "Nature and Variability of Mercury's Sodium Atmosphere." *Nature* 323 (1986): 696–699.

Stone, E. C., and E. D. Miner "The Voyager 2 Encounter with the Neptunian System." *Science* 246 (1989): 1417–1421.

Trafton, L. M. "Pluto's Atmosphere near Perihelion." *Geophysical Research Letter* 16 (1989): 1213–1216.

Wayne, R. P. *Chemistry of Atmospheres.* Oxford, 1986.

KATHRYN P. SHAH

PLANETARY BOUNDARY LAYER. The atmosphere near the Earth's surface is almost always turbulent; that is, the air is continually undergoing chaotic motions in addition to whatever mean wind is present. We observe this in the random patterns that become visible as the wind blows across a corn or wheat field, or in the oscillations of a wind vane, the fluctuations and dispersion of a smoke plume, or the bumpiness of an airplane ride just after takeoff or just before landing. These random motions efficiently mix energy and substances that are introduced into the atmosphere near the surface. We call this turbulent layer the *planetary boundary layer*—the lowest few hundred meters of the atmosphere—because it is the part of the atmosphere that interacts directly with the Earth's surface and responds quickly to changes in surface properties. In contrast, air through the rest of the troposphere above the boundary layer is only intermittently turbulent. [See Chaos; Turbulence.]

The boundary layer is also the part of the atmosphere within which almost all human activity takes place. It is the conduit through which trace atmospheric constituents are transported to and from the surface. We depend on the diffusive properties of this turbulent layer to dilute pollutants released into the atmosphere. Pollen and seeds released by many plant species, fungal and plant spores, and pathogens are effectively dispersed by boundary-layer turbulence. Soaring birds such as hawks and gulls, as well as sailplanes and hang-gliders, depend on the vertical motions (updrafts) generated in this layer.

The energy sources of this turbulent motion are the heating of Earth's surface by the Sun and the frictional drag that the Earth's surface exerts on the mean wind flow. The boundary layer typically extends above the surface over land to a height from several hundred meters to several kilometers in the daytime, and up to several hundred meters at night. Over the ocean, the height is more constant—typically around 1 kilometer.

The diurnal variation in boundary-layer depth over land is a result of solar heating of the Earth's surface during the day and infrared cooling of the surface at night. In the course of a typical diurnal cycle on a clear day, the surface becomes warmer than the overlying air within a few hours after sunrise; conduction and reradiation warm the lowest layer of air so that it is warmer and less dense than the air higher up. This relatively warm air rises and thereby converts its potential energy to kinetic energy in the form of turbulent fluctuations, which deepen the boundary layer through the morning. By midafternoon the boundary layer reaches a plateau at one to several kilometers altitude. Clouds may form at the top of the boundary layer if there is sufficient moisture and if the boundary layer grows to sufficient depth. [See Convection; Energetics; Longwave Radiation.]

In late afternoon, the turbulence decreases as the difference between the ground and air temperature decreases. In the evening, infrared emission from the Earth's surface typically cools the surface so that it is colder than the overlying air. Consequently the air immediately above the surface becomes cooler and denser than air higher up. This acts to suppress turbulence generated by the frictional slowing of the mean wind by the surface. As a result, the daytime boundary layer collapses and a much shallower nocturnal boundary layer forms, which may be only intermittently turbulent through the night.

This diurnal cycle is strongly dependent on surface characteristics. A bare dry surface, for example, will become warmer than a moist or vegetated surface. Therefore, turbulence is likely to be stronger and the boundary layer deeper over a bare dry surface. The moisture given off by vegetation will probably cause clouds to form at lower heights than over a bare dry surface. If these phenomena occur over a large enough area, there may also be significant differences in mean temperature, humidity, and cloudiness.

Over the ocean, the diurnal cycle is quite different and often scarcely noticeable. Because the ocean has a much larger heat capacity and heat conductivity than land, its surface temperature is hardly perturbed by the daily solar cycle. More importantly, in clear daytime conditions the boundary-layer air is heated directly by the Sun. Over the ocean this typically more than compensates for long-wave (infrared) radiational cooling. In this case, the boundary layer is more stable (has smaller buoyancy flux and less cloudiness) in the daytime than at night. [See Boundary Fluxes.]

The turbulence generated in the daytime boundary layer mixes the boundary-layer air so efficiently that only around 1 hour is required to mix a constituent released to the atmosphere at the surface throughout the layer. At the top, the mixing is capped by a sharp temperature increase with height, called a *temperature inversion*. This inversion results from a balance between the erosive effects of turbulent gusts or eddies impinging on the top and the amount of energy required to pull the relatively warm and buoyant overlying air into the boundary layer, where it can then be mixed by turbulence. This temperature difference across the top can vary from a few tenths of a degree to 10°C or more over a vertical distance of a few tens of meters.

Potential temperature is a useful variable for evaluating whether a layer of air will generate or suppress turbulence, or whether an overlying layer can be entrained (engulfed) into a turbulent boundary layer. If the potential temperature increases with height (as in a nocturnal boundary layer over land), mixing the air in the layer will require that energy be added to the volume of air. This is known as *stable stratification*. As a result, in the nocturnal boundary layer the turbulence intensity normally decreases with height; the turbulence generated by shear decreases rapidly with height and may even become intermittent, especially in the upper part of the layer. Because of this, the nocturnal mixing process here is slower and less efficient, and the top of the boundary layer is less well defined than in the daytime boundary layer. [See Potential Temperature.]

If the potential temperature decreases with height (as in the lower part of a clear-air daytime boundary layer over land), mixing the air will generate more turbulence; this is known as *buoyancy-generated turbulence*, and the boundary layer is classified as *unstable*. Buoyancy-generated vertical velocity turbulence increases in intensity with height in the lower third of the boundary layer as the warmer, more buoyant air accelerates as it rises. Above this level, the turbulence slowly decreases with height as the warm updraft loses energy by mixing with its environment. If the potential temperature is constant with height, a parcel of air displaced vertically will be at the same temperature (and thus the same density) as its environment, and the total energy in the layer will not change.

A boundary layer with constant potential temperature throughout, including at the surface, is called a *neutral boundary layer*. This is a useful simplification (even though it is seldom met in actual conditions), because the turbulence then depends only on the roughness of the surface and the mean wind.

One aspect of boundary-layer flow that may at first seem surprising is the rotation of the wind with height even when the horizontal pressure gradient—the force that drives the mean wind—is constant with height. This is due to the rotation of the Earth and is known as the *Coriolis effect*. Above the boundary layer the wind is assumed to be in *geostrophic balance*, which means that the flow is perpendicular to the pressure gradient. Within the boundary layer, surface friction acts as an additional force to oppose the flow, decreasing with height above the surface. In response to this additional force, the wind must rotate toward lower pressure. The result is that the wind profile rotates clockwise in the Northern Hemisphere and counterclockwise in the Southern Hemisphere (looking from above), and it increases in magnitude with increasing height as the frictional force decreases. At temperate latitudes, the angle of rotation ranges from a few degrees up to 30° or more depending on latitude, surface roughness, and boundary-layer stability. This rotation was first analyzed by V. W. Ekman in 1905 to describe the rotation of ocean surface currents with depth. [See Coriolis Force.]

Topography has an impact on the boundary layer in two major ways. First, features such as hills and valleys, as well as buildings, embankments, and other structures, increase the effective roughness of the surface, which increases frictional drag and consequently the production of turbulence by wind shear near the surface. Second, terrain variations—particularly on scales of a few kilometers or larger—can cause channeling of the flow (as in a mountain valley) and, in combination with heating and cooling of the surface, gravitationally forced flows. Heating or cooling of a sloping surface causes horizontal density gradients in the overlying air, resulting in an upslope flow over heated sloped surfaces and a downslope flow over cooled ones. [See Small-Scale or Local Winds.]

Above the boundary layer, the potential temperature always increases with height, so that the air is turbulent for only short periods of time in very limited regions. The time scale for vertical mixing is highly variable but is measured in days, because it is dependent mostly on the frequency and intensity of cumulus convection.

The vertical velocity fluctuations in the middle of the boundary layer that are generated by daytime surface heating have a typical magnitude of several meters per second, and a horizontal scale roughly the same as the depth of the boundary layer (several hundred meters to several kilometers). By contrast, within a few meters of the surface, the magnitude of the vertical velocity fluctuations is only a few tens of centimeters per second and the horizontal scale a few tens of meters. This is why birds cannot soar near the surface except where cliffs produce orographical lift; they must ordinarily be 100 meters or more above the surface before the updraft velocities are large enough to overcome their fall velocity.

Fair-weather cumulus or stratocumulus clouds often form at the top of the daytime boundary layer in the afternoon over land, or at any time of day over the ocean. This is the cumulative result of two important processes. First, vertical mixing of boundary-layer air results in a nearly constant potential temperature and constant water vapor mixing ratio (the ratio of the mass of water vapor to the mass of dry air in a specified volume) throughout the boundary layer. Because air pressure decreases with height, constant potential temperature means that the actual air temperature decreases roughly 10°C with height; consequently, the saturation mixing ratio (the value of the mixing ratio of saturated air at a particular temperature and pressure) decreases rapidly with height. For example, if the temperature and relative humidity near the surface are 25°C and 50 percent respectively, the air in a well-mixed boundary layer would become saturated, and clouds could form at a height of about 1,400 meters.

Second, daytime surface heating evaporates water from moist and vegetated surfaces. This surface flux of water vapor can enhance the boundary-layer mixing ratio, although it is often at least partially compensated for (and sometimes overwhelmed by) the incorporation of relatively dry overlying air into the boundary layer by turbulent mixing at the top. (An additional complication is the possibility of horizontal transport of moister or drier air into the region.) The interplay between these competing processes determines whether or not clouds will form on a particular day.

The type of cloud that forms at the top of the boundary layer depends on the magnitude of the jump in temperature across the top, the rate of increase in potential temperature with height above the jump level, and the humidity of the overlying air. If the jump and the increase in potential temperature with height are small, and the humidity is high enough that the clouds do not evaporate immediately after they form and mix with overlying air, the clouds can grow vertically as cumulus clouds. If there is sufficient moisture and a small enough increase in potential temperature with height, some cumulus clouds may continue to grow into cumulonimbus clouds. Cumulus clouds, in turn, modify the boundary layer by inducing circulation patterns in the boundary-layer flow through pressure perturbations, by shading the surface and thus altering solar heating, and by producing precipitation that evaporates before reaching the ground. Evaporating precipitation can cool the air column beneath the cloud sufficiently that outflow air currents of several meters per second or more occur. Precipitation that reaches the ground cools and moistens the surface.

If, on the other hand, the jump and increase in potential temperature with height are large, the clouds may be unable to penetrate through the top of the layer. They then spread out laterally just below the top, forming stratocumulus and stratus clouds. These clouds also modify the boundary layer by reflecting solar radiation, absorbing and emitting infrared radiation, and producing precipitation. Of special importance for climate is the formation of stratiform clouds over the ocean. Marine stratus clouds are persistent and extensive and have a large impact on the Earth's radiation budget; they reflect much more solar radiation than the ocean but emit a similar amount of longwave radiation because their temperature is close to the ocean's temperature. It has been estimated that a mere 4 percent increase in the area of the globe covered by marine stratus clouds would be sufficient to offset the 2° to 3°C predicted rise in global temperature that would be caused by a doubling of carbon dioxide in the atmosphere.

[*See also* Atmospheric Structure; Clouds; *and* Greenhouse Effect.]

BIBLIOGRAPHY

Arya, S. P. S. *Introduction to Micrometeorology*. San Diego, Calif., 1988.

Brown, R. A. *Fluid Mechanics of the Atmosphere*. San Diego, Calif., 1991.

Garratt, J. R. *The Atmospheric Boundary Layer*. Cambridge, 1992.

Haugen, D. A., ed. *Workshop on Micrometeorology*. Boston, 1973.

Kaimal, J. C., and J. J. Finnigan. *Atmospheric Boundary Layer Flows: Their Structure and Measurement*. Oxford, 1992.

Lenschow, D. H., ed. *Probing the Atmospheric Boundary Layer*. Boston, 1986.

Nieuwstadt, F. T. M., and H. van Dop. *Atmospheric Turbulence and Air Pollution Modeling*. Dordrecht, Netherlands, 1982.

Sorbjan, Z. *Structure of the Atmospheric Boundary Layer*. Englewood Cliffs, N.J., 1989.

Stull, R. B. *An Introduction to Boundary Layer Meteorology*. Dordrecht, Netherlands, 1988.

DONALD H. LENSCHOW

PLANETARY ORBITAL PARAMETERS. *See* Orbital Parameters and Equations.

PLANETARY ORBITAL VARIATIONS. *See* Orbital Variations.

PLANTS. *See* Vegetation and Climate.

POETRY. *See* Literature and Weather.

POLAR CELL. *See* Meridional Circulations.

POLAR CLIMATE. *See* Antarctica; Arctic; Climatic Zones; *and* North Pole.

POLAR FRONT. *See* Cold Fronts.

POLAR LOWS. Since the pioneering work of Jule Charney in 1947, the development of extratropical cyclones has generally been associated with the formation of baroclinic waves in the west-wind belt. Other types of cyclogenesis were seldom considered, although it was suspected—even before the era of meteorological satellites, and mainly among Scandinavian and British mete-

orologists—that depressions other than the well-known polar-front lows existed in the high latitudes. These depressions were of a relatively small horizontal scale, so they often went undetected by the synoptic meteorological observing network as they traversed the Norwegian Sea or the northeastern North Atlantic before reaching the coasts. [*See* Cyclogenesis; Fronts; Low Pressure.]

Satellite observations over more than 2 decades have confirmed this suspicion. Small-scale cyclonic disturbances with a horizontal scale ranging from about 200 kilometers (occasionally even less) to about 800 kilometers frequently develop in the cold air masses poleward of the polar front.

These so-called *polar lows*, which develop only over the sea in the cold season, are often associated with severe weather, including heavy hail and snow showers as well as strong winds. These inherent qualities are reflected in the definition of a polar low as a small but fairly intense maritime cyclone that forms poleward of the main baroclinic zone (the polar front). The horizontal scale of a polar low is approximately between 200 and 800 kilometers, and the surface winds are around gale force or greater. The surface analysis shown in Figure 1 illustrates

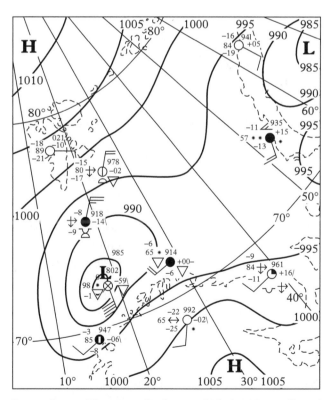

POLAR LOWS. Figure 1. *Surface analysis for 00 coordinated universal time, 14 December 1982, showing a polar low between Svalbard and northern Norway.*

the small horizontal scale of the polar low and the relatively strong winds around its center.

Polar lows may develop quickly—almost explosively—in as little as 3 to 6 hours, and the wind and associated high seas may also arise with little warning. The development of polar lows is difficult to forecast; their sudden appearance and occasional severe weather have accounted for great loss of life at sea, even though the wind strength associated with most polar lows seldom exceeds 30 meters per second. In the early 1980s attention was focused on polar lows in connection with growing offshore oil-drilling along the Norwegian coast as far north as 70° north latitude. The Norwegian Sea is a favorite region of genesis for polar lows, so the problem of forecasting polar lows for the offshore industry became an important meteorological challenge. This led to the Norwegian Polar Lows Project (1983–1986), which generated much basic knowledge about polar lows. [See Cyclones, article on Explosive Cyclones.]

Distribution. Polar lows invariably form over the sea. The best-known area of genesis is the region comprising the Greenland Sea, the Norwegian Sea, and the Barents Sea, where relatively high sea-surface temperatures may occur close to the edges of land or sea ice.

Polar lows may, however, develop at many other places around the world where cold arctic air masses flow out over an open, relatively warm sea. For example, they have been observed almost everywhere along the Greenland coast and over the Labrador Sea; an example of a Labrador Sea polar low is shown in Figure 2. Polar lows form also in the Davis Strait (between Greenland and Baffin Island) and in Baffin Bay; they have even been observed a few times over the Beaufort Sea north of Alaska during the short period of the year when this region is ice-free. They can be observed at many places in the northern Pacific, as well as over adjacent waters such as the Gulf of Alaska and the Japan Sea.

Relatively little is known about the occurrence of polar lows in the Southern Hemisphere. Important geographical differences between the high-latitude regions in the two hemispheres make it difficult to apply results from the Arctic directly to the Antarctic. So far, most interest has been focused on possible developments in the coastal regions around Antarctica, but very few polar lows seem to form close to the coasts or ice edges. Polar lows have, however, been observed to occur quite frequently over the Tasman Sea near New Zealand.

Timing. Polar lows are basically cold-season phenomena that occur when huge outbreaks of polar or arctic air from the continents (or from the ice-covered Arctic

Ocean) can move out over relatively warm water. Statistics on the frequency of polar-low formation as a function of the time of the year, from records for the Norwegian Barents Sea region and for the Gulf of Alaska, show a maximum from November through March, varying from around 2 to 5 polar lows per month in each region. Polar lows also develop quite frequently in midwinter over the Japan Sea and parts of the Pacific.

Mechanism of Formation. The question of the basic mechanism responsible for the formation of polar lows gave rise to much debate in the early 1980s. One group of meteorologists claimed that polar lows, although much smaller than typical extratropical cyclones, were nevertheless of basically the same structure—that is, small baroclinic waves. Another viewpoint held that polar lows were more like small tropical cyclones. Indeed, some polar lows are strikingly similar in appearance to small tropical cyclones. It is well established, however, that in order to develop, tropical cyclones require that the temperature of the underlying sea surface be at least 26°C—far higher than the sea-surface temperatures at the high latitudes where polar lows form. Despite this difference, Scandinavian meteorologists demonstrated that the same basic mechanism responsible for the development of tropical cyclones might work even over much colder high-latitude seas. The latter type of development is typically observed deep within polar air masses, far away from the polar front. Closer to the polar front, so-called baroclinic polar lows may form, smaller but otherwise similar to much larger extratropical cyclones. These baroclinic systems include characteristic comma-shaped clouds. Comma clouds do not invariably accompany strong polar-air outbreaks, and some controversy still remains as to whether comma clouds should be considered an indicator of polar lows. Baroclinic polar lows may also form along the ice edges in the polar regions where strong temperature contrasts develop between the air over the relatively warm sea and that over ice or snow.

Regardless of the basic mechanism, polar lows generally need some kind of triggering mechanism to get started. This triggering mechanism is almost always a preexisting disturbance in the form of an upper-level, small-scale cold trough. The combination of the ascending air ahead of the trough and the cold upper-level temperature provides the ideal environment for the development of a polar low. [See Convection.]

Over the years, the problem of which mechanism is the dominant one has been resolved by saying that many polar lows seem to develop through a combination of both mechanisms. Polar lows that initially form close to

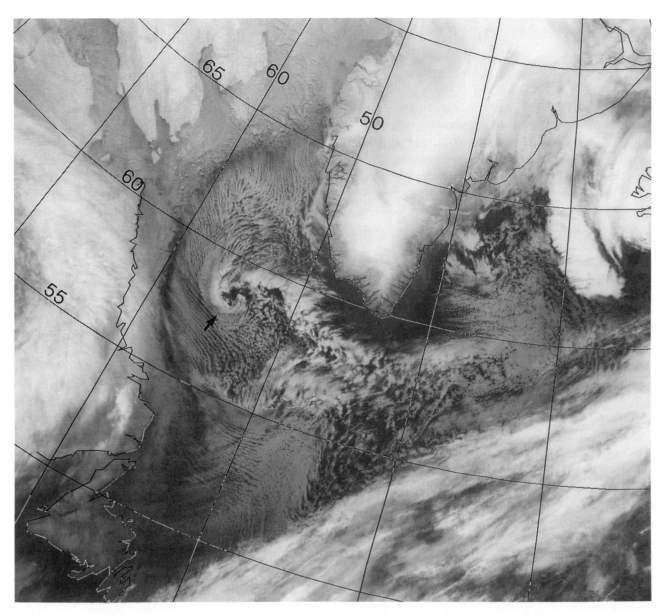

POLAR LOWS. Figure 2. *NOAA-11 satellite image (infrared, Channel 4), 1543 coordinated universal time, 28 January 1989, showing a polar low (arrowed) over the Labrador Sea.* The polar low is seen as a tight vortex of fairly deep convective clouds (the white tones in the picture indicate cold, high cloud tops, and the grayer tones lower, warmer clouds). The lines of clouds (called *cloud streets*) around the center are roughly parallel to the wind direction near the sea surface, where wind velocity reaches about 25 meters per second or more. Note the small horizontal scale of the polar low. (Photography courtesy of Department of Electrical Engineering and Electronics, University of Dundee.)

an ice edge often start their development as shallow baroclinic waves; later, when the low has moved away from the coast toward a warmer sea surface, convection may take over as the primary forcing mechanism.

[*See also* Heat Low; *and* Icelandic Low.]

BIBLIOGRAPHY

Twitchell, P.F., E.A. Rasmussen, and K.L. Davidson, eds. *Polar and Arctic Lows.* Hampton, Va., 1989.

Rasmussen, E. A., J. Turner, and P. F. Twitchell. "Report of a Workshop on Applications of New Forms of Satellite Data in

Polar Low Research." *Bulletin of the American Meteorological Society* 74.6 (June 1993): 1057–1073.

ERIK A. RASMUSSEN

POLLUTION.

Air quality has been an increasingly serious issue in many parts of the world for several decades. The greenhouse effect, acid rain, and smoggy skies over large regions are familiar examples of the adverse effects associated with airborne chemicals. Recognition of the harmful effects of air pollution has resulted in increased public interest in understanding its social context as well as its physical nature, and in developing effective strategies to mitigate its impacts.

Definition. Clean air is composed of chemicals that occur naturally. Oxygen and nitrogen make up the bulk of Earth's air. Clean air also contains varying amounts of water vapor and trace gases such as helium and carbon dioxide. This air is referred to as *clean* because it does not contain harmful levels of chemicals or harmful chemicals that adversely affect living things.

Air pollution occurs as a result of the addition of noxious and often unpleasant gases and aerosols (liquid droplets, solid particles, and mixtures of liquids and solids suspended in air). Highly polluted air is often referred to as *smog*, whose key components include ozone, acids, and aerosols. Ozone, in the *troposphere* (the first 10 kilometers of atmosphere above Earth's surface), is a major gaseous component of smog in the first few kilometers above the ground; however, stratospheric ozone produced high above the Earth's surface forms a protective shield against harmful incoming ultraviolet radiation. Some gases that are normal components of clean air—for example, carbon dioxide—become dangerous when their concentration (the amount of chemical in the air) becomes much higher than normal.

Effects. Polluted air can harm people, plants, animals, and materials. Some gases and aerosols have a direct harmful impact; others cause damage indirectly by forming other harmful gases and aerosols in the air or by altering the amount or character of incoming solar radiation. Changes in incoming radiation can then affect temperature and rainfall at the Earth's surface. Acid precipitation produced by air pollutants has been implicated in the destruction of forests, crops, aquatic life, and materials. Aerosols interact with solar radiation to degrade visibility. Atmospheric pollution can be involved in short-term severe weather changes as well as in long-term climatic changes that harm humans and the environment.

Some gases that are not very reactive near the surface of the Earth—including methane, nitrous oxide, carbon dioxide, and chlorofluorocarbons (CFCs)—become chemically reactive in the *stratosphere* (over 10 kilometers above the Earth), leading to a disruption of the protective ozone layer. Ozone in the stratosphere shields living things from solar ultraviolet radiation, which can damage DNA. Ozone in the stratosphere and aerosols produced by sulfur dioxide from volcanoes and industrial sulfides influence the Earth's climate by absorbing or scattering solar radiation. [*See* Ozone.]

Characteristics. Air pollution problems occur on three scales. *Local* problems are caused by chemicals that have an immediate effect on nearby people and environments. On the *regional* scale (on the order of 10 to 1,000 kilometers around the sources of pollution), damage results mainly from acids, oxidants, and aerosols formed in the air through chemical reactions involving the emitted chemicals. On the *global* scale (the entire Earth), problems are associated mainly with chemicals that are not very reactive in the troposphere and, as a result, are effective beyond local and regional scales. Some affect radiation balance directly. Others are transported up to the stratosphere, where they participate in chemical reactions that can result in changes in the radiation balance. These changes, in turn, can affect temperature and rainfall all over the Earth, effects often referred to as *climate change*. In addition, reactions involving some of the same chemicals in the stratosphere affect the protective ozone layer.

Brown clouds, acid rain, and ozone holes are familiar examples of the many ill effects associated with pollutants. The phrase *brown clouds* suggests the visual impact of air pollution. In some cities, the air is so polluted that the air appears to be a brown haze on most days. Pollution haze is also a regional effect, contributing to degradation of scenic vistas in many national parks in the United States.

Acid rain results when pollution chemicals are taken up by clouds and fall in solution in precipitation, potentially harming trees, crops, and fish among other life forms, as well as those that depend on them for food, shelter, and so on. In addition, building materials and outdoor sculpture can be damaged by acid precipitation, as can natural rock and mineral formations. Pollution chemicals can also be deposited on the ground in dry form, with similar effects.

The depletion or reduction of the ozone layer in the stratosphere, at an altitude of about 10 kilometers or more, has led to the appearance of ozone holes. Ozone in the upper atmosphere protects living things from ultraviolet (UV) radiation. When the ozone layer is depleted, more UV can reach the Earth's surface. One major threat to human health from additional UV is skin cancer, especially lethal melanomas. [See Ozone Hole.]

Another problem associated with changes in chemicals in the stratosphere is the *greenhouse effect*. This effect occurs when the amounts of atmospheric chemicals that absorb radiation increase, resulting in changes in temperature and rainfall at the Earth's surface. One major greenhouse gas is carbon dioxide; others include methane, nitrous oxide, and chlorofluorocarbons (CFCs, especially Freon). Increases in these gases are believed to cause general climatic warming by blocking the emission of infrared radiation to space. Conversely, aerosols have a cooling effect. [See Greenhouse Effect.]

Smog, tropospheric ozone, haze-causing aerosols, and acid rain are all derived from the same set of sulfur, nitrogen and carbon compounds. Many of these harmful compounds are emitted directly into the air as a result of human activities, mainly fossil-fuel combustion, and natural processes associated largely with soils and plants. Many air pollutants, including ozone, acids, and secondary aerosols, are produced in atmospheric reactions.

Until the early-to-mid 1800s, levels of harmful chemicals in the air were quite low, and the sources were mainly natural. As people operated factories and used automobiles, the levels of pollution increased. Most anthropogenic pollution is associated with the burning of fossil fuels—as coal, to generate electricity or factories, and gasoline to power vehicles. This produces sulfur oxides, nitrogen oxides, and volatile organic compounds, as well as small particles such as soot and fly ash.

The effects of air pollution can extend far away from its sources. Gaseous chemicals emitted from tall stacks often travel long distances. Pollutants emitted near the ground, such as carbon monoxide from automobiles and wood-burning, affect people close to the sources. Gases that are not chemically reactive in the lower atmosphere do their damage only after years. For example, it takes about 10 years for Freon (the widely used refrigerant) to reach the stratosphere, where it becomes chemically reactive, affecting the ozone layer and greenhouse warming.

The relationships between the sources and effects of air pollutants depend on wind characteristics and the amount of sunlight present, as well as on their interactions with other pollutants. For example, when the wind is not blowing, local concentrations of pollutants may be

higher. Reactions tend to occur more quickly, however, when the Sun's energy is available.

Managing Air Pollution. Over the past several decades, air pollution has had increasingly widespread and serious impacts on the natural environment and human society. Expanding populations and industrial growth have made it inevitable that foul air would move beyond city limits, cross national boundaries, create international tensions, and present new difficulties for those attempting to control its harmful effects. Controlling certain pollutants or finding the best general solutions may be hard, but it is a goal that the people of the Earth must try to achieve.

In order to establish cause-and-effect relationships, it is important to understand the links between emissions, atmospheric processes, optical characteristics, and the human and environmental responses to changes in air quality. Such knowledge is obtained through measurements and modeling of the human-environment system. Modeling provides a framework for organizing important processes and relationships so that future as well as current conditions can be examined by simulating the systems with computers. Measurements of emissions, atmospheric conditions, and other environmental factors provide the necessary information for model execution and evaluation. [See Models and Modeling.]

Models vary in complexity, from simple empirical descriptions of current situations to complex deterministic models based on fundamental principles. The most comprehensive models are the most credible, but such complex models place large demands on computing resources. Use of a suite of models of varying complexity can provide a feasible and defensible approach. One recent example is the use of multiple model frameworks for addressing the issue of acid rain in the National Acid Precipitation Assessment Program (Renné, 1990).

It is essential to recognize that policies to reduce chemical emissions can involve major social, economic, and political considerations. Many sources of air pollution are directly connected to fundamental activities in our industrial society. Automobile use and energy production are among the major sources, so strategies to reduce emissions can involve changes in lifestyle as well as technological modifications. Direct and indirect social costs resulting from the adoption of such strategies could range from the inconvenience of reduced use of personal automobiles to higher energy bills.

The history of efforts to control atmospheric pollution has been characterized by awareness and action on scales from local to global. As discussed in detail by Firor and Rhodes (1990), control of certain pollutants in the atmosphere is an achievable but very difficult goal. The biggest problem now on the international agenda—climate change—demands negotiations even more complex than those required for regulation of CFCs. Progress cannot be made without the participation of developing countries, as solutions to many air pollution problems seem to require that these societies find a path to prosperity that does not involve increased use of fossil fuels.

[*See also* Aerosols; Coevolution of Climate and Life; Ecology; Environmental Economics; Health; *and* Population Growth and the Environment.]

BIBLIOGRAPHY

Chang, J. S., et al. *The Regional Acid Deposition Model and Engineering Model.* State-of-Science/Technology Report 4, National Acid Precipitation Assessment Program. Washington, D.C., 1990.

Clark W. C., and R. E. Munn, eds. *Sustainable Development of the Biosphere.* Cambridge, 1986.

Firor, J. W., and S. L. Rhodes "Political and Legislative Landmarks in Air Pollution Control." In *Global Atmospheric Chemical Change*, edited by W. T. Sturges and C. W. Hewitt. New York., 1990.

Gilpin, A. *Air Pollution.* Queensland, Australia, 1978.

Holgate, M. W., M. Kassas, and G. F. White. *The World Environment, 1972–1982.* Dublin, 1982.

Middleton, P., T. R. Stewart, and J. Leary "On the Use of Human Judgment and Physical/Chemical Measurements in Visual Air Quality Management." *Journal of the Air Pollution Control Association* 35 (1985): 11–18.

Myers, N., ed. *Gaia: An Atlas of Planet Management.* Garden City, N.Y. 1984.

Renné, D. S., ed. "Models Planned for Use in the NAPAP Integrated Assessment." In *Report of the National Acid Precipitation Assessment Program (NAPAP).* Washington, D. C., 1990.

Rhodes, S. L., and P. Middleton. "The Complex Challenge of Controlling Acid Rain." *Environment* 25 (1983): 33–38.

Scientific American. *Managing Planet Earth.* Special issue, vol. 261 (September 1989).

PAULETTE MIDDLETON

POPULATION GROWTH AND THE ENVIRONMENT.

As the twentieth century draws to a close, humanity faces daunting prospects for supporting its rapidly growing population without seriously undermining Earth's life-support systems. Human activities are responsible for rates of land degradation, freshwater depletion, biodiversity loss, and disruption of global biogeochemical cycles that are unprecedented in human history. Moreover, the planet appears destined to ex-

perience another doubling (or more) of the human population before growth halts. Few issues are as complex and important, or as politically and emotionally charged, as the impact of population size and growth on human and environmental well-being.

In 1993, this set of circumstances prompted more than 1,600 of the world's most distinguished scientists, including a majority of the living Nobel laureates, to issue a warning to humanity. It stated, in part:

> The Earth is finite. Its ability to provide for growing numbers of people is finite. And we are fast approaching many of the Earth's limits. . . . Pressures resulting from unrestrained population growth put demands on the natural world that can overwhelm any efforts to achieve a sustainable future. . . . No more than one or a few decades remain before the chance to avert the threats we now confront will be lost and the prospects for humanity immeasurably diminished. . . . A great change in our stewardship of the earth and the life on it is required, if vast human misery is to be avoided and our global home on this planet is not to be irretrievably mutilated.

This was followed a year later by a similar statement issued by the world's scientific academies (Population Summit, 1994). Although some details may be disputed, there is little disagreement in the international scientific community regarding the overall seriousness of the threat posed by continuation of present patterns of population growth and resource utilization.

The possibility that *Homo sapiens* would attain the status of a dominant global force would have seemed remote throughout most of human history. The total population at the dawn of the agricultural revolution, around 8000 B.C.E., is thought to have numbered about 5 million people. This early population estimate is based on knowledge of the environmental conditions suited to human occupation, the extent of appropriate habitat, and an estimate of population density derived from densities of modern gatherer-hunter societies. By the time of Christ, the population is thought to have grown to roughly 250 million, inferred from archeological remains and population densities of present agricultural societies.

By 1650, the population had increased to about 500 million. A population size of 1 billion was attained by around 1800. Whereas it took virtually all of human history to grow to 1 billion, subsequent billions have been added in ever shorter periods. The population doubling time (the time required for a population to double in size) was 130 years to the second billion in 1930, 30 years to the third billion in 1960, 15 years to the fourth billion in 1975, and 12 years to the fifth billion in 1987; it is projected to take a mere 11 years to reach the sixth billion in 1998.

The growth of the global human population as depicted in Figure 1 conceals numerous local population increases and crashes caused by various combinations of epidemic disease, famine, and warfare. The one interruption of the steady increase in human numbers discernible on the graph represents the toll of the bubonic plague (Black Death), which killed an estimated 25 percent of the inhabitants of Europe between 1348 and 1350. Not visible, for example, is the impact of the famine in 1769–1770 in India, in which 3 million people perished. The worldwide influenza epidemic of 1917–1918 killed 20 million people in one year. Even today, about 10 million people are estimated to die of hunger and hunger-related disease each year.

Nonetheless, human population size is projected by demographers to continue to increase rapidly over the coming decades. United Nations agencies estimate the mid-1994 global population of 5.607 billion people will grow to 8.4 billion by 2025 and to 10 billion by around 2050, eventually leveling off at around 11.6 billion. Low-fertility and high-fertility variants of this projection place the global population in 2050 anywhere between 7.9 and 11.9 billion.

The world population growth rate declined in the 1990s to 1.6 percent per year, down from 1.7 percent per year in the 1980s and the lowest recorded since World War II. However, the number of people added to the planet each year will continue to increase, from 93 million in 1992 to a peak of about 98 million between 1995 and 2000.

The HIV/AIDS epidemic is not now expected to influence these projections greatly, although its ultimate effect is difficult to evaluate—in part because its impact extends beyond the death toll to potentially serious and wide-reaching social disruption. By the turn of the century, the World Health Organization estimates that 40 million people will be infected, resulting in perhaps 1 million deaths per year, roughly the same as for malaria.

Projections of human population growth vary primarily as a function of the rate of change in fertility patterns, measured as the total fertility rate (TFR). Mortality rates are assumed to remain low and independent of resource constraints or general environmental deterioration. The TFR is the average number of children per woman, calculated under the assumption that prevailing age-specific fertility rates remain constant throughout a woman's reproductive years. The global TFR in 1994 was 3.2. Assuming achievement of replacement fertility—a TFR of about 2.1, to account for prereproductive mortality—by 2010 yields an ultimate population size of about 7.8 billion in 2060. In contrast, assuming that replacement fertility is not achieved until 2065 yields a global population size in 2100 of 14.2 billion that is still growing.

This growth in population size that occurs after achievement of replacement fertility is called *demographic momentum*. In populations with age compositions skewed toward young people, population growth will continue

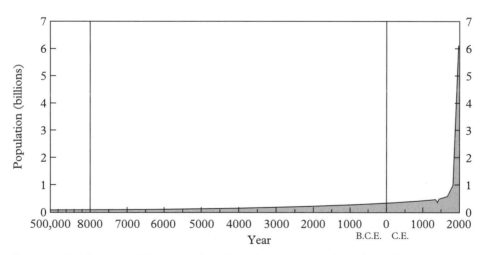

POPULATION GROWTH. Figure 1. *Growth of the human population over the past 500,000 years.* Depiction of the preagricultural period on the same scale as used for the rest of the plot would require extending its axis about 18 feet to the left. (Adapted from Ehrlich et al., 1977).

for one life expectancy, or about 70 years, after reaching replacement fertility. Worldwide, the proportion of the population under the age of 15 is 33 percent; at the extremes, it is 50 percent in several sub-Saharan African nations; it is no more than about 20 percent in most industrialized nations. In India, where 36 percent of the population is under age 15, a drop in TFR from 3.6 to 2.2 by 2025 would still involve growth of the 1994 population of 912 million to more than 2 billion people around 2100.

There is considerable variation among nations in all demographic parameters. In 1994, the combined populations of less-developed nations (excluding China) were growing at a rate of 2.2 percent per year, with an associated doubling time of 31 years. (Projected doubling times assume a constant growth rate). In contrast, the annual population growth rate of industrialized nations was 0.3 percent; their mean doubling time was 264 years. The developing countries' contribution to global population growth, having increased from 77 percent in 1950 to 93 percent in 1990, will increase further to 95 percent by the end of the century. Progress in reducing fertility rates has been slower than hoped for generally, as reflected in repeated upward revisions of United Nations projections.

The demographic transition was originally proposed to describe a transition between two theoretical equilibrium states—from one characterized by high fertility and mortality rates, regulated by nature, to one of low fertility and mortality rates, controlled by human behavior. This transition, purportedly exemplified by the European experience, involved reductions in mortality (mostly achieved through basic sanitation) followed by reductions in fertility at the time of the Industrial Revolution. This led to the idea that "development is the best contraceptive."

Several problems in the widespread application of this model have emerged. First, there is evidence that early human societies did control their fertility through cultural practices of abstinence and a variety of primitive contraceptives. Second, there exists no natural tendency (or feedback mechanism) whereby birth rates exactly balance death rates; rather, social control over birth rates will always be required if a stationary population is desired. Third, experience—even in Europe—often does not follow the model. For example, in France declines in birth rates preceded the start of industrialization by decades.

In parts of Africa, reductions in mortality were followed by an *increase* in fertility rates in the 1960s. The fertility declines in several developing nations (for example, Costa Rica, Thailand, South Korea, and India) have stopped, at least temporarily, at above-replacement plateaus. This model has slowly given way to a more thorough understanding of fertility patterns, based on the influence of cultural, economic, and environmental factors on individual or household incentives for childbearing.

Nonetheless, demographic statistics alone give a misleading impression of the population problem because of vast regional differences in environmental impact. The impact, I, of any population on the environment can be expressed as a product of three characteristics: the population's size, P; its affluence or per capita consumption, A; and the environmental damage, T, inflicted by the technologies used to supply each unit of consumption.

These factors are not independent. For example, T varies as a nonlinear function of P, A, and rates of change in both of these. This dependence is evident in the influence of population density and economic activity on the choice of local and regional energy-supply technologies and on land-management practices. Per capita impact is generally higher in very poor societies and especially in affluent societies.

Because of the difficulty in estimating the A and T factors in isolation, per capita energy use is sometimes employed as an imperfect surrogate for their product. Using that crude measure, and dividing the rich and poor nations at a per capita gross national product of $4,000 (1990 dollars), each inhabitant of a rich nation does roughly 7.5 times more damage to Earth's life-support systems than does an inhabitant of a poor one. At the extremes, the impact of an average citizen of the United States is roughly 30 times larger than that of a typical inhabitant of, say, Sudan. The U.S. population is the world's third largest in size and has one of the highest per capita impacts, giving it the largest share of responsibility for global environmental deterioration.

The *IPAT* identity makes explicit the multiplier effect of each factor on environmental impact. John Holdren has calculated the share of population in the growth of total energy consumption; over the period 1890–1990, it was 49 percent. Holding per capita impact constant, the population multiplier alone could account for a future doubling of emissions of carbon dioxide from fossil fuel combustion. To the extent that emissions of other greenhouse gases result from activities related to population size (such as agriculture, the source of a large fraction of methane and nitrous oxide emissions), population growth may exert a major influence on the buildup of greenhouse gases.

Another measure of the impact of the global population on the environment is the fraction of the terrestrial net primary productivity (NPP, the basic energy supply of all terrestrial animals) directly consumed, coopted, or eliminated by human activity. This figure is now an estimated 40 percent. Of course, the most accessible and desirable sources of NPP have been exploited first; supplying NPP to the projected additional 5 billion human beings will be considerably more difficult and destructive than meeting present demands.

Whether 10 billion people or more can be supplied with the basic material ingredients of human well-being without causing irreversible deterioration of the planet's life support systems is a matter of great concern and debate. The biophysical carrying capacity of the planet is defined as the maximum population size that could be supported under given technological capabilities, sustainably (without interruption, weakening, or loss of valued qualities). The present population has exceeded carrying capacity by the simple standard that it is being maintained only through the exhaustion and dispersion of a one-time inheritance of natural capital. The rapid depletion of essential and largely nonsubstitutable resources, coupled with deleterious changes in atmospheric quality, indicates that the human enterprise is reducing future potential biophysical carrying capacities. Important interdisciplinary efforts are now focused on crafting public policy that makes short-term, individual incentives consistent with long-term, societal well-being. Economic policy instruments, in particular, have the potential to induce the development and widespread adoption of less environmentally damaging technologies and behaviors.

Indeed, capitalizing on human behavioral flexibility and ingenuity is critical to transforming global society into a sustainable enterprise. The recent work of the Oxford economist Partha Dasgupta and others posits that the paramount need is for greater socioeconomic equity at all levels of organization—between the sexes, between households, between rural and urban regions, and between nations. Greater equity is required not only to interrupt vicious cycles of poverty and environmental deterioration, but also to facilitate cooperation in forging international agreements on atmospheric quality (e.g., emissions of ozone-destroying compounds and greenhouse gases) and to protect numerous other internationally controlled resources.

[See also Health; Environmental Economics; and Pollution.]

BIBLIOGRAPHY

Daily, G. "Population, Sustainability, and Earth's Carrying Capacity." *BioScience* 42 (1992): 761–771.
Dasgupta, P. *An Inquiry into Well-being and Destitution.* Oxford, 1993.
Ehrlich, P., A. Ehrlich, and J. Holdren. *Ecoscience: Population, Resources, Environment.* San Francisco, 1977.
Population Summit. *Population—The Complex Reality: A Report of the Population Summit of the World's Scientific Academies.* Golden, Colo. 1994.
United Nations Population Fund. *Population, Resources, and the Environment.* London, 1991.

GRETCHEN C. DAILY

POTENTIAL ENERGY. *See* Energetics; General Circulation; *and* Meridional Circulations.

POTENTIAL TEMPERATURE. Temperature is not conserved as air parcels (although it is often convenient to think of the atmosphere divided into infinitesimal, idealized, individual packets or parcels of air rather than molecules) that undergo adiabatic expansion or compression (*adiabatic* refers to processes where there is no gain or loss of heat). When an air parcel rises it expands, due to lower pressures, and will cool. When an air parcel sinks, it is compressed, due to higher pressures, and warms. This is a direct result of the ideal gas law, which states that the temperature of a gas is proportional to its pressure. Instead, it is useful to express temperature in a form in which it is considered to be conserved under adiabatic processes called *potential temperature*.

Potential temperature is the temperature that any parcel of air would have if it were brought adiabatically to the reference pressure, usually chosen to be 1,000 millibars (which is approximately the surface of the Earth). The expression for potential temperature is derived from the first law of thermodynamics for adiabatic conditions. The equation for potential temperature is:

$$\theta = T(P_0/P)^{R/C_p}$$

where θ is the potential temperature, T is the temperature, P_0 is standard pressure (1,000 millibars), P is the pressure of the gas, R is the gas constant (for dry air, this is 287 joules per degrees Celsius per kilogram), C_p is the specific heat at constant pressure (1,004 joules per degrees Celsius per kilogram). The equation for potential temperature is also known as Poisson's equation.

The gradient, or change of potential temperature with respect to height, $d\theta/dz$ (z is the vertical coordinate in feet or meters) determines the static stability of the atmospheric column (a column of air in the vertical direction, usually thought to consist of an infinitesimal area). *Static stability* refers to the ability of air parcels to accelerate freely in the vertical direction if they become displaced. If the vertical gradient of potential temperature is positive, the atmospheric column is said to be *statically stable,* and air parcels, if they are displaced, will return to their original point prior to being displaced. If the vertical gradient of potential temperature is exactly zero, the atmospheric column is said to be *neutrally stable,* and if an air parcel is displaced, it will remain in the new location it achieves after being displaced. If the vertical gradient of potential temperature is negative, the atmospheric column is said to be *statically unstable,* and air parcels rise freely after being displaced, usually giving rise to convection and thunderstorms.

If an air parcel is or becomes saturated while moving vertically, it no longer undergoes adiabatic compression or expansion, and potential temperature is no longer conserved. This is caused by some additional heating, from latent heat release during condensation. Another quantity referred to as the *equivalent potential temperature* is conserved. The equivalent potential temperature, θ_e, is defined as:

$$\theta_e = \theta \exp(Lw_s/c_pT)$$

where L is the latent heat of condensation, w_s is the saturation mixing ratio (the ratio of the mass of water vapor to the mass of the remaining dry air, if the air were saturated). The equivalent potential temperature is the temperature of an air parcel after all its water vapor has been condensed out and it is taken down to 1,000 millibars (the reference pressure). The equivalent potential temperature is higher than the potential temperature because it includes the heating from condensation.

[*See also* Adiabatic Processes; Conditional and Convective Instability; Convection; Latent Heat; Static Stability; *and* Temperature.]

BIBLIOGRAPHY

Gedzelman, S. D. *The Science and Wonders of the Atmosphere.* New York, 1980.

Wallace, J. M., and P. V. Hobbs. *Atmospheric Science: An Introductory Survey.* New York, 1977.

JUDAH COHEN

POTENTIAL VORTICITY.

POTENTIAL VORTICITY. [*This article treats a technical aspect of climate and weather studies; it is intended for readers at an advanced level.*]

A conservation law is a statement that a particular property of a fluid remains constant, moving with the flow. Such laws play a central role in fluid dynamics because they present constraints on motion and often simplify the interpretation of fluid behavior. Fluid (or atmospheric) conservation of a quantity known as potential vorticity (PV) was developed by Carl-Gustaf Rossby in the late 1930s and refined by Hans Ertel a few years later. Vorticity describes the rotation or the shear within a fluid—shear being the variation of velocity, orthogonal to the direction of flow. Atmospheric phenomena such as cyclones and fronts can be identified as areas of large vorticity.

Potential vorticity was defined by Rossby as the vorticity that a fluid element, bounded above and below by material surfaces, would have if it were moved to a standard latitude and its mass per unit area brought to a standard value. This may be restated symbolically as PV = (absolute vorticity) $\times \Delta_0/\Delta$, where Δ is the mass per unit area between the bounding surfaces and Δ_0 is a constant reference value of Δ. In the simple case of a homogeneous fluid (uniform density), PV = (absolute vorticity) $\times D_0/D$, where D is the depth of the fluid and D_0 a standard depth. (See Table 1.)

Conservation of PV in a homogeneous fluid with a free surface may be derived by considering the basic equations of motion in such a system (without friction).

POTENTIAL VORTICITY. Table 1. *List of Symbols*

v	horizontal velocity vector
$\hat{k}$	local unit-vector pointing vertically
∇	gradient operator
Δ^2	horizontal Laplacian operator
ρ	density
p'	pressure deviation from a horizontal mean
t	time
g	gravitational acceleration
θ	potential temperature
q	potential vorticity
η	absolute vorticity
f	Coriolis parameter
β	gradient of planetary vorticity
$(\ )_0$	subscript denoting a constant
N	Brunt-Vaisala frequency
R	gas constant for dry air
C_p	specific heat of air
Ψ	geostrophic stream function

$$\frac{du}{dt} - fv = -g\frac{\partial H}{\partial x} \qquad (1)$$

$$\frac{dv}{dt} + fu = -g\frac{\partial H}{\partial y} \qquad (2)$$

$$\frac{dH}{dt} = -H\left(\frac{\partial u}{\partial x} + \frac{\partial v}{\partial y}\right) \qquad (3)$$

Here, u and v are velocities in the x and y directions, H is the depth of the fluid, g is gravity and f is the Coriolis parameter (the local rotation rate about the Earth's axis). Equations 1 and 2 are the expression of Newton's law of motion, where the right-hand side is the pressure-gradient force, and terms involving f describe apparent accelerations in the rotating coordinate frame of the Earth. Equation 3 describes conservation of mass.

If we form the vorticity equation from Equations 1 and 2 by operating with $\partial/\partial x$ on Equation 2, and $\partial/\partial y$ on Equation 1 and subtracting, we may write,

$$\frac{d\eta}{dt} = -\eta\left(\frac{\partial u}{\partial x} + \frac{\partial v}{\partial y}\right), \qquad (4)$$

where $\eta \equiv (f + \partial v/\partial x - \partial u/\partial y)$ is the absolute vorticity. This states that absolute rotation of a fluid element can be altered only if there is convergence or divergence. From Equation 3 it is apparent that changes in fluid depth can occur only under this same condition, so the divergence and convergence in a shallow fluid provide the interplay between fluid depth and fluid rotation. If the divergence term is eliminated between Equations 3 and 4, a single equation obtains for a conserved variable, the potential vorticity,

$$\frac{1}{\eta}\frac{d\eta}{dt} - \frac{1}{H}\frac{dH}{dt} = 0.$$

or

$$\frac{d}{dt}(\eta/H) \equiv \frac{dq}{dt} = 0. \qquad (5)$$

As the layer of fluid deepens following a fluid element, its absolute vorticity must increase, and vice versa. If the latitude of the fluid element does not change, a deepening fluid will spin cyclonically (counterclockwise in the Northern Hemisphere).

The homogeneous-fluid model has some application to the ocean, but little application to the atmosphere because the atmosphere's density decreases rapidly with height. In particular, the atmosphere is generally stratified, meaning that fluid elements will have densities that differ from their environment if they are moved vertically, unlike a homogeneous fluid layer. A more realistic view of the atmosphere is therefore one with a continuum of stacked fluid layers, each with a different effective density, delineated by interfaces above and below that move with the fluid. That is, mass is not transported from one layer to another, so that the total mass within each layer is constant, even though the top and bottom of each layer are free to move independently. For the atmosphere, this geometry may be achieved by transforming the equations of motion into a vertical coordinate system that is a function of a conserved variable.

For unheated (adiabatic) flows, potential temperature is a conserved variable and can be used as a vertical coordinate because it generally increases monotonically with height. Potential temperature, $\theta = T(p/p_0)^{R/C_p}$, is temperature weighted by a function of pressure in which p_0 is a constant reference pressure and R and C_p are the gas constant and specific heat for dry air, respectively. In general, θ is a more useful measure of the internal energy of an air parcel than is temperature because it accounts for the pronounced expansion and compression air undergoes when raised or lowered due to the rapid decrease of pressure with height. With θ as a vertical coordinate, layers of fluid are bounded above and below by material surfaces. The analogy with Equations 1 and 5 is then exact, such that *isentropic* absolute vorticity—$\eta_\theta = f + \partial v/\partial x|_\theta - \partial u/\partial y|_\theta$, with the subscript θ implying derivatives are taken with θ constant—changes proportionately with layer depth (Δ), and the ratio η_θ/Δ is constant. The ratio of isentropic absolute vorticity to layer thickness is the essence of Rossby-Ertel PV. If we assume that isentropic surfaces are relatively flat, η_θ is very similar to conventional absolute vorticity. The quasi-geostrophic equations, commonly applied to the study of cyclones and larger-scale weather systems, make this assumption and represent an approximate form of conservation of PV on isentropic surfaces. These equations are discussed below.

The basic assumptions necessary to derive a conservation relation for PV—frictionless and adiabatic motion—are often violated in the atmosphere. When fluid is externally heated, θ changes following a fluid element, meaning that there is fluid moving through the isentropic surfaces. Therefore there can be a net transport of mass across isentropic surfaces and PV may no longer be conserved. Broadly speaking, if the heating increases upward, a layer loses mass (Δ decreases) and the PV increases. If heating decreases upward, a layer gains mass and the PV decreases. The vertical distribution of heating is very important for determining the resulting rearrangement of PV. When solar heating acts on the Earth's surface, it is most intense near the ground and decreases upward, decreasing the PV over much of the lower troposphere. Heating from water vapor condensation in clouds typically becomes maximal in the middle troposphere, resulting in larger values of PV in the lower troposphere and lower values in the upper troposphere.

Quasi-geostrophic Potential Vorticity. Because of the complexity of the full equations of motion, considerable effort has been made to derive simpler equations, valid under specific approximations. The primary complication lies in the fact that the full equations permit sound waves, gravity waves, and Rossby waves. Even though conservation of PV is a property of adiabatic, inviscid evolution of the primitive equations, the velocity, temperature, and pressure fields are not related in any simple way. It is desirable to filter the full equations to remove motions occurring on space and time scales outside the range of interest, and therefore to constrain the types of motion a system may exhibit.

For example, if we consider motions that vary on time scales that are long compared to the time scale $1/f$ (a few hours), and if we consider a geometry in which isentropes are relatively flat, then the conservation relation for PV may be written,

$$
\frac{d_g}{dt}\left[\zeta_g + f_0 + \beta(y - y_0) + \frac{f_0}{\bar{\rho}}\frac{\partial}{\partial_z}\left(\frac{\bar{\rho}}{\frac{d\Theta}{dz}}\theta'\right)\right]
$$

$$
\equiv \frac{d_g q_g}{dt} = 0. \tag{6}
$$

In Equation 6, Θ and $\bar{\rho}$ are functions of z alone and represent a reference atmospheric state that can exist in the absence of motion. The quantity $f_0 + \beta(y - y_0)$ is a Taylor series expansion of the Coriolis parameter about latitude y_0. The subscript g on the total derivative denotes that only the geostrophic wind, defined below, is acting to advect the quantity in brackets. The quantity q_g, known as the *pseudo-potential vorticity* is conserved following the geostrophic wind, assuming an absence of heating and friction. Pseudo-PV is nearly, but not exactly, Rossby-Ertel PV linearized about the reference state mentioned above. The static-stability term in Equa-

tion 6, involving the vertical derivative of θ', is directly related to variations in layer thickness relative to a reference isentropic state. On flat, horizontal boundaries, conservation of potential temperature results in $d_g\theta/dt = 0$ and completes the system.

Equation 6 is the essence of quasi-geostrophy (QG) and represents a closed dynamical system because velocity and temperature are both obtained from the pressure field alone. It turns out that one can obtain the pressure field from q_g through the relations

$$
q_g = \nabla^2\Psi + f_0 + \beta(y - y_0) + \frac{f_0^2}{\bar{\rho}}\frac{\partial}{\partial z}\left(\frac{\bar{\rho}}{N^2}\frac{\partial\Psi}{\partial z}\right). \tag{7}
$$

and the geostrophic wind,

$$
v_g = f_0^{-1}\hat{k}\times\nabla(p'/\bar{\rho}) \equiv \hat{k}\times\nabla\Psi,
$$

where p' is the pressure deviation from the reference state. Given q_g one can solve for the streamfunction Ψ, obtain the velocity v_g, and use the velocity to move the PV to a new position. With a new distribution of PV, one obtains a new velocity and repeats the process. In this way, the system is advanced in time. This is a significant simplification over the full equations of motion, because momentum, temperature and mass have been effectively combined into a single equation for a single dependent variable.

Practical applications of Equation 6 coupled with Equation 7 to the study of mid-latitude weather systems are discussed later. However, it is perhaps helpful to mention a few aspects of Equation 7 first. If we consider that f, $\bar{\rho}$ and N^2 are constant, then f_0 can be removed from Equation 7, and we have

$$
q_g' = \nabla^2\Psi' + \frac{f_0^2}{N^2}\frac{\partial^2\Psi'}{\partial z^2}, \tag{8}
$$

where primes denote departures from an areal average. In this case q_g' is still conserved, but the relation between q_g and Ψ now resembles the classical Poisson equation in three dimensions. Because Equation 8 is linear, one can find the geostrophic stream function associated with any distribution of q_g' by summing the solutions to individual pieces of the total distribution. If we consider such a piece to be isolated at a location (x_0, y_0, z_0), then in an unbounded domain, the stream function associated with that piece should decay in all directions like $1/r$, where $r^2 = (x - x_0)^2 + (y - y_0)^2 + [N/f_0(z - z_0)]^2$. The pressure is low and the circulation cyclonic if $q_g' > 0$, but if $q_g' <$

0, the pressure is locally high and the circulation is anticyclonic. In this solution, the ratio of the horizontal and vertical decay scales is given by N/f_0, meaning that if N is small, or f_0 is large, the circulation about a given piece of q'_g penetrates relatively more in the vertical. For disturbances that are wavelike in the horizontal, the longer the wavelength, the greater the vertical penetration for a given amplitude. Because of the mathematical form of Equation 8, the effect of a localized distribution of q'_g on the wind and temperature fields can extend to large distances, albeit in a decaying manner. This has an important implication for atmospheric cyclogenesis.

Application to Cyclogenesis. Throughout the past century, a significant portion of the literature on atmospheric dynamics has been devoted to the study of the development of baroclinic waves or extratropical cyclones. The work of Jule Charney (1947) and Eric Eady (1948) first used the essence of QG to extract analytic solutions for disturbances growing in a westerly current whose velocity increased with height. The early theory generally accounted for many of the observed growth and structural characteristics of atmospheric disturbances on scales of a few thousand kilometers. Perhaps the greatest shortcoming of this theory and many later contributions by other researchers was the neglect of clouds and precipitation, as well as a general underestimate of the growth rates of observed cyclones. These factors are related, as shown by the fact that computer-simulated cyclones generally grow faster when one incorporates the effects of heat, released as water vapor condenses in clouds. [See Cyclones.]

In the original theories of Charney and Eady, the background current on which the cyclones grew was unstable, meaning that any initial disturbance placed in the flow would grow. The ultimate structure and rate of growth of this disturbance, known as a normal mode, were completely determined by the character of the background flow. A more recent, alternative explanation advanced most notably by Brian Farrell is that cyclogenesis is a fundamentally transient process and depends more on the structure and amplitude of an initial disturbance than on the stability character of the mean flow. In contrast with modes, characterized by a fixed spatial structure, the structure evolves in transient growth. In unstable flows, modes will generally be excited by transients and may dominate the longtime behavior (after a few days). In cases with no unstable modes, neutral modes may be excited and grow for a short time, sometimes with growth rates greatly exceeding those of modes in unstable

flows. Elements of this theory have been offered as an explanation for observed cases of extremely rapid cyclogenesis.

For simplicity, we will use the model described by Eady as the framework for discussing phase-locked (modal) and transient baroclinic growth. This model consists of a flow with constant westerly shear (westerlies increasing with height) bounded by two rigid horizontal boundaries. The stratification N^2 and the density $\bar{\rho}$ are constant. For quasi-geostrophic motions, q'_g is conserved and Equation 8 relates the PV to the streamfunction. This means that PV perturbations in the interior cannot grow (since the model is adiabatic). If we start with $q'_g = 0$, it remains zero. Therefore, all the interesting dynamics involve the time-dependent condition

$$\left(\frac{\partial}{\partial t} + \bar{u}\frac{\partial}{\partial x}\right)\theta' = v'\frac{\partial\bar{\theta}}{\partial y}, \qquad (9)$$

applied at each horizontal boundary.

Equations 8 and 9 admit a wave solution on each boundary (considered in isolation) whose effective restoring force is the meridional temperature gradient. Such waves are called *Rossby edge-waves*, where edge-waves are boundary-trapped disturbances. The wave at the lower boundary propagates eastward (relative to the surface flow), and the wave at the lid propagates westward. These waves are very similar to Rossby waves in the interior, which rely on variations in planetary vorticity (β) or other mean gradients of PV, and can be thought of in a similar way. Localized warm surface air is associated with low pressure and cyclonic flow and is effectively a positive perturbation of PV. Cold air at the surface features high pressure and anticyclonic flow, analogous to a negative PV perturbation, provided we identify the variations of θ near the ground as effective PV variations. [See Rossby Waves.]

Baroclinic instability occurs in this model when the waves at each boundary amplify the temperature perturbations at the opposite boundary. For this to happen, the disturbance penetration depth ($H = NL/f_0$) must be at least comparable to the distance between the boundaries. Thus if the disturbance horizontal scale L is too short, or if the boundaries are too far apart, unstable modes do not exist. Mutual amplification can occur if the disturbance pressure field tilts westward with height. This phase displacement allows the velocity and temperature perturbations to be partly in phase at each boundary. For a pure linear mode, the structure does not evolve. Therefore, the boundary waves must interact not only to amplify but to alter each other's propagation enough that the structure is fixed. For short waves, upstream propagation must be increased to counter the tendency of the mean shear advection. Because long waves propagate faster, each wave must slow the propagation of the other to maintain a fixed structure.

The concept of interacting boundary waves also applies to situations where no unstable modes exist. Even if the interaction is insufficient to maintain a fixed structure, transient growth can occur as neutral edge-waves pass by each other. The mechanism of growth is similar, in that the wave at one boundary makes the other grow. In the limiting case that we remove the temperature gradients from the boundary, but still move disturbances at each boundary relative to each other, transient growth and decay occurs completely from interference effects. This is the other end of the development spectrum. Development in the atmosphere is usually a combination of these scenarios, here termed *mutual intensification* and *superposition*, respectively.

Recent effort has examined the usefulness of thinking about observed cyclogenesis as the interaction of PV perturbations at the tropopause and at the ground. The best-known work in this area occurred in a collaboration by Brian Hoskins, Michael McIntyre, and Andrew Robertson in the mid-1980s, who themselves brought to light some of the pioneering studies by Ernst Kleinschmidt in the 1950s. Kleinschmidt's work was the first significant attempt to use the properties of conservation and invertibility (the ability to calculate velocity and temperature given only the PV) together to diagnose atmospheric flows, but his efforts went largely unnoticed. The idea that the tropopause and the ground are the loci of important atmospheric disturbances was couched in terms of forecast rules for predicting cyclogenesis by Sverre Petterssen in the 1950s. However, it has not been until recently that PV concepts have been applied to the study of cyclogenesis to try to determine useful theoretical models of development, and quantitatively to deduce the importance latent heat release, evaporation of rain, boundary-layer mixing, and other processes for cyclone behavior.

A cross-section of the time-averaged PV (Figure 1) shows that the strongest gradients of PV along isentropic surfaces occur at the tropopause. The other region of importance is the large horizontal temperature gradient at the surface in mid latitudes. These are the regions that most easily support wave propagation and disturbance amplification. Rossby waves will not propagate without

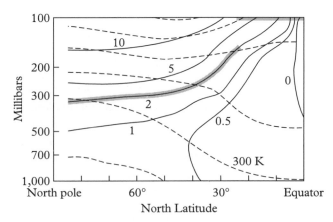

POTENTIAL VORTICITY. Figure 1. *The climatology of potential vorticity (PV) and potential temperature θ for Northern Hemisphere winter. Isentropes are dashed and drawn in intervals of 30 K; PV has units of 10^{-6} square meters-kelvin per kilogram per second ≡ 1 PV unit. Contours of PV are 0, 0.5, 1, 2, 5, and 10 PV units, with the contour at 2 PV units stippled to indicate the approximate position of the dynamical tropopause.* (From Hoskins, 1990).

a PV gradient to provide a restoring force, and baroclinic development is strongly aided by the presence of a PV gradient that may be deformed to produce growing PV perturbations. In QG, it is the horizontal gradient of q_g; more generally, it is the gradient of Rossby-Ertel PV on an isentropic surface that is important. Upper-level waves are generally visible as excursions of the strong, tropopause PV gradient from its mean position; such disturbances can achieve large amplitudes, with associated wind perturbations up to 50 meters per second. Excursions of the surface baroclinic zone from its mean position may be interpreted as Rossby edge-waves and are manifested as warm lows and cold highs on surface weather maps.

An example of one such intense upper-level disturbance is shown in Figure 2a. The PV perturbation is mainly the result of an extreme southward meander of the tropopause from its mean latitude of 40° north latitude. This feature is moving eastward in the upper-level westerlies as it encounters a nearly stationary zone of large temperature gradient north of the Gulf of Mexico. The region of large PV is surrounded by cyclonic circulation, which penetrates to the ground and later induces cyclogenesis within this temperature gradient. For illustrative purposes we have extracted, by computational methods, the wind field near the ground that is associated only with the downward-penetrating upper-tropospheric PV perturbation. Although the velocity field associated with this upper perturbation decays down-

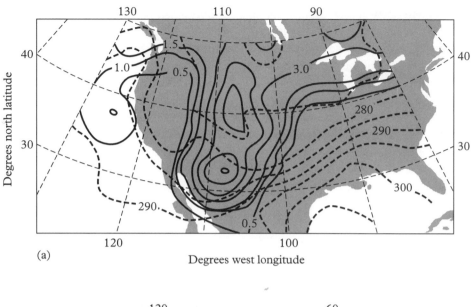

(a)

Degrees west longitude

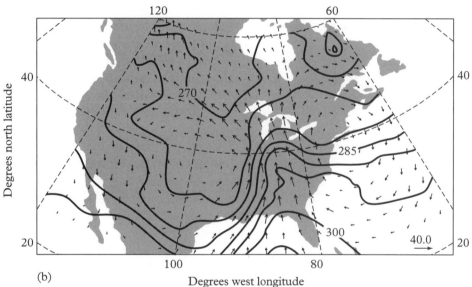

(b)

Degrees west longitude

POTENTIAL VORTICITY. Figure 2. *Overview of cyclogenesis over the central United States on 14–15 December 1987.* (a) PV on the 315 K isentropic surface (solid), with contours of 0.5, 1, 1.5, 3, 4.5, 6 and 7.5 PVU, and θ (dashed) averaged between 1,000 millibars and 850 millibars for 1200 coordinated universal time (UTC) 14 December; (b) θ averaged between 1,000 millibars and 850 millibars at 1200 UTC 15 December and 925 millibars nondivergent winds associated with the upper-level PV perturbation depicted in (a). Vector scale is at lower right in meters per second. The contour interval for θ is 5 K in both (a) and (b).

ward, its magnitude near the surface still exceeds 20 meters per second in places. Warm air is drawn northward by the strong, low-level southerly winds between the upper trough (high PV) and downstream ridge (low PV). The remainder of the surface wind field (not shown) shows cyclonic winds around the locally warm surface air (where the baroclinic zone bulges northward)

and anticyclonic flow about the region where the baroclinic zone has pushed southward. When these two wind fields are put together, the cyclone's circulation turns out to be roughly equally partitioned between the surface winds associated with the downward-penetrating upper-level disturbance, and the local temperature perturbations which the upper disturbance creates.

In addition, the southerly flow ahead of the upper trough ascends along the isentropes that slope upward and northward within this region, and heavy precipitation is produced just to the north of the developing surface cyclone. Development of the surface cyclone is rapid, occurring over a 12-hour period. It is an example in which both the nature of the initial disturbance (the upper-level trough) and the background state (the pronounced thermal gradient north of the Gulf of Mexico) are important in cyclogenesis.

[See also Vorticity.]

BIBLIOGRAPHY

Eady, E. T. "Long Waves and Cyclone Waves." *Tellus* 1 (1949): 33–52.

Farrell, B. F. "Modal and Non-modal Baroclinic Waves." *Journal of the Atmospheric Sciences* 41 (1984): 668–673.

Hoskins, B. J., M. E. McIntyre, and A. W. Robertson. "On the Use and Significance of Isentropic Potential Vorticity Maps." *Quarterly Journal of the Royal Meteorological Society* 111 (1985): 877–946.

Hoskins, B. J., M. E. McIntyre, and A. W. Robertson. "Theory of Extratropical Cyclones." In *Extratropical Cyclones*, edited by C. Newton and E. O. Holopainen. Boston, 1990.

Pedlosky, J. *Geophysical Fluid Dynamics*, New York, 1982.

Petterssen, S. *Weather Analysis and Forecasting.* 2d ed. New York, 1956.

Rossby, C. G. "Planetary Flows in the Atmosphere." *Quarterly Journal of the Royal Meteorological Society* 66 (1940): Suppl., 68–87.

CHRISTOPHER A. DAVIS

PRECESSION CYCLE. *See* Orbital Variations.

PRECIPITATION. When water vapor in the atmosphere condenses, it changes state from a gas to either a liquid (cloud droplets) or a solid (ice crystals). Initially, very small droplets or crystals are produced that have extremely slow rates of fall (for example, cloud droplets fall at a rate of approximately 6 millimeters per second). As they grow in size through either collision and coalescence (cloud droplets) or the Bergeron process (ice crystals), their speed of fall increases. Water drops or ice particles that eventually reach the ground are collectively called *precipitation*. A more general term, *hydrometeor*, is used to denote these and any other product of the condensation of atmospheric water vapor. Table 1 lists and describes common forms of hydrometeors. [See Condensation.]

Condensation. Precipitation occurs when moisture in the atmosphere condenses, and condensation can occur only when the relative humidity within the cloud exceeds 100 percent. As the air continues to decrease in temperature—through rising motions or radiational cooling or by coming into contact with a cooler surface—the relative humidity in the cloud eventually exceeds 100 percent, a condition called *supersaturation*. If the air temperature is below freezing, ice crystals begin to form around atmospheric aerosol particles. At a given temperature, the saturation vapor pressure with respect to ice is smaller than the saturation vapor pressure with respect to water. Therefore, ice crystals begin to grow at the expense of water droplets. Precipitation occurs when the ice crystals become large enough to fall by gravity overcoming the upward force of rising air currents. This mechanism, called the *Bergeron process*, is responsible for condensation (which may eventually become precipitation) when the air temperature is below freezing.

When the air temperature is above freezing, condensation occurs through the collision and coalescence of cloud droplets. Atmospheric aerosols again act as a surface around which water vapor can condense when supersaturation occurs in the atmosphere, producing extremely small droplets of water with radii of approximately 10 microns, called *cloud droplets*. As these cloud droplets collide with one another, they can merge, or coalesce, to form larger drops. As the size of a drop increases, its falling velocity increases, and the drop begins to fall as precipitation. A typical raindrop is composed from the coalescence of approximately 1 million cloud droplets.

Forms of Precipitation. Rain, the most common form of precipitation, is defined as liquid water falling with an intensity exceeding 1 millimeter of water depth per hour, with drop diameters greater than 0.5 millimeter. Raindrops average about 1 to 2 millimeters in diameter but can be as large as 7 millimeters. Light rain known as *drizzle* consists of smaller water drops (less than 0.5 millimeters in diameter) that fall with lower intensities. Raindrops are larger but fewer in number, while drizzle is associated with smaller and more numerous drops. Although both rain and drizzle are forms of liquid precipitation, they may originate as ice crystals which subsequently melt when they fall through air that is above freezing. Indeed, most raindrops from thunderstorms initially grow from ice crystals. [See Rain.]

Solid precipitation takes a variety of forms, the most common of which is snow. *Snow* is defined as white or translucent ice particles organized into a crystalline

PRECIPITATION. Table 1. *Some common forms of hydrometeors*

Liquid Precipitation	
Drizzle	Liquid water droplets with diameters less than 0.5 millimeter that fall with accumulation rates of not more than 1 millimeter per hour.
Rain	Liquid water droplets with diameters that exceed 0.5 millimeter with accumulation rates greater than 1 millimeter per hour.
Solid Precipitation	
Graupel	Conical or round ice particles, white and opaque in color, approximately 2 to 5 millimeters in diameter that grow through aggregation with other ice particles and exhibit no crystalline structure. Also called snow pellets.
Hail	Hard pellets or irregularly shaped lumps of ice with a diameter exceeding 5 millimeters formed with concentric layers of clear and milky-white ice.
Sleet	Small translucent ice particles formed either by rain freezing as it falls through an atmospheric layer that is below freezing or snow melting briefly and then refreezing before reaching the ground. In Great Britain, a mixture of rain and snow.
Snow	White or translucent ice particles organized into a crystalline structure usually resembling flakes.
Snow Grains	Very small, white, opaque ice particles with diameters less than 2 millimeters.
Other Hydrometeors	
Dew	Water condensing directly on the ground without first producing drops. Occurs when the ground temperature is below the dewpoint of the surrounding atmosphere.
Frost	Ice sublimating directly on the ground. Frozen equivalent of dew. Sometimes called hoar frost.
Glaze	A clear smooth coating of ice on exposed objects formed by the freezing of rain or drizzle.
Rime	A white granular deposit of ice formed by rapid freezing of supercooled water droplets coming into contact with an exposed object.
Virga	Water drops or ice particles falling from the base of clouds that evaporate before reaching the earth.
White Dew	Dew that freezes subsequent to its formation. More opaque than frost.

Source: From Mather (1974)

structure resembling flakes. As small ice crystals grow through the Bergeron process, additional moisture changes directly from a gas to a solid without passing through the liquid phase, building the ice crystal into a larger structure. *Graupel* (snow pellets) consists of round or conical white, opaque ice particles, approximately 2 to 5 millimeters in diameter, which grow through aggregation with other ice particles. Unlike snow, graupel does not exhibit a crystalline structure. Very small white ice particles less than 2 millimeters in diameter are called *snow grains* and are the solid equivalent of drizzle. [*See* Snow.]

Rain may freeze as it falls through an atmospheric layer that is below freezing. Snow also may melt briefly and then refreeze before reaching the ground. Solid precipitation produced under these circumstances is usually called *sleet* and is comprised of small translucent ice particles. In Great Britain, however, the term *sleet* usually denotes a mixture of rain and snow. [*See* Sleet.]

When supercooled water droplets (liquid water at a temperature below freezing) come into contact with an exposed object, they rapidly freeze on its surface, producing a white granular deposit of ice called *rime*. When rain or drizzle falls on an exposed object at a temperature below freezing, a clear smooth coating of ice called *glaze* or *freezing rain* is created. [*See* Freezing Rain.] Rime is produced when the water drops are relatively small, the precipitation intensity is relatively slow, strong supercooling occurs, and the latent heat of fusion—released when the water freezes—is dissipated rather rapidly. Because the air molecules dissolved in the water do not have time to escape, the ice takes on a white or milky color. Glaze, by contrast, is associated with larger water drops and greater intensities, weak supercooling, and a slow dissipation of the latent heat of fusion. The dissolved air molecules have time to escape as glaze forms, giving it a clear appearance. Severe damage can be caused by the accumulation of glaze if the ice becomes thick enough to weigh down and break tree limbs and power lines.

Snow, graupel, snow grains, sleet, rime, and glaze are largely wintertime phenomena, but solid precipitation can also occur at other times of the year. In cumulonimbus clouds, strong updrafts and a considerable amount of supercooled water often exist. When liquid water drops are lifted above the freezing level by the strong updrafts, or when water drops fall through a freezing layer, they may freeze and subsequently grow through collisions with supercooled cloud droplets as they fall. Additional strong updrafts may keep aloft these hailstones—hard pellets or irregularly shaped lumps of ice with a diameter exceeding 5 millimeters—causing them to grow in size. Hail often grows to a diameter of 10 millimeters but may exceed 100 millimeters. Obviously, large hail can cause considerable damage. [*See* Hail.]

Water can also condense directly on the ground without first producing drops in the atmosphere. This occurs when the ground temperature is below the dew point of the surrounding atmosphere. If the ground temperature is above freezing, the water condenses as a liquid and is called *dew*; frost occurs when the ground temperature is below freezing. Dew that subsequently freezes is termed *white dew* and is more opaque than frost. Although dew and frost are not really forms of precipitation, they are hydrometeors produced by the condensation of atmospheric moisture. [*See* Dew; Frost.]

A dry layer of air occasionally exists beneath the base of a cloud. As precipitation begins to occur, the water drops or ice particles falling through this dry layer may begin to evaporate. If the precipitation evaporates completely before reaching the Earth, a visible *rain shaft* is produced that does not extend to the ground. This phenomenon, called *virga* or *fallstreaks*, is not really precipitation, although the processes required for precipitation, condensation, and the growth of water drops have all occurred.

Acid Precipitation. Carbon dioxide dissolves readily in water both in the atmosphere and in lakes and oceans, producing a relatively weak acid called *carbonic acid* (H_2CO_3). In the absence of any pollution in the atmosphere, rain and other forms of precipitation are slightly acidic, with a pH of 5.6 (pure water has a pH of 7.0).

The pH of precipitation, however, can also be influenced by the presence of other compounds in the atmosphere, particularly aerosols, because both of the condensation mechanisms (the Bergeron process and collision-coalescence) require aerosols to provide a surface on which condensation can begin. Many aerosols occur naturally, such as those arising from blowing dust and soil, sea spray (salt), forest fires, and volcanic activity. Others are *anthropogenic* (resulting from human activities), particularly from the burning of fossil fuels and industrial emissions.

Of all the atmospheric aerosols, oxides of sulfur and nitrogen have the most dramatic effect on the pH of precipitation. Sulfur dioxide (SO_2) is produced by coal-burning power plants and in other industrial emissions, as well as naturally by volcanic eruptions and as a byproduct of some biological activities. When it reacts with water, sulfur dioxide forms a strong acid called *sulfuric acid* (H_2SO_4). Nitrogen oxides (NO_2, NO) are largely produced by internal-combustion engines as well as from burning coal, industrial emissions, and certain biological activities. Like sulfur dioxide, nitrogen oxides also form a strong acid, *nitric acid* (HNO_3), when they react with water. Levels of sulfur dioxide emission are about twice those of nitrogen oxide emission.

When sulfur dioxide or nitrogen oxides react with ice crystals or cloud droplets to form acids, the pH of the ice or water is lowered. The term *acid precipitation*—actually a misnomer, because rainfall is naturally acidic—is used to describe any precipitation that has a pH lower than 5.6. Over large regions that are heavily populated and highly industrialized (such as Europe, or the eastern United States and Canada), the pH of precipitation is typically about 4.5, or more than 10 times as acidic as normal rainwater (pH is a logarithmic scale). Rainfall with a pH as low as 1.5 has been measured, and fogs in Los Angeles often have a pH less than 3.0. By comparison, vinegar has a pH of 2.6, and lemon juice a pH of 2.1. [*See* Acid Rain; Aerosols; Pollution.]

Precipitation and the Global Hydrologic Cycle. Precipitation is an essential component of the global hydrologic cycle. Estimates of annual precipitation averaged over the globe give a value around 1,134 millimeters (44.6 inches), with a probable error of 5 to 10 percent. Precipitation in lower latitudes is largely driven by the circulation of the Hadley cells. Precipitation averaged by latitude is greatest from 10° north latitude to 10° south, owing to the rising air along the Intertropical Convergence Zone (see Figure 1). Conversely, precipitation decreases over the descending arm of the Hadley cell,

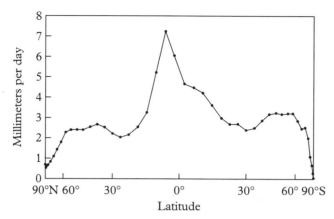

PRECIPITATION. Figure 1. *Annual precipitation rate averaged by latitude in millimeters per day.* (From data given by Legates, 1987.)

between 10° and 30° latitude in each hemisphere. Precipitation is driven by the interactions of air masses and fronts in mid latitudes and decreases poleward of 60° north and south latitudes owing to the decreased water vapor capacity of cold air. [*See* General Circulation; Hydrological Cycle.]

Precipitation also exhibits considerable spatial variability. Over land, precipitation is estimated to average 820 millimeters (32.3 inches) per year, and 1,251 millimeters (49.3 inches) over oceanic regions. The mean annual distribution of precipitation (shown in Figure 2) indicates that precipitation is greatest between 0° and 10° north latitude along the Intertropical Convergence Zone (ITCZ). Precipitation rates also exceed 4 millimeters per day over central South America, Madagascar, the western South Pacific Ocean in an area called the South Pacific Convergence Zone, and much of Southeast Asia. In the tropics, precipitation is greatly affected by the sea-surface temperature (SST) of the oceans. Equatorial precipitation is affected by the seasonal migration of the ITCZ, caused by the seasonal changes in solar heating due to the tilt of the Earth. Over the tropical oceans, convection is caused by changes in oceanic SSTs resulting from the solar heating and wind-induced upwelling of deeper and colder oceanic waters. The eastward propagation of a warm pool of water in the equatorial South Pacific and the concomitant changes in atmospheric circulation that occur during El Niño events is largely responsible for the offshore migration of precipitation

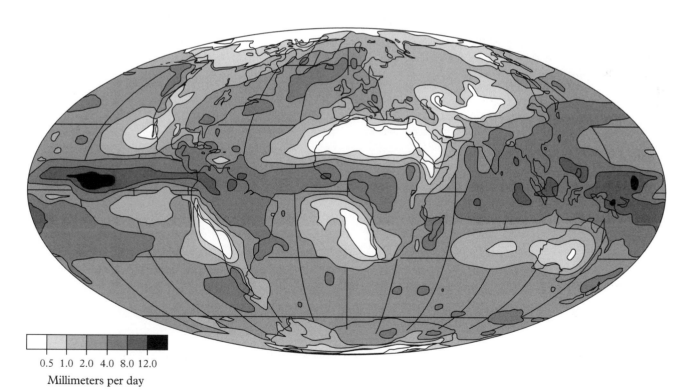

PRECIPITATION. Figure 2. *Mean annual precipitation for the globe.* (From data given by Legates, 1987.)

systems and the failure of the summer monsoons of the Indian subcontinent. Moreover, the clouds associated with precipitation shield the ocean from the solar radiation that helps to suppress SSTs and thus limits the convection, which, in turn, decreases precipitation.

In the middle latitudes, heavy precipitation occurs in marine West Coast climates (the northwestern United States, southern Chile, and western New Zealand) and over the warm Gulf Stream and Kuroshio current. By contrast, precipitation is extremely low over northern Africa (the Sahara), central Asia (the Gobi), southwestern North America, the Australian interior, western South America (the Atacama Desert), and southwestern Africa (the Kalahari). These regions are associated with cold offshore ocean surface currents. Orographic influences occur where high ranges—such as the Alps, Andes, Atlas, Himalayas, or Rocky Mountains—intercept and block moist air. Thus global precipitation patterns are determined largely by the global circulation, surface ocean currents, and topography.

[See also Instrumentation.]

BIBLIOGRAPHY

Legates, D. R. "A Climatology of Global Precipitation." *Publications in Climatology* 40 (1987).
Legates, D. R. "A High-Resolution Climatology of Gage-Corrected, Global Precipitation." In *Precipitation Measurement*, edited by B. Sevruk. Zurich, 1989, pp. 519–526.
Legates, D. R. "Global and Terrestrial Precipitation: A Comparative Assessment of Existing Climatologies." *International Journal of Climatology* 15 (1995): 237–258.
Legates, D. R., and C. J. Willmott. "Mean Seasonal and Spatial Variability in Gauge-corrected, Global Precipitation." *International Journal of Climatology* 10 (1990): 111–127.
Mather, J. R. *Climatology: Fundamentals and Applications.* New York, 1974.
Wallace, J. M., and P. V. Hobbs. *Atmospheric Science: An Introductory Survey.* New York, 1977.

DAVID R. LEGATES

PREDICTABILITY. *See* Scientific Methods.

PREDICTION. *See* Numerical Weather Prediction; *and* Weather Forecasting.

PRESSURE. *See* Barometric Pressure; High Pressure; *and* Low Pressure.

PRIMITIVE EQUATIONS. *[This article treats a technical aspect of climate and weather studies; it is intended for readers at an advanced level.]*

Advanced numerical models of Earth's atmosphere for weather prediction and climate studies now adopt the system of physical relationships that consists at least of the primitive equations of motion and the equation of mass continuity, in addition to the equations of thermodynamics and state. Primitive equations of motion are derived from the Eulerian equations of motion through the various simplifications discussed below.

One unique aspect of the atmospheric motions is that the motion of air parcels is subjected to an "apparent" force (called the *Coriolis force*) due to the rotation of the Earth, as well as the gravitational force of the Earth. [*See* Coriolis Force; Gravity.] Let us adopt the spherical coordinate system of longitude λ, latitude ϕ, and radial distance r from the center of the globe that is rotating with it about an axis through the poles with a constant angular velocity Ω. The Eulerian equations of a moving air parcel are expressed by:

$$\frac{du}{dt} + \frac{uw}{r} - \frac{uv}{r}\tan\phi - fv + \hat{f}w = \\ -\frac{1}{\rho r\cos\phi}\frac{\partial p}{\partial\lambda} - \frac{1}{r\cos\phi}\frac{\partial\Phi}{\partial\lambda} + F_\lambda, \quad (1)$$

$$\frac{dv}{dt} + \frac{vw}{r} + \frac{u^2}{r}\tan\phi + fu = \\ -\frac{1}{\rho r}\frac{\partial p}{\partial\phi} - \frac{1}{r}\frac{\partial\Phi}{\partial\phi} + F_\phi, \quad (2)$$

$$\frac{dw}{dt} - \left(\frac{u^2+v^2}{r}\right) - \hat{f}u = -\frac{1}{\rho}\frac{\partial p}{\partial r} - \frac{\partial\Phi}{\partial r} + F_r, \quad (3)$$

where

$$f = 2\Omega\sin\phi, \qquad \hat{f} = 2\Omega\cos\phi,$$

and the total derivative $\frac{d}{dt}$ with respect to time t is

$$\frac{d}{dt} = \frac{\partial}{\partial t} + \frac{u}{r\cos\phi}\frac{\partial}{\partial\lambda} + \frac{v}{r}\frac{\partial}{\partial\phi} + w\frac{\partial}{\partial r} \quad (4)$$

in which u, v, and w are the velocity components given by

$$u = r\cos\phi\frac{d\lambda}{dt}, \qquad v = r\frac{d\phi}{dt}, \qquad w = \frac{dr}{dt}. \quad (5)$$

In Equations 1 through 3, p and ρ denote the pressure and the density of air, respectively; F_λ, F_ϕ, and F_r represent frictional components. The quantity Φ, called *geopotential*, is the sum of the Earth's gravitational potential $\Phi^\star$ and centrifugal potential $-1/2\,(\Omega r \cos \phi)^2$. The second and third terms in Equations 1 and 2 and the second term in Equation 3 are apparent acceleration terms due to the curvature of the coordinate system. Terms fv and $\hat{f}w$ in Equation 1, fu in Equation 2, and $\hat{f}u$ in Equation 3 are also apparent acceleration (Coriolis) terms, resulting from the rotation of the coordinate system.

The equations of motion deal with the relationship between the acceleration of an air parcel and the total applied forces, including the pressure gradient force. The rate of change of air density per unit of time is determined by the conservation equation of air mass, written in the form

$$\frac{\partial \rho}{\partial t} + \frac{1}{r \cos \phi}\left[\frac{\partial(\rho u)}{\partial \lambda} + \frac{\partial}{\partial \phi}(\rho v \cos \phi)\right]$$
$$+ \frac{\partial(\rho r^2 w)}{r^2 \partial r} = 0. \quad (6)$$

The equations of motion relate the dynamics of flow to the pressure and density fields. The equation of mass continuity determines the time rate of change in the density field in terms of the kinematics of flow. We now need a relationship between the time rates of change of pressure and density; this is given by the first law of thermodynamics. The atmosphere is considered to be an ideal gas, so internal energy is a function only of temperature and not of density. We can also assume for an ideal gas that the specific heat at constant volume C_v is constant. For an ideal gas or a mixture of ideal gases, the equation of state defines temperature T in relation to pressure p and density ρ:

$$p = R\rho T, \quad (7)$$

where R denotes the specific gas constant for the particular ideal gas under consideration and relates to C_v through $C_p = R + C_v$, which defines the specific heat at constant pressure C_p. Now, the first law of thermodynamics may be expressed by

$$C_p \frac{dT}{dt} - \left(\frac{1}{\rho}\right)\left(\frac{dp}{dt}\right) = Q, \quad (8)$$

where Q denotes the rate of heating per unit mass per unit time.

The vertical extent of the relevant part of the atmosphere is approximately 100 kilometers above the Earth's surface. This means that the thickness of the atmosphere is very small compared to the Earth's radius. Consequently, the magnitude of the vertical velocity is expected to be much smaller than that of the horizontal velocity for large-scale motions. We also assume that the surface of constant apparent gravitational potential Φ is approximated by a sphere, so Φ depends only on the altitude z relative to the Earth's radius a, which is treated as a constant. By incorporating this idealization of Φ and the shallowness approximation of the atmosphere, we can simplify the equations of motion (Equations 1 through 3) and the equation of mass continuity (Equation 6) in the following form suitable for atmospheric dynamics:

$$\frac{du}{dt} - \left(f + \frac{u \tan \phi}{a}\right)v = -\frac{1}{\rho a \cos \phi}\frac{\partial p}{\partial \lambda} + F_\lambda, \quad (9)$$

$$\frac{dv}{dt} + \left(f + \frac{u \tan \phi}{a}\right)u = -\frac{1}{\rho a}\frac{\partial p}{\partial \phi} + F_\phi, \quad (10)$$

$$\frac{dw}{dt} = -\frac{1}{\rho}\frac{\partial p}{\partial z} - g + F_z, \quad (11)$$

$$\frac{\partial \rho}{\partial t} + \frac{1}{a \cos \phi}\left[\frac{\partial(\rho u)}{\partial \lambda} + \frac{\partial}{\partial \phi}(\rho v \cos \phi)\right]$$
$$+ \frac{\partial(\rho w)}{\partial z} = 0, \quad (12)$$

where f is called the *Coriolis parameter*,

$$\frac{d}{dt} = \frac{\partial}{\partial t} + \frac{u}{a \cos \phi}\frac{\partial}{\partial \lambda} + \frac{v}{a}\frac{\partial}{\partial \phi} + w\frac{\partial}{\partial z} \quad (13)$$

and

$$u = a \cos \phi \frac{d\lambda}{dt}, \quad v = a\frac{d\phi}{dt}, \quad w = \frac{dz}{dt}. \quad (14)$$

In Equation 11, g denotes the Earth's gravitational acceleration, and we use $g = 9.8$ meters per second per second. The error committed by assuming the value of g as constant is about 3 percent at $z = 100$ kilometers.

Equations 7 through 12 constitute a dynamical model of the atmosphere applicable to such mesoscale motions as severe storms and squall lines, as well as to global-scale motions. In this system, λ, ϕ, z, and t are coordinate variables, and u, v, w, ρ, and p are prognostic variables, because there are corresponding equations for the time rate of change of these variables. However, temperature T may be considered as a diagnostic variable, because once p and ρ are known, T can be calculated from Equation 7, which contains no time derivative. The frictional terms F_λ, F_ϕ, and F_z and the heating term Q must be expressed in terms of prognostic and diagnostic variables in order to complete the system. Although the role of water vapor in the atmosphere is important and we should be concerned with predicting it, we will not here discuss the physical aspect of thermal forcing and frictional terms.

The characteristics of atmospheric motions can be examined from the solutions of a linearized version of the basic equations (Equations 7 through 12). Linearization is a mathematical technique used to simplify the basic equations by considering small-amplitude motions (called *perturbations*) around a simple specified state, such as the atmosphere at rest with a basic state temperature that depends only on z and neglects the frictional and heating terms altogether. The assumption of small-amplitude motions permits us to neglect all product terms involving two or more perturbed variables. Solutions to such a linearized system are referred to as *free oscillations*. The characteristics of free oscillations in the resulting linearized system are called *normal modes*.

The normal modes of the basic equations consist of acoustic waves due to the compressibility of air, inertia-gravity waves due to the combination of Coriolis and gravitational forces, and low-frequency planetary waves that arise from the latitudinal variation of the Coriolis parameter. Although the presence of low-frequency planetary waves in the atmosphere was known from the work of P. Laplace, M. Margules, and S. S. Hough in the nineteenth century, the meteorological implication of planetary waves was not clarified until Carl-Gustaf Rossby (1898–1957) and his collaborators began to examine daily synoptic weather charts of the Northern Hemisphere in the late 1930s. Large-scale weather changes in the mid latitudes are primarily controlled by mid-tropospheric planetary waves, with wavelengths of several thousand kilometers, in the westerlies. These are now referred to as *Rossby waves*. [See Rossby Waves.]

The evolution of atmospheric flow patterns may be predicted by integrating the basic equations with respect to time starting from initial conditions. Since the basic equations are nonlinear and not amenable to solution by analytical methods, we must resort to the numerical approach. When the system of basic equations is solved by a popular numerical algorithm, called the *finite-difference explicit method*, there is a numerical constraint, called a *computational stability condition*, on the selection of the value of time step Δt in order to execute computing schemes stably. This constraint states that Δt must satisfy the condition $C_m \Delta t / \Delta s < 1$, where Δs represents one of the space increments in the three-dimensional grid of prognostic and diagnostic (dependent) variables, and C_m denotes the maximum characteristic speed of waves. In the atmospheric system, C_m is the speed of acoustic waves, approximately 300 meters per second. If we choose a representative horizontal grid increment of $\Delta s = 300$ kilometers, which is appropriate for large-scale motions, we find that $\Delta t < 1,000$ seconds. However, sound waves in this system propagate not only in the horizontal but also in the vertical. If we choose $\Delta s = 3$ kilometers in the vertical, we find that $\Delta t < 10$ seconds. The use of such a small time step for the numerical prediction of weather patterns for several days and the numerical simulation of climate for many years is impractical even with present-day computing capability. A customary approach to solving this difficulty is to modify the atmospheric equations so that the vertical propagation of sound waves is suppressed.

For large-scale motions of the atmosphere, the horizontal extent of motion is much larger than the vertical, and we can show by scale analysis that the vertical acceleration $dw \div dt$ and the frictional term F_z can be neglected in Equation 11. This approximation leads to the following important relationship, called the *hydrostatic equilibrium*:

$$\frac{\partial p}{\partial z} = -\rho g. \tag{15}$$

This states that the pressure at any point in the atmosphere is equal to the static weight of the column of air above it. [See Barometric Pressure.]

The system of equations of horizontal motion (Equations 9 and 10), hydrostatic equilibrium (Equation 15), mass continuity (Equation 12), thermodynamics (Equation 8), and the ideal gas law (Equation 7) is referred to as the *hydrostatic prediction model*, or *primitive equations*. The hydrostatic assumption modifies the basic atmospheric prediction system in such a way as to eliminate the vertical propagation of sound waves.

When the prediction of weather patterns by means of high-speed electronic computers became possible in the early 1950s, their computing capability was too small to perform the numerical integration of the hydrostatic prediction model that was originally proposed in 1922 by Lewis Fry Richardson (1881–1953). Changes in large-scale weather patterns are mostly associated with Rossby waves that have a characteristic speed of propagation on the order of 10 meters per second rather than with inertia-gravity waves, which may have a much greater propagation speed — as large as 300 meters per second. Therefore, if the presence of inertia-gravity waves is eliminated from the hydrostatic prediction model, one can use the time step of 1 hour or more for the space increment of $\Delta s = 300$ kilometers. Clearly the use of a 1-hour time step economizes the labor of numerical calculation. Filtering of inertia-gravity waves from the hydrostatic prediction model was achieved by approximating the horizontal acceleration terms in the equations of motion by the acceleration of the geostrophic wind (the resulting wind in the state of large-scale motions in which the pressure force, directed at right angles to the isobars from high to low pressure, is balanced by the Coriolis force). Note that the geostrophic relationship is used only in horizontal acceleration terms in the hydrostatic prediction model, and no approximation is introduced elsewhere. Such a filtered model is customarily called the *quasi-geostrophic prediction model.* [*See* Numerical Weather Prediction.]

During the 1950s various prediction models were developed in many parts of the world by adopting the quasi-geostrophic approximation and operationally testing the models on a daily basis, producing forecasting skills comparable in accuracy to those made subjectively by experienced forecasters. However, researchers began to notice systematic errors in the forecasts produced by quasi-geostrophic models. For example, mid-tropospheric long waves in the westerlies tended to move too fast westward in the forecasts. Also, the forecast models failed to predict the development of frontal discontinuities. Researchers found that these errors were attributable to the quasi-geostrophic approximation used in the forecast models, and that the use of the original, unmodified hydrostatic model (primitive equations) would do much to correct the defects. The nomenclature of primitive equations was introduced by Jule Charney (1917–1981), who produced the first successful numerical prediction using an electronic computer, to distinguish the primitive equation models from the quasi-geostrophic models.

One important aspect of the primitive equation model is that the vertical velocity w is no longer a prognostic variable. We must calculate w diagnostically from the condition that the calculations of $\partial p/\partial t$ from Equation 8 and $\partial p/\partial t$ from Equation 12 must always satisfy the hydrostatic equation (Equation 15). This was first pointed out by Richardson in 1922. Richardson's proposal of the hydrostatic prediction model was not moti-

vated by his intention to eliminate the vertical propagation of sound waves; however, he realized that the vertical velocity w was too small to calculate from the vertical dynamical equation (Equation 11), because in his words, "dw/dt would have to be found from a tiny difference between two large terms [$\partial p/\partial z$ and ρg] in the vertical dynamical equation." Although Richardson first formulated a concrete model of numerical weather prediction, the concept of the computational stability condition discussed earlier did not appear until 1928, when R. Courant, K. O. Friedrichs, and H. Lewy discovered it.

Richardson's foresight lay in the fact that he considered the height of an isobaric surface z as a dependent variable in place of the pressure p through the hydrostatic equation and presented transformation formulas, showing how to derive the prediction system in pressure coordinates. Although Richardson did not present the explicit form, he mentioned that "this system readily yields elegant approximations." He might have known that the continuity equation (Equation 12) in the p-system is transformed to the following diagnostic form

$$\frac{\partial \omega}{\partial p} + \frac{1}{a \cos \phi}\left[\frac{\partial u}{\partial \lambda} + \frac{\partial}{\partial \phi}(v \cos \phi)\right]_p = 0, \quad (16)$$

where ω is a measure of vertical velocity $\omega = \dfrac{dp}{dt}$ with

$$\frac{d}{dt} = \left(\frac{\partial}{\partial t}\right)_p + \frac{u}{a \cos \phi}\left(\frac{\partial}{\partial \lambda}\right)_p$$
$$+ \frac{v}{a}\left(\frac{\partial}{\partial \phi}\right)_p + \omega \frac{\partial}{\partial p}. \quad (17)$$

The subscript p in Equations 16 and 17 indicates that the respective differentiations are carried out on the isobaric surfaces.

The primitive equations in pressure coordinates now consist of the horizontal equations of motion (Equations 9 and 10) with replacement of the pressure gradient terms by $-g\,(a \cos \phi)^{-1}\,\partial z/\partial \lambda$ and $-g\,a^{-1}\,\partial z/\partial \phi$, the hydrostatic equation, the continuity equation (Equation 16), and the equations of thermodynamics (Equation 8) and state (Equation 7) with the definition of the total derivative (Equation 17).

In predicting the atmospheric flow with a numerical model, it is necessary to set up the upper and lower boundary conditions. As the lower boundary condition, it is straightforward to assume that there is no mass transport through the surface of the Earth. For the upper boundary condition, the requirement may differ depending on the specific vertical coordinate to be chosen. It is customary to assume that the upper and lower boundary conditions ensure the global conservation of mass and total energy without a source and sink of energy in the system.

The development of primitive equation models, written in various vertical coordinates, took place actively during the 1960s and 1970s in conjunction with efforts to study the general circulation of the atmosphere for climate studies and to extend the period of weather forecasting from short-range (2 to 3 days) to medium-range (7 to 10 days). Most of the shortcomings associated with the quasi-geostrophic prediction models of the 1950s were eliminated by the adoption of primitive equation models. One serious flaw in the quasi-geostrophic models is their inability to deal accurately with the atmospheric circulations in the tropics, where the Coriolis force is weak in comparison with the mid latitudes and, therefore, the geostrophic approximation is a poor assumption. For medium-range weather forecasts and climate simulations, it is crucial to include the effects of tropical circulations correctly. For this reason advanced numerical models for weather prediction and climate studies now adopt primitive equations.

Obviously, the role of inertia-gravity motions is essential for describing the global circulations; however, the use of primitive equations for weather forecasting puts an extra burden on researchers in regard to the preparation of initial conditions. In the atmosphere the magnitudes of sound waves and high-frequency inertia-gravity motions are small, but such motions may overwhelm low-frequency meteorologically significant motions in the prediction models. Therefore special care is necessary to prepare the initial conditions such that large-amplitude high-frequency motions (known as *noise*) are suppressed initially and remain small during the period of time integrations. The process of adjusting the input data to suppress large-amplitude noise (left in the input data owing to observational and data analysis errors) is called *initialization*. In recent years several successful initialization schemes have been proposed, including one called *nonlinear normal mode initialization* (NNMI). The input data can be represented in terms of the normal modes of primitive equation models, thereby finding how large high-frequency wave components are in the input data. If we completely eliminate the high-frequency components from the initial conditions, then some high-frequency components may reappear owing to nonlinear contributions of dynamical terms in the time changes of the high-frequency components. The essence of NNMI

is to leave a small amount of high-frequency components initially, so they do not grow during the course of time integrations.

It took scientists almost half a century since Richardson first formulated the primitive equation model for "weather prediction by numerical process" to accomplish his dream. That is, in his words, "perhaps some day in the dim future it will be possible to advance the computations faster than the weather advances at a cost less than the saving to mankind due to the information gained." Several major conceptual and engineering developments were necessary to produce today's successful weather forecasts and climate simulations. It is clear by now that the adjective *primitive* means "first of its kind" or "original", rather than implying a lack of sophistication.

BIBLIOGRAPHY

Baer, F., and J. J. Tribbia. "On Complete Filtering of Gravity Modes through Nonlinear Initialization." *Monthly Weather Review* 105 (1977): 1536–1539.

Charney, J. "The Use of the Primitive Equations of Motion in Numerical Weather Predictions." *Tellus* 7 (1955): 22–26.

Daley, R. *Atmospheric Data Analysis.* Cambridge, 1991.

Kasahara, A. "Various Vertical Coordinate Systems Used for Numerical Weather Prediction." *Monthly Weather Review* 102 (1974): 509–522. Corrigendum, 103 (1975): 664.

Machenhauer, B. "On the Dynamics of Gravity Oscillations in a Shallow Water Model, with Application to Normal Mode Initialization." *Contributions in Atmospheric Physics* 50 (1977): 253–271.

Monin, A. S., and A. M. Obukhov. "A Note on General Classification of Motions in a Baroclinic Atmosphere." *Tellus* 11 (1959): 159–162.

Phillips, N. A. "Principles of Large-scale Numerical Weather Prediction." In *Dynamic Meteorology*, edited by P. Morel. Dordrecht, Holland, 1987, pp. 3–96.

Richardson, L. F. *Weather Prediction by Numerical Process.* Cambridge, 1922; reprinted with an introduction by Sydney Chapman, New York, 1965.

AKIRA KASAHARA

PRIMITIVE EQUATIONS MODELS. *See* Models and Modeling.

PROFESSIONAL SOCIETIES. The primary scientific and professional society for atmospheric scientists in the United States is the American Meteorological Society, which has more than 10,000 members. Its primary headquarters are at 45 Beacon Street, Boston, Massachusetts 02108–3693.

The objectives of the Society are the development and dissemination of knowledge of the atmospheric and related oceanic and hydrologic sciences and the advancement of their professional applications. Numerous programs are designed to meet these objectives.

Among them are the publication of several scientific journals and books, the conduct of scientific meetings and short courses, and the certification of consulting meteorologists and radio and television weathercasters. Of particular interest to students are publications on career opportunities, a biennial volume on the offerings of various colleges and universities, and a monthly listing of employment opportunities. In recent years the society has initiated major efforts in pre-college (kindergarten through twelfth grade) education and in awarding supported fellowships and undergraduate scholarships.

Membership is open to all; there is a grade of Associate Member for those who are interested in the goals of the Society but not educationally qualified for full membership. There is also a grade of Student Member for those enrolled at least half-time at an accredited institution of higher learning. All receive the official organ, the *Bulletin of the American Meteorological Society*.

Since its founding in 1919 the Society has sought to serve both scholars and practitioners of atmospheric science. As the scope of the discipline has broadened, so have the interests and activities of the Society. Most recently, with the emergence of global change as a major concern, the Society has become increasingly interdisciplinary.

The other major professional society for meteorologists and climatologists in the United States is the American Geophysical Union, headquartered at 2000 Florida Avenue NW, Washington, D.C. 20009. It is a scientific society, which covers all Earth, solar, and planetary sciences, publishing periodicals and organizing meetings. One of its sections, about one-fifth the size of the American Meteorological Society, is devoted to the atmospheric sciences. While the American Meteorological Society does issue statements from time to time, usually intended to enlighten the public on such topics as the dangers of hurricanes or the limitations of long-range weather forecasting, the American Geophysical Union has been somewhat more active in advocacy on public issues.

The expansion of atmospheric science practice has also led to the formation of other groups with limited or special interests in one or another phase of the field. There

exist or have existed, for example, the National Weather Association, intended primarily for weather forecasters; the Weather Modification Association, for individuals and businesses interested in this area; and the Council of Industrial Meteorology, for individuals and firms providing consulting services to commerce and industry.

Outside the United States, there exist numerous independent meteorological professional societies. Though there is no formal coordination among these organizations, most of which are in nations with well-developed meteorological services and education, informal relations are common. For example, the Chinese Meteorological Society and the American Meteorological Society were among the earliest sponsors of exchange visits after the two nations were opened to each other. Many individuals hold membership in more than one society, and the American Meteorological Society usually holds at least one conference annually in joint sponsorship with one of the other societies.

[*See also* Weather Services.]

WERNER A. BAUM

PROVERBS.

For centuries, people throughout the world have summarized their observations and experiences of weather in easily memorable proverbs. To this day the acute farmer or sailor relies on such formulaic bits of weather wisdom to plan various agricultural tasks or to navigate a ship. The ordinary citizen, whether residing in a rural or urban setting, also recalls traditional proverbs to forecast the weather. While weather proverbs lack the scientific precision of modern meteorology, they do contain the collective wisdom of generations of people who have depended on knowing at least to some degree of certainty what the coming weather might be. There have been those who have relegated weather proverbs to the status of mere beliefs or superstitions lacking any kind of scientific reliability. In recent years, however, numerous scholarly treatises by meteorologists have appeared in defense of the truth element in many weather proverbs.

Proverbs have been defined as concise statements of apparent truths with currency among the folk—that is, among any identifiable group of people. More elaborately stated, proverbs are short, generally known sentences that contain wisdom, truths, morals, and traditional views in a metaphorical, fixed, and memorizable form and that are handed down orally from generation to generation. Weather proverbs are a subgenre of this larger class of proverbial wisdom. Their major function is to predict the weather so that people can plan the daily affairs of life without too many climatic uncertainties or surprises. They are based on keen observation and scrutiny of natural phenomena by experienced people who formulated and promulgated their wisdom in proverbial form. Because weather proverbs usually contain prognostic statements, they have also been called *predictive sayings, weather rules, farmers' rules,* and *weather signs.* Their intent is to establish a causal or logical relationship between two natural events that will lead to a reasonable statement concerning the weather of the next hour, day, week, month, or even year.

Structurally, weather proverbs follow a distinct pattern that can be summarized with the formula "If A, then B"; each proverbial text consists of a clear antecedent in the first part and a consequent in the second part. The well-known proverb "April showers bring May flowers" clearly illustrates this common structure. "April showers" is element A, and the ensuing "May flowers" is element B. By looking at this structure in a bit more detail, it can also be stated that the binary form of weather proverbs includes element A in the form of a weather sign and a date as well as element B stating a prognostication and its particular date. Thus the proverb "Green Christmas, white Easter" basically states that if there is a thaw in late December (early winter in northern latitudes), there will be a severe snowfall in spring. It should be noted that many weather proverbs do not necessarily express a precise time element in the second part, especially if the texts refer to agricultural cycles. The proverb "A wet March makes a sad harvest" is a case in point, but it does, of course, indirectly project from the month of March to harvest time in the summer and fall.

While weather proverbs contain plenty of colorful images or metaphors, they usually do not convey a figurative meaning, as most other proverbs do. In fact, some scholars prefer not to use the genre name *proverb* for weather signs at all, arguing that bona fide proverbs take on various meanings depending on the context in which they appear. A good example for such a figurative proverb with multiple semantic possibilities is "A rolling stone gathers no moss," which might imply that mobility avoids stagnancy (gathering moss), or that mobility never establishes roots and wealth (moss = money). Predictive sayings, on the other hand, are usually interpreted literally; they have but one distinct meaning. Thus "A snow year, a rich year" means only that a blanket of snow in the winter will act as a ground cover, preventing the soil from freezing too deeply, and will later on provide needed moisture when plants begin to grow in spring.

Here there is no metaphorical shift of meaning; the weather proverb means just what it says and no more.

Yet it would be wrong to deny the thousands of predictive sayings any proverbial status merely because they are more literal than other proverbs. It has even been argued that they are nothing but superstitions in proverbial form—but then, there are also plenty of common proverbs that are not at all figurative—for example, "Honesty is the best policy." In addition, some of the most popular weather proverbs are figurative as well. Such texts as "One swallow does not a summer make" and "Lightning never strikes twice in the same place" can take on different semantic shades in various situations. This is also true of the most frequently used English weather proverb, "Make hay while the sun shines," which, in addition to its solid advice to farmers, is used to encourage entrepreneurs of whatever kind to take advantage of a good situation. There appears to be no sound reason to reclassify weather proverbs as proverbial superstitions—that is, merely as superstitions expressed in proverbial language using rhyme ("A sunshiny shower never lasts half an hour"), parallelism ("Frost year, fruit year"), personification ("Winter eats what summer gets"), or other literary conventions. Such scholarly genre differentiation ignores the close connections and overlap of proverbs per se and weather proverbs. In addition, it is known, for example, that native informants in both oral societies of Africa and urban environments of the United States consider such texts as "Rain before seven, clear before eleven" as traditional folk proverbs and not as superstitions.

Weather proverbs can be classified into various groups according to their content or referents. Collections range from limited regional listings to major national or international compendia of several thousand proverbs, of which the three best comparative reference works are Otto von Reinsberg-Düringsfeld, *Das Wetter im Sprichwort* (1864; rpt. Leipzig, 1976); Reverend C. Swainson, *A Handbook of Weather Folk-Lore* (1873; rpt. Detroit, 1974); and Alexis Yermoloff, *Die landwirtschaftliche Volksweisheit* (Leipzig, 1905). Most collectors have grouped their massive materials according to proverbs relating to the year, seasons, individual months, days of the week, sun, moon, stars, rainbow, fog, clouds, frost, snow, rain, thunder and lighting, wind, animals, plants, and saints' days. The following examples from different countries illustrate at least a few representative categories, indicating at times the broad geographical and linguistic distribution of some weather proverbs by citing identical texts in various languages.

- *Year.* "A snow year, a rich year" (English). "Anné neigeuse, anné fructueuse" (French). "Schnee Jahr——reich Jahr" (German).
- *Seasons.* "A cold spring kills the roses" (Arabic). "Mild winter—dry spring" (French). "When the butterfly comes, the summer comes also" (Native American). "Warm fall—long winter" (German).
- *Months.* "March windy, April rainy, clear and fair May will be" (English). "Mars venteux et avril pluvieux font le mai gai et gracieux" (French). "Märzenwind, Aprilregen verheissen im Mai grossen Segen" (German).
- *Clouds.* "Mackerel scales and mare's tails make lofty ships carry low sails" (English). "Black clouds bring rain" (Persian). "Every cloud has a silver lining" (English). "Clouds are the sign of rain" (African).
- *Wind.* "When there is wind, it is always cold" (Italian). "Every wind has its weather" (American). "Much wind, little rain" (Dutch). "Let the sail come down if the wind becomes strong" (Filipino).
- *Animals.* "When dogs bark at the moon, a strong frost will follow" (German). "When ducks are driving through the burn, that night the weather takes a turn" (Scottish). "When swallows touch the water as they fly, rain approaches" (English).
- *Plants.* "Many acorns bring a strong winter" (German). "Onion's skin very thin, mild winter's coming in; onion's skin thick and rough, coming winter cold and rough" (English). "Ill weeds are not hurt by frost" (Portuguese).

The origin, history, and dissemination of weather proverbs are quite complex, and each individual text deserves a special historical study. Many of them, having been transmitted orally from generation to generation, go back to preliterate times. They were recorded in classical antiquity by Aristotle in *Meteorologica* and by his disciple Theophrastus in *On Weather Signs* and *On Winds.* The wisdom literature of the Bible also contains early references to weather proverbs, notably "When it is evening, you say, 'It will be fair weather, for the sky is red.' And in the morning, 'It will be stormy today, for the sky is red and threatening'" (Mt. 16.2–3). Through translations of the Bible this proverb has become current in many languages, albeit in many variants predicting the weather by these signs of the sky for the sailor, shepherd, traveler, and others. A few examples in English are the following:

> Evening red and morning gray
> Will send the sailor [or traveler] on his way,
> But evening gray and morning red
> Will bring rain down upon his head.

> When the evening is gray and morning is red,
> The sailor is sure to have a wet head.

> Red sky at night, sailor's [or shepherd's] delight;
> Red sky in the morning, sailors take warning.

The last example is also known in a variant form in which "red sky" is replaced by "rainbow."

Modern meteorologists have gone to some lengths to prove scientifically that there is a kernel of truth in the weather proverbs about a red sky. In simple terms, a red or pink hue in the evening sky is a result of dry dust particles, indicating fair weather ahead. On the other hand, a gray evening sky means that the atmosphere is heavy with water droplets that will probably fall the next day as the weather system moves in from the west.

Other proverbs were coined as people observed certain weather signs locally and regionally, resulting in such specific proverbs as "Snow on Mount Mansfield [highest mountain in Vermont] and in six weeks the valley will be white" and "In New England we have nine months of winter and three months of damned poor sledding." Many older English weather proverbs have been registered in John Claridge's many editions of *The Shepherd of Banbury's Rules to Judge of the Changes of the Weather* (London, 1748), which itself goes back to *The Shepherd's Legacy* (1670).

Various farmers' almanacs, notably Benjamin Franklin's *Poor Richard's Almanac* (1732–1757), helped to spread traditional weather wisdom in the New World, and immigrants from many lands brought their native weather proverbs along as well.

A subclass of weather proverbs are those attached to saints' days, a phenomenon that spread throughout the Christian world. The following examples should be taken with a grain of salt; not even the originators of such proverbial wisdom meant their sayings to be taken as gospel truth. All that these country folk did was to summarize a general weather or agricultural time sequence by attaching its beginning to a memorable saint's day and adding a reasonable prognosis. "He who shears his sheep before St. Servatius [13 May] loves more his wool than his sheep" makes sense in an environment in which sudden cold weather might still set in during the middle of May. However, the observation that "If St. Martin's Day [11 November] is dry and cold, the winter will not be long lasting" is at least questionable, although chance might have it that in a certain region a dry and cold day on 11 November was in fact followed by a relatively short winter. It should be remembered that people took these proverbs with them when they moved about, and clearly these sayings lost validity when they were transported to a region with different climate and weather conditions.

Other weather proverbs are attached to church festivals such as Candlemas Day (2 February), commemorating the purification of Mary and the presentation of Christ in the Temple. Because of the date involved, most of these proverbs contain predictions of spring, as does the British text

> If Candlemas Day be mild and gay,
> Go saddle your horses and buy them hay;
> If Candlemas Day be stormy and black,
> It carries the winter away on its back.

A Massachusetts variant of this text cites the same double message a bit less metaphorically as

> If Candlemas Day is fair and bright,
> Winter will take another fight;
> If Candlemas Day bring storm and rain,
> Winter is gone and will not come again.

Even more unscientific or based on mere folkloric superstition are the numerous proverbial interpretations of the meteorological phenomenon of the Sun shining during a rainfall. Several scholars have studied the folk beliefs attached to this particular weather situation; Matti Kuusi of Finland published a 420-page book entitled *Regen bei Sonnenschein* (Helsinki, 1957) in which he lists and discusses the weather proverb "When it rains and the sun shines, the devil is beating his grandmother" and its many variants from all around the world. Clearly these texts are folk beliefs couched in proverbial language, but this study also shows the complexity of the origin, history, dissemination, and meaning of weather lore. Similar studies can and certainly should be undertaken for other weather proverbs as well.

A few international variants in English translation must suffice to illustrate the widespread interest in the coincidence of sunshine and rain. Most variants follow the binary pattern "If A, then B": element A is a statement that says in general "When it rains and the sun shines"; element B is an explanatory comment such as "foxes are on a marriage parade" (Japanese), "the devil is getting married" (Bulgarian), "the devil is beating his wife" (Hungarian), "witches are doing their wash" (Polish), "the gypsies are washing their children" (Finnish), "a tailor is going to hell" (Danish), "mushrooms are growing" (Russian), "good weather is coming" (German), "husband and wife are quarreling" (Vietnamese), "the sheep get scared" (Spanish), and so on. Many of these variants are current throughout Europe and other parts of the world, while others are known only locally. They all show a fascination with this peculiar weather sign, which has led to some truly astounding folkloric explanations. Belief in superstitions is decreasing today, but that does not mean that these proverbial statements are not uttered anymore, even if only to comment in jest on a bizarre natural phenomenon.

Such proverbial superstitions are the exact opposite to the modern science of meteorology. They lack any validity and appear nonsensical to the modern educated person who has acquired a certain meteorological sophistication from the modern mass media and their involved weather reports. Nevertheless, the general proverb "Some are weather-wise, and some are otherwise" continues to have truth value today. Some people do indeed know more about natural weather signs than others, and they certainly continue to employ those weather proverbs that make sense even in the modern world. After all, many of them are based on ancient empiricism, and some of these basic weather truths have not changed and will not change in the foreseeable future.

Even uneducated folk are well aware of the obvious shortcomings of some weather proverbs, claiming with considerable proverbial irony and insight that "All signs fail in dry weather." There are even such parodies as the German "Wenn der Hahn kräht auf dem Mist, ändert sich das Wetter oder bleibt wie's ist" (When the rooster crows on the dunghill, the weather will change or stay as it is). Other humorous reactions to the human preoccupation with the weather are such gems as "Change of weather is a discourse of fools," from the early seventeenth century, and the American standard "Everybody talks about the weather, but nobody does anything about it," from the turn of the twentieth century. In the long run, those who delight in citing weather proverbs as well as those who pride themselves on being meteorological scientists will all have to agree with the medieval English proverb that simply recommends "Take the weather as it comes." Weather prediction will likely always be a proverbial mixture of truth and superstition, of sense and nonsense, and of science and folklore.

[*See also* Weather Lore.]

BIBLIOGRAPHY

Arora, S. L. "Weather Proverbs: Some 'Folk' Views." *Proverbium* 8 (1991): 1–17.

Brunt, D. "Meteorology and Weather Lore." *Folklore* 57 (1946): 66–74.

Delsol, P. *La météorologie populaire.* Montreal, 1973.

Dundes, A. "On Whether Weather 'Proverbs' Are Proverbs." *Proverbium* 1 (1984): 39–46. Reprinted in Alan Dundes, *Folklore Matters* (Knoxville, 1989), pp. 92–97.

Dunwoody, H. H. C. *Weather Proverbs*. Signal Service Notes IX. Washington, D.C., 1883.

Freier, G. D. *Weather Proverbs*. Tucson, 1992.

Garriott, E. B. *Weather Folk-Lore and Local Weather Signs*. Washington, D.C., 1903; Detroit, 1971.

Hauser, A. *Bauernregeln: Eine schweizerische Sammlung mit Erläuterungen*. Zurich, 1973.

Humphreys, W. J. *Weather Proverbs and Paradoxes*. Baltimore, 1923.

Inwards, R. *Weather Lore: A Collection of Proverbs, Sayings, and Rules Concerning the Weather* (1869). 4th ed. edited by E. L. Hawke for the Royal Meteorological Society. London, 1950.

Lee, A. *Weather Wisdom: Facts and Folklore of Weather Forecasting* (1976). Rev. ed. Chicago, 1990.

Malberg, H. *Bauernregeln: Ihre Deutung aus meteorologischer Sicht*. Berlin, 1989.

Page, R. *Weather Forecasting: The Country Way*. London, 1977, 1981.

Roucy, F. de. *Dictons populaires sur le temps, ou Recueil de proverbes météorologiques de la France*. Paris, 1877.

Ting, N. "Chinese Weather Proverbs." *Proverbium*, no. 18 (1972): 649–655.

Wolfgang Mieder

PSEUDOADIABATIC CHART. *See* Atmospheric Structure.

PSYCHROMETER. *See* Instrumentation.

Q–R

QUASI-BIENNIAL OSCILLATION. In 1908, the German meteorologist Arthur Berson found the first evidence of the phenomenon now called the *Quasi-Biennial Oscillation* (QBO). During an expedition to Central Africa, Berson launched a series of balloons and thus discovered that there were westerly winds (which come from the east) at about 20 kilometers altitude. At that time, permanent easterly winds (which come from the west) were assumed to exist in the equatorial stratosphere, a theory confirmed by the observation of westward-moving aerosol clouds after the 1883 eruption of Krakatau, a volcanic island of Indonesia. After the establishment of a global, regularly reporting radiosonde observation network, R. J. Reed in 1960 and R. G. Veryard and R. A. Ebdon in 1961 independently identified the alternation of easterly and westerly winds in the equatorial stratosphere, with an average period of 27 months. Today, the QBO is regarded as the main dynamic feature in the lower and middle tropical stratosphere. [*See* Tropical Circulations.]

Structure. A time-height section of monthly mean zonal wind near the equator (Figure 1) shows the following properties of the QBO:

- Alternating easterly and westerly wind regimes propagate downward with time.
- Westerlies move down faster and more regularly than easterlies.
- The downward propagation of easterly winds is often delayed between 30 and 50 hectopascals (a measure of pressure).
- Easterlies are generally stronger (at around 30 to 35 meters per second) than westerlies (around 15 to 20 meters per second).
- Maximum amplitudes of both phases typically occur near 20 hectopascals.
- The average period is about 27 months.
- Both period and amplitude vary considerably from cycle to cycle.

As the last point indicates, the QBO is not a periodic oscillation, but is instead characterized by period lengths from less than 2 years to nearly 3 years. The variability of the QBO period is related to the seasonal cycle; at 50 hectopascals, the onset of the easterly wind usually takes place during the Northern Hemisphere late spring, summer, or autumn, but only rarely during winter. Once the transition to easterlies has taken place, the next westerly phase is initiated in the upper stratosphere. This "phase-locking" of the QBO to the annual cycle is obviously connected with the delay of easterlies between 30 and 50 hectopascals, which is thought to be the most important element of the observed variability.

The QBO, as described so far, is confined to tropical latitudes; its meridional extent at 30 hectopascals is between 20° south and north latitudes. Comparisons of data sets from several equatorial radiosonde stations indicate that the QBO is a zonally highly symmetric phenomenon above the equator. Longitudinal asymmetries, however, probably occur off the equator, especially during transitions between easterly and westerly winds.

The vertical wind shears associated with these transitions between easterly and westerly winds are also present in the vicinity of the equator. According to the thermal wind relation, these shears require a warm (or cold) temperature anomaly near and above the equator. The range of temperature over the QBO is on the order of 3 to 5 K. Because the temperature change is associated with the vertical wind shear rather than with the zonal wind itself, it precedes the alternation of the zonal wind by about one-quarter of a cycle.

The temperature anomaly due to the QBO would be damped by radiative processes within a few weeks; thus, a secondary adiabatic meridional circulation is required for its maintenance. Above the equator, the warm shear zone of downward-propagating westerlies is accompanied by descending air, while ascending motion prevails during the onset of the easterlies. The meridional circulation cells extend toward 15° to 20° north and south lat-

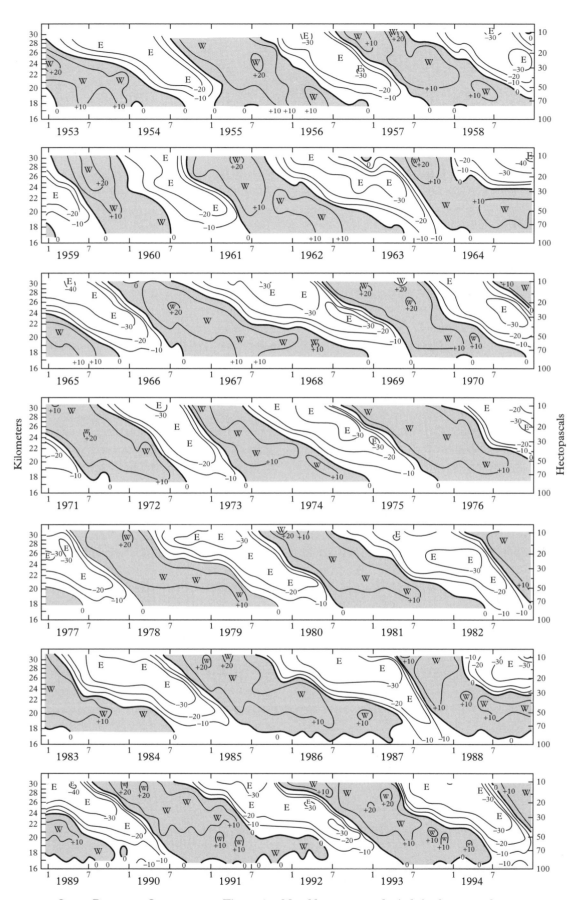

QUASI-BIENNIAL OSCILLATION. Figure 1. *Monthly mean zonal winds in the stratosphere at equatorial stations (meters per second).* Westerlies are shaded. the data set is combined from observations at several radiosonde stations: Canton Island in the central Pacific Ocean (1953–1967); Gan/Maldives in the Indian Ocean (1967–1975); Singapore (since 1976). (Updated from Naujokat, 1986).

itudes, propagating downward along with the shear zones.

The QBO-induced meridional circulation affects the distribution of trace gases in the tropical and subtropical stratosphere. Clear QBO signals have been detected in all trace gases that have sufficiently long lifetimes to be at least partially controlled by dynamics. Prominent examples are volcanic aerosols and ozone—although the phase relation between ozone and the QBO is complicated; photochemistry is another important influence on ozone in the lower tropical stratosphere.

Theory. The QBO is one of the finest examples of wave-mean-flow interaction. The theory, developed by J. R. Holton and R. S. Lindzen (1972), is widely accepted as the basic explanation of the generating mechanism of the QBO. The general concept is as follows:

Vertically propagating traveling planetary waves confined to the tropics are damped during their upward propagation. The waves' horizontal momentum is deposited into the mean flow, accelerating the zonal wind. The waves that Holton and Lindzen had in mind were eastward-propagating Kelvin waves and westward-propagating mixed Rossby-gravity waves. The maximum acceleration of the mean flow takes place slightly below the maximum of the zonal wind, yielding a downward propagation of the wind regime in question. [*See* Dynamics; Turbulence.]

Because of their vertical propagation properties, tropical Kelvin waves (forcing westerly winds) are dissipated most strongly within westerly wind regimes. The waves believed to be responsible for the easterly winds—mixed Rossby-gravity waves, in Holton and Lindzen's theory—can pass through the same wind regime rather unaffected and reach the upper stratosphere, where they initiate the succeeding easterly QBO phase. With easterlies rather than westerlies prevailing in the lower stratosphere, the same arguments apply, but with a reversal of the roles of the involved waves.

A detailed mathematical discussion of the properties of tropical planetary waves as well as of the driving mechanism can be found in Andrews et al. (1987). A review of the development of the theory of the QBO is given by Lindzen (1987).

Global Effects. The QBO is certainly the dominating dynamical signal in the tropical stratosphere. After appropriate filtering, however, variations on the time scale of the QBO can be found in many meteorological time series. Examples are surface pressure variations in the tropics and values of temperature and ozone column even in high latitudes.

Some investigations have addressed the effects of the QBO on trace gases, most notably on ozone in mid and high latitudes. In general, however, details of the dynamical coupling between the QBO and atmospheric phenomena in higher latitudes are not yet completely understood.

Of special interest in regard to the circulation in the stratosphere is a possible link between the QBO and the dynamics of the arctic polar vortex during winter—specifically, the occurrence of sudden midwinter warmings. This was first proposed by Holton and Tan (1980), who also gave observational evidence for their theory.

Sudden warmings are driven by extratropical planetary waves propagating upward from the troposphere into the stratosphere. According to the Charney-Drazin theorem, these waves can propagate vertically into the mid-stratosphere as long as (not too strong) westerly winds prevail, but they cannot propagate into regions of easterlies. The Northern Hemisphere stratosphere in winter, therefore, can be viewed as a "waveguide" for planetary waves. Its southern boundary is, at least in linear theory, given by the line of zero mean zonal wind. Consequently, during QBO easterlies, when the zero wind line shifts northward from the equator, the waveguide becomes narrower, and the waves are focused toward the polar cap. In such winters the polar vortex should therefore be more perturbed, weaker, and warmer than during years with a westerly QBO.

The theoretical considerations of Holton and Tan are quite convincing, but in practice, there is a major drawback. Roughly 60 percent of the observed winters indeed occur as Holton and Tan's theory predicts. The other 40 percent, however, behave in just the opposite way: cold, unperturbed stratospheric winter conditions during years with easterly QBO, and disturbed, warm conditions with westerly QBO.

Labitzke (1982, 1987) has discovered that Northern Hemisphere winters can be classified in a statistically highly significant manner by grouping them not only by the phase of the tropical QBO, but also by the state of the 11-year sunspot cycle. This is illustrated in Figure 2, which shows correlations between 30 hectopascal geopotential heights and the solar flux at a wavelength of 10.7 centimeters for the Northern Hemisphere during February. (The solar cycle is usually measured in terms of sunspot numbers. It is, however, accompanied by small variations—which are in phase with the sunspot number—in the entire electromagnetic spectrum. Here, the 10.7 centimeter solar flux is used as an indicator of solar activity.) Figure 2 also shows negative correlations

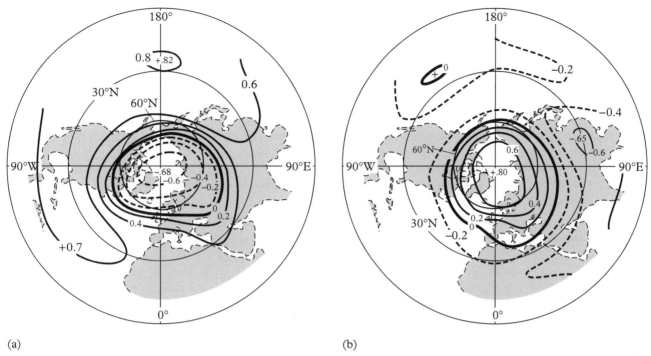

(a)   (b)

QUASI-BIENNIAL OSCILLATION. Figure 2. *Correlations between the solar cycle (measured by the intensity of the 10.7 centimeter solar flux and the geopotential height of the 30 hectopascal level over the Arctic in February for (a) the east years and (b) the west years of the QBO.* Negative correlations indicate low values of 30 hectopascal geopotential heights (weak and warm polar vortex) during maxima of solar activity, and a strong, cold polar vortex during periods of low solar activity. The number of winters involved in each calcualtion is 16 for (a) and 20 for (b). Easterly and westerly QBO years were distinguished according to the mean of equatorial zonal winds at 45 hectopascals in January and February.

between the 30 hectopascal height and the solar cycle above the northern polar region during QBO easterlies, indicating a deep and strong polar vortex (low values of 30 hectopascal heights) during maxima of solar activity. For westerly QBO phases, the map shows positive correlations, representing a weak polar vortex (high geopotential heights) during the solar maximum. During minima in solar activity, the 30 hectopascal layer is located lower down; the vortex, accordingly, is stronger and colder.

Thus, the effect of the tropical QBO onto high latitudes is modulated by some Ten-to-Twelve-Year Oscillation (TTO). The interaction between the dynamics in the tropics and in high latitudes is in good agreement with the Holton and Tan (1980) theory during the solar minima. But during maxima of the solar activity, the relation between QBO and the polar vortex in February is reversed. Later, Labitzke and van Loon found the TTO not just in the wintertime arctic stratosphere but also in both the entire Northern Hemisphere over the whole year

and in the middle and upper troposphere. A recent review of the TTO is given by van Loon and Labitzke (1994).

Unanswered Questions. Although the existence of the TTO is now fairly well established by observations (and even used in operational weather forecasting), it must be stressed that no physical mechanism is yet known to explain the influence of the 11-year solar cycle on the Earth's stratosphere and troposphere. The TTO's locking in phase with the 11-year solar variation over nearly four recorded cycles might, therefore, be only coincidental.

As mentioned above, the various dynamical feedbacks between the tropical QBO and the atmosphere at higher latitudes, as well as its interaction with other mechanisms of interannual variability such as El Niño and the TTO, need intensive investigation to be completely understood. This is even more important because the dynamic variability of the stratosphere affects the natural variability of all chemical constituents, such as ozone.

The tropical QBO itself is not fully understood. Until now, no general circulation model has been able to simulate a realistic QBO. This may be due to excessively coarse vertical resolution, which limits the model's ability properly to simulate the vertical propagation of tropical waves. Even highly simplified high-resolution models, however, experience serious difficulties when applied to the generation of the QBO. In particular, the waves proposed by Holton and Lindzen (1972) do not seem to sufficiently account for all the required QBO forcing. Other waves are probably involved; recent observational and numerical studies suggest that tropical gravity waves may be the missing source of forcing. Just which waves contribute what amount of forcing still needs to be determined.

Finally, although some annual and interannual variability of the wave forcing is certainly to be expected, it may not be sufficient to explain the phase-locking of QBO easterlies to the annual cycle. The observed variability of individual QBO cycles, therefore, is another open question.

[*See also* Conditional and Convective Instability; *and* Stratosphere.]

BIBLIOGRAPHY

Andrews, D. G., J. R. Holton, and C. B. Leovy. *Middle Atmosphere Dynamics.* London, 1987.

Holton, J. R., and R. S. Lindzen. "An Updated Theory of the Quasi-Biennial Oscillation of the Tropical Stratosphere," *Journal of Atmospheric Sciences* 29 (1972): 1076–1086.

Holton, J. R., and H. C. Tan. "The Influence of the Equatorial Quasi-Biennial Oscillation on the Global Circulation at 50 mb." *Journal of Atmospheric Sciences* 37 (1980): 2200–2208.

Labitzke, K. "On the Interannual Variability of the Middle Stratosphere during Northern Winter." *Journal of the Meteorological Society of Japan* 60 (1982): 124–139.

Labitzke, K. "Sunspots, the QBO, and the Stratospheric Temperature in the North Polar Region." *Geophysical Research Letter* 14 (1987): 535–537.

Lindzen, R. S. "On the Development of the Theory of the QBO." *Bulletin of the American Meteorological Society* 68 (1987): 329–337.

Naujokat, B. "An Update of the Observed Quasi-Biennial Oscillation of the Stratospheric Winds over the Tropics." *Journal of Atmospheric Sciences* 43 (1986): 1873–1877.

Reed, R. J. et al., "Evidence of a Downward-Propagating, Annual Wind Reversal in the Equatorial Stratosphere." *Journal of Geophysical Research* 66 (1961): 813–817.

van Loon, H., and K. Labitzke. "The 10–12-Year Atmospheric Oscillation." *Meteorologische Zeitschrift* n.s. 3 (1994): 259–266.

Veryard, R. G., and R. A. Ebdon. "Fluctuations in Tropical Stratosphere Winds." *Meteorological Magazine* 90 (1961): 125–143.

K. LABITZKE, B. NAUJOKAT, and C. MARQUARDT

QUASI-GEOSTROPHIC THEORY. *See* Vorticity.

RADAR. The word *radar* is an acronym derived from "radio detection and ranging." A transmitter broadcasts a focused beam of electromagnetic waves through an antenna. The receipt of the beam's returning backscattered energy is used to identify a discrete target, such as an aircraft, or many distributed targets, such as raindrops. Doppler radar uses the Doppler frequency shift to assess the scatterer's component of motion toward or away from the radar. The signal returned to these radars is used to identify storms and to characterize their severity and amount of rainfall.

Radar was developed during World War II to identify and track warships and aircraft. It was greatly advanced by the British through the development of the cavity magnetron, a device used to increase the power of the transmitted signal. The use of radar was a crucial advantage for the Allies during the Second World War; for example, it played an integral role in the destruction of one-third of the German submarine fleet during March through June 1943. Similar contributions have been repeated in other applications in other war theaters since then.

The role of radar in meteorological studies began shortly after World War II. Initial observations were descriptive. Only recently has it become possible to infer the dynamical and microphysical properties of convection directly from radar observations. This opens a new era of investigation in convection dynamics.

Operating Principles. Most weather radars transmit electromagnetic energy, a form of radiation with much longer wavelengths than those of visible light. The microwave frequencies most often used are in the 3-centimeter (X-band) to 10-centimeter (S-band) wavelength bands. The designation of frequencies by letters of the alphabet was a product of wartime security when radar was being developed.

Electromagnetic waves can be described by their amplitude, phase, and polarization. Weather radar is based on the interaction of the transmitted electromagnetic waves with hydrometers (precipitation drops). A small fraction of the energy that is incident on a raindrop is scattered, some of it back in the direction of the antenna used for both transmitting and receiving. [*See* Longwave Radiation.]

There are many radar designs, and only the main elements are described here and illustrated in Figure 1. A

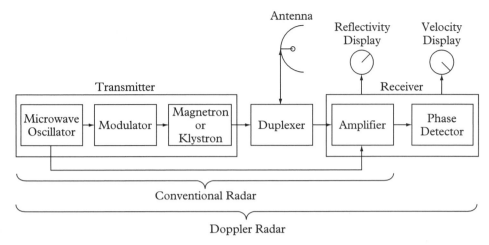

RADAR. Figure 1. *The main elements of conventional radar and Doppler radar designs.*

transmitter produces power at a known frequency in the microwave oscillator. This continuous signal is formed into pulses in the modulator. Most radars use either a *magnetron* (a device that uses strong magnetic fields and cavities to amplify a signal) or a *klystron* (a large "traditional" vacuum tube) to amplify the pulse to be transmitted. The advantage of the magnetron is that it is much lighter and less costly; however, until recently klystrons were necessary to preserve the radar phase coherence (or the same phase) over many pulses. The principal difference between conventional (incoherent) radar and Doppler radar is that the conventional radar does not provide phase information; it only measures the amplitude of the received power. Knowing the phase of the transmitted pulse is essential to estimating the velocity in a Doppler radar. Everyone is familiar with the example of Doppler shift that is observed when a train whistle's pitch (frequency) decreases as the train passes the observer. The degree of frequency change is a measure of the speed of the train. Through the phase information provided by the Doppler radar, it is possible to determine the speed of the particles (usually rain) in the direction toward or away from the radar. Some important weather hazards, such as tornados, are associated with characteristic patterns in the winds that can be perceived through the use a Doppler radar scanning through a storm.

Most meteorological radars use a sequence of short pulses, transmitting a burst of energy for about 10^{-6} second (1 microsecond), every 10^{-3} second. The interval between pulses is the *interpulse period,* and the reciprocal is the *pulse repetition frequency* or simply the PRF (here, 1,000 per second). Both these parameters may vary widely.

RADAR. Figure 2. *Tower and radome for the WSR-88D located in Norman, Oklahoma.* The antenna is 30 feet (about 9 meters) in diameter and is housed inside the spherical radome.

In either case, the high-power pulse goes through the *T/R* (transmit, receive) *switch,* or *duplexer.* The purpose of this switch is to protect the receiver from damage by the high-powered transmitted pulse. It connects the antenna to the transmitter when it is "on," and to the receiver when it is "off."

An antenna directs the transmitted waves in a beam, usually about 1 degeee wide, thereby increasing the power density by a factor of 10^4 or so while indicating direction. Additionally, the antenna receives the fraction of the power backscattered from the target. The receiver detects, amplifies, and converts the reflected microwave signal into a low frequency, in which it is eventually electronically processed and displayed.

To determine range, the return signals are sampled periodically in bins (small volumes along the beam in the direction the antenna is pointing) called *gates.* Because it takes time for signals to travel out to distant targets and back, the interval of time between the transmission of the pulse and the reception of the echo is used to determine distance.

Uses of Reflectivity Information. Reflectivity information is used for tracking storms, identifying them by characteristic structures or features, and estimating rainfall based on the strength of the returned signal. Tracking storms is straightforward—generally, just a linear extrapolation from past to future time. [*For an example of a storm being tracked, see Figure 1 in* Squall Lines.]

An example of a storm structure easily identified by radar is the hook echo often found with tornadic and other severe thunderstorms. This reflectivity feature is formed by the advection of precipitation particles by the rotating flow in which the tornado is embedded. Other storms show a distinctly linear echo pattern, or some other structure that indicates a possible threat or probable evolution.

The power backscattered to the radar is approximately proportional to the sixth power of the drop diameter, giving the largest contribution to the signal from the largest drops, which also contribute the most to the mass of rainwater. If we know the received power and assume a drop-size distribution and the drops' fall speeds, we can calculate the rain rate and, over time, the amount of rainfall. Among other applications, this is obviously useful in anticipating possible floods.

Uses of Doppler Velocity Information. The Doppler theorum relates the observed velocity (V) to the frequency shift (f_d) and wavelength of the transmitted signal (λ) by $V = -f_d\lambda/2$. Thus, the larger the velocity is, the greater the frequency shift. When we know the wavelength of the transmitted wave and the frequency shift of the received wave, we can then calculate the velocity of

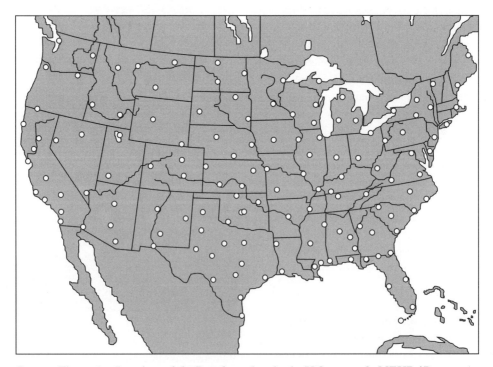

RADAR. Figure 3. *Locations of the Doppler radars in the U.S. network (NEXRAD system).*

the rain—but only the component of its motion toward or away from the radar.

Many important features of storms have characteristic velocity patterns that make the mapping of this variable important. For example, the winds that usually swirl counterclockwise around a tornado, approach the radar on the left side, as seen when looking at the tornado, and move rapidly away on the right side. A hurricane shows the same pattern on a much larger scale. A microburst, or strong low-level wind, is observed to show strong flow toward the radar on the side closest to the radar, with receding winds behind the microburst. Sometimes, for example, the "hook" pattern associated with tornados is not visible, but the characteristic velocity pattern nearly always is. Early observations of these and other hazards allow for more timely warnings.

National Doppler Radar Networks. Over the course of the 1990s, the United States will be covered by a new network of Doppler radars, the NEXRAD (NEXt generation weather RADars) system. These S-band Doppler radars, designated WSR-88D (for Weather Surveillance Radar, 1988, Doppler), have a beamwidth of less than 1 degree, thereby replacing non-Doppler radars with a beamwidth of nearly 3 degrees. This radar is pictured in Figure 2. Data on reflectivity, velocity, and spectrum width are all computed and displayed. These new radars will significantly improve identification of severe weather hazards and estimation of rainfall. The high-quality digital database will be of great value for establishing a national archive of rainfall and storms. The locations of the radars in the network are shown in Figure 3.

BIBLIOGRAPHY

Atlas, D., ed. *Radar in Meteorology.* Boston, 1990.
Battan, L. *Radar Observations of the Atmosphere.* Chicago, 1973.
Doviak, R. J., and D. S. Zrnic. *Doppler Radar and Weather Observations.* Orlando, Fla., 1984.
Ray, P. S., ed. *Mesoscale Meteorology and Forecasting.* Boston, 1986.
Skolnik, M., ed. *Radar Handbook.* New York, 1990.

PETER SAWIN RAY

RADAR METEOROLOGY. *See* Meteorology.

RADIATION.

[This article treats a technical aspect of climate and weather studies; some of it is intended for readers at an advanced level.]

This article focuses on *electromagnetic radiation,* although the term *radiation* is also used in physics in several ways: it categorizes particles from cosmic sources (cosmic radiation); it describes types of emissions from paticle interactions and decay; and it describes types of emissions from radioactive decay (alpha radiation, beta radiation, and gamma radiation). As used hear, the term *radiation* describes either the emission of electromagnetic radiation from a source or the propagation of these rays through a medium. The classical (nonquantum) view of electromagnetic radiation was completed as an electromagnetic theory in 1862 by James Clerk Maxwell and published in 1873 (in his *A Treatise on Electricity and Magnetism*). He stated that electromagnetic radiation consists of energy propagated as *electromagnetic waves* by oscillating electric and magnetic fields at right angles to each other. The radiation is emitted by electrically charged entities called *dipoles.*

The distribution of different wavelengths of radiative energy (electromagnetic radiation) is called the *electromagnetic spectrum.* The electromagnetic spectrum spans a great range of wavelengths, from the very short (10^{-14} meters) to the very long (10^8 meters). These range from (the shortest to longest) X-rays through light waves to radio waves: hard X-rays (those emitted as gamma radiation), soft X-rays, ultraviolet light (vacuum-ultraviolet, extreme-ultraviolet, far-ultraviolet, and near-ultraviolet), visible light (violet, indigo, blue, green, yellow, orange, and red), infrared radiation (near-infrared, middle-infrared, and far-infrared), microwaves, and radio waves. All these types of radiation are emitted to a greater or lesser extent by cosmic bodies, including the Sun and the Earth; however, only the energy in the solar spectrum, important for climate processes, is shown in Figure 1a (energy amplitudes at shorter or longer wavelengths could be depicted here only on a logarithmic scale). Rays of short wavelength are generally blocked by the Earth's atmosphere, and only radiation of wavelengths greater than about 280 nanometers (1 nanometer equals 10^{-9} meters) reaches the surface, as shown in Figure 1b. Radio waves emitted by radio sources in deep space reach the surface of the Earth with negligible absorption. At wavelengths below about 1 millimeter, radiation is largely absorbed in the atmosphere; there are few "windows" for radiation to reach the Earth's surface until the middle-infrared where the so-called *atmospheric window* between 8 and 13 microns occurs (1 micron equals 10^{-6} meters). The chief absorbers of radiation with wavelengths of between 1 millimeter and 10 microns are water vapor and clouds; carbon dioxide also absorbs radiation at about 15 microns. [*See* Longwave Radiation; Shortwave Radiation.]

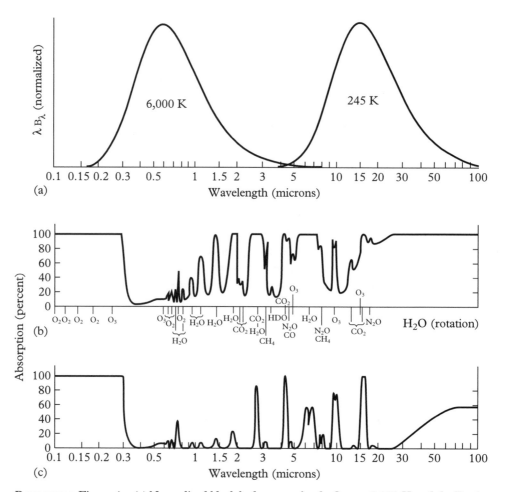

RADIATION. Figure 1. *(a) Normalized black body curves for the Sun at 6,000 K and the Earth at 245 K. (b) Atmospheric gaseous absorption for a solar beam reaching ground level. (c) The same for a solar beam reaching an altitude of 11 kilometers.*

The solar radiation reaching the Earth is effectively black body radiation at the Sun's temperature of 6,000 K, with a peak in emissions at about 600 nanometers in the visible part of the spectrum. The planetary radiation is emitted from the Earth at temperatures between about 200 and 300 K, which in the black body spectrum peak at about 15 microns and 9 microns respectively. It is of interest from the point of view of measuring and calculating radiation fluxes that, as shown by the normalized black body curves in Figure 1a, the solar and planetary spectra overlap little, particularly when we compare reflected and diffuse solar radiation fluxes with emitted and scattered planetary infrared radiation fluxes. [*See* Black Body and Grey Body Radiation.]

Other radiation also reaches Earth, in the form of charged particles (ions—the nuclei of many elements) from cosmic sources. This is called *cosmic radiation* (also known as *cosmic rays*). This radiation arrives at the top of the Earth's atmosphere with very high energy. Some of these charged particles penetrate the atmosphere, reaching the surface of the Earth unaffected; however, most collide with nitrogen and oxygen atoms, producing what are called *secondary cosmic rays*. These include altered and spallated nuclei and various elementary particles such as neutrons, mesons (pions), muons, electrons, positrons, and photons. Some of these particles undergo further tranformation (decay) before a fraction of them reach the Earth's surface.

The well-established classical treatment of radiation as electromagnetic waves is most appropriate when we consider long radio waves and far-infrared waves. For shorter wavelengths, quantum theory as developed by Max Planck in 1900 and formulated in his equations tells us that radiation can also be described as discrete units of energy, the elementary particles called *photons*. This description does not contradict that of wave theory; for example, the statistical distribution of photons arriving in optical interference patterns can be described by the wave theory, even though the energy is actually incident as photons.

The energy (E) of a photon is a function of the frequency of the radiation v:

$$E = \hbar v = \frac{\hbar c}{\lambda}$$

where λ is the wavelength, c is the velocity of the electromagnetic radiation (commonly called the *speed of light*), and $\hbar$ is Planck's constant. Often one is dealing with a large number of photons arriving in a short time, and the radiation can be treated physically on average as a wave motion. However, in the visible and ultraviolet region of the spectrum, weak sources of light can be detected by the arrival of single photons; photon-counting techniques are then used to measure intensity.

The main laws of black body radiation are as follows. First, the Stefan-Boltzmann law gives E in terms of T:

$$E = \sigma T^4$$

where E is the total energy emitted by a black body, integrated over all wavelengths, at an absolute temperature T, and σ is the Stefan-Boltzmann constant.

The energy density E_λ of radiation from a black body in a wavelength interval $d\lambda$ is given by:

$$E_\lambda = \frac{8\pi\hbar c}{\lambda^2(e^{\hbar c/kt} - 1)}$$

where k is the Boltzmann constant. The resultant curve E_λ increases to a peak at some wavelength λ_{max} given by Wien's law:

$$\lambda_{max} = \frac{a}{T}$$

where a is a constant. A radiation spectrum does not always exhibit black body characteristics, particularly when black body radiation interacts with a gaseous, liquid, or solid medium, as explained in the following sections.

Interaction of Radiation with the Medium. The velocity of electromagnetic radiation in a vacuum is 299,792,458 meters per second (186,282 miles per second), independent of wavelength. Light travels slower in typical media consisting of gases, solids, liquids, or combinations of these, however. Furthermore, in absorbing media, some of the photons are converted into infrared (heat) energy and thus removed from the beam. When

the speed of the light in a medium is slowed due to changes in the refractive index, refraction, reflection, and scattering can also occur. [*See* Scattering.]

These phenomena can generally be explained by the wave theory of light, except for the case of absorption in gases. In reflection and refraction, the change from one medium to another causes sources of secondary waves to spread out in all directions at the junction of the two media. Plotting the back waves in the first medium and their subsequent wavefronts leads to the law of reflection: The incident and reflected angles to the normal are equal. In the secondary medium, the waves have different velocity, and the secondary wavefront moves out at a different angle (i_2) according to the formula,

$$n_1 \sin i_1 = n_2 \sin i_2$$

where n_1 and n_2 are the respective refractive indices of the two media.

In a medium of continuously changing density, such as the atmosphere, radiation can be bent (refracted) in a different direction, causing phenomena such as mirages. The incidence of electromagnetic waves onto particles of various shapes results in fairly complex patterns of reflection, refraction, and diffraction, together treated as *scattering*. The removal of energy from a narrow beam of rays passing through a medium, through either scattering or absorption, is known as *extinction* or *attenuation*. Attenuation through a narrow slab dz (Figure 2) can be described by the Bouguet-Lambert law:

$$dI = \sigma I dz$$

where dI is the attenuation of a beam through a slab of thickness dz on which is incident a beam of energy I, and σ is the extinction coefficient. As energy can be removed

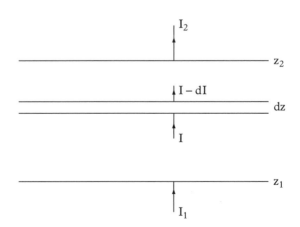

RADIATION. Figure 2. *Schematic of a ray passing through a medium of extinction coefficient* σ.

from the beam by both absorption and scattering, σ can be written as $\sigma_{abs} + \sigma_{sca}$, where the two subscripts define the absorption coefficient (abs) and the scattering coefficient (sca) respectively. Integration of this equation through a slab of finite thickness from z_1 to z_2 yields:

$$I_2 = I_1 \exp \{-\sigma(z_2 - z_1)\}. \tag{1}$$

Radiative Transfer. Although radiation can be scattered out of a light beam (and lost, if the beam is isolated), with an extended medium and a wide light beam—as from the Sun—radiation can also be scattered *into* the beam from nearby rays. Furthermore, in the case of an absorbing medium, energy can be emitted from the beam as well as absorbed. The full equation of radiative transfer and its integral properties is therefore too complex to present here.

Polarization of Radiation. As we have seen, radiation propagates as an electromagnetic wave, and the wave properties of the radiation are described by the oscillating electric field. If we consider one small wave packet, the radiation will oscillate in one plane; such radiation is said to be *plane polarized*. If, however, we consider many wave packets emitted or scattered from a source, each packet can have its own plane of polarization—or *electric vector*—oriented in a random direction. There is then no preferred plane of polarization, and the radiation is said to be *unpolarized*. Natural sources of radiation such as the Sun and planets generally emit unpolarized light; certain artificial sources, such as lasers or radio transmitters, can emit polarized light. It is also possible to polarize a beam of light by causing it to pass through a material which selectively passes only one plane of polarization, either in transmission or in reflection. For example, if light is reflected from a transparent glass plate at an angle of incidence known as the *polarizing angle,* then the reflected light will be plane polarized in a direction perpendicular to the plane of incidence. Materials that polarize radiation in transmission are known as *polaroids*.

Molecules and aerosol particles can scatter unpolarized light selectively, causing *partial polarization*. The extent to which visible light from the Sun is polarized depends on the solar altitude and the angle of scattering. Light scattered from surfaces such as roads or cars can also be partially polarized; the use of polaroids in sunglasses can reduce the glare from these sources.

Associated with polarization is coherence. Imagine a continuous train of waves of a given constant frequency stretching away infinitely and a second similar train of waves; the two wave trains are said to be *coherent*. Even if there is a phase difference between them, this phase difference is constant. If two such wave trains are brought together, they will interfere with each other. If they are in phase, they are said to interfere *constructively,* and the amplitudes will be additive; if they are 180° out of phase, the resultant disturbance will be zero. A good example of coherent radiation is the train of waves from a radio transmitter emitting waves at a fixed frequency. Laser light can also possess good coherence properties. Coherent radiation is by its very nature plane polarized.

Interaction of Radiation with Gases. Radiation can be either scattered or absorbed by gaseous molecules. Scattering can be explained by either the electromagnetic theory [*see* Scattering] or by quantum theory in terms of an *elastic collision* between a photon and a molecule. However, absorption can only be explained adequately in terms of the quantum theory. The energy of a molecule is *quantized,* that is, it exists in discrete values or *levels* that are a property of the molecule. If a photon possesses an energy that is exactly equal to the difference between two adjacent molecular energy levels, then the molecule can be excited from the lower to the higher level, the photon being absorbed into the molecule. The molecule is subjected to numerous collisions with adjacent molecules, which has the effect of smearing out the energy levels allowing a narrow range of photon energies to be absorbed but with decreasing efficiencies from the central wavelength. This effect is called *collisional broadening* or *pressure broadening* and increases with gaseous pressure. After transmission through a gas, the black body spectrum can thus have discrete (narrow) gaps in the energy spectrum known as *absorption lines*. The widths of these lines decrease with altitude in the atmosphere as the atmospheric pressure decreases.

The excitation of molecules occurs in several ways and at different wavelengths, or frequencies. The energy E of an isolated molecule can be written as

$$E = E_e + E_v + E_r + E_t$$

where the terms stand, respectively, for *e*lectronic, *v*ibrational, *r*otational, and *t*ranslational energy. The first three terms in the equation are quantized. For a given molecule, there is a characteristic series of energy levels between which transitions can occur, resulting in a *band* of absorption frequencies at regular intervals.

Electronic transitions, which involve the most energy, occur in the ultraviolet and visible spectra. Vibrational energy transitions are usually in the intermediate infrared region, while rotational energy changes occur in the far infrared and microwave spectra. However, molecules usually exhibit complex band spectra that involve both vibrational and rotational energy changes, leading to *vibration rotation bands*. The types of spectra describe the molecular motions. The individual atoms in a molecule may vibrate characteristically with respect to one another, or alternatively, whole molecules may revolve around different axes. *Polyatomic molecules* composed of three or more atoms generally exhibit strong infrared spectra; typical atmospheric molecules of this type include H_2O, CO_2, NO_2, NH_4, and SO_2. *Diatomic molecules* of two atoms—for example, N_2 or O_2—tend to have only electronic spectra, although some magnetic dipole transitions occur at much lower frequencies. Figure 1c shows characteristic absorption spectra for certain atmospheric components at ground level and at 11 kilometers above the surface. The effects of decreasing atmospheric pressure on the widths of the absorption bands are obvious.

In addition to band spectra, continuous absorption is observed from the near to the far infrared. This continuum has been the subject of much theoretical and experimental work, particularly in the atmospheric window, (8 to 13 micron wavelengths) where the water vapor continuum causes appreciable absorption of importance to the Earth's radiation balance. The absorption coefficient here is proportional to the square of the water vapor pressure. Since 1970, it has been believed that this absorption is due to *dimer* molecules, pairs of molecules that bind together for a short time; more recently, however, a perhaps more satisfactory theory has been proposed. Strong water vapour absorption lines, having their central wavelengths some distance away from the window, can be sufficiently broadened so that the line "wings" can cause significant absorption in the window.

Transfer through Liquids. The only important natural absorber in the atmosphere is water, along with compounds of water such as sulphuric acid. The most important reservoirs of water from the standpoint of radiation are clouds. Like gases, water both absorbs and transmits radiation, depending strongly on the wavelength of the radiation. Because of the strong interactions between the closely spaced molecules of liquid water, absorption here is much more intense than in an equivalent volume of water vapor. In addition, these strong interactions result in a continuum of absorption throughout the near and far infrared spectral regions. In the near-infrared and visible wavelengths, however, water becomes almost transparent to radiation; strong absorption again occurs in the far-ultraviolet region.

The absorption coefficient of a substance is determined by its refractive index—in particular, by what is defined as the "imaginary" part of the refractive index.

It is so defined because the theory of complex numbers is used as a convenient mathematical framework in the treatment of the electromagnetic theory of radiation, particularly where absorption occurs.

The complex refractive index can be written

$$n = n_1 - in_2$$

where $i = (-1)^{1/2}$ and n_2 is a complex number. The distance through which radiation can travel before a certain fraction is absorbed can be obtained from Equation 1. If we make:

$$\sigma(z_1 - z_2) = 1, \quad \text{then} \quad \frac{I_2}{I_1} = \frac{1}{e} \approx 0.37. \quad (2)$$

Table 1 shows the distance travelled through water at different wavelengths before $I_2 = 0.37\, I_1$.

Water clouds consist of very small drops—about 1 to 10 microns in radius in nonraining clouds, but up to several millimeters in raining clouds. Over this size range, absorption is still very strong in the infrared, as shown in Table 1, and cloud drops absorb strongly. By contrast, cloud drops are very transparent in the visible region, although they still scatter the radiation strongly.

Transfer through Solids. In many solids in the natural state, absorption is very strong in the infrared; it can also be strong in the visible, depending on the apparent color of the object. When absorption is very strong—for example, when visible light falls on a shiny metal object—much of the radiation is reflected rather than absorbed. Selective absorption and reflection cause the color variations in terrestrial surfaces.

A medium can either reflect, absorb or transmit radiation. That is,

$$R + \alpha + T = 1$$

where α is the absorptance, R the reflectivity, and T the transmittance. According to Kirchhoff's law, the emis-

RADIATION. Table 1. *The distance $(z_1 - z_2)$ traveled in water to obtain $1/e$ attenuation (Equation 2).*

Wavelength (microns)	$(z_1 - z_2)$ (centimeters)
0.2	7.1×10^1
0.5	4×10^3
1.0	3.3×10^0
2.0	1.7×10^{-2}
5.0	3.3×10^{-3}
10.0	1.6×10^{-3}
20.0	3.1×10^{-4}

sivity ϵ of a body is equal to its absorptance. If the medium is strongly absorbing so that $T = 0$, then

$$\epsilon = I - R.$$

Many natural surfaces have a high emissivity and therefore a low reflectance in the thermal infrared. For water or land surfaces, the emissivity varies from about 0.94 to 0.99, depending on wavelength as well. The absorption of visible radiation by water is small, in which case $R = 1 - T$. Water surfaces have a reflection value of about 0.05, with the rest of the light being transmitted.

Solar reflectivity from land surfaces is quite variable, ranging from about 0.05 over dark forests, 0.3 over light grasses and deserts, to 0.8 or 0.9 over snow-covered surfaces. The reflection is usually spectrally selective, with fractional transmission, absorption, and reflection. The characteristic green color of vegetation is a result of selective absorption by chlorophyll outside the green wavelengths. The light is reflected and scattered at the leaf's cell boundaries and appears green in both reflection and transmission. At wavelengths beyond 0.65 microns, reflection becomes strong, indicating strong absorption in the leaf.

Ice crystals in high cirrus clouds are strong absorbers of infrared radiation with absorption intensities rather similar to water drops. Because of this, and because of their cold temperatures, cirrus clouds have a significant effect on climate.

Radiation and Climate. The flows of radiation to and from the Earth–atmosphere system largely control the planetary temperature. The Sun's rays incident on the Earth are the primary source of the energy that drives the climate system. Incoming solar radiation is balanced by outgoing infrared radiation, according to Planck's law. The balance between these radiation flows determines the mean planetary temperature, T_p. Figure 3 illustrates this process schematically. T_p is very low—only $-18°C$. The Earth's surface temperature, T_s, is much higher at about $15°C$. The difference is caused by the so-called *greenhouse effect*. Infrared radiation emitted at the Earth's surface is absorbed by carbon dioxide, water, and other greenhouse gases, to be emitted higher in the atmosphere at, on average, the planetary temperature T_p. Simultaneously, the greenhouse gases also emit infrared radiation back to the surface, causing T_s to increase above T_p. [*See* Greenhouse Effect.]

Another major modulator of radiation flows is atmospheric clouds, which produce large uncertainties in the study of climate. Clouds affect both incoming solar radiation and outgoing infrared radiation. They reflect

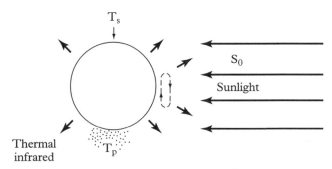

RADIATION. Figure 3. *Schematic of the radiation balance of the Earth.* S_0 is the solar constant. The mean planetary temperature T_p is $-18°C$. The mean surface temperature T_s is $15°C$. Atmospheric gases (such as carbon dioxide and water vapor) increase T_s. Clouds on average decrease T_s.

greater amounts of solar radiation than does a cloudless atmosphere, decreasing the radiation reaching the surface and consequently lowering the surface and planetary temperatures. They also strongly absorb infrared radiation emitted from the surface and reemit it at the lower cloud temperature, causing additional planetary heating. On average, the cooling effect is greater than the heating caused by clouds; it is estimated that a cloudless Earth would have a surface temperature of about $30°C$. The relationship between clouds and global warming is complex and poorly understood, because it is not known how different kinds of clouds would respond to warming.

[*See also* Clouds; *and* Global Warming.]

BIBLIOGRAPHY

Goody, R. M., and Y. L. Young. *Atmospheric Radiation: Theoretical Basis.* New York, 1989.
Kerker, M. *Electromagnetic Scattering.* New York, 1963.
Liou, Kuo-nan. *Radiation and Cloud Processes in the Atmosphere.* New York, 1992.
Paltridge, G. W., and C. M. R. Platt. *Radiative Processes in Meteorology and Climatology.* Amsterdam, 1976.

C. MARTIN R. PLATT

RADIOSONDES. *See* Instrumentation.

RAIN. The form of precipitation classified as *rain* consists of falling drops of liquid water with a spherical equivalent diameter of at least 0.5 millimeters. Drops smaller than this fall very slowly and are called *drizzle*. Even smaller drops of less than 0.1 millimeter diameter are termed *cloud droplets;* these make up the visible clouds. Typical raindrops, about 2 millimeters in diameter, are around 100 times larger in diameter and about 1 million times larger in volume than cloud droplets, usually about 0.02 millimeters in diameter. In a typical cloud, there are millions of cloud droplets for every raindrop. [See Clouds; Precipitation.]

Rain usually forms in regions of clouds where there is rising motion. In order for rain to descend to the Earth's surface, the velocity of the drops (called their *fall speed*) must exceed that of the rising air; otherwise, the drops are borne upward. Fall speed increases as drop size increases. Cloud droplets fall very slowly and so are effectively suspended in the atmosphere, maintained aloft by small turbulent fluctuations until they evaporate. Raindrops fall at between 2 meters per second (for drops 0.5 millimeters in diameter) and 10 meters per second (for 5-millimeter drops); the vertical motions in rain clouds are generally at least this strong. Drizzle falls at speeds between 0.5 and 2 meters per second, so this form of precipitation is associated with clouds in which vertical motion is weak.

The shape of raindrops is determined by several factors. Surface tension, the electrostatic force that holds a drop together, dominates and tends to make the drop spherical. Aerodynamic forces induced by the air flowing past a drop interact with surface tension. Thus, most falling raindrops exhibit a slight flattening of the leading (bottom) face; this flattening becomes more pronounced as the size of the drop increases and the drop falls faster.

The formation of rain is not completely understood. In principle, the same processes that produce cloud droplets and ice crystals—nucleation, the solute effect, and condensation—can also produce raindrops. In practice, however, this appears to be a path nature follows only rarely. To produce a 2-millimeter raindrop by condensation alone would require many hours or even days under normal atmospheric conditions, yet rain is often observed falling from a growing cumulus cloud only 20 to 30 minutes after the initial appearance of the visible cloud. This puzzle is solved when we realize that a relatively large (relatively faster falling) drop can overtake and capture smaller droplets that have lesser fall speeds. If only a few droplets in a cloud become large enough to fall, they can sweep up the smaller droplets in their paths. This is the essence of the process called *collision and coalescence.*

Collision and coalescence, however, is not very efficient, for two reasons. First, a falling drop can capture only those drops that are just slightly smaller than itself.

This is true because air, a slightly viscous gas, must flow around the falling drop. Because of inertia, drops much smaller than the falling drop are carried along with the flow of air and thus never touch the large drop's surface. Second, even when a falling drop does collide with another drop, coalescence does not necessarily occur. A very thin residual film of air can keep the two drops apart, so that the smaller drop rolls around the larger one and is left behind. For the two drops to coalesce, this film of air must drain away; this will occur only if the two drops are in contact for a sufficient time, and thus they must be of similar size so that their rates of fall are similar. In addition, the surface tension of both drops must be overcome. Electrostatic effects appear to be important in triggering the final merger, but exactly how this occurs is still an object of active research.

The efficiency of collision and coalescence is enhanced by the aerodynamic instability of large raindrops. As a falling raindrop grows larger, forces induced by the air flowing past the drop progressively flatten the drop's leading face. When the spherical diameter of the drop grows to approximately 5 millimeters, these forces greatly distort the drop's shape and then tear it into many smaller drops. These smaller drops immediately begin to grow again by collision and coalescence. This process leads to a *cascade effect,* in which a few large drops from high in the cloud eventually produce a great many raindrops falling from the cloud base.

In order for the sequence just described to begin, a few cloud droplets must grow to around 40 millimeters in diameter. Under normal atmospheric conditions, condensation rarely produces droplets much larger than 20 millimeters in diameter during the time intervals over which rain is observed to form. How, then, is this gap bridged to create the few very large droplets needed? It appears that these can be created by several different processes, depending on the dynamics of different types of clouds and the differing locations in which these cloud types form. We will examine two processes, one associated with continental clouds and the other with marine clouds in the tropics.

In clouds over continents, the high concentration of very small atmosphere-borne particles, called *condensation nuclei,* permits the rapid formation of many cloud droplets from the available water vapor. The interior of the cloud quickly becomes occupied by a large number of droplets of roughly equal size, all in quasi-equilibrium with the remaining water vapor. Few if any of these droplets grow large enough to fall with respect to the other droplets. Observational evidence suggests that if the temperature is above freezing throughout a continental cloud, so that no ice crystals form, it is unlikely that the cloud will produce any rain.

The requisite ice crystals can form directly by condensation on aerosols called *freezing nuclei* or *ice nuclei,* which are particles such as kaolinite whose structure mimics the shape of ice crystals. Alternatively, the ice crystals can form when supercooled cloud droplets (water that remains liquid below the freezing point, 0°C) freeze, often as a result of coming into contact with an ice nucleus or ice crystal. Observations of supercooled droplets in the process of freezing show that the surface of the droplet freezes first to produce a thin shell of ice. Then, as the interior water begins to freeze and expand, this shell shatters into many small ice crystals.

As water vapor in the cloud migrates to condense on the forming ice crystals, liquid droplets in the cloud evaporate. This occurs because the saturation vapor pressure over ice is less than that over water. Thus, the ice crystals grow at the expense of the liquid droplets. When this happens, the cloud is said to *glaciate.*

In a cloud on the verge of glaciating, water droplets are always much more numerous than ice crystals or ice nuclei. Nonetheless, through the cascade effect resulting from the freezing and shattering of water droplets—constantly providing more ice crystals—and the rapid migration of water vapor from evaporating droplets to ice crystals, the subfreezing portion of a cloud can glaciate in a few minutes. As additional water vapor is carried upward in the cloud by rising air, the crystals grow in the upper part of the cloud to form snowflakes. Because of their aerodynamics, snowflakes can be held aloft by very weak upward air motion, so they grow quite large before they begin to descend through the cloud. The flakes eventually fall through the freezing level into warmer air, where they melt into relatively large drops, which in turn initiate collision and coalescence.

The atmosphere over the tropical oceans contains much smaller concentrations of condensation nuclei than that over continents. As a consequence, droplets in tropical clouds are far less numerous than in similar continental clouds, reducing competition for the available water vapor. This allows a few droplets to grow large directly by condensation. In addition, some of the nuclei that are present here are likely to be relatively large salt crystals derived from ocean spray. Condensation on these can produce a few more large droplets. Even though they are rare relative to the number of cloud droplets present, these few large droplets appear to be enough to initiate (or *seed*) the process of collision and condensation.

Once raindrops fall out of the cloud into the unsaturated air below it, they begin to evaporate. Consequently, the drops that reach the ground are somewhat smaller than when they left the cloud base. Because the fall speed of the drops decreases as their size decreases, the drops slow down as they evaporate, lagging farther and farther behind the moving cloud. In this way, raindrops can give rise to drizzle at the surface. Drizzle can also form directly, usually from thin, low-lying stratus clouds in which the drops have no opportunity to grow larger. If rain falls from a cloud with a very high base, or if the air between the cloud base and the surface is very dry, the drops may evaporate completely before reaching the ground. This phenomenon, often visible as a long plume streaming out below a cloud, is termed *virga.*

[*See also* Condensation; Freezing Rain; Hydrological Cycle; Lake Effect Storms; *and* Religion and Weather.]

BIBLIOGRAPHY

Byers, H. R. *Elements of Cloud Physics.* 2d ed. Chicago, 1973.
Fletcher, N. H. *The Physics of Rainclouds.* Cambridge, 1969.
Pruppacher, H. R., and J. D. Klett. *Microphysics of Clouds and Precipitation.* Boston, 1980.

JOHN T. SNOW

RAINBOWS. *See* Religion and Weather; *and* Scattering.

RAINFORESTS. The rainforest is one of the major terrestrial biomes, or life zones, of the Earth. The geographic distribution of rainforests, is determined primarily by climate; the crucial climatic elements are mild to hot temperatures and abundant, relatively constant rainfall. Although there are some tracts of rainforest in temperate regions, the most extensive and biologically rich are the *tropical rainforests,* which typify this biome in the popular mind.

The first written account of tropical rainforests was that of Christopher Colombus, who described the forests of Hispaniola (presently Haiti and the Dominican Republic) to the king and queen of Spain around the end of the fifteenth century. The term *tropical rainforest* was coined by the German botanist A. F. W. Schimper in 1898. His original diagnosis of such forests fits the concept most modern ecologists accept. This can be stated as follows: evergreen forests, *hygrophilous* (wetness-related) in character, with trees at least 30 meters tall as the predominant life form; rich in thick-stemmed *lianas* (woody climbing plants), palms, and woody as well as herbaceous *epiphytes* (plants that grow on other plants, including orchids and other flowering plants, ferns, algae, and mosses). The trunks of the trees are usually straight and slender; their bases commonly present *buttresses*—flange-like outgrowths—a distinctive character of these forests. The rainforest understory consists of shrubs, herbaceous plants, sapling and seedling trees, and frequently palms.

Tropical rainforests form a belt roughly following the equator, but its northern and southern limits are highly irregular. In some areas, such as Africa, the latitudinal limit is more restricted than the geographic tropics, while in others it approaches (e.g., in Mexico), or even exceeds (e.g., in Brazil) the boundaries of the geographic tropics. Rainforests are also irregular in altitudinal distribution. In general, the latitudinal boundaries are determined by precipitation, and the altitudinal ones mainly by temperature. More specifically, in the area occupied by tropical rainforests, the mean annual temperature is about 25°C, but there are localities with considerably lower average minimum monthly temperatures; however, temperatures in the latter areas never fall to the freezing point. In typical rainforest localities, annual rainfall is close to 2,000 millimeters, but the upper limit varies considerably, with some areas approaching 8,000 millimeters. Although it is popularly believed that rainforests experience precipitation throughout the year, most of them show some degree of seasonality; the length of the dry season—the number of months with less than 100 millimeters of rainfall—determines the variation in appearance observed in rainforests, mostly as a result of differences in the luxuriance, size, and proportion of deciduous trees.

Land area with climates suitable for this biome is roughly 1,600 million hectares, but about 10 percent of this is thought originally to have been covered by other types of vegetation, mainly because of soil or other ecological factors. Thus, the original extent of the world's rainforests was around 1,440 million hectares. The three major tracts were in central Africa (325.8 million hectares), Asia and the Pacific (391.5 million hectares), and Latin America (722.7 million hectares).

The most outstanding feature of tropical rainforests is their biological diversity. It is estimated that no less than 50 percent of all species of plants and close to 80 percent of insect species live in tropical rainforests, even though they cover less than 2 percent of the globe. Trees, birds, mammals, reptiles, and butterflies all have their greatest known local species richness in Amazonian tropical rainforests. Great Britain has slightly more than 1,800 species of flowering plants, but a similar number can be found in an area several thousand times smaller in the tropical rainforests of Veracruz, Mexico. The world record for concentration of tree species is held by Amazonian Ecuador, with 473 species in 1 hectare, considering only trees greater than 10 centimeters in girth. A total of 43 species of ants were found in a single tree in the Peruvian Amazon—a number equivalent to all the species known from the British Isles.

With such biological diversity, tropical rainforests constitute a reservoir of natural products of great importance to humankind—not only wood products, but also other resources. For example, it is estimated that approximately 1,400 species of tropical rainforest plants may contain drugs potentially useful in treatment of various types of cancer, yet only 1 percent of the known species have been studied for their anticancer properties. Another important consideration is new foodstuffs, such as the species of perennial corn recently discovered in Central America, which may promise revolutionary changes in food production. In addition, these forests are highly valuable for the ecosystem services they provide, including, erosion control, carbon sequestration, and local climate regulation. In regard to the last, some studies in the Amazon Basin suggest that in this area approximately half of the local rainfall is derived from local evapotranspiration; therefore, a major modification of the forest cover could have a significant climatic impact. Similarly, it is estimated that up to 30 percent of carbon dioxide emission to the atmosphere results from tropical deforestation. [*See* Deforestation; Greenhouse Effect.]

Despite their great biological diversity, ecological services, and applied value, tropical rainforests are being dramatically reduced. Recent estimates of annual deforestation rates for Africa, Asia and the Pacific, and Latin America are 1.5, 3.7, and 5.3 million hectares respectively, or 10.5 million hectares for the whole rainforest biome. This destruction results from conversion to grassland for cattle grazing, uncontrolled logging, and overexploitation of several rainforest elements. Climate change can profoundly affect tropical rainforests, given their fragility and the restricted range of temperature and precipitation regimes to which their species are adapted.

Ecologists also recognize the existence of *temperate rainforests*. These rainforests have a more restricted distribution in three geographical areas: the northwestern coast of North America, southeastern Australia, and southern South America. Annual precipitation is on the order of 2,000 to 3,500 millimeters (but never as high as in some tropical counterparts); this is augmented by condensation of water from dense coastal fogs. The proximity to the coastline regulates the temperature so that seasonal fluctuations are relatively narrow, with cool summers and mild winters, though in contrast to the tropics, freezing temperatures do occur. The physiognomy of these forests is characterized by the dominant trees, which are large (sometimes larger than in tropical rainforests) and evergreen. As in the tropics, these forests are rich in epiphytes, though here this exuberance is due to the number of plants (largely mosses, lichens, and ferns) rather than to the richness of species. The understory vegetation includes abundant ferns, herbs, shrubs, and saplings. There is a rich associated fauna, insects and birds, followed by mammals; however, the biological richness, in general, is considerably lower than in tropical rainforests. These temperate forests are seriously threatened by human activities, mostly by logging, because they are one of the most important producers of lumber and pulpwood worldwide. Climate change models suggest that these forests are likely to be more seriously affected by global warming than are tropical rainforests.

[*See also* Ecology.]

BIBLIOGRAPHY

Gentry, A. H. "Tropical Forest Biodiversity: Distributional Patterns and Their Conservational Significance." *Oikos* 63 (1992): 19–28.
Kricher, J. C. *A Neotropical Companion.* Princeton, N.J., 1989.
Richards, P. W. *The Tropical Rain Forest.* Cambridge, 1976.
Valencia, R., H. Baslev, and G. Paz y Miño. "High Tree Alpha-Diversity in Amazonian Ecuador." *Biodiversity and Conservation* 3 (1994): 21–28.
Whitmore, T. C. *Tropical Rain Forests of the Far East.* 2d ed. Oxford, 1984.

RODOLFO DIRZO

RAIN GAUGE. *See* Instrumentation.

RAINSHADOW. *See* Orographic Effects.

RECREATION AND CLIMATE. Recreation, sports, and tourism are activities that we pursue in our leisure time to make our lives more interesting and enjoyable. Activities as simple as strolling through a park or skipping rocks on a quiet pond, or as complex as hang-gliding or hot-air ballooning, fall within this definition. All types of team or individual sports are also included. Tourism involves visiting new places and, perhaps, participating in new activities. This requires time and transportation and therefore has traditionally been restricted to wealthier people of advanced civilizations. Tourism in the form of leisurely travel has taken place for thousands of years, but rapid and inexpensive transportation systems developed in recent decades have made tourism feasible for millions across the globe.

Because most recreation and sports are pursued out of doors, and tourism requires moving about, these activities are directly influenced by the prevailing local climate and by day-to-day fluctuations in the weather. The very activities that are available are determined directly by climate and geography. People who live by bodies of water swim, sail, fish, and explore in their leisure time; people who live near mountains hike, climb, and ski; in cold climates, they skate or cross-country ski; in urban areas, residents may prefer team or individual sports and watching professional sporting events. Vigorous activities are most often pursued in cool climates and during cool times of year, while recreation may be slower-paced in tropical areas. If they can afford it, people who wish to participate in activities that are not available or popular where they live may travel elsewhere to do so.

The preferred season of the year for many sporting events has evolved as a direct response to the climate. Ice hockey is still played primarily in the winter, slow-paced sports like baseball and golf in the heat of summer, when water sports are also most popular and enjoyable. Activities that require great exertion—such as soccer, distance running, or football—are best suited for cool weather. Such activities as sailboarding and kite-flying depend on wind characteristics. Hot-air ballooning requires light winds and a cool, stable atmosphere, while glider pilots seek out convective updrafts.

Weather conditions also affect how well athletes perform. Marathon runners hope for cool, humid weather for best performance; ski racers know that fresh snow slows their speeds, as does temperature near the freezing point or far below freezing. Wind affects ski jumping and track and field events so much that wind speeds are measured and recorded along with athletic performance, and limits are specified for the determination of records. In team sports, extreme weather such as snow, heavy rain, strong winds, or extreme temperatures may help weaker teams overcome the competition. Some field sports continue despite adverse weather, but others are usually suspended until weather conditions improve and fields dry.

Spectator comfort is a function of weather conditions. In dry and mild climates like southern California, weather conditions rarely interfere with sports competitions at any time of the year. Spectator discomfort can be a major problem, however, in higher-latitude cities; sports fans in Minnesota, for example, have been much more comfortable, and therefore more likely to pay money to attend games, since a domed stadium was built there. In Houston, Texas, where heat and humidity are oppressive half the year, the Astrodome provides year-round comfort.

Even indoor recreation is indirectly influenced by weather conditions. Participation in activities such as bowling, indoor swimming, squash, racquetball, and basketball are greatest during sustained periods of cold, cloudy weather. Participation in these indoor activities declines as outdoor weather conditions improve. Health and fitness clubs, which have become an integral part of sports and recreation in the United States since the late 1970s, originally flourished because of a demand for indoor exercise opportunities. These clubs have recently had to develop more outdoor opportunities to remain economically viable throughout the year.

The relationship of weather and climate to tourism has several different dimensions. Some of the motivation for travel and tourism is the desire to escape undesirable weather at home for more comfortable and enjoyable conditions. Tourists from hot, humid areas flock to the mountains, beaches, or higher latitudes during the summer months; during winter people from cold climates often visit tropical areas. As a result, many tourist destinations have distinct seasonal cycles in their local economies. Efforts have been made to lengthen the tourist seasons by adding attractions and broadening activities. The explosive growth in popularity in winter recreation and skiing since the 1960s has helped stabilize many local tourism-based economies. Colorado, for example, traditionally derived most of its tourism revenues from summer and fall visitors. Now, with more than a dozen destination downhill ski areas, winter recreation has become the primary tourist season for much of the state.

Part of the enjoyment in travel and tourism is derived from experiencing at first hand the elements of the weather that define a region's climate. Being brushed by a hurricane while visiting the Bahamas, baking in the dry heat of Death Valley, dodging hailstones from a vicious thunderstorm in Kansas, creeping through dense fog in the California redwoods, or waking to a surprise snow in the Rocky Mountains in July—these experiences all tend to enhance, not diminish, the pleasure of being a tourist.

NOLAN J. DOESKEN

RELATIVE HUMIDITY. *See* Humidity.

RELIGION AND WEATHER. In the history of religion many cultures have demonstrated a worldview in which human beings, nature, and the divine all interact, for good or ill. Humans can have divine characteristics, such as great wisdom or power, and gods are often perceived in human terms. In the same way, animals are frequently viewed as having both human and divine characteristics, and humans often seek to assume the characteristics of certain animals in order to participate in their powers, especially powers of fertility.

For example, in many ancient cultures bulls were linked with thunder (because of the noise they make when running) as well as with rain (because of their sexual prowess, since semen is linked with rain). Somewhat like bulls, who are perceived as great but hard-to-control sources of power, the divinities who control the weather are believed to be sources of both fertility (the right amount of rain at the right time) and destruction (through flooding or drought), and some are identified with bulls (e.g., Indra, Shiva, Zeus, Baal).

Weather is a vivid, all-encompassing, and often spontaneous manifestation of these ideas. Believing themselves to be subject to gods with unlimited meteorological powers—which are frequently invested in mythical animals such as dragons (particularly by East Asians) and thunderbirds (particularly Native Americans)—humans attempt to influence these gods through understanding, propitiating, and/or threatening them. In this powerfully charged and symbolically rich worldview, human beings achieve a sense of being part of a larger whole; they transcend their mundane concerns and attempt to arbitrate their fates.

Divine Powers. One telling example of the effect of the natural environment on theological and religious life can be seen in the contrasting pantheons of ancient Egypt and Mesopotamia. Ancient Egypt was almost rainless and depended on the annual flooding of the Nile to irrigate its crops. This was a predictable and mostly measurable phenomenon. The north wind provided a cooling breeze that made life possible. Mesopotamia, on the other hand, was subject to the unpredictable and powerful rising of the Tigris and Euphrates rivers, which could flood the fields. Scorching winds blew in off the desert and alternated with torrential rains. Nature in Mesopotamia made humans conscious of their insignificance, while nature in Egypt seemed to work harmoniously with human efforts to contain it. As a consequence, the Egyptian gods of nature were essentially benign. In Mesopotamia these gods were fierce and punitive. For example, the defeat of the ancient city of Ur by the Elamites was understood to be due to a storm sent by Enlil, the Mesopotamian god of the wind. This was a force so powerful that it alone was believed to keep the sky and Earth separated. Many of Enlil's characteristics and insignia were passed on to the Babylonian deity Marduk, who became king of the gods after using his meteorological powers as god of the storm to defeat Tiamat, the primeval sea. The account of this cosmic battle is told in the Babylonian creation story, the *Enuma Elish*, dating from about the twelfth century B.C.E.

In the Hebrew Bible (Old Testament), one of the many instances when God (Yahweh) displayed his control of weather and other natural phenomena is recorded in the account (Ex. 9–10) of Moses's confrontation with the magicians of Pharaoh, when Yahweh caused hail to fall on the Egyptians but not on the Israelites, caused a wind to bring locusts, and blotted out the light of the Sun for three days. The ancient Hebrews believed that Yahweh spoke and acted through thunder and lightning, either warning them or punishing their transgressions of his law. Although Psalms 147 and 148 praise his timely sending of rain, snow, hail, frost, wind, and sunshine, Yahweh has all the unpredictability of a god of the storm.

Jesus was also believed to have meteorological power: he calmed the winds and a stormy sea (Mt. 8.23–27, Mk. 4.35–41, Lk. 8.22–25) and showed his transcendence of nature quite literally by walking upon the sea (Mk. 6.45–52). Christian saints also share in this power. It is said, for example, that Saint Genovefa (423–502) prayed for relief when rain threatened while she and her neighbors were harvesting their crops; her prayer was answered, and only her part of the field remained dry. With similar effects, Saints Radegund (525–587) and Gertrude of Nivelles (628–658) calmed stormy seas, as did the Virgin Mary, who became the protector of sailors. In fact, Mary's control over weather could be quite specific, as when she appeared in Rome after the victory of the Christians against the Turks at Lepanto in 1571 requesting that a church be built. She caused snow to fall just on the spot where she wanted the church and nowhere else.

Theological interpretations of natural events and conditions have continued in modern times among Christians of many nations. In the Netherlands, for example, weather and geography seem to have played a significant role in shaping Dutch self-understanding. Simon Schama (1988) presents a fascinating account of this thesis, calling particular attention to the theological implications of Dutch efforts to reclaim land from the sea and God's use of storm and flood both to punish them for transgressions and to protect them against political enemies.

In the Islamic tradition, Muslim saints are also believed to participate in controlling the weather and other natural forces. The twelfth-century Persian mystic Habib al-Ajami, for example, is said to have calmed winds and walked on water. In general, Muslim saints are believed to be so close to Allah that they have access to his mercy and are therefore able to make the rains fall. Villagers in Turkey and Iran evoke this belief in their word for rain, *raḥmat*, meaning "mercy."

Like many deities, divinities, and saints, the Buddha, too, had to show his mastery of weather. His meteorological power was clearly demonstrated when he did battle with the demon Mara, who sent lightning and storm in order to prevent the Buddha's enlightenment.

Rain. Religious rituals designed to cause rain are related to what Sir James Frazer (1922) called "imitative magic," and such rituals reveal the arbitrariness of a distinction between magic and religion. The rain dances of various Native American societies offer a good example—the dancers may imitate rain by sprinkling water on the ground, use bull-roarers to simulate the sound of thunder, make paintings of rain clouds, and so on. In this way they are imitating the effect they desire while also acknowledging the power of their gods to give or withhold such benefits.

Australian Aborigines believe clouds are bodies that release rain because of their ceremonies. Consequently, the male rainmakers are bled, and their blood is sprinkled over the rest of the men of the tribe. Significantly, it is male, not female, blood that is used, and male participants must avoid contact with their wives until the rains come. Similar beliefs existed among the Javanese and ancient Babylonians. The association of rain with maleness continues, for example, among the Australians, who attribute rain-giving powers to the foreskins removed during circumcision and preserved, carefully out of the sight of women, in the event of drought. Given rain's association with semen, male chastity is recommended as a way of avoiding unwanted rain. Conversely, in some cultures (in ancient Europe, India, and sub-Saharan Africa), women play a role in bringing rain by stimulating male desire by showing their nude bodies or singing ribald songs to the male rain gods. Some Native American peoples, however, distinguish between male (pelting) rain and female (soft) rain. In parts of Europe until fairly recent times, when rain was desired, water was sprinkled over people of both sexes dressed in leaves (representing the crops that depend on rain) or was used to soak puppets wearing leaves.

One way to prevent rain, in order to avoid flooding or perhaps to send a drought to enemies, is to use the element opposite to water—fire. Here one either sets wood on fire or heats up stones, which are then placed outside in the rain. The way these objects make the rain sizzle and disappear imitates the desired drying up of rain, serving as a sign to the rain, or rain gods, that the people do not want any more. Similarly, when Southeast Asians want to change the weather they expose statues of their gods to the elements; setting them out in the sun is thought to bring rain, while setting them out in the rain will bring sunshine. European Christians have done the same thing with images of their saints.

Various aspects of the natural world, such as animals and plants, are associated with rain around the globe. Some examples for rain are snakes, frogs, salamanders, turtles, fish, lizards, and spiders, as well as various birds that fly up to the clouds and are thus capable of carrying messages to the gods. All these animals are used in rituals to bring rain. For example, one Native American group would confine frogs in hot places in the hope that their croaking would make the rain god relent and send rain. Plants that live in water, such as lotuses, are thought by many different peoples to contain the power to make rain, and rain rituals are often performed near bodies of water, or even springs or wells.

It is interesting to compare evidence of rain rituals among tribal peoples with the biblical account in 1 Kings 18, where it is told that Yahweh caused a drought to punish the faithless Hebrews and that the prophet Elijah performed an elaborate ceremony (and contest with the priests of Baal) involving fire and water. This ceremony was designed to prove the superiority of his god and to bring rain. The Hebrews were a people living in an arid land and much in need of water, which they believed Yahweh could provide (Pss. 65.19–23, 134.5–7, 147.8; Joel 2.23; Amos 4.7–8; Zec. 10.1–2). Further, they believed that drought was God's punishment for sin (Jer.

31.3, 14.7; Hg. 1.11). To Westerners this seems theologically sound, even sophisticated, especially when compared to the magical rituals of tribal peoples, yet the story of Elijah contains many of the same magical elements. He sacrifices a bull, builds an altar of twelve stones (representing the twelve tribes of Israel), soaks the bull being offered and the altar three times with water, and then calls upon Yahweh to make it all burn. Yahweh does, and soon after it begins to rain, ending the drought.

Rainmakers are known under a wide variety of names by the scholars who study them: sorcerers, magicians, medicine men, shamans, and wizards. These names often reveal a European and American prejudice about the faiths of non-European peoples. When Christian priests or ministers are called upon to bless the fields, to ask their god for timely rains, they are not really behaving any differently from tribal rainmakers.

Rainbows. The rainbow is an almost universal sign of a bridge linking this world and the divine world. Shamans in many parts of the world (e.g., Siberia, Japan, Australia, Polynesia, North and South America) are believed to climb up to heaven on it. It is the calm after the storm that indicates the gods are once again at peace with humankind. In the Hebrew Bible it is told that Yahweh created the rainbow to be a reminder of his covenant with Noah and the rest of humankind that never again would there be total destruction through flood (Gn. 9.13–17). In myths of the ancient Greeks the rainbow is the goddess Iris, a messenger of the gods. Some Native Americans have seen it as the road of the dead (Catawba and Tlingit), but it has almost as many meanings as there are tribes: it is the path of the Cloud People and the kachinas (Hopi); of the gods (Navajo); to another world (North Pacific Coast, Iroquois); it prevents rain from falling (Yuchi); to point at it is dangerous (Hopi, Northern California), and so on.

Many peoples have thought of the rainbow as a snake: native peoples of both South and North America, Australian Aborigines, West Africans, and the ancient Persians. In China it is a dragon, a symbol of the ubiquitous rain goddesses, especially Yao Chi; and rainbows are associated with the feminine, with *yin*. The ancient Chinese saw dragons–rainbows–women–rain–fertility–moon–death on the same continuum. Their water goddesses were great seductresses, much like the Lorelei of German legend, and the subject of many erotic tales. In ancient times, the rain goddesses were ritually married to the king in order to channel their fertility. Perhaps most curiously, in Romania rainbows are associated with sex changes; anyone passing underneath one will be transformed, woman into man, man into woman. This adds a particular challenge to those seeking the legendary pot of gold.

In Tibet rainbows are generally perceived to be an auspicious sign. They are particularly associated with *ḍākiṇīs*, divine women who act as guides to enlightenment, and advanced yogis, female and male, are said to possess a rainbow body. Tibetan (Tantric) Buddhism continues to use meteorological imagery in its most ubiquitous symbol, the *dorje* (Sanskrit *vajra*), the "thunderbolt." This symbolizes the invincible force of transcendental reality that shatters illusion and empowers the process of enlightenment. Indeed, Tantric Buddhism is often referred to as the path of the thunderbolt, Vajrayāna.

Thunder and Lightning. The main companions of rain, thunder and lightning, have been deified and/or personified all over the world. They are often the voice and manifestation of a great sky god, although sometimes they are separate divinities, as in the native traditions of North and South America, and sometimes they are byproducts of actions of the gods. In Norse mythology, for example, the ruling deities Odin and Thor each have specialized functions as storm gods and different kinds of power. Odin is the divine sovereign, while Thor is the god of tempests and combat, whose atmospheric battles with his famous hammer result in thunder, lightning, and beneficial rain.

Together, thunder and lightning have been understood by many peoples (e.g., the ancient Babylonians, Hebrews, Hindus, Germans, Greeks, Romans, Native Americans) as the weapon of the gods, which can be used either to warn or to punish. As such, wherever lightning strikes, that place often becomes a sacred spot. Tribal peoples, such as the Buriat of Siberia, may believe that a person struck by lightning has been chosen as a shaman. Lightning is thus often portrayed on the ritual clothing of a Siberian shaman, and the wooden parts of the all-important shaman's drum are in some areas chosen from a tree that has been struck by lightning.

Wind. The intangibility, yet power, of wind has suggested the presence of divine or semidivine beings to many peoples. The Israelites believed that Yahweh began creation with a wind moving across the waters (Gn. 1.2), and the ancient Celtic belief that wind is the special province of fairies is still held by some in Ireland today: When the wind is gentle, fairies are passing by in it; when fierce, it is their weapon. Most peoples have imagined more than one wind living together, contained in a cave (northwestern Native Americans) or a bag (ancient Greeks,

Native Americans), or one wind residing in each of the four cardinal directions. Each wind has its own distinct personality and stratagems for getting free to blow as it will.

The culture of ancient Greece provides many examples of the power of the winds. The Athenians established a state cult for the north wind, Boreas, in gratitude for his destruction of a large part of an attacking Persian fleet. Homer tells us that Odysseus was given a magical bag containing the winds by the wind god Aeolus to guide him and his ship safely home, but while he slept his men undid the bag and their ship was tempest-tossed. Throughout the Aegean world it was commonly believed that the winds could be magically bound by three knots, as in a handkerchief. These were sold to sailors who untied them when they needed a wind to fill their sails.

Whirlwinds are considered dangerous or even evil by many people, especially Native Americans. Some believe they transport the spirits of the dead and/or witches; therefore to be caught in one, or even to dream of one, predicts sickness and death. The power of a whirlwind can, however, sometimes be benevolent; in the biblical book of Exodus (13.21–22, 33.9–10), Yahweh appears as a pillar of cloud, possibly a whirlwind, to guide Moses and the children of Israel on their journey to the Promised Land.

Perhaps the most dramatic blending of meteorology, divinity, and humanity centers on the terrifying and destructive typhoons (great wind) of the China Seas. Traditional Japanese believe that the *kami* ("divinities"), especially the spirits of past heroes, protect Japan (as they did in 1281 in the form of a typhoon that drove back the huge Mongol fleet attempting to conquer Japan). They referred to this as a *kamikaze* ("divine wind"). During World War II Japanese pilots transformed themselves into *kami* by having funeral ceremonies before they flew off on suicide missions. They, too, were known as *kamikaze*, the wind of the gods.

Cyclones, tornadoes, hurricanes, and other winds of destructive force have been represented as mythical monsters around the world. The Seneca Indians, for example, believed hurricanes were the activities of a monstrous bear, while the English word *hurricane* may come from the name of Hurucán, the terrible wind god of the Quiché Indians of Mesoamerica.

Seasonal Ceremonies. Terrestrial ideas about gender have universally been projected on the natural phenomena of the seasons, which have everywhere been personified, for example, as Old Man Winter, or as Primavera and numerous other goddesses of spring. This is a survival of ancient religious ideas going back to Mesopotamia and ancient Egypt and possibly even earlier cultures. There are, in fact, two related ideas. First, this is the celebration of a sacred marriage between a goddess and god at the New Year, traditionally when crops are sown, and the ritual enactment of the divine marriage to assure the fertility of the fields. This means that rain must be plentiful in the coming year, but not too much, that sunshine must be equally moderate, and that storms must be mild. Seasonal rituals are an attempt to harness cosmic energies in order to get what is needed from them without being overwhelmed by them. This idea of a sacred marriage between goddess and god is enacted by their human devotees through a general orgy that takes place in the freshly planted fields. Human sexual activity is believed to stimulate the divine marriage so that the Earth (female) is productively related to the sky (male) in the form of rain (semen). Such rituals were performed by rural people in pagan Europe (and even after the spread of Christianity), in ancient Greece and Rome, and among innumerable tribal peoples.

The second idea is that of a ritual combat between summer and winter, or more to the point, between rainfall and drought, enacted by people all over the world. It is perhaps most familiar to Westerners in the English Mummers' Play, or Saint George Play, a survival of an ancient drama in which characters slain in combat are afterward resuscitated. Although the characters and action of the play vary in different localities, the central incident is doubtless connected with the celebration of the death of the year and its resurrection in the spring.

Theodor Gaster (1950) has formulated four stages to seasonal rites that take place over several days: those of mortification, purgation, invigoration, and jubilation. Both the sacred marriage and the ritual combat are rites of the third stage, invigoration. Preceding this stage are rites of mortification and purgation, which include periods of fasting, such as the Christian Lent or the Islamic Ramadan, and the removal of communal evil through a scapegoat, bonfire, or the modern ritual of noisemaking on New Year's Eve. After the sacred marriage or ritual combat, the final rites of jubilation center on a communal feast to which the ancestral dead and the gods are invited. This feast represents the plenty that all three—the living, the dead, and the divine—share in; but there is only plenty if all three do their parts, and the purpose of these rites is to remind all the participants of their proper roles.

Conclusion. The religions of the world are rich in meteorological divinities; religious experts, whether priests or shamans, are expected to exercise control over the weather by propitiating deities, mastering the elements through their spiritual powers, or by at least being able to foretell the weather. In the public's attendance on television and radio weathermen, it is not so very far removed from people who have placed their faith in religious experts; repeated failures of accurate forecasting do nothing to stay the belief in their powers.

[*See also* Drama, Dance, and Weather; Literature and Weather; Music and Weather; *and* Weather Lore.]

BIBLIOGRAPHY

Dumézil, G. *Gods of the Ancient Northmen* (1959). Berkeley and Los Angeles, 1977.

Eliade, M. *Patterns in Comparative Religion* (1958). Translated by Rosemary Sheed. New York, 1974.

Eliade, M., et al., eds. *The Encyclopedia of Religion*. New York, 1987. Includes numerous articles that explain relationships of religion, climate, and weather; see especially "Ecology" by Åke Hultkrantz, "Flood, The" by Jean Rudhardt, "Meteorological Beings" by Peter C. Chemery, "Rain" by Ann Dunigan, and "Seasonal Ceremonies" by Theodor H. Gaster.

Frankfort, H., et al. *Before Philosophy: The Intellectual Adventure of Ancient Man* (1946). Baltimore, 1949.

Frazer, J. G. *The Golden Bough.* Abr. ed. (1922). New York, 1963.

Gaster, T. H. *Thespis: Ritual, Myth, and Drama in the Ancient Near East* (1950). New York, 1977.

James, E. O. *The Worship of the Sky God.* London, 1963.

Leach, M., and J. Fried, eds. *Funk and Wagnalls Standard Dictionary of Folklore, Mythology, and Legend* (1949). 1-vol. ed. San Francisco, 1984.

Schafer, E. *The Divine Woman: Dragon Ladies and Rain Maidens* (1973). San Francisco, 1980.

Schama, S. "Moral Geography." In *The Embarrassment of Riches: An Interpretation of Dutch Culture in the Golden Age*, by Simon Schama, chap. 1. Berkeley and Los Angeles, 1988.

SERINITY YOUNG

REPLICABILITY. *See* Scientific Methods.

RICHARDSON, LEWIS FRY (1881–1953), British mathematician and meteorologist. The visionary scientist Lewis Fry Richardson tried in the 1920s to forecast weather by using numerical techniques to solve equations. He fell short of his goal because computers had not yet been invented; as a result, no one studied the techniques that he was pioneering. Born in 1881, Richardson was an imaginative thinker who won recognition for the theories he promoted in a number of disparate fields, including mathematics, meteorology, peace research, and psychology.

A committed Quaker, Richardson joined the Friends Ambulance Unit in World War I and spent two years in France. He often attempted to converse in Esperanto with captured German prisoners; the other ambulance drivers suspected that he was trying to get hold of German gas shells so that he could analyze their contents and neutralize them.

After working as a chemist in industry and experimenting with echo-sounding (following the sinking of the *Titanic* in 1912), Richardson became superintendent of Eskdalemuir Observatory in Scotland. He began to develop his ideas about numerical weather prediction by applying differential equations and using weather maps to track depressions and weather systems. This led him to the study of atmospheric turbulence; he tried to measure upper winds by shooting small spheres the size of peas and cherries vertically into the air.

Richardson's seminal study, *Weather Prediction by Numerical Process*, was first published in 1922 and was reissued in 1965. He is probably best known as the originator of the *Richardson number*, a mathematical criterion for predicting whether atmospheric turbulence is likely to increase or decrease. He was also a scientific dreamer who proposed a futuristic "forecast factory."

Richardson's nephew, the renowned actor Sir Ralph Richardson, was not the only dramatist in the family. Long before the advent of electronic computers, Lewis Richardson had envisioned a room full of human calculators who would solve equations in record time in a breathtaking attempt to predict the world's weather. In a large theatrical hall, he pictured walls painted to form a map of the globe, with England in the gallery, Australia on the dress circle, and the Antarctic in the orchestra pit. He wrote:

> A myriad computers [human calculators] are at work upon the weather of the part of the map where each sits. . . . Numerous little "night signs" display the instantaneous values so that neighbouring computers can read them. . . . Four senior clerks in the central pulpit are collecting the future weather as fast as it is being computed, and dispatching it by pneumatic carrier to a quiet room. There it will be coded and telephoned to the radio transmitting station. . . . In a basement an enthusiast is observing eddies in the liquid lining of a huge spinning bowl, but so far the arithmetic proves the better way. . . . Outside are playing fields, houses, mountains and lakes, for it was thought that those who compute the weather should breathe of it freely. (*Weather Prediction by Numerical Process*, 1922, pp. 219–220)

Richardson was pursuing his numerical methods of weather prediction at the same time that Vilhelm Bjerknes was formulating his less quantitative air-mass analyses. Unfortuately, Richardson's attempts to forecast the weather by a much scaled-down version of his forecast factory failed because the mathematical techniques he used to fashion approximate solutions to complex differential equations will work only under carefully chosen conditions, of which he was unaware. Decades later, mathematicians and scientists would create a new discipline known as *numerical analysis*, using Richardson's techniques in conjunction with sophisticated computers. That is how the large-scale, long-term weather forecasts are made today, and Richardson's techniques underlie projections of climate change in the coming century.

[*See also* Numerical Methods; Numerical Weather Prediction; *and* Turbulence.]

BIBLIOGRAPHY

Ashford, O. M. *Prophet or Professor? The Life and Work of Lewis Fry Richardson.* Bristol and Boston, 1985.
Schneider, S. H., and R. Londer. *The Coevolution of Climate and Life.* San Francisco, 1984.

DIANE MANUEL

ROSSBY, CARL-GUSTAF (1898–1957), Swedish meteorologist. Carl-Gustaf Rossby is identified with several concepts that bear his name: the *Rossby diagram*, used to identify air masses; the *Rossby waves*, huge horizontal waves found in the eastward drift of mid-latitude, high-altitude air around the Earth; and the *Rossby equation*, which determines the speed of those waves.

Born in Stockholm in 1898, Rossby developed an early interest in meteorology at the Geophysical Institute at Bergen, Norway, under the tutelage of Vilhelm Bjerknes. In later years Rossby would further develop Bjerknes's ideas about using air masses for weather forecasting. [*See the biography of the Bjerkneses.*]

After overseeing several expeditions for the Swedish Meteorological and Hydrological Service, Rossby was awarded a year-long fellowship at the U.S. Weather Bureau. This marked the beginning of a 20-year stay in the United States, during which he became an American citizen, established a weather service for civil aviation in California, taught at the Massachusetts Institute of Technology and the University of Chicago, tried to set up a worldwide forecasting service, and reorganized the American Meteorological Society.

In the 1930s, while he was at MIT, Rossby began to theorize that by studying air pressure, temperature, and other variables from a high altitude, meteorologists could obtain new information about the atmosphere's large-scale circulation. He hired a small airplane to make daily weather observations from a Boston airport. He then began to see an emerging pattern of huge horizontal waves. These Rossby waves, as they have become known, were elements of circumpolar vortices. [See Rossby Waves.]

Using an equation for the fluid flow of air on the rotating Earth, Rossby was able to predict conditions for the onset of waviness in the circumpolar winds, thus partially explaining the shifting of the atmospheric waves. Because large-scale weather phenomena depend on this shifting, his equation made it theoretically possible to forecast evolving large-scale weather patterns by starting with the existing flow pattern and applying certain dynamical equations.

Rossby's achievements were considered a breakthrough for the science of meteorology. He also studied ocean currents and developed the basic concepts of the jet stream, which he is credited with naming. With the advent of electronic computers, Rossby's calculations and formulas could be applied to the development of sophisticated forecasting techniques.

[See also Jet Stream; and Numerical Weather Prediction.]

BIBLIOGRAPHY

Gillispie, C. C., ed. Dictionary of Scientific Biography. Vol. 11. New York, 1975.
Schneider, S. H., and R. Londer. The Coevolution of Climate and Life. San Francisco, 1984.

DIANE MANUEL

ROSSBY WAVES. *[This article treats a technical aspect of climate and weather studies; it is intended for readers at an advanced level.]*

The Rossby wave is a mathematical paradigm for understanding atmospheric disturbances of planetary scale. Generally, flow patterns of larger spatial scales tend to be associated with slower motions; thus Rossby waves are characterized by long periods. They are named after Carl-Gustaf Rossby, who, during the 1930s and 1940s, described them mathematically and showed their relevance for understanding the kinematics of disturbances appearing on weather charts of the upper troposphere. [See the biography of Rossby.]

All types of oscillatory motions that develop around a stable equilibrium state, represent the response to a restoring force produced by a departure from that equilibrium. The atmosphere can support a wide array of wave phenomena, including sound waves, gravity or buoyancy waves, and Rossby waves. Sound waves have periods on the order of minutes and arise owing to the compressibility of the gaseous atmosphere. Gravity waves are an oscillatory response to a perturbation acted on by gravitational forces and have periods of minutes to hours. By contrast, Rossby waves exist because of the rotational effects induced by the spinning planet on atmospheric motions; their periods are measured in days. Because meteorological phenomena unfold on time scales similar to those of Rossby waves, the latter have become widely used in interpreting the results of both short-range and long-range weather forecasts.

Planetary waves were first studied in the 1890s by Max Margules and Sydney Samuel Hough independently as solutions to Laplace's tidal equations. In the late eighteenth century, Laplace developed a version of the shallow-water equations for his studies of the Earth's oceanic tides. One hundred twenty years later, Margules and Hough showed that the tidal equations possessed two classes of solutions—one corresponding to high-frequency gravity waves, and a second containing low-frequency planetary waves (now called *Rossby waves*). As such, the solutions describe the possible global, free oscillations of the Earth's atmosphere. However, Rossby was interested in deciphering the nature of the wavy perturbations encircling the globe at middle and upper tropospheric levels that control the progression of synoptic weather systems in middle latitudes. (The *troposphere* is Earth's lowest atmospheric layer, from the surface to about 10 kilometers.) Rossby's insight was to realize the meteorological importance of a mathematical theorem, originated by German physicist Hermann Helmholtz (1821–1894) and used in 1913 by Augustus Edward Hough Love in the solution of the shallow-water equations. The procedure effectively decomposes the flow into divergent and rotational components. In the limit of slow planetary rotation, these components correspond, respectively, to solutions of the first and second classes described above. Also, in the same limit, the shallow-water equations simplify to the barotropic vorticity equation. The latter formed the basis of Rossby's work on the rotational modes of motion characterizing the weather. [See Vorticity.]

Mathematical Description. It is a characteristic of observations that atmospheric flows on the terrestrial scale are in an approximate geostrophic equilibrium. In addition, they are almost divergence-free. Rossby waves are viewed as perturbations on this balanced state. Rigorously, the mathematical development would use the primitive equations (which represent an atmosphere in hydrostatic balance) in spherical coordinates on a rotating Earth. However, for the sake of simplicity and to emphasize the dynamics of Rossby waves in contrast to other oscillatory motions, the simplest prototype model capable of generating Rossby waves will be derived. [*See* Dynamics; Primitive Equations.]

Consider an idealized atmosphere that is homogeneous, isothermal, incompressible, inviscid, and initially at rest. The horizontal momentum equations (taken from the set of primitive equations)—in which the spherical coordinates of longitude and latitude have been replaced, respectively, by eastward and northward cartesian coordinates (x, y)—are:

$$\frac{\partial u}{\partial t} + u\frac{\partial u}{\partial x} + v\frac{\partial u}{\partial y} + w\frac{\partial u}{\partial z} - fv + \frac{\partial \Phi}{\partial x} = 0$$

$$\frac{\partial v}{\partial t} + u\frac{\partial v}{\partial x} + v\frac{\partial v}{\partial y} + w\frac{\partial v}{\partial z} + fu + \frac{\partial \Phi}{\partial y} = 0 \tag{1}$$

where z is the vertical coordinate, t indicates time, (u, v, w) are the eastward, northward, and vertical velocity components respectively, f is the Coriolis parameter ($= 2 \cdot \Omega \cdot \sin \varphi = 2 \cdot$ *Earth's rotation rate* $\cdot$ *sine of latitude*, a measure of planetary vorticity), and Φ is the geopotential. The focus will be on linear Rossby waves; therefore, all terms that are quadratic in the velocity fields are neglected from Equations 1 to obtain:

$$\frac{\partial u}{\partial t} - fv + \frac{\partial \Phi}{\partial x} = 0$$

$$\frac{\partial v}{\partial t} + fu + \frac{\partial \Phi}{\partial y} = 0. \tag{2}$$

In the steady-state regime the fields are in geostrophic equilibrium. This results in the following approximation to the planetary atmospheric flow:

$$fv_g = \frac{\partial \Phi}{\partial x}$$

$$fu_g = -\frac{\partial \Phi}{\partial y}. \tag{3}$$

Small, quasi-geostrophic departures from the equilibrium described by Equations 3 are supported by Equations 2. The relationship among the various wind variables is

$$u \equiv u_g + u'$$

$$v \equiv v_g + v' \tag{4}$$

with $|u'| < |u_g|$, $|v'| < |v_g|$ and primed quantities representing the departures from equilibrium. In addition, if

the horizontal excursions of air parcels are small compared to the radius of the Earth, the Coriolis parameter can be reduced to

$$f = 2\Omega \sin \varphi \approx f_o + \beta y \tag{5}$$

which is a linear approximation (called the *β-plane approximation*) to the $\sin \varphi$ function centered around latitude φ_0. The different parameters are $\Omega = 7.292 \times 10^{-5}$ per second, $f_0 = 2\Omega \sin \varphi_0$, $\beta = 2\Omega/a \cos \varphi_0$ and $a = 6.37 \times 10^6$ meters, which is the Earth's radius. Substituting Equations 4 and 5 into Equations 2 gives the following set of perturbative equations:

$$\frac{\partial u'}{\partial t} - f_0 v' - \beta y v' = 0$$

$$\frac{\partial v'}{\partial t} + f_0 u' + \beta y u' = 0. \tag{6}$$

For large-scale winds that are quasi-nondivergent (that is, $\partial u/\partial x + \partial v/\partial y \approx 0$), it becomes plausible to reduce the set of Equations 2 to only one equation governing the vorticity of the flow. It is observed that the latter captures the dominant modes of wave activity in extratropical latitudes. A simple operation on Equations 6 yields

$$\frac{\partial\left(\frac{\partial v'}{\partial x} - \frac{\partial u'}{\partial y}\right)}{\partial t} + f_0\left(\frac{\partial u'}{\partial x} + \frac{\partial v'}{\partial y}\right)$$
$$+ \beta y\left(\frac{\partial u'}{\partial x} + \frac{\partial v'}{\partial y}\right) + \beta v' = 0 \tag{7}$$

where the relative vorticity of a flow is defined as $\partial v/\partial x - \partial u/\partial y$. Since the perturbative velocities are almost divergence free, that is,

$$\frac{\partial u'}{\partial x} + \frac{\partial v'}{\partial y} \approx 0 \tag{8}$$

Equation 7 reduces to

$$\frac{\partial\left(\frac{\partial v'}{\partial x} - \frac{\partial u'}{\partial y}\right)}{\partial t} + \beta v' \approx 0. \tag{9}$$

Equation 9 is called the *unforced, quasi-geostrophic, barotropic vorticity equation*. The simplest solution to this differential equation corresponds to a (plane) wave—the Rossby wave. It is a consequence of the latitudinal variation of the Coriolis parameter. Because the motion is horizontal and the idealized atmosphere is incompressible, the wave is transverse: air parcels will oscillate in a direction perpendicular to the movement of the wave.

Now, assume that the motion of the wave is along a parallel of latitude (zonal displacement), while the induced oscillations (represented by the primed variables) occur along the meridians of longitude. Mathematically, this is described by

$$u' = 0$$
$$v' = v'_0 \cos(kx - \omega t). \tag{10}$$

Incorporating Equations 10 into Equation 9 yields a dispersion relation relating ω ($= 2\pi \cdot frequency$) to the wavenumber k ($= 2\pi/wavelength = 2\pi/\lambda$):

$$\omega = -\frac{\beta}{k}. \tag{11}$$

From Equation 11, the zonal phase velocity c_p (defined as $c_p \equiv \omega/k$) of this wave, which is the speed at which its crests or troughs are moving, can be derived:

$$c_p \equiv \frac{\omega}{k} = -\frac{\beta}{k^2}. \tag{12}$$

The dependence on k indicates that Rossby waves are dispersive. This means that if a packet of many waves, each having a different wavelength, is moving in some direction, it will spread out in time as each wave propagates with a different speed. The minus sign indicates that Rossby waves more westward. Rossby's original *trough formula* ($c_p = U - \beta/k^2$) was not obtained for an atmosphere at rest (with $U \equiv 0$ meters per second), but rather for waves superimposed on an eastward current (for which $U > 0$). In this case, it is possible that a wave of a given wavenumber k moving westward with a velocity β/k^2 that exactly cancels that of the basic eastward flow U, appears stationary ($c_p = 0$) with respect to an observer on Earth. All other waves (whether they are superimposed on a background current flowing eastward or westward) are denoted as *traveling waves* because of the propagating character of their troughs (with respect to a fixed observer on Earth). It is a simple matter to calculate which component of the wave spectrum will be stationary. For this purpose, Rossby's original trough formula is rewritten using the fact that the wavelength of a zonal wave along the latitude circle φ_0 is given by $\lambda = (2\pi a \cos \varphi_0)/n$. Here, wave n corresponds to an alternate succession of n highs (crests) and n lows (troughs) along the parallel of latitude. The stationary character ($c_p = 0$) of the wave implies that

$$U = \frac{\beta}{k^2} = \frac{\beta\lambda^2}{(2\pi)^2} = \frac{\beta a^2 \cos^2 \varphi_0}{n^2} \tag{13}$$

or

$$n = \left(\frac{2\Omega a}{U} \cos^3 \varphi_0\right)^{1/2} \tag{14}$$

For winds typical of the upper troposphere over middle latitudes, U is on the order of 25 meters per second. Hence, waves 3 or 4 would appear to be stationary. Longer waves (wavenumbers 1 or 2) would travel slowly westward, while shorter waves (wavenumbers > 4) would be advected eastward by the zonal flow. For an observer on Earth, the short waves appears to propagate eastward but slower than the background current. In temperate latitudes, these short waves are conspicuous features of daily weather charts of the middle and upper troposphere.

Clearly, more than one wave component can be present simultaneously. This situation is described as a *wave packet*. A simple mathematical representation for such an occurrence is given by a plane wave moving at speed c_p and characterized by a modulated envelope which itself is moving at the group velocity c_g. In this case, Equations 10 is replaced by

$$u' = 0$$
$$v' = v'_0(x - c_g t) \cos k(x - c_p t) \tag{15}$$

where v' is now ascribed the modulating amplitude $v'_0 (x - c_g t)$ which is a function of space and time. The character of the envelope is such that the phase of (the argument of the function) $v'_0 (x - c_g t)$ varies slowly compared to that of the plane wave. An expression for the speed c_g of longitudinal displacement of $v'_0 (x - c_g t)$ is obtained by substituting Equations 15 into Equation 9:

$$c_g = -\frac{\omega}{k} = \frac{\beta}{k^2}. \tag{16}$$

Since waves with modulated amplitude carry information (similar to those emitted by AM radio stations), the group velocity of the wave packet describes the propagation of its energy. In the case explored here, an atmosphere initially at rest, the group velocity is equal in magnitude but opposite in direction to the zonal phase velocity and therefore is toward the east.

When the β-plane approximation and the requirement that motions be parallel to latitude circles are relaxed, it becomes necessary to consider the field of Rossby waves over the entire horizontal extent of the atmosphere. Rossby waves can propagate in all directions except along the meridians. If a wave is generated with its crests and troughs oriented in an east–west direction, the restoring force on these (owing to the latitudinal variation of planetary vorticity) disappears, and the wave cannot move.

At this point, it becomes useful to draw a parallel between the propagation of atmospheric planetary waves and that of light rays or waves. In geometrical optics, the concept of the index of refraction is used to infer the path followed by rays of light. A ball thrown on a surface characterized by a topographic relief of mountains and valleys rolls under the influence of gravity, avoiding mountains and curving toward valleys. The light ray propagates in the same way, curving away from low values (mountains) toward high values (valleys) of the refractive index. In the case of planetary Rossby waves, the index of refraction consists of a combination of terms: the latitudinal gradient of planetary vorticity (similar to the β term defined previously), and the various directional gradients of the different shears of the background flow upon which the waves are superimposed. For perturbations of any zonal wavenumber, the atmosphere tends on average to exhibit low values of the index toward the polar regions, with high values toward the subtropical regions. Therefore, a wave excited somewhere in the extratropics will refract away from the polar latitudes and propagate in the direction of the tropics.

Physical Description. The conspicuous westward propagation of a Rossby wave, with respect to the background flow on which it is superimposed, can easily be understood in physical terms. Consider a basic state consisting of an atmosphere in purely zonal motion, with the magnitude of the flow decreasing poleward; an atmosphere at rest or one characterized by constant zonal motion could have been used just as well. The vorticity field describing the initial atmospheric state is composed of planetary and relative vorticity parts. The former corresponds to the Coriolis term $f = 2 \Omega \sin \varphi$, while the latter is equal to $-du(y)/dy$. For the situation at hand, the magnitude of both terms increases poleward. Hence, the vorticity field of the undisturbed initial state has contours parallel to latitude circles with contour values increasing poleward. Now, suppose that in a zonal band between two parallels of latitude, the contours are slightly deformed into a wavy pattern (Figure 1). These waves representing the total flow, can be viewed as the superposition on the initial vorticity field of a longitudinal sequence of vorticity anomalies of alternating sign. Since the magnitude of the vorticity contours increases poleward, positive (or cyclonic) anomalies correspond to the troughs, while negative (or anticyclonic) ones are associated with the crests of the disturbance. In the Northern

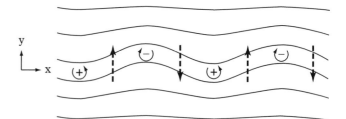

ROSSBY WAVES. Figure 1. *Depiction of an Idealized Rossby wave.* For simplicity, the wave is created parallel to latitude circles. The x and y arrows point eastward and northward respectively. Thin solid lines represent contours of total vorticity (the sum of planetary vorticity, relative vorticity, and vorticity anomalies). The + and − signs indicate the centers of the vorticity anomalies, while the curved arrows depict the anomalous motions associated with the anomalies. The heavy, dashed arrows indicate the direction of the velocity field, induced by the vorticity anomalies, that causes the westward migration of the wave pattern.

Hemisphere, a cyclonic vortex describes counterclockwise motion, while an anticyclonic one displays clockwise circulation. (The reverse is true in the Southern Hemisphere.) Hence, the particular spatial arrangement of superposed vorticity anomalies will generate, in between the troughs and crests, a field of wind anomalies oriented in a north–south direction. Northward components of velocity will be located to the west of the crests, and southward components to their east. The induced anomalous wind field dictates the direction in which the wave pattern migrates. Westward of crests the wind will displace the vorticity contours northward, while displacing them southward to the west of troughs, inducing the wave pattern to propagate westward relative to the background flow.

Because pure Rossby waves are characterized by nondivergent motions, they tend to occur above the planetary boundary layer corresponding to the lowest kilometer of the atmosphere. Their appearance in the free atmosphere is due to a variety of generating mechanisms. One possible source is the phenomenon of *flow instability*. Random departures from geostrophic equilibrium, resulting in freely traveling Rossby waves, can be created through baroclinic instability (converting available potential energy of the basic zonal current into wave kinetic energy) or through barotropic instability (converting kinetic energy associated with horizontal shears in the velocity of the zonal flow into wave kinetic energy).

Additionally, departures from geostrophic equilibrium can be directly forced. Flow over bottom relief, or a distribution of thermal sources embedded in a current, will under certain conditions produce Rossby waves. For instance, the temperature anomalies that appear at the surface of the tropical Pacific Ocean during El Niño events are sometimes strong enough to set up Rossby wave trains that carry the influence of the tropical phenomenon to the southern latitudes of the North American continent.

The prototype model developed above describes linear dynamics. Eventually, the disturbances evolve and attain large amplitudes. At this point, nonlinear effects, resulting in coupling among wave components, transfer energy among various wavenumbers. This is also seen as a way to induce Rossby waves of large scale. Part of the group of waves of planetary scale can be maintained in this manner by a spectrum of smaller-scale disturbances. In turn, the broad-scale flow pattern of the planetary-scale waves, which is likely to contain stationary components, steers the transient cyclone eddies composing the group of small-scale disturbances.

But do these mathematical entities really represent the physical undulations displayed on daily and temporally averaged hemispheric weather charts of the middle and upper troposphere? The many possible generating mechanisms are continuously operating in the troposphere. They generate a mixture of Rossby waves with various temporal and spatial scales. Their simultaneous presence has impeded detailed comparison of observed tropospheric planetary waves with their theoretical counterparts. Interestingly, even though Rossby's original intent was to apply the trough formula to diagnostic analyses of mean charts, in practice it is more useful for daily forecasting charts.

The situation is somewhat different in the stratosphere. Assuming that tropospheric sources are responsible for the excitation of stratospheric planetary waves, theoretical studies have shown that during upward propagation, all but the largest-scale waves are filtered out before they reach the tropopause (the transition layer between the troposphere and stratosphere). In fact, observed stratospheric disturbances exhibit simpler structural patterns, denoting the presence of fewer wave components in the flow. Current theories also indicate that upward-propagating planetary waves play a crucial role in the dynamics of the stratospheric phenomenon known as *sudden warming*.

From a global perspective, Rossby waves participate in the momentum and heat balances that maintain the Earth's planetary wind system. In their large-amplitude stage the waves tend to acquire, on a horizontal plane, a southwest-to-northeast tilt. This is characteristic of waves propagating energy southward. However, the transport of angular momentum is poleward, because air parcels moving northward through the tilted waves are associated with a greater eastward wind component than are southward-moving parcels. In the vertical plane, the phase lines of baroclinic Rossby waves tilt westward with height. This implies an upward flux of energy together with a northward eddy heat transport. This heat flux, however, drives secondary circulations in the latitudinal-vertical plane that cancel the transport of momentum by the waves. This cancellation by the induced secondary circulations can occur only in an inviscid and adiabatic environment—and when the stationary and traveling Rossby waves have a temporally-fixed amplitude. In all other situations, the planetary waves interact with and force the global atmospheric wind system.

[*See also* Stratosphere.]

BIBLIOGRAPHY

Dickinson, R. E. "Rossby Waves: Long-Period Oscillations of Oceans and Atmospheres." *Annual Review of Fluid Mechanics* 10 (1978): 159–195.

Gill, A. E. *Atmosphere-Ocean Dynamics.* New York, 1982.

Holton, J. R. *An Introduction to Dynamic Meteorology.* 3rd ed. San Diego, Calif., 1992.

Palmén, E., and C. W. Newton. *Atmospheric Circulation Systems: Their Structure and Physical Interpretation.* New York, 1969.

Pedlosky, J. *Geophysical Fluid Dynamics.* 2nd ed. New York, 1987.

Platzman, G. W. "The Rossby Wave." *Quarterly Journal of the Royal Meteorological Society* 94 (1968): 225–248.

LIONEL PANDOLFO

RUNOFF. The term *runoff* denotes water that originates as precipitation and ultimately reaches stream channels or joins surface water bodies. In the context of the hydrological cycle, *runoff* refers to the movement of water between the time that precipitation reaches the ground and the time of the water's return to the atmosphere through evapotranspiration or its eventual discharge into the oceans or landlocked lakes through stream-channel flow. The term *runoff* can refer specifically to water that flows over the land surface directly into stream channels without ever infiltrating the ground; it can also have a more general use, as *total runoff,* to describe all water that eventually becomes streamflow, even though some of it may move in pathways below the surface as delayed subsurface runoff or as groundwater runoff before it discharges into streams. [*See* Ground Water; Hydrological Cycle.]

When precipitation is delivered to the ground at a rate faster than it can infiltrate the soil surface, runoff moves directly over the land surface toward stream channels in a process known as *Hortonian overland flow,* named after R. E. Horton, who first described the process. Runoff is also generated at the surface as *saturation overland flow,* or *saturation excess runoff,* when infiltration is impeded because the soil is saturated and cannot absorb any more water. Areas contributing saturation overland flow grow larger during storms as infiltration raises shallow water tables to the surface. *Subsurface stormflow,* also referred to as *throughflow,* occurs during storm events when water infiltrates the soil and very quickly moves laterally through the soil via interconnected large pores or by flow in a temporarily saturated zone, soon reaching stream channels. These direct runoff processes combine with the delayed contribution from deeper in the soil and from groundwater to yield the total runoff that appears in streams and rivers. The degree to which runoff is generated by these different processes varies geographically and seasonally owing to the nature of the land surface, the subsurface, and climatic factors.

Land surface and subsurface factors that influence runoff include the type and thickness of vegetation cover, the porosity and permeability of the soil and geologic substrata, the type of land use, and the topography. The type of vegetation cover found in different climatic zones influences the amount and timing of runoff. During a given precipitation event, a portion of the falling rain or snow is intercepted by vegetation and can be evaporated and returned to the atmosphere immediately, thereby reducing the total amount of possible runoff. This process is most effective during short-lived, low-intensity rain showers and in forested regions. Vegetation cover also has great influence on the time it takes for runoff to concentrate in stream channels. Trees, shrubs, and grasses increase infiltration to varying degrees by intercepting precipitation and allowing it to be funneled along foliage, stems, and trunks directly into the soil. Extensive root networks also augment infiltration by enhancing the permeability of the soil. The influence of vegetation cover is apparent when we compare runoff generation in arid and humid regions. An intense rainstorm tends to produce surface runoff almost immediately in an arid region with sparse vegetation cover. A similar rainstorm in a well-vegetated humid region produces less direct runoff and instead delivers a more sustained flow to stream channels over time, owing to delayed runoff from subsurface contributions. [*See* Vegetation and Climate.]

The properties of porosity and permeability in an unsaturated soil are important factors for the generation of runoff. They determine the soil's capacity to absorb the water that infiltrates the surface as well as the amount of percolation through the soil to the groundwater table at deeper levels. As water percolates to progressively greater depths in the subsurface, the time until it reemerges as stream channel discharge increases. It is this contribution of delayed runoff from the subsurface, also called *baseflow,* that sustains water levels in stream channels during long rainless periods and droughts. Land use can affect runoff by altering the permeability of the ground surface—for example, by urbanization or agricultural activity—or by concentrating runoff into channelized flow in artificial drainage networks.

The natural shape of the land surface and topography are also important in the runoff process. A portion of precipitation collects temporarily in depressions on the ground surface, especially in flat or gently sloping terrain. This can slow the runoff process or reduce the total amount of runoff if evaporation occurs from the surface of the stored water. In contrast, steeply sloping topography enhances runoff. Other topographic factors that influence the amount and type of runoff are slope length and the concavity or convexity of the slope.

Climatic factors that affect runoff include the magnitude and intensity of rainfall, the degree of saturation of the soil from previous precipitation events, and, in seasonally cold climates, the amount of snow accumulation, rapidity of snow melt, and depth of frozen soil. Extreme rainfall, such as might occur during a heavy thunderstorm, can be so intense that the precipitation is delivered to the ground surface faster than it can infiltrate, thereby generating Hortonian overland flow even where soils are permeable or unsaturated. This occurrence is common in arid and semiarid regions. The duration of a rainstorm and the amount of antecedent precipitation determines the degree of soil saturation. In humid regions, characterized by frequent rainstorms of long duration, saturation overland flow and subsurface storm flow are the dominant processes. In cold climates, an entire season's snow accumulation can be released gradually as runoff during the snowmelt season, or rapidly during an abrupt thaw. At such times, the existence of frozen ground, which prevents infiltration, is another determinant of the rate at which runoff occurs.

In addition to runoff processes observed at the basin or catchment scale, runoff is also analyzed at larger scales and modeled in the context of energy and moisture balances in global or regional climate models. In water-balance climatology, runoff can be derived from the surplus component of a water balance. The surplus is the amount of water that remains when precipitation exceeds evapotranspiration. In general circulation models, the runoff portion of the hydrologic cycle is often modeled by considering the soil as a bucket that fills when precipitation exceeds evaporation. When the bucket is completely filled, excess water becomes runoff. In models with more detailed parameterization, the runoff process is portrayed in a layered soil scheme that more accurately simulates the pathways of water at the catchment scale.

The annual global runoff to the oceans from streams and rivers is estimated to be about 43,500 cubic kilometers per year. An additional amount is contributed to the oceans directly by underground water and glacial runoff to yield a total annual global runoff estimate of about 47,000 cubic kilometers per year. In volume, river flows represent less than 0.0002 percent of the total water reserves on Earth. The annual volume of flow in the Mississippi River Basin is about 580 cubic kilometers; the flow volume in the largest river on Earth, the Amazon, is about 6,900 cubic kilometers annually. This contrasts with an annual flow volume of greater than 900,000 cubic kilometers in a major ocean current, such as the Gulf Stream, and the estimated movement of water back into the atmosphere as evaporation from both land and ocean,

at a flow rate of 577,000 cubic kilometers per year. Although small in comparison to the amount of water in other phases of the hydrological cycle, runoff is of critical importance to humanity because of its role in water supply, irrigation, industry, power generation, and navigation.

BIBLIOGRAPHY

Dunne, T., and L. B. Leopold. *Water in Environmental Planning.* San Francisco, 1978.

Famiglietti, J. S., and E. F. Wood. "Evapotranspiration and Runoff from Large Land Areas: Land Surface Hydrology for Atmospheric General Circulation Models." In *Land Surface–Atmosphere Interactions for Climate Modeling,* edited by E. F. Wood. Dordrecht, 1991, pp. 179–204.

Horton, R. E. "The Role of Infiltration in the Hydrological Cycle." *Transactions of the American Geophysical Union* 14 (1933): 446–460.

Korzun, V. I., ed. UNESCO. *World Water Balance and Water Resources of the Earth.* UNESCO Paris, 1978.

L'vovich, M. I., and G. F. White. "Use and Transformation of Terrestrial Water Systems." In *The Earth as Transformed by Human Action,* edited by B. L. Turner II et al. Cambridge, 1990, pp. 235–252.

Maidment, D. R. *Handbook of Hydrology.* New York, 1993.

Shiklomanou, I. "World Fresh Water Resources." In *Water in Crisis: A Guide to the World's Fresh Water Resources,* edited by P. H. Gleick, 1993, pp. 14–24.

KATHERINE K. HIRSCHBOECK

S

SATELLITE INSTRUMENTATION AND IMAGERY. It was not until astronauts on the way to the Moon in 1968 looked back and saw the Earth as a "big blue marble" that humans first literally viewed the Earth as a globe. As suitable as human eyesight is for appreciating the esthetic qualities of our planet, we must depend on instruments and sensors to measure and analyze the planet's physical phenomena, processes, and resources. Perhaps the only way to obtain a truly global view is with the perspective gained by placing Earth-pointing instruments on satellites in Earth orbit. Sensors on satellites have been used since the 1960s for this purpose; with time, we have become much more sophisticated in our technology and science, to the point that we are now involved in planning and implementing a truly global-scale examination of the Earth, using remote sensing and in situ measurements.

Satellite instruments fall into several broad classes according to the part of the electromagnetic (EM) spectrum they are designed to measure, the nature of their electromechanical components, and the nature of the measurements to be taken. All remote sensing depends on the differential response of physical phenomena or objects to regions of the EM spectrum. Mainly the photons (elementary particles, quanta of EM radiation) measured are from incoming solar radiation; in other cases, infrared or microwave emission from the Earth–atmosphere system is the source. All objects absorb or reflect light to a greater or lesser degree depending on the part of the EM spectrum being examined. As a result, an object has a characteristic absorption, transmission, or reflection spectrum. When a sensor is close to the object under examination, there is little need to correct for disturbances to the photons impinging on the instrument or sensor; however, in satellite-based remote sensing there is significant distance between the sensor and the target. Between the sensor and target are often several hundred miles of atmosphere, which contain many emitting, reflecting, and absorbing species—such as gases, particulate matter, and thermal regimes—that may affect the desired signal. As a result, it is often necessary to measure the characteristics of the atmosphere in addition to those of the Earth-level target, in order to subtract contributions to the signal from these spurious signals; thus it is necessary to infer properties of the intervening atmosphere. Figure 1 shows the differences in spectral distribution between solar irradiance at the top of the atmosphere and irradiance reaching sea level after passage through a scattering and absorbing atmosphere containing aerosol particles, water vapor (H_2O), oxygen (O_2), ozone (O_3), and carbon dioxide (CO_2). Each of these constituents affects the amount of radiation that is transmitted through the atmosphere at a given wavelength.

One type of instrument useful for this purpose is the *sounder,* an instrument that measures some process or phenomenon in the atmosphere through a column of air between the instrument on a satellite and the ground. Sounders often work in the microwave and infrared regions of the EM spectrum; they are useful for determining vertical profiles of temperature and humidity, as well as cloud characteristics, based on the spectral emission properties of a column of the atmosphere. These measurements are useful in their own right and are often subjects of intense scrutiny by scientists interested in atmospheric phenomena. Sounders can help in understanding the roles that greenhouse gases play in the global climate system and in monitoring climate variations and trends. They are important to investigators who wish to examine ground-level or sea-level phenomena because the data they provide may be used to remove any contributions to signals received by other instruments whose targets are at ground level. This is an excellent example of *instrument synergy,* an approach in which two or more instruments are used in concert to generate data and information that otherwise would not be available.

Instruments that measure in the microwave region of the EM spectrum are often used to determine surface wind speed and direction. These radar-based *scatterometers* can provide all-weather measurements of near-

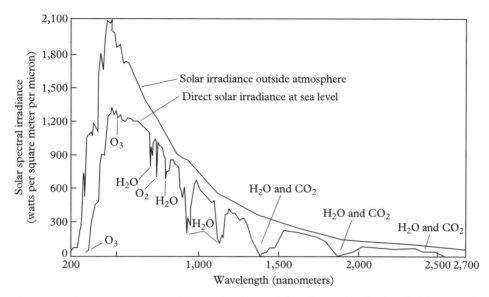

SATELLITE INSTRUMENTATION. Figure 1. *Spectral irradiance distribution incident on the Earth from the Sun.* The upper curve is the observed solar irradiance at the top of the atmosphere, and the lower curve is the solar irradiance at sea level.

surface vector winds over ice-free global oceans. Microwaves are well suited for this purpose because cloud water is transparent to microwaves in the regions where these instruments work. Also, microwave-based *limb sounders* have measured thermal emission from the atmosphere between the Earth's surface and free space by looking through the atmosphere at an angle roughly tangential to the surface. Microwave limb sounders allow measurements of the chemistry of the upper troposphere and lower stratosphere (at about 5 to 15 kilometers above Earth's surface), including concentrations of water vapor, ozone, sulfur dioxide, chlorine monoxide, bromine monoxide, hydrochloric acid, hydroxyl radical, nitrous oxide, and carbon monoxide—important gases in understanding phenomena such as the greenhouse effect, radiative forcing of climate change, and ozone depletion.

Longwave and shortwave radiation may be used to investigate the role of clouds in modifying and regulating the Earth's energy balance. Cloud and radiation flux measurements are essential for models of oceanic and atmospheric energetics and are also useful in long-range weather forecasting. Among the geophysical parameters that such data will generate are areal coverage of clouds, cloud top altitude and liquid water content, cloud droplet size, and cloud optical depth. Radiation parameters include fluxes at the top of the atmosphere, shortwave flux at the Earth's surface, and estimates of surface temperature, albedo, and emissivity.

In addition to generating these and other geophysical parameters, direct imaging has long played a large role in understanding the Earth system. It will continue to do so as the nations of the Earth embark on a multidecade, global-scale examination of our planet as participants in the Global Change Research Program. Several imaging instruments are currently under development for launch in the era of the U.S. National Aeronautics and Space Administration's (NASA's) Earth Observing System (EOS), beginning in 1998.

The Advanced Spaceborne Thermal Emission and Reflection Radiometer (ASTER) is being built by the Japanese Ministry of International Trade and Industry for flight on a U.S.-provided platform, known as EOS AM-1. ASTER will measure photons in both the visible and the thermal wavelength regions and will provide high spatial resolution images of the land surface, water, ice, and clouds. Surface mapping, soil and geologic studies, volcano monitoring, and surface temperature, emissivity, and reflectivity will be addressed. The data will be used to investigate land-use patterns and vegetation distribution, to study coral reefs and glaciers, and to generate digital elevation models. This should lead to better understanding of the physical processes that affect climate change. ASTER is an example of a class of instruments known as *multispectral imagers,* in this case used for both reflected and emitted radiation. Derived from a long line of similar instruments, it will image surface features at a scale comparable to the physical phenomena being measured—as small as 15 meters.

Multispectral imagers on spacecraft such as the Landsat series have long provided useful information to Earth scientists. Landsat images such as those in Figure 2 have become familiar because of their many practical applications. The images shown depict the dramatic contrast in water levels of the Missouri, Mississippi, and Illinois rivers in a drought year (1988) and a flood year (1993).

Another EOS-era imager is the Moderate Resolution Imaging Spectroradiometer (MODIS). MODIS has been described as the quintessential EOS instrument because it provides state-of-the-art measurement tools, has a long heritage of related instrumentation, and will make use of several on-board calibration methodologies. It will measure biological and physical processes on a global basis every 1 to 2 days. Its data will enhance our understanding of global dynamics and processes occurring on the Earth's surface and in the lower atmosphere. It will provide atmospheric, oceanic, and land surface measurements of cloud cover and associated properties, aerosol particle concentration and optical properties, sea surface temperature and chlorophyll, land cover changes, land surface temperature, and vegetation properties. Its most notable feature is its ability to use 36 spectral bands with high spectral resolution. It combines some of the best features of such proven successes as the Advanced Very High Resolution Radiometer (AVHRR), High-resolution Infrared Radiation Sounder (HIRS), Landsat Thematic Mapper (TM), and the Coastal Zone Color Scanner (CZCS), flown in 1978 on the *Nimbus 7* satellite. MODIS data will be calibrated with several onboard methodologies, including measurements of free-space, black body, and internal calibrators on every scan, and the system will also provide periodic views of the Sun and Moon. When used in conjunction with the sounders described above, MODIS will indeed provide a model for integrated, synergistic measurement of Earth system processes. Figure 3 is a schematic cutaway drawing of

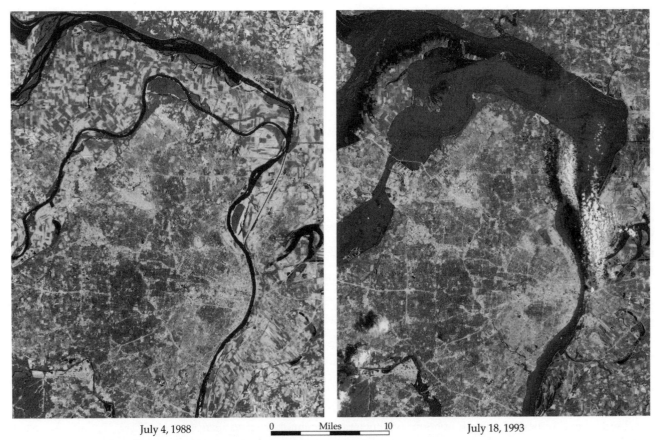

July 4, 1988   0 ———— Miles ———— 10   July 18, 1993

SATELLITE INSTRUMENTATION. Figure 2. *Landsat Thematic Mapper (TM) imagery acquired 4 July 1988 (left) and 18 July 1993 (right).* (Landsat imagery courtesy of the Earth Observation Satellite Company, Landover, Maryland.)

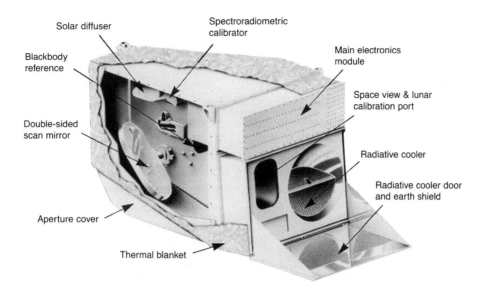

Solar diffuser
Spectroradiometric calibrator
Blackbody reference
Main electronics module
Space view & lunar calibration port
Double-sided scan mirror
Radiative cooler
Radiative cooler door and earth shield
Aperture cover
Thermal blanket

SATELLITE INSTRUMENTATION. Figure 3. *Cutaway drawing of the Moderate Resolution Imaging Spectroradiometer (MODIS) satellite sensor.* (Courtesy of NASA Goddard Space Flight Center.)

the MODIS satellite sensor, showing the positions of the various onboard calibrators; the instrument is approximately 160 centimeters long, 100 centimeters wide, and 100 centimeters deep, weighing about 250 kilograms.

Regardless of how the photons from any target are measured, the raw data must be converted into a form useful by scientists. In the case of the EOS-era instruments, such as ASTER and MODIS, described above, data will be sent—either by direct broadcast or by transmission through NASA's Tracking and Data Relay Satellite System (TDRSS)—to ground stations. There the data will be corrected based on instrument calibration, merged with spacecraft and instrument "housekeeping" data so scientists will know when, where, and under what conditions the data were obtained, and then used as input to computational algorithms that will convert the data into geophysical parameters such as temperature, humidity, or wind speed and direction. These geophysical parameters may be used as further input to generate mathematical models, such as general circulation models of the Earth system, which may eventually be used for predictive purposes. These model-based predictions, when compared with observations, may then be used to refine the models further, as well as to support policy and decision making in the effort to manage the resources of our planet.

[*See also* Floods; Instrumentation; *and* Radiation.]

BIBLIOGRAPHY

Asrar, G., and R. Greenstone, eds. *EOS Reference Handbook.* NASA NP-215. Greenbelt, Maryland, 1995.
Gille, J. C., and G. Visconti, eds. *The Use of EOS for Studies of Atmospheric Physics.* New York, 1992.

MICHAEL D. KING and MITCHELL K. HOBISH

SATELLITE METEOROLOGY. *See* Meteorology.

SATELLITE OBSERVATION. It is no coincidence that the rapid increase in studies of the Earth's climate during the three decades since 1960 paralleled the new era of weather and climate observation from near-Earth satellites. Satellite observations of the "blue planet" have focused scientific attention on the complex and deeply intertwined atmosphere, hydrosphere, and biosphere of Earth. We call this the *climate system.*

Furthermore, Earth-orbiting satellites can detect ample evidence of human presence on Earth. Evidence of urban sprawl, deforestation, and pollution of the atmosphere and oceans are painfully evident from unmanned weather and climate satellites such as the TIROS, NIMBUS, NOAA, DMSP, Earth Radiation Budget, and Upper Atmosphere Research satellites. Satellite-mounted instruments probe our climate system in spectral radiation wavelengths often unseen by human eyes. They measure reflected solar radiation and radiation emitted from the Earth. Gases, aerosols, and clouds in the atmosphere and the surfaces of the land, oceans, ice, and snow selectively send radiation "signals" to the satellites depending on the physical and electrical properties of the "target." In this way they have detected ozone fluctuations, rainfall variations, ocean currents, temperature and moisture in the atmosphere, ice and snow amounts, the life cycles of storms, and many other physical aspects of our climate.

Perhaps most importantly, the satellite observations of recent decades and those planned for the next 10 to 20 years provide a seamless, unifying view of many key climatic variables. No longer do the distances between surface weather stations cause great concern. No longer are the ocean regions or vast, sparsely populated continental areas regions of undocumented weather and climate. Our climate system is viewed from space as a single entity.

The satellite data complement other land-surface, balloon, aircraft, and ship observations very well. Within the assimilated climate data set, they have helped scientists develop new understanding of the physical bases for climate change, the natural variability of climate, and possible anthropogenic (human-caused) effects on climate.

The first sketchy weather and climate observations from space were provided by *Explorer 7* (1959) and TIROS 1 (1960). Today, these pioneering experiments have been extended by a regular series of satellites operated by the United States, Russia, Japan, the European Community, and other governments. In 1990 at least 10 satellites were measuring some part of the climate system on a daily basis. More than a terabit of data (10^{12} bits = 1 terabit) is sent from these satellites each day to various data centers. Climate analysts, climate modelers, and scientists from numerous disciplines and countries large and small, all share these data under the auspices of the World Climate Programme. This worldwide effort in sharing data and discussion is fostered by the United Nations and by worldwide scientific organizations.

Since the 1960s, the principal results derived from satellites and related to climate have been as follows:

- We have measured the total energy reaching Earth from our Sun at 1,367 watts per square meter, with small variations about the mean. This is the principal energy source for our climate system.
- We have measured the energy exchange between Earth and space—both the incoming solar energy and the infrared heat energy emitted to space. We found that the planet is warmer and darker than previously thought, and that the atmosphere and oceans transport 40 percent more energy between the hot tropical regions and both cold poles than was previously believed.
- We have mapped sea-surface temperatures from space in great detail and have observed important regional climate variations (such as El Niño and the Southern Oscillation) related to air–sea interactions.
- We now know, after several decades of measurement, much more about the natural variability of the occurrence, height, and types of clouds on Earth. These water and ice clouds are the principal modulators of energy exchange from Earth to space and greatly affect regional climates.
- We are beginning to measure all of the variable greenhouse gases—water vapor, ozone, methane, and others—in the atmosphere by remote sensing from satellites. We have atmospheric temperature data accumulated over many years by satellite remote sensing and use it to search for evidence of global warming or cooling.
- We map the daily variation of ozone and the "holes" in that life-sheltering gas layer which have been recently discovered. We also monitor the spread and effects of volcanic aerosol clouds.
- We map the extent and properties of sea ice in both polar regions because it is important in several climatic prossesses.

Finally, one of the major contributions of satellite observations to better understanding of climate has been the use of the data to help improve global climate models. We depend on these very large, complex computer models to help answer many questions about possible climate change. Satellite observations of today's climate provide important benchmarks for the models, which must be able to replicate today's climate properly before they can predict possible changes or help explain variability in the past. [*See* Models and Modeling.]

Looking to the future, the spacefaring nations and groups of scientists have made plans for increasingly sophisticated satellite observations of climate. First, they will continue to extend the time series of well-measured variables. This is very important for the study and understanding of climate. In addition, they will reduce measurement uncertainty and provide data on previously unmeasured climate parameters, such as the water content of clouds and a measurement of cloud bases. They also hope to obtain better information about ice packs, the chemical composition of aerosols, wind at the ocean surface, and the variable moisture in land surfaces. In the United States, NASA calls this measurement program "Mission to Planet Earth." With continued worldwide cooperation, more secrets of the Earth's climate will be discovered, using continued satellite observations as one of our most important scientific tools.

BIBLIOGRAPHY

Kidder, S., and T. Vonder Haar. *Satellite Meteorology: An Introduction.* San Diego, Calif., 1995.
Rao, P. K., S. J. Holmes, R. K. Anderson, J. S. Winston, and P. E. Lehr, eds. *Weather Satellites: Systems, Data, and Environmental Applications.* Boston, 1990.
Stephens, G. *Remote Sensing of the Lower Atmosphere: An Introduction.* New York, 1994.

THOMAS H. VONDER HAAR

SAVANNA CLIMATE. The savanna is an environment with a vegetation cover of grasses, trees, and shrubs in varying proportions. Generally the savanna marks a transition between humid and dry climates—that is, between forest and a subtropical or mid-latitude desert. The only common element of the physical environment of all savannas is a highly seasonal rainfall regime, with a wet season during the high-Sun period (summer) and a dry season during the cooler, low-Sun months. This is the main feature of the savanna climate. During the dry season there is a deficiency of available soil moisture, its degree being dependent on the severity and length of the dry season and on soil characteristics. This period of moisture shortage is the prime factor creating savannas;

however, other aspects of the physical environment, such as soils, nutrients, drainage, geology, and environmental history, also play roles. Some savannas are produced by human activity rather than by climate; an example is savannas produced by fires set to clear the land at the end of the dry season.

Savannas cover about 23 million square kilometers, occupying about 20 percent of the Earth's land surface—about 65 percent of Africa, 60 percent of Australia, 45 percent of South America, and 10 percent of India and Southeast Asia. They are found in relatively low latitudes (Figure 1). Savannas are not found in Europe or North America because the required climate with dry, mild winters and wet summers is absent from these continents.

Various types of savannas are distinguished on the basis of the dominance of trees or grasses and the size and spacing of trees and shrubs. The dominant type in a given environment depends not only on moisture availability but also on the nutrient status of soils and on the long-term environmental history of the region. The various savannas of the Earth are diverse in plant species and in structure, but all share characteristics that allow them to tolerate the seasonal drought. All likewise exhibit a distinctive seasonal cycle of growth in response to the seasonal availability of moisture.

In general, two classes of savannas—called *wet* and *dry*—can be distinguished on the basis of climate. The wetter savannas consist primarily of trees spaced relatively closely; the drier ones consist mainly of low trees and shrubs, well spaced in continuous grass cover. In most regions, a continuum exists along the rainfall gradient, with a gradual transition from wet savanna woodlands to low tree and shrub savannas (the latter maybe termed *thorntree–tall grass savannas,* or *steppe,* or other

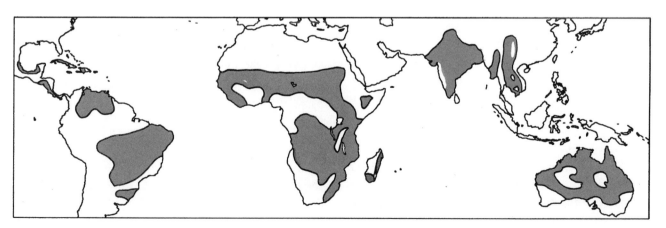

SAVANNA CLIMATE. Figure 1. *Global distribution of savannas.*

local names). The savanna woodlands correspond roughly to the wet-dry tropical climates; and the low tree and shrub savannas correspond to either semiarid or semidesert tropical climates.

Mean annual rainfall in the world's savannas ranges from about 250 millimeters in those along desert margins to 1,500 millimeters or more in the savanna woodlands along the forest margins. Transition between the low tree and shrub savanna and woodland occurs at around 500 millimeters. The dry seasons are of varying length and intensity, ranging from 3 or 4 months in the wetter savannas to 8 or 9 months in the drier ones. Most of the rainfall (as much as 50 to 80 percent) comes as a small number of brief but intense showers. Light rainfall frequently occurs on some days within the rainy season, but dry spells of 1 to several days are common.

The causes of the seasonal rainfall regime and the dryness of the cool season are so diverse for the various savannas that few generalizations can be made. In some cases—such as northern Australia, most of Africa, and parts of Asia—the seasonal cycle is due to the north-south displacement of the wind and pressure systems that form the general atmospheric circulation. The dry season occurs when the subtropical highs prevail, and the rainy season occurs when these move toward higher latitudes and are displaced by the Intertropical Convergence Zone, which brings convective rainfall. In other savannas—such as those in the high latitudes of southern Africa, in southern Australia, and in much of South America—additional and complex factors play a role.

The savannas occupy relatively low latitudes, so insolation (sunlight) and temperature are high during most of the year, favoring photosynthesis and growth. Mean maximum temperatures of the warmest months are in the range of 30° to 35°C near the desert margin and 25° to 30°C at the forest boundary. Temperatures of the cooler months are on the order of 13° to 18°C and 8° to 13°C along these two boundaries. In many savannas the highest temperatures occur just before the start of the rainy season, when the sun is high in the sky and the sparse vegetation cover of the dry season and dry soils promote intense heating. The annual range of temperatures in most savannas is relatively small—on the order of 3° to 10°C in Africa and South America, but as high as 15° to 20°C in Australia and Asia. The daily range can be quite large, especially during the dry season and near the desert margins.

Because of the high insolation, water demand (potential evapotranspiration or PET) is on the order of 1,000 to 1,500 millimeters annually, exceeding 2,000 millime-

ters in some areas. Because rainfall occurs during the warm season, much of it is lost through evaporation and is unavailable for plant growth. The length of the dry season varies among the world's savannas, but in all of them the moisture deficiency during this season is sufficient to produce stress in plants and limit their growth. In savannas with a wet–dry tropical climate, rainfall exceeds PET during at least several months, and there is a substantial water surplus (an excess of rainfall over evapotranspiration and soil moisture storage) during the wet season, generally 10 centimeters or more. During the dry season, water deficits (the amount by which potential evapotranspiration exceeds rainfall plus soil moisture storage) are on the order of 20 centimeters or more. Soil moisture storage generally exceeds 20 centimeters in 5 months or fewer. In the dry savannas, rainfall may exceed PET in only 1 or 2 months and may suffice only to saturate the soil, producing no water surplus.

Another climatic characteristic common to most savannas is the high year-to-year variability of rainfall. In areas where mean annual rainfall is on the order of 500 millimeters, amounts in individual years may vary from 100 to 1,000 millimeters. Where the mean is as high as 1,500 millimeters annual totals vary roughly from 500 to 2,500 millimeters. As a consequence, the savanna regions are prone to severe drought and to longer-term climatic fluctuations. The droughts of the African savannas, particularly the Sahelian region just south of the Sahara, are well known for their severity and long duration. For the Sahel as a whole, mean annual rainfall was twice as great during the 1950s as during the 1970s and 1980s. These fluctuations produce severe stress on agriculture and require a flexible system that can sustain such a range of conditions.

Longer and more extreme climatic fluctuations have influenced the savannas in the past. Few generalizations can be made, however, because the climatic histories of the various savannas are quite diverse. In many a much wetter period occurred toward the end of the nineteenth century, when rainfall in the savannas of Australia and much of Africa averaged about 25 percent above the twentieth-century mean. Long, severe dry episodes of the past have forced migration to wetter regions; such changes in the West African savannas a few centuries ago may have contributed to the downfall of major centers of civilization, including the Songhai and Mali empires near the fabled city of Tombouctou (Timbuktu).

On a much longer time scale, the extent of the savannas has changed markedly. Around the last ice age maximum, about 18,000 years ago, the low-latitude deserts encroached onto many of the savannas, contracting them or displacing them equatorward. After the ice age ended, the low-latitude deserts retreated and were largely replaced by savannas. About 5,000 years ago this change was so marked that the Sahara, today the world's largest desert, occupied only small hyperarid core regions of its current expanse.

The genera of plants in savannas are many, but several are common to a number of regions. Typical are the *Acacia* species, of which hundreds exist, with compound leaves having numerous small leaflets. Many savanna trees, such as *Combretum* and *Terminalia*, are broadleafed and rather large. Many species are deciduous, losing their leaves during the dry season. A characteristic savanna tree of Africa is the odd-shaped baobab, which stores water in a massive trunk. In some savannas, succulents such as cacti and euphorbias grow. Low shrubs and thickets, where evapotranspiration is reduced, are also common. Herbs and grasses may be annuals, growing only during the wet season, or perennials that grow year-round but may retreat beneath the surface during dry periods.

In the drier savannas the trees are typically of low or medium height, with a relatively thin canopy which often spreads laterally toward its top. The trees and shrubs are heavily grazed, so many species have evolved structures that protect them from foragers. Some have poisonous sap or fruits. Thorns—common in acacias—may be long and numerous in the young trees most accessible to grazing animals but become smaller and less dense as the trees grow.

The savannas are home to a multitude of grazing animals and their predators. In Africa, buffalo, zebra, giraffes, elephants, and various species of antelope feed on the vegetation, in turn providing food for jackals, hyenas, lions, and other carnivores. As the savannas are converted to croplands and human settlement expands, these large animal populations are rapidly diminishing.

The peoples who reside in the savannas have to adapt to the alternation between seasons of intense rainfall and high atmospheric humidity, and seasons of moisture deficiency and a scarcity of resources. They are generally sedentary during all or most of the year. For protection from the heat and moisture, their dwellings are usually small structures with adobe walls and thatched, slanted roofs. In some regions the adobe walls are replaced by thatched matting, which contracts during the dry season to allow breezes to penetrate.

The savanna peoples traditionally practice both sedentary farming and pastoralism (herding grazing animals such as sheep and goats). The drier savannas support mainly pastoralism; and their crops require irrigation. Ingenious strategies, such as constructing small mounds to catch runoff down slopes, may be utilized to gather water. In many savanna regions the farmers move into the drier parts when the grass thrives during the rainy season and retreat to the woodlands to tend crops during the drier months. During the twentieth century about 10

SAVANNA CLIMATE. Figure 2. *Typical savanna landscape of Namibia in southern Africa.*

to 15 percent of the savanna environment worldwide has been replaced by cropland. Crops that do well here include corn, millet, sorghum, and peanuts; in many savannas rice is grown, usually along floodplains where it can be irrigated.

The human occupants of the savanna have transformed them over time. Woody plants are cleared to make room for crops or gathered for fuel. At the end of the dry season, both natural and farmed savanna land is often burned to stimulate new growth. These grassland fires can be so extensive and intense that they can be detected from satellites. The constituents of the smoke alter the chemical composition of the lower atmosphere; some of the gases produced include carbon dioxide, carbon monoxide, hydrocarbons, methane, and various oxides of nitrogen.

In some savanna regions human impact has been so great that the land is said to be *desertified*. This term, which has no precise definition, is used to describe land that has been degraded and its productivity reduced. Soil and nutrients are eroded away; and water supplies diminish. When a region is simultaneously devastated by drought, the result can be badly deteriorated land. In some cases the area may not be able to recover its fertility without many years of careful management.

[*See also* Climatic Zones; *and* Desertification.]

BIBLIOGRAPHY

Bourliére, F., ed. "Tropical Savannas." *Ecosystems of the World.* Vol. 13. Amsterdam, 1983.
Cole, M. M. *The Savannas.* London, 1986.
Cole, M. M. "The Savannas." *Progress in Physical Geography* 11 (1987): 334–356.
Strahler, S. N., and A. H. Strahler. *Elements of Physical Geography.* 4th ed. New York, 1989.
Trewartha, G. T., and L. H. Horn. *An Introduction to Climate.* 5th ed. New York, 1980.

SHARON E. NICHOLSON

SCALES. *[This article treats a technical aspect of climate and weather studies; it is intended for readers at an advanced level.]*

Atmospheric phenomena and features have a very wide range of both size and duration. The space scales used in describing these phenomena are determined by their typical size (or *wavelength*), and the time scales by their typical lifetime (or *period*). As illustrated in Figure 1, atmospheric features range from turbulence on a space scale of meters and a time scale of seconds, up to jet streams on a space scale of thousands of kilometers and a time scale of months.

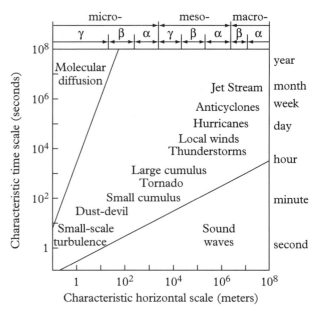

SCALES. Figure 1. *Scale definitions and different processes with characteristic time and horizontal scales.* (Adapted from Orlanski, 1975, and Oke, 1987.)

Because atmospheric phenomena occur in continua, classifying them is problematic with regard to scale limits; however, the classification suggested by Orlanski (1975) is widely accepted. Here the characteristic horizontal distance scale is divided into a macroscale including phenomena that extend over scales larger than 2,000 kilometers, a microscale including features smaller than 2 kilometers, and a mesoscale, which includes all atmospheric phenomena with scales between the macroscale and the microscale. As shown in Figure 1, these scales are further subdivided into macro-α (larger than 10,000 kilometers), macro-β (2,000–10,000 kilometers), meso-α (200–2,000 kilometers), meso-β (20–200 kilometers), meso-γ (2–20 kilometers), micro-α (200 meters–2 kilometers), micro-β (20–200 meters), and micro-γ (smaller than 20 meters).

Classifications according to time scales are more ambiguous and less frequently used. In this context, *macroscale* usually refers to atmospheric features with time scales of several days or longer, and *microscale* refers to phenomena with time scales on the order of minutes. Using the classification suggested by Orlanski (1975), meso-α scale corresponds to time scales of 1 day to 1 week, meso-β scale to several hours to 1 day, and meso-γ scale to one-half hour to several hours.

Thus, tidal and baroclinic waves are considered as macro-α and macro-β scale features, respectively. Frontal systems and hurricanes having horizontal scales of 200 to 2,000 kilometers and time scales of 1 day to 1 week are classified as meso-α scale systems. The nocturnal low-level jet, squall lines, inertial waves, cloud clusters, breezes, and other thermally induced circulations having horizontal scales on the order of 20 to 200 kilometers and time scales on the order of several hours to 1 day are classified as meso-β scale systems. Meso-γ scale systems include thunderstorms, internal gravity waves, clear air turbulence, and urban effects. Finally, tornadoes and deep convection are classified as micro-α scale, thermals and dust devils as micro-β scale, and plumes as micro-γ scale processes.

The fundamentals of scale analysis, also called *scaling*, are introduced below. Some aspects of the subgrid scale problem in atmospheric modeling are discussed following this.

Scale Analysis. The method of scale analysis, or scaling, involves estimation of the order of magnitude of the individual terms in the governing equations for a particular type of motion through the use of representative values of the dependent variables and constants that make up these terms. Based on such calculation, small terms can be eliminated. This has the advantages of simplifying the mathematics that describe the particular motion and of eliminating types of motions that are not relevant to the process of interest.

Because the space and time scales of the various atmospheric features can be associated with their motion, one can use the different terms of the equation of motion for scaling purposes. Using Einstein's summation notation, this equation, which is derived from Newton's second law, is given as:

$$\underbrace{\frac{\partial u_i}{\partial t}}_{\text{I}} = \underbrace{-u_j \frac{\partial u_i}{\partial x_j}}_{\text{II}} \underbrace{-\delta_{i3}g}_{\text{III}} \underbrace{-2\epsilon_{ijk}\Omega_j u_k}_{\text{IV}} \underbrace{-\frac{1}{\rho}\frac{\partial p}{\partial x_i}}_{\text{V}} + \underbrace{\nu \frac{\partial^2 u_i}{\partial x_j^2}}_{\text{VI}} \quad (1)$$

where u_i represents the three Cartesian wind components ($u_1 = u$ for eastward-moving; $u_2 = v$ for northward-moving; and $u_3 = w$ for upward-moving); t is time; x_i represents the three Cartesian directions ($x_1 = x$ for eastward; $x_2 = y$ for northward; and $x_3 = z$ for upward); δ_{ij} is the Kronecker Delta; g is the gravitational acceleration (approximately 10 meters per second per second); ϵ_{ijk} is the Alternating Unit Tensor; Ω_j represents the three Cartesian components of the angular velocity vector of the Earth's rotation; ρ is the air density; p is the pressure; and ν is the air kinematic viscosity (approximately 1.5×10^{-5} square meters per second).

In this equation, Term I represents the storage of momentum (inertia); Term II describes the advection; Term III allows gravity to act vertically; Term IV describes the influence of the Earth's rotation (the Coriolis effect); Term V describes pressure-gradient forces; and Term VI represents the influence of viscous stress (i.e., molecular interactions). We define the following characteristic scales of the field variables:

- U — horizontal velocity scale;
- W — vertical velocity scale;
- L — length scale;
- D — depth scale;
- $\dfrac{\Delta P}{\rho}$ — pressure fluctuation scale;
- $\dfrac{L}{U}$ — horizontal time scale; and
- $\dfrac{D}{W}$ — vertical time scale.

If latitude $\phi_0 = 45°$ is considered and the notation $f_0 = 2\Omega \sin \phi_0 = 2\Omega \cos \phi_0$ is introduced, characteristic scales of the different terms of this equation can be defined:

- $\left|\dfrac{\partial u}{\partial t}\right| \sim \left|\dfrac{\partial v}{\partial t}\right| \sim \left|u\dfrac{\partial u}{\partial x}\right| \sim \left|v\dfrac{\partial u}{\partial y}\right| \sim \left|u\dfrac{\partial v}{\partial x}\right| \sim \left|v\dfrac{\partial v}{\partial y}\right| \sim \dfrac{U^2}{L}$;

- $\left|\dfrac{\partial w}{\partial t}\right| \sim \left|w\dfrac{\partial w}{\partial z}\right| \sim \dfrac{W^2}{D}$;

- $\left|w\dfrac{\partial u}{\partial z}\right| \sim \left|w\dfrac{\partial v}{\partial z}\right| \sim \dfrac{UW}{D}$;

- $\left|u\dfrac{\partial w}{\partial x}\right| \sim \left|v\dfrac{\partial w}{\partial y}\right| \sim \dfrac{UW}{L}$;

- $2v\Omega \sin \phi \sim 2u\Omega \sin \phi \sim f_0 U$;

- $2w\Omega \cos \phi \sim f_0 W$;

- $\left|\dfrac{1}{\rho}\dfrac{\partial p}{\partial x}\right| \sim \left|\dfrac{1}{\rho}\dfrac{\partial p}{\partial y}\right| \sim \dfrac{1}{\rho}\dfrac{\Delta P}{L}$;

- $\left|\dfrac{1}{\rho}\dfrac{\partial v}{\partial z}\right| \sim \dfrac{1}{\rho}\dfrac{\Delta P}{D}$;

- $\left|\nu\dfrac{\partial^2 u}{\partial x^2}\right| \sim \left|\nu\dfrac{\partial^2 u}{\partial y^2}\right| \sim \left|\nu\dfrac{\partial^2 v}{\partial x^2}\right| \sim \left|\nu\dfrac{\partial^2 v}{\partial y^2}\right| \sim \nu\dfrac{U}{L^2}$;

- $\left|\nu\dfrac{\partial^2 u}{\partial z^2}\right| \sim \left|\nu\dfrac{\partial^2 v}{\partial z^2}\right| \sim \nu\dfrac{U}{D^2}$;

- $\left|\nu\dfrac{\partial^2 w}{\partial x^2}\right| \sim \left|\nu\dfrac{\partial^2 w}{\partial y^2}\right| \sim \nu\dfrac{W}{L^2}$;

- $\left|\nu\dfrac{\partial^2 w}{\partial z^2}\right| \sim \nu\dfrac{W}{D^2}$;

where ϕ is the latitude.

Based on observations for mid-latitude synoptic systems, for example, typical values for these characteristic scales are $U \sim 10$ meters per second, $W \sim 10^{-2}$ meters per second, $L \sim 10^6$ meters, $D \sim 10^4$ meters, $\Delta P/\rho \sim 10^3$ square meters per second per second, $L/U \sim 10^5$ seconds, and $f_0 \sim 10^{-4}$ per second. Table 1 shows the corresponding characteristic values for the various scaled terms of the equation of motion.

Clearly, for such systems, the Coriolis term associated with the horizontal wind and the pressure gradient term in the equations of the horizontal components of the wind are in approximate balance and are at least one order of magnitude larger than the other terms (note, too, that the viscous stress terms are several orders of magnitude smaller than the other terms). Therefore, retaining only these two terms in the equations that describe the horizontal wind (that is, $\partial u/\partial t$ and $\partial v/\partial t$ derived from Equation 1) as a first approximation, the geostrophic wind components are obtained:

$$v \approx \frac{1}{\rho f} \frac{\partial p}{\partial x}; \qquad u \approx \frac{1}{\rho f} \frac{\partial p}{\partial y}$$

where $f = 2\Omega \sin \phi$ is called the *Coriolis parameter*. It is important to note that these relations are diagnostic—that

SCALES. Table 1. *Scale of the various terms of the equation of motion and their characteristic magnitudes for a typical synoptic system*

Scaled Term	Magnitude (meters per second per second)
$\dfrac{U^2}{L}$	10^{-4}
$\dfrac{W^2}{D}$	10^{-8}
$\dfrac{UW}{D}$	10^{-5}
$\dfrac{UW}{L}$	10^{-7}
$f_0 U$	10^{-3}
$f_0 W$	10^{-6}
$\dfrac{1}{\rho} \dfrac{\Delta P}{L}$	10^{-3}
$\dfrac{1}{\rho} \dfrac{\Delta P}{D}$	10
$\nu \dfrac{U}{L^2}$	10^{-16}
$\nu \dfrac{U}{D^2}$	10^{-12}
$\nu \dfrac{W}{L^2}$	10^{-19}
$\nu \dfrac{W}{D^2}$	10^{-15}

is, they contain no reference to time and consequently cannot be used to predict the evolution of the velocity field.

To obtain a prediction equation it is necessary to retain the acceleration (expressed by du_i/dt, which is the sum of Term I and Term II in Equation 1). The resulting approximate horizontal motion equations are:

$$\frac{du}{dt} = fv - \frac{1}{\rho} \frac{\partial p}{\partial x} \qquad (2)$$

$$\frac{dv}{dt} = -fu - \frac{1}{\rho} \frac{\partial p}{\partial y}. \qquad (3)$$

The scale analysis summarized in Table 1 shows that the acceleration terms are about one order of magnitude smaller than the Coriolis and pressure gradient terms. This makes actual applications of these equations in weather forecasting difficult, because acceleration is given by the small difference between the two large terms.

A convenient measure of the magnitude of the acceleration compared to the Coriolis term may be obtained by forming the ratio of the characteristic scales for the horizontal acceleration and the Coriolis terms:

$$\frac{U^2/L}{f_0 U}.$$

This ratio, a nondimensional number called the *Rossby number*, is designated by

$$Ro = \frac{U}{f_0 L}.$$

Thus, the smallness of the Rossby number is a measure of the validity of the geostrophic approximation. For the typical synoptic conditions given above, $Ro \equiv 0.1$. For a mesoscale system with a characteristic length scale of 100 kilometers and a similar horizontal velocity scale, $Ro \equiv 1$, emphasizing that acceleration is on the same order of magnitude as the Coriolis term. For a microscale system with a characteristic length scale of 1 kilometer and a similar horizontal velocity scale, $Ro \equiv 100$, and the Coriolis term, which is two orders of magnitude smaller than the acceleration, can be ignored.

The *Reynolds number*, which expresses the magnitude of the viscous stress terms compared to the advective terms, is another helpful parameter for scale analysis. Designating the length and velocity scale by S and V, respectively (that is, $S \sim L \sim D$, and $V \sim U \sim W$), the Reynolds number is defined as:

$$Re = \frac{vs}{\nu}.$$

When Re is much greater than 1, changes in motion by advection are much more important than the dissipation of velocity by molecular interactions. Under this condition, the flow is said to be *turbulent,* and transfers of all properties of the air (such as heat or water vapor) are performed through the movement of air from one point to another. By contrast, when Re is much less than 1, the molecular dissipation of velocity dominates, and the flow is said to be *laminar;* under this condition, properties of the air are transferred on the molecular scale.

For the typical synoptic conditions given above, $Re > 10^7$. For mesoscale systems, this number is also very large ($Re \approx 10^6$). Clearly, for such systems, viscous stress can be ignored. Only near the ground do molecular transfers become important. For instance, for a microscale system with $D = 10^{-4}$ meters and $W = 1.5 \times 10^{-2}$ meters per second, the Reynolds number is 0.1, and the movement of air is laminar.

Table 1 indicates that in the vertical direction the gravity and pressure gradient terms are dominant in synoptic systems. Thus, this scale analysis demonstrates that, to a high degree of accuracy, the pressure field is in hydrostatic equilibrium; that is, the pressure at any point is simply equal to the weight of a unit cross-section column of air above that point. At such scales, the vertical equation of motion can be reduced to

$$\frac{\partial p}{\partial z} = -\rho g. \qquad (4)$$

This is called the *hydrostatic equation.* In utilizing this relationship it must be emphasized that the results of the scale analysis imply only that the magnitude of the vertical acceleration is much less than the magnitude of the pressure gradient term—not that the magnitude of the vertical acceleration is identically zero. Although this relation is valid for most meso-α and even meso-β scale systems, it is not appropriate for smaller-scale systems.

Subgrid-Scale Processes. At any location and time, the atmosphere is characterized by a set of conservation principles applying to mass, heat, motion, water, and other gaseous and aerosol materials, which form a coupled set of relations that must be satisfied simultaneously. This set of relations consists of nonlinear partial differential equations, which have no known analytical solution. Therefore, only a simplified set of relations can be solved analytically. As discussed above, scale analysis can be used for simplifying these equations.

An alternative method consists of converting these equations into a different but related set of equations which require only a finite number of operations for their solution: for example, a differential equation might be replaced by a finite difference equation. This process requires that continuous variables (such as the temperature and pressure fields) be reduced to a finite set of values at given locations (usually a rectangular grid on the horizontal plane, with multiple levels in the vertical) and given times. Computers are used to solve this new set of equations. [*See* Numerical Methods.]

Because of limited computational resources, only a limited number of operations can be carried out within the available time period. Accordingly, the size of the grid and the length of the prediction are defined. For instance, current general circulation models (GCMs) used in climate studies typically represent the Earth's surface with a horizontal resolution (the distance between two neighboring grid points) of 2 to 8 degrees of latitude and longitude, and the vertical structure of the atmosphere as having 2 to 19 layers. A typical mesoscale model covers an area of 10^4 to 10^6 square kilometers with a horizontal resolution of 5 to 50 kilometers. In the vertical direction, such models use 5 to 30 layers to describe the structure of the first 3 to 15 kilometers of the atmosphere. Large-eddy simulation (LES) models, which are typically used to study microscale processes at a horizontal resolution of 100 to 200 meters, cannot presently be used to simulate an area larger than a few square kilometers. [*See* Models and Modeling.]

Consequently, all atmospheric numerical models are formulated on the concept that any atmospheric variable, ϕ, can be separated into a resolved (mean) variable part, $\overline{\phi}$, and an unresolved (perturbation) part, or *subgrid-scale* part, ϕ':

$$\phi = \overline{\phi} + \phi' \qquad (5)$$

Accordingly, the equation of motion (Equation 1) can be written:

$$
\begin{aligned}
\frac{\partial \overline{u}_i}{\partial t} + \frac{\partial u'_i}{\partial t} = &-\overline{u}_j \frac{\partial \overline{u}_i}{\partial x_j} - \overline{u}_j \frac{\partial u'_i}{\partial x_j} - u'_j \frac{\partial \overline{u}_i}{\partial x_j} \\
&- u'_j \frac{\partial u'_i}{\partial x_j} - \delta_{i3g} - 2\epsilon_{ijk}\Omega_j \overline{u}_k \\
&- 2\epsilon_{ijk}\Omega_j u'_k - \frac{1}{\overline{\rho} + \rho'} \frac{\partial \overline{p}}{\partial x_i} \\
&- \frac{1}{\overline{\rho} + \rho'} \frac{\partial p'}{\partial x_i} + \nu \frac{\partial^2 \overline{u}_i}{\partial x_j^2} + \nu \frac{\partial^2 u'_i}{\partial x_j^2}
\end{aligned}
\qquad (6)
$$

When we average this equation while applying Reynolds averaging rules, and using the Boussinesq approximation and the continuity equation for motion perturbations, we are left with a prognostic equation for the resolved (mean) wind:

$$\frac{\partial \overline{u}_i}{\partial t} = -\overline{u}_j \frac{\partial \overline{u}_i}{\partial x_j} - \delta_{i3g} - 2\epsilon_{ijk}\Omega_j \overline{u}_k$$
$$- \frac{1}{\overline{\rho}} \frac{\partial \overline{p}}{\partial x_i} + \nu \frac{\partial^2 \overline{u}_i}{\partial x_j^2} - \frac{\partial \overline{u_j' u_i'}}{\partial x_j} \qquad (7)$$

which is mathematically similar to Equation 1, except for the last term on the right-hand side of the equation, which involves correlations of the velocity perturbations (covariance). This term represents the divergence of subgrid-scale fluxes.

This additional term, which adds another set of unknowns in the set of equations, may or may not significantly affect the resolved flow. For example, Deardorff (1974) showed that for an LES model with a horizontal resolution of 125 meters, the resolvable fluxes were much more important than the subgrid-scale fluxes. By contrast, current GCMs resolve only macroscale features, and all mesoscale and microscale processes known to have an important impact on the resolved features need to be included within the subgrid-scale fluxes.

Prognostic equations can be developed for subgrid-scale fluxes. Unfortunately, they include triple correlation terms, which are new unknowns. When equations are developed for these new unknowns (changing them to known variables), one discovers quadruple correlation terms, and so on *ad infinitum*. Thus, for any finite set of those equations, the description of subgrid-scale fluxes is not closed. This is known as the *closure problem* and remains one of the unsolved problems of classical (non-quantum) physics.

One approach to making the mathematical-statistical description of these terms tractable is to use only a finite number of equations, and then to approximate the remaining unknowns in terms of known quantities. Such closure approximations are named by the highest-order prognostic equations that are retained. For example, if the double correlation terms in Equation 7 are approximated and only prognostic equations for the mean quantities are retained (as is done in current GCMs), we have a first-order closure.

Regardless of which order of closure is used, there are unknown terms that must be *parameterized* as a function of known quantities and parameters. A known quantity is any quantity for which a prognostic or diagnostic equation is retained. A parameter is usually a constant, the value of which is determined empirically. By definition, a parameterization is an approximation to nature. In other words, the true (natural) equation describing a value is replaced with some artificially constructed approximation. Sometimes parameterizations are employed because the true physics is not yet understood; sometimes, the physics is known but is too complicated to use for a particular application, given cost or computer limitations.

Parameterization of subgrid-scale processes is rarely perfect and clearly requires different degrees of complexity in different numerical models. The goal is that it will be adequate for the application at hand.

Important among the various processes that need to be parameterized in macroscale models are clouds and their interactions with radiation and precipitation. This task is further complicated by the redistribution of energy and water at the ground surface, and subsequent interactions with the biosphere at various scales (not necessarily similar to the scales of the atmospheric processes).

Summary. Atmospheric features can be classified according to their characteristic space and time scales. Although there is no universal classification, a reasonable consensus exists on the general definition of macro-, meso-, and microscales. The atmosphere is described by a set of conservation principles represented by a set of complex mathematical equations which has no known analytical solution. Therefore, to predict changes in the atmosphere, this set of equations may be simplified based on a scale analysis of the various terms of these equations, which is specific to the scale of motion of interest; alternatively, the equations may be converted into a different but related set of equations, which is then solved numerically using computers. Because of computer limits, these equations can be solved only for a limited number of grid points and time intervals. Atmospheric quantities resolved at the grid resolution of a model may be considerably affected by unresolved (subgrid-scale) processes. These subgrid-scale processes must be parameterized, especially in models that resolve only macroscale features and, therefore, do not resolve mesoscale and microscale processes. The development of adequate parameterizations for such models represents one of the greatest challenges for the scientific community.

BIBLIOGRAPHY

Avissar, R., and F. Chen. "Development and Analysis of Prognostic Equations for Mesoscale Kinetic Energy and Mesoscale

(Subgrid-scale) Fluxes for Large-scale Atmospheric Models." *Journal of the Atmospheric Sciences* 50 (1993): 3751–3774.

Deardorff, J. W. "Three-dimensional Numerical Study of the Height and Mean Structure of a Heated Planetary Boundary Layer." *Boundary-Layer Meteorology* 7 (1974) 81–106.

Louis, J.-F. "Parametrisation in Weather Prediction Models." *Rivista di Meteorologia Aeronautica* 42 (1982): 219–255.

Oke, T. R. *Boundary Layer Climates.* 2d ed. London and New York, 1987.

Orlanski, I. "A Rational Subdivision of Scales for Atmospheric Processes." *Bulletin of the American Meteorological Society* 56 (1975): 527–530.

RONI AVISSAR

SCATTERING. Radiation interacts with atmospheric molecules and/or aerosol particles and is then propagated away at an angle different to the original direction of incidence. Scattering occurs in the atmosphere when light is refracted, reflected, or diffracted by molecules or small particles. The treatment of scattering of radiation on particles depends on the size parameter χ where

$$\chi = \frac{\pi d}{\lambda} \qquad (1)$$

and d is the particle diameter and λ is the wavelength of exciting radiation. For particles such as molecules, or fine aerosols, the particle diameter d is much smaller than λ and therefore χ is much less than 1. Molecules act as single electric dipoles that are excited by the incident radiation and then reradiate the energy in all directions around the direction of incidence. This type of interaction of light with molecules is known as *Rayleigh scattering*, after the English physicist Lord Rayleigh (1842–1919). The phenomenon is also known as the *Tyndall effect,* after the physicist John Tyndall (1820–1893) who first observed it experimentally. He also found that aerosols much smaller than the wavelength of the scattered radiation can act in the same way as molecules.

Clouds and hazes have particle sizes that are much closer to, comparable with, or even larger than the exciting wavelength. That is, χ varies from somewhat less than to much greater than 1. For spherical particles, such as cloud water drops and some aerosols, the scattering properties were determined by G. Mie in 1908. When the size parameter becomes large and d is much greater than λ, Mie theory becomes complicated, with many terms needed in the mathematical solution. For cloud drops and some ice crystal forms, Mie theory can be replaced by tracing many ray paths through the particle using standard theories of reflection, refraction, and diffraction. [*See* Radiation.] This is accomplished using sophisticated ray-tracing computations.

Patterns that describe the intensity of scattered radiation and its angular dependence are called *scattering phase functions*. These patterns are continuous with scattering angle and describe the probability of a photon being scattered at a given angle to the original direction. The shape of a phase function depends on the aerosol particle size parameter χ.

As the size parameter increases, the scattering pattern becomes more forward peaked; that is, there is more radiation scattered into the forward than the back hemisphere. Diffraction, although involving no refraction or reflection at the particle, is an important factor in scattering. Because some light is blocked by the aerosol particle, light waves from the side spread out beyond the particle and form an interference pattern around the forward direction with an angular width $\delta\theta$, which is inversely proportional to particle size. In fact, for a circular particle of diameter d,

$$\delta\theta = \frac{1.22\lambda}{d} \qquad (2)$$

where angle $\delta\theta$ is in radians, or from Equation 1,

$$\delta\theta = \frac{1.22}{\pi\chi}. \qquad (3)$$

The diffraction effect presents a well-known anomaly in that the particle scattering cross-section is twice the geometrical cross-section, at least when χ is much greater than 1, one unit for diffraction and one for reflection and refraction.

The phase functions for Rayleigh scattering are shown in Figure 1, and those for two different values of χ for

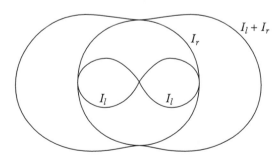

SCATTERING . Figure 1. *Intensity distribution with scattering angle in Rayleigh scattering.* Inner curves relate to separate polarized components, outer curve to total intensity. I_r —Scattering out of plane including electric vector and incident beam. I_l —Scattering in the above plane. (After Twomey, 1977.)

Mie scattering in Figure 2. The direction of the incident radiation is to the left. The scattering is symmetrical for any angle about that direction, as expected. The three-dimensional distribution is obtained by "spinning" the diagram around the axis of incidence. It can be seen in Figure 2 that the phase function for larger size parameters has many peaks and troughs, particularly in the back hemisphere, as well as a forward diffraction peak. By contrast, the function for Rayleigh scattering is symmetrical in the forward and back direction. Figure 1 also shows the scattering patterns for two directions of polarization, discussed below. As the polar diagrams in Figures 2a and 2b are plotted on a logarithmic scale, it can be seen that even for a size parameter of about 2, the scattering in the forward direction is about 10 times the value in the back direction.

Polarization. In the electromagnetic theory of radiation, the amplitude of light is described by the electric vector. Wave packets of radiation in which the electric vectors are all propagating in one plane are said to be *polarized*. The intensity of scattering of polarized radiation—that is, the phase function—depends on the direction of scattering as well as on the size parameter. In the case of Rayleigh scattering, the incident field causes the electric dipole to oscillate in the same direction as the incident electric vector. The intensity of light scattered at right angles to the plane defined by the incident radiation and electric vector is the same at every angle, whereas light scattered in directions in the *same* plane falls to zero at 90° scattering angle. As an example of these effects, solar radiation is unpolarized, but if the scattered sky radiation is viewed near 90° to the solar beam, the solar radiation is quite polarized. The polarization patterns for Mie scattering are more complex. [*See* Radiation.]

Wavelength Dependence. The total scattering intensity, when integrated through all scattering angles—that is, the *scattering cross-section*—is also dependent on the size parameter χ. For Rayleigh scattering, the intensity is found to vary as λ^{-4}. This implies that radiation at the "red" end of the spectrum, at about 700 microns, is scattered with one-sixteenth the intensity of that in the ultraviolet at 350 microns. This phenomenon can account for the blue of the sky, which is visible only through scattered radiation. The blue effect is enhanced by the high solar intensity at "blue" wavelengths. Conversely, viewing the setting run through a long path of atmosphere gives the sun a red appearance. The wavelength dependence also implies that the scattering of ultraviolet light is intense; thus, the potential for sunburn from a clear blue sky is considerable, even if the direct sunlight is obstructed by buildings, sunshades, or isolated clouds.

The scattering intensity from aerosel particles also depends on the wavelength λ or size parameter χ. For values of χ up to about 1 (for particles less than about 1 micron in diameter in the solar spectrum), the scattering intensity increases with the size parameter and then levels off. In the case of a given particle size and a variable wavelength, scattering intensity varies from λ^{-4} for "Rayleigh size" particles to λ^{-1}, or even λ^{0} for other particle size distributions. In the case of large cloud drops of 10 microns in diameter, the scattering becomes rather independent of wavelength. This can explain why the sky is milky when it is very hazy, grading to very white in the case of clouds. In the case of deep clouds, the darkness of the cloud base is due to more radiation being scattered back to space than is transmitted, through the action of *multiple scattering*. [*See* Haze.]

The intensity of scattering is controlled by a quantity called the *scattering efficiency*. This efficiency describes how much radiation is scattered compared to that from the particle cross-sectional area. When the size parameter is much less than 1, then the particle removes radiation from a much smaller cross-sectional area compared to the particle area; that is, the scattering efficiency is much

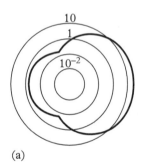

(a)

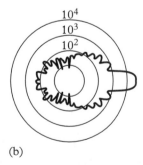

(b)

SCATTERING . Figure 2. *Scattering polar diagrams for two values of the size parameter* χ. (a) χ = 2.28; (b) χ = 15.99.

less than 1. For particles of size parameter of unity or greater, the efficiency tends to a value of 2. This apparent *extinction anomaly* is due to the effects of diffraction.

Absorption. Aerosol particles absorb as well as scatter radiation, and this absorption can be strong. It can be selective, covering only a band of wavelengths, or it can cover most of the visible region. The latter is exemplified by soot particles. Large soot particles produced on glass held above a candle flame look black, meaning that they absorb strongly at all wavelengths. Thus a cloud of smoke from a forest or other fire can appear quite grey or even black. This means that some of the scattered radiation is reabsorbed by the particles. Sometimes, polluted haze layers appear red-brown, owing to selective absorption at shorter wavelengths, generally thought to be due to pollutant molecules.

Multiple Scattering. Light scattered by aerosols, molecules, or clouds can suffer several subsequent scatterings during its passage through the atmosphere. Multiple scattering can cause the emergent radiation from a water cloud to be quite diffuse; that is, the cloud looks the same regardless of the observer's viewing angle. It also has the effect, for a deep water cloud, of scattering a large fraction of the incident solar radiation up through the cloud top and back to space. This fraction of radiation (or *albedo*) can have values as high as 0.7 or 0.8. [*See* Albedo.]

Scattering Phenomena in the Atmosphere. Scattering of radiation by atmospheric particles manifests itself in spectacular ways; for example, brilliant sunsets, rainbows, and the arcs and halos observed in clouds of ice crystals. Spectacular sunsets have occurred for several years after major volcanic eruptions, the latest being Mt. Pinatubo in June 1991 in the Philippines, the largest eruption this century. A veil of sulfate particles became diffused around the world, high in the stratosphere, some 25 kilometers above the ground. After sunset, radiation scattered from such a veil can cause brilliant sunsets. They are red in color because the blue light has been scattered out by Rayleigh scattering.

The familiar phenomenon of the rainbow was first explained by René Descartes in 1637. Rainbows are caused by refraction and reflection of sunlight in raindrops. The light is refracted into the drop, reflected once at the opposite face, and then refracted back out again. If an observer is facing directly away from the Sun, the rainbow is seen at angles of 42° from the horizontal line made by the observer's shadow—the *antisolar* direction. The rainbow exhibits different colors at slightly different

angles because the different wavelengths in white sunlight are refracted by slightly different angles: from 42° for red light to 40° for violet.

Another quite familiar phenomenon is a halo around the Sun at 22° to the solar direction and occurring when sunlight passes through high cirrus ice clouds. A common ice crystal shape in these clouds is hexagonal, and the crystals can grow quite long in the third dimension (that is, they are then shaped like pencils). The sunlight is refracted through two crystal faces successively to give the 22° angle. If the ice crystals are tumbling down at random angles to the horizontal then a complete ring will occur around the Sun. More complicated arcs and Sun pillars occur from similar crystals when they become oriented with their long axes in the horizontal. This can occur when the air is calm and when the crystals have grown to a specific size, about 0.1 millimeter.

[*See also* Aerosols.]

BIBLIOGRAPHY

Coulson, K. L. *Polarization and Intensity of Light in the Atmosphere.* Hampton, Va., 1988.
Deirmendjian, D. *Electromagnetic Scattering by Spherical Polydispersions.* New York, 1969.
Greenler, R. *Rainbows, Halos and Glories.* New York, 1980.
Hansen, J. E., and L. D. Travis. "Light Scattering in Planetary Atmosphere." *Space Science Reviews* 16 (1974): 527–610.
Kerker, M. *The Scattering of Light.* New York, 1969.
Longhurst, R. S. *Geometrical and Physical Optics.* London and New York, 1957.
Meinel, A., and M. Meinel. *Sunsets, Twilights, and Evening Skies.* New York, 1983.
Twomey, S. *Atmospheric Aerosols.* Amsterdam, 1977.

C. MARTIN R. PLATT

SCIENTIFIC METHODS. Physicist Niels Bohr (1885–1962) wrote, "The aim of Science is to extend our experience, and reduce it to order." Scientific methods guide the arts of scientific inquiry, evolving in the process. They are intended to provide fruitful and reliable procedures of investigation and argument.

In the early history of Western science and philosophy a major emphasis was given to systematic ordering and the extension of common-sense knowledge. This included knowledge of Earth and sky, the seasons, states of matter, attributes of life, and the practical arts. There developed methods of deduction, induction, definition, classification, measurement, and geometry. These early reductions to order led to such broad, unifying theories as Aristotle's biology or the atomic theory of Leucippus and Democritus. Descended from the practical arts, the earliest laboratories—such as that of Archimedes—were devoted first to mathematical physics.

Descriptive astronomy began in very ancient times and was transformed by the Greeks into quasi-realistic geometrical models that embodied the most essential parameters of planetary motion. The heliocentric transformations of Nicolaus Copernicus and Johannes Kepler produced a redefinition of these parameters adequate for Isaac Newton's explanation of the motion of the planets, all within his unified science of mechanics and gravitation. This synthesis brought together the traditions of observational astronomy and the laboratory.

Basic to the experimental tradition is the recognition that mechanical contrivance is often necessary to satisfy conditions under which hypotheses can be tested empirically. So Evangelista Torricelli hypothesized that the air we breathe was a gaseous terrestrial "ocean"; its pressure would balance liquid columns, evacuated above, exerting the same pressure. Predicting the height of a column of mercury from that of water lifted as high as possible by a suction pump, he confirmed his hypothesis: Atmosphere is not some weightless matter pervading space, as Aristotle had supposed and common sense often still does. Blaise Pascal showed later that the height of this mercury "baro-meter" decreased with altitude, thus finding the empirical basis for the Law of Atmospheres.

Such logic is central to all scientific methods; some testable hypothesis is formulated that would reduce to order some phenomenon that otherwise appears anomalous. Results of tests may—strongly or weakly—support the hypothesis or oppose it. The aim always is to minimize the risk of rejecting what is true, or accepting what is false.

But how are fruitful hypotheses generated? Torricelli's came from the Copernican conclusion that the Earth speeds through space, surely not through air. The atmosphere, then, moves with us. Contradicting common sense, Aristotle, and Christian dogma, this idea needed powerful support. The image of an open universe gave it life, but not proof. Torricelli's weighing machine and Pascal's use of it, like the telescope of Galileo Galilei, provided dramatic confirmation. It then also gave new impetus to the investigation of the gaseous state of matter and to atmospheric science. Such important testable hypotheses are often invented, as this was, under the sway of far-reaching beliefs too general themselves to be immediately testable.

The impressive evolution of physics and chemistry has often led philosophers to the belief that the only final aim of science is to formulate *covering laws*—unconditional, invariant relationships among essential variables that explain phenomena and make possible their prediction.

Natural history, however, has quite different aims. Hypotheses about the past require confirmation or rejection by phenomena observable in the present. Such investigations may make full use of known laws, but they must also rely on *uniformitarian* assumptions: Essential causes of change in the past must be the same as those still operative in the present. This guidance, however, evolves along with historical knowledge that may qualify it. Comets and asteroids are permanent members of the solar system, but their impact with Earth may have caused catastrophic changes in biological history.

The computer brings important extensions of scientific methods. Among these is the possibility of constructing working models of systems far more complex than the planetary models of early astronomy. With computer capabilities one can construct real-time models that use both established theory and new hypotheses. One then can "play the game" repeatedly, with variations of the hypotheses, to see how well the models agree with available evidence. One may study the gravitational dynamics of hypothetical star clusters, too complex for analytical methods, or the operation of a large urban hospital. Computerized weather and climate modeling has become essential, radically extending and redefining traditional research agenda. [*See* Models and Modeling.]

Complex systems, with interconnected components that are themselves complex systems, may sometimes be represented in first approximations by models that describe these components in simple holistic form. Philosophical *holism* maintains, and *reductionism* denies, that such components may contribute to the system some properties not derivable from their own detailed description. Philosophical debates aside, there can be important differences between studying a subsystem independently of the part it plays in the larger system and seeking to define just the part it plays in the whole system, without regard for its inner complexities.

Thus, thunderstorms play an import part in energy transport within the atmosphere but are too small in scale and too complex for inclusion within a climate model. Their effect is *parametrized* by greatly simplified assumptions. In different computer runs of the model, these assumptions can be varied somewhat in order to test the model's sensitivity to such simplification. This is a holistic strategy. Climate modelers first hoped, likewise, for a relatively simple account of the effect of the ocean circulation, asking for a more holistic account than the oceanographers themselves were accustomed to look for. Since the mass of the ocean water is about two orders of magnitude greater than that of the atmosphere, the climatic influence of the oceans and their complex circulation hardly could be something merely added on to an otherwise purely atmospheric model. Climatology becomes increasingly a joint enterprise.

The study of the Earth's climate can be a link in a range of increasingly important studies. Climate history is part of geological history, especially that of the biosphere. Its future changes may depend, as well, upon human behavior. Much as Torricelli's hypothesis and experiments were embedded in the wider world view of Galileo, so the scientific study of climate gains meaning as part of a wider and richer view of the history and future of Earth.

The evolution of the biosphere has always been linked to that of climate; humans may well be responsible for further changes. Scientific methods needed for such investigations include not only those of climate modeling, but also the analogies of history, of what can be learned from our growing knowledge of past climatic changes and their effects on the biosphere.

BIBLIOGRAPHY

Dewey, J. "The Logic of Scientific Method." Part 4 of *Logic: The Theory of Inquiry*. New York, 1938.

Goldberg, L. P. *Interconnectedness in Nature and in Science: The Case of Climate and Climate Modeling*. Cooperative Thesis No. 121, University of Colorado and National Center for Atmospheric Research. Boulder. Colo., 1989.

Laudan, L. *A Source-book in Scientific Method: Galileo to Mach*. Cambridge, Mass., 1968.

Laudan. L. "Theories of Scientific Method from Plato to Mach: A Bibliographical Review." *History of Science* 7 (1968): 1–63.

Lovelock, J. *The Ages of Gaia: A Biography of Our Living Earth*. New York, 1988.

Lucretius. *De Rerum Natura*. English translation by W. D. Rouse. Boston, 1982.

Mayr, E. *Toward a New Philosophy of Biology: Observations of an Evolutionists*. Cambridge, Mass., 1988.

Nagel, E. *The Structure of Science: Problems in the Logic of Scientific Explanation*. New York, 1961.

Popper, K. R. *Objective Knowledge: An Evolutionary Approach*. Oxford, 1972.

Schneider, S. H., and R. Londer. "The Twin Earth: Computer Models of Weather and Climate." In *The Coevolution of Climate and Life*, edited by Stephen H. Schneider and Randi Londer. San Francisco, 1984, pp. 205–221.

Segre, M. *In the Wake of Galileo*. New Brunswick, N.J., 1991.

Suppe, F. ed. *The Structure of Scientific Theories*, 2d ed. Urbana, Ill. 1977.

Toulmin, S. *Human Understanding*. Vol. 1. Princeton, N.J., 1972.

DAVID HAWKINS

SCULPTURE. *See* Painting, Sculpture, and Weather.

SEA-AIR INTERACTIONS. *See* El Niño, La Niña, and the Southern Oscillation; Maritime Climate; *and* Oceans.

SEA BREEZES. *See* Small-Scale or Local Winds.

SEA ICE is ice formed on oceans, seas, and bays through the freezing of sea water. Although it is sometimes attached to the shore and relatively immobile, sea ice is more often freely floating. In locations experiencing very calm weather, the ice at times forms as a vast, solid sheet, but such a sheet rarely remains intact for long, because buffeting by winds and waves tends to break it up. Hence, a field of sea ice usually consists of numerous individual ice floes and smaller pieces. Figure 1 shows a Landsat satellite image of sea ice near the continent of Antarctica; the larger floes are about 50 kilometers long.

Sea ice is distinguished from icebergs in that it is formed in the sea rather than entering the sea by breaking off from land ice. As a result, sea ice floes tend to be much thinner and smaller than icebergs. Typical ice floe thicknesses are 0.5 to 4.0 meters, and typical ice floe diameters range from 1 meter to many kilometers.

Formation and Types. The determining factors in the formation of sea ice are that the water reaches its freezing point and that existing conditions allow ice crystals to form. Calm conditions are conducive to ice formation, as are shallow water depths. The freezing point of sea water depends on its salinity, with higher salinities requiring lower temperatures before freezing occurs. Sea water salinities are typically about 34 parts per thousand, at which value the freezing point is $-1.8°C$, compared to $0°C$ for fresh water. As sea ice forms, much of the salt content of the freezing water is released to the ocean below, but some is trapped in *brine pockets,* which tend to migrate downward through the ice and into the underlying water as the ice ages. As a result, the salinity of the ice generally decreases with time, unless the ice gains additional salt by being flooded with sea water, for example, from wave action during a storm. Sea-ice salinities range from about 2 parts per thousand for old ice to about 15–20 parts per thousand for new ice formed under particularly cold conditions, which cause rapid sealing of the brine pockets.

Sea ice can take many forms, depending not only on atmospheric and oceanic conditions at the time of its formation but also on conditions later in its evolution. Some

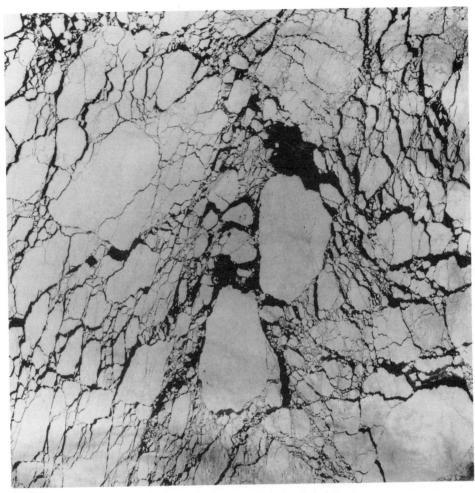

SEA ICE. Figure 1. *Landsat image of sea ice near Antarctica.* Image size is approximately 185 kilometers by 185 kilometers.

common ice types include *frazil ice,* consisting of fine ice particles suspended in water; *grease ice,* a soupy surface layer of ice crystals; *nilas,* a thin elastic sheet 0.01 to 0.10 meter in thickness; *pancake ice,* circular pieces of ice up to 0.1 meter thick and 0.3 to 3.0 meters in diameter; *young ice,* ice floes that have thickened to 0.1 to 0.3 meter; *first-year ice,* floes thicker than 0.3 meter that have not yet survived a summer melt period; and *multiyear ice,* floes that have survived at least one summer melt. Ice that is not freely floating but is attached to the shore or to the sea bed is called *shorefast ice.* Frazil ice, especially common in Antarctica, tends to appear in the first stages of ice growth, grease ice under agitated water conditions, and nilas under calm conditions. Pancake ice often has raised rims all along its edges, resulting from the pressure of the pancake floes against one another. Multiyear ice tends to have particularly low salinity, not only because

of its age but also because the very process of surviving a summer melt period tends to decrease the brine content of the ice, as the meltwater from the surface washes through the ice, flushing out some of the brine pockets.

Role of Sea Ice in the Climate System. The presence of sea ice influences the adjacent oceans and atmosphere in several important ways, especially after the ice has progressed beyond the frazil and grease ice stages to form a more solid interface. Sea ice of more than a few centimeters in thickness is a strong insulator, greatly restricting the amount of heat transferred between the ocean below it and the atmosphere above. This can be especially significant in winter, when the water temperature is very close to the freezing point but the air temperature is considerably lower, providing a sharp temperature contrast that encourages heat flow from the water to the air. Under such conditions, the heat transfer from the ocean to the atmosphere can be considerable (500 to 600 watts per square meter) in open spaces between ice floes or in the absence of ice. However, if the ocean and atmosphere are separated by ice a meter or more thick, very little heat (perhaps 10 to 20 watts per square meter) is conducted through the ice and transferred into the atmosphere. Ice thinner than 1 meter allows more heat through, but not as much as the heat flux under ice-free conditions. Hence, in wintertime the presence of sea ice leads to more heat being retained in the oceans and less being transferred to the atmosphere.

Sea ice also lessens the transfer of water vapor to the atmosphere, because the rate of evaporation from the ocean, during which liquid water is converted to water vapor, is much greater than the rate of sublimation from ice floes, during which ice is converted to vapor. Thus the presence of ice tends to reduce the amount of water vapor in the atmosphere, and consequently there is less precipitation over the ocean and nearby land areas.

Another transfer decreased by the presence of sea ice is the transfer of momentum (mass times velocity) from the atmosphere to the water. Surface-level winds tend to transfer some of their momentum to the water, resulting in waves. In the presence of ice, however, part of the momentum from the winds gets transferred to the ice floes, leaving less to be transferred to the water and hence dampening the wave motion. Anyone on board a ship in the polar oceans during a period of strong winds can appreciate the result; the waters outside the ice pack are almost always far rougher than the waters within it.

Sea ice has a much higher albedo than water does, meaning that a much larger percentage of incident solar radiation gets reflected off a sea-ice surface than off a water surface. With sea ice, generally well over half the incident solar radiation is reflected; and if fresh snow covers the ice, the portion reflected can exceed 90 percent. The portion reflected off ocean water tends to be only 5 to 15 percent. Most of the reflected energy travels through the atmosphere and leaves the Earth-atmosphere system, so the existence of a sea-ice cover means that less solar energy gets absorbed within the system than would be the case without the ice. Thus, if the ice cover begins to retreat substantially, year by year, more solar radiation will be absorbed, warming will occur (barring possible counteracting influences) and the ice could be expected to retreat even more, in what is termed a *positive feedback*, because the initial change is accentuated rather than dampened.

In each of these effects, sea ice is known to influence the local atmosphere and ocean; however, its influence

on the atmosphere and ocean distant from the polar regions is less certain. There is, however, one sea-ice process known to have influences extending throughout the deep waters of the global oceans. This process is the previously mentioned rejection of salt to the underlying water as sea ice forms and ages. This salt rejection increases the density of the underlying water and therefore increases the chance that the surface water will become denser than the water beneath, encouraging convective overturning. The overturning is occasionally deep enough to reach the ocean bottom; once it does so, the newly generated bottom water becomes incorporated in the larger-scale bottom-water circulation. Although measurements are scant, it is believed that a substantial portion of the bottom water of the global oceans was generated through this process of overturning in the region of the sea-ice cover.

Distribution and Extent. The many local impacts of sea ice are important in a global context because of the large areal expanse of the sea-ice cover (Figure 2). In the Northern Hemisphere, even at the end of the summer melt period, the ice covers approximately 9×10^6 square kilometers; in winter, it covers almost the entire Arctic Ocean, Kara Sea, Canadian Archipelago, Baffin Bay, and Hudson Bay, as well as large portions of the Sea of Okhotsk, Bering Sea, Barents Sea, and Greenland Sea. Altogether, the Northern Hemisphere winter ice extends over approximately 16×10^6 square kilometers, greatly exceeding the area of the United States (9.5×10^6 square kilometers) or of Canada (10.0×10^6 square kilometers).

In the Southern Hemisphere, the central polar region is covered by the continent of Antarctica, preventing sea ice from extending to the highest latitudes (80° to 90° latitude) as it does in the Northern Hemisphere. As a result, the extent of summertime ice in the Southern Hemisphere is much less, covering only 4×10^6 square kilometers and confined largely to two regions—the western Weddell Sea and the Bellingshausen and Amundsen seas. In wintertime, however, the sea ice extends far out from the Antarctic continent at almost all longitudes, covering a total ocean area of approximately 19×10^6 square kilometers, even larger than the wintertime ice coverage in the Northern Hemisphere.

The Arctic sea-ice cover generally reaches its maximum extent in March and its minimum extent in September; the Antarctic ice generally reaches its maximum extent in September and its minimum extent in February. One attempted explanation for the more rapid decay of the ice in the Antarctic is the large upward flux of heat from the southern ocean. The ice covers of the two hem-

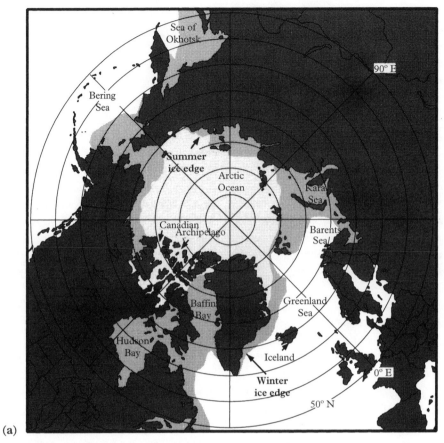

(a)

(b)

ispheres also differ in other ways. For example, the ice in the central Arctic tends to be thicker and older than the ice in the Antarctic and tends to be more extensively covered by meltponds (ponds of meltwater on the ice surface) in summer. Additionally, the Antarctic ice is not as constrained geographically as the Arctic ice and can therefore expand outward more readily in response to divergent wind forcing and thermodynamic growth.

When the ice extents in the Northern and Southern hemispheres are added, the total sea ice coverage of the global oceans at any time is between about 20×10^6 and 30×10^6 square kilometers. The average global sea-ice area, approximately 25×10^6 square kilometers, exceeds the area of North America.

Ice Motion. Sea ice moves in response to a variety of factors, including winds, waves, currents, and collisions between ice floes. Ice-floe speeds vary considerably depending on the forces acting on the ice, but speeds up to 15 kilometers per day are not unusual. Ice can also remain nearly stationary, either under very calm conditions or in regions of highly compact ice.

Observations of the Arctic ice over the past century have revealed two major components in its long-term mean drift: a clockwise gyre termed the Beaufort Gyre, centered at about 80° north latitude, 155° west longitude and a Transpolar Drift Stream in which ice moves away from the Siberian coast, across the pole or near to it, and out of the Arctic through the Fram Strait (80° north latitude, 5° west longitude) between Greenland and Svalbard. Since the late 1970s, buoys placed on the ice and tracked remotely have added many details to the knowledge of the ice drift patterns, including such deviations from the long-term mean as a recurring late-summertime reversal in the direction of flow in the Beaufort Gyre, at least partly in response to changed atmospheric forcing.

Less is known of Antarctic ice drift than of Arctic ice drift. However, considerable evidence exists for a basic clockwise gyre in the Weddell Sea; and satellite data as well as buoys placed on the Antarctic ice are expected to reveal within the next several years many details of Antarctic ice motions.

Research. Traditionally, most knowledge of sea ice has come from ground-based observations; but since the 1960s, much has also been learned from satellite observations and numerical modeling.

Beginning with ground-based observations, the indigenous people of the Far North have probably had a vast store of information about sea ice for many centuries. Most of that information is not incorporated into the scientific literature, although there exists an extremely valuable, several-hundred-year record of sea-ice incidence along the coast of Iceland from the Norse settlers there. Others who encountered polar sea ice before the twentieth century were generally explorers seeking new land masses, the North Pole, or passages through the ice to the Far East, or whalers or sealers taking advantage of the high biological productivity of polar waters. A major early highlight in the scientific study of sea ice was the 1893–1896 expedition, led by Fridtjof Nansen and Otto Sverdrup, on the ship *Fram* in the central Arctic basin, with many measurements made of the ice, the atmosphere, and the ocean.

Since the *Fram* expedition there have been hundreds of scientific expeditions to the polar ice, some based on ships, some on aircraft, and some with field parties directly on the ice. These studies have revealed details regarding the structure, velocity, thickness, temperature, salinity, and roughness of the ice, snow thickness and meltponding on the ice, biological activity within and near the ice, and various meteorological and oceanic forcings. Such ground-based studies continue to yield valuable new information of the traditional type; since the 1960s, however, another primary purpose of many polar expeditions is the validation of the data now being obtained remotely by orbiting satellite instruments.

The remoteness of the polar regions and the generally harsh conditions within them make it prohibitively difficult and expensive to obtain large-scale information about the ice cover through ground-based observations alone. However, satellite technology developed in the second half of the twentieth century has made possible frequent, large-scale data coverage, and this has spawned or contributed to considerable sea-ice research.

Landsat images (for example, see Figure 1) can be used to study the sizes, shapes, and distributions of sea ice floes. When the same ice floe is identified on images taken at different times, scientists can examine structural changes and can calculate the average velocity of the floe over the intervening period. However, Landsat images— along with all other imagery based on visible radiation—

SEA ICE. FIGURE 2. *Typical spatial coverage of sea ice at the summer minimum and the winter maximum, for (a) the north polar region and (b) the south polar region.* The sea ice distributions are determined from the satellite passive-microwave data of the *Nimbus 7* Scanning Multichannel Microwave Radiometer (SMMR).

have two major shortcomings: they are unobtainable during darkness, and they can be badly obscured by cloud cover. Both these difficulties have been eliminated by using microwave radiation instead of visible radiation.

Both active-microwave and passive-microwave instruments have been used for sea-ice studies. The active instruments send a signal out and receive it back; whereas the passive instruments simply receive radiation emitted from elsewhere (and then record or transmit the data). Current satellite active-microwave data, from Synthetic Aperture Radar (SAR), have resolutions of 10 to 100 meters, comparable to the 80-meter resolution of Landsat data, and they can be used for the same types of detailed studies as the Landsat data. Satellite passive-microwave data have a much coarser resolution of 20 to 150 kilometers and cannot be used to identify individual ice floes or to perform comparably detailed local studies. Because of the sharp contrast in microwave emissions of ice and water, however, passive-microwave data can be used to calculate the percent areal coverage of ice (the ice *concentration*) and to reveal the large-scale distribution of sea ice throughout the polar regions every few days. This large-scale monitoring, particularly relevant to climate research, is not yet possible with the higher-resolution data.

Satellite passive-microwave data have been collected for most of the period since December 1972 and have allowed documentation of the day-by-day and year-by-year changes in sea-ice distributions. Many sizeable open-water features (called *polynyas*) within both polar ice covers have been revealed and monitored, some persisting through entire seasons and others opening and closing repeatedly on short timescales. Available data exhibit interannual variations since 1972 of about 12 percent in the yearly maximum extent of the Antarctic ice and about 8 percent in the yearly maximum extent of the Arctic ice, with neither hemisphere exhibiting a strong long-term trend toward either increasing or decreasing ice coverage. Individual regions in both hemispheres show far greater percentage variations, though again without convincing long-term trends. For example, the Sea of Okhotsk had maximum ice extents of 1.2×10^6 square kilometers in 1979, 0.65×10^6 square kilometers in 1984, and 1.0×10^6 square kilometers in 1985.

Necessary and valuable as ground-based and satellite observations are, they alone can never provide a complete understanding of the ice cover or explain sea-ice phenomena. For complete understanding, a theoretical context is essential. One technique that has proven extremely useful in this regard is numerical modeling. The spatial

distribution and thickness of the sea-ice covers of both polar regions have been modeled numerically by several research groups, and the resultant sea-ice models have been used both independently and in conjunction with atmospheric and oceanic models. [*See* Models and Modeling.]

Model calculations of the freezing and melting of sea ice are based on energy balances at the top and bottom surfaces of the ice. At the top surface of the ice, the fluxes incorporated in the energy balances include incoming solar radiation, incoming longwave radiation from the atmosphere, outgoing longwave radiation from the ice, sensible and latent heat fluxes to the atmosphere, the conductive flux through the ice, and the energy flux from melting on the ice surface. At the bottom of the ice, the fluxes incorporated are the heat flux from the water underneath the ice, the conductive flux through the ice, and the energy flux from melting or freezing at the bottom surface. Model calculations of ice motions are based on balancing the forces acting on the ice: the air stress from above, transferring momentum from the air to the ice; the water stress from below, transferring momentum from the water to the ice; the Coriolis force owing to the rotation of the Earth; the force owing to the local tilt of the ocean surface; and the stresses internal to the ice pack, for example, from collisions among ice floes.

Numerical models have been used to help understand large-scale aspects of the ice cover, such as the basic seasonal cycle, as well as individual features observed within the ice cover, such as a very large polynya that occurred in the Weddell Sea in the 1970s. They have also been used to simulate conditions that have not yet occurred, such as the effect on the ice of possible future global warming or global cooling. When used in conjunction with models of the atmosphere and oceans, the ice models have demonstrated the significant impact that the ice cover has on local atmospheric and oceanic conditions. Together, these various studies have added considerably to the understanding of the polar sea ice covers and their role in the global climate system.

[*See also* Antarctica; Antarctic Ice Sheet; Arctic; Greenland Ice Sheet; *and* Icebergs.]

BIBLIOGRAPHY

Barry, R. G. "Arctic Ocean Ice and Climate: Perspectives on a Century of Polar Research." *Annals of the Association of American Geographers* 73 (1983): 485–501.
Bourke, R. H., and R. P. Garrett. "Sea Ice Thickness Distribution in the Arctic Ocean." *Cold Regions Science and Technology* 13 (1987): 259–280.
Gloersen, P., et al. *Arctic and Antarctic Sea Ice, 1978–1987: Satellite*

Passive-Microwave Observations and Analysis. NASA SP-511. Washington, D.C., 1992.

Gordon, A. L., and J. C. Comiso. "Polynyas in the Southern Ocean." *Scientific American* 256 (June 1988): 90–97.

Hall, D. K., and J. Martinec. *Remote Sensing of Ice and Snow.* London, 1985.

Massom, R. A. "The Biological Significance of Open Water within the Sea Ice Covers of the Polar Regions." *Endeavour* 12 (1988): 21–27.

Nansen, F. *Farthest North: The Norwegian Polar Expedition, 1893–1896.* 2 vols. Westminster, 1897.

Parkinson, C. L. "On the Value of Long-term Satellite Passive Microwave Data Sets for Sea Ice/Climate Studies." *GeoJournal* 18 (1989): 9–20.

Steffen, K. *Atlas of the Sea Ice Types, Deformation Processes and Openings in the Ice.* Zurich, 1986.

Washington, W. M., and C. L. Parkinson. *An Introduction to Three-Dimensional Climate Modeling.* Mill Valley, Calif., 1986.

CLAIRE L. PARKINSON

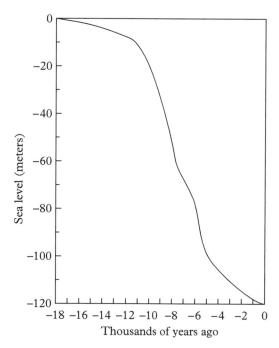

SEA LEVEL. Figure 1. *The sea-level record from Barbados, derived by dating of coral samples found at several depths. It covers the last period of deglaciation. (From Fairbanks, 1989.)*

SEA LEVEL. As measured at tide gauge stations all around the world, sea level varies on a wide range of time scales. The tides are the most obvious manifestation of sea level change, but short-term air pressure, wind set-up (wind forces acting on the sea surface that pile up water against the coasts), and dynamical adjustment of the sea surface to changing currents also cause variations on short time scales. Such changes may be as large as several meters during a storm surge in a shallow sea. Devastating tsunamis (waves caused by earthquakes) produce even larger changes, albeit for a short time.

On longer-term time scales, geophysical effects become more important. On time scales larger than a decade, ocean volume may change so much that it becomes detectable in local records from stations in tectonically stable regions. Going farther back in time, thousands to 2 million years, the large changes associated with the Pleistocene glacial cycles dominate the picture. These changes, of the order of 100 meters worldwide, were mainly associated with the advance and retreat of the big ice sheets in the Northern Hemisphere. On even longer time scales, millions to billions of years, tectonics and sedimentation of ocean basins are the most important factors.

Figure 1 shows a reconstruction of the rise in sea level during the last 17,000 years—the time during which the large ice-age ice sheets of the Northern Hemisphere melted. It is based on data derived by R. G. Fairbanks from carbon-14 dating of samples from coral reefs off Barbados. The species *Acropora palmata* is especially suitable for this type of analysis, because it grows fast and

only in the upper few meters of the sea, permitting an interpretation of the depth of a sample in terms of relative sea level. In addition, the Barbados region was only marginally affected by vertical movements of isostatic effects associated with the decaying ice sheets. This record is probably the most reliable one available today.

Aside from these events, sea-level change generally results from shifts in the water budget of the Earth, and from changes in the configuration of ocean basins and coastlines. The water budget can be formulated as:

$$M_{oc} + M_{at} + M_{ice} + M_{gw} + M_{la} \approx \text{constant.}$$

M_{oc} is the amount of water in the ocean, M_{at} the amount in the atmosphere (mainly in the form of water vapor), M_{ice} the amount stored as land ice, M_{gw} the amount of ground water, and M_{la} the amount of water in lakes. The sum of all these components should remain almost constant; small changes may occur because of the formation of water in chemical reactions, important only on very long geological time scales. For practical reasons, sea ice and floating ice shelves are included in M_{oc}, because melting does not affect sea level directly. Table 1 gives a crude estimate of the components of this water balance (Warrick and Oerlemans, 1990; SMIC Report, 1971).

In terms of the water budget, during the deglaciation process the exchange was mainly between the ocean and

SEA LEVEL. Table 1. *Amounts of mass (in 10^{18} kilograms) in the different components of the Earth's water budget (the third column indicates the uncertainty of these estimates)*

M_{oc}	1,370	5%
M_{at}	0.013	15%
M_{ice}	29	10%
M_{gw}	64	100%
M_{la}	0.023	10%

land ice reservoirs. During deglaciation, the rate of sea-level rise may have been as large as 1 centimeter per year. This seems to assign an upper limit to sea-level changes associated with natural processes, as opposed to anthropogenic (humanly caused) effects.

Most instrumental records of sea-level change are not very long. Many started only in the twentieth century, but there are a few old records, including the Amsterdam record (1682–1928, calibrated extension from overlap with nearby stations) and the Stockholm record (1780 to present). Annual mean values from these records are shown in Figure 2, together with a global mean sea-level record derived by Gornitz and Lebedeff (1987).

Figure 2 illustrates some of the problems encountered in the analysis of instrumental sea-level records. First, gaps in the records occur. This is often associated with changes in observational procedure, or even movement of the gauge to another location. The largest difficulties arise, however, when one tries to derive a global figure

from local records. For example, the records of Amsterdam and Stockholm show little resemblance. In this case the reason is well known: glacial rebound of the Earth's crust in Scandinavia, following the last deglaciation, is still going on, and relative sea level is dropping. In the case of Amsterdam, there is a small effect of isostatic sinking and compaction of the soil. Accounting for such effects with high precision will soon be possible, through newly developed geodetic techniques employing quasar interferometry and geolitic satellites. It is not well understood why the interannual variability in the Stockholm record is larger than in the Amsterdam record.

Various methods have been applied to existing records to derive a figure for global mean sea-level rise over the last 100 years. The number of records available ranges from 30 to 100, depending on the criteria set for length, continuity, reliability, and isostatic/tectonic stability of the location. One possibility is to discard all records from unstable regions and to compose a weighted average based on the remaining stations. The disadvantage is obvious: much information is lost, and the remaining stations are unevenly distributed over the globe. Another method is to try to correct for local isostatic/tectonic movements. In many cases, it is possible to determine local uplift from elevated shorelines dated with radioactive methods. The estimate of global sea-level rise shown in Figure 2 is based on the latter procedure (Gornitz and Lebedeff, 1987).

In spite of the problems just discussed, evidence is

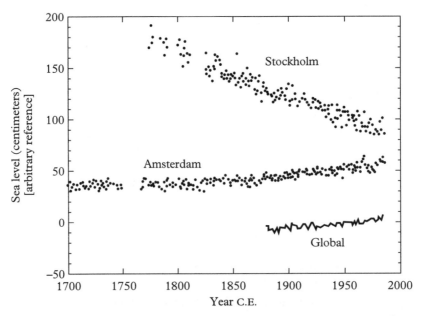

SEA LEVEL. Figure 2. *Two local sea-level records and an estimate of the change in global mean sea level.* (From Gornitz and Lebedeff, 1987.)

becoming convincing that global sea level has indeed been rising over the past hundred years. Almost all estimates of the amount of rise are in the 10 to 20 centimeters per century range (Warrick and Oerlemans, 1990, give a review). Why sea level is rising at this rate is unclear. The worldwide retreat of mountain glaciers since the end of the nineteenth century is without doubt a real feature, but estimates show that this cannot have contributed more than a few centimeters (Meier, 1984). Thermal expansion can provide another explanation for the observed sea-level rise.

Understanding this requires the introduction of the *steric effect* on global mean sea level. Ignoring changes in the geometry of the oceans for the moment, it is the *volume* of ocean water, rather than its mass, that determines global mean sea level. Thus, changes in sea level are the result of changes in total mass of the oceans, coupled with changes in mean water density. Wigley and Raper (1987) suggest, on the basis of a calculation with a very simple ocean model, that the ocean warming of 0.5 K over the past 100 years may have caused a thermal expansion effect of 7 centimeters. There is considerable uncertainty in this estimate, however. The rate at which heat goes into the ocean is an important parameter, but modeling it is very difficult. The expansion coefficient of sea water depends strongly on temperature: a 100-meter-thick layer of ocean water (of average salinity) at a temperature of 0°C will expand about 0.5 centimeters in response to a 1°C warming; for the same layer at 25°C, this figure is approximately 3 centimeters. This implies that a useful hindcast of sea-level change owing to thermal expansion requires the *spatial* pattern of temperature change as input to an ocean model with spatial resolution. It is not clear, however, that the quality of input data is sufficient for such an analysis. In summary, it is likely that thermal expansion has contributed in a positive sense to sea-level rise, and the contribution is about 5 centimeters.

The large ice sheets of Greenland and Antarctica remain the major unknown factors. In spite of increasing research efforts, it is not yet possible to determine whether these ice sheets are growing, shrinking, or in close equilibrium with the present climate. Long time scales are involved in the dynamics of ice sheets, and it seems likely that, at least during the Pleistocene, the Antarctic and Greenland ice sheets were hardly ever in real equilibrium. The currently available data sets on the mass budget do not allow us to detect an imbalance smaller than about 25 percent.

These uncertainties and inadequacies all contribute to one observation. At this time, the observed sea-level rise cannot be explained, in the sense that the shift in the water budget responsible for it cannot be identified with confidence.

Sea-level rise may accelerate during global warming. In view of the important consequences this may have for coastal defense and water management in many regions, it is not surprising that attempts have been made to project future sea levels (Hoffman et al., 1983; Robin, 1986; Oerlemans, 1989; Warrick and Oerlemans, 1990). Although the diagnosis of the present state is poor, as discussed above, this does not necessarily imply that estimates of the sensitivity of sea level to climate warming cannot be made. Great caution is required, however.

Most studies agree on the following. It is likely that in the case of a climate warming, glaciers and small ice caps will shrink considerably. The increase of melting on the Greenland ice sheet will be larger than a possible compensating increase of snow accumulation. Total snow accumulation on the Antarctic ice sheet will increase. Ice-flow instabilities, leading to rapid discharge of ice into the ocean, may occur at some places in response to increasing ocean water temperature. Finally thermal expansion will probably make the largest contribution to sea level rise on the 100-year time scale.

Quantification of the various effects has been done in different ways, which are not reviewed here. It is instructive, however, to provide a feeling for the uncertainties involved. Figure 3 summarizes the various contributions. The vertical bars indicate a 95-percent interval—that is,

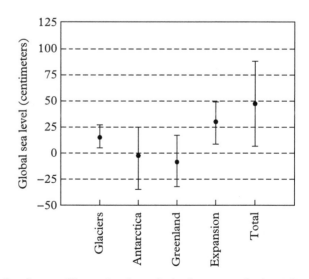

SEA LEVEL. Figure 3. *A synthesis of recent projections of sea level in the year 2100 under the influence of predicted greenhouse warming.* The error bars refer to a 95-percent probability interval.

the probability of actual effects being in the range indicated is 0.95. The figure makes it clear that the largest contributions (best estimates) are expected to come from thermal expansion and melting of glaciers. Contributions from the major ice sheets of Greenland and Antarctica seem small, but these have the largest share in the overall uncertainty.

[*See also* Antarctic Ice Sheet; *and* Ice Age Sea Level.]

BIBLIOGRAPHY

Fairbanks, R. G. "A 17,000 Year Glacial-Eustatic Sea-Level Record: Influence of Glacial Melting on the Younger Dryas Event and Deep Ocean Circulation." *Nature* 342 (1989): 637–641.

Gornitz, V., and S. Lebedeff. "Global Sea-Level Changes during the Past Century." In *Sea-Level Change and Coastal Evolution,* edited by D. Nummedal et al., 1987.

Hoffman, J. S., J. B. Wells, and J. G. Titus. *Projecting Future Sea Level Rise: Methodology, Estimates to the year 2100 and Research Needs.* U.S. Environmental Protection Agency. Washington, D.C., 1983.

Meier, M. F. "Contribution of Small Glaciers to Global Sea Level." *Science* 226 (1984): 1418–1421.

Oerlemans, J. "A Projection of Future Sea Level." *Climatic Change* 15 (1989): 151–174.

Robin, G. de Q. "Changing the Sea Level." In *The Greenhouse Effect: Climatic Change and Ecosystems,* edited by B. Bolin et al. New York, 1986, pp. 323–359.

SMIC-Report. *Inadvertent Climate Modification.* Cambridge, Mass., 1971.

Warrick, R. A., and J. Oerlemans. "Sea Level Rise." In *Climate Change: The IPCC Scientific Assessment.* Cambridge, 1990, pp. 257–281.

Wigley, T. M. L., and S. C. B. Raper. "Thermal Expansion of Sea Water Associated with Global Warming." *Nature* 330 (1987): 127–131.

JOHANNES OERLEMANS

SEASONS. The word *season* comes from the Latin word *satio,* meaning "time of sowing." In the context of climate and weather, *season* refers to a period of time, often on the order of several months, that is characterized by reasonably similar or coherent weather patterns. People who live in Earth's temperate zones traditionally think of four seasons of the year, each defined astronomically by specific events in the planet's orbit around the Sun and culturally by the cycle of plant growth. The discussion that follows deals with the astronomical seasons directly controlling annual cycles in both weather and climate.

Recognizing that the seasons should define periods with relatively homogeneous weather patterns, some scientists have abandoned the astronomical definitions and have elected to define seasons based entirely on actual weather events. These seasonal definitions are often based on such things as the length of time between frosts, wet versus dry times of year, tornado probabilities, or periods of frequent snowfall. The techniques used to define these seasons range from subjective decisions made by an individual interested in a particular climate-related phenomenon to highly complex, multivariate statistical procedures applied to large matrices of many interrelated climate variables. Alternate definitions of the seasons are especially useful in parts of the world other than the northern temperate zone, such as the tropics of Australia.

Causes of Seasonality. Most of us are familiar with the concept of the seasons; from early childhood we become aware of the weather associated with winter, spring, summer, and fall. Later we learn that in some parts of the world winter can last months without any sunlight, that summer can produce 24 hours of continuous daylight, while other regions have no seasonal shifts, and that the seasons are reversed in the Northern and Southern Hemispheres. We discover that much of the biosphere seems conscious of seasonal cycles; indeed, the evolution of almost all life on the planet is closely related to the procession of the seasons. Scientists dealing with weather and climate must learn about the fundamental causes of these seasonal patterns.

Seasonal patterns are caused by the tilt of the Earth's axis and the orbit of the Earth around the Sun. The Earth travels around the Sun in a slightly elliptical orbit with the Sun at one focus of this ellipse. The Earth is farthest from the Sun—approximately 152 million kilometers (94.5 million miles)—on 4 July; this date is called the *aphelion.* The *perihelion,* or closest approach, occurs on 3 January, when the Earth is only 147 million kilometers (91.5 million miles) from the Sun. Because the Sun's output remains relatively fixed from day to day (± 1.5 percent), the Earth receives about 3 percent more energy than the annual daily average from the Sun on the perihelion and about 3 percent less on the aphelion. Thus this ellipticity in the orbit cannot be the fundamental cause of the seasons in the Northern Hemisphere.

The single greatest cause of the observed seasons is the relative tilt of the Earth's axis. With respect to the Sun, the Earth rotates on its axis, on average, once every 24 hours. The result of this rotation is the day/night pattern that has such an obvious effect on daily weather patterns. This axis of rotation is not vertical with respect to the

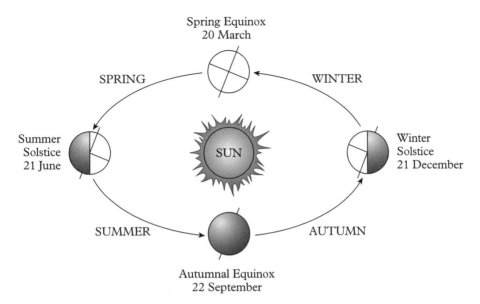

SEASONS. Figure 1. *Revolution of the Earth around the Sun (Northern Hemisphere).* In the Southern Hemisphere the winter solstice is about 21 June and the summer solstice is about 21 December.

plane of the Earth's orbit; rather, it is tilted at an angle of approximately 66.55° with respect to the ecliptic plane, or 23.5° from a vertical from the plane.

Very importantly, the Earth does not wobble much on its axis. The position of the axis on any given day is almost exactly parallel to the axis position on any other day of the year, a characteristic called *parallelism.* The Earth's axis in the Northern Hemisphere points to a position in the sky very close to the star Polaris, and owing

to parallelism, this "North Star" is in approximately the same position throughout the year.

As seen in Figure 1, at some times of the year the axis of the Northern Hemisphere leans toward the Sun while the axis of the Southern Hemisphere leans away from it. At other times of the year the opposite is true. This leaning of the axis with respect to the Sun is the ultimate cause of the seasons. Figure 2 shows the Earth on the day when the axis in the Northern Hemisphere leans the most

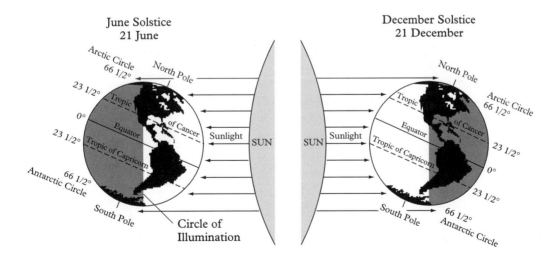

SEASONS. Figure 2. *Characteristics of incoming solar radiation at the solstices.*

toward the Sun. This day, called the *solstice*, occurs about 21 June. As seen in Figure 2, the rays of the noonday Sun then strike the Earth perpendicularly at 23.5° north latitude (the tropic of Cancer). The latitude where the perpendicular rays strike is called the *latitude of declination*. In the absence of clouds this latitude would receive the most intense sunlight. As one moves away from 23.5° north latitude, the angle of the Sun's rays is reduced by 1° for every 1° of latitudinal movement; for example, New York City is 18° north of the tropic of Cancer, so the noonday Sun there is 72° above the southern horizon on the date of the summer solstice, less intense than at the tropic of Cancer. Sydney, Australia, is 57° south of the tropic of Cancer—the noonday Sun in Sydney on 21 June is only 33° above the northern horizon. Latitudinal differences in the intensity of sunlight are one reason why the Northern Hemisphere on this date has warm summer conditions and the Southern Hemisphere has cooler, winter conditions.

Several other interesting things can be observed in Figure 2. Notice that as the Earth rotates through the summer solstice, the North Pole receives continuous sunlight, as does the entire area from the Arctic Circle to the pole. In contrast, the area between the Antarctic Circle and the South Pole is in constant darkness at this time.

As the Earth continues in its orbit for the following 6 months, the situation completely reverses. Near 21 December another solstice is reached when the most direct rays of the Sun strike the Earth at the tropic of Capricorn, the Antarctic area receives 24 hours of daylight, and the Arctic region is in total darkness.

The solstice actually occurs when the perpendicular rays of the Sun strike the tropic of Cancer or tropic of Capricorn, respectively. Throughout the rest of the year, the perpendicular rays of the Sun fall at different points between these latitudes. The "figure 8" found on many globes, called an *analemma* (Figure 3), shows the latitudes of declination where these perpendicular rays strike on each day of the year. On two of these days, the *equinoxes*, the noonday Sun is directly overhead at the equator. On these equinox days, near 20 March and 22 September, all places on the Earth receive 12 hours of daylight and 12 hours of darkness.

In most weather and climate applications, the seasons are defined by these astronomically determined dates. In the Northern Hemisphere the spring season begins at the vernal or spring equinox in March and extends until the summer solstice in June. During this season the point of most intense sunlight is moving from the equator to 23.5°

north latitude and the days in the Northern Hemisphere are lengthening; in the Southern Hemisphere, this period is marked by increasingly less intense sunlight and shorter days. From the solstice in June until the equinox in September, the perpendicular rays of the Sun migrate from the tropic of Cancer back to the equator. During this Northern Hemisphere summer season, the days shorten, while in this Southern Hemisphere winter season, the days lengthen. By the September equinox, all points on the earth have returned to 12 hours of daylight and 12 hours of darkness. From the equinox in September to the solstice in December, the most intense sunshine moves from the equator to the tropic of Capricorn. In the Northern Hemisphere, the sunshine gets progressively less intense and the days get shorter; conversely, this is the Southern Hemisphere's spring, when the days are getting longer and the Sun's energy is becoming more intense. By the December solstice, the perpendicular rays have reached the tropic of Capricorn. From the December solstice to the March equinox, the direct rays of the

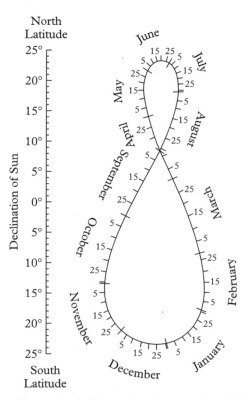

SEASONS. Figure 3. *Analemma showing latitudes of declination on dates throughout the year.* To find the declination of the Sun, trace a perpendicular line from the desired date on the analemma to the vertical declination measure. For example, 21 June has a declination of 23.5° north latitude.

Sun move from the tropic of Capricorn back to the equator. By the March equinox, the entire seasonal cycle has been completed.

Variations in the Seasons. The length of the present-day astronomical seasons varies from 89.0 days for the Northern Hemisphere winter season (December solstice to March equinox) to 93.6 days for the summer season (June solstice to September equinox). Over long periods of time the length of the seasons is known to vary owing to fluctuations in the orbital parameters. For example, approximately 100,000 years ago, the Northern Hemisphere summer season (or Southern Hemisphere winter season) was fully 8.8 days shorter than it is today, and the Northern Hemispheric winter was 8.8 days longer, because of a change in the orbital path around the Sun. These fluctuations in the length of the astronomically determined seasons certainly had an impact on the overall climate of the hemispheres. In addition to changes in the length of specific seasons, the dates of the aphelion and perihelion have changed over time, and the distance between Earth and Sun on these dates also varied. Finally, despite its short-term parallelism, the Earth wobbles slightly on its axis at times, changing the 23.5° tilt with respect to a perpendicular off the ecliptic plane. All these changes in orbital characteristics combine to create differing durations and intensities of the basic seasonal weather and climate patterns over very long periods of time. Many atmospheric scientists believe that these changes in the orbital characteristics of the Earth were responsible for the major ice ages during the past 1 million years.

Consequences of the Seasonal Cycle. The changing position of the perpendicular rays of the Sun profoundly influences the general circulation of the Earth's atmosphere, which in turns alters weather patterns throughout the world. During the equinox periods, the most intense energy from the sun arrives near the equator. In response to this heating, air rises and the intertropical convergence zone of cloudiness and intense rainfall forms around the world near the equator. The subtropical high pressure systems are located between 20° and 30° in both hemispheres, and the subpolar lows and associated upper-level jet stream are located near 60° in both hemispheres.

As the perpendicular rays migrate into the Northern Hemisphere between March and June, the general circulation features also move in a northward direction. During the Northern Hemisphere's summer season, the intertropical convergence is well into the Northern Hemisphere, with greatest northward movement over most land areas. The subtropical high moves northward, and the Northern Hemisphere's subpolar jet stream is far to the north of its mean position. Conversely, in the Southern Hemisphere, the subtropical high pressure cells, the circumpolar vortex, and all other circulation features expand northward.

This change in circulation in the atmosphere is associated with similar changes in oceanic circulation. The seasons cause oceanic currents to move into new positions, and upwelling zones appear in different locations; these changes in oceanic circulation provide further feedback to the atmosphere, creating highly seasonalized climate events. The Asian monsoon is an outstanding example of a strongly seasonal phenomenon that results from closely interrelated oceanic and atmospheric adjustments to the migration of solar energy over the surface of the Earth.

Conclusions. A season is a period of time, generally a few months in length, when the weather and climate of a location, region, or hemisphere are relatively homogeneous in terms of some prescribed definition of atmospheric conditions, such as temperature. precipitation, or storminess. The change from one season to the next is fundamentally caused by the tilt of the Earth on its axis and the resulting annual migration of the most intense rays of the Sun. The atmospheric and oceanic circulation systems respond to this movement of energy. Understanding the underlying causes of these seasonal patterns in weather and climate provides a solid foundation for further exploring the complexities of the atmospheric system.

[*See also* Orbital Variations; *and* Religion and Weather.]

BIBLIOGRAPHY

Barry, R. G., and A. H. Perry. *Synoptic Climatology.* London, 1973.

Bryson, R. A., and J. F. Lahey. *The March of the Seasons.* Madison, Wis., 1958.

Hays, J. D., J. Imbrie, and N. J. Shackleton. "Variations in the Earth's Orbit: Pacemaker of the Ice Ages." *Science* 194 (1976): 1121–1132.

Kukla, G., A. Berger, R. Lotti, and J. Brown. "Orbital Signature of Interglacials." *Nature* 290 (1981): 295–300.

Prescott, J. A., and J. A. Collins. "The Lag of Temperature behind Solar Radiation." *Quarterly Journal of the Royal Meteorological Society* 77 (1951): 121–126.

Strahler, A. N. *The Earth Sciences.* New York, 1971.

Trenberth, K. A. "What are the Seasons?" *Bulletin of the American Meteorological Society* 64 (1983): 1276–1282.

ROBERT C. BALLING, JR.

SEA WEATHER OBSERVATIONS AND PREDICTIONS. *See* Marine Weather Observations and Predictions.

SEMI-GEOSTROPHIC THEORY. *See* Vorticity.

SENSIBLE HEAT refers, generally speaking, to the heat we can feel. A parcel of air at 20°C has more sensible heat than one at 18°C. Sensible heat is a form of energy and is usually measured in units called *joules*. It is an informal way of referring to the thermodynamic state function more technically known as *enthalpy;* sometimes, however, *sensible heat* is used as a synonym for *dry static energy,* which also includes the energy associated with a parcel's altitude—atmospheric scientists are not entirely consistent in how they define and use this term.

Only changes in sensible heat are important in physical calculations, rather than its absolute magnitude. This limited usage avoids the many difficulties associated with defining and determining the absolute internal energy of a substance. [*See* Enthalpy; Thermodynamics.]

Consistent with the intuitive notion that sensible heat is something we can feel, the change in the sensible heat of a substance (ΔH) is the change in temperature (ΔT) multiplied by the specific heat for that substance (c_p)—assuming constant pressure and no phase changes. This change is expressed by the equation

$$\Delta H = c_p \Delta T.$$

The specific heat (c_p) of dry air is approximately 1,000 joules per kilogram for each kelvin; that of liquid water is approximately 4,200 joules per kilogram per kelvin.

The term *sensible heat* is used mainly in the context of a *flux* or *transport,* rather than by itself. In the atmospheric sciences, it is used in contrast to *latent heat* and *radiant heat.* Latent heat is the heat needed to transform water into vapor from either its liquid or ice phase. Radiant heat, or radiation, is electromagnetic energy, such as sunlight or the infrared radiation emitted from the Earth. Solar heating is the radiant heat flux from the Sun to the Earth. [*See* Latent Heat; Radiation.]

Within both the atmosphere and oceans, there is a large heat transport from the tropics toward the poles, driven by the large difference in solar heating between these regions. Much of this transport is a sensible heat flux. For the average temperature of the Earth to remain nearly constant—as it does—the infrared radiation from the Earth to space must balance the solar heating, globally totaled. Locally, however, these two radiant heat fluxes do not balance. If it were not for the large horizontal heat flux, the tropics would be much hotter and the poles much colder than they are. On average, the currents of air and water move warm parcels from the tropics toward the poles and cooler parcels from the polar regions toward the equator. The total amount of the sensible heat flux in the atmosphere at 40° north latitude (where the meridional fluxes are near their maxima) is roughly 2.5×10^{15} joules per second (1 joule per second = 1 watt); at this latitude the oceanic sensible heat flux is about 1×10^{15} watts, and the atmospheric latent heat flux is about 2×10^{15} watts. For comparison, the electrical generating capacity of all of the world's power plants is only about 3×10^{12} watts.

There is also a large vertical heat transport in the atmosphere, because most of the incoming solar radiation goes through the atmosphere and is absorbed at the Earth's surface. Very close to the surface, in the lowest few centimeters, there is a large sensible heat flux from diffusion. Just above that diffusive layer, heating of the air near the Earth's surface drives convection, which leads to large sensible and latent heat fluxes in the vertical. These sensible heat fluxes can be attributed to the rising of warmer air and the sinking of cooler air. Vertical fluxes of both sensible and latent heat in the atmosphere vary enormously from place to place and time to time, from tens to hundreds of watts per square meter.

[*See also* Convection; *and* General Circulation.]

BIBLIOGRAPHY

Gill, A. E. *Atmosphere-Ocean Dynamics.* New York, 1982.
Michaud, R., and J. Derome. "On the Meridional Transport of Energy in the Atmosphere and Ocean as Derived from Six Years of ECMWF Analysis." *Tellus* 43A (1991): 1–14.

MARK DAVID HANDEL

SEVERE WEATHER. The term *severe weather* probably summons different images for different people; however, there is a strict definition of a severe thunderstorm that is used by the U.S. National Weather Service (NWS). To be classified as severe, a thunderstorm must produce at least one of the following: hail reaching a diameter of 1.9 centimeters (0.75 inches) or greater; wind gusts of at least 26 meters per second (58 miles per hour); or a tornado.

Because of the impact of severe weather on human life, researchers and forecasters expend great efforts to increase the understanding and improve the prediction of severe weather. Early efforts included the 1949 Thunderstorm Project and the 1950 Tornado Project, which led to establishment of the Severe Local Storms (SELS) unit in Kansas City, Missouri, where severe weather watches, forecasts of events up to 6 hours ahead, are still issued today. More focused research on severe weather came after the formation of the National Severe Storms Laboratory (NSSL) in Norman, Oklahoma, in 1964, and of programs at the National Center for Atmospheric Research (NCAR) in Boulder, Colorado, established in 1960.

Property and crop damage from hailstorms and strong winds can be extremely high. Damage from a single hailstorm has exceeded 300 million dollars, but deaths are rare. Tornadoes are responsible for a number of deaths and injuries each year, in addition to devastation of property. Tornado deaths in the United States averaged between 100 and 150 per year in the 1960s, but during most of the 1980s they declined to about 5 per year, in part due to improvements in NWS watches and warnings, as well as public preparedness and awareness. [*See* Hailstorms; Tornadoes.]

In a broad sense, the factors that support severe thunderstorms can be thought of as extreme forms of the same conditions that produce ordinary thunderstorms. For a thunderstorm to form, rising air must be warm and moist enough to produce a cloud that then continues to grow. To determine whether air can rise through the troposphere (the atmospheric layer closest to Earth's surface) and produce a thunderstorm, meteorologists examine the above-surface atmospheric structure by plotting the temperature, dewpoint, and wind speed and direction at various levels through the troposphere, as measured by a radiosonde. Radiosondes are carried aloft by balloons launched every 12 hours from a number of sites around the world. Though accurate, the depiction of the atmosphere derived from radiosonde data suffers from its coarse resolution in both time and space. Meteorologists continue to search for new means of remotely sensing the atmosphere; for example, automated wind profilers accurately measure winds through the troposphere every 6 minutes. Other profiling instruments resolve the temperature structure. More complete coverage of the globe is offered by satellite-borne instruments positioned far above the Earth's surface, although their accuracy and methods of dealing with cloud cover need further refinement to be as accurate as radiosonde measurements. Instrumented commercial aircraft sending wind, temperature, and moisture data in real-time are yet another source of information. The search for improved methods

is important, because significant differences in the above-surface moisture and temperature can develop in the intervals between radiosonde sites and launch times. [*See* Instrumentation; Thunderstorms.]

Despite the faults of radiosonde data, it remains an important source of information for the severe-storms forecaster. The data for a number of stations are plotted and analyzed, both manually and by computer, at various *significant levels* in the vertical. Forecasters also examine the radiosonde data for individual stations, using a pseudoadiabatic chart, referred to as a *sounding*. The most common method of analyzing a sounding is called the *parcel method*. Parcels or thermal bubbles of rising air, perhaps originating in the planetary boundary layer and rising as a result of the heating of Earth's surface by solar radiation, are represented on the pseudoadiabatic chart. The first problem is to determine whether enough moisture is present for the water vapor in the rising air to reach its saturation point and condense to form a cumulus cloud, which occurs at an altitude called the *lifting condensation level* (LCL). If the updraft air can remain buoyant, or warmer than the surrounding environment, the air will continue ascending and the cloud will grow. The pseudoadiabatic chart is used to determine if such buoyancy exists; the amount of buoyancy can easily be calculated with a small computer, the result being known as the *convective available potential energy* (CAPE) or *positive energy*. The parcel method neglects both mixing or entrainment of drier environmental air into the rising updraft, and the weight of condensed water being supported by the updraft, both of which tend to decrease the rising motion. In fact, mixing may destroy many incipient cumulus clouds before they have an opportunity to grow. [*See* Clouds.]

Eventually, the rising air in a cumulus cloud that grows to the towering cumulus stage reaches a level where it is cooler than the surrounding environment, though perhaps not until it reaches the warmer stratosphere, where the temperature increases or remains constant with height. The altitude where the rising parcel's temperature equals that in the environment is known as the *equilibrium level*; above this, the air is no longer buoyant and so must sink. As more and more rising air arrives at the equilibrium level, it spreads out and forms an anvil-shaped structure on top of the cloud, making the towering cumulus cloud a cumulonimbus cloud, the type most often associated with thunderstorms. Because the rising air in the cloud updraft can have a significant amount of momentum, it can overshoot the equilibrium level. The amount of overshooting, which can be measured via infrared satellite imagery, sometimes indicates storm severity, because it is proportional to the magnitude of the storm updraft.

Stronger updrafts are associated with more severe storms, in part because a strong updraft is capable of supporting the weight of hailstones long enough for them to become quite large. In addition to hail, severe surface winds can occur when strong downdrafts are associated with falling precipitation. The amount of CAPE is directly proportional to the strength of the updraft, so its calculation is an important part of forecasting severe weather. Before the advent of high-speed computers, meteorologists used a number of *stability indices* as quick substitutes for the more complicated calculation of CAPE. (Some of the indices also try to summarize characteristics of the vertical windshears, which is also important to severe weather environments.) Table 1 lists some of the better-known stability indices. Although the actual positive area or CAPE is a more complete indication of the energy available for thunderstorm development, some of these indices were developed after case studies of severe weather outbreaks and can be useful to forecasters in recognizing areas threatened by severe weather.

The hot, humid conditions common in summer are favorable for a warm LCL temperature and often for high values of CAPE, so it may be puzzling that there are not more thunderstorms and severe weather. The reason is that a layer of warmer air, called an *inversion*, often exists above the surface and prevents heated air rising from the surface from continuing to rise. In summer the inversion can be quite strong. It often lies between 2 and 4 kilometers above sea level, and can be the result of sinking air beneath a strong ridge of high pressure. It is not uncommon for inversions to be so strong that even the hottest days are not sufficient to create rising thermals that can break through the warm layer aloft. Meteorologists often refer to a thermodynamic sounding with such a warm layer aloft as being *capped*. The strength of the cap can be calculated; the result is the amount of energy necessary to overcome the inversion, called the *negative energy* or *convective inhibition* (CIN). Usually the CIN is far smaller than the CAPE, but even a small amount of CIN can effectively prevent convection. It is apparent from examining a plotted sounding, however, that having a cap or inversion increases the CAPE. Indeed, almost all soundings associated with severe weather outbreaks have strong capping inversions. Determining whether the cap can be broken and its energy released into a severe thunderstorm is one of the major problems facing severe-weather forecasters.

How can the inversion cap be overcome? This may happen if the air below is heated sufficiently so that the

SEVERE WEATHER. Table 1. *Stability indices*

Stability Index	Brief Description*	Typical Severe Weather Values
Lifted Index (LI)	$T_x - T_{parcel}$ (x = 500 millibars)	≤ -3
Modified Lifted Index (MLI)	$T_x - T_{parcel}$ (x = 700, 400, or 300 millibars)	≤ -3
Showalter Index (SI)	Like LI, but parcel begins at 850 millibars instead of the surface	≤ -2
K Index (K)	Uses temperature difference between 850 and 500 millibars combined with moisture at 850 and 700 millibars	$\geq$ about 30
Modified K	Like K but uses moisture from surface-850 for 850 millibars	$\geq$ about 40
Total Totals (TT)	$(T_{850} + T_{d850}) - 2(T_{500})$	$\geq$ about 48
Severe Weather Threat (SWEAT)	TT combined with an 850- and 500-millibar wind velocity $[12(T_{d850}) + 20(TT - 49) + 2(W_{850}) + W_{500}]$	$\geq$ about 300

*T = temperature: 850 millibars, about 1.5 kilometers above sea level; 500 millibars, about 5.5 kilometers above sea level; 400 millibars, about 7.0 kilometers above sea level; 300 millibars, about 9.0 kilometers above sea level. T_{parcel} is typically computed by mixing moisture in the lowest 50 to 100 millibars to determine the representative moisture, combined with an expected surface temperature, then lifting the parcel on the pseudoadiabatic chart to the desired level. T_d = dew point temperature. W_x = wind speed at level x.

rising parcels become warmer than the temperature of the air in the inversion. Alternatively, the inversion itself can weaken over time, either because cooler air moves in, or because the layer of air containing the inversion is lifted and therefore cooled. Such lifting can occur ahead of an approaching synopotic-scale storm system, perhaps over a period of many hours, during which the air below may continue to warm. Such situations are often associated with extensive outbreaks of severe weather.

Yet another way to break through the capping inversion is to physically force the air upward with enough positive energy to overcome the negative energy. Such forced lifting is often accomplished ahead of an approaching cold front, where the forced lifting of the warm and unstable air often results in a line (or several lines) of storms, called *squall lines*. Multiple squall lines can occur ahead of a cold front as a result of more complicated upward forcing, for example, by gravity waves. The huge outbreak of 148 tornadoes on 3–4 April 1974 in the central United States occurred in several squall lines ahead of a very strong cold front. In addition to cold fronts, other types of boundaries where winds are converging and forcing air to rise include thunderstorm gust fronts and sea and lake breezes. A particularly potent area known as *Tornado Alley* exists in Oklahoma and northern Texas where strong storms often develop along a boundary known as the *dry line,* so named because it forms at the leading edge of dry westerly downslope winds off the higher terrain of New Mexico. East of the dryline, strong southerly winds can bring high amounts of lower-level moisture northward from the Gulf of Mexico, creating ideal conditions for severe thunderstorms to form, if the forcing along the dry line, perhaps in combination with an approaching synoptic-scale disturbance, can overcome the capping inversion.

The strength and direction of the wind from the surface through the troposphere, called *wind shear,* also influences thunderstorm development and structure. When the winds throughout the troposphere are very light—as they often are in the summertime—the thunderstorm updraft is vertically erect, and the weight of the supported precipitation eventually falls directly down on it, replacing the updraft with strong downdraft and ending the storm. If the atmosphere has vertical wind shear, where the horizontal winds increase with height, then the developing storm becomes tilted in the vertical, so that precipitation does not fall directly onto the lower-level updraft. Instead, the updraft and downdraft remain separate, and the storm can persist longer. Thunderstorms

that form in an environment of high CAPE and strong wind shear can last for hours; called *supercell storms,* these almost always produce severe weather.

Another process that occurs when storms grow in the presence of vertical wind shear is the transfer of the horizontal vorticity in the environment into vertical vorticity, or circulation about a vertical axis, which causes the thunderstorm to rotate. If this rotation is strong enough, the storm is said to possess a *mesocyclone*. A persistent mesocyclone is a defining characteristic of a supercell storm. Nearly all strong tornadoes are preceded by the development of a mesocyclone in the middle levels (approximately 3 to 7 kilometers above ground), sometimes by 30 to 60 minutes. Because Doppler weather radars can measure the air flow as well as the precipitation within a thunderstorm, and hence can detect a mesocyclone, the NWS has been replacing their older radars with Doppler radars, a process that should improve severe weather warnings.

Only about half of the storms that have mesocyclones go on to produce tornadoes, although a much larger percentage of them are associated with severe wind or hail. The steps by which a supercell storm with a mesocyclone forms a tornado are not completely understood yet. Trained storm-spotters, usually combining volunteers and law-enforcement personnel, provide visual observations to supplement Doppler radar and satellite information, helping the NWS forecasters to issue warnings.

Recent studies in eastern Colorado and elsewhere have shown that tornadoes can form from storms that do not possess a mesocyclone, apparently by a mechanism similar to that in a waterspout (see Figure 1). These non-supercell storms apparently obtain the required lower-level vertical vorticity needed to produce a tornado from the interaction of the developing storm with a low-level boundary or boundaries. In such cases, the thunderstorm itself is usually caused by the forced rising motion associated with a boundary, or with the collision of two or more boundaries. Such boundaries may be thunderstorm gust fronts, or they may exist where winds converge as a result of topography or air mass differences. In such a scenario the vorticity, already aligned in the vertical, is stretched beneath the rapidly growing storm updraft to produce a tornado. These types of storms occur in vertical wind shears that are often quite weak, unlike the strong wind shear needed to support a supercell storm. Fortunately, most of the tornadoes produced by non-supercell storms are not strong, although they can cause significant damage and may represent a significant pro-

SEVERE WEATHER. Figure 1. *Photograph of a non-mesocyclone tornado about 4:20 P.M. mountain daylight time on 15 June 1988, as it approached the Denver, Colorado, NWS Forecast Office.* The tornado is on the ground, about 2 kilometers to the southwest, although the visible condensation funnel does not extend fully to the ground. To the left of the tornado is an automated atmospheric wind profiler, while to the right is the building from which the radiosondes are launched. This tornado was on the ground for about 20 minutes and temporarily forced the evacuation of the Stapleton International Airport Control Tower, closing the airport.

portion of all tornadoes. A topic of ongoing research is whether this mechanism of producing tornadoes can also occur with supercell storms, which could result in the production of a strong tornado and answer the question of how tornadoes form in some supercell storms.

[*See also* Cyclones; Floods; *and* Hurricanes.]

BIBLIOGRAPHY

Ahrens, C. D. *Meteorology Today: An Introduction to Weather, Climate, and the Environment.* 4th ed. New York, 1990.
Anthes, R. A., H. A. Panofsky, J. J. Cahir, and A. Rango. *The Atmosphere.* Columbus, Ohio, 1975.
Farrand, J. *Weather.* New York, 1990.
Galway, J. G. "Early Severe Thunderstorm Forecasting and Research by the United States Weather Bureau." *Weather and Forecasting* 7 (1992): 564–587.
Johns, R. H., and C. A. Doswell. "Severe Local Storms Forecasting." *Weather and Forecasting* 7 (1992): 588–612.
Kessler, E., ed. *Thunderstorm Morphology and Dynamics.* 2d ed. Norman, Okla., 1986.
Ludlum, D. M. *The Audubon Society Field Guide to North American Weather.* New York, 1991.
Milner, S. "NEXRAD-The Coming Revolution in Radar Storm Detection and Warning." *Weatherwise* 39 (1986): 72–85.
Ostby, F. P. "Operations of the National Severe Storms Forecast Center." *Weather and Forecasting* 7 (1992): 546–563.
Riehl, H. *Introduction to the Atmosphere.* New York, 1972.

EDWARD J. SZOKE

SHALLOW-WATER EQUATIONS. *See* Models and Modeling.

SHEAR INSTABILITY. An instability of a flow in which different layers of fluid move at different speeds is termed *shear instability*. It is observed in the laboratory, in the oceans, in the flow of air around a moving object such as a car, and in the atmosphere. It is often manifest as waves that grow quickly and then break up. The German physicist and physiologist Hermann von Helmholtz first recognized the nature of shear instability in 1868, and the British physicist and mathematician William Thomson, later Lord Kelvin, presented the first detailed theory of it in 1871. Helmholtz later applied the theory to billow clouds. In the 1950s the name *Kelvin-Helmholtz instability* was given to shear instability.

In most parts of the world shear instability can be easily seen as billow clouds, although it also occurs in clear air, where it is invisible. The structure of the unstable flow may be complicated, but its essence is illustrated in Figure 1. In the shear, the upper air moves at a different speed from that of the lower air. The shear may be created by tilting of a horizontal stream, as when air flows over a mountain. The inherent instability of this sheared horizontal flow has led to the formation of the pattern of the streamlines, resembling a long row of cats' eyes, which is depicted. Thus there are parallel rolls perpendicular to the plane of Figure 1; of course, in practice they may be observed from a different direction. The wavelength—the distance between adjacent "eyes"—varies considerably from instance to instance of shear instability in the atmosphere, but is typically in the range of 40 to 100 meters. The rolls usually stretch several wavelengths perpendicular to the plane of flow shown. The whole pattern is carried along by the wind. It grows and may soon break up, typically being of only a few minutes' duration. The breakup resembles the breaking of a steep wave on the ocean's surface.

The part of the process, if any, that is visible depends on the humidity, temperature, and pressure of the air at the time. As air rises, its pressure decreases and it cools adiabatically; as it falls, the processes are usually reversed. This may lead to condensation of water vapor during rising and to evaporation during falling, with formation of cloud at the top of each wave, so that a row of *billow clouds* may be seen. Figure 1 shows schematically some of the possible positions of clouds according to the prevailing humidity and other factors. Figure 2 is a photo-graph of some billow clouds; note that the waves have reached the stage of breaking and, by chance, have been caught by the camera end-on to the rolls. (Other kinds of clouds also occur in rows, and not every row of clouds is due to shear instability.) Billows and other kinds of rows of cloud are sometimes called a *mackerel sky,* after the pattern of bars on the back of the North Atlantic fish.

Shear instability occurs at all levels in the troposphere (the atmospheric layer closest to Earth's surface). It may be severe and lead to the development of turbulence as the waves break. This can disturb an aircraft as it rises and falls rapidly with the waves; it has caused injury to a few passengers who have been thrown to the ceilings of aircraft. It is especially dangerous when the pilot cannot see signs of the instability to get advance warning, a phenomenon called *clear-air turbulence.* Radar can help to detect clear-air turbulence.

The instability arises when a flow has large enough shear—for example, when the speed of the wind increases rapidly enough with height. In 1880 the British physicist Lord Rayleigh treated the instability of a general parallel shear flow mathematically—with, say, horizontal velocity U at height z, in which case the shear is the velocity gradient dU/dz. Rayleigh found that the flow is stable unless the vertical gradient of the shear vanishes at some level—that is, unless $d^2U/dz^2 = 0$ for some value of z. The mechanism of instability is quite complicated, but it can be explained in terms of the dynamics of vorticity. As a rough and ready rule, one may say that the greater the shear in a flow, the more likely it is to be unstable, although the overall structure of the flow may also be important. This tendency to instability is countered in practice by the buoyancy of the atmosphere. If the temperature increases rapidly enough with height, the den-

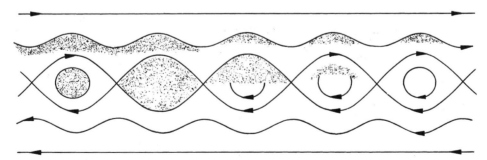

SHEAR INSTABILITY. Figure 1. *A sketch of the cat's eye pattern of shear instability.* The view is an elevation, an end-on view of the long rolls. Some possible positions of clouds are shaded; in actual occurrence, the clouds are usually similar in each "eye" for a given row of billow clouds.

SHEAR INSTABILITY. Figure 2. *A row of billow clouds (Kelvin-Helmholtz waves) near Pueblo, Colorado.* (Photo by George E. Arnott.)

sity of the air decreases with height, so that light air lies above heavy air and stabilizes the shear flow; then gravity acts to pull down the heavy air, allowing the light air up, and thereby inhibits instability. Lewis Richardson formulated and quantified this physical idea in 1920. His and subsequent work has led to a sufficient criterion for the *stability* of a shear flow in the form $Ri > 0.25$ everywhere, the local Richardson number being defined as the dimensionless quantity

$$Ri = \frac{g}{T}\left(\frac{dT}{dz} + \Gamma\right) \Bigg/ \left(\frac{dU}{dz}\right)^2;$$

here g denotes the acceleration due to gravity, T the absolute temperature, Γ the adiabatic lapse rate, and dU/dz the shear as before. (In fact, in the troposphere $\Gamma = g/c_p$ is approximately 10 K per kilometer, where c_p is the specific heat of air at constant pressure.) The criterion expresses the requirement that the stabilizing effect of the buoyancy $g(dT/dz + \Gamma)/T$ is sufficiently strong relative to the destabilizing effect of the shear $(dU/dz)^2$ to prevent instability.

Shear instability has been characterized here as an occasional phenomenon on a scale much smaller than the planetary. Nonetheless, it is more fundamental to atmospheric dynamics than it may appear. Shear instability and the turbulence into which it often develops are very widespread. They transfer energy and momentum from larger to smaller scales of atmospheric motion. In this way they play a significant role in the energy budget of the atmosphere, participating in the chain of phenomena whereby the kinetic energy put into the atmosphere by the Sun's radiation is broken down from a planetary scale to smaller and smaller scales until it is finally dissipated as heat by molecular action.

[*See also* Aviation; Conditional and Convective Instability; Dynamics; Gravitational Instability; Inertial Instability; *and* Vorticity.]

BIBLIOGRAPHY

Colson, D. "Wave-Cloud Formation." *Weatherwise* 7 (1954): 34–35.
Drazin, P. G., and W. H. Reid. *Hydrodynamic Stability.* Cambridge, 1981.

PHILIP G. DRAZIN

SHELTER. *See* Human Life and Climate.

SHORT-RANGE WEATHER FORECASTING. *See* Weather Forecasting.

SHORTWAVE RADIATION. The Sun is the prime mover behind the climate of its planet Earth. The radiation emitted by the Sun that is incident (falls) on the Earth is called *shortwave radiation.* The Sun emits radiation approximately as a black body at a temperature of 5,800 K. This radiation covers a range of wavelengths from about 0.1 microns on the short wavelength (high-energy) end of the spectrum to about 4 microns on the long wavelength (low-energy) end. This shortwave radiation spectrum can be conveniently divided into three portions: (1) the *ultraviolet* (UV), including all radiation with a wavelength shorter than about 0.35 microns; (2) the *visible,* including all radiation with wavelengths between 0.35 and 0.8 microns; and (3) the *near-infrared* (NIR), including all solar radiation at wavelengths greater than 0.8 microns (Figure 1). The total shortwave energy incident at the orbital radius of Earth on an annual average is 1,370 watts per square meter (1 watt = 1 joule per second). Approximately half of this energy occurs at wavelengths in the UV and visible portions of the spectrum, and the other half in the NIR range. The wavelength of peak energy emission is roughly 0.6 microns (green light). [*See* Black Body and Grey Body Radiation; Sun.]

On a global and annual average, 30 percent of the incident solar radiation is reflected back to space by a combination of clouds, molecules, aerosol particles, and the surface, with clouds the dominant reflector. About two-thirds of the remaining 70 percent is absorbed by the surface, and about one-third is absorbed directly in the

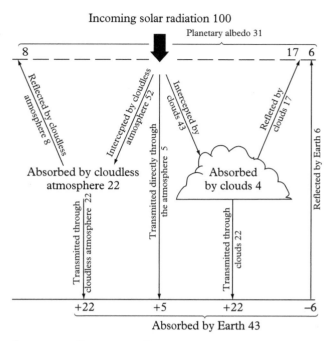

SHORTWAVE RADIATION. Figure 2. *The radiation balance of the Earth-atmosphere.* The numbers indicate the percent of incoming solar radiation reflected, intercepted, absorbed, or transmitted. (From Liou, 1980).

atmosphere by water vapor (Figure 2). Most of the radiation absorbed by the surface is absorbed by the upper few meters to tens of meters of ocean water in the tropics and subtropics. The absorbed solar energy maintains the temperature of the Earth, and the differential absorption between equatorial and polar regions, which heats the former more than the latter, is the primary driver for both atmospheric and oceanic circulations.

Ultraviolet Radiation. In collisions, UV radiation (highly energetic radiation) can break the chemical bonds that bind atoms into molecules; it can also remove electrons from atoms, thus forming charged atoms called *ions.* When very high-energy UV radiation encounters atoms at the upper edge of the Earth's atmosphere, about 100 kilometers above the Earth's surface, it ionizes these atoms, forming the *ionosphere.* [*See* Ionosphere.] Slightly less energetic UV radiation penetrates lower into the atmosphere to about 50 kilometers altitude, where it dissociates, or breaks apart, oxygen molecules (O_2) to form single oxygen atoms (O). These oxygen atoms combine with other O_2 molecules to form O_3, *ozone.* The bulk of the ozone lies in a layer between 25 and 50 kilometers above the surface. Still less-energetic UV radiation penetrates down to this ozone layer and dissociates an O_3

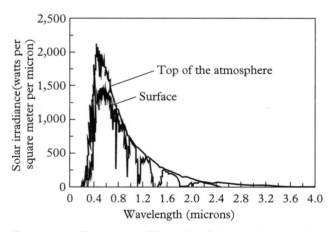

SHORTWAVE RADIATION. Figure 1. *Solar irradiance with a spectral resolution of 20 per centimeter at the top of the atmosphere and at the Earth's surface (with a 60° solar zenith angle) in a clear atmosphere without aerosols or clouds.* Based on the LOWTRAN 7 program. (From Liou, 1992.)

molecule into a single oxygen atom, O, and an O_2 molecule. It is this dissociation of O_3, resulting in the depletion of UV radiation through absorption, that protects surface-dwelling plants and animals from the potentially harmful effects of UV radiation. Virtually no radiation with wavelengths shorter than 0.32 microns penetrates to the surface, owing to this absorption by O_3 and O_2. [*See* Ozone.]

Some portion of the UV energy absorbed by O_3 also goes into heating the atmospheric layer between 20 and 50 kilometers. This heating stabilizes this layer because it becomes warmer than the underlying air. (Since warm air is less dense than cold air, warm air over cold air is stable.) This layer of the atmosphere is called the *stratosphere* because it is stratified into stable layers by solar heating. [*See* Stratosphere.]

Visible Radiation. Human vision is sensitive to light between about 0.4 and 0.7 microns in wavelength. Here, the term *visible* is used to refer to wavelengths between 0.35 and 0.8 microns—because all these wavelengths exhibit the same characteristics with regard to the atmosphere. The clear atmosphere is basically transparent to all shortwave radiation in the visible region; that is, visible radiation is not absorbed during transit through the clear atmosphere. It can be scattered by the molecules that comprise the atmosphere, however.

Molecular scattering, often called *Rayleigh scatter* after the eminent British scientist Lord Rayleigh, is the physical process responsible for much of the color, and associated beauty, in our atmosphere. Because molecules are very small relative to the wavelength of light, they are poor scatterers of light. In addition, as the wavelength of light increases, molecules become increasingly ineffective scatterers of the light. Mathematically, the scattering efficiency of molecules decreases as $\lambda^{(-4)}$ where λ is the wavelength of the light being scattered. This means that molecules scatter blue light ($\lambda = 0.4$ microns) much more efficiently than red light ($\lambda = 0.7$ microns). From the power law relationship, we can see that the ratio of scattering efficiency is $(0.7/0.4)^{(-4)}$, about a factor of 9. [*See* Scattering.]

It is this factor of 9 that is responsible for the blue color of the sky. Visible radiation is a mixture of monochromatic (single-color) radiation ranging from blue to red. As this radiation passes through the atmosphere, some of it is scattered by molecules. Because blue wavelength radiation is nine times more likely to be scattered than red wavelength radiation, much more of the blue is scattered in all directions out of the direct solar radiation beam (the radiation coming directly from the Sun to the ground). This radiation is scattered again and again by molecules, much as balls are scattered by a pinball machine, until all sense of the original direction of entry to the atmosphere is lost. Thus, an observer on the ground looking up at the clear sky sees blue wavelength radiation in every direction. The factor of 9 is also responsible for the red color of sunrise and sunset. Because our atmosphere is spherical, the path of the direct solar beam through the atmosphere is much longer when the Sun is low in the sky than when it is overhead. A longer path means more atmospheric molecules, and thus a higher probability of the radiation being scattered. By the time the direct beam traverses this long atmospheric path when the Sun is low in the sky, all the blue and green radiation has been removed from the beam, and we see a red Sun at sunrise or sunset.

Visible radiation is also scattered by small aerosol particles that are present everywhere in the atmosphere. The size of these particles tends to increase with available moisture; on dry days, the particles are small and relatively inefficient scatterers. To an observer, the dry atmosphere appears clean and one can see long distances. On very humid days, the particles grow perhaps an order of magnitude larger in diameter and become much more efficient scatterers of radiation at all wavelengths. The atmosphere appears hazy, and visibility is reduced to a few kilometers or less. In this case, the actual number of particles has not changed; rather, the particles have increased in size, and their resultant impact on visible radiation has also increased. Because these particles scatter all visible radiation with approximately equal efficiency, hazy skies appear to be white, a blend of all the colors. If the particles grow even larger because the atmosphere reaches saturation (the atmospheric water vapor begins to condense onto the particles), the haze can become fog. In this case the particles become very efficient scatterers, and visibility may be reduced to a few meters to tens of meters. The interaction between shortwave radiation and clouds is discussed in more detail below. [*See* Aerosols.]

Explosive volcanic eruptions inject large amounts of sulfur dioxide into the lower stratosphere; this combines with water vapor to form a haze of small sulfuric acid droplets. This particle haze has a typical lifetime of 1 or 2 years. During this time, it increases the reflection of solar radiation slightly and thereby decreases the temperature of the surface and lower atmosphere. Observations following the June 1991 eruption of Mt. Pinatubo in the Philippines suggest that the particle haze from moderately large eruptions can reduce the surface temperature by a few degrees Celsius in the year or two after the eruption. [*See* Volcanoes.]

In regard to energy, the combined effect of molecular and particle scatter is to reflect a small portion of the incoming visible radiation back to space. About 8 percent of the total incoming shortwave radiation is reflected back to space as a result of molecular and particle scattering. A substantially higher percentage is scattered but still finds its way down to the surface as *diffuse,* as opposed to *direct beam,* shortwave radiation. Thus, most of the incident visible radiation finds its way to the surface of the planet, where most of it is absorbed and contributes to heating our world.

Near-Infrared Radiation. This is shortwave radiation at wavelengths greater than about 0.8 microns. Roughly half of the total shortwave energy is incident in this wavelength regime. Because the wavelengths of NIR radiation are longer than those of visible radiation, molecular and particle scatter is of little importance in this spectral region. Thus, we would expect NIR radiation to be transmitted directly through the atmosphere to the surface; however, this is not the case, because of the absorption by water vapor.

Water vapor is the dominant molecular absorber in the atmosphere. It is a very efficient absorber of NIR radiation in a group of spectral bands between 0.94 and 3.2 microns. NIR radiation at these and surrounding wavelengths is absorbed as it travels through the atmosphere. Because most of the atmospheric water vapor is contained in the lowermost 1 or 2 kilometers of the atmosphere, most of the absorption of NIR radiation also occurs in this region. The absorbed energy is used to heat the lower atmosphere directly. Although some NIR radiation is transmitted to the surface and heats it, most of it is absorbed in the lower atmosphere. Vegetation reflects more NIR than visible radiation, a fact that makes it possible to map vegetated areas by remote sensing techniques using satellites.

Clouds. The interaction of shortwave radiation with clouds is important, interesting, and complicated. If one looks at our planet from outer space, one of the dominant impressions is the presence of bright clouds. These clouds are bright because they reflect a substantial portion of the incident shortwave radiation. This reflectivity can range from a few percent for thin cirrus (ice) clouds to perhaps 50 to 60 percent for overcast stratus and stratocumulus clouds, to as much as 80 or 90 percent for deep cumulonimbus clouds. Because most of this radiation would otherwise be absorbed in the atmosphere or at the surface, clouds play an extremely important role in

the energetics of the Earth. Clouds are composed of liquid water droplets, ice crystals, or a combination of both. Both droplets and ice crystals are large relative to the wavelength of shortwave radiation, so they are efficient scatterers and scatter all shortwave radiation equally well, independent of wavelength. Because clouds are efficient scatterers, they rapidly convert direct beam solar radiation to a diffuse field—that is, one in which radiation is transmitted roughly equally in all directions. Even relatively thin stratus clouds only a few hundred meters in depth can completely obscure the solar disk, which occurs when the direct solar beam is entirely scattered. Fortunately, radiation scattered by large droplets or crystals tends to continue in the same general direction (the "forward" direction) in which it was traveling before the scattering event. Thus, much of the radiation that was incident downward continues downward through thin-to-moderately thick clouds. This scattered field makes such clouds appear white when viewed from below. If it were not for this forward scattering, our world would be very dark whenever clouds were present. When clouds become very thick, as in the case of cumulonimbus or thick stratus, most of the incident solar radiation is reflected, and the clouds appear gray or dark gray from below. Observations show that as much as 90 percent of the solar radiation that would reach the ground under clear skies can be reflected by thick clouds.

The scattering characteristics of clouds are determined by the total amount of water or ice in a vertical column (called the *liquid water* or *ice water path*) and by the distribution of the droplet or ice crystal size. For ice, the shape of the crystals is also important. These factors in turn depend on the state of the atmosphere (temperature, humidity, and wind fields) in which the clouds form, so understanding the interactions among the atmospheric state, clouds, and the solar radiation field is exceedingly complicated. Understanding and modeling these interactions is one of the major challenges in atmospheric science. [*See* Cloud Physics.]

The Surface. Because the atmosphere is largely transparent at visible wavelengths, the majority of the solar radiation incident at the top of the atmosphere reaches the Earth's surface. The Earth's surface is about 70 percent ocean, and water is a poor reflector of solar radiation, so most of this incident radiation is absorbed in the top few meters or tens of meters of the ocean. This energy heats the oceanic mixed layer, which in turn heats the atmosphere, partly by direct contact but primarily by the evaporation of water. Evaporation requires energy, and that energy subsequently heats the atmosphere when condensation occurs in clouds.

The fate of solar radiation incident on land surfaces is slightly different. Vegetated surfaces reflect about 10 percent of the incident radiation and absorb the remainder. The absorbed solar radiation is largely used to fuel photosynthesis, or plant growth. In the process of photosynthesis, plants give up water vapor to the atmosphere, so some absorbed solar energy again heats the atmosphere via condensation of water in clouds. A smaller fraction of the solar radiation directly heats the vegetative surface. On nonvegetated surfaces such as deserts, the bulk of the solar radiation directly heats the surface, which in turn directly heats the overlying atmosphere. Because nonvegetated surfaces are generally dry, no evaporation can occur, no clouds form, and the land surface becomes very warm from direct solar heating.

[*See also* Clouds and Radiation; Longwave Radiation; *and* Radiation.]

BIBLIOGRAPHY

Liou, K. N. *An Introduction to Atmospheric Radiation.* New York, 1980, p. 328.
Liou, K. N. *Radiation and Cloud Processes in the Atmosphere.* New York, 1992, p. 103.

THOMAS P. ACKERMAN

is in the process of changing from snow to rain or vice versa. It usually occurs in overrunning precipitation during winter.

Sleet is a result of snow or rain falling through a warm air layer (temperature above freezing) which overlies a cold layer (temperature below freezing) (see Figure 1). Overrunning precipitation occurs when warm air from the south overrides cold air in place or moving equatorward. If the temperature remains below freezing throughout the atmospheric column, then the precipitation falls as snow. However, if a layer at temperatures above freezing becomes sandwiched between layers of subfreezing temperatures above and below it, the snow will first melt in the above-freezing layer and then refreeze as sleet before reaching the ground. If the lower subfreezing layer is particularly shallow, the rain will not have time to refreeze before reaching the ground and instead will refreeze upon contact with the ground; this in known as *freezing rain.* Sleet is not the same thing as *hail,* which is primarily a warm-weather phenomenon.

[*See also* Freezing Rain; Hail; Hydrological Cycle; Precipitation; Rain; *and* Snow.]

BIBLIOGRAPHY

Gedzelman, S. D. *The Science and Wonders of the Atmosphere.* New York, 1980, pp. 190–191.

JUDAH COHEN

SLEET. The form of precipitation called *sleet* is transitory between snow and rain, occurring as precipitation

SMALL SCALE. *See* Scales.

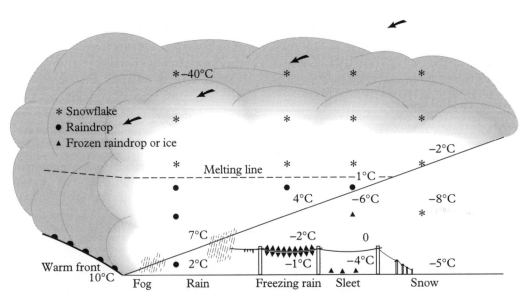

SLEET. Figure 1. *Meteorological conditions leading to freezing rain and sleet.* (From Gedzelman, 1980.)

SMALL-SCALE   METEOROLOGY.   *See* Meteorology.

SMALL-SCALE OR LOCAL WINDS. Atmospheric circulations may range in size from a small gust or swirl of wind to the global circulation cells that give rise to the trade winds and the westerlies. The scale of motion considered in this article may be termed *small-scale* or *local scale*—the scale of motion associated with winds within the planetary boundary layer (the lowest few hundred meters of the atmosphere); these are the various kinds of winds we experience daily. Surface winds of this scale are often shaped by local conditions such as topography or location.

Winds of this scale may be classified into five main groups: *diurnal* winds, including winds associated with the diurnal cycle; *jet-effect* winds, including those that are strongly influenced by local topography; *antitriptic* winds, arising from pressure and thermal gradients, such as land and sea breezes or chinook and foehn winds, or from gravity, such as fall winds; local winds created by local heating or instability caused by the overrunning of cold air, such as dust storms or haboobs; and winds created by strong pressure gradients over a relatively small area, by the uninterrupted flow over a level surface, or by these conditions in combination—such as the desert khamsin wind or winds associated with blizzard events. Small-scale or local winds may produce significant landforms, particularly in desert environments. They can also pose problems for human activities and engineering.

In the classification described below, general meteorological terminology is used. However, as is evident in many of the examples cited, a wide variety of interesting and colorful names exist for specific local winds. A partial list of local wind names and where they are found is given in Table 1.

The first class of local winds in this scheme is the diurnal winds. The local surface winds in many places often undergo variations in strength during the course of a day-night cycle. Under normal circumstances, wind velocities tend to be weakest near dawn as a result of radiative cooling of the surface, which often creates low-level inversions. An inversion of this type isolates the surface winds from the winds of higher velocity in the upper atmosphere above. Because the atmosphere higher above the surface is not so strongly influenced by friction, it can maintain high velocities. The morning surface winds, isolated from these upper winds, are weaker than average. Conversely, local surface winds are often strongest in the afternoon, when mixing between the faster-flowing upper atmosphere and the surface is maximal owing to thermal heating of the surface. [*See* Diurnal Cycle.]

The second type of local circulation includes winds that are strongly influenced by variations in local topography. For example, local topography can intensify the general wind flow by channeling air through some long configuration such as a narrow mountain pass or canyon. The resulting local-scale wind is significantly faster than the general regional wind pattern. Such winds are referred to as *jet-effect winds* or, sometimes, as *canyon winds* or *mountain-gap winds*. Specific examples include the düsenwind (Dardanelles and the Aegean Sea region), the tehuantepecer (Central America), the kossava (east central Europe), and the Wasatch (western North America).

A third type may be termed *antitriptic winds*. Local pressure differences and thermal effects are the primary causative mechanisms of this group, which includes land and sea breezes, mountain and valley winds, many fall winds, and adiabatic winds. Sea and land breezes are created by the thermal differences between land and water. Because of their differing heat capacities, land surfaces heat up and cool down much faster than water bodies do. This creates thermal gradients along coasts that lead to the formation of small-scale circulations. During the day the hotter land surface produces vertical uplift, and the cooler water surface experiences descending air. In addition, the thermal gradient and vertical air movement generate a pressure gradient such that lower pressure exists over the land and higher pressure over the water. Consequently, a horizontal wind at the surface develops from the sea to the land—a sea breeze. At night the situation is reversed and the flow is from the cooler land to the warmer water, in the form of a land breeze.

Valley upslope and downslope winds are also created through diurnal temperature and pressure gradients. On clear nights, loss of longwave radiation from high mountain ridges produces marked cooling of the ridge and surrounding air. The valley, however, is protected from excessive cooling by the radiation exchanges between the valley walls and floor. Consequently, the cooler, denser air at the top sinks to the floor. Such a wind is termed *katabatic*. The daytime equivalent of the katabatic wind is the *anabatic* wind. It forms when mountain ridges receive the maximum sunlight and thus maximum heating, while the valley, shaded by its walls, receives less. The warmer mountain ridges produce rising air and lower pressure; the cooler valley air then flows up the valley toward the ridges.

SMALL-SCALE WINDS. Table 1. *Some local wind names, with region of origin and brief description*

Name	Region	Description
afghanets	Uzbekistan	southerly wind from Afghanistan
almwind	Poland	foehn wind
angin	Malay Peninsula	land-sea breeze
aperwind	Alps	warm wind
aracaty	Ceará, Brazil	northeast wind
austru	Romania	east or southeast foehn wind
autan	south-central Gascony	strong southeast wind
ban-gull	Scotland	summer sea breeze
barat	Indonesia	strong northeast or west wind
barinés	eastern Venezuela	westerly winds
belat	Yemen	dusty seaward northeast wind
berg wind	South Africa	hot dry squally foehn wind
blizzard	Europe; North America	70 kph, reduced visibility due to snow
bora	Dalmatian coast, Adriatic	cold fall wind
bordelais	Quercy, southwestern France	west wind
breva, tivano	Lake Como, Italy	valley winds
buran	Siberia, Russia	northeast wind, blizzard
caracenet	eastern Pyrenees	cold violent gorge wind
charduy	Guayaquil, Ecuador	cool descending wind
chergui	Morocco, North Africa	east or southeast wind
chili	Tunisia	warm dry wind, sirocco
chocolatta north	West Indies	northwest gale
cierzo	Ebro Valley, Spain	mistral wind
collado	Gulf of California	north or northeast winds
criador	northern Spain	rain-bearing wind
dadur	Ganges Valley, India	down-valley wind
düsenwind	Dardanelles	east-northeast mountain-gap wind
elvegust	Norwegian fjords	cold descending squall
embata	Canary Islands	onshore southwest wind
etesian	Aegean Sea region	summer northerly winds
euraquilo	Yemen, Near East	northeast, north-northeast stormy
furiani	Po River, Italy	strong southwest wind
gallego	Spain; Portugal	cold northerly wind
gending	northern plains, Java	foehn wind
gharra	Libya	northeast squalls
ghibli	North Africa	desert foehn wind
gregale	Malta, Mediterranean Sea	strong northeast wind
guxen	Swiss Alps	cold wind
haboob	Sudan	sandstorm or duststorm
harmattan	West Africa	northeast or east dry wind
helm wind	northern England	cold northeasterly down-valley wind
jauk	Austria	foehn wind
junk wind	Vietnam; China; Japan	south or southeast monsoon wind
junta	Andes	mountain-gap wind
kachchan	Sri Lanka	hot, dry, west/southwest foehn wind
kal baisakhi	Bengal	short-lived squall

SMALL-SCALE WINDS. Table 1. *(Continued)*

Name	Region	Description
karaburan	Central Asia	strong summer northeast wind
karema wind	Lake Tanganyika, Africa	strong east wind
karif	Somalia	southwest wind, large land breeze
kaus	Persian Gulf	strong southeasterly wind
khamsin	Egypt	southerly desert wind
kloof wind	Simons Bay, South Africa	cold southwest wind
knik wind	Alaska	strong southeast wind
koembang	Java	dry southeast or south foehn wind
kona	Hawaii	southwest rain-bringing wind
kossava	Eastern Europe	cold east or southeast, jet-effect wind
lagnkisau	Indonesia	foehn-like day winds
leveche	Spain	sirocco wind, hot southerly wind
libeccio, afer	Italy	southwest winds
llebetjado	Catalonia, Spain	hot wind descending from Pyrenees
Maloja wind	Switzerland	reverse mountain valley wind
mamtele	Sicily	light northwest wind, *mistral* type
marin	French Mediterranean	warm moist southeast wind
medina	Cádiz, Spain	winter land wind
meltém	Bulgarian coast	northeast or east summer wind
mistral	France	northwest, fall and jet effect wind
n'aschi	Persian Gulf	northeasterly winter wind
nevada	Ecuador	cold glacier wind
nor'easter	eastern North America	northeast wind
orsure	Gulf of Lions, France	north to northeast wind
ouari	Djibouti, Africa	south wind
pampero	Argentina; Uruguay	cold south or southwest wind
panas oetara	Indonesia	warm dry north wind
ponente	southern coast of France	west wind (land breeze)
poniente	Strait of Gibraltar	west wind
pontias, solore	southeastern France	mountain winds
poriza	Black Sea region	strong northeast winds
puelche	Andes	foehn wind of west coast
purga	Siberia, Russia	similar to blizzard or *buran*
raffiche	Mediterranean Sea region	mountain winds
reshabar	Iraq	"black wind," northeast wind
Santa Ana	California	foehn-like mountain-gap hot wind
scharnitzer	Tyrol	cold northerly wind
seistan	Iran; Afghanistan	northwest monsoon wind
shamal	Iraq	northwest wind
siffanto	Adriatic Sea region	southwest wind
simoom	Middle East	dry dust-laden desert wind
sirocco	North Africa	warm south or southeast wind
solano	southeastern coast of Spain	southeasterly or easterly wind
solarie	central and southern France	easterly wind
sukovei	Russia	"dry wind," dry easterly wind
sur	Brazil	cold wind
suracon	Bolivia	cold rainy wind *(Table 1 continues over.)*

SMALL-SCALE WINDS. Table 1. *(Continued)*

Name	Region	Description
tarantata	Mediterranean Sea region	northwest wind
tehuantepecer	Central America	north or north-northeast wind
terral levante	Spain; Brazil	land breeze
terrenho	India	cold dry land wind
tongara	Makasar Strait	southeast wind
touriello	Ariège Valley, France	south wind of foehn type
vardar	Greece	cold fall wind
vaudaire	Lake Geneva	south foehn wind
virazon	Andes	southwesterly sea breeze
yamase	Senriku district, Japan	onshore easterly wind
zonda	Argentina	hot foehn wind

Related to the katabatic winds is the *foehn* or *chinook* wind. A foehn wind is a warm dry wind found on the lee side of a mountain range. The warmth and dryness of the air arises from adiabatic compression as the air descends the mountain slopes, so many scientists refer to these winds as *adiabatic* winds. Foehn winds are characteristic of nearly all mountain areas. They are normally associated with larger-scale synoptic patterns, in that they can occur only when the large-scale circulation is sufficiently strong and deep to force air completely across a major mountain range in a short period of time. The precise strength and character of a given foehn wind, however, varies widely around the world; it depends on many factors, such as topography, the strength of the regional circulation across the mountain, the amount of moisture lost through precipitation on the windward side, and the conditions prior to the onset of the foehn.

The term *foehn* is derived from Latin *favonius* "favoring wind" and refers to the adiabatic winds of the European Alps, where it is particularly common. In North America, the most common term for such winds is *chinook* after the Chinook Indians of the Pacific Northwest. When a chinook occurs after a spell of intense cold, the temperature may rise by 20° to 40°F (11° to 22C°) in 15 minutes. For example, at Havre, Montana, a rise from 11°F to 42°F has occurred within as little as 3 minutes. A foot of snow or more may melt in just a few hours under chinook conditions. In other alpine regions the foehn has other local names: *ijuka* in Carinthia, Austria; *halny wiatr* in Poland; *austru* in Romania; and *favogn* in Switzerland. In New Zealand a foehn blowing from the Southern Alps onto the Canterbury plains is called a *Canterbury northwester.*

Fall or *drainage winds* are created by the gravity-drawn flow of cold air from higher to lower regions. They are characterized by cold, dense air that has collected in winter over a high plateau or high interior valley. When regional weather conditions are favorable, some of this cold air spills over low divides or through passes into the adjacent lowlands as a very strong cold wind. A fall wind differs from a foehn in that the air is initially cold enough that it remains relatively cold despite adiabatic warming during descent. Fall winds occur in many mountainous regions of the world and go by various local names. The cold, dry *mistral* of the Rhone valley in southern France is a well-known example. Other fall winds may be found on the coast of Norway and for some distance inland. They are also well developed on the northern coast of the Aegean Sea, often descending at gale force and extending several miles out to sea. At Rio de Janeiro, Brazil, the descending fall winds from the northwest are called *terre altos*. In the Antarctic, fall winds off the inland ice produce violent blizzards.

The fourth type is characterized by local heating or instability created by the overrunning of cold air. This type of wind is often associated with convective instability, and subsequent air-mass or convective thunderstorms. Examples include various types of dust storms or sandstorms, including the lens-shaped *haboob* common in the desert regions of North Africa and southwestern North America. A recently discovered example of instability-generated local wind is the *microburst*, a downburst of air formed by a convective thunderstorm. The winds of a microburst may exceed speeds of 60 kilometers per hour (100 miles per hour). Such wind shears pose severe problems to landing aircraft and may lead to crashes.

The final class of local winds contains those created by

a strong pressure gradient over a relatively small area, uninterrupted flow over a level surface, or both conditions in combination. This includes regional-scale desert winds such as the khamsin or sirocco, as well as polar winds associated with blizzard, buran, or purga events.

Many of the Earth's features have been shaped by wind processes. Such geologic features are termed *eolian landforms* after Aeolus, the Greek god of wind. The majority of eolian landforms are found in desert environments where other erosional processes may not be very active. Examples include crescent and parabolic dunes, yardangs (elongate ridge formations), and deposits of loess (windblown silt).

Research has also been conducted on changes in local winds in the long-term past. Geologic formations known as *ventifacts* and *dreikanters* have been created through abrasion of rock by sand-carrying winds. Analyses of fossilized dune fields, such as those in western North America, have also led to new insights about changes in local wind circulations. Microparticle analyses of arctic and antarctic ice cores may also define past changes in local and regional circulations, particularly for the polar latitudes.

Small-scale or local winds have a direct impact on many human activities. Microbursts and the winds associated with thunderstorms pose serious concerns to the aviation industry. [*See* Aviation.] Sudden dust storms in desert environments have caused death and injury on highways. Chinook or foehn winds in many alpine regions of the world can cause rapid melting of snow, contributing to avalanches and flash flooding. And architects have begun to consider the impact on building design of air flow within the urban landscape. Such interrelationships between human activity and the climatology of local winds will undoubtedly be the focus of more research in the future.

BIBLIOGRAPHY

Brazel, A. J. "Blowing Dust and Highways: The Case of Arizona, U.S.A." In *Highway Meteorology*, edited by A. H. Perry and L. J. Symons. London, 1991, pp. 131–161.

Defant, F. "Local Winds." In *Compendium of Meteorology*, edited by T. F. Malone. Boston, 1951, pp. 655–672.

Forrester, F. "Winds of the World." *Weatherwise* 35.5 (1982): 204–210.

Henderson-Sellars, A., and P. J. Robinson. *Contemporary Climatology*. Essex, U.K., 1986, pp. 296–310.

Huschke, R. E., ed. *Glossary of Meteorology*. Boston, 1959.

Ritter, D. F. *Process Geomorphology*. Dubuque, Iowa, 1978, pp. 309–344.

RANDALL S. CERVENY

SNOW. Crystals nucleated on condensation nuclei grow by vapor condensation supplied by the supercooled water droplets (liquid water below 0°C) in clouds. The crystals then fall from the sky in many different forms which, once on the ground, undergo continuous changes in their important characteristics. These changes, usually called *metamorphism*, account for snow's widely varying behavior and complicate its characterization. Snow is always white and cold, but its albedo, strength, density, grain size, grain shape, insulating quality, hardness, and impurity content vary widely with time and location. Thus snow's properties and characteristics make it an interesting material to study, and its common occurrence on the surface of the Earth makes it an important material to understand.

Snow is very active thermodynamically because of the large surface area of the many crystals and because its absolute temperature is at or proportionately close to the melting temperature. Ice crystals in clouds grow rapidly with intricate shapes, but on the ground they undergo rapid changes to minimize the Gibbs free energy. [*See* Thermodynamics.] In addition, there are fundamental differences between wet and dry snow—that is, snow with or without any liquid water. The most common form of dry snow (Figure 1) contains individual crystals bonded together by a process known as *sintering*, the welding together of grains by necks or bonds. The most common form of wet snow (Figure 2) is a complicated collection of individual crystals bonded together in clusters containing liquid water in pores in the interior of the cluster and in crevices. The liquid content of freely drain-

SNOW. Figure 1. *Ice crystals in dry snow are usually well rounded and bonded together to form a lattice.* These are 0.1 to 1.0 millimeter in size.

SNOW. Figure 2. *Ice crystals in wet snow form into clusters with liquid water in the crevices and pores in the interior.* The individual crystals are 0.5 to 1.0 millimeter in size.

ing wet snow is generally only 3 to 6 percent of the total volume, but, as shown in the figures, a small amount of liquid makes a large difference. In addition to these forms of snow on the ground, different grain sizes and shapes arise under specific surface conditions, such as wind, sun, rain, melt-freeze, and low temperatures. Within the snowpack specific types of layers develop, ranging from weak horizons of depth hoar crystals—depth hoar consists of large, faceted grains that form weak layers from lack of sintering—or slush layers to strong, icy layers.

The snow surface is constantly changing because of constant changes in the driving forces of temperature, humidity, wind, and solar radiation. Nevertheless, we can generalize about one of its most important properties, the albedo or reflectivity. Albedo is of considerable interest because it is very high when the snow is fresh but drops rapidly thereafter. Given the large areal extent of snow during winter in the Northern Hemisphere, placing a white material over the surface with an albedo of 80 percent clearly reduces the absorption of solar energy at the Earth's surface. However, as soon as the snow is deposited, it tends to densify and metamorphose into larger grains. These processes are accelerated at higher temperatures or if the snow is wet, and both of these normal aging processes cause a rapid decrease in albedo. In addition, the accumulation of contaminants, especially dark materials such as soot, has a dramatic effect on the albedo. There is a natural feedback in this aging process, because a lower albedo increases solar radiation absorption, which warms the snow surface, causing faster aging and lowering the albedo. [*See* Albedo.]

Snow melts in earnest when the incoming energies at the surface are much larger than the outgoing energies. Melt water percolates into the snow, causing an increase in density to about one-third that of water. The snow is still quite porous and permeable, and water flows easily downward unless interrupted by internal layering. These layers may be buried wind or sun crusts, or they may arise from sequences of melting and freezing events. Snow is a water reservoir that liquefies and supplies melt water at a time of year when it is beneficial to life. For example, melt water from high mountain areas is the water supply in many parts of the world that would not otherwise have a water source during the summer months. For both water supply and flood forecasting purposes, it is important to know how much water is going to be available throughout the melt season. Efficient ways of making such an inventory are still being developed by remote-sensing specialists. For example, radar systems can be used to survey large areas of snow, but the interpretation of the radar signals requires considerable knowledge of the electrical properties of snow in both its wet and dry states. The presence of internal layers consisting of a variety of grain types and liquid water contents greatly complicates these interpretations. [*See* Hydrological Cycle.]

The internal layers in snow also complicate the prediction of avalanches because weak layers develop on mountain slopes without giving any outward indication of the danger. In particular, the growth of depth hoar crystals during periods of cold weather, the burial of surface layers without good bonding to adjacent layers, and the formation of wind slabs are known to lead to instability. [*See* Avalanches.]

The chemistry of snow is of considerable interest because of the so-called *acid shock* that often accompanies the initial snowmelt. Snow crystals nucleate on atmospheric impurities and gather other foreign materials as they fall. The impurity load is increased by fallout while the snow is on the ground; as the snow metamorphoses, all the impurities tend to accumulate on the crystals' surfaces because foreign substances are rejected by growing ice crystals. Thus, even though the snow may be fairly pure as a whole, the soluble impurities tend to be available for removal with the first fraction of melt water. This frequently leads to a distinct increase in acidity when the first meltwater passes into streams.

Many other properties of snow are of interest. For example, dry snow is a good thermal insulator and protects plants and animals living beneath it during the winter. Snow quantity and quality are important in the skiing industry, and snow removal is an important process in many inhabited areas. Thus snow has many benefits and poses many problems, partly because it is one of the most common materials on the surface of the Earth.

[*See also* Precipitation; Sleet; *and* Snow Cover.]

BIBLIOGRAPHY

Colbeck, S. C. "The Layered Character of Snow Covers." *Reviews of Geophysics* 29 (February 1991): 81–96.

Jones, H. G., and W. J. Orville-Thomas, eds. *Seasonal Snowcovers: Physics, Chemistry and Hydrology.* Dordrecht, 1987.

Langham, E. J. "Physics and Properties of Snowcover." In *Handbook of Snow*, edited by D. M. Gray and D. H. Male. Toronto, 1981, pp. 275–337.

Male, D. H. "The Seasonal Snowcover." In *Dynamics of Snow and Ice Masses*, edited by S. C. Colbeck. New York, 1980 pp. 305–395.

SAMUEL C. COLBECK

SNOW COVER. A blanket of snow accumulates on the ground from falling and drifting snow particles. Snow covers much of the Earth during the winter months, especially in the Northern Hemisphere, where most of the land mass exists (Figure 1) and the polar ocean is ice-covered. Its direct influence on people is most important in the temperate regions of the Northern Hemisphere, where the populations affected by snow are largest, but it is also important as an insulating cover in the polar and subpolar areas and as the source of mass for the large ice sheets. Its insulating qualities reduce frost penetration into the ground, but its high reflectivity (albedo) reduces solar warming and could intensify the winter accordingly. However, because the presence of snow cover affects the circulation of the atmosphere, its effect on climate and weather is not as straightforward as the increase in albedo might suggest. Otherwise, the simple feedback mechanism of a higher albedo causing lower temperatures, causing the snow cover to grow, and in turn causing a higher albedo, might lead to a much cooler planet—even to ice ages. [*See* Albedo; Glaciers; Ice Ages; Little Ice Ages.]

The snow cover forms by the accumulation of falling snow in distinctive layers arising from separate snowstorms and wind events. These layers can often be traced throughout the entire winter, as shown in Figure 2. Each layer can retain or develop certain characteristic features that sometimes control the behavior of the entire cover; for example, avalanches tend to be released off buried layers of low strength.

At higher latitudes or elevations, it is more likely that the snow cover will build throughout the colder months to form a continuous winter blanket. When melting

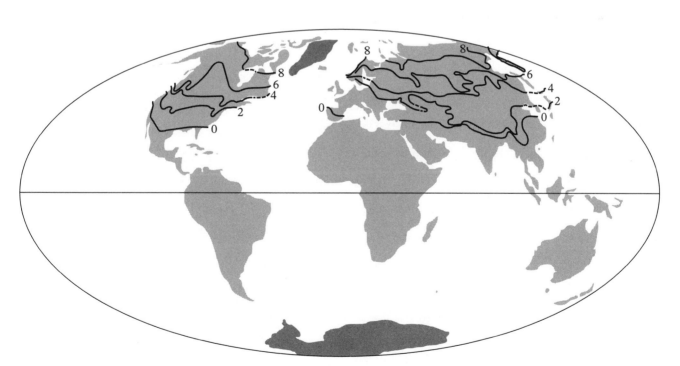

SNOW COVER. Figure 1. *Typical duration of the snow cover in months.* The major ice sheets are shaded darkly.

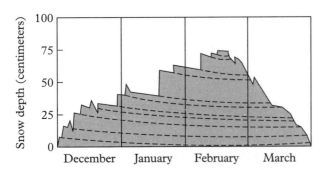

SNOW COVER. Figure 2. *Snow covers build up slowly through the colder months and then melt away rapidly during warmer weather.* The covers usually consist of distinctive layers arising from separate snowstorms that often persist until meltout.

begins, the snow usually melts away relatively quickly, as shown in Figure 2. The melting occurs in response to a change in the balance of energy inputs and outputs, including solar energy input, longwave radiation loss to the sky, and latent and sensible heat transfer to or from the air. In the Northern Hemisphere, air temperature and solar input increase most rapidly around the vernal equinox (21 March), when a wave of melting travels northward across the continents and upward to the higher elevations. As the snow cover melts over large, flat areas like the Missouri River Basin, snow melt water travels down the rivers and can produce large floods. The deep snow covers of the higher mountains take months to melt and are thus important sources of summer water for many arid regions.

Thus the insulating but reflective blanket of snow has a major impact over large areas during much of the year, even when its own mass of water is only a fraction of the total annual precipitation. It can be simultaneously a climate moderator, water reservoir, recreational medium, transportation nightmare, and mountain menace.

BIBLIOGRAPHY

Berry, M. O. "Snow and Climate." In *Handbook of Snow*, edited by D. M. Gray and D. H. Male. New York, 1981, pp. 32–59.

Colbeck, S. C. "The Layered Character of Snow Covers." *Reviews of Geophysics* 29 (February 1991): 81–96.

Male, D. H. "The Seasonal Snowcover." In *Dynamics of Snow and Ice Masses*, edited by S. C. Colbeck. New York, 1980, pp. 305–395.

SAMUEL C. COLBECK

SNOWSTORMS. The definition of a snowstorm varies greatly from region to region. Snowstorms can consist of wet, heavy snow or very dry, powdery snow. They can also occur over a wide range of temperatures, from relatively warm temperatures at or even above freezing to extremely cold temperatures below −18°C (0°F). The storm may have winds up to hurricane force, or no wind at all. In fact, a blizzard can consist of wind-blown fallen snow when no precipitation is actually occurring. There are no standards or categories set for snowstorms similar to those for other severe weather events, such as tornadoes or hurricanes. One useful general definition of a snowstorm is whether it is a "plowable snow," or about 10 centimeters of accumulation. The National Weather Service issues a heavy snow warning when at least 10 centimeters of snow is expected within a 12-hour period, or at least 15 centimeters within a 24-hour period.

The most severe type of snowstorm is called a *blizzard*. A blizzard is defined by three criteria: significant accumulation—usually 25 centimeters or more—or blowing of snow that reduces visibility to less than 400 meters (one-quarter mile); sustained winds of at least 56 kilometers (35 miles) per hour; and temperatures of −7°C (20°F) or less (this last criterion has recently been dropped in some locations).

Snowstorms can affect large areas and populations. For example, during the "Storm of the Century" of 12–14 March 1993, snow at least 30 centimeters (1 foot) deep covered much of the ground from the Appalachians to the Atlantic coast, continuously from Alabama to Maine. The effects of a snowstorm last longer than those of other types of severe weather: snow cover can remain for an entire winter season, and flooding from snowmelt can occur weeks or months after a particular snowstorm.

It has been hypothesized, but not proven, that one snowstorm can result in another snowstorm through positive feedback mechanisms with the atmosphere, a cause-and-effect relationship not associated with other types of severe weather. The snow cover produced by a snowstorm reinforces the cold air and also helps to maintain a trough of low pressure in the upper atmosphere. Both of these synoptic conditions are conducive to more snowstorms (Namias, 1962).

There is no structural difference between snowstorms and rainstorms; in fact, much of the precipitation that reaches the surface as rain begins as snow in clouds and melts in the atmosphere. The only difference is in the temperature gradient of the atmosphere through which the precipitation falls. In a snowstorm, the atmospheric column is generally at or below freezing throughout, while in a rain, sleet, or ice storm, the temperature is above freezing somewhere between the cloud where the precipitation forms and the ground. In rare cases, a heavy snowstorm can occur when the temperature near the surface is slightly above freezing. In the early-season snowstorm in central New England on 4 October 1987, up to 60 centimeters (2 feet) of snow fell with temperatures a few degrees above freezing.

Snow can occur anywhere on the globe—over the oceans, on land, at the equator, or at the poles. Obviously, the more frequently temperatures remain below freezing in a certain location, the greater is the likelihood of snowstorms; therefore, in general, snowstorm frequency increases proportionally to distance away from the equator, and on the eastern sides of continents proportionally to the distance away from the coast. During winter, oceans tend to be warmer than land surfaces, and so the ratio of snow to rain also tends to increase with distance away from the coast. Temperatures decrease and precipitation increases with height, so snowfall increases at higher elevations. Even though snow never falls in the tropics near sea level, it can fall at any time of the year in the highest elevations near the equator.

Temperature is not the only factor limiting snowstorms; moisture availability is also important. Near the poles the temperature remains below freezing year round, yet it hardly snows because the atmosphere is so dry. When it does snow, some of the snow can last for hundreds of thousands of years as glacial ice.

Snow can result from several synoptic-scale or mesoscale features. Even hurricanes have produced snow over New England in their dying stages. Most snowstorms result from one of four different meteorological events: the advection of warm air forced to rise up and over cold, denser air already in place, so that water vapor condenses and falls as snow; cold air passing over relatively warmer water, the origin of *lake effect snow;* air cooled due to rising terrain, resulting in *upslope snow;* and extratropical cyclones. Many snowstorms result from a combination of two or even of all three such events, especially the most spectacular snowfalls. The snowiest cities in the United States are situated in places where a significant portion of the annual snowfall is a product of lake effect or upslope snowfall. Cities with heavy lake effect snowfall include Syracuse and Buffalo, New York; Cleveland, Ohio; and Salt Lake City, Utah. Cities with increased snowfall owing to their high elevations include Denver, Colorado, and Flagstaff, Arizona.

In certain geographical regions, very localized snowstorms can occur as a result of strong winds passing over large bodies of water. The region most noted for these lake effect snows is that just east of the Great Lakes. On the eastern shores of the Great Lakes half of the annual snowfall may be due to lake effect snow. Other locations experiencing lake effect snow include the Salt Lake Valley in northern Utah and even Cape Cod, where a bay of the Atlantic Ocean produces a similar effect. [*See* Lake Effect Storms.]

Lake effect snows result when cold arctic air passes over relatively large warm lake waters. The temperature difference between the water and the air causes great quantities of heat and moisture to be transferred from the lake to the atmosphere. This process destabilizes the atmosphere, forming clouds and eventually snow squalls downwind of the lakes. Lake effect snow squalls are very similar to summertime thunderstorms in that both originate from cumulus type clouds rather than the stratus clouds more commonly associated with snowstorms. Also needed for heavy snow squalls are low-level convergence, upper-level divergence, positive vorticity advection, and cold upper-level temperatures.

Heavy amounts of snow can fall in a short period of time, but heavy snowfall is usually very localized. It is not uncommon for certain sites to receive snowfalls measured in feet while other sites just a few miles away have no snow at all. These narrow bands of snowfall can extend for hundreds of miles or more, especially over rising terrain. The mountains of Vermont and West Virginia receive large amounts of lake effect snow even though they are relatively far from the Great Lakes. One famous lake effect blizzard paralyzed the Buffalo, New York, area on 28 January through 1 February 1977. Snow on the ground approached 100 centimeters (40 inches), which combined with snow blown off the frozen lake to create drifts many feet high.

Snowfall resulting from air being forced to rise and cool by rising terrain is known as *upslope snow.* Whenever moist air is forced to rise, whether over a mountain or a gently sloping plain, the air cools to the saturation point and clouds form, followed by rain or snow. Once the air passes over the highest point, it begins to sink or subside, quickly drying out. This process creates large differences in precipitation between the windward and leeward sides of mountains. The region of North America that receives the greatest upslope snowfalls is the western slope of the major north–south mountain ranges of the West Coast. Single storm totals here can exceed 3 meters (10 feet), and single-season snow totals can exceed 25 meters (1,000 inches). Just east of these mountains lie deserts where the air subsides and little snow or rain falls.

The majority of snowstorms are associated with extratropical cyclones, which form in environments of large temperature gradients. Large temperature gradients can render the atmosphere unstable, an atmospheric state or condition known as *baroclinic instability.* The other condition conducive to large-scale instability is wind shear, an atmospheric condition known as *barotropic instability.* Both of these types of instability in the atmosphere can form a wave or an eddy, more commonly known as a storm. The baroclinic wave that develops exists at many levels in the atmosphere. The reflection of the wave near the Earth's surface is the storm itself and is identified on the weather maps as an *L.* The reflection of the wave near the jet stream (about 10 kilometers above the surface) is known as an *upper level low* or *trough* and is generally west of the surface storm during development. Other preconditions for the strongest snowstorms include low-level convergence, upper-air divergence, strong positive vorticity advection, jet streaks, and large quantities of sensible and latent heat release into the atmosphere. [*See* Baroclinic Instability; Cyclogenesis.]

When a storm is forming in an environment of large temperature gradients, to the south and east of the storm there is warm air, and to the north and west of the storm there is cold air. As the storm forms, it spins counterclockwise, causing the warm air to move north and to rise over the colder air near the ground. This region of the storm, where warmer air from the south is forced to rise above colder air in place to the north of the storm, is where snowstorms occur. In the strongest storms the warm air can be forced to rise violently, causing thunder and lightning; when these accompany snowfall, they are called *thunder snows.* In the warmer regions of the storm, or where the cold air is eroded by the warm air, rain occurs instead of snow. If the cold air is only partially eroded, sleet or freezing rain may occur. [*See* Precipitation.]

In the eastern United States most snowstorms are called *northeasters* because the general wind direction during the snowstorm is from the northeast. Throughout the country, most snowstorms are accompanied by easterly winds, because most of the precipitation associated with an extratropical cyclone is located in the northern half of the storm—especially in its northeastern quadrant—where the winds are generally east to northeast. In the low-elevation, coastal northeastern United States, a storm or cyclone is usually not sufficient in itself to produce a snowstorm. A strong high pressure originating to the north is also needed to maintain a northeasterly flow of air and to supply cold air throughout the duration of the storm. [*See* Cold Air Damming.]

Over the centuries, many great snowstorms have become part of meteorological lore. The "Blizzard of '88," which occurred on 11–14 March 1888, produced blizzard conditions in the northern mid-Atlantic states and central and western New England. Many locations in New York and New England received record snowfalls which still stand today. Maximum snow totals approached 4 to 5 feet in the upper Hudson Valley. In New York City more than 200 people died (400 people died over the entire Northeastern region), making the "Blizzard of '88" the worst natural disaster in the city's history. New York Harbor was completely frozen over, and people were able to walk across the East River between Brooklyn and Manhattan.

Another "Blizzard of '88" occurred on 11–13 January 1888 in the northern Plains (Montana, North and South Dakota, and Minnesota) and is considered to be the worst storm ever recorded there. During the "Great Snow of 1717," a series of 4 snowstorms from 27 February to 7 March 1717, snow totals are thought to have exceeded 3 feet in eastern Massachusetts. The "Great Atlantic Snowstorm" of 14 to 16 January 1831 probably produced the most snow over the largest area of any storm that century. The "Great Eastern Blizzard" or "Blizzard of '99" on 11–14 February 1899 blanketed the entire East Coast from central Florida to Maine, with up to 3 feet of snow in New Jersey.

The "Blizzard of '58" on 16–17 February 1958 and the "Storm of the Century" on 12–14 March 1993 were the greatest East Coast snowstorms of the twentieth century. Both storms produced heavy snowfalls of at least 25 centimeters (10 inches) in every state from Alabama to Maine. The "Blizzard of '93" produced several feet of snow the entire length of Appalachian Mountains, along with killer tornadoes, straight-line wind gusts over 100 miles per hour, record low sea-level pressures, and record cold temperatures (some for the entire month of March).

Every year many snowstorms occur in North America, especially in the regions downwind of the Great Lakes and in the highest elevations of the mountain chains on both coasts. Some lake effect snowstorms have produced as much as 5 feet of snow in 24 hours. The greatest 24-hour snowfall on record for the United States was 76 inches in Silver Lake, Colorado, on 14–15 April 1921. The greatest single-storm snow total was 189 inches in Mt. Shasta, California, on 13–19 February 1959. However these snowstorms were very localized and occurred in relatively unpopulated areas.

Unlike other types of severe weather, snowstorms have certain beneficial effects. Winter sports is a big industry, especially in mountainous states such as Colorado and Vermont. Winter snowpack usually translates into spring and summer runoff used for agriculture and urban water supply, even in areas that never see snow. Most of the water supply in the western United States, and in many other arid parts of the world, comes from mountain snowmelt.

Blizzards are probably most common in the United States, Canada, and Russia (where they are known as pugas); however, severe snowstorms do occur in other populated regions of the world. The worst snowstorm to affect England in recent history occurred 19–20 January 1881. Snowfall measured up to 1 meter, and wind gusts up to 35 meters per second (80 miles per hour) whipped the snow into drifts 3 to 6 meters high. Transportation was disrupted and 100 to 200 people died because of the storm. This century, the winters of 1947 and 1963 were the harshest. The winter season of 1946–1947 climaxed with a snowstorm on 6 March 1947. Much of England was covered by 30 to 65 centimeters of snow, and in Wales total snow depths reached 1.5 to 2 meters. In the winter season of 1962–1963, snow was recorded on the ground for 66 continuous days. The worst storm of the season occurred on 29 December 1962. In London, 30 centimeters of snow fell on top of an existing 30 centimeters of snow. More recently one of the worst snowstorms ever to affect continental Europe occurred 25 December 1978. In northern Germany and southern Scandinavia, including the cities of Hamburg and Copenhagen, 60 centimeters to 1 meter of snow fell, and winds of 22 to 27 meters per second (50 to 60 miles per hour) caused drifts 6 to 7.5 meters high. January 1985 brought unusually heavy snowfall to the Mediterranean countries of Europe, and Turkey was especially hard hit.

[*See also* Storms.]

BIBLIOGRAPHY

Braham, R. R., and M. J. Dungey. "Quantitative Estimates of the Effect of Lake Michigan on Snowfall." *Journal of Climate and Applied Meteorology* 23 (1984): 939–949.

Gedzelman, S. D. *The Science and Wonders of the Atmosphere.* New York, 1980.

Kocin, P., and L. W. Uccellini. *Snowstorms along the Northeastern Coast of the United States: 1955 to 1985.* Boston, 1990.

Ludlum, D. M. *Early American Winters 1604–1820.* Boston, 1966.

Ludlum, D. M. *Early American Winters 1821–1870.* Boston, 1968.

Namias, J. "Influences of Abnormal Heat Sources and Sinks on Atmospheric Behavior." In *Proceedings of the International Symposium on Numerical Weather Prediction.* Tokyo, 1960.

JUDAH COHEN

SOILS. The word *soil* conveys a wide array of meanings, ranging from "dirt" to "the thin veneer on which all life depends." Many people's notion of soil reflects their particular use of it—for example, growing plants or building on it. Here we will discuss soil as a living system that forms as a result of physical, chemical, and biological processes modifying the land surface. In turn, the soil

mediates processes crucial for supporting life in other components of the Earth system.

Soils preserve a "memory" of their condition of formation, which is reflected in their appearance. This appearance or *morphology*—including color, arrangement, and size of particles—is highly variable among soils and differs even among layers at various depths within a single soil profile. The layers that form in soils are referred to as *horizons;* they develop more or less parallel to the land surface in response to soil-forming processes that exert their strongest influence near the soil surface.

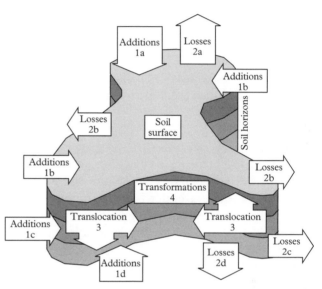

SOILS. Figure 1. *Processes affecting the soil profile through time.* (1) *Additions:* (a) to the soil surface (epipedon); including solar energy, rainwater and dissolved substances, atmospheric gases and dust, airborne organisms, organic matter, and artificial additives (e.g., fertilizers). (b) Lateral input—soil material, water and dissolved substances, migrating and transported organisms. (c) Lateral input within the profile—oblique throughflow in sloping conditions, water plus dissolved and suspended organic and inorganic substances. (d) Upward input—capillary movements of water and dissolved substances from groundwater, translocations by organisms. (2) *Losses:* (a) from the soil surface, including radiation energy, evapotranspirated water, gases (e.g., N through denitrification, or C through CO_2 oxidation), artificial losses through harvesting, fire. (b) Lateral runoff and mass movement—soil material, water and dissolved substances, wind transport, migrating and transported organisms. (c) Lateral drainage within the profile—throughflow of water, dissolved and suspended substances. (d) Eluviation out of the profile—water plus dissolved and suspended organic and inorganic material, or (more rarely) small organisms. (3) *Translocations:* gas exchanges, diffusion within the soil, eluviation and illuviation, materials moved by animals, plants, or microorganisms, uptake of substance by organisms. (4) *Transformations:* weathering and decomposition reactions, synthesis of organic and inorganic compounds. (See Furley and Newey, 1983.)

Soil-formation processes can be characterized as transformations and translocations within the soil, or as additions to and losses from the soil (Figure 1). *Transformations* are processes occurring in the intact soil—for example, the conversion of roots into decomposing organic matter—and *translocations* are movements of material within the soil. Soil diversity is apparent over a range of spatial scales from millimeters to thousands of kilometers. Unravelling the conditions of soil formation and developing schemes for classification, mapping, and use present major challenges for soil scientists.

Soil Formation and Diversity. In the late-nineteenth century the Russian scientist V. V. Dokuchaev developed a paradigm for explaining soil diversity, which was later formalized by Hans Jenny as the *five factors of soil formation* (Buol et al., 1989). The state of soil (S) at a given point in time is described as a function (f) of climate (c), organisms (o), relief or shape of the land (r), parent material (p), and time (t):

$$S = f(c, o, r, p, t).$$

Climate. The influence of climate on soil formation is pervasive and integrating. Temperature and precipitation control both the growth of organisms (the accumulation of biomass) and the transformation of minerals in the soil. Climate is a major factor influencing the distribution and productivity of ecological communities, which in turn affect soil formation.

Chemical reactions are estimated to increase by a factor of two to three for every 10°C rise in soil temperature. Attributes of the soil landscape, such as shape, moisture content, surface soil color, and vegetation architecture, influence the absorption of sunlight and thus soil heating, so that for many landscapes, a simple correspondence between soil and atmospheric temperature is not apparent.

Precipitation that enters the soil is the principal mediator of changes that occur at various depths within the soil profile. Local influences, such as vegetation, topography, and the structure and composition of soil particles, control the entry and movement of water in soil. Changes in water content and state result in shrinking, swelling, freezing, and thawing of the soil, which together with soil organic matter and clay contribute to the development of distinct aggregations or linkages among individual soil particles. The magnitude and frequency of these hydrological stresses decrease with depth and are typically reflected in an increase in size and a decrease in complexity of soil aggregates with depth. Ions and particulate matter can be transported in soil water. Climate exerts a strong control on the fate of soil water and its constituents. In humid-soil landscapes, where precipitation exceeds water loss from plants and soil (evapotranspiration), soluble salts may be leached from the soil to the ground water, with less soluble ions and the smallest clay-size particles accumulating in the subsoil. In arid-soil landscapes, by contrast, water can move from greater soil depths toward the surface in response to the drying caused by excessive evapotranspiration; this results in the concentration of soluble salts near the soil surface. In both cases, hydrological processes result in vertical changes of soil attributes that are reflected in soil morphology.

An understanding of past climates is crucial for identifying soil-forming processes. The advance and retreat of ice sheets and associated global climatic change during the past 2 to 3 million years have had a profound influence on soils. Many soils cannot be sensibly interpreted by reference to their contemporary conditions of formation. For example, the occurrence of red, strongly weathered soils in the southern Saharan and Australian deserts suggests that these regions experienced much wetter climates in the past. Some landscapes have been continuously exposed to weathering, soil formation, and erosion for a very long time; portions of the Northern Territory, Australia, are believed to have been exposed since the Cambrian period, about 560 million years ago. [*See* Paleoclimates.]

Parent material. The initial materials from which a soil has formed are called its *parent materials*. They may include consolidated igneous, sedimentary, and metamorphic rocks; decaying plant and animal material, in which organic soils such as peat and muck form; and transported materials such as glacial drift, river sediment, dune sand, or volcanic ash. The chemistry and mineralogy of the parent material have a limited but significant influence on subsequent soil formation. This influence diminishes with weathering, which tends to remove all but the most stable minerals. Climate and vegetation strongly influence rates and pathways of soil development; thus basalt, a fine-grained igneous rock, can be transformed under humid conditions into a red soil composed largely of resistant iron and aluminum oxides and clay minerals that show little or no volume increase when the soil is wet. In semiarid landscapes, however, basalt can be transformed into dark soils with a high content of clay minerals that increase their volume with addition of moisture.

Organisms. Plants and animals influence many facets of soil formation by modifying soil climate, providing biomass for incorporation into soil, mixing soil by digging or root disturbance, and nutrient cycling. The bulk of the biomass that accumulates on the soil surface as litter, or that is incorporated into the soil and transformed by animals and microorganisms into soil organic matter (SOM), is derived from higher plants. Local to global-scale variation in the amounts of SOM can be broadly explained by differences in climate and ecological communities. Mineralogy also influences SOM accumulation because clay-size minerals form more stable organo-mineral complexes than do larger soil particles. Increasing temperature speeds biomass oxidation, while high near-surface humidity and saturated soil conditions retard oxidation. The interplay of these climate-related variables is well correlated with the spatial distribution of SOM both locally and globally.

Ecological communities show considerable variation in the composition of the litter they produce and in the manner in which they distribute biomass above and below ground. Coniferous vegetation stores the bulk of its biomass above ground and produces litter with a low nitrogen content, which retards the rate of its decomposition by microorganisms. Soils forming under coniferous vegetation typically have a large accumulation of litter above the soil surface and comparatively small amounts of SOM. In contrast, prairie ecosystems store the bulk of their biomass below ground, and their litter has a high nitrogen content. The litter and fine plant roots, which turn over rapidly, result in deep SOM accumulations. Tropical rainforests produce abundant litter that is typically oxidized rapidly, so that little of it is stabilized into SOM.

Relief. The relief or topography of the land is the product of geological processes, erosion, and deposition, and thus should be viewed as a changing condition that is continually influencing soil formation. Topography affects the fate of precipitation reaching the land surface by modifying the climate and controlling surface water flow and infiltration. Locally, relationships between soil and topography are useful for deciphering soil patterns, because morphological variation is strongly linked to hydrologic processes; for example, vegetation patterns reflect the availability of water, and texture and particle size distribution affect water flow. As such, an understanding of hydrology provides a useful link between factors of soil formation and the processes that result in soil diversity. [*See* Hydrological Cycle.]

Time. It is clear that soil development is complicated by interdependency among the five factors discussed here; for example, climate influences organisms and is in turn influenced by topography. Although time may be viewed as an independent variable, soils are always disturbed during the course of their development, and it can be difficult to assign a definite start of formation to any but the youngest soils. Nonetheless, a *factorial* view, with an emphasis on deciphering processes, provides a useful framework for estimating soil-formation rates.

Soil Classification. Numerous soil classification systems have been developed, most of them national rather than international in scope. The UNESCO–FAO (United Nations Educational, Scientific, and Cultural Organization-Food and Agricultural Organization) System and Soil Taxonomy are the most widely used international systems. In this article the Soil Taxonomy is used to illustrate global soil distribution and the principles of classification (see Figure 2).

The Soil Taxonomy has a six-level hierarchy, composed of 11 orders and progressively more suborders, great groups, subgroups, families, and series. More than 12,000 series are recognized in the United States. The system is based on differentiating among measurable and observable soil properties. Soil *orders* are distinguished largely by differences in horizons that correspond with differences in the relative influence of soil-formation processes. Soil *series* use combinations of many properties to distinguish among soils locally, including differences in the size of surface particles.

Histosols are soils formed in organic material. Distinct parent material also distinguishes *andisols* (formed in volcanic material) and *vertisols* (swelling clay soils of semi-arid regions) from other orders. Dry climates separate *aridisols* from other orders, but within climatically defined arid regions other orders, such as *entisols,* may occur. Entisols are very young soils with alteration confined to the surface; *inceptisols* exhibit early stages of subsoil development in addition to surface alteration.

The remaining orders reflect broad differences in climate and vegetation. *Mollisols* occur mainly in grassland ecosystems of temperate zones and are distinguished by their thick, dark surface horizons rich in organic matter. Agricultural ecosystems based upon mollisols are among the most productive in the world. *Alfisols* are common on glaciated, deciduous-forest landscapes of humid, cool temperate regions. Alfisols have thinner, lighter-colored surfaces than mollisols, they have appreciable accumulations of translocated clay in their subsoils, and they are generally less agriculturally productive than mollisols. *Spodosols* are associated mostly with cool, humid climates and vegetation that produces acid litter strongly resistant to decomposition (e.g., conifers). Spodosols are charac-

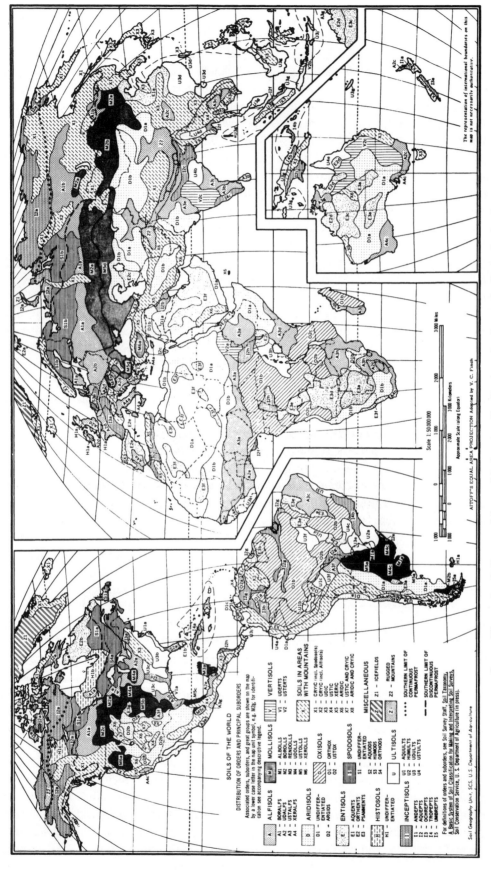

SOILS. Figure 2. *Soil map of the world, showing distribution of orders and principal suborders.* (Provided by the Soil Geography Unit, Soil Conservation Service, U.S. Department of Agriculture.)

terized by subsurface horizons rich in iron, aluminum, and organic matter, derived from intense weathering of surface horizons by organic acids. Climatic constraints and low fertility make spodosols poorly suited for intensive agriculture. *Ultisols* and *oxisols* are found on nonglaciated landscapes largely in humid, warm temperate or tropical climates and are considered to be old, strongly weathered soils. Ultisols have subsurface horizons rich in translocated clay and are more weathered than alfisols. Oxisols are composed largely of resistant minerals and show limited development of horizons. Both ultisols and oxisols have inherently low fertility but physical properties favorable for storing water. Clearing of forested ultisols and oxisols (as in tropical rain forests) typically results in major loss of nutrients from the soil ecosystem because the bulk of the nutrients is stored above ground. This loss of nutrients presents a particular challenge for agricultural management in the tropics.

Climate, Soils, and Humans. The study of soils that formed on landscapes of the past provides insight into the biological and chemical evolution of the Earth's surface and its atmosphere. *Paleosols* (ancient soils) are abundant in the geological record; types 3.1 billion years old have been identified. Like modern soils, paleosols are products of the conditions under which they formed, and, if well preserved, they can be sensitive indicators of former environments. The iron chemistry and mineralogy of paleosols provide important clues about the evolution of the atmosphere by serving as a measure of changing levels of primordial atmospheric oxygen.

The impact of prehistoric and historic human societies on the landscape is reasonably well documented. The transformation of alfisols to spodosols by soil acidification during prehistoric times in western portions of Europe has been linked to human manipulation of the landscape. The clearing of deciduous woodland in northeastern England, followed by the invasion of acid-litter-producing heath vegetation, is suggested as the primary cause of acidification, in preference to other causes such as climatic change. The rise and fall of civilizations have also been interpreted from the perspective of utilization and eventual degradation of soil and water resources. The irrigation-based society of the Sumerians around the Tigris and Euphrates floodplains of Mesopotamia gave rise to the impressive first civilization, about 5,500 years ago. Surplus agricultural production was an important source of wealth and allowed many people to engage in activities beyond agricultural toil. The decline of Mesopotamian agriculture and the civilizations it supported has been traced in part to siltation of the irrigation systems and salinization of the soils.

Soil continues to have a profound impact on the interplay between humans and their environment. Successful agriculture is still a foundation for successful societies. Humans have a major impact on soil through manipulation of agricultural, urbanized, and forested lands; these altered soils then influence the climate.

Soils support vegetation, and both can profoundly affect the climate by altering the surface albedo—the fraction of incoming sunlight that is reflected from the surface—and by influencing the amount of evapotranspiration that can be sustained. Evapotranspiration is a cooling process; therefore, deep soils with a high water-holding capacity can supply large amounts of water to deep-rooted plants and maintain evapotranspiration during periods without precipitation, thus moderating the climate of a region. The grain-growing region of the midwestern United States and Canada is such a region. Humans can alter vegetation types through agriculture, forestry, grazing, and horticulture, which also affect the duration of cover and the soil depth. This in turn influences albedo and the amount of water available for evapotranspiration.

Humans have altered the depth and quality of soils through practices that expose the surface to water and wind erosion. Erosion is a natural process of soil formation and has been a major influence on the landforms of the world today. However, unnecessary exposure of soil surfaces to wind and water erosion displaces the most fertile topsoil from agricultural regions to streams and lakes, where the nutrients it contains disrupt ecological balance. Conservation practices that preserve live vegetation and residue cover help to maintain the productivity of the soil and to reduce pollution of water bodies. [*See* Erosion and Weathering.]

Wind of sufficient speed moves and sifts soils, with the smallest particles being carried the longest distance. Recent satellite images of dust clouds reveal extensive areas of wind erosion in Africa, Asia, Australia, South America, and North America. The "Dust Bowl" years of the 1930s left vivid memories of the effects of wind and loose soil. [*See* Dust.] Further, the geologic record shows extensive deposits of wind-blown soil, called *loess*. These soils can be agriculturally very productive but remain susceptible to wind erosion if soil conservation is not practiced by maintaining a cover of residue or vegetation or by contour plowing. Fertile soils may depart from careless managers and be lost to future generations, but humans who sustain this invaluable natural resource can maintain its life-giving potential almost indefinitely.

[*See also* Agriculture and Climate.]

BIBLIOGRAPHY

Birkeland, P. W. *Soils and Geomorphology.* London, 1984.

Buol, S. W., F. D. Hole, and R. J. McCracken. *Soil Genesis and Classification.* 3d ed. Ames, Iowa, 1989, pp. 10–12.

Dudal, R. *Key to the Soil Units for the Soil Map of the World.* Vol. 1. *Legend.* UNESCO-FAO. Rome, 1974.

Foth, H. D. *Fundamentals of Soil Science.* New York, 1990.

Furley, P. A., and W. W. Newey. *Geography of the Biosphere: An Introduction to the Nature, Distribution and Evolution of the World's Life Zones.* London, 1983.

Hillel, D. *Out of the Earth: Civilization and the Life of Soil.* New York, 1991.

Holmes, A. *Principles of Physical Geology.* 3d ed. New York, 1978.

Hyams, E. *Soil and Civilization.* New York, 1976.

Simonson, R. W. "Outline of a Generalized Theory of Soil Genesis." *Soil Science Society of America Proceedings* 23 (1959): 152–156.

Skidmore, E. L. "Soil and Water Management and Conservation: Wind Erosion." In *Handbook of Soils and Climate in Agriculture,* edited by V. J. Kilmer. Boca Raton, Fla., 1982, pp. 55–72.

Soil Survey Staff. *Soil Taxonomy: A Basic System of Soil Classification for Making and Interpreting Soil Surveys.* USDA-SCS Agricultural Handbook 436. Washington, D.C., 1975.

KEVIN MCSWEENEY and JOHN M. NORMAN

SOLAR ACTIVITY.   *See* Sun.

SOLAR ELECTROMAGNETIC RADIA-TION. *See* Shortwave Radiation; *and* Sun.

SOLAR WIND. *[This article treats a technical aspect of climate and weather studies; some of it is intended for readers at an advanced level.]*

The term *solar wind* refers to a stream of high-energy particles, mostly protons and electrons, with ions of many elements, traveling out from the Sun at speeds ranging from 300 to 1,000 kilometers per second (200 to 600 miles per second). It results from the continuous, high-velocity expansion of the tenuous, outer layers of the Sun's atmosphere into space. In order to understand the origin and attributes of this outflow of gas from the Sun, it is first necessary to consider some of the physical processes affecting the structure of that portion of the solar envelope above the Sun's visible surface (the *photosphere*). Here we adopt the following values for the solar mass ($M\odot$), radius ($R\odot$), and bolometric luminosity ($L\odot$, the rate at which radiative energy is emitted at all wavelengths from the entire solar surface): $M\odot = 1.989 \times 10^{33}$ grams, $R\odot = 6.96 \times 10^{10}$ centimeters, and $L\odot = 3.82 \times 10^{33}$ ergs per second.

The Sun's radiative energy output $L\odot$ derives principally from the continuous fusing of hydrogen into helium via nuclear fusion reactions (the so-called *proton-proton chain*) deep within the core. The energy liberated in this way is transported outward by diffusion, and, in the outermost layers of the solar interior, by the mass motions associated with convection. Virtually all of this energy flux ultimately emerges from the photosphere in the form of photons (elementary particles, the quanta of electromagnetic radiation), most of which subsequently escape into space. However, a small portion of the energy generated within the Sun is converted from radiation to other forms as a result of its interaction with matter. For example, because the material located at radial distances r (measured from the center of the Sun) in the range $0.75R\odot \le r \le R\odot$ is very opaque to the passage of radiant energy, it is subject to convective instability. As a byproduct of the convective fluid motions within this region, a flux of mechanical energy in the form of compressive wave modes is produced. The interaction between convection and differential rotation in this portion of the Sun's interior is also thought to sustain the solar magnetic field. The heating which arises from the dissipation of these nonradiative forms of energy is believed to be responsible for causing the temperature of the gas in the solar atmosphere to increase with height above the photosphere. Observations of the quiet solar atmosphere indicate that over the height interval between the visible surface and an altitude of approximately $0.1R\odot$, the temperature increases from a photospheric value of about 6,000 K to values in excess of 10^6 K. Over the same height interval, the total hydrogen number density decreases from about 10^{17} particles per cubic centimeter (in mostly neutral form) at the photosphere to fewer than 10^8 particles per cubic centimeter (in nearly fully ionized form) at greater elevations. The rarefied, high-temperature layers in the Sun's outer atmosphere constitute the *solar corona*, the source region of the solar wind. [*See* Sun.]

The existence of the solar wind was suspected for many years before it was directly observed by means of instruments on satellites in orbit around the Earth. During the first half of the twentieth century, investigators suggested that such diverse phenomena as auroral displays, geomagnetic storms, the modulation of cosmic rays, and the orientation and acceleration of comet tails were due in part to charged "corpuscular radiation" emitted by the Sun. In 1958, E. N. Parker demonstrated that the outward expansion of the Sun's atmosphere to form the solar wind was an inescapable consequence of the high temperature of the corona and the physical state of the local interstellar medium. He reasoned that on the basis of its inferred conditions of temperature and density, the coronal gas should be fully ionized, and that the transport of heat within it should take place primarily by means of thermal conduction. The latter process is a macroscopic manifestation of collisional encounters among electrons of the coronal plasma. If, in addition, the corona is presumed to exist in a state of static equilibrium, then at each height, the inward-directed gravitational force exerted on the gas must be balanced by the outward-directed force arising from the gradient in thermal pressure. By using the mathematical expressions of these physical statements concerning momentum and energy balance, Parker was able to calculate the gas pressure distribution throughout the corona. He found that although the pressure initially decreased with increasing height in the corona, it tended toward a finite value at very large heights. The calculated magnitude of this asymptotic coronal pressure was much larger than reasonable estimates for the magnitude of the confining pressure provided by the gaseous component of the interstellar medium beyond the boundaries of the solar system. Hence, since it appeared unlikely that the extended solar corona could be contained by any external pressure source, Parker concluded that the assumption of hydrostatic equilibrium was invalid, and that the corona must instead expand dynamically.

Parker also constructed models for the stationary expansion of the solar corona, assuming that the outflow was solely in the radial direction and describable by the hydrodynamic equations expressing conservation of mass, momentum, and energy for the gas. Physical considerations led him to suppose that the wind should begin with very small velocities in the corona, and that it should be characterized by a gas pressure which tended toward zero at a great distance from the Sun. He found that the solution which satisfied these boundary conditions showed that the solar wind was smoothly accelerated by the thermal pressure gradient force, from velocities less than the local speed of sound in the corona to a constant, supersonic velocity far from the Sun. His predictions about the existence and properties of the solar wind were subsequently confirmed in the early 1960s, when the first generation of satellite-borne experiments detected the presence of a continuous supersonic flow of plasma from the Sun in interplanetary space. After more than 30 years

of observations, average values for the basic parameters of the quiet solar wind have been established. The primary chemical constituents of the solar wind are hydrogen and helium, which, given the ionization state of the flow, are present in the form of protons, helium nuclei, and electrons. On average, the abundance of helium relative to hydrogen is about 4 percent by number. Heavier species are present in the flow as trace constituents, with abundances more than an order of magnitude less than that of helium. At a distance of 1 astronomical unit (AU, defined to be the mean Sun-to-Earth distance, approximately equal to $215R\odot$), the radial component (the velocity component in the radial direction along a radius from the Sun's center) of the quiet solar wind flow velocity is typically in the range 300 to 350 kilometers per second, and the proton number density is between 8 and 12 particles per cubic centimeter. For comparison, the speed of sound in the outflow at 1 AU, assuming a gas temperature of 10^5 K, is about 40 kilometers per second. The corresponding flux densities of particles and flow kinetic energy are 2 to 4×10^8 particles per square centimeter per second and 0.2–0.4 ergs per square centimeter per second, respectively. As is evident from these numbers, the solar wind at the Earth's orbit is not only quite supersonic but also of extremely low density. Indeed, a cylindrical column with a cross-sectional area of 1 square centimeter extending from the coronal base of the wind to the Earth contains less than 10^{-5} grams of material. If the measured values of the mass and kinetic energy flux densities are assumed to be representative of the wind emanating from the corona at all heliographic latitudes, then the Sun is estimated to be losing mass at the rate of about 10^{12} grams per second (that is, about 10^{-14} $M\odot$ per year), and losing energy at the rate of almost 10^{27} ergs per second (about 10^{-7} $L\odot$) due to coronal expansion.

The discussion thus far has neglected the effects the Sun's magnetic field might have on the dynamical state of the solar wind. It was noted earlier that the hot, fully ionized coronal gas has a large thermal conductivity. The same microscopic collisional processes that contribute to such efficient heat transport also enable the coronal plasma to have a high electrical conductivity. Consider a perfectly electrically conducting fluid containing a magnetic field. In such a situation, the fluid can move freely along the lines of force associated with the magnetic field. However, fluid motions in a direction perpendicular to the lines of force cause the magnetic field lines to be carried along with the flow. Hence, to the extent to which the electrical conductivity of the coronal gas is large enough that little field diffusion occurs, it is expected that the outward motion of the solar wind will transport the Sun's magnetic field into space. If the corona expanded only in the radial direction, the outflow would include a similarly directed magnetic field. But because the Sun rotates (with a period of about 27 days, as viewed from the Earth), there is also a nonradial (azimuthal, in a spherical coordinate system whose polar axis coincides with the Sun's rotation axis) component of the solar wind flow velocity. As a result, the lines of force describing the extension of the Sun's magnetic field into interplanetary space have a spiral shape.

The preceding analysis omits some important consequences of the solar rotation and magnetic field for the motion of the solar wind. In particular, the assumption that magnetic field lines are passively advected by the outflow is not entirely correct. In fact, the field exerts a force on the highly conducting wind material in which it is embedded. This force (called the *Lorentz force*) is proportional to the vector product of the electric current density associated with the nonradial component of the field and the magnetic field itself. At large heliocentric distances where the field has a substantial azimuthal component, the magnetic force acts so as to accelerate the gas in the radial direction. However, because the magnetic field in the distant solar wind is quite weak (the measured field strength in the flow at 1 AU is typically about 5×10^{-5} gauss), the outward expansion speed is enhanced by only a few percent by this extra, nonthermal force. Alternatively, near the coronal base of the wind, the vector magnetic field is primarily radially directed, so that the Lorentz force acts in the azimuthal direction (in the sense of the Sun's rotation). Since the magnetic field is considerably stronger closer to the Sun (coronal field strengths are, on average, about 1 gauss), the magnetic force can exert a correspondingly greater influence on the dynamics of the wind there. Specifically, as a result of this interaction between the solar wind plasma and the coronal magnetic field, an approximate state of corotation of the flow with the Sun is established. That is, the rotational motion of the gas in the wind is described by an angular velocity, which is nearly equal to that of the footpoints of magnetic fieldlines in the surface layers of the Sun. (*Footpoints* refers to apparent points of emergence of the magnetic field lines from the visible, photospheric surface of the Sun.) This state persists from the base of the wind out to a distance of about 20 to 30 $R\odot$, beyond which point the magnetic field and attendant azimuthal Lorentz force are too weak to continue to enforce such motion. Note, however, that the existence of a non-radial component of the solar wind flow velocity is an indication that, in addition to transporting mass, momentum, and energy, the expanding, magnetized corona also removes angular momentum from the Sun. Because the azimuthal motion of the wind is coupled to the solar rotation by the coronal magnetic field in the manner just described, the angular momentum flux in the wind is larger than that transported by the gaseous component of the flow alone. Measured values of the nonradial components of the gas velocity and interplanetary magnetic field at 1 AU suggest that the magnitude of the torque exerted on the Sun by the solar wind could be as large as 10^{31} dynes per centimeter. The angular momentum of the Sun is estimated to be about 1.9×10^{48} grams per square centimeter per second, assuming that the Sun rotates uniformly with an angular frequency of 3×10^{-6} radians per second. It therefore follows that at the present rate of removal, the solar angular momentum would be reduced by a factor of e ($= 2.71828...$) within 6×10^9 years, a period less than the Sun's *main sequence lifetime*, which is the period of time during which a star's luminosity derives from the fusion of hydrogen to form helium. For a star like the Sun, this period lasts about 10^{10} years.

This picture of the solar wind is one of a steady, structureless outflow emanating from a spherically symmetric corona. While such a description may be appropriate in some averaged sense, in reality the wind is frequently observed to contain persistent, large-scale structures and to be variable (both spatially and temporally) on a variety of scales. The earliest in situ measurements of solar wind plasma properties showed that components of the flow velocity and magnetic field underwent considerable variation on time scales ranging from a few minutes to a few hours. Subsequent analyses of data obtained over extended periods of time have shown that much of this fluctuating behavior is due to the presence of propagating hydromagnetic waves which probably originate near the coronal base of the wind. The solar wind, as observed at 1 AU, also contains features that reflect the structural nonuniformity of its source regions in the corona. In recent years, X-ray images of the Sun have revealed that much of the corona is quite nonhomogeneous, consisting of numerous looplike magnetic structures containing high-temperature gas in their uppermost regions and rooted at their bases in the underlying photosphere. Alternatively, other portions of the corona are magnetically open, containing weak, divergent fields that extend outward into interplanetary space. These so-called *coronal hole* regions are characterized by particle densities somewhat lower (by about a factor of 3) than that of the average solar corona; they have been identified as the sources of recurrent, high-speed streams in the solar wind. High-speed wind streams have limited solar longitudinal extent and are recognizable by the occurrence of flow velocities substantially in excess of those found in the quiet solar wind—often as high as 700 to 800 kilometers per second. Over the lifetime of a coronal hole, streams are observed to reappear with a period which is about equal to the 27-day synodic solar rotation period, and these streams are causally associated with recurrent geomagnetic activity. At present, the physics of high-speed streams is incompletely understood. Although it is known that the thermal pressure gradient force alone is incapable of producing such high-speed flow, the source of the requisite additional acceleration remains unidentified.

A final topic for consideration is the nature of the interaction between the solar wind and the Earth. It is because of this interaction that the Earth's dipolar magnetic field does not occupy a volume of infinite extent but is instead confined to a finite region of space called the *magnetosphere*. Solar wind plasma that is incident on the magnetospheric cavity must be deflected around it, in much the same way that water is deflected around an obstacle in a stream. In this case, however, the solar wind flow is supersonic and cannot be turned by the gas dynamic pressure gradient force. Rather, a detached bow shock wave forms in the wind upstream from the magnetospheric boundary (termed the *magnetopause*), a flow feature similar to the shock wave that forms in front of an aircraft traveling at supersonic speed. Upon passing through the bow shock, the wind is decelerated and deflected around the periphery of the magnetosphere. The plasma flow in this *magnetosheath* region is denser and hotter than unshocked solar wind and acts to compress magnetic field lines on the Earth's dayside while extending them into an elongated magnetic tail on the nightside. When the interplanetary magnetic field has a polarity which is opposite to that of the terrestrial field, its advection by the flow into the dayside magnetopause can result in the merging of solar and terrestrial field lines through the process of *magnetic reconnection*. The field lines so joined are swept back across the polar caps by the solar wind and into the magnetotail, whereupon field lines originating at opposite polar caps can again reconnect to become part of the closed terrestrial field. Associated with this magnetic flux transfer process is the occurrence of magnetospheric substorms, with the auroral displays and related geomagnetic activity they engen-

der. The amount of energy involved in interactions between solar wind and upper atmospheric is quite small in comparison to that involved in driving the Earth's climate. Hence, the extent to which the solar wind can influence conditions at lower levels in the Earth's atmosphere is at present unclear.

[*See also* Magnetosphere.]

BIBLIOGRAPHY

Brandt, J. C. *Introduction to the Solar Wind.* San Francisco, 1970.
Crooker, N. U., and G. L. Siscoe. "The Effect of the Solar Wind on the Terrestrial Environment." In *Physics of the Sun,* edited by P. A. Sturrock et al. Dordrecht, 1986, vol. 3, pp. 193–249.
Holzer, T. E. "The Solar Wind and Related Astrophysical Phenomena." In *Solar System Plasma Physics,* edited by E. N. Parker et al. New York, 1979, vol. 1, pp. 101–176.
Hundhausen, A. J. *Coronal Expansion and Solar Wind.* New York, 1972.
Leer, E., T. E. Holzer, and T. Flå. "Acceleration of the Solar Wind." *Space Science Reviews* 33 (1982): 161–200.
Parker, E. N. *Interplanetary Dynamical Processes.* New York, 1963.
Russell, C. T. "The Magnetosphere." In *The Solar Wind and the Earth,* edited by S. I. Akasofu and Y. Kamide. Dordrecht, 1987, pp. 71–100.
Schwenn, R. "The Average Solar Wind in the Inner Heliosphere: Structures and Slow Variations." In *Solar Wind Five,* edited by M. Neugebauer. NASA Conference Publication 2280. Washington, D.C., 1983, pp. 489–507.
Vasyliunas, V. M. "Large-scale Morphology of the Magnetosphere." In *Solar-Terrestrial Physics,* edited by R. L. Carovillano and J. M. Forbes. Dordrecht, 1983, pp. 243–254.
Withbroe, G. L. "The Temperature Structure, Mass and Energy Flow in the Corona and Inner Solar Wind." *Astrophysical Journal* 325, (1988): 442–467.
Withbroe, G. L., and R. W. Noyes. "Mass and Energy Flow in the Solar Chromosphere and Corona." *Annual Reviews of Astronomy and Astrophysics* 15 (1977): 363–387.

K. B. MacGregor

SOLSTICE. *See* Orbital Parameters and Equations.

SOUTH AMERICA. The Earth's fourth largest continent, South America, comprises 12 percent (17.8 million square kilometers) of the global land mass and 6 percent (290 million) of its human population. Situated in the Western Hemisphere between 12° north latitude and 55° south latitude, its 32,000-kilometer shoreline is lapped by the Caribbean Sea to the north, the Atlantic Ocean to the east, the Pacific to the west, and the Drake Passage to the south. South America has climates ranging from tropical rainforest to arid desert. The Andes Mountains extend more than 8,900 kilometers, from Tierra del

Fuego in the far south of the continent to the northern-most coast of Venezuela, including Mount Aconcagua, at 6,960 meters the highest peak in the Western Hemisphere. The mighty Amazon River annually discharges 6,300 cubic kilometers of water, transporting 900 million tons of sediment. In addition to this physiography, distinctive oceanographic phenomena—including El Niño, La Niña, cyclones, anticyclones, and storms in the "roaring forties" near the Straits of Magellan—help to produce the continent's diverse climates.

Climates. The great variety of physical geographic features in South America—such as its latitudinal (10° north to 55° south) and altitudinal (6,960 meters) extents, its tapered shape south of 20°, and the imposing Andes mountain range that divide it into east and west—contribute to a wide range of differences in weather. Climatic conditions range from steaming tropical rain forests in the Amazon to arid desert-like coasts in Peru and northern Chile. In summer, the high plateau of western Bolivia and southern Peru frequently experiences violent thunderstorms accompanied by rain, hail, and snow. Trade winds blowing from northeastern Brazil to the Gran Chaco region of Argentina cause a tropical wet climate, while those blowing over the southwestern strip of

the continent cause a very unpleasant and oppressive marine climate (Rumney, 1968).

Although most of South America receives ample rainfall, the annual precipitation ranges from 0.76 millimeter at the seaport city of San Marcos de Arica in extreme northern Chile to 2,700 millimeters in Iquitos in northeastern Peru. The Chocó region of Colombia experiences rain 300 days a year, totaling 885 millimeters. Four areas that receive an average annual rainfall of 200 centimeters are coastal French Guiana and Surinam, the Amazon River Basin, southwestern Chile, and the coasts of Colombia and northern Ecuador.

South America enjoys a great variety of climates, with the lowest temperatures (−33°C) in far southern Argentina and the hottest (43°C) in the Gran Chaco region of Argentina. On the Pacific coast of Colombia, in the Amazon Basin, and on part of the Brazilian coast the average temperature is fairly stable (30°C), with a limited monthly and annual variation of 3°C. In the Orinoco Basin, the Brazilian Highlands, and part of western Ecuador the temperatures are moderately high (18°C), with small annual variations; however, the daily averages range between 6°C and 17°C. Temperature differences are pronounced south of the Tropic of Capricorn (Par-

SOUTH AMERICA. Table 1. *South American climates and representative areas*

Climate	Characteristics	Region
Oceanic Moist (sub-polar)	Rainy, high latitude, winter very cold, short summer	Southern Andes, Falkland Islands, southern Patagonia, Puerto Gallegos
Subtropical Dry	Winter cool, summer warm, rain throughout the year	Chile, Evanjelistas, Puerto Montt
Subtropical Prairie (steppe)	Winter cool and dry, summer warm with early rain maximum	The Pampas, Entre Rios Country, Uruguay, southern coast of Brazil
Subtropical Moist (humid temperate highlands)	Winter warm and wet, summer hot and wet	Brazilian central plateau, Sao Paulo, Curitiba
Mediterranean	Winter mild with moderate rain, summer warm and dry	Middle Chile, Santiago, Coquimbo.
Semi Arid Tropical (trade wind coasts)	Winter cool with slight rain, summer hot and dry	Venezuela, coastal Colombia, south Argentina
Desert	No regular rainfall, considerable temperature range	Western Argentina, southwest Bolivia, San Juan
Dry Tropical Wet (low latitude highlands)	Winter hot and dry, summer hot and wet	Northern and central Andes, Bogota, Quito, La Paz, Cochabamba
Wet Tropical	Hot, year round rainfall	Amazonia, Guyana, Georgetown, Manaos, Para, Iquitos
Mountain Highlands	Extremes of temperature with snow and ice	High Andes, Chaco

aguay, Bolivia, Brazil, Argentina, and Chile). In the north on the Atlantic coast, temperatures range between 17°C and 25°C. Central Chile enjoys a Mediterranean climate of about 21°C. In Patagonia, because of the Andes that divide it, differences in temperature are more pronounced from east to west than from north to south. In the east, the annual temperature range is 20°C, the largest in South America.

With certain regional modifications, South America's climates generally conform to latitudinal zones: north of 20° south latitude, the climate is equatorial; from 20° to 30° south it is tropical; from 30° to 40° subtropical; from 40° to 50° warm temperate; and south of 50° cold temperate. Nevertheless, ten climatic types are distinguishable (Table 1 and Figure 1).

Winds associated with the anticyclone regions of the Atlantic and Pacific Oceans influence the climate of South America. In the Northern Hemisphere, trade winds blow from northeastern Brazil, while in the Southern Hemisphere they blow from the southeast. In addition, the presence of a high-pressure region off the western coast results in westerly winds blowing toward the continent. Consequently, two low-pressure regions develop; one covers an extensive area from northeastern Brazil to the Gran Chaco region of Argentina, and the other lies in the South Pole region. The lofty Andes bar the cool westerlies from the continental interior. Atmospheric inversions produce a 400-meter thick cloud along the Pacific coast, which blocks sunshine over Peru for about 6 months a year, making this region the cloudiest desert in the world. In contrast, persistent trade winds converge over the Amazon from May to September, producing torrential rains.

Two strong rotational ocean currents (called *gyres*)—one in the western Atlantic and the other in the southeastern Pacific—affect the coninent's climate. The South Pacific anticyclonic gyre forces the northward-flowing Peru Current (also called the Humboldt Current), which skirts the coast of Peru and Ecuador. As the winds blow parallel to the coast, cold waters well up from the depths

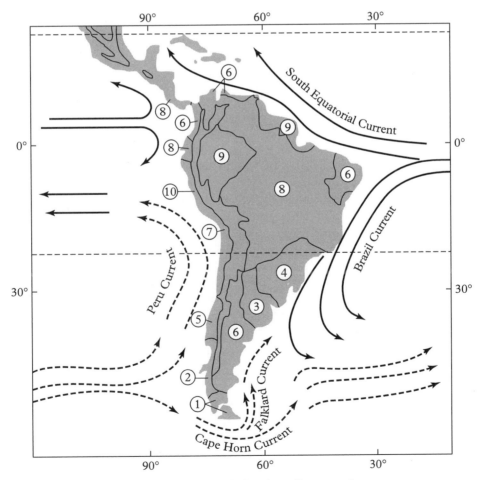

SOUTH AMERICA. Figure 1. *South American climates and ocean currents.*

of the Pacific. When the westward trades diminish, the warm surface waters that have accumulated in the western Pacific flow toward this coast and an anomalously warm season with heavy rainfall ensues. This phenomenon, known in Spanish as *El Niño* (the child) because it occurs around Christmas, causes marked changes in marine productivity. El Niño occurs about once every 3 to 7 years in response to the interannual wind fluctuations termed the *Southern Oscillation*. The cold and dry phase complementary to El Niño is referred to as *La Niña* (the girl). Recent major El Niños occurred in 1957–1958, 1982–1983, and 1991–1993. Researchers have developed dynamic models employing data on sea surface temperature and surface winds which have successfully predicted the occurrence of El Niño. [*See* El Niño, La Niña, and the Southern Oscillation.]

Climatic Changes. Reconstruction of South America's recent paleoclimatic history relies on historical records, dendrochronology (analysis of annual tree-rings), ice cores from glaciers, and variations in planktonic foraminiferal shells found in sediments. These sources reveal a cooling in the Northern Hemisphere during the Holocene, at about 2,500 years ago, with concurrent warming, low lake levels, and degeneration of the Bolivian rainforests in the Southern Hemisphere. The forests revived some 500 years later (Servant, 1984). Cores from the Quelccaya ice cap in the Peruvian Andes show three major paleoclimatic episodes (Jones, 1990): a cool period in the ninth to tenth centuries, the Medieval Warm Period in the twelfth century, and the Little Ice Age also documented for Europe between the fourteenth and the nineteenth centuries. Between the Medieval Warm Period and the Little Ice Age the mean annual temperatures differed by 1°C.

Nearly 30 percent of the increase in global atmospheric carbon dioxide since the eighteenth century has been attributed to tropical deforestation and burning (Kuo et al., 1990). It is not certain, however, to what extent the 0.4°C warming since 1940 in the Southern Hemisphere is related to deforestation in the Amazon Basin, which comprises about half of the Earth's tropical forest but is being rapidly destroyed. The area cleared in 1985 totaled 89,000 square kilometers, and the area disturbed by human activities was three times that large. Computer simulations of Amazon deforestation predict a 20 percent decrease in rainfall, warming of the local climate, and reduction in evapotranspiration (Lean and Warrilow, 1989). Although major uncertainties exist in the global climatic models of the Intergovernmental Panel on Climatic Change (IPCC), which are largely focused on other regions, IPCC studies predicted that total defor-

estation of the Amazon basin could irreversibly reduce rainfall locally, leading to a decline in vegetative cover and an increase of several degrees in mean temperature. It is not clear whether the intensity and the frequency of El Niño events might change as a result of this warming (Folland et al., 1990). Under the projected conditions of global warming, agricultural production in Argentina and Brazil (the regions considered most vulnerable to climatic changes) is expected to increase by 25 percent (Parry, 1990).

Vegetation. More than 25 families of plants, represented by at least 3,500 genera, are native to South America. The northern rainforests are renowned for their biological diversity. The Amazon Basin alone supports 20 percent of the world's 10,000 vascular plant species (Bennett, 1992).

The rainforests support about 2,500 species of hardwood trees, some taller than 100 meters. Spiny palms, orchids, and epiphytes (plants that grow on other plants or rocks and trap nutrients and rainwater) are innumerable in the tropical climate of Colombia, Venezuela, the Amazon Basin, and northern Ecuador. On the humid Amazonian side of the Andes, the subtropical rainforest extends up to 2,000 meters elevation, and epiphytes thrive at 4,200 meters. Thick forests interspersed with savannas extend from the southern edge of the Amazon rainforest to northern Argentina and upper Uruguay, and are characterized by Parana pine, hardwood trees, and the South American hollies. In parts of Peru, northern Chile, northeastern Colombia, and southwestern Ecuador that experience alternating wet and dry seasons, woody shrubs predominate.

Tropical deciduous forests of moderate height occur in regions of Venezuela, Colombia, and Brazil that experience a prolonged dry season. In northeastern Brazil, periods of drought foster the growth of palms, cacti, and other xerophytes (drought-tolerant plants). The arid climate of the western coastal deserts and the eastern slope of the southern Andes also support xerophytes. Many broad-leaved evergreens and conifers, with an understory of bamboos and ferns, are found in southern Chile where rainfall is heavy. At high altitudes, typical steppe and alpine tundra plants survive near areas of permanent snow.

Agriculture. Savanna grasslands, such as the plains and delta of the Orinoco Basin in Venezuela, are among South America's most productive areas; these, with the rainforests, sustain much of the continent's economic development. Some of the indigenous peoples of South America had considerable skill in irrigation, agriculture, and terracing. By 1530, the Spanish and Portuguese had colonized the coasts of much of the continent. Large estates produced sugar in northeastern Brazil, sugar and cotton in the Chancay Valley of Peru, and cacao (the source of chocolate) in the Guayes Basin and Ecuador. After 1850, coffee plantations flourished in subtropical eastern Brazil, as did livestock and cereal-grain cultivation in the temperate climates of Uruguay and the plains of eastern Argentina. In the latter region, known as the *pampas,* alfalfa-growing provides cattle fodder. The pampas are now also being planted with exotic timber trees, including pines, eucalypts, oaks, and poplars.

In the Amazon Basin, tapping wild rubber trees for their sap was a booming industry from the mid-nineteenth century until World War I. Since the 1960s, foreign investors have been encouraged to establish agribusiness in the Amazon Basin, concentrating on crops grown for export. Many of the crops grown are indigenous species, while others, such as coffee, have been introduced.

The Amazonian Indians have traditionally used native medicinal plants to treat ailments. Pharmaceutical companies are interested in discovering therapeutic phytochemicals such as quinine; however, cooperation with the native peoples is crucial to such discovery and to the sustainable development of the rainforests.

Wildlife. Like Australia, South America was biologically isolated from the rest of the world during the Tertiary period and has many endemic animals. The hoglike tapir of the Amazon, an animal the size of a pony, is the largest land mammal. The Amazon forest is also home to jaguars dwelling in the shadows, a vast array of insects, the capybara (the world's largest rodent), armadillos, sloths, the anaconda, and the New World monkeys. The forests also support parrots, flamingos, woodpeckers, egrets, tinamous, toucans, and barbets—birds well known for their brilliant plumage. The Amazon River is home to the manatee, a large aquatic herbivore, as well as turtles, crocodiles, caymans, electric eels, and about 3,000 fish species, including the razor-toothed piranha.

In the Argentinian pampas, large herbivores are limited to guanaco (*Llama guanacoe,* resembling deer but related to camels) and the Pampa deer *(Blastoceros campestris).* Vicuna and guanaco live in the high Andes. The alpaca and llama, are domesticated for their fine wool and for carrying loads, respectively. Other Andean mammals include chinchillas, guinea pigs, foxes, and pumas. The condor with a wing span of 3 meters is a notable bird of the peaks.

The Peruvian and Chilean coasts have few terrestrial animals, however the enormous plankton production associated with the Peru Current supports abundant fish life. Sea lions, birds, penguins, gulls, terns, shearwaters, petrels, cormorants and gannets thrive on this fishery. Huge catches of anchovy are a major constituent of the valuable fishing industry. Flocks of guanay, variegated booby, and brown pelican nest on the small coastal islands, and their droppings form *guano,* a traditional source of fertilizer.

[*See also* Cores; Deforestation; Evapotranspiration; Little Ice Ages; Paleoclimates; Rainforests; *and* Tree-Ring Analysis.]

BIBLIOGRAPHY

Bennett, B. C. "Plants and People of the Amazonian Rainforests." *Bioscience* 42 (1992): 599–607.

Folland, C. K., T. R. Carl, and K. Ya. Vinnikov. "Observed Climatic Variations and Change." In *Climatic Change: The IPCC Scientific Assessment,* edited by J. T. Houghton, G. J. Jenkins, and J. J. Ephraums. Cambridge, 1990, pp. 194–238.

Kuo, C., C. Lindberg, and D. J. Thomson. "Coherence Established between Atmospheric Carbon Dioxide and Global Temperature." *Nature* 343 (1990): 709–714.

Jones, P. D. "The Climate of the Past 1000 Years." *Endeavour* n.s. 14 (1990): 129–136.

Lean, J., and D. A. Warrilow. "Simulation of the Regional Climatic Impact of Amazon Deforestation." *Nature* 342 (1989): 411–413.

Mitchell, J. F. B., S. Manabe, T. Tokioka, and A. Meleshko. "Equilibrium Climate Change and Its Implications for the Future." In *Climatic Change: The IPCC Scientific Assessment,* edited by J. T. Houghton, G. J. Jenkins, and J. J. Ephraums. Cambridge, 1990, pp. 131–172.

Parry, M. *Climatic Change and World Agriculture.* London, 1990.

Philander, S. G. *El Niño, La Niña and the Southern Oscillation.* New York, 1990.

Philander, S. G. "El Niño." *Oceanus* (1992): 56–61.

Rind, D., C. Rosenzweig, and R. Goldberg. "Modelling the Hydrological Cycles in Assessments of Climatic Change." *Nature* 358 (1992): 119–122.

Rumney, G. R. *Climatology and the World's Climates.* New York, 1968.

Schwerdfeger, W. "Climates of Central and South America." In *World Survey of Climatology,* edited by H. E. Landsberg. New York, 1976, vol. 12, pp. 1–12.

Servant, M. "Climatic Variations in the Low Continental Latitudes During the Last 30,000 Years." In *Climatic Changes on a Yearly to Millennial Basis,* edited by N. A. Morner and W. Karlen. Dordrecht, Netherlands, 1984.

D. V. SUBBA RAO

SOUTHERN HEMISPHERE. *See* General Circulation.

SOUTHERN OSCILLATION. *See* El Niño, La Niña, and the Southern Oscillation.

SOUTH POLE. *See* Antarctica.

SPORTS. *See* Recreation and Climate.

SPRING. *See* Seasons.

SQUALL LINES. Convective systems organized in a linear fashion are termed *squall lines*. These storm systems are important in the United States because much of the country's precipitation falls from squall lines. In addition, they can harbor severe weather hazards such as tornados, large hail, and damaging winds. Squall lines have long been recognized as an important form of convection and have been studied for decades (see Hane, 1986 and Ray, 1990 for a review of past studies). These studies have relied heavily on the use of radar observations; an example of a radar-depicted squall line is shown in Figure 1. [*See* Radar.]

Formation. Squall lines usually form in the warm sector of a synoptic front—that is, between the cold front (usually oriented north–south) and the warm front (usually oriented east–west). Almost all thunderstorms require at least a conditionally unstable environment as well as a mechanism to release the instability. [*See* Convection.] The initiation may occur in response to a boundary at the surface (for example, a cold front or dry line) or in response to dynamical lifting from an upper-level disturbance. Such a mechanism, acting on a boundary of potentially unstable air, can produce a line or a complex of thunderstorms. Once initiated, storms can be sustained by their interaction with the environment. Some squall lines also form as a result of successive splitting of individual storms, eventually forming a line of storms. Others form as the result of continuous new storm formation on the south side of a line of storms, constantly extending the line of storms to the south. Often the individual cells within the line move in a northerly direction as the line of storms moves to the east. This produces an apparent cell motion to the northeast.

Structure. A squall line is defined, in practice, as any continuous or broken line of thunderstorms. Their distinguishing characteristics include the continuity and geometric relationship of the radar echoes and the persistence of the precipitation. Soon after formation, the line is characterized by a thin line of intense rain. As the system matures, a broad area of more uniform and light-to-moderate rain forms to the rear. The strongest rain and most active convection is usually on the line's most easterly side. Sometimes the line is made up of distinct individual cells; at other times, the pattern is nearly uniform along the line.

The flow features that are perpendicular to the leading edge of a squall line are shown in Figure 2. From the leading edge through the stratiform region (one of light uniform rain), front-to-back flow dominates in the lowest 6 kilometers. The convective region has forward flow above 6 kilometers. A region of weak back-to-front flow is sometimes observed between 3 and 8 kilometers at the far western edge of the squall line. Usually the strongest convection is at the leading edge, which is also the easternmost edge for middle-latitude squall lines (and it is westernmost in the tropics). This is also the region of inflow. Widespread descent at low levels exists throughout the remainder of the precipitation area. Except for cases exhibiting strong rear inflow that penetrates forward to the convective line, the flow at all levels behind the vigorous convection at the system's leading edge is

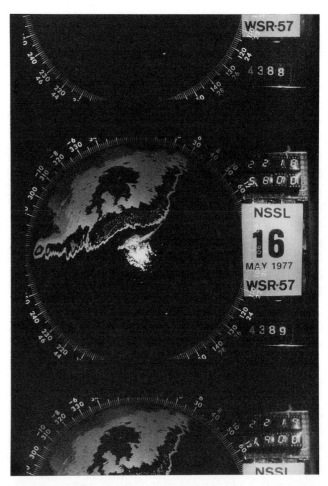

SQUALL LINES. Figure 1. *Radar image of a squall line in Oklahoma, U.S.A.* Echoes are from rainfall, with contours of increasing rainfall near the system's leading edge. Ahead of the squall line, a gust front is visible.

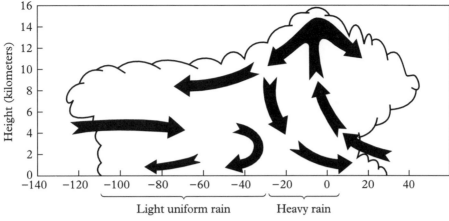

SQUALL LINES. Figure 2. *Schematic of size and airflow within a squall line.* This is a representative vertical cross section, perpendicular to the direction of movement.

usually away from the most vigorous convection. The principal updraft slopes toward the rear, but the degree of slope varies both among systems and within the same system at different times. Over a portion of the stratiform region, a radar bright-band—a level of enhanced radar reflectivity—is often observed about 4 kilometers above the ground. This band of enhanced radar return is due to liquid water coating large ice crystals or hailstones, which appear as very large water drops to the radar. Generally, there is weak rising motion above the level of the bright band, but its spatial uniformity and strength vary among systems, and perhaps during the lifetime of a given system.

Dynamics. Rotunno et al. (1988) and Weisman et al. (1990), among others, have examined some characteristics of the environment that are vital to the formation and long life of squall lines. They found that low-level shear (change of wind speed and direction with height) favors the development of long-lived linear systems, but that the magnitude of the shear required depends on the properties of the cold pool generated by the squall line, which in turn depends on the temperature structure of the atmosphere with height. Two types of long-lived squall lines are possible. One is a line of ordinary cumulonimbi that grows and decays, the other is a line of more nearly steady-state supercells. The former occurs when the shear of the environmental wind at low levels is strong and directed perpendicular to the line. The latter occurs when the shear in the environmental wind is strong and deep and directed at an angle to the line. This allows the three-dimensional circulation to develop with each cell, without interference from neighboring cells.

Without a cold pool, the updraft will tilt toward the back of the squall line. Ensuing precipitation will form a cold pool through evaporation, and the boundary thus produced will trigger new cell development. The strength of the new development depends critically on the shear profile. The optimal profile is a function of the cold pool. In time, the increasing negative buoyancy in the cold pool's circulation makes it progressively difficult for new cells to develop, and the system slowly decays.

There is a critical balance between the speed of the squall line and the generation of the cool pool, or *gust front.* If the pool moves too far ahead of the squall line, it weakens and nothing happens, or else cells form ahead of the existing squall line, choking off its moisture supply. If there is an insufficient cold pool, then the squall line rains out without regenerating itself. In the optimal condition, the speed of the storm and the production of the cold pool are balanced with the inflow to produce a downdraft that is not over the updraft, one that lifts inflowing environmental air (as it rides up over the heavier cold downdraft-produced cold pool) into the storm.

[*See also* Fronts; *and* Gust Front.]

BIBLIOGRAPHY

Hane, C. E. "Extratropical Squall Lines and Rain Bands." In *Mesoscale Meteorology and Forecasting,* edited by Peter S. Ray. Boston, 1986, pp. 359–389.

Ray, P. S. "Convective Dynamics." In *Radar in Meteorology,* edited by David Atlas. Boston, 1990, pp. 348–390.

Rotunno, R., J. B. Klemp, and M. L. Weisman. "A Theory for Strong, Long-lived Squall Lines." *Journal of Atmospheric Science* 4 (1988): 453–485.

Weisman, M. L., and J. B. Klemp "Characteristics of Isolated Convective Storms." In *Mesoscale Meteorology,* edited by Peter S. Ray. Boston, 1986, pp. 331–389.

Weisman, M. L., J. B. Klemp, and R. Rotunno. "The Structure and Evolution of Numerically Simulated Squall Lines." *Journal of Atmospheric Science* 45 (1990): 1990–2013.

PETER SAWIN RAY

STABILITY INDICES. *See* Severe Weather.

STATIC STABILITY. Generally, the atmosphere is stratified or statically stable; that is, buoyancy increases and density decreases with height. In a stratified atmosphere, vertical overturning is inhibited by gravity, which acts as a restoring force. If an air parcel at a given level of the atmosphere is displaced upward it will find itself in an environment where it is heavier than its surroundings, and it will return to its original level; the reverse is also true.

The density of the atmosphere is determined by its temperature. However, temperature is not conserved in vertical motions (air cools due to adiabatic cooling or the decrease in pressure), so temperature alone does not determine buoyancy. To determine buoyancy, we must also employ a lapse rate, as discussed below. Potential temperature is conserved in vertical motions; the change of potential temperature with height—$d\theta/dz$—in the atmosphere determines static stability. If the change in potential temperature with height is positive, then the atmosphere is said to be *statically stable*. Under positive static stability, gravity acts as a restoring force and a vertically displaced air parcel will return to its original level. If the change in potential temperature with height is exactly zero, the atmosphere is said to be *neutrally stable*, and a displaced air parcel will remain at the level to which it has been displaced. If the change in potential temperature with height is negative, the atmosphere is said to be *statically unstable*, and a vertically displaced air parcel will accelerate in its direction of displacement. Eventually, convective overturning will return the atmosphere to a statically stable state.

In the atmosphere, the change in temperature with respect to height—dT/dz—is known as the *atmospheric lapse rate*. When an air parcel is displaced vertically, it changes temperature based on the *dry adiabatic lapse rate* (9.8°C per kilometer). If the atmospheric lapse rate is greater than the dry adiabatic lapse rate, it is referred to as *super-adiabatic*, and displaced air parcels will accelerate freely. However, the atmospheric lapse rate is almost always more gradual or less than the dry adiabatic lapse rate. Therefore, even if a relatively warm air parcel begins

to rise, it will cool at a quicker lapse rate than the background atmosphere, will soon be at a lower temperature than its surroundings, and will cease to rise. Thus an air parcel is almost always stable when rising according to the dry lapse rate. However, once an air parcel becomes saturated (100 percent relative humidity), it no longer rises at the dry atmospheric lapse rate, but rather at the *moist* or *saturated* adiabatic lapse rate (which can vary between 4° and 7°C per kilometer), which takes into account heating due to the condensation of moisture (known as *latent heating*). Thus, if the atmospheric lapse rate lies between the dry and moist adiabatic lapse rates and an air parcel begins to rise, it will cool at a faster rate than the background atmosphere. If the air parcel cools to its saturation point (the *lifting condensation level* which is defined as the level to which an air parcel becomes lifted before it become saturated), it will then continue to cool as it rises according to the *saturated lapse rate*, which is slower than the dry adiabatic rate applicable to the background atmosphere. If the air parcel can reach a level where it is warmer than the surrounding atmosphere (known as the *level of free convection*), it will achieve static instability and actually accelerate upward. This process of convection is known as *conditional instability*.

[*See also* Adiabatic Processes; Convection; Latent Heat; *and* Potential Temperature.]

BIBLIOGRAPHY

Gedzelman, S. D. *The Science and Wonders of the Atmosphere.* New York, 1980.
Holton, J. R. *An Introduction to Dynamic Meteorology.* 2d ed. New York, 1979.
Wallace, J. M., and P. V. Hobbs. *Atmospheric Science: An Introductory Survey.* New York, 1977.

JUDAH COHEN

STATISTICAL METHODS. With the advent of high-resolution numerical weather prediction models and the increasing speed of computers and concomitant increases in accuracy and range of weather forecasts, it would seem that statistical methods of weather prediction would soon become redundant. Unfortunately, it appears that skillful forecasts of weather patterns a month or more ahead are out of the question. The reason for the inability of computer models to predict the weather on these longer time scales is the nonlinearity of the equations governing the evolution of the Earth's weather, which generate chaotic behavior. Although it appears that chaotic behavior represents a fundamental obstacle to long-range prediction of instantaneous weather patterns, it should be noted that gross features of the climate, such as monthly mean temperatures and rainfall, may nonetheless be predictable, at least in a certain statistical sense. [*See* Chaos; Numerical Weather Prediction; Weather Forecasting.]

The statistical methods used in the long-range prediction of a specific weather variable for the period of a month, a season, or a year assume that, although the variability of weather from day to day during some period in the future must be regarded as unpredictable, the *average character* of the weather during that period is at least potentially predictable. The average character of the weather is expressed in relation to the normal climatologically expected value for a particular weather variable for that period. Thus, in statistical weather prediction, one does not expect to provide a detailed weather forecast for each day of the coming summer, but rather to say whether it is likely that the summer is, on average, going to be hotter or cooler than usual. For example, as a result of the effect of the El Niño-Southern Oscillation, the average December-January-February temperatures in New Orleans can be predicted with some degree of skill from the sea-level pressure in Darwin, Australia, during the previous June-July-August. This is the simplest possible statistical weather prediction model. [*See* El Niño, La Niña, and the Southern Oscillation.]

In general, when making a long-range weather forecast we may believe that several variables are going to contribute to the average character of the weather. For example, current models for the Indian monsoon use not only the state of the El Niño-Southern Oscillation but also upper air flow over India and surface conditions in southern Asia. When many variables are involved in the prediction we can use linear statistical models. These models are based on the idea that some weather variable $y(j)$, the *predictand*, which is the deviation of a seasonal average from its mean climatological state in the year j, can be predicted using a weighted sum of other variables $x_i(j)$, the predictors, from preceding seasons. The mathematical expression of such a model is

$$\hat{y}(j) = \sum_{i=1}^{Q} z_i x_i(j)$$

where $\hat{y}(j)$ is the model prediction of the weather variable, the weighting coefficients are z_i, and Q is the total number of predictors used. Before such a model is used we must decide which predictors to use and estimate their weighting coefficients.

The usual method for estimating the weighting coefficients z_i is to find those coefficients that provide a least mean square best fit between the predicted and observed values of the weather variable for previous years. That is, we minimize the error, E, which is defined to be

$$E = \frac{1}{N} \sum_{j=1}^{N} [y(j) - \hat{y}(j)]^2$$

where N is the number of years, in our predictions of historic data. This criteria cannot however be used in deciding how many predictors Q to use, because, provided we use predictors that vary relatively independently of one another, the error will always tend to decrease as we add more predictors, and there is therefore no minimum in the error as we increase Q. We might consider that having lots of predictors, and hence a low error in our fit to the historical data, would be beneficial; however, historical data will contain noise that is caused by the chaotic and unpredictable variability of the weather. Therefore, if we have too many predictors, many of them will be fitting the noise in the historical data and will tend to cause our predictions to be noisy and unreliable. This leads to a tradeoff between including extra variables in the set of predictors, for which the weighting coefficients z_i are dominated by these spurious variations, and decreasing the error in the prediction of historical data. It is this tradeoff that determines which predictors should be included in a statistical model.

One can quantitatively evaluate this tradeoff by estimating the probability that the true value of an estimated weighting coefficient is different from zero. This estimate uses the fact that the difference between the true weighting coefficients and the estimated weighting coefficients may be considered to be a multivariate Gaussian random process. If there is not a strong probability that a weighting coefficient is really different from zero, then the associated predictor should be excluded. This procedure is termed *analyzing the statistical significance of the model coefficients*. An alternative approach to determining what predictors should be included employs information theoretic concepts and was originally proposed by H. Akaike. In this method the set of predictors is found that minimizes Akaike's information criteria (AIC), which is defined as follows:

$$\text{AIC} = N\log(E) + Q.$$

Initially, as the number of predictors is increased, the error (and AIC) decreases rapidly, but as more predictors are added, Q increases more rapidly than the log of the error decreases, leading to a minimum in AIC for a particular number of predictors.

Once we have established which variables should be included in the statistical model, we must decide how accurate, and therefore how effective, our predictions are. One measure of this is the *skill* (represented S), defined to be

$$S(t) = 1 - \frac{E(t)}{\langle y^0(t)y^0(t)\rangle}$$

If the skill is 100 percent, then there is no residual error, and all of the variance of $y_j^0(t)$ is explained by the model forecast, while 0 percent skill indicates no explanatory power for the forecast. Confidence limits for $S(t)$ can be constructed from the distribution of $\hat{Z}^i$ discussed above. If the skill is not significantly different from 0 percent, then the forecast model does not provide useful forecasts. Perhaps the most reliable method for validating any statistical forecast model is to estimate the model coefficients using one set of data and then to test the efficacy of the model predictions on an independent data set. The skill obtained on the original data is then termed the *hindcast skill,* and the skill obtained with the independent data is termed the *forecast skill.* If the forecast skill is insignificant, then the model forecasts are not useful. A salutary lesson on the use of statistical models is provided by the fact that prior to 1980, the operational long-range weather forecast models used by most of world's major weather forecasting centers showed little skill.

BIBLIOGRAPHY

Nicholls, N. "Long-range Weather Forecasting: Value, Status and Prospects." *Review of Geophysics and Space Physics* 18 (1980): 771–788.
Priestley, M. B. *Spectral Analysis and Time Series.* San Diego, Calif., 1981.

BRIAN CAIRNS

STORMS. A storm can be defined as a disturbed state of the atmosphere, the opposite of what we would call *calm.* Storms are a natural part of the environment, arising as a consequence of solar heating and the Earth's topography and rotation. Humans tend to categorize storms according to what we see as the weather's most damaging or impressive aspect; hence, we refer to snowstorms, thunderstorms, ice storms, hailstorms, windstorms, and so on. Other types of storms have special or local names, such as tornadoes, hurricanes, or blizzards. The latter events often involve several different aspects of weather. For example, the definition of a blizzard involves a set of threshold values for snowfall rate, wind speed, and perhaps temperature. A tornado is a special kind of windstorm associated with thunderstorms.

From the point of view of a meteorologist, the notion of a storm takes on a different meaning. The meteorologist wants to understand the *causes* of the snow, wind, hail, and other events. Disturbed weather of all sorts tends to be associated with regions of relatively low atmospheric pressure, whereas the weather associated with relatively high pressure is typically calm and undisturbed. Because of the way wind and pressure are related, regions of low pressure are almost always characterized by winds rotating cyclonically (counterclockwise in the Northern Hemisphere, clockwise in the Southern Hemisphere) around the center of the low pressure. Thus, disturbed weather of all sorts occurs primarily in association with *cyclones.* Cyclones come in all sizes: the largest span thousands of kilometers and last for days or even weeks; the smallest are only a few kilometers across and last for a few hours at most. There are three classes of cyclones considered to be most important in connection with storms, as discussed below. [*See* Cyclones.]

Extratropical Cyclones. Outside the tropics, the most important storm makers are those with characteristic sizes of several thousand kilometers, called *extratropical cyclones.* These cyclones, which persist for several days or occasionally longer, are the traveling centers of low pressure seen on television and newspaper weather maps. Extratropical cyclones in middle latitudes are the atmosphere's primary large-scale response to the unequal distribution of solar heating between the equatorial regions and the poles. The tropics are much warmer than the polar regions, and when the temperature difference becomes too large, the atmosphere exports the excess heat from the tropics to the poles by means of these large cyclones.

In this heat-transfer process, many of the important weather events we call storms are created—rain, snow, ice, wind, and hail. In a sense, all the stormy weather events that affect us are simply side-effects of the large-scale heat transfer from the equator to the poles. The extratropical cyclone is itself a storm, of course, but its actions create conditions that make many other types of storms possible. For example, thunderstorms arise when conditional instability in the presence of water vapor in the air is triggered by a process of strong lifting. An extratropical cyclone can create such conditions by its poleward transport of warm, moist air at low levels, while the frontal boundaries associated with the moving air masses in the cyclone can provide lifting to initiate thunderstorms. Water falling as snow in a blizzard probably originated in the tropics and was brought poleward in the flow ahead of a cyclone.

As the extratropical cyclone develops, it can produce very strong winds over a large, deep region. To obtain a feeling for the energies involved in an extratropical cyclonic storm, assume that such a storm covers an area of about 1 million square kilometers and that the average windspeed within the cyclone is about 10 meters per second (roughly 20 miles per hour). Just to maintain those winds within the volume of a single storm takes the energy of around a hundred 1-megaton thermonuclear bombs every second. There are several such cyclones occurring worldwide at any moment. The winds themselves can be damaging to structures and vegetation, and they are important in creating the potential for other types of damaging storms. Severe thunderstorms, including tornadic storms, usually arise in the high-energy environment associated with an extratropical cyclone. In the winter, moisture can fall as rain in one part of the extratropical cyclone, ice in another part, sleet in another, and snow in yet another. As an extratropical storm moves by, an observer may see several different types of storms in succession: a severe thunderstorm, hail, a tornado, a rainstorm, an ice storm, and a snowstorm, all within 24 hours. Figure 1 provides an enhanced satellite portrait of such a complex storm.

The damage and loss of life associated with extratropical cyclones can be quite extensive. A notable example was the major winter storm that occurred over the eastern United States on 13–14 March 1993. This event produced severe thunderstorms in the southeastern United States and heavy snowfalls over a swath covering the middle Atlantic states and New England, as well as wind gusts exceeding hurricane force. Another well-known storm surprised the United Kingdom on 15–16 October 1987, with devastating winds. Such storms are almost always characterized by a phase of very rapid development, with accelerating winds and rapidly falling pressure.

Tropical Cyclones. The second major type of storm-producing cyclone is the tropical variety. Tropical cyclones have different names in different parts of the world, including *typhoon, hurricane,* and *cyclone.* All such storms originate in the tropics from an initial disturbance characterized by a cluster of thunderstorms. These thunderstorms become organized into a cyclonic disturbance with a warm core. Tropical cyclones characteristically span hundreds of kilometers and last from several to 10 days or more. Whereas extratropical cyclones gain energy from vertical wind shear (that is, winds changing speed and direction with height) ultimately related to the equatorial-polar temperature contrast, tropical cyclones are dissipated by vertical wind shear and develop only in the relatively weak shear of the tropics. Tropical cyclones develop almost exclusively over warm ocean waters and dissipate rapidly after making landfall or when they travel over cold water. A major energy source for tropical cyclones lies in the warm surface waters of the ocean.

Tropical cyclones come in a variety of sizes and intensities. At their strongest, they are potentially the most damaging storms produced by the atmosphere, because they produce violent winds—sometimes approaching the speeds seen in tornadoes—over an area of hundreds of square kilometers, often combined with torrential rainfall and sometimes with a *storm surge.* The storm surge is an increase in the height of the sea caused by the reduced atmospheric pressure of the disturbance and the force of the wind, combined with ocean-floor topography. This increase in sea height is enhanced by storm winds, so that low-lying areas where tropical cyclones make landfall can be inundated with rising seawater, at the same time as they are experiencing destructive winds and torrential rainfall. Storm surges occasionally coincide with the high tide phase of the normal tidal cycle; and when this happens, destruction and casualties can be especially great. Tropical cyclones have been responsible for tremendous property damage, as has been seen with hurricanes Hugo (South Carolina, 22 September 1989) and Andrew (24 August 1992 in Florida and 26 August in Louisiana) in the United States. Devastation in tropical cyclones can be intense, as with Cyclone Tracy (Darwin, Australia on 25 December 1974), with many casualties; reports of tens of thousands of deaths have been associated with tropical cyclones in the Bay of Bengal, near India.

Tropical cyclones also can produce local conditions favoring the development of tornadoes within them, typically after a storm makes landfall. Tornadoes within tropical cyclones usually are not as violent as those produced in other circumstances, but they can produce local swaths of particularly intense damage within the area affected by the parent cyclone.

Dissipating hurricanes may contribute to further damage from torrential rainfall long after their winds have diminished below hurricane force. In some arid parts of the world, such as the American Southwest, the occasional hurricane's rainfall may be a significant portion of the total precipitation in the region over a decade. Not all of the effects of such storms are damaging, however;

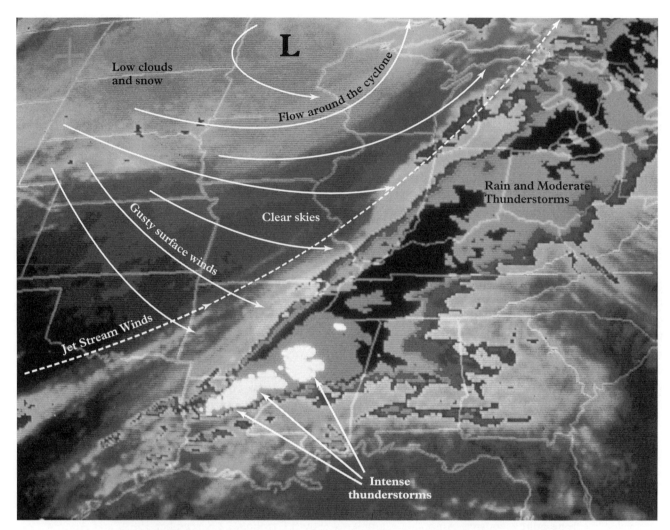

STORMS. Figure 1. *An example of the distribution of various types of stormy weather within an extratropical cyclone.* The center of the large-scale cyclone is denoted with an "L," and various types of storms, including thunderstorms, rainstorms, and snowstorms are indicated on an infrared satellite image. Streamlines of windflow at selected places at the surface and along the jet stream aloft are also indicated. On the satellite image, cloud-top temperatures are shown with various shades of gray; the image has been artificially enhanced by changing the shading abruptly at certain specified temperatures. The very coldest, and therefore the highest, clouds are indicated by the lightest gray shades.

the rainfall from a tropical storm in the fall can provide relief after a summer's drought.

Other Important Cyclones. Over the polar seas, another type of intense cyclone, called a *polar low,* can occur. These small systems form in airstreams flowing off the poles over relatively warm sea surfaces. Such a polar airstream occurs in the wake of an extratropical cyclone, after cold frontal passage. As with all other cyclones, the sizes and intensities of polar lows vary, but they characteristically are less than about 100 kilometers in diameter. These polar lows can produce blinding,

heavy snowfalls with near-hurricane force winds and sometimes have embedded thunderstorms. They appear to have some features in common with tropical cyclones, including an energy source from relatively warm ocean surface temperatures. The name *arctic hurricanes* has been applied to them. Aside from the danger they represent to shipping and aircraft, their impact when making landfall can be both surprising and devastating. Scandinavia and the countries bordering the Baltic Sea are often affected, because polar lows are especially common in air flowing off Greenland. [*See* Polar Lows.]

Another type of cyclone, called a *mesocyclone,* arises in association with thunderstorms. In some special situations, a large portion of a single thunderstorm can be in cyclonic rotation. A thunderstorm that has this characteristic is called a *supercell thunderstorm* and is almost always accompanied by some sort of severe local storm activity, such as tornadoes, damaging winds, or large hailstones. Supercells can also produce heavy rain and abundant lightning. Not all such severe local storm events are associated with mesocyclones, but most mesocyclones include at least one form of hazardous weather. Mesocyclones are characteristically less than about 10 kilometers wide and can last for several hours. [*See* Thunderstorms.]

When thunderstorms develop in groups, they can form a system called a *mesoscale convective system.* It is not uncommon for such a system to develop a cyclonic disturbance in association with it. Such disturbances generally exist above the Earth's surface, so their presence can be rather difficult to observe. On some occasions, however, the disturbance can persist after the thunderstorm activity dissipates; such remnant cyclonic disturbance has occasionally been found to be associated with subsequent redevelopment of new thunderstorms and the organization of those new thunderstorms into another mesoscale convective system. This formation and dissipation of deep, moist convection in association with a traveling cyclone can continue for several days. Much remains to be learned about cyclones that develop in conjunction with thunderstorms.

Observation, Technology, and Forecasting. Our ability to observe the weather is a key factor in recognizing and forecasting storms. Meteorological satellites and special radars have increased knowledge about storms significantly. Weather satellites have been especially important in the detection and prediction of storms coming onshore from the ocean. A few decades ago, tropical and extratropical storms could make landfall with little or no warning, resulting in many casualties. Current and planned observing systems will continue to improve our capability to detect and forecast such evens.

As we learn more about storms and their causes, we should be able to use that knowledge to make better forecasts, including forecasts by computer models of the atmosphere. [*See* Models and Modeling.] With the rapid development of computer technology, computer-based models of atmospheric behavior are becoming increasingly capable of predicting many of the processes that lead to storms. However, most storm phenomena are nonlinear, which means that the phenomena are not the simple sum of their components. [*See* Chaos.] This imposes an inherent limit to the predictability of storms. Thus the future of storm forecasting will not depend entirely on computer models.

[*See also* Convection; Cumulonimbus Clouds; Dust Storms; Hailstorms; Hurricanes; Ice Storms; Severe Weather; Snowstorms; *and* Tornadoes.]

BIBLIOGRAPHY

Anthes, R. A. *Tropical Cyclones: Their Evolution, Structure and Effects.* Boston, 1982.
Cotton, W. R. *Storms.* Fort Collins, Colo., 1990.
Frazier, K. *The Violent Face of Nature.* New York, 1979.
Houze, R. A., Jr., and P. V. Hobbs. "Organization and Structure of Precipitating Cloud Systems." *Advances in Geophysics,* 24 (1982): 225–315.
Ludlam, F. H. *Clouds and Storms.* University Park, Pa., 1980.

CHARLES A. DOSWELL III

STRATOCUMULUS CLOUDS. The term stratocumulus is a compound of the words *stratus* (Latin for "layer") and *cumulus* (Latin for "pile"). It denotes an extended layer of cumuliform clouds. Stratocumulus *decks* occur at low altitudes (usually between 1,500 and 3,000 meters) and can cover areas of several hundred thousand square kilometers. They are observed primarily in subtropical marine environments or, more frequently, in the regions off the western coasts of the major continents. Stratocumulus clouds consist primarily of water droplets. Their liquid water content ranges from a few tenths of a gram to about 1 gram per cubic meter; the radii of their water particles are primarily between 8 and 16 microns. Stratocumulus clouds produce only a little precipitation, mainly in the form of drizzle that often evaporates before it reaches the ground. Their large vertical extents and relatively high optical thicknesses make them efficient reflectors of solar radiation and therefore important factors in the radiative budget of the planet. [*See* Clouds and Radiation; Cumulus Clouds.]

Formation. Marine stratocumulus decks, particularly those off the western coast of North America, have been a subject of intense study since the 1920s. Several ideas have been proposed to explain the formation of these decks. In most cases where stratocumulus clouds are present, the large-scale cloud field consists of a solid layer of stratus clouds near the coastline, which breaks into a stratocumulus deck farther away from the coast. The stratus clouds are often capped by a temperature inversion which inhibits their vertical growth. The prevailing mechanism in the breakup of the stratus layer and the formation of the stratocumulus clouds is associated with the entrainment of dry air into the top of the stratus layer. This entrainment happens because the cloud top is cooled as the cloud radiates heat to space and as cloud droplets evaporate when turbulent motions mix them with drier air. When the cloud top cools, a convectively unstable situation is created, with colder, drier air located above warmer, more humid air. The colder, heavier air falls downward and is entrained into the body of the cloud. This produces discontinuities in the cloud layer and eventually breaks up the stratus layer into the smaller elements that form the stratocumulus field. The conditions under which the top of the cloud layer becomes unstable to the entrainment of dry air from above (known as conditions for *cloud top entrainment instability*) are the subject of intense research in the field of cloud physics. [*See* Cloud Physics.]

Stratocumulus clouds produce little precipitation, for two reasons. First, stratocumulus clouds are warm clouds that do not contain ice crystals in large concentrations. This means that in these clouds the process in which ice particles serve as nuclei that collect water vapor and form large snowflakes does not operate. Second, stratocumulus clouds have small liquid water contents. This means that the coalescence process, in which large droplets grow to precipitation size by collecting smaller ones

as they fall through the cloud, also does not operate efficiently in these clouds. The water droplets in stratocumulus clouds grow primarily by condensation, a slower process. Thus, the droplets remain small and only occasionally fall in the form of drizzle, which often evaporates below the cloud and does not reach the ground. When this happens, the area below the cloud is cooled by the evaporation. Then a stable layer is created with warmer air in the cloud and colder air below it, which tends to inhibit transport of moisture from the Earth's surface to the cloud. The cutting off of the stratocumulus deck from its moisture source tends to dissipate the cloud layer. [*See* Precipitation.]

Radiative Impact. The radiative impact of a cloud is determined by the relative strengths of two opposing effects, the solar albedo (reflectivity) effect and the thermal greenhouse effect. The solar albedo is a cooling effect, because the clouds reflect solar radiation and prevent it from reaching the surface of the planet. The greenhouse effect causes warming as the clouds trap thermal radiation emitted by the surface of the planet and prevent it from escaping to outer space. The amount of solar radiation reflected by a cloud depends primarily on the cloud *optical thickness*, which is directly proportional to the physical extent and water density of the cloud, and inversely proportional to the average size of the cloud particles. Optically thick clouds reflect more solar radia-

STRATOCUMULUS CLOUDS. Figure 1. *Stratocumulus clouds over Kenora, Ontario, Canada.* (Photo copyright Ronald Holle.)

tion than optically thin ones. The amount of thermal radiation trapped by a cloud also depends on its optical thickness as well as on the altitude of the top of the cloud. High, colder clouds trap thermal radiation more effectively than low, warmer clouds.

Stratocumulus clouds have high optical thicknesses and large horizontal extents, so they are efficient reflectors of solar radiation. At the same time, they have low cloud tops that are not significantly colder than the sea surface, so they are inefficient in trapping thermal radiation. Therefore, they exert a net cooling effect on the Earth's climate. Stratiform water clouds cover about 45 percent of the ocean surface and account for about one-third of the total planetary albedo. Model calculations show that it would take only a 4 percent increase in the amount of marine stratocumulus cover to counteract the warming that would result from a doubling in atmospheric carbon dioxide (CO_2) concentration. This emphasizes the great importance of stratocumulus clouds in the determination of the Earth's radiative budget, and therefore of the Earth's climate.

[See also Clouds.]

BIBLIOGRAPHY

Ramanathan, V., et al. "Cloud-Radiative Forcing and Climate: Results from the Earth Radiation Budget Experiment." *Science* 243 (1989): 57–63.
Rogers, R. R., and M. K. Yau. *A Short Course in Cloud Physics.* 3d ed. Oxford, 1989.
Rossow, W. B., and A. A. Lacis. "Global, Seasonal Cloud Variations from Satellite Radiance Measurements. Part II: Cloud Properties and Radiative Effects." *Journal of Climate* 3 (1990): 1204–1213.
Wallace, J. M., and P. V. Hobbs. *Atmospheric Science: An Introductory Survey.* Orlando, Fla., 1977.

GEORGE TSELIOUDIS

STRATOSPHERE. The Earth's atmosphere is conventionally divided into layers based on the average vertical variation of temperature. The layer between approximately 10 and 50 kilometers of altitude, known as the *stratosphere*, is characterized by a more or less continuous increase of temperature with height. Below the stratosphere lies the *troposphere*, where the temperature decreases sharply with height up to the *tropopause*, which forms the lower boundary of the stratosphere. The altitude of the tropopause varies with latitude; it is near 16 to 17 kilometers in the tropics and at 7 to 8 kilometers in the polar regions. Above the tropopause the temperature generally increases—very slightly at first, and then much

more rapidly—until it reaches a maximum at the *stratopause*, the upper boundary of the stratosphere. At altitudes above 50 kilometers the temperature decreases again up to a height of 85 kilometers, the *mesopause* level, and then begins a very rapid increase to the outer edge of the atmosphere. Figure 1 shows this structure. The vertical increase of temperature that defines the stratosphere is the result of heating by the absorption of ultraviolet solar radiation by ozone (O_3), the triatomic form of oxygen. In addition to playing a central role in the thermal budget of the stratosphere, ozone absorbs ultraviolet radiation that would be lethal to life on the surface of the Earth. [See Atmospheric Structure.]

The existence of the stratosphere was unknown until 1902, when a series of balloon observations by Léon Teisserenc de Bort demonstrated that the temperature decrease with altitude typical of the troposphere ceased above 11 kilometers. Scientific understanding of the stratosphere increased rapidly after World War II, when balloon observations reaching the lower stratosphere became possible, and accelerated during the International Geophysical Year (1957–1958). In the 1990s, a network of meteorological stations provides routine measurements of such variables as temperature, pressure, and wind velocity up to altitudes of about 30 kilometers. Certain observations of chemical composition, including ozone concentrations, are also made occasionally from balloons. Observations at altitudes above 30 kilometers have been made since the 1960s using instruments carried aloft by meteorological rockets. The development of satellite observing systems in the 1970s revolutionized scientific understanding of the stratosphere by providing a global view of many aspects of its meteorology and chemical composition.

Radiative Processes and Thermal Structure. Ozone plays a central role in the thermodynamics of the stratosphere. It is a strong absorber of ultraviolet radiation at wavelengths between 200 and 350 nanometers, a region of the spectrum that contains the Hartley and Huggings bands, and contributes most of the heating in the upper stratosphere. Near the stratopause in summer, ozone heating rates can exceed 15°C per day. At lower altitudes ozone absorbs ultraviolet radiation in the Chappuis bands, between 44 and 80 nanometers. The resulting heating rates are much smaller than those near the stratopause, but they are important because they constitute one of the major heating sources for the lower stratosphere. Ozone is also an efficient absorber and emitter of infrared radiation at wavelengths near 9.6 microns. Molecular oxygen (O_2) absorbs ultraviolet radi-

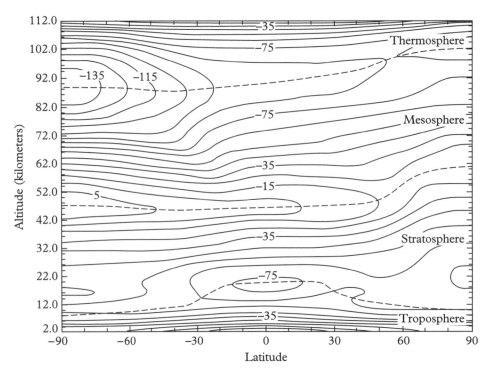

STRATOSPHERE. Figure 1. *Longitudinally averaged temperature distribution of the Earth's atmosphere.* Contours are isolines of temperature (°C). The boundaries of the troposphere, stratosphere, mesosphere, and thermosphere are denoted by broken lines.

ation, but its contribution to the heat budget of the stratosphere is insignificant. More importantly, dissociation of molecular oxygen produces atomic oxygen (O), which leads to the formation of ozone. [*See* Ozone; Shortwave Radiation.]

The thermal budget of the stratosphere is also influenced by the presence of several other constituents, including nitrogen dioxide, carbon dioxide, and water vapor. Nitrogen dioxide is an important heating source in the lower stratosphere. Carbon dioxide and water vapor contribute to the radiative budget by absorbing and emitting infrared radiation, principally at wavelengths near 15 and 6.3 microns, respectively. Chlorofluorocarbons (CFCs), which have increased rapidly in the atmosphere as a result of industrial activity, are very efficient absorbers and emitters of radiation in certain regions of the infrared spectrum where other radiatively important gases have a minor influence. Two other stratospheric trace gases, methane and nitrous oxide, have both natural and anthropogenic sources; they also contribute to infrared cooling and heating at wavelengths unaffected by carbon dioxide, ozone, or water vapor.

The temperature distribution of the stratosphere represents a balance among heating by absorption of ultra-violet radiation, cooling through infrared emission, and transport of heat by the atmospheric circulation. Throughout much of the stratosphere, transport effects are of secondary importance. Thus a calculation of the stratospheric temperature distribution that ignored transport effects would yield a result not much different from that shown in Figure 1 below 50 kilometers, with the important exceptions of the tropical upper stratosphere and the high-latitude stratosphere in winter. Outside these regions, net stratospheric heating rates (solar heating plus infrared cooling) are generally less than 1°C per day. By contrast, at high latitudes in winter an imbalance of cooling over heating of more than 5°C per day is commonly found, implying that transport processes play a significant role in balancing the radiative budget.

Stratospheric Dynamics. The stratosphere possesses a vigorous wind system which, in an average sense, is directed predominantly in the east–west direction. This is not to say that north–south winds are unimportant, but rather that an average taken with respect to longitude yields a large east–west resultant, with much weaker motions in the north–south and vertical directions. It might be thought that, if the atmosphere is heated in the summer hemisphere and cooled in the winter hemisphere, a compensating circulation would be set up, with air rising in the summer hemisphere and descending in the winter hemisphere. However, such motions, commonly referred to as *mean meridional circulations*, are relatively weak in a rotating atmosphere whose angular momentum must be conserved. This constraint manifests itself as an "apparent" force, the *Coriolis force*, which deflects north–south motions toward an east–west direction. On the other hand, strong east–west motions are possible because the Coriolis force acting on them can be balanced by the north–south pressure gradients set up by atmospheric heat sources. This situation, known as *geostrophic balance*, prevails throughout much of the stratosphere. Because geostrophic balance supresses mean meridional circulations, the extratropical temperature distribution tends toward radiative equilibrium. The longitudinally averaged circulation can depart from geostrophic balance if sources of east–west momentum are present in the atmosphere. Various types of waves can transport and deposit momentum in the stratosphere, enhancing the mean meridional circulation and driving the temperature distribution away from radiative equilibrium. [See Coriolis Force; Dynamics; Meridional Circulations.]

Planetary waves are Rossby waves of dimensions comparable to the circumference of the Earth. Planetary Rossby waves are the result of the conservation of angular momentum in a rotating atmosphere. Propagation of these waves into the stratosphere is possible only during winter, when the stratospheric winds blow from west to east. When propagation is possible, the waves can reach very large amplitudes, especially in the Northern Hemisphere, where jet streams blowing across mountainous terrain are an efficient forcing mechanism for the waves. [See Rossby Waves.]

Rapid amplification of planetary waves can disturb the dynamics of the winter stratosphere so as to produce a reversal of the average west-to-east flow, and consequently, of the north–south pressure and temperature gradients. At high latitudes, the temperature can increase by as much as 50°C in a few days during one of these wave events—a phenomenon so striking that it has become known as a *sudden stratospheric warming*. Sudden warmings also bring about large, rapid changes in the abundance of stratospheric trace species, including ozone. Even when planetary waves are not amplifying rapidly, they are continuously being dissipated by processes such as radiative damping and certain types of instabilities. Wave dissipation modifies the longitudinally averaged circulation and temperature distribution of the

stratosphere, and it is responsible for the considerable departures from radiative equilibrium found at higher latitudes in winter.

In the tropics the weakness of the Coriolis force makes possible complex interactions between wave motions and the wind system of the tropical stratosphere. In particular, the tropical atmosphere supports certain "equatorially trapped" waves that propagate vertically and influence the circulation of the stratosphere. The most important large-scale waves in the tropics are known as *Kelvin waves* and *Rossby-gravity waves*. Both types result from the combined effects of the gravitational and Coriolis forces. These waves, as well as smaller-scale gravity waves (of longitudinal wavelengths of several hundred to a few thousand kilometers), are believed to be excited by latent heat release in tropical convective cloud systems. The dissipation of vertically propagating tropical waves gives rise to the quasi-biennial and semiannual oscillations of the tropical stratosphere.

The *quasi-biennial oscillation* (QBO) consists of alternating regimes of eastward and westward winds that propagate downward at the rate of about 1 kilometer per month. Eastward and westward regimes alternate, with a mean period of some 27 months (hence the name, meaning "about once every two years") at altitudes between about 16 and 30 kilometers. The eastward and westward phases of the oscillation are thought to be caused by deposition of momentum by Kelvin waves and Rossby-gravity waves, respectively, although equatorially trapped gravity waves may also play a role. Quasi-biennial variability in longitudinally averaged winds and in the abundance of certain minor chemical species is also observed at high latitudes. Its connection with the tropical QBO is not altogether clear because, on theoretical grounds, the direct effect of the tropical QBO should not extend beyond the subtropics. Quasi-biennial variability at high latitudes appears to be mediated by planetary waves, whose propagation is influenced by the wind regime in the tropics. The cause of the high-latitude QBO is of considerable theoretical interest because the oscillation manifests itself in many meteorological fields, including longitudinally averaged wind and temperature, and total ozone abundance.

The *semiannual oscillation* (SAO) is the dominant mode of seasonal variability in the upper stratosphere, above about 35 kilometers. In common with the QBO, it is characterized by the appearance of alternating regimes of eastward and westward winds, the former appearing near the equinoxes and the latter at the solstices. The eastward regime appears to be driven by planetary-scale

Kelvin waves and smaller-scale gravity waves; the westward regime is thought to be driven by the transport of westward momentum from the summer hemisphere by the mean meridional circulation. This momentum transport is a consequence of the seasonal cycle in the stratosphere; it is evidently responsible for the fact that the SAO stays in phase with the annual cycle (as opposed to the QBO, whose period can vary by several months about the mean of 27 months). Recent studies suggest that redistribution of momentum in the tropics by inertial instabilities also plays a role in enhancing westward winds at the solstices. The SAO in the upper stratosphere is known to be related to a similar oscillation at much greater heights, near the top of the mesosphere. The stratospheric and mesospheric SAOs are out of phase, which suggests that selective transmission of tropical waves through the wind system produced by the stratospheric oscillation determines the evolution of its mesospheric counterpart.

Stratospheric Composition The major constituent gases in the stratosphere, as in the troposphere, are molecular nitrogen (N_2) and molecular oxygen (O_2). With argon, these account for about 99 percent of the gas molecules in the stratosphere. The remaining 1 percent are the so-called minor, or trace, constituents, including ozone (O_3), atomic oxygen (O), carbon dioxide (CO_2), nitrous oxide (N_2O), methane (CH_4), water vapor (H_2O), and the oxides of nitrogen, hydrogen, chlorine, and bromine. Water vapor, abundant in the troposphere, is present only in trace amounts in the stratosphere. The dryness of the stratosphere is due to freezing and precipitation in the tropical cloud systems that are the major means of injecting tropospheric air into the stratosphere. Several industrial compounds, most notably the chlorofluorocarbons (CFCs), are found in the stratosphere in relatively high concentrations. CFCs are important because they are the major source of stratospheric chlorine, which plays an important role in the ozone budget. The abundance of certain species, such as N_2O, CH_4, and CO_2, has also increased as a result of human activities.

Although the oxides of nitrogen, hydrogen, chlorine, and bromine are present in very small amounts (a few molecules per 10^9 molecules of air), they are of great importance to stratospheric photochemistry because of their ability to convert ozone into molecular oxygen in catalytic cycles. Catalytic cycles are very efficient sinks of ozone because the species that remove ozone are not themselves destroyed, but are instead continuously recycled. [See Ozone.]

In addition to the minor gaseous constituents, dust and other minute solid particles are found in the stratosphere. In winter, temperatures over the polar caps can become low enough to promote the formation of so-called *polar stratospheric clouds*, which are composed of mixtures of water and nitric acid or, if the temperature drops below about −85°C, of water ice. Chemical reactions occurring on polar stratospheric cloud particles, which transform the chlorine contained in "reservoir" species such as chlorine nitrate ($ClNO_3$) and hydrochloric acid (HCl) into forms capable of catalyzing ozone, have been implicated in the recurrent destruction of ozone during the Antarctic spring. While polar stratospheric clouds are confined to high latitudes during the winter season, a layer of aerosol particles (the *Junge layer*) is found near 20 to 25 kilometers throughout the year. This aerosol layer is composed mainly of sulfuric acid droplets derived from sulfur-containing gases of tropospheric origin. The aerosol concentration can increase by factors of 10 to 100 after volcanic eruptions, when it can have a significant impact on stratospheric temperatures. Very large eruptions can also reduce significantly the amount of solar radiation reaching the Earth's surface and produce detectable decreases in tropospheric temperatures. Recent studies suggest that reactions occurring on stratospheric aerosols can liberate chlorine from reservoir species, much as reactions occur in polar stratospheric clouds in winter. After large volcanic eruptions, when stratospheric aerosols are very abundant, enough active chlorine may be released to cause significant ozone depletion at middle and high latitudes. [See Aerosols.]

The abundance of the major constituents in the stratosphere is essentially constant, but that of most minor species varies substantially with altitude and latitude; it is determined by the interplay between photochemical processes and transport by the stratospheric circulation. Thus an understanding of both atmospheric photochemistry and dynamics is required to explain the distribution of minor species in the stratosphere. Detailed discussion of this problem is beyond the scope of this article, but some idea of its complexity can be gained from a brief examination of the stratospheric ozone budget.

Figure 2 shows the distribution of ozone in the stratosphere near the December solstice, as obtained using a computer model of the stratosphere. Superimposed on the ozone distribution is a second set of contours represented by broken lines which indicate the photochemical lifetime of ozone—that is, the time scale on which ozone is produced or destroyed by photochemical processes. The direction of the stratospheric mean meridional cir-

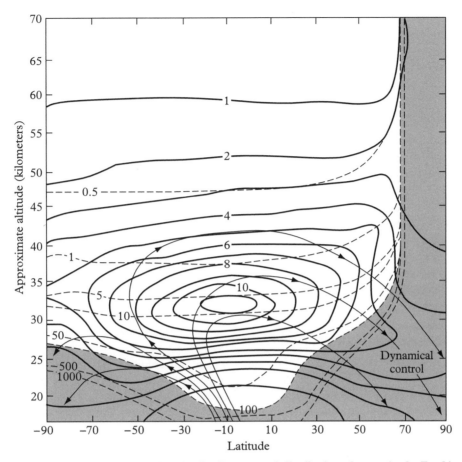

STRATOSPHERE. Figure 2. *Longitudinally averaged distribution of ozone in the Earth's atmosphere in winter for the Northern Hemisphere.* The solid contours are isolines of mixing ratio in parts per million by volume; the broken contours indicate the photochemical lifetime of ozone in days. The direction of the atmospheric mean meridional circulation is denoted by the arrows. The shaded region (labeled "Dynamical Control") includes polar night and the lower stratosphere, where photochemistry is slow and chemical composition is determined by dynamical processes such as the meridional circulation.

culation is indicated schematically by arrows. During most of the year the mean meridional circulation consists of two equator-to-pole cells in the lower stratosphere, and a summer-pole-to-winter-pole cell in the upper stratosphere.

The ozone distribution displays maximum mixing ratios of over 10 parts per million by volume (10 molecules of ozone per 10^6 molecules of air in a reference volume) in the tropical middle stratosphere, where ozone is produced rapidly by a series of reactions that begin with the photolysis of molecular oxygen. In the production region the lifetime is so short that ozone abundance is essentially determined by photochemistry. Note, however, that substantial ozone mixing ratios are found at higher latitudes and lower altitudes, in regions where ozone is produced very slowly if at all. The presence of

ozone at these locations is due to the stratospheric circulation, which generally transports ozone poleward and downward from the production region in the tropical middle stratosphere. Because the photochemical lifetime of ozone is usually long at high latitudes and low altitudes, the ozone transported by atmospheric motions can accumulate there without appreciable loss. As a consequence, the total ozone content of the atmosphere—the so-called *ozone column amount*—is usually larger at middle and high latitudes than it is in the tropics.

The appearance of the Antarctic "ozone hole" in the 1980s is an important exception to this pattern. This departure from normal high-latitude behavior is now understood to be caused by rapid destruction of ozone during the Antarctic spring by chlorine liberated from reservoir species by reactions on polar stratospheric clouds. In effect, the presence of active chlorine in the Antarctic lower stratosphere reduces the photochemical lifetime of ozone there to about 1 month and overwhelms the effect of the relatively slow atmospheric transport processes. The fact that no ozone hole has developed in the Arctic can also be understood in terms of the stratospheric circulation, which is known to be much stronger in the northern winter than in the southern winter. Among other effects, the stronger circulation in Northern Hemisphere winter leads to a warmer lower stratosphere, where conditions are not conducive to the survival of polar stratospheric clouds until spring; therefore, much less reactive chlorine is available to destroy ozone during the northern spring.

[*See also* Ozone Hole.]

BIBLIOGRAPHY

Andrews, D. G., J. R. Holton, and C. B. Leovy. *Middle Atmosphere Dynamics.* New York, 1987.

Brasseur, G. P., and S. Solomon. *Aeronomy of the Middle Atmosphere.* Dordrecht, Holland, 1986.

Garcia, R. R. "On the Mean Meridional Circulation of the Middle Atmosphere." *Journal of Atmospheric Science* 44 (1987): 3599–3609.

Garcia, R. R. "Causes of Ozone Depletion." *Physics World* 7.4 (1994): 49–55.

Hamill, P., and O. B. Toon. "Polar Stratospheric Clouds and the Ozone Hole." *Physics Today* 44 (1991): 34–42.

Holton, J. R. *The Dynamic Meteorology of the Stratosphere and Mesosphere.* Boston 1975.

McIntyre, M. E., and T. E. Palmer. "Breaking Planetary Waves in the Stratosphere." *Nature* 305 (1983): 593–600.

Salby, M. L., and R. R. Garcia. "Dynamical Perturbations to the Ozone Layer." *Physics Today* 43 (1990): 38–46.

Solomon, S. "Progress towards a Quantitative Understanding of Antarctic Ozone Depletion." *Nature* 347 (1990): 347–354.

ROLANDO R. GARCIA

STRATUS CLOUDS. The name *stratus* comes from the Latin word for "layer." In contrast with cumulus clouds, which have small diameters and short lifetimes, stratus clouds form in layers that cover large horizontal areas and last for extended periods. The mechanism of their formation is the forced lifting of stable air. When the lifting takes place on a large horizontal scale, the cooling of the moist air condenses the water in it and produces the large cloud layers collectively known as *stratiform clouds*. These layers can form at various altitudes, from near the ground to near the tropopause (the transition between the lowest atmospheric layer, the troposphere, and the stratosphere). The term *stratus* is tradi-

tionally reserved for the lowest-altitude stratiform clouds. Specific terms for middle- and high-altitude stratiform clouds include the word *stratus* preceded by a prefix characterizing the altitude, shape, or rainmaking potential of the cloud—for example, *altostratus,* with a prefix meaning "high." Depending on their altitude and their temperature (which usually decreases with altitude), stratiform clouds consist of either water or ice particles. The concentration of these particles in the cloud also varies significantly with height. Because of their large horizontal extents, stratus clouds play a very important role in the radiative budget of the planet, with their influence varying according to the altitude of the cloud tops and the optical depth of the clouds. The different types of stratiform clouds are described below in terms of their formation, composition, rainmaking potential, and radiative properties.

Types. The primary mechanism by which stratus clouds form is the ascent of air associated with the development of cyclones. In the boundary region between warm and cold air masses, a large horizontal mass of warm air is lifted over denser cold air, forming the cloud layers associated with the warm and cold sectors of a frontal zone. To the observer on the ground, the approach of a front is signaled by the approach of high cirrus clouds. Cirrus clouds (the name comes from the Latin word for "filament") form in relatively large layers, but, unlike stratus clouds, the layers are discontinuous. They are composed entirely of ice crystals. The passage of the cirrus cloud layer is followed by the approach of a more uniform cloud deck formed by very thin, high clouds known as *cirrostratus*. Cirrostratus clouds form a thin, transparent layer in which fine streaks may be observed. They are located at high altitudes, usually above 8,000 meters. The cloud layer is so thin that at times the clouds are invisible; more often, however, the hexagonal prisms of ice that form the cirrostratus layer refract sunlight and form the optical phenomenon known as a *halo*. Halos appear as concentric rings around the Sun or the Moon, changing color from yellow to green, white, and finally blue from center to outer edge. The appearance of halos has traditionally been considered an omen of stormy weather. Cirrostratus clouds do not produce precipitation makers, so they do not have a significant effect on local weather.

The cirrostratus clouds are usually followed by a thicker, lower layer of *altostratus* clouds. This layer can be 2 or 3 kilometers thick, so that the Sun can barely be seen through it. The altostratus layer is located at a height of about 5 to 6 kilometers above the surface and can

STRATUS CLOUDS. Figure 1. *Stratus clouds over Boulder, Colorado, viewed looking southwest.* (Photo copyright Ronald Holle.)

cover several hundred thousand square kilometers. The upper part of an altostratus cloud consists of ice particles, but the lower part is usually warm enough to include supercooled liquid water droplets. Precipitation is infrequent in altostratus clouds; however, when ice crystals from the upper cloud layers fall into the lower layers, they can produce light snow or drizzle through the ice crystal process, explained in more detail below. Altostratus clouds grow thicker as their base moves lower. Below the base, small fragments of black clouds of irregular form and ragged contour can be observed. Halos are rarely observed in altostratus clouds, but one or more sequences of brilliantly colored rings centered around the Sun or the Moon are sometimes seen. These rings, known as *coronas,* are formed by the diffraction of light by small water droplets.

As an altostratus cloud continues to thicken and its base extends closer to the ground, it forms a dark gray layer through which the Sun or the Moon is no longer visible. This layer, which begins to produce rain or snow, is called a *nimbostratus cloud* (the prefix *nimbo-* is generally used to identify precipitating clouds). Nimbostratus clouds are a primary source of precipitation in the atmosphere. Because these clouds can extend several kilometers vertically, (their bases are usually at around 2,000 meters), they often consist of ice particles and supercooled water droplets in their upper, colder layers, and

water particles in the lower, warmer layers. (In general, clouds warmer than $-10°C$ consist primarily of water, clouds between $-10°C$ and $-20°C$ consist of a mixture of supercooled water droplets and ice crystals, and clouds colder than $-20°C$ consist primarily of ice.)

The composition of the upper layers of nimbostratus clouds is ideal for the formation of precipitation through the ice crystal process. This process is based on the fact that the saturation vapor pressure over water is greater than the saturation vapor pressure over ice for the same subfreezing temperatures. This means that when the air is saturated with respect to water, it is supersaturated with respect to ice. This condition favors the growth of ice crystals at the expense of water droplets, because the evaporation from the liquid droplets provides a source of water vapor for the ice crystals. The ice crystals grow larger and eventually fall to the ground as precipitation. The precipitation is in the form of snow when the air between the cloud layer and the ground is sufficiently cold, or in the form of rain when the temperature exceeds $0°C$ at some level inside or below the cloud. The melting level is often visible in the cloud: snow scatters more sunlight than rain, so the area above the melting line is darker than that below it. Most of the rain that reaches the earth originates as snow inside the cloud layer. Even in nimbostratus clouds that consist only of water, precipitation can occur through the process of coalescence (the collection of smaller droplets by larger ones as the larger ones fall through the cloud). As the air underneath the cloud is moistened by rain, fragments of low-level clouds known as *pannus* or *scud* are often observed below nimbostratus layers. [*See* Rain.]

It is not unusual, particularly in the winter, for a nimbostratus cloud to be followed by a low, gray uniform layer of *stratus clouds*. The primary mechanism of formation of stratus clouds, however, is the lifting of fog, which occurs when a mass of relatively warm, humid air passes over a cold ocean or land surface—or, in coastal regions, when two air masses of widely different temperature mix. Stratus clouds are uniform in their properties; they form low in the atmosphere (usually below 2,000 meters) and can cover large horizontal areas. They consist of water droplets, having water contents typically of between 0.05 to 0.1 grams per cubic meter. Stratus clouds can produce long-lasting precipitation in the form of drizzle, because the water droplets that form these clouds are relatively small, and the clouds are too warm to initiate the ice crystal process. In precipitating stratus clouds, particle radii range between 10 and 12 microns, while in nonprecipitating ones they range between 5 and 8 microns. Stratus clouds are often capped by a temperature inversion that inhibits their vertical growth. When stratus clouds break up, they form large layers of stratocumulus clouds. [*See* Stratocumulus Clouds.]

Radiative Impact. The effect of a cloud on the Earth's radiation budget is determined by the relative strengths of two opposing effects; the solar albedo effect and the thermal greenhouse effect. The solar albedo is a cooling effect, because the clouds reflect solar radiation and prevent it from reaching the surface of the planet. The thermal greenhouse is a warming effect, because the clouds trap thermal infrared radiation emitted by the surface of the planet and prevent it from escaping to outer space. The amount of solar radiation reflected by a cloud depends primarily on the cloud optical thickness, which in turn is directly proportional to the physical extent and water density of the cloud, and inversely proportional to the size of the cloud particles. Optically thick clouds reflect more solar radiation than optically thin ones. The amount of thermal radiation trapped by a cloud depends on both its optical thickness and on the altitude of the cloud top. High clouds trap thermal radiation more effectively than low clouds. [*See* Clouds and Radiation.]

The different types of stratiform clouds have different radiative effects depending on the altitude of the cloud layer. High stratiform clouds—cirrostratus and altostratus—have small optical thickness and reflect small amounts of solar radiation. At the same time, they have very high, cold cloud tops and are very effective in trapping thermal radiation. Therefore, these cloud types produce much stronger thermal greenhouse effects than solar albedo effects, resulting in a net warming of the planet. Middle stratiform clouds, or nimbostratus, have large optical thicknesses and reflect large amounts of solar radiation, but they also have high, cold cloud tops that effectively trap thermal radiation. The net effect of nimbostratus varies for individual cloud layers, but overall it is considered negligible for the radiative balance of the planet. The low stratiform clouds, or stratus, have significant optical thicknesses to reflect large amounts of solar radiation, their warm cloud tops have little ability to trap thermal radiation. Therefore, they produce a net cooling effect on the planet. Low stratus clouds, with their large horizontal extents and their reflection of solar radiation, are very influential on the radiative balance of the planet.

[*See also* Cloud Climatology; *and* Clouds.]

BIBLIOGRAPHY

Ramanathan, V., et al. "Cloud-Radiative Forcing and Climate: Results from the Earth Radiation Budget Experiment." *Science* 243 (1989): 57–63.

Rogers, R. R., and M. K. Yau. *A Short Course in Cloud Physics*. 3d ed. Oxford, 1989.

Rossow, W. B., and A. A. Lacis. "Global, Seasonal Cloud Variations from Satellite Radiance Measurements. Part II: Cloud Properties and Radiative Effects." *Journal of Climate* 3 (1990): 1204–1213.

Wallace, J. M., and P. V. Hobbs. *Atmospheric Science: An Introductory Survey*. Orlando, Fla., 1977.

GEORGE TSELIOUDIS

SUBSIDENCE. In meteorology, the word *subsidence* is used to denote sinking motions of all types. A gas cools when it expands and heats up when it is compressed; for example, a bicycle pump gets hot from rapid compression of air, and a cylinder of compressed carbon dioxide cools rapidly when discharged. This effect is known as *adiabatic* heating (in the case of compression) or cooling (in the case of expansion). Thus all subsiding air in the atmosphere experiences an adiabatic warming effect, and all rising air an adiabatic cooling effect. Subsidence also carries cold dry air from aloft to the lower troposphere, thus lowering the relative humidity of the air below and suppressing the rising of air in the form of convection—the process that initiates thunderstorms. For these reasons, subsidence is often associated with high surface pressure and clear sky conditions. [*See* Adiabatic Processes; Convection.]

Subsidence motions occur on all scales. On the global scale, subsidence is most predominant in the subtropical latitudes, coinciding with the sinking branches of the Hadley circulation. These include subtropical high-pressure zones over the ocean and over continental deserts. The deserts are generally hotter than their surroundings, being dominated by subsidence, which maintains dry, clear sky conditions. Because warm desert air is forced to descend, the desert circulation represents a *thermally indirect circulation,* as opposed to a *thermally direct circulation* where warm air rises and cold air sinks.

In the tropics, rising motions are concentrated in regions of heavy precipitation, whereas subsidence is widespread elsewhere. The most extensive region of subsidence is found over the equatorial region of the eastern Pacific, particularly along the coast of South America. The equatorial eastern Atlantic is also a region of strong subsidence. These equatorial tropical oceanic subsidence zones form the subsiding branches of the so-called *Walker circulation.* In these regions, the sea surface temperature is relatively cold. Although the air near the surface contains abundant moisture, the air above it is capped by a strong inversion created by the subsidence, which virtually blocks the occurrence of deep convection and thunderstorms. [*See* General Circulation.]

Subsidence also plays an important role in maintaining the vertical and horizontal thermal conditions needed for the development of mid-latitude synoptic weather systems. Finally, subsidence motions within convective clouds, known as *downdrafts,* are crucial in maintaining the moisture budget within cumulus convection, initiating new convective cells and linking these to the large-scale environment. These downdrafts are often felt at the ground as a gust outbreak in front of a thunderstorm.

[*See also* Thunderstorms.]

BIBLIOGRAPHY

Cotton, William, and R. A. Anthes. *Storm and Cloud Dynamics.* New York, 1989.
Wallace, J. M., and Peter V. Hobbs. *Atmospheric Science: An Introductory Survey.* New York, 1977.

WILLIAM K.-M. LAU

SUBTROPICAL CLIMATE.

The subtropics extend from the tropics of Cancer (23°27′ north latitude) and Capricorn (23°27′ south latitude) to about 40° north and south latitude, respectively. These regions are dom-

inated atmospherically by the subtropical ridges situated between the tropical easterlies and mid-latitude westerlies. The ridges are a series of *anticyclones,* elongated in an east–west direction, with their areas of highest pressure over the eastern parts of the subtropical oceans. In the centers of the anticyclones, winds are light and variable, air subsides and diverges, and skies are clear. [*See* Anticyclones.]

There is a marked contrast between the eastern and western ends of the anticyclones. The eastern part is characterized by strong subsidence and a dry climate, while the western part has little or no subsidence and is wetter. The strong subsidence in the east produces a marked temperature inversion, above which the air is dry. The inversion and dryness result in meager rainfall along the eastern oceans and the adjacent western coastal margins of continents. The western side of a subtropical oceanic anticyclone, by contrast, is usually characterized by moist, unstable air and abundant rainfall. The eastern continental regions adjacent to the western sides of the oceanic anticyclones have more humid climates than the western sides of the continents. In winter the subtropical ridges weaken and move equatorward, allowing westerlies and embedded depressions to bring rain to both coasts. Thus the western coasts of the continents receive winter rainfall (except in the equatorial edges of the subtropics) and experience dry summers owing to the stronger ridge in this season. These are called the *dry-summer subtropics.* The eastern coasts have rainfall throughout the year, with no distinct dry season, and are called the *humid subtropics.* The continental interiors of the subtropics are generally arid or semiarid because of their distance from the oceanic moisture source.

The dry-summer subtropics are situated along western coasts. This climate is characteristic of the Mediterranean Sea, so it is popularly termed a *Mediterranean climate.* Temperatures are moderate, rain rarely falls in summer, and there is plenty of sunshine. During summer the weather is dominated by the sinking and equatorward-flowing air from the subtropical anticyclones over the ocean to the west. This brings stable air, clear skies, and high temperatures. Where coasts are influenced by cool ocean currents (notably in California and Chile), summer temperatures are lower and marine fogs and stratus clouds are common. During winter, the subtropical ridge moves equatorward and weakens, allowing the mid-latitude westerlies to increase their influence. These winds bring upslope winds, frontal activity, and rain. The moderating influence of the oceans keeps winter temperatures mild. Frosts are infrequent, except well inland or at high elevations. The annual temperature variation is small because of the marine influence. Strong and persistent temperature inversions often trap pollutants, leading to chronic air pollution in some cities of the dry-summer subtropics.

The humid subtropics occupy similar latitudes on the eastern sides of continents. Here the air from the subtropical highs comes from the tropics and is warm and moist. This results in showery weather in summer, so there is no distinct dry summer as is seen on the western coasts. In some parts of the humid subtropics, decaying tropical cyclones can bring heavy rains as they move poleward. The summer months are normally hot, with a relatively small diurnal range because of the humid atmosphere and cloudiness. Winter temperatures are in general mild, but there are important regional differences. Subtropical China and the southeastern United States, because of the presence of a severe continental climate on their poleward edge, have colder winters than the humid subtropics of the Southern Hemisphere, all of which lack large continental masses on their poleward edge. As a result, winter frosts are a prominent feature in the humid subtropics of the Northern Hemisphere but are rare in those of the Southern Hemisphere. The equatorward shift of the subtropical ridge in winter allows disturbances and frontal activity in the westerlies to bring rains to these regions.

Many parts of the continental subtropics have arid or semiarid climates. These occupy the interiors of the continents and the northern parts of the subtropical western coasts. The eastern subtropical oceans also receive little rainfall because of the subsidence and stability of the eastern portions of the anticyclones. Mid-latitude depressions rarely penetrate far enough equatorward to affect these regions, and the tropical systems also usually remain equatorward. In the continental interior, the air is dry because of its distance from the oceans, so skies are clear, leading to cold nights and hot days. The extreme temperatures produce unstable, windy conditions. The lack of vegetation allows the wind to pick up and carry sand. The relative humidity in these regions is very low, resulting in very high evaporation rates. When rainfall does occur in these regions, it can be intense.

The relatively low rainfall and long dry summers of the dry-summer subtropics contribute to the dominance of scrubby brush vegetation cover, in some regions referred to as *chapparal* or *maquis*. Toward the end of the winter rains, plants grow and flower in an intense burst. The humid subtropics, with their heavier and more consistent rains, are densely forested. Both these relatively mild cli-

mates are ideal for human settlement, although air pollution can be a problem for the dry-summer subtropical cities. The arid or semiarid centers of the subtropical continents can support pastoral industries, provided that water storage facilities are built. The natural vegetation is characterized by succulents and other drought-tolerant plants; trees are rare.

Many parts of the subtropics are affected by the phenomenon known as El Niño-Southern Oscillation (ENSO). This increases the variability of rainfall. Occasional very heavy rains occur in the subtropical deserts affected by the phenomenon, such as those of coastal Chile and inland Australia. Most of the other nonarid subtropical regions are also affected, in various ways, by the ENSO. For example, the humid subtropics of the southeastern United States tend to be cool and wet during an El Niño event. The humid subtropics of Argentina also tend to be wet, while the semiarid inland region of Australia and the humid subtropics of its eastern coast tend to be drier than normal. Inland semiarid Australia usually receives good rains only during the other extreme of the phenomenon, the La Niña episode.

[*See also* Climatic Zones; *and* El Niño, La Niño, and the Southern Oscillation.]

BIBLIOGRAPHY

Gedzelman, S. D. *The Science and Wonders of the Atmosphere.* New York, 1980.

Kiladis, G. N., and H. F. Diaz. "Global Climatic Anomalies Associated with Extremes in the Southern Oscillation." *Journal of Climate* 2 (1989): 1069–1090.

Trewartha, G. T., and L. H. Horn. *An Introduction to Climate.* 5th ed. New York, 1980.

NEVILLE NICHOLLS

SUBTROPICAL HIGH. *See* Teleconnections.

SUMMER. *See* Seasons.

SUMMER SOLSTICE. *See* Orbital Parameters and Equations.

SUN. The Sun is classified as a typical G2 star (G2 indicates a specific spectral type, based on calcium emission lines and other characteristics) with an age of nearly 4.5 billion years. It is a *second-generation* star. This means that it is composed not only of the elements produced by the big bang (hydrogen and helium) but also of the heavier elements that can be produced only by nucleosynthesis in stars (e.g., carbon, iron, oxygen, and others) plus the heavy elements that can only be produced in supernovas (all the radioactive elements above atomic number 83, such as uranium and thorium). The elemental composition of the Sun is 94 percent hydrogen, 5.9 percent helium, and 0.1 percent of other elements; these include, in decreasing proportion, oxygen, carbon, nitrogen, silicon, magnesium, neon, iron, sulfur, and many others.

From the astronomical point of view, the Sun is a very ordinary star. To paraphrase a remark by physicist Robert Leighton, "If the Sun did not possess a magnetic field, it would be as uninteresting as most astronomers think it is!" We are fortunate in this, because if the Sun underwent violent fluctuations, it could not have provided a safe haven for the evolution of life, which required stability over hundreds of millions of years. Nevertheless, the Sun does exhibit interesting though modest variations—a hot corona with a temperature of millions of degrees, solar flares, and sunspots. The Sun displays a wide range of exciting astrophysical phenomena at a distance as near as an observer would wish to go.

The Sun appears to be an important contributor to natural climatic variability. To answer current questions about anthropogenic (human-caused) climate change, we must be able to distinguish these against the backdrop of naturally occurring variability in our climate system. For example, at present, we are uncertain whether industrial doubling of atmospheric carbon dioxide (CO_2) will result in an increase of 1.5°C or 4.5°C in mean global temperature. Although these temperature changes seem small, they can have tremendous global impact on the survival of species and on many other aspects of life. For example, the "Little Ice Age" (ca. 1600–1850 C.E.) affected the whole planet, but the average global temperature changed by less than 1°C. [See Little Ice Ages.]

One of the most interesting puzzles concerning past climate involves solar luminosity. From astronomical studies of the interiors of stars, we know that since its beginning, the Sun has gradually become more luminous by about 30 percent. If the Sun were fainter in the past by even a few percent, then the Earth would have been covered by ice, from which it should not have recovered, because the ice, with its high albedo, would have reflected the solar radiation back into space. Although a number of suggestions have been advanced to solve the "faint young Sun" problem—for example, volcanic aerosols covering the ice, early oceans moderating the climate, or higher CO_2 levels—we are not certain how Earth escaped being a barren planet.

Natural climatic variability is thought to arise from a number of sources. Some of the most important are volcanic influences, inherent solar variations, changes in surface albedo, changes in oceanic heat transport, orbital forcing, and the chaotic nature of the climate system.

Some researchers believe that the Sun plays a major role in natural climatic variations, while others consider its role to be negligible. In favor of the former view, solar activity has been increasing this century in a manner that appears to match the global temperature record better than does CO_2 forcing. The Sun certainly had a more prominent role in past climate change than did human activities, but there is also disagreement about the extent of the difference. To understand the Sun's relevance to climate studies, let us consider how the Sun as a star provides energy for the Earth, and then return to its variations and their terrestrial impact.

What allows the Sun to generate vast amounts of energy for billions of years, and still continue to shine? Before the development of nuclear physics, energy for the Sun's radiation was thought to come from gravitational energy, through the solar material's self-attraction. However, this process would allow the Sun to shine for only about 30 million years before its energy was depleted. For the Sun to shine longer, stronger energy sources are needed. We now know that many nuclear reactions occur within the solar interior in a *nuclear reaction zone* (see Figure 1), in which 4 hydrogen nuclei fuse into 1 helium nucleus. Because the 4 hydrogen nuclei have more mass than the helium nucleus, a mass deficit results, and this mass is converted into energy according to Einstein's famous (1905) formula, $E = mc^2$.

This energy, produced near the Sun's center, leads to a high core temperature of 15 million K, and this energy exits at the surface of the Sun. The energy is transported from the interior, first by radiation and then, in the outer layers of the Sun, by convection. Figure 1 shows these different layers, as well as the temperature and density structure of the Sun. The energy is deposited in the surface layers of the Sun (the *photosphere*) at 5,800 K. From here the energy is finally radiated into space, and a small fraction bathes the Earth with heat and light.

The energy output of the Sun is variable. The turbulence in the Sun's outer layers, the convection zone, creates a magnetic field resulting in sunspots, flares, and other types of magnetic activity, as well as the *solar cycle*. Solar cycles are the periods of relative activity and inactivity, occurring with approximately an 11-year period. In addition to the 11-year variation, there is also variability of longer duration, such as the *Gleissberg cycle* with

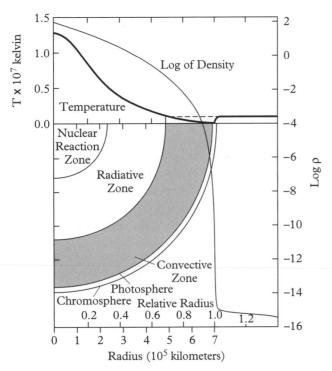

SUN. Figure 1. *Solar structure.* Shown against the structure of the Sun (in the lower portion of the figure) are plotted the log of the density (light line, right) and temperature (heavy line, left). The nuclear reaction zone is also shown, as well as the radiative zone, which allows the energy to be transported by radiation. Farther out is the convective zone, where the energy is transported by bulk gas motions. The photosphere and chromosphere are the outer layers where the radiant energy leaves the Sun.

time scale variations of the order of 100 years. It is these long-period variations that make the Sun an excellent candidate for influencing climate on long time scales. Other terrestrial variations (e.g., volcanic aerosols) may cause irregular, temporary disturbances of climate, but many successive eruptions would be needed to drive the climate system with a long enough time scale to provide a lasting effect.

Let us now consider how the Sun's radiative output varies. We refer to the average energy radiated to Earth as the Sun's *total irradiance,* or *solar constant.* This is a misnomer: although long thought to be constant, this quantity is now known to vary on time scales from days to decades. The value of the solar constant is near 1,367 watts per square meter. Variations in this value are associated with solar *active regions,* most notably sunspots. Sunspots occur when the surface magnetic field of the Sun gains sufficient strength to offset the convective heat flow from the Sun's interior. Sunspots are cooler than the

rest of the Sun's surface by about 1,500 K and therefore appear darker. As a result, when an active region is centrally located on the face of the solar disk toward the Earth, there is a reduction in the energy radiated by the Sun to the Earth. This reduction in energy, which has been observed by satellites in space, can amount to a decrease of as much as 0.3 percent. By itself, this is unlikely to affect climate because of the short time period of reduced energy; the Sun rotates completely in about 27 Earth days.

Surprisingly, at the height of the solar cycle (sunspot maximum), when sunspots are most numerous on the solar disk, the Sun shines with a greater intensity than it does when there are fewer sunspots (solar activity minimum). This is another poorly understood enigma related to the interior of the Sun. We do, however, see some extra energy leaving the Sun's surface at sunspot maximum, in areas called *faculae* (Latin for "torches"). These are bright areas that surround sunspots on the outer edges of active regions. It is a mystery how and why the energy variations of sunspots and faculae occur; nevertheless, these features play an important role in the variation of the solar constant.

Perhaps even more important than variations in solar constant are variations in *spectral irradiance.* The Sun's short wavelength radiation, in the ultraviolet (UV) regions of the spectrum, displays variations ranging from a few percent to a factor of 10 throughout the solar cycle. These variations may have important implications for climate change because they significantly affect the thinnest, most sensitive layers of the Earth's atmosphere, including the protective ozone layer. These are important areas for future research.

The longer-term effects of solar activity upon the solar constant are even less understood. Nevertheless, some information about the record of past solar activity can be inferred from cosmogenic isotopes (beryllium-10, oxygen-18, carbon-14, and others) stored in ice cores and tree rings. These suggest that the solar activity record appears often to correlate directly with the temperature record of the Earth. When solar activity is higher than normal, the Earth is warmer than normal.

During the period 1600 to 1850, there was a "Little Ice Age," notably in Europe, but to a lesser extent throughout the Northern Hemisphere. The coldest temperatures coincided with the so-called *Maunder minimum* on the Sun, a period when there were few sunspots. In the eleventh and twelfth centuries, the *medieval maximum* in solar activity corresponded to the *medieval optimum* in climate, when global warming occurred to such an extent some of Greenland was colonized by the Norse. As solar activity declined, so did the global temperature, and the Norse were forced to retreat from their northern outpost. More recently, solar activity dipped at the end of the eighteenth century and the early years of the nineteenth (the *modern* or *Dalton minimum* in solar activity), and this era was also unusually cold. The twentieth century has been marked by generally increasing levels of solar activity. Cycle #19, which peaked in 1958, was the largest maximum recorded since Galileo first observed sunspots in 1610, and cycle #22 appears to have been the second-largest. The global temperature record matches this general tendency, making it difficult to disentangle the effects of solar activity from those of greenhouse gases.

[*See also* Global Warming; Ice Ages; Shortwave Radiation; *and* Solar Winds.]

BIBLIOGRAPHY

Herman J. R., and R. A. Goldberg. *Sun, Weather and Climate.* NASA report SP-426, Washington, D.C., 1978.
Lean, J., et al. *Solar Influences on Global Change.* Washington, D.C., 1994.
White, O. R., *The Solar Output and Its Variation.* Boulder, Colo., 1977.

KENNETH H. SCHATTEN and RICHARD A. GOLDBERG

SUNSPOTS. *See* Sun.

SYNOPTIC METEOROLOGY. *See* Meteorology.

SYNOPTIC SCALE. *See* Scales.

T

TAIGA CLIMATE. The climate of northern Russia, interior Alaska, and much of Canada, Norway, Sweden, and Finland is characterized by long, cold winters with short days and short, warm summers with long days. The needleleaf evergreen, needleleaf deciduous, and broadleaf deciduous tree species that dominate the vegetation in these regions are collectively called the *boreal forest* or *taiga*; some authors use the Russian word *taiga* to mean exclusively needleleaf vegetation. The boreal forest is the northernmost forest in the Northern Hemisphere, lying just south of the treeless arctic tundra. Temperate forests and grassland lie to the south. The exact definition of the boreal forest varies greatly among authors—some exclude the broadleaf deciduous tree species; the definition of the treeline separating tundra from forest is the subject of considerable debate; and there is a large transition zone into the temperate vegetation to the south. In general, however, the boreal forest occupies some 12 million square kilometers, ranging in some locations as far north as 70° north latitude and as far south as 50° north latitude, making it one of the most extensive natural vegetation types on the Earth (Figure 1).

The climate of the boreal forest varies greatly depending on geographic location. Two features common to all regions, however, are the strong seasonal variation between summer and winter and the cold mean annual air temperature. The seasonal temperature variation is caused to a large degree by the strong seasonal variation in incoming solar radiation, which increases with higher latitude. For example, daylength at 65° north latitude varies from a minimum of less than 4 hours in late December to a maximum of 22 hours in late June. In contrast, at 50° north latitude the minimum and maximum are 8 hours and 16 hours respectively. Cold mean annual air temperatures at high latitudes result in part from decreased annual solar radiation received at the surface. Low solar angles cause insolation to be received obliquely, so that it is spread over a large horizontal area, and create a long path length through the atmosphere over which solar radiation is attenuated. With the decreased solar radiation, the annual net radiation (the net energy at the surface) also decreases with higher latitude.

Large seasonal and interannual variations in air temperature are distinct features of the continental regions of interior Alaska and eastern Siberia, which are largely insulated from moderating oceanic air masses. Intraannual variation in mean monthly air temperature is on the order of 40° to 60°C (for example, −24°C in January to 15°C in July at Fairbanks, Alaska); an annual range in extreme air temperature of 100°C is common (for example, a low of −60° to 70°C and a high of 30° to 40°C). Continental regions also have pronounced interannual temperature variation. For example, in the 16-year period from 1971 to 1986, mean monthly January air temperature at Fairbanks, Alaska, ranged from −35° to −8°C. In eastern Canada and the northern European boreal forest region west of the Ural Mountains, where an oceanic climate prevails, intraannual and interannual temperature fluctuations are not so extreme.

Annual precipitation increases eastward across North America from a relative dry climate (less than 40 centimeters and often less than 20 centimeters) in interior Alaska and western Canada, where high mountain ranges restrict the inland penetration of moist air, to a moister climate in eastern Canada. In Eurasia, annual precipitation is higher west of the Ural Mountains than in the east. Precipitation is especially low (10 to 20 centimeters) in northeastern Siberia.

Permafrost, or perennially frozen soil, is an important feature of the boreal forest (Figure 1). In most boreal regions where permafrost is found (Alaska, northern Canada, northern Russia, and Siberia), it is spatially discontinuous—it occurs where local microclimatic conditions are favorable. In this region, locally cold air temperature, low solar radiation (as on north-facing slopes), dense vegetation, wet soils, and the presence of a thick organic layer (moss, lichen, or peat) on the soil surface

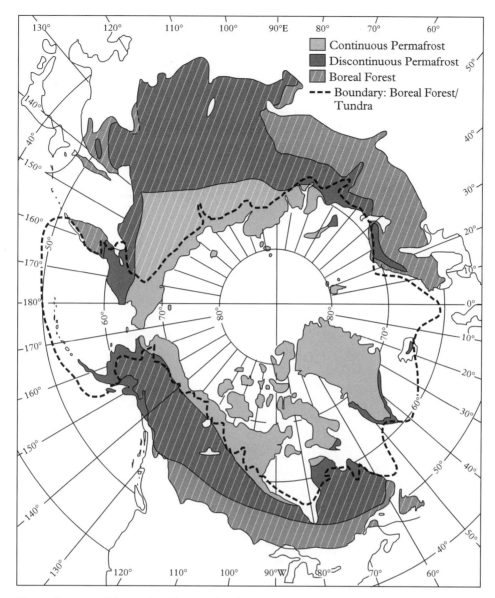

TAIGA CLIMATE. Figure 1. *Circumpolar distribution of the boreal forest and permafrost.* Asia and Europe are in the upper part of the figure, and North America is in the lower part. (Adapted from the *Canadian Journal of Forest Research* 1983, vol. 13, p. 696).

can create conditions favorable for permafrost. Permafrost is spatially continuous only in the vast region of northeastern Siberia. The presence of permafrost is ecologically important because the cold soil temperatures reduce physiological processes such as photosynthesis, respiration, water absorption, and translocation of assimilated carbon, resulting in slow tree growth. Soil temperature rather than air temperature defines the growing season for much of the boreal forest. Permafrost also acts as an impervious layer that impedes soil water drainage, cre-

ating cold, wet soils. Cold soil temperatures also reduce soil carbon decomposition and nutrient mineralization, leading to nutrient-poor soils, which further reduce tree growth, and to the accumulation of a thick peat layer on the soil surface.

People have adapted remarkably well to living in the extreme conditions of the boreal climate. Although the severe winter cold is physically dangerous, the short winter days pose a more serious psychological challenge; numerous studies have linked this to depression.

From an engineering point of view, the cold winter temperatures and permafrost are important challenges. Special care must be given when building on permafrost to account for frost heaves and to prevent melting of the permafrost and subsequent subsidence. Despite the low productivity of many forest types, forest harvesting is common throughout the circumpolar boreal forest.

The floristic diversity of the boreal forest is low compared to other biomes, and the woody vegetation is limited to a few species of needleleaf trees (firs, larches, spruces, and pines), and broadleaf deciduous trees and shrubs (alders, birches, poplars, and willows) that have adapted to the cold climate. Mosses and lichens are common ground cover. These trees form four major vegetation types—closed forest, woodlands, bogs, and forest-tundra ecotone. In general, there is a long-term trend of degenerating environmental conditions including increased thickness of the forest floor organic mat, decreased soil temperature and nutrient availability, poor soil drainage, increased carbon storage; this decline is interrupted only by recurring disturbance such as fires.

Closed forest vegetation is typically dominated by a single tree species. Broadleaf deciduous species and pines are frequently found in areas recently disturbed by fires, floods, or insects because their growth patterns allow for massive regeneration and rapid growth in the high light regimes of recently disturbed sites. In the absence of recurring disturbances, these species give way to slower-growing, longer-lived spruce and fir trees, often with a moss ground cover. In spruce-feathermoss communities, this moss organic layer is 10 to 30 centimeters thick. In extreme cases—for example, spruce growing on permafrost in interior Alaska—the thick organic mat accounts for 80 to 90 percent of aboveground biomass, and annual moss production is twice annual tree foliage production. Stands greater than 300 years old are extremely rare because of the high frequency of insect attacks, fire, and flooding.

Woodlands are characterized by low productivity, low standing biomass, and wide spacing among trees, resulting in an open forest canopy. In the vast continuous permafrost region of Siberia, where mean annual air temperature is as low as −10°C and soils thaw only to depths of 10 to 50 centimeters, deciduous larch trees form unproductive woodlands. Lichens are a conspicuous component of the boreal forest, often growing in open spruce, pine, or larch woodlands where they cover the ground surface between trees. These lichen woodlands form on dry, sandy, nutrient-poor soils and are most abundant in the northern part of the boreal forest.

Both extensive forested and nonforested peatlands are found throughout the boreal forest. Plant growth is extremely slow in the wet, acidic, nutrient-poor soils. However, peatlands store vast quantities of carbon because of the extremely slow decomposition of organic material here.

The forest-tundra vegetation forms a broad transition zone between the boreal forest and tundra. This vegetation is composed of scattered or isolated trees, often shrubby in form, intermixed with tundra plants.

Climate is thought to be the overriding factor determining the biogeographic distribution of the boreal forest and vegetation types within it because of the strong correlation with many climatic variables (for example, air temperature, precipitation, or net radiation). In Canada, the northern limit to the boreal forest is correlated with the position of the July 13°C isotherm, with local departures attributable to montane or oceanic influences. The southern limit correlates with the July 18°C isotherm. Certain airmass characteristics also coincide with the northern and southern limits of the boreal forest. In much of Canada, the northern treeline is marked by the modal July position of the front that separates continental arctic cool, dry air and maritime Pacific warmer, moister air masses, though this relationship is not clear in northeastern Canada. In the south, the boreal forest is bounded by the winter position of the arctic front. Similar relationships have been found in Eurasia.

The current boreal forest reflects the floristic response to changes in climate over the past 18,000 years. Between 18,000 and 15,000 years ago, when the influences of the last glaciation were strongest, the tree line separating forest from tundra vegetation was much farther south. Trees common to the boreal forest today grew to the south of the ice sheet, but in association with other species not found in the modern boreal forest. For example, in eastern North America spruce is thought to have formed a parkland consisting of widely scattered trees with abundant herbs. As the ice sheet began to retreat with a warmer climate 12,000 to 9,000 years ago, these tree species began to migrate northward; however, they did so in an individualistic manner, depending on their different life history characteristics, rather than as a cohesive unit. In Canada, the treeline reached its maximum northern extent approximately 5,000 to 7,000 years ago, during a period of greater summer solar radiation than that of today.

Historically, climate has been perceived as a controller of the biogeographic distribution of the boreal forest. However, in recent years this view has given way to a more interdisciplinary perspective in which the boreal forest is seen as an important component of the climate system.

Climate model simulations indicate that the warming caused by increased concentrations of greenhouse gases in the atmosphere is most pronounced at high latitudes in the Northern Hemisphere. There is much concern about the accumulation of carbon in cold, wet, nutrient-poor boreal soils and the possibility that this carbon will be released to the atmosphere with a warmer, drier climate, resulting in a positive feedback with the atmosphere. [See Greenhouse Effect.]

Furthermore, by influencing the absorption of solar radiation by the land surface and the partitioning of net radiation into sensible and latent heat, the boreal forest significantly affects climate. Observational studies have shown that although snow reflects more solar radiation than foliage, the presence of trees masks this reflectance, so that in snow-covered areas the albedo of a forested site is much less than that of a nonforested site. Observational studies at the forest-tundra treeline also indicate greater sensible heat flux (surface-to-air heat exchange) and less latent heat flux (evaporation) for a forest site compared to tundra. [See Albedo.]

Several climate model experiments have shown that by absorbing more solar radiation during the winter than nonforested sites, the boreal forest warms the climate, compared to simulations in which the boreal forest is replaced with bare ground or tundra. This warming extends into the summer because the colder air temperatures caused by removal of the boreal forest reduce sea-surface temperatures in Arctic regions. This introduces a thermal inertia which inhibits summer warming. In addition, the extent of sea ice increases, reinforcing the cooling because of the higher ocean albedos.

These effects of boreal forest on climate clearly suggest a coupled climate-vegetation system in which the biogeographical distribution of the boreal forest both affects and is affected by climate. For example, the position of the forest-tundra treeline moved northward in response to the last deglaciation. Climate model experiments show that the northward migration of the treeline in response to the climate warming caused additional climate warming. It is likely that the northward extension of the boreal forest into the tundra in response to future climate warming will also produce additional warming.

[See also Deforestation; and Global Warming.]

BIBLIOGRAPHY

Bonan, G. B., D. Pollard, and S. L. Thompson. "Effects of Boreal Forest Vegetation on Global Climate." Nature 359 (1992): 716–718.
Bonan, G. B., and H. H. Shugart. "Environmental Factors and Ecological Processes in Boreal Forests." Annual Review of Ecology and Systematics 20 (1989): 1–28.
Bryson, R. A. "Air Masses, Streamlines, and the Boreal Forest." Geographical Bulletin 8 (1966): 228–269.
Foley, J. A., J. E. Kutzbach, M. T. Coe, and S. Levis. "Feedbacks between Climate and Boreal Forests during the Holocene Epoch." Nature 371 (1994): 52–54.
Larsen, J. A. The Boreal Ecosystem. New York, 1980.
Ritchie, J. C. Postglacial Vegetation of Canada. Cambridge, 1987.
Shugart, H. H., R. Leemans, and G. B. Bonan. A Systems Analysis of the Global Boreal Forest. Cambridge, 1992.
Van Cleve, K., et al. Forest Ecosystems in the Alaskan Taiga. New York, 1986.
Wright, H. E., Jr., et al. Global Climates since the Last Glacial Maximum. Minneapolis, 1993.

GORDON B. BONAN

TELECONNECTIONS. The concept of teleconnections is important and useful for describing and understanding weather and climate patterns on the Earth. Literally, the word means "connections across space." More specifically, in the context of weather and climate teleconnections are correlations of atmospheric and oceanic variables—such as temperature, pressure, and wind—in one region with those in another. A numerical value assigned to the correlation indicates the degree to which fluctuations (deviations from the mean or average conditions) of the variables at different locations are related to each other. Values range from magnitudes of 0 for no correlation to 1 for perfect correlation.

Teleconnections are most evident for variables averaged over broad regions and long time periods. Teleconnections can extend many thousands of kilometers across the Earth and involve fluctuations that may persist for months, seasons, or even years; thus they refer directly to climate conditions. Teleconnections have been found for temperature, pressure, precipitation, and wind flow conditions extending from one ocean basin to another, between the tropics and middle latitudes, and around the equatorial region. For example, fluctuations from long-term averages in monthly averaged sea surface temperature in the eastern tropical Pacific Ocean region correlate to the fluctuations from long-term averages in the pressure field and air flow conditions over the northern Rocky Mountains in North America. Teleconnections clearly have global dimensions.

Teleconnections were identified as early as 1897 by scientists making careful statistical analyses of weather observations from different parts of the world. By comparing long series of observations from India and Australia, it was found that periods of drought over India corresponded to periods of drought over Australia. It was also discovered that seasonally or annually averaged surface air pressure in the Australian and Indonesian regions tended to vary inversely to the pressure in the eastern half of the Pacific Ocean, especially south of the equator; specifically, when the pressure in one area was above its long-term mean, it was below its mean value in the other. This seesaw relationship in pressure is now known as the Southern Oscillation. [*See* El Niño, La Niña, and the Southern Oscillation.]

Basic Principles. Teleconnections depend on the physical interconnectedness of the atmospheric and oceanic systems. A wide range of circulation and weather systems provides the physical interconnections for the atmosphere on all scales. Systems with global horizontal dimensions include the low-level trade winds, which flow from the subtropics toward the equator, and the upper-level west-to-east air motions in the middle latitudes. Systems with much smaller horizontal dimensions include thunderstorms and clouds. Time durations in the global-scale systems range from months to seasons, and those for the smaller-scale phenomena from minutes to hours.

With such an array of organized systems, it is no surprise that changes in atmospheric variables in one region would correlate with changes in others. The distance over which the correlations are found depends on the spatial dimensions of the associated atmospheric and oceanic phenomena. Teleconnections are enhanced when they are related to fluctuations in phenomena that are nearly stationary in space and persistent in time, because teleconnections are defined for fixed geographical positions.

The large-scale atmospheric circulation has many phenomena that are nearly stationary and persistent. The uneven heat input from solar radiation across the Earth causes polar regions to be generally cooler than equatorial regions, giving rise to large-scale and seasonally varying air motions such as the trade winds and westerly upper-level flow mentioned above. Temperature contrasts between land and sea, as well as mountains, cause large-scale east-west variations in these flow patterns, which also vary slowly over time. The Asian monsoon is a good example. Near the surface, the monsoon air flows onto the Asian continent from the surrounding oceans in summer and reverses in winter. At higher levels in the atmosphere, the monsoon circulations produce wavy patterns in the west-to-east flow in the middle latitudes. Such wave flow is characteristic of the airflow around the world; it is due in part to combined effects of all continents and oceans. Physical processes related to the wave patterns themselves in the upper-atmospheric air flow can enhance the persistence of some of the fluctuations in large-scale flow features. [*See* General Circulation.]

Systematic variations in atmospheric pressure accompany these global flow fields. One example is the Siberian High, a persistent surface high-pressure area that exists in winter over Asia. Another is the Bermuda High, a region of high pressure over the western part of the north Atlantic Ocean that is most pronounced in summer. Persistent low-pressure areas include the Icelandic Low, centered in the north Atlantic near Iceland, and the Aleutian Low centered in the north Pacific south of the Aleutian Islands. These are more pronounced in winter than in summer. [*See* High Pressure; Icelandic Low; Low Pressure.]

The existence of nearly stationary large-scale atmospheric systems is conducive to global and climate-related teleconnections. The nearly stationary wave structure and circulations in the flow field set the pattern for how changes in atmospheric conditions in one region —for example, an increase in heating—affect conditions in another region. The vibrations in the string of a violin or guitar provide a simple illustration. The characteristics of the string vibration pattern depend on the tension and length of the string and the amount of excitation applied. The string vibrations have a well-defined structure, being zero at specified end points and, under most conditions, having a maximum in the middle. This pattern holds as the intensity of vibration is changed, resulting in a direct correlation in the amplitude of the vibrations between any two positions on the string.

The string analogy is only approximate; the atmospheric situation is much more complex. Atmospheric phenomena are influenced by a large number of physical processes, including heating owing to condensation of water vapor, solar heating, cloud effects on radiation energy transfer, topographic influences, effects of smaller-scale phenomena, and distortions from displacements caused by air motions. These factors may greatly affect the correlation of fluctuations between regions. The physics of teleconnections is very complex, and much remains to be learned.

Interactions between the ocean and atmosphere have an important role in teleconnections. The ocean has a much larger heat capacity than the atmosphere above it; therefore, values greater or less than average for ocean temperature at the surface can persist and cause pronounced changes in the atmospheric conditions leading to teleconnections. Influence of the ocean fluctuations on fluctuations in the atmosphere is particularly pronounced in the tropics because of the enhancement of concurrent fluctuations in water vapor when temperatures are high.

Observed Teleconnections. A large number of teleconnections have been found through statistical analysis of observations. Many of these have been described as direct or inverse correlations of fluctuations in atmospheric and oceanic conditions between two regions. Table 1 lists a number of examples. Other teleconnections involve correlations among conditions in more than two regions. Continuing studies may add new teleconnections to previously identified ones in the same regions. This has produced a truly global network. Teleconnections have been found for regions all over the world, but there are locations referred to as *centers of action* where they are more prevalent and where the correlated fluctuations tend to be largest.

Some teleconnections have the largest correlation values when fluctuations of the atmosphere or ocean are compared with fluctuations at another region at a different time. This relationship may assist in weather prediction at the place where the conditions occur later. Such teleconnections exist, for example, between sea surface temperature over the Indian and tropical Pacific oceans and summer monson rainfall over India; between sea surface temperature in the Newfoundland area of the north Atlantic and surface air pressure over western Europe one month later in wintertime; and between ocean surface temperature in the tropical central and east Pacific and rainfall over China or the southeastern United States in a later season. In general, such findings have not led to much enhancement in predictions, except involving ocean temperature conditions in the tropical Pacific Ocean; here the temperature variation is commonly associated with an important predictable cycle, the El Niño-Southern Oscillation.

The best-defined and globally most widespread teleconnections are those associated with the El Niño-Southern Oscillation (ENSO). This consists of a long-term (typically 2 to 5 years) fluctuation in ocean surface temperature in the tropical Pacific Ocean (the El Niño),

TELECONNECTIONS. Table 1. *Examples of teleconnections involving two regions* *

Variable 1	Variable 2
Surface pressure in Indonesia	Surface pressure in the eastern Pacific
Precipitation over Australia	Precipitation over India
Ocean temperature in the eastern tropical Pacific	Upper-level air pressure over the northern Rocky Mountains
Ocean temperature in the eastern tropical Pacific	Rainfall in the southeastern United States
Surface temperature in Greenland	Surface temperature in northern Europe
Ocean temperature in the central tropical Pacific	Ocean temperature in the tropical Indian Ocean
Low-level east-west air flow over the Indian Ocean	Low-level and east-west airflow over the tropical Pacific Ocean
Ocean temperature in the eastern tropical Pacific	Strength of upper-level westerly wind flow in the North Pacific
Rainfall in northeast Brazil	Ocean surface temperature in the eastern tropical Pacific
Rainfall in the sub-Saharan area of Africa	Surface pressure difference between the Indonesian and tropical east Pacific areas
Upper-level pressure in the subtropical regions of the West Atlantic	Upper-level pressure in the polar region of the north Atlantic
Upper-level pressure in the western subtropical Pacific area	Upper-level pressure in the north Pacific area

*Correlations may be either positive or negative for the variables listed. In all cases, the variables are for monthly or seasonally averaged times and for a broad area in the regions listed.

combined with concurrent fluctuations in atmospheric pressure and east–west air motions over the Pacific Ocean, Indian Ocean, and Australian areas (the Southern Oscillation). ENSO involves a cyclic interaction between the ocean and atmosphere. A change in ocean surface temperature leads to variations in the supply of moisture to the atmosphere. The moisture change in turn leads to variation in heating of the atmosphere through condensation of water vapor. The subsequent changes in atmospheric pressure and air flow produce variations in ocean currents and in the ocean surface temperature.

Global teleconnections result from ENSO because its large spatial scale and long-term variations in tropical Pacific atmospheric conditions affect the atmospheric conditions elsewhere. Teleconnections related to ENSO (those in Table 1 that include tropical Pacific Ocean temperature) extend to many regions in the tropics and subtropics, and even to temperate zones. Effects on rainfall are particularly important and are found in Australia, sub-Saharan Africa, northeastern Brazil, India, China, and the southeastern United States. In these areas a wet or a dry season can mean serious floods or droughts, which can be devastating, especially if the region's agricultural system is marginal.

The teleconnections related to ENSO also extend into higher latitudes. A well-documented example is the Pacific-North American (PNA) teleconnection. In this, sea-surface temperature variations in the tropical central eastern Pacific correlate with upper-level pressure variations in the North Pacific, western Canada, and the northern Rocky Mountain region, as well as the southeastern United States. When ocean temperatures are above normal in the tropical east Pacific (an El Niño condition, or warm phase of ENSO), air pressures will be below normal in the north Pacific, above normal in the western Canada and the northern Rocky Mountain area, and below normal in the southeastern United States. This results in changes in weather systems, air flow, temperature, and rainfall over a large portion of the eastern north Pacific and North America. Over North America these atmospheric variables are sensitive to other influences in addition to ENSO, so a simple correlation with ENSO conditions in the tropics may not be readily apparent in the observations.

Teleconnections describe variations in climatic conditions that are correlated between one region and another. They are most evident in the time-averaged conditions over large areas for atmospheric and oceanic variables such as sea surface temperature, surface pressure, high-level and low-level wind flow, pressure at upper atmospheric levels, and precipitation. The associated anomalies in temperature and rainfall are significant aspects of the total climatological picture and can have great impact on human lives.

[*See also the biography of the Bjerkneses.*]

BIBLIOGRAPHY

Glanz, M. N., R. W. Katz, and N. Nicholls, eds. *Teleconnections Linking Worldwide Climate Anomalies.* Cambridge, 1991.

Hoskins, B. J., and R. P. Pearce, eds. *Large-scale Dynamical Processes in the Atmosphere.* London, 1983.

Leathers D. J., B. Yarnal, and M. A. Palecki. "The Pacific/North American Teleconnection Pattern and United States Climate, Part I: Regional Temperature and Precipitation Associations." *Journal of Climate* 4 (1991): 517–528.

Leathers, D. J., and M. A. Palecki. "The Pacific/North American Teleconnection Pattern and United States Climate, Part II: Temporal Characteristics and Index Specification." *Journal of Climate* 4 (1991): 707–716.

DAVID D. HOUGHTON

TELEVISION METEOROLOGY. *See* Meteorology; *and* News Media.

TEMPERATE CLIMATE. The word *temperate*, meaning mild or moderate, characterizes a type of climate with a generally well-defined changing of the seasons and whose temperatures during the warmest months are suitable for the growing of food crops. The representative range of latitudes associated with this type of climate extends from about 25° latitude (near the latitude of the tropics of Cancer and Capricorn in the Northern and Southern hemispheres, respectively) to around 50° (the approximate latitude of the 10°C annual mean sea-surface temperature isotherm. Hence, temperate climates are often referred to as *middle latitude climates*.

A principal characteristic of temperate climates is the changing amounts of solar energy received during the course of a year. The intensity of radiation received at the earth's surface at a given time depends on the solar declination, which varies with the geographic latitude. As the Earth orbits the Sun, the tilt of its axis of rotation against the plane of its orbit (called the *obliquity*) results in changing amounts of solar radiation (*insolation*) and gives rise to the annual insolation cycle, and thus to the seasons. The changes in intensity and duration of sunlight, coupled with the effects of atmospheric transmissivity and path length, combine to produce large changes in daily solar radiation receipts between the dates of the solstices. For example, at 40° north latitude, daily solar insolation at the end of December is only about 30 percent of that received in late June.

The Earth's temperate climates can be placed in four broad categories: warm and cool maritime climates, and warm and cool continental climates. Because of water's high heat capacity, large bodies of water exert a moderating effect on climate. Continental climates typically differ from maritime climates in having a much greater amplitude (difference between extremes) of the annual cycle of monthly mean temperature, as well as a more pronounced diurnal temperature range at comparable latitudes. The 48 contiguous United States lie almost entirely within the northern temperate zone. The southern half of Australia lies in the southern temperate zone, along with much of Chile, Argentina, and Uruguay in South America. In contrast only a relatively small fraction of the African land mass is situated within the temperate zone. Land areas in the parts of temperate zones nearer the poles may experience climates warmer or colder than would be expected by their latitude, depending on their proximity to the ocean and on whether cold or warm ocean currents predominate. Much of Europe lies at latitudes north of the contiguous United States but enjoys a milder climate because of the warming influence of the Gulf Stream. In contrast, much of the large Russian land mass, though situated in roughly the same latitude as western Europe, experiences more intense cold and shorter growing seasons, because of its distance from the Gulf Stream.

Figure 1 illustrates the differences in the annual cycle of mean monthly temperature for representative stations in three latitudinal belts across the contiguous United States. The first set of stations are situated along the southern tier of states at approximately 32° north latitude; they are representative of the southern part of the northern temperate zone. Two maritime stations—San Diego, California, moderated by relatively cool ocean temperatures, and Savannah, Georgia, modified by a warmer ocean—are contrasted with the more continental, warmer climate of Dallas, Texas. At San Diego the amplitude of the annual cycle of monthly mean temperature is only 8.6°C (15.4°F), whereas in Dallas, it is 23.5°C (42.3°F), nearly three times as large; at Savannah, the annual monthly temperature range is intermediate at 17.8°C (32°F).

To represent the middle section of the United States at approximately 40° north latitude—two maritime (San Francisco, California, and Dover, Delaware), and two continental sites (Denver, Colorado, and St. Louis, Mis-

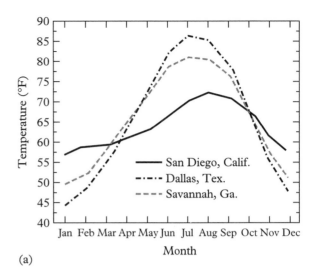

(a)

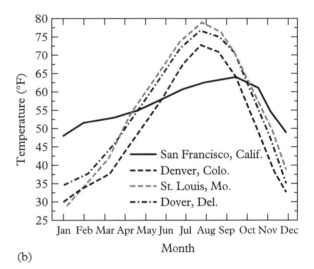

(b)

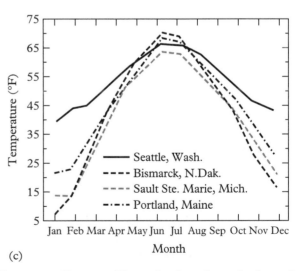

(c)

TEMPERATE CLIMATE. Figure 1. *Annual march of monthly mean temperature (in degrees Fahrenheit) for selected cities in the United States.*

souri)—are also compared in Figure 1. The annual monthly temperature range for San Francisco is 8.6°C (15.4°F), identical to that at San Diego, but at Dover the annual range is somewhat greater than at Savannah (23.7°C or 42.6°F). At St. Louis the range is much greater (27.8°C or 50.1°F), while at Denver, because its high elevation lowers the summer temperatures, it is 24.3°C (43.8°F).

Of the sites near the border with Canada, Seattle, Washington, is also representative of a cool maritime climate, with an annual temperature range of 14.3°C (25.8°F). Bismarck, North Dakota, represents the interior continental region, with winter temperatures considerably below freezing and the largest annual temperature range of these locations at 35.4°C (63.7°F). The proximity of Sault St. Marie, Michigan, to several of the Great Lakes considerably modifies its climate, which would otherwise exhibit an annual temperature curve much like that of Bismarck. The annual range at Sault Ste. Marie (27.9°C or 50.2°F) is nearly the same as that of St. Louis 840 kilometers to the south. Note, however, that Sault Ste. Marie's spring warming is delayed compared to Bismarck or St. Louis, a result of residual lake ice and slowly warming lake surface temperature. Portland, Maine, illustrates the climate of New England maritime areas. Here, the annual temperature range (25.9°C or 46.6°F) is larger than at Denver, despite Portland's coastal location, and temperatures are also slow to rise in the spring. The monthly temperature change between January and April in Bismarck is 19.9°C (35.8°F), three and one-half times that of Seattle (5.8°C or 10.4°F) and about twice that of Portland, Maine (11.8°C or 21.3°F).

A characteristic feature of the general atmospheric circulation in the middle latitudes is the relatively rapid succession of weather systems, which in both hemispheres are embedded in the belt of mean tropospheric westerly winds (Houghton 1985; Oliver and Fairbridge 1987). The belt of mean westerly winds is maintained by the mean meridional (north–south) temperature gradient, which arises from the uneven latitudinal distribution of incident solar radiation.

Strong meridional pressure differences develop in response to this differential heating and cooling of the atmosphere. In mid-latitudes at a given elevation, the mean pressure gradient exerts a poleward force on the atmosphere. Air moving poleward in response to this pressure gradient will, because of the Earth's rotation, be deflected toward the east. This force, known as the *Coriolis effect*, results in the occurrence of generally westerly (west-to-east) winds in temperate latitudes, and instabil-

ities are generated in the westerlies as a result of spatial wind and density gradients. The results are large eddies (circulation cells), manifested at Earth's surface as low and high pressure weather systems; these are often associated respectively with disturbed or fair weather.

In the Northern Hemisphere the westerlies undergo strong seasonal variations in response to the changes in insolation. The greater proportion of land mass to ocean in northern high latitudes means that during the months with long nights and low sun angles, cold air masses are generated in the continental interiors. In the same way, summer warming of the air over the continental land masses is amplified. The situation is different in the Southern Hemisphere, where more land lies at tropical and subtropical latitudes than in the higher latitudes (except for Antarctica). This leads to lower seasonal temperature variability over the southern continents.

Another weather characteristic of the climate in temperate latitudes is the passage of frontal systems. A weather front represents the boundary between air masses with differing values of certain characteristics, such as temperature and humidity. Thermal gradients, caused by differential heating and cooling during the course of the year, generate potential energy, which is converted into kinetic wind energy by the traveling weather systems (as shown on a typical weather map). The storm tracks typically move equatorward during winter and poleward during summer. In the Northern Hemisphere, the mean tracks are located farthest south during February and March, and farthest north in August and September.

As noted above, a very important control on the climate of a given location is its distance from open ocean areas, along with such land features as surface elevation and the surrounding topographic relief. In coastal areas along the western continental margins of the Americas and the Mediterranean Sea, monthly precipitation exhibits a distinct winter maximum and summer minimum. The entire West Coast of the contiguous United States experiences this type of climate. At the eastern margins of the continents precipitation tends to be more evenly distributed throughout the year; in interior continental regions, precipitation generally exhibits a summer maximum.

A seasonal feature of temperate climates is the occurrence during the coldest months of the year of snowstorms, which often paralyze transportation and commerce over large areas and inflict severe economic losses to farmers and ranchers. Particularly in areas of the central and southeastern United States during the spring months, warm moist air out of the Gulf of Mexico often clashes with cold air from the north to produce severe thunderstorms and tornadoes. Tornadoes are intense vortices that can attain wind speeds of 320 kilometers per hour, causing destruction of building structures and occasionally loss of life. Along the eastern continental margins, large tropical storms—known as *hurricanes* in the Atlantic and *typhoons* in the western Pacific—sometimes hit land with disastrous effect. Fortunately, improved prediction and communications have lowered the loss of human life to these climate hazards, although inflation-adjusted property losses have increased significantly.

Temperate climates contribute most of the world's grain, much of it grown without irrigation. A characteristic feature of middle latitude climates is the recurrent development of major drought episodes. Although difficult to define precisely, in general, drought can be understood as a period of prolonged precipitation deficiency, below-normal streamflow, and soil moisture deficits. The United States has experienced a number of regional droughts, notably in the 1930s. Of the climatological features characteristic of temperate latitudes, drought may perhaps have the greatest impact on society (Warrick, 1980; Wilhite et al., 1987; Riebsame et al., 1991). Drought typically affects a large area and may persist up to several years. The western United States is a semiarid region that depends on winter precipitation, falling mostly as snow, to provide the spring and summer runoff necessary for irrigation. Without an extensive water storage system, much of the region west of the Rocky Mountains could not support its current level of population and economic activity.

Vegetation zones can provide an integrative measure of climate. The most abundant vegetation types of temperate zone climates include deciduous forests, found in areas with adequate year-round moisture; grasslands and shrublands, present in the continental interior; and evergreen forests, which tend to predominate in cooler environments and montane parts of the region. Pine forests are present in warmer, more humid areas of the temperate zone on poor, shallow soils. Deserts, which have been estimated to cover as much as a quarter of the world's land, are typically found along the tropical margins of the temperate zone. For example, because of its extensive land mass and remoteness from oceanic moisture sources, large parts of Asia receive relatively low amounts of precipitation and are classified as arid.

The temperate regions of the Earth have long been home to a majority of the world's population, but tremendous population growth in tropical regions during the past century has nearly closed the gap. Temperate agriculture provides much of the world's supply of food grains, as a combination of rich soils and the application of technological tools have greatly increased crop yields. Nevertheless, the prospect of global climate changes resulting from human activities suggests that the generally stable climate that has nurtured modern societies may not remain stable in the future.

[*See also* Climatic Zones.]

BIBLIOGRAPHY

Crowe, P. R. *Concepts in Climatology*. London, 1971.
Houghton, D. D. ed. *Handbook of Applied Meteorology*. New York, 1985.
Oliver, J. E., and R. W. Fairbridge, eds. *Encyclopedia of Climatology*. New York, 1987.
Riebsame, W. E., S. A. Changnon, and T. R. Karl. *Drought and Natural Resources Management*. Boulder, Colo. 1991.
Warrick, R. A. "Drought in the Great Plains: A Case Study of Research on Climate and Society." In *Climatic Constraints and Human Activities*, edited by J. Ausubel and A. K. Biswas. Oxford, 1980, pp. 93–123.
Wilhite, D. A., W. E. Easterling, and D. A. Wood, eds. *Planning for Drought: Toward a Reduction of Societal Vulnerability*. Boulder, Colo., 1987.

HENRY F. DIAZ

TEMPERATURE. Temperature is a measure of the relative warmth or coolness of an object. This in turn is determined by the speed of motion of the molecules that compose the object. All substances—gas, liquid, or solid—are composed of molecules. These molecules are always in motion, though in the case of a solid the motion is restricted to vibrations of the molecules themselves. The temperature of an object is related to the average kinetic energy (equal to one-half times the mass times the speed of motion squared) of the molecules composing the object, taken with respect to the center of mass of the object. The greater the kinetic energy, the higher the temperature.

The fact that the kinetic energy is calculated with respect to the center of mass of the object is important. This says that the speed of motion of the object itself is irrelevant to the temperature. For example, suppose that we have two identical metal balls, except that ball *A* is hot and at rest, and ball *B* is cold and thrown so that it is moving very fast. Since ball *A* is at a higher temperature than ball *B*, the molecules composing it are moving much faster (with respect to the center of the ball itself) than are the molecules in ball *B*. The part of the movement of the molecules that is due to the movement of the ball itself is unrelated to the temperature of the ball.

Measurement. Temperature is measured with an instrument called a *thermometer,* a device that depends on the heating and consequent expansion of a given material. Although there are a variety of different types of thermometers, common atmospheric thermometers use either mercury or alcohol inside a graduated glass envelope; both mercury and alcohol are substances that tend to expand and contract freely, and to remain unfrozen, as the temperature varies through the range experienced near the surface of the Earth.

In order to determine a temperature scale, two reference temperatures are needed. The temperature difference between them is then divided into a certain number of units called *degrees.* These are the units in which the scale of the thermometer is calibrated.

Temperature Scales. There are presently three different temperature scales that are widely used: *Fahrenheit, Celsius* (formerly called *centigrade,* from its 100-degree scale), and *Kelvin.* A fourth, *Reaumur,* is used in a few parts of Europe (its gradations range from 0°, freezing, to 80°, boiling point of water, and it was devised in 1730 by and named for René Antoine Ferchault de Réaumur, 1683–1757). The Fahrenheit temperature scale is widely used in the United States; the Celsius scale is used throughout most of the rest of the world, and the Kelvin scale is commonly used by scientists.

The Fahrenheit scale was developed from 1708 to 1724 by the physicist Gabriel Fahrenheit (1686–1736). His two reference temperatures were the *ice point* (the freezing point of pure water at a pressure of one standard atmosphere) and the *boiling point.* These were assigned the values of 32 degrees and 212 degrees (written 32° and 212°), respectively. The origin of these reference values is interesting. Fahrenheit noted that temperatures lower than the ice point could be achieved in mixtures of crushed ice and various salts. He believed that the lowest temperature that could be achieved in this manner was a limiting value and therefore assigned a temperature of 0° to it. Given his decision to have 180 scale divisions between the ice and boiling points, he found that this zero point was 32 scale divisions below the temperature of the ice point—hence his use of 32° and 212° as the two reference values.

The Celsius temperature scale is currently much more widely used than the Fahrenheit temperature scale. This scale was developed by Swedish astronomer Anders Celsius (1701–1744) in 1741. Here again the two reference temperatures are taken to be the freezing and boiling points of water (at a pressure of one standard atmosphere, defined as the average atmospheric pressure at sea level), but now these are assigned values of 0° and 100° respectively. Thus there are 100 degrees (scale divisions) between these two values, hence the original designation *centigrade temperature scale.* It is now known as the Celsius temperature scale in honor of its developer. At the Ninth General Conference on Weights and Measures held in 1948, the designation *degrees centigrade* was replaced by the designation *degrees Celsius.*

The temperature scale most often used by scientists is the Kelvin scale, introduced in 1848 by British scientist William Thomson, Lord Kelvin (1824–1907). Its scale division is the same as that of the Celsius temperature scale. In the Kelvin temperature scale, however, the boiling point of water is 373 K, the freezing point is 273 K, and the zero value is set to the theoretical lowest temperature that can exist. We have already noted that temperature is determined by the speed of motion of the molecules composing an object; thus, the lowest possible temperature is that at which all molecular motion ceases. This is referred to as *absolute zero* and is indicated in the following way: 0 K; $-273.15°$C; and $-459.67°$F. Because this is the beginning point of the Kelvin scale, this scale is known as an *absolute temperature scale* and was originally called the *absolute scale,* with symbol A.

Figure 1 shows how the Celsius, Kelvin, and Fahrenheit temperature scales compare to one another. To convert between these temperature scales, one can either read the corresponding temperature from the adjacent scale, or use the following formulas:

$$°C = \tfrac{5}{9}(°F - 32)$$
$$°F = \tfrac{9}{5}°C + 32$$
$$K = °C + 273.15.$$

For example, to convert degrees Fahrenheit to degrees Celsius, subtract 32 degrees from the Fahrenheit temperature and then multiply by 5/9.

Laws of Heat Flow. Temperature and heat are not identical. While temperature tells us how hot or cold an object is, *heat* is a form of energy that is transferred from one object to another because of the temperature difference between them. In the atmosphere heat can be transferred by conduction, convection, and radiation.

Heat conduction. This refers to the transfer of heat from molecule to molecule within a substance, or from one substance to another, by means of internal molecular activity. As an example, consider a situation where a long, thin and straight piece of metal is held horizontally with

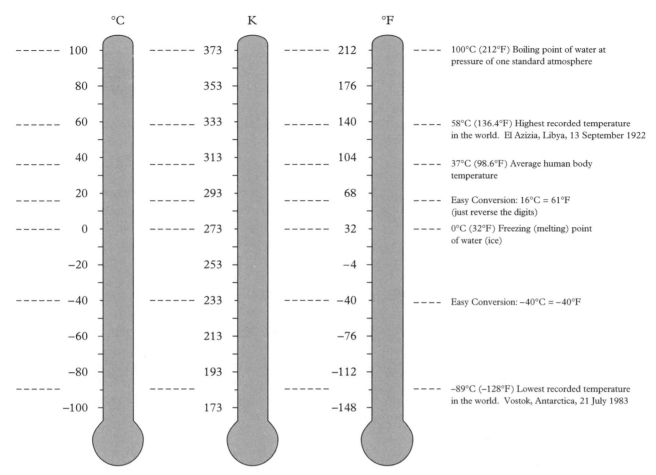

°C   K   °F

------ 100   ------ 373   ------ 212   ---- 100°C (212°F) Boiling point of water at
                                                      pressure of one standard atmosphere

80   353   176

------ 60   ------ 333   ------ 140   ---- 58°C (136.4°F) Highest recorded temperature
                                                      in the world. El Azizia, Libya, 13 September 1922

------ 40   ------ 313   ------ 104   ---- 37°C (98.6°F) Average human body
                                                      temperature

------ 20   ------ 293   ------ 68   ---- Easy Conversion: 16°C = 61°F
                                                      (just reverse the digits)

------ 0   ------ 273   ------ 32   ---- 0°C (32°F) Freezing (melting) point
                                                      of water (ice)

-20   253   -4

------ -40   ------ 233   ------ -40   ---- Easy Conversion: -40°C = -40°F

-60   213   -76

-80   193   -112

------   ------   ------   ---- -89°C (-128°F) Lowest recorded temperature
                                                      in the world. Vostok, Antarctica, 21 July 1983
-100   173   -148

TEMPERATURE. Figure 1. *Comparison of Celsius, Kelvin, and Fahrenheit temperature scales.*

one end just above the flame of a candle. As the molecules in the part of the metal that is directly above the flame absorb heat from it, they vibrate faster. These molecules then collide with neighboring molecules, causing them also to vibrate faster, and so on. Eventually the molecules at the other end of the metal vibrate faster, which in turn causes the molecules in the fingers of the person holding the metal to vibrate faster. Heat is now being transferred from the metal to the person's finger, and both now feel hot. The heat transfer from one end of the metal to the other and from the metal to the fingers occurs by conduction. Heat that is transferred by conduction always flows from higher-temperature to lower-temperature regions. Generally, the larger the temperature difference, the greater the rate of heat transfer.

The rate of heat transfer by conduction also depends on the type of material. Some materials are much better conductors of heat than others. Metals, for example, tend to be good conductors of heat, while air is a very poor conductor of heat. Consider a piece of wood and a metal pipe, both at room temperature. The metal will feel much colder when touched than will the wood. Room temperature is lower than the temperature of a person's hand, so heat will flow by conduction from the hand to the object at room temperature held in the hand. Thus the object feels cool to the touch because it causes loss of heat from the hand. However, metal is a much better conductor of heat than wood, so the metal pipe feels cooler to the touch than does the piece of wood, even though both are exactly the same temperature. It can therefore be difficult to judge the relative temperatures of objects of different composition by how they feel when touched.

Heat convection. Air is a very poor conductor of heat. Even if the ground becomes very hot, for example, only a very shallow layer of air, perhaps a few centimeters thick, will be warmed as a result of heat conduction from the ground. However, the atmosphere can transfer heat rapidly from one region to another. Heat transfer within a fluid or a gas that occurs as a result of movement of the fluid or gas itself is called *heat convection*. Note that although heat transfer by conduction can take place in a solid, heat transfer by convection cannot. In order for convective heat transfer to occur, it must be possible for motion in the form of currents to occur within the substance. By definition, such motion within a solid is not possible.

In meteorology, the term *convection* is applied only to motions that are predominantly vertical; sometimes it is further restricted to upward motion, with the word *subsidence* used to refer to downward vertical motion. The analogous term for horizontal motion is *advection*. Vertical motions can be either free or forced. The former, referred to as *free convection,* occurs whenever air is locally heated to the point that it becomes warmer than the surrounding air. If this occurs, it will be positively buoyant and will rise much as a hot-air balloon does. Since the atmospheric pressure decreases with increasing height, as air rises it expands and cools. However, the structure of the atmosphere itself is generally such that the temperature decreases with height. As long as the rising heated air stays warmer than the air surrounding it at the same height, it will continue to rise. In this manner, heat is transferred by convection from lower to higher in the atmosphere. If the rising air is sufficiently moist, puffy clouds called *cumulus clouds* may form. In much the same manner, if air away from the surface becomes colder than the air surrounding it at the same height, it will sink. As it sinks, the pressure on it increases and it warms up. However, the temperature of the surrounding atmosphere also generally increases with decreasing height. As long as the sinking air stays colder than the surrounding air at the same height, it will continue to sink. [*See* Convection.]

Forced convection occurs when air rises or sinks without initially being warmer or colder than the surrounding air at the same height (positively or negatively buoyant, respectively). Such vertical motion can occur, for example, when air is forced upward by flowing along upsloping terrain (termed *orographic lifting*), when air is lifted by an atmospheric front *(frontal lifting),* or when air is forced to rise or sink as a result of convergence or divergence of the wind flow.

As noted above, heat *advection* refers to atmospheric heat transfer by horizontal motion. Such heat transfer occurs if there is wind and if the wind is blowing in a direction in which there is a local change in temperature.

Radiation. All objects at a temperature greater than absolute zero emit radiation, energy carried in the form of electromagnetic waves. These waves may travel either through empty space (a vacuum) or through a material medium. *Visible radiation* (light) represents a particular range of wavelengths of such electromagnetic waves. All types of electromagnetic radiation are reflected and absorbed in basically the same manner as visible light. [*See* Longwave Radiation; Radiation; Shortwave Radiation.]

The wavelengths of radiation and the total amount of radiation emitted by an object depend primarily on the object's temperature. The higher the temperature, the shorter the wavelength at which peak emission of radiation occurs and the greater the total amount of radiation emitted. The Sun therefore emits far more radiation than does the Earth. The wavelength of maximum radiation emission by the Sun's surface (which is at a temperature of almost 6,000 K) is right in the middle of the visible part of the spectrum. The Earth, on the other hand, has an average surface temperature of only 288 K. This results in maximum radiation emission at a much longer wavelength than that of visible light; this radiation is referred to as *infrared radiation.*

All objects absorb as well as emit radiation. If an object absorbs more radiation than it emits, it gets warmer; if it emits more radiation than it absorbs, it gets colder. If the rates of emission and absorption are equal, it remains at constant temperature. The rate at which an object emits or absorbs radiation is determined not only by the temperature of the object (as discussed above) but also by its surface characteristics. Some objects are good absorbers and emitters of only certain wavelengths of radiation. For example, fresh snow is a very poor absorber of visible light (most of the visible light that strikes it is reflected), but it is a very good absorber (and emitter) of infrared radiation.

We can summarize the net effect of these processes on temperature with the following equation:

Change of temperature at a point in space	=	advection (horizontal of nearby air of different temperature)	+	cooling or warming due to convection or subsidence
		+ temperature change due to conduction	+	temperature change due to radiation

There is one other type of process that can cause a significant local change in temperature within the atmosphere: phase changes of water. When water changes from vapor to liquid (condensation) or from liquid to solid (freezing), heat is released to the environment (the surrounding air), which can cause warming of the air. When water changes phase in the reverse direction—from liquid to vapor (evaporation) or from liquid to solid (melting)—heat is taken from the environment, which can result in cooling of the air. If any of these processes is occurring, the resulting temperature change would need to be added to the right-hand side of the equation.

Importance of Temperature for Biological and Human Activities. Temperature significantly affects all biological systems. Most chemical reactions are temperature-dependent. As metabolic processes are in fact complex combinations of chemical reactions, changes in the temperature can have significant consequences. Of course, many other factors—availability of water and nutrients, the degree of windiness, or the amount of sunshine, for example—also affect biological systems.

Most biological processes, such as photosynthesis in plants or digestion in animals, occur more quickly at higher temperatures. However, some of the essential compounds involved in these processes tend to become altered at high temperatures such that the processes no longer readily occur. The biological processes essential to the existence of most organisms function most efficiently at some particular intermediate temperature appropriate to the typical environment inhabited by the organism. For mammals and birds, this optimum temperature is the regulated body temperature, which is in the range of 30°C to 40°C. Warm-blooded animals can survive within a wide range of temperatures. Cold-blooded animals, and plants, are much more affected by changes in environmental temperature and are thus potentially more susceptible to climatic changes.

Human activity is impaired when the temperature is either excessively high or excessively low. In determining the deleterious effects of high temperatures on humans, one has to consider both the amount of water vapor in the air (the humidity) and the wind speed in addition to the air temperature itself. The human body is cooled by a combination of conduction, radiation, and evaporation of moisture from the sweat glands. The lower the humidity, the more readily sweat can evaporate and thereby cool the body. The higher the wind speed, the more air that is brought into contact with the body, allowing more evaporation and greater conductive heat loss from the

skin. Hence high air temperatures are most unpleasant when the humidity is high and the wind speed is low.

At low temperatures, maximum human discomfort occurs when the humidity is high and the wind speed is high. When the humidity is high, the skin is more likely to remain damp, and thus the conductive heat loss is likely to be greater; the conductive heat loss will also become greater as the wind speed increases. The type and amount of clothing worn can significantly affect the degree of discomfort and impairment produced when temperatures are extreme.

[*See also the biography of Fahrenheit*; Instrumentation; Potential Temperature; *and* Water Vapor.]

BIBLIOGRAPHY

Ahrens, C. D. *Meteorology Today: An Introduction to Weather, Climate and the Environment.* 4th ed. New York, 1991.
Anthes, R. A., J. J. Cahir, A. B. Fraser, and H. A. Panofsky. *The Atmosphere.* 3rd ed. Columbus, Ohio, 1981.
Fairbridge, R. W., ed. *The Encyclopedia of Atmospheric Sciences and Astrogeology.* New York, 1967.
Ford, M. J. *The Changing Climate: Responses of the Natural Flora and Fauna.* London, 1982.
Huschke, R. E., ed. *Glossary of Meteorology.* Boston, 1959.
Ludlum, D. M. *The Audubon Society Field Guide to North American Weather.* New York, 1991.
Wallace, J. M., and P. V. Hobbs. *Atmospheric Science: An Introductory Survey.* New York, 1977.

WARREN BLIER

THEATRICAL PERFORMANCE. *See* Drama, Dance, and Weather.

THERMAL CELLS. *See* Meridional Circulations.

THERMOCLINE. *See* El Niño, La Niña, and the Southern Oscillation; Maritime Climate; *and* Oceans.

THERMODYNAMICS. The science of thermodynamics is concerned with the equilibrium states of systems and the changes in system properties after transformations from one state to another. A *system* is a well-defined aggregate of matter. An example of a *thermodynamic system* is one that is thermally insulated so that no heat flows into or out of it, from or to its environment; alternatively, a thermodynamic system can be one that is in contact with a heat reservoir, so that its temperature is held fixed. These are only two of many ways in which a system can interact with its environment. The *state* of the system is defined by the values of its *state variables*. For a gas, these state variables are the mass of each chemical constituent, temperature, pressure, and volume. The state of a system may be changed by allowing a net amount of heat or mass to flow into the system; in addition, we may do work on the system—for example, by compressing it with external force. After the net flows of energy and mass have ceased, the system will come to a new thermodynamic state characterized by a new set of values of the state variables. A special case of changes of state are *reversible* changes. These take place quasi-statically, and each infinitesimal change in one direction can be reversed, returning the system to the original state. Each step along the path of a reversible process is itself occupying an equilibrium state; thus it can be defined continuously along the path by the usual state variables. *Irreversible* changes often occur in nature. These may take place along paths that may not have equilibrium states corresponding to each instantaneous stage.

One important state variable is the *phase* of a substance—solid, liquid, or gas. A thermodynamic system may comprise a mixture of phases. Important energy releases occur if the fraction of the mass of a system in a less-condensed phase goes into a more condensed phase. The energy exchange is called *latent heat*. The release and absorption of latent heat play crucial roles in the weather and climate system, because water occurs naturally in all three of its states (as solid ice, liquid water, and gaseous water vapor) at or near the Earth's surface.

First Law of Thermodynamics. The first law deals with the changes in state after a transition from one thermodynamic state to another. The transition may be along either a reversible or an irreversible path. The essence of the first law is that there exists a quantity called the *internal energy*, which is a function only of the state of the system. The internal energy change during a transition is given by the sum of the work done on the system by external agents plus the heat given to the system during the transition. For example, an ideal gas (most gases are ideal at sufficiently low pressures; air at ordinary pressures satisfies the conditions) in a cylinder being compressed an infinitesimal amount by a piston exhibits a change in its internal energy $dU = dQ - p dV$, where dU is the infinitesimal change in internal energy, dQ is the infinitesimal amount of heat entering the gas from the walls of the cylinder, p is the pressure of the gas, and dV is the infinitesimal change in the volume occupied by the gas. The term $-p dV$ is the work performed on the gas

by the piston. It represents a force applied through a distance—the classical definition of *work*. The first law includes the statement of the conservation of energy, but it actually goes beyond this principle because of the claim that there exists a function U that is only a function of the thermodynamic state.

Two important types of transition are the *adiabatic* and *isothermal transitions*. An adiabatic transition is one in which the system is thermally isolated from its surroundings. In other words, the quantity dQ vanishes during the transition. In this case, the change in internal energy consists solely of the amount of work done on the system. An isothermal transition is one conducted at constant temperature. This case requires constant contact with a heat reservoir, defined to be of such large mass that it can transfer heat to the system without its own temperature changing. While rarely realized perfectly, these types of transition are often good approximations of some of the important transitions that occur in weather and climate.

Second Law of Thermodynamics. The second law of thermodynamics introduces a new function of the state variables, called *entropy*. The change in entropy dS is given by dQ^{rev}/T, where T is the absolute temperature, and dQ^{rev} is the heat given to the system computed along a reversible path. While the transition in question may be irreversible, the computation of change in entropy must be conducted along an artificially constructed reversible path. Spontaneous irreversible changes in the state of a system always result in changes in the entropy of the system and its neighboring systems (together called the *universe*) that are greater than or equal to zero. Entropy is sometimes considered a measure of a system's disorder. When unconstrained, nature always moves irreversibly toward states of the universe with higher entropy or disorder. An example is when a warm mass is placed in contact with a cool one. The final state is with both at the same intermediate temperature. The orderly state is the one with the isolated masses. Nature always proceeds to the less ordered state, which has higher entropy. [*See* Entropy.]

Thermodynamics deals with the macroscopic laws of matter. By contrast, *statistical mechanics* is the branch of theoretical physics and chemistry dealing with the microscopic laws governing the motions of the individual atoms and molecules making up the bulk of solids, liquids, or gases. A major goal of statistical mechanics is to derive the laws of thermodynamics from first principles. This goal has now been satisfactorily accomplished, but its possibility was the subject of heated scientific debate around the turn of the twentieth century. While the macroscopic science of thermodynamics is most directly applied to problems in science and engineering, statistical mechanics provides insight; occasionally, it allows us to evaluate some of the macroscopic properties observed in the bulk matter from microscopic principles—for example, relating the molar heat capacity (heat in mole units) of an ideal gas to molecular properties.

The Earth's atmosphere is composed mostly of nitrogen and oxygen, with small, variable amounts of water vapor. For most thermodynamic problems, the gases comprising the atmosphere can be considered locally as an ideal gas. The balance between the gravitational force drawing the gas toward the Earth's surface and the vertical force on a slab of air owing to a negative vertical gradient in the pressure leads to the barometric law, which states that the atmospheric pressure falls off in nearly exponential fashion in the vertical direction. It is a good approximation to consider the pressure at a particular altitude to be equivalent to the weight of air above a horizontal plane of unit area. An important consequence of the laws of thermodynamics is that for an ideal gas, the internal energy of a given air mass is simply proportional to its temperature and is independent of the other state variables. [*See* Barometric Pressure.]

Small masses of air in the atmosphere, called *parcels,* can often be considered individually as thermodynamic systems for the investigation of many problems. For example, lifting a parcel from one elevation to another results in the parcel's expansion owing to the decreased pressure aloft. Pressures inside the parcel adjust rapidly to become equal to the pressures outside (called the *environmental pressure*). If the parcel is lifted rapidly, the amount of heat exchanged between the environment and the interior of the parcel will be negligible; hence, we can consider the process to be adiabatic. During such adiabatic lifting, the parcel expands and in the process does positive work on the environment. This work expended by the parcel on the environment leads to a decrease of the parcel's temperature. In fact, for a dry parcel, the decrease is fixed at nearly 10°C per kilometer of lifting. This is called the *adiabatic lapse rate.* If the air has water vapor in it, as the parcel is lifted it may reach the dew point, and the water it contains will begin to condense into fog or cloud droplets. As parcels are mechanically forced upward by turbulence or passage of the air flow up an incline, we can expect clouds to form at a certain altitude, called the *lifting condensation level.* This effect accounts for the fact that cloud bottoms appear to have a uniform elevation even on a partly cloudy day. [*See* Adiabatic Processes.]

Coexistence of the Phases of Water. At temperatures between 0°C and 100°C at ordinary pressures, water can exist in equilibrium between its liquid and gaseous phases. Imagine a cylinder in which a puddle of water exists at the bottom and vapor exists above. (The presence of air above makes very little difference to this example.) The equilibrium vapor pressure over the flat surface of the water is known as the *saturation vapor pressure.* It is a strong function of the ambient temperature, roughly doubling for every 10°C. In nature, unsaturated air often passes over wet surfaces. Because in such an air mass the saturation vapor pressure exceeds the existing vapor pressure (the ratio of the existing vapor pressure to the saturation vapor pressure is called the *relative humidity*), evaporation will occur at a rate proportional to the difference between the saturation and actual vapor pressures. Such evaporation processes are much more influential in the tropics than they are in the higher latitudes because of their strong dependence on temperature.

The interior of clouds is always at or very near 100 percent local humidity. If, through mechanically forced lifting of parcels, the humidity becomes slighter supersaturated, the droplets in the cloud will grow. (As parcels within a cloud mix with dry air, however, cloud droplets shrink.) As droplets grow, there is a release of latent heat into the surrounding air. This represents the freeing of heat that was removed from the wet surface when the water was originally evaporated. The evaporation of water at one location and the resulting cooling at that place, and the subsequent recondensation of the same vapor and accompanying release of latent heat somewhere else, constitute a transport of heat from one geographical location to another. This transport of heat by the latent heating mechanism accounts for about half of the heat transported in the atmosphere from equatorial regions toward the poles. A good example is afforded by the trade winds, shallow winds only a few kilometers deep flowing along tropical surfaces toward the equator. As they do so, they pass over warm oceanic surfaces, picking up large amounts of moisture. When the parcels arrive at the Intertropical Convergence Zone, the air is lifted along an irregular narrow band girdling the Earth near the equator. The lifting occurs in tall towers of clouds in which the vapor is at last condensed into droplets. The droplets grow in the cloud towers and eventually become so large that they fall out as torrential rains. The net energetic effect is that the sea-surface waters in the tropics are cooled by evaporation, and the transported heat is released in high altitudes in the tropics, fueling the buoyancy of the air rising in the towers and accelerating the tropical large-scale overturning circulation. [*See* Tropical Circulations.]

Great volumes of water are also stored in the *cryosphere,* the term applied to the total global ice contained in clouds, sea ice, and various forms of land-based ice. The existence of the cryosphere affects climate in two important ways. On land or sea it forms a surface that is highly reflective to sunlight, tending to make the surface air cooler than it would be over a dark surface. Second, the cryosphere tends to buffer changes in temperature due to the latent heat absorption needed to melt ice or the latent heat released in the freezing process. Near the freezing point the latent heat effect tends to delay the thermal response forced by heating rate changes.

[*See also* Cryosphere; Diabatic Processes; Latent Heat.]

BIBLIOGRAPHY

Wallace, J. M., and P. V. Hobbs. *Atmospheric Science: An Introductory Survey.* New York, 1977.
Zemansky, M. W., and R. H. Dittman. *Heat and Thermodynamics: An Intermediate Textbook.* 6th ed. New York, 1981.

GERALD R. NORTH

THERMOMETER. *See* Instrumentation; *and* Temperature.

THERMOSPHERE. The thermosphere is a spherical shell of the Earth's atmosphere lying between 80 and 500 kilometers altitude. This highly rarefied region near the edge of space is strongly influenced by two variable components of the Sun's output—the solar ultraviolet (UV) radiation, with wavelengths between 5 and 200 nanometers, and the plasma outflow from the Sun, called the *solar wind,* which streams outward past the Earth. All of the solar UV radiation with wavelengths between 5 and 200 nanometers is absorbed by the gases of the thermosphere. The physical and chemical processes driven by this absorption establish the basic chemical, thermal and, dynamic structure of the thermosphere.

The composition of the thermosphere's neutral gas differs considerably from that in the lower atmosphere. The main constituents of the lower atmosphere are molecular nitrogen (N_2) and oxygen (O_2). The solar UV radiation in the thermosphere begins to dissociate N_2 and O_2 into their atomic components N and O. Furthermore, at these altitudes the air is so thin that atomic N and O cannot chemically recombine back to molecular N_2 and O_2; turbulence is also lacking, and the atmosphere approaches a diffusive equilibrium where constituents settle according to their molecular weights. Molecular N_2 and O_2 dominate the composition of atmosphere below 140 kilometers. Above this altitude, atomic O is the major constituent up to about 500 kilometers. Atomic hydrogen (H) released during the UV dissociation of water vapor (H_2O) dominates above 500 kilometers in the region called the *exosphere.* Temperature increases dramatically with altitude within the thermosphere, from values near 200 K at 80 kilometers to 1,000 to 2,000 K near 500 kilometers. This temperature increase occurs because solar UV radiation is absorbed in the thermosphere, but no effective infrared-active gas species exists to cool the thermosphere by reradiating this energy to space. Instead, heat must be thermally conducted downward to the base of the thermosphere, near 100 kilometers, where infrared-active molecular species can radiate this energy back to space. This downward heat conduction requires that the temperature increase with altitude, accounting for the high temperature that occurs in the thermosphere.

A large amount of solar wind energy is also deposited in the thermosphere at high magnetic latitudes through physical and chemical processes associated with the aurora borealis in the Northern Hemisphere and the aurora australis in the Southern. These competing forms of variable energy deposition make the thermosphere a dynamically active region with large variations in temperature, composition, and winds about its basic state, which is established mainly by solar radiative processes. Many satellites orbit in the upper thermosphere, where they experience an atmospheric drag that eventually brings them down to burn up in the lower atmosphere. Satellites at a given altitude experience variations in atmospheric drag as the atmosphere expands and contracts in response to atmospheric heating, caused by changing solar UV radiation and auroral energy input. This variability makes accurate prediction of satellite drag difficult. [*See* Aurora Borealis; Solar Wind.]

Embedded within the thermosphere is the *ionosphere,* a weakly ionized plasma that reflects certain radio waves; the bouncing of radio waves off the ionosphere has been used for long-distance communication since the beginning of the twentieth century. The ionosphere is formed by solar UV radiation that is energetic enough to ionize the gases of the thermosphere, those wavelengths between 5 and 122 nanometers. The ions and electrons

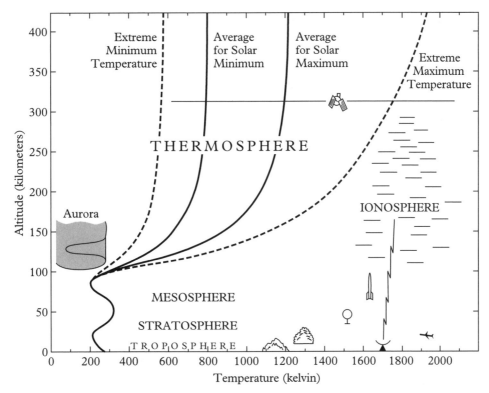

THERMOSPHERE. Figure 1. *Schematic illustrating atmospheric regions.* Thermosphere and ionosphere exist above 80 kilometers altitude. The thermospheric temperature profile illustrates response to solar UV radiation output over a solar cycle. Cold thermospheric temperatures occur when the Sun is quiet with no sunspots, and warm thermospheric temperatures when the Sun is active with many sunspots.

formed by this process are confined by the Earth's magnetic field, and the neutral-gas winds flowing in the thermosphere experience a collisional drag because these ions and electrons are not free to move with the wind. The meteorology of the thermosphere therefore differs from the familiar weather that we experience near the Earth's surface. The fluid motions are governed by the same equations as those used by meteorologists studying weather systems in the lower atmosphere; however, there are notable differences between the upper and lower atmospheres. In the troposphere (the lowest 10 kilometers of Earth's atmosphere), the temperature decreases with altitude, whereas in the thermosphere it increases with altitude. This difference makes the thermosphere more stable than the troposphere. Two additional forces are also important at thermospheric heights: the viscous force and the ion drag force. Molecular viscosity is small in the lower atmosphere, but within the thermosphere it increases exponentially with altitude by many orders of magnitude, as atmospheric density decreases. Viscosity primarily transfers momentum between various altitudes,

smoothing out vertical gradients in wind velocity. Above 300 kilometers, the atmospheric motion is nearly constant with height. [*See* Ionosphere; Thermodynamics.]

The ion drag force is due to collisions between charged particles and neutral particles and results in an effective drag on the neutral winds. Plasma processes in the ionosphere also create strong electric fields—especially at high magnetic latitudes in the auroral zones—and these electric fields can accelerate the plasma relative to the neutral gases. Under these conditions, the ion drag force becomes a source of collisional momentum that has been observed to accelerate the neutral winds from about 200 to 1,000 meters per second during strong auroral activity.

These differences in the dynamics of the thermosphere and the lower atmosphere make the former region interesting for geophysical fluid studies. Atmospheric scientists studying the Earth's upper atmosphere have to consider the interacting properties of the neutral gases in the thermosphere and the plasma of the ionosphere. General circulation models of the coupled thermosphere and ionosphere system, with electrodynamic feedbacks, are being used to study such topics as how the upper atmosphere and ionosphere respond to solar and auroral variability; how waves, tides and the other disturbances generated in the lower atmosphere and propagating upward affect the thermosphere; how the thermosphere and ionosphere will respond to global change and the trace gases released by man's activities; and how the great variability of the coupled thermosphere and ionosphere system affects human activities in space.

[See also Atmospheric Structure.]

BIBLIOGRAPHY

Kelley, M. C. *The Earth's Ionosphere*. San Diego, Calif., 1989.

Rees, D., ed. *Cospar International Reference Atmosphere, 1986, Part I: Thermosphere Models*. Oxford, 1988.

Rees, M. H. *Physics and Chemistry of the Upper Atmosphere*. Cambridge, 1989.

Roble, R. G. "The Earth's Thermosphere." In *The Solar Wind and the Earth*, edited by S.-I. Akasofu and Y. Kamide. Norwell, Mass., 1987, pp. 245–264.

RAYMOND G. ROBLE

THORNTHWAITE, CHARLES WARREN

(1899–1963), American climatologist and geographer. Thornthwaite ranks even now among the world's most cited climatologists, mainly for his concepts of potential evapotranspiration and the climatic water budget. Making use of *potential evapotranspiration* (the climatic demand for water) and *measural precipitation* (the water supply), he developed a procedure for solving the mass water budget for any location. Called the *climatic water budget* approach, this method provides quantitative information on water stored in the upper soil layers, on water surplus or recharge to the water table, and on streamflow. It also makes possible practical solutions to such problems as trafficability (vehicle and human movement over unpaved surfaces), irrigation scheduling, leaching rates of materials through soils, and the hydroclimatic effects of human alterations of the environment. New applications of the climatic water budget to environmental problems are continually emerging. Thornthwaite's water budget and his practical approaches to both basic and applied studies in climatology, as well as to the training of climatologists, were instrumental in shaping the development of the field of climatology after World War II. [See Evapotranspiration; Hydrological Cycle.]

Thornthwaite was born near Pinconning in central Michigan, the son of a farmer and harnessmaker. He worked his way through Central Michigan Normal School, graduating with a teacher's certificate in 1922, and began teaching high-school science. Summer course work in geography at the University of Michigan led him to pursue graduate study in the field; he entered the doctoral program in geography at the University of California, Berkeley, in 1923, receiving his Ph.D. in 1929. After completing his coursework in 1927, Thornthwaite had accepted a teaching position in a new geography program at the University of Oklahoma.

Thornthwaite immediately began work on what was to become one of his most significant contributions to climatology—the development of a new and more rational classification of world climates. His first version, published in 1931, attempted a new expression for the moisture factor in climate—that is, the relation between the supply of water by precipitation and the climatic demand for water. Thornthwaite was satisfied neither with the way earlier climatologists expressed the relative moistness of a climate nor with his own 1931 attempt. His concepts of potential evapotranspiration and a procedure to estimate it were first published in 1944; his climatic water budget research, first published in 1946, received international exposure in his seminal 1948 paper, "An Approach toward a Rational Classification of Climates."

Thornthwaite left Oklahoma in 1934 to enter government service. He became chief of the new Climatic and Physiographic Division of the Soil Conservation Service in the U.S. Department of Agriculture in 1935. He remained in that position until 1946, initiating a wide variety of research studies involving the influence of climate on soil erosion, measurement of evapotranspiration, evaluation of the water requirements of crops, improvement of scientific irrigation, and development of sensitive microclimatic instruments to obtain more detailed observations of the heat and moisture fluxes near the Earth's surface. In 1939, he and his associates undertook a year-long study of evapotranspiration from a grassed field in Arlington, Virginia; they determined the vertical flux of humidity near the ground and the horizontal movement of moisture from the area by the wind—a remarkable achievement in view of the limited capabilities of their instrumentation.

Thornthwaite took a leave of absence from government service in 1946 to explore opportunities as a consulting agricultural climatologist at Seabrook Farms, in southern New Jersey, a major grower and packager of frozen vegetables. During the next 2 years, he completed his work on the climatic water budget and produced his new climatic classification of 1948; he developed a significant improvement in the field of phenology (the relation of climate to periodic biologic activity), permitting the scheduling of planting and harvesting of crops to regulate the flow of fresh vegetables to the processing plant; he used his water budget approach to develop a simplified bookkeeping approach to scientific irrigation; he began work on a multiyear research study, "Microclimatology of the Surface Layers of the Atmosphere," that was supported by the U.S. Air Force; and he established the Laboratory of Climatology, associated with the Johns Hopkins University. With the founding of the laboratory, he was able to fulfill his long-expressed goal of creating a center for training students in climatology and undertaking significant basic and applied research. The success of these activities encouraged him to resign from government service and to begin a new career as a private consulting climatologist.

Scientists from all parts of the world came to visit the Laboratory of Climatology, to participate in its various research studies, and to lecture to students. During the 1950s, Thornthwaite published significant research reports on the movement of vehicles and humans over unpaved fields, the global water budget, the leaching of radioactive strontium through the soil, the measurement of vertical wind movement, and microclimatic conditions near Point Barrow, Alaska. Laboratory personnel under his direction developed a range of sensitive micrometeorologic instruments, including matched anemometers for profile studies, sensitive dew-point hygrometers, a net radiometer, soil heat flux transducers, and a vertical anemometer. Three patents were issued for instrument developments completed at the laboratory. He formed C. W. Thornthwaite Associates in 1952 to undertake consulting activities primarily concerned with disposal of waste effluent from food processing plants, agricultural climatology, hydroclimatic problems of nuclear reactors, and the sale of microclimatic instruments.

In 1951 Thornthwaite was elected the first president of the United Nations World Meteorologic Organization, Commission for Climatology, and reelected in 1953. He became perhaps the best-known climatologist of his day, and his reputation was international. Honors he received included the Cullum Medal of the American Geographical Society, the Outstanding Achievement Award from the Association of American Geographers (he was honorary president in 1960–1961), and an honorary doctorate from Central Michigan University. He died of cancer at the height of his career in 1963.

BIBLIOGRAPHY

Thornthwaite, C. W. "The Climates of North America According to a New Classification." *Geography Review* 21 (1931): 633–655.
Thornthwaite, C. W. "The Moisture Factor in Climate." *Transactions, American Geophysical Union* 27 (1946): 41–48.
Thornthwaite, C. W. "An Approach toward a Rational Classification of Climate." *Annals, Association of American Geographers* 38 (1948): 55–94.
Thornthwaite, C. W. "The Task Ahead." *Annals, Association of American Geographers* 51 (1961): 345–356.
Thornthwaite, C. W., and B. G. Holzman. *Measurement of Evapotranspiration from Land and Water Surfaces.* U.S. Department of Agriculture Technical Bulletin 817. Washington, D.C., 1942.
Thornthwaite, C. W., H. G. Wilm, et al. "Report of the Committee on Transpiration and Evapotranspiration 1943–44." *Transactions, American Geophysical Union* 26 (1945): 686–693.

JOHN R. MATHER

THUNDER. *See* Lightning.

THUNDERSTORMS. In some storms, convection produces lightning and thunder. Although storm electrification can occur in a variety of circumstances, almost all thunderstorms develop under atmospheric conditions of low static stability, with abundant heat and moisture at low levels. In such a situation, termed *conditional instability*, if low-level air is lifted by some atmospheric process, such as a front, to its lifting condensation level and beyond, it can reach a level (called the *level of free convection*) above which it becomes warmer than the air surrounding it. Like a hot air balloon, this air is positively buoyant and can continue to rise without further energy being supplied to it. Buoyancy continues to accelerate the air upward until once again it becomes cooler than its surroundings. The point where the rising air's temperature again equals that of its surroundings after having been positively buoyant is called the *equilibrium level*, above which its rising motion is slowed and eventually stopped by negative buoyancy. As the rising motion of the air is stopped, the air settles back down to the equilibrium level, spreading out as it does so. The vertical motion within a thunderstorm is depicted in Figure 1. [*See* Conditional and Convective Instability.]

The visible manifestation of this process is a *cumulonimbus cloud*, often called a *thunderhead*. [*See* Cumulonimbus Clouds.] Such clouds are not always visible to observers on the ground because they can be obscured by other clouds and atmospheric haze, but their presence

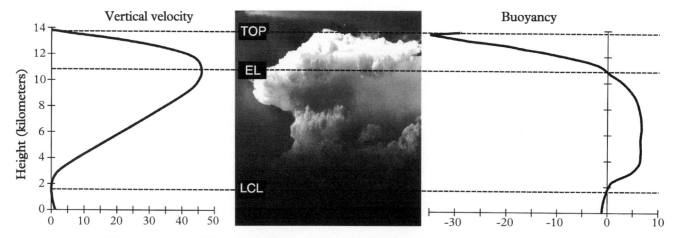

THUNDERSTORMS. Figure 1. *Schematic profiles of a thunderstorm.* Schematic vertical profiles of the vertical motion (left) and the buoyancy (right) associated with an actual thunderstorm (center). On the diagrams, the lifting condensation level is denoted by LCL and the equilibrium level by EL. Values shown for buoyancy and vertical motion are only qualitative and do not necessarily reflect the actual values for the thunderstorm shown.

is made evident by lightning and thunder. Because such clouds can be very deep, with tops as high as 16 to 20 kilometers, they absorb a large proportion of the sunlight, so an observer sees a dark cloud base. The crisply defined, cauliflower-like part of a visible cumulonimbus corresponds to the rising air, or *updraft*, within the thunderstorm. Cloud material accumulates at the equilibrium level, forming the spreading cumulonimbus *anvil.* As air rises through the equilibrium level before settling back into the anvil, it can form a cloud dome above the anvil level, called an *overshooting top.* The rise and fall of overshooting tops is associated with intermittent rising surges of low-level air. Each such surge constitutes a thunderstorm *cell*, and the typical thunderstorm is made up of several such cells.

The fact that thunderstorms are made up of cells was discovered during the Thunderstorm Project, carried out in Florida and Ohio shortly after World War II. The importance of thunderstorms to aviation had become especially apparent during the war, so a major research program was initiated under the direction of Horace R. Byers and Roscoe R. Braham. By using a newly developed observing tool, radar, as well as extensive surface and aircraft observations, the researchers documented the cellular nature of thunderstorms for the first time.

Thunderstorm cells have a distinct life cycle, illustrated in Figure 2. The cell begins with a *towering cumulus stage*, when the cell is dominated by updrafts and any precipitation that develops is retained aloft. As the precipitation falls to the Earth, its drag on the air and the cooling from

its evaporation combine to develop a downdraft that eventually reaches the ground and spreads out as an *outflow*, with a *gust front* at its leading edge. As this outflow develops, the cell is made up of both updrafts and downdrafts and the cell is in its *mature stage.* During the mature stage the cell attains its greatest strength, with the strongest updraft and downdraft, the heaviest precipitation, and the most frequent lightning. Finally, the cell's updraft diminishes, and it becomes dominated by downdraft as the remaining precipitation falls out. The cell has now entered the *dissipating stage*, and all the effects associated with it gradually wane. A typical cell completes its life cycle in about 20 to 40 minutes, corresponding to the time it takes a parcel of air to rise from near the surface to the storm top. At any given time, a thunderstorm usually contains several cells in different stages of their life cycles.

About 90 percent of all thunderstorms are benign, producing beneficial rains and few other effects. Some thunderstorms produce little or no rain, but their lightning can set destructive fires. Only about 10 percent of thunderstorms become severe, producing hazardous weather that may include large hailstones 2 or more centimeters in diameter, damaging surface winds of at least 25 meters per second, or tornadoes. Although not officially classified as severe events, large amounts of small hail, heavy rains that create flash floods, and frequent cloud-to-ground lightning flashes can also be serious hazards. Any thunderstorm can be quite dangerous to aviation, especially when aircraft encounter intense localized down-

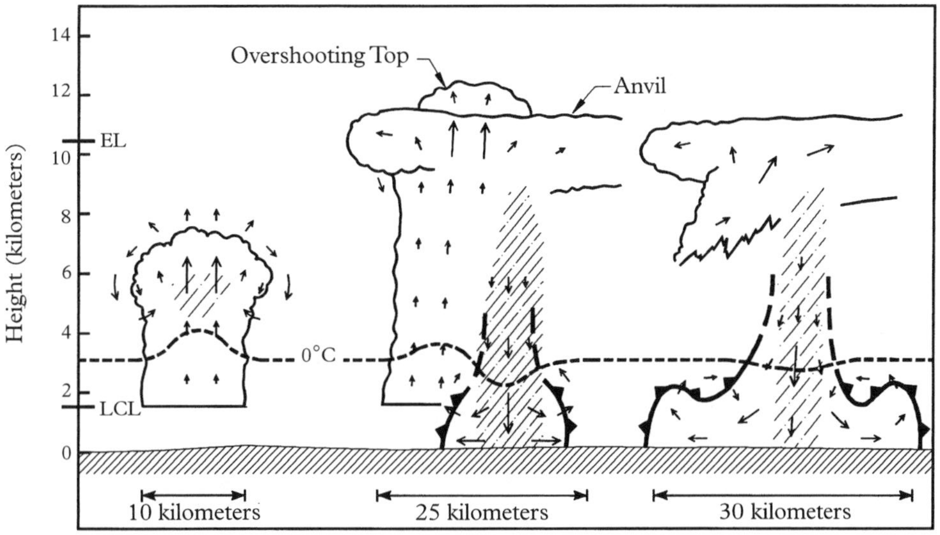

THUNDERSTORMS. Figure 2. *Stages in the life cycle of a single thunderstorm cell.* The figure shows air motions (arrows), precipitation (hatching), cloud outlines, and the freezing level (dashed line, labeled 0°C). From left to right: towering cumulus stage, mature stage, and dissipating stage. The gust front is denoted by a thick line with cold front symbols.

drafts called *microbursts* during take-off or landing. The occurrence of microbursts, even in relatively benign-appearing thunderstorms, has been recognized only recently, in part through the pioneering studies of T. Theodore Fujita. [*See* Aviation; Weather Hazards.]

The most characteristic property of thunderstorms is the lightning that produces the thunder. Lightning represents an electrical discharge between regions of different electrical charge—a giant spark. Surprisingly, relatively little is known for certain about how cumulonimbus clouds become electrified, but laboratory experiments suggest that air's breakdown voltage (the voltage necessary to allow the charged regions to equalize with a spark between them) is measured in millions of volts per meter. Most lightning flashes begin and end within the cloud, but some extend between the cloud and the ground. Of those reaching the ground, the great majority lower negative charge to the Earth, but a small fraction lower positive charge to the ground. It is not well understood why a few flashes bring down positive charge, but it appears that at least some severe storms have an unusually high proportion of positive flashes; thus this is an important topic of current research. Recently, cloud-to-ground lightning-flash detection networks have become widespread throughout the United States, and it is becoming possible to develop better knowledge of the distribution of thunderstorms. [*See* Atmospheric Electricity; Global Electric Circuit; Lightning.]

Present knowledge about the distribution of thunderstorms has been developed from surface observations, radar data, and satellite images. The implementation of new radars that include Doppler capability, permitting them to sense air motion in thunderstorms is an important recent development that should eventually lead to improved understanding and prediction of thunderstorms. [*See* Radar.]

Generally, thunderstorms are most frequent during the afternoon and early evening, when the Sun's radiation has warmed the low levels and the atmosphere's conditional instability is the greatest; however, over a large section of the central North American plains in the summer, thunderstorms are most frequent at night. Many of these nocturnal thunderstorms actually develop during the afternoon and persist well into the night, aided by a low-level poleward wind speed maximum that tends to develop soon after dark over the plains east of the Rocky Mountains. This low-level wind speed maximum brings warm, moist air into the thunderstorms, prolonging their existence into the nighttime hours and sometimes into the following morning.

Thunderstorms usually are more frequent and intense over continental land masses than over the open oceans. The main reason for the concentration of thunderstorms over land rather than water is that ground surfaces warm more rapidly than do most water surfaces, given equal amounts of sunshine. This is the result of the high heat capacity of water—that is, its temperature changes relatively slowly as heat is added. The heated land surface transfers its heat to the low-level air, producing conditional instability more readily than does a water surface. One exception to this occurs over warm ocean-surface currents such as the Gulf Stream. Another exception is associated with the frequent thunderstorms that develop in association with the Intertropical Convergence Zone, which lies generally near the equator. This atmospheric convergence zone lifts the warm, moist air of the tropics and produces thunderstorms over both land and sea. The thunderstorms that form over the ocean in this zone are usually in clusters associated with westward-moving disturbances. Occasionally, these disturbances can organize into tropical cyclones.

Just as many thunderstorms tend to follow the daily solar heating cycle, their overall timing reflects the seasonal cycles. In many parts of the United States, conditional instability is at its highest during the warmest part of the year, but during the summer, the lifting processes needed to initiate thunderstorms are often missing. The mechanisms that initiate thunderstorms are usually associated with the traveling weather systems that tend to follow the jet streams of middle latitudes, which move far poleward during the summer. Important exceptions to this general rule occur in Florida and southern Texas, where sea and land breeze fronts may initiate thunderstorms nearly every afternoon. The mountain-valley circulations in the western United States act similarly to set off nearly daily thunderstorms. This is why thunderstorm frequency maps show Florida and Colorado with the most frequent occurrences of thunderstorms, generally in the summer. Apart from these exceptions, thunderstorms are most common in late spring, when conditional instability is high and traveling weather systems are still relatively strong; these conditions are also important ingredients for creating severe thunderstorms. The peak in thunderstorm frequency tends to move poleward during the spring and, to a lesser extent, equatorward in the fall. In order to understand this asymmetry in the distribution of thunderstorms between spring and fall, the role of thunderstorms in the atmosphere must be considered.

As a manifestation of convection, a thunderstorm can be thought of as a giant heat engine, transporting latent and sensible heat upward to alleviate conditional and convective instability. The heated air at low levels rises in updrafts and is replaced by downdraft air. That is cool and relatively dry. Although the thunderstorm updraft does not cover a large area, its rising motion is on the order of 10 meters per second or more, much faster than the larger but weaker vertical motions of the large-scale traveling weather disturbance. Therefore, thunderstorms can rapidly process large amounts of air and carry aloft large amounts of moisture in a short time. A single updraft averaging 10 meters per second in vertical speed, and covering an area of 100 square kilometers (a circular updraft about 5.6 kilometers in radius), is lifting about 1 million kilograms of air per second. If that air has a water vapor content of 10 grams of water vapor per kilogram of air, that means a vertical transport of 10 million grams (or 10,000 kilograms) of water vapor per second. For the most part, the energy required to perform this feat is derived from latent heat release associated with the condensation of the transported water vapor.

We can get a feeling for how much energy this represents if we assume that this hypothetical storm lasts for 2,000 seconds (about 30 minutes), so that 20 million kilograms of water is processed. If all that water vapor condenses, this produces roughly 1.2×10^{13} calories of released latent heat. The explosion of 1 kiloton of TNT corresponds to about 10^{12} calories, so the thunderstorm releases heat roughly equivalent to a 12-kiloton bomb (about half the yield of the first atomic explosion). Of course, this comparison is somewhat misleading, because the thunderstorm spreads its energy release out over 2,000 seconds, whereas a bomb releases its heat in a tiny fraction of one second. Nonetheless, the comparison at least suggests the enormous energies involved in even a relatively weak thunderstorm. Severe storms can be substantially more energetic.

The net effect of thunderstorms in the overall heat balance of the planet is quite significant. They are important in transferring the heat received daily from the Sun upward into the atmosphere, especially during the warm season in each hemisphere, and in the tropics during much of the year. If thunderstorms act to transfer heat upward, thereby mitigating conditional and convective instability, then the asymmetry between spring and fall, mentioned earlier, can be easily explained. During the spring, the upper atmosphere tends to be relatively cold, in part because the increasing solar heat input at the surface has not yet been transferred upward. Thunderstorms play a major role in that transfer process. Solar heating warms the low levels increasingly during the spring, so, because the upper atmosphere is still relatively cold, conditional instability is high. In addition, the strong weather systems of spring produce conditions very favorable for thunderstorms, bringing moisture in at low levels and providing mechanisms to initiate thunderstorms. In the fall, however, thunderstorms have been operating during the spring and summer to reduce instability. Thus fall conditions are generally more stable than in spring and summer, and fall thunderstorms usually are correspondingly less frequent and intense. Naturally, exceptions to this occur occasionally—the atmosphere does not know anything about the calendar, and if sufficient conditional instability is produced, then thunderstorms will develop to alleviate it.

About 1,500 thunderstorms are occurring worldwide, on the average, at any given time. The equilibrium level in some thunderstorms (often the more severe ones) can be in the stratosphere, so a few thunderstorms deposit considerable amounts of water into the stratosphere in the form of ice. These stratospheric ice clouds can be an important factor in the atmosphere's radiation and heat balance. Not only are thunderstorms important in the heat and water balance of the planet; their lightning also acts to fix atmospheric nitrogen into forms that can be used by plants. The lightning also maintains the Earth's overall negative charge with respect to the atmosphere—the so-called *fair-weather electric field*—because the vast majority of cloud-to-ground lightning flashes lower negative charge to ground. On the whole, despite the damage they sometimes cause, thunderstorms are an important and mostly beneficial component of the processes sustaining life on the planet.

[*See also* Meteorology; *and* Storms.]

BIBLIOGRAPHY

Anthes, R. A., H. A. Panofsky, J. J. Cahir, and A. Rango. *The Atmosphere,* Columbus, Ohio, 1975.
Atkinson, B. W. *Meso-scale Atmospheric Circulations,* London, 1981.
Cotton, W. R. *Storms.* Fort Collins, Colo., 1990.
Johns, R. H., and C. A. Doswell III. "Severe Local Storms Forecasting." *Weather and Forecasting* 7 (1992): 588–612.
Kessler, E., ed. *The Thunderstorm in Human Affairs.* 2d ed. Norman, Okla., 1983.
Ludlam, F. H. *Clouds and Storms.* University Park, Pa., 1980.
Ray, P. S., ed. *Mesoscale Meteorology and Forecasting.* Boston, 1986.
Uman, M. A. *Understand Lightning.* Carnegie, Pa., 1971.

CHARLES A. DOSWELL III

TIDES. The term *tide* denotes the periodic motions of the ocean, atmosphere or solid Earth that are directly forced by the motions of the Earth and other astronomical bodies, notably the Sun and the Moon. These astronomical bodies can influence the Earth both by gravitational attraction (important for ocean tides) and by radiative heating (important for the solar atmospheric

tide). The tidal variations of ocean-surface height can be observed easily at almost any coastal location on the open ocean. The influence of the Moon and Sun on the ocean tides has been appreciated since classical antiquity. Sea level usually reaches its maximum (called *high tide*) and minimum *(low tide)* twice each lunar day, during the interval between one appearance of the Moon at its highest point in the sky and the next such appearance (on average, a lunar day is about 24.85 hours). The amplitude of the fluctuations in sea level between high and low tides at any location depends on the relative positions of the Sun, Earth and Moon. The largest tidal fluctuations *(spring tide)* occur when the three bodies are most nearly aligned.

The basic astronomical explanation for all these facts is straightforward. Figure 1 is a schematic picture of the orbital motions of the Moon and the Earth. From the terrestrial viewpoint, it appears that the Moon orbits around the Earth, but in fact, the Earth and the Moon both orbit about the center of mass of the sum of the two

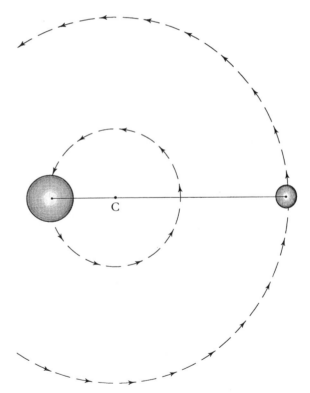

TIDES. Figure 1. *A view of the motions of the Earth and Moon, looking downward at the North Pole of the Earth.* Both the Moon and Earth orbit about the center of mass marked *C*. The geometry is depicted here in very schematic form: the actual diameters of the orbits of the Earth and Moon are much smaller than shown, and the orbits are depicted as circles, whereas in fact they are elliptical.

bodies (marked C in the figure). In the frame of reference of the Earth, there is a balance of the centrifugal force pushing outward from point C and the gravitational attraction of the Moon pulling toward point C. This balance is exact only at the center of the Earth. The centrifugal force increases with increasing distance from C, and the lunar gravitational attraction decreases with distance. Thus, on the side of the Earth closer to the Moon, the gravitational attraction is stronger than the centrifugal force, leading to a net attraction toward the Moon. On the opposite side of the Earth, the outward centrifugal force is stronger than the gravitational force. A fluid, such as the water in the ocean, might be expected to respond by bulging up on the surface of the Earth closest to the Moon and also on the exactly opposite surface.

These bulges resulting from the gravitational and centrifugal forces would result in a high tide twice each lunar day as the Earth rotates—once when the Moon is directly overhead *(lunar noon),* and once exactly half a lunar day later *(lunar midnight).* This theoretical prediction is called the *equilibrium tide.* The actual high tide is observed to occur at different lunar times depending on location; it generally occurs sometime after lunar noon and midnight. The imbalance of gravitational and centrifugal forces does indeed create a forcing that pushes the water upward twice each day. Rather than being static, however, the response is best visualized as a global-scale ocean wave propagating around the world. The wave is affected by friction at the ocean floor and by its encounter with coastlines. These effects lead to significant geographical variation in the observed tidal phases and amplitudes.

The same mechanism is applicable to the effects of the solar gravitational field. We would expect two high/low tide cycles per solar day (24 hours). The actual gravitational tidal force of the Sun turns out to be, on average, only about 47 percent of that of the Moon. This explains the dominance of lunar tides in ocean height measurements. The solar tides are not negligible, however, and when the Sun, Earth, and Moon are most nearly aligned (close to new moon or full moon), the two gravitational fields act together, and we see the greatest difference between high and low tides. Thus, the spring tide is observed twice each lunar month.

Lunar and solar gravitational effects also influence the atmosphere. The simplest way to observe these effects is with barometric measurements at the surface. The lunar *atmospheric tide* is much too small to see in a simple plot of the pressure through a single day. This tide can be detected by averaging long periods of hourly pressure data segregated by local lunar time. For example, a long-term average of the pressure at times when the Moon is near its zenith differs from an average over times when the Moon is on the horizon. Statistical analysis of long records of barometric observations has been carried out at many locations throughout the world. The greatest peak-to-peak amplitude of the twice-daily lunar fluctuations is 20 pascals. This maximum is observed near the equator and drops rapidly with increasing latitude.

The most striking aspect of the tidal oscillations in barometric data is the presence of a very strong solar oscillation. Near the equator, the peak-to-peak amplitude of the twice-daily (or *semidiurnal*) solar oscillation is nearly 300 pascals. This is strong enough that, at low latitudes, a simple plot of observed pressure at hourly intervals is almost always dominated by this solar semidiurnal oscillation. An example is shown in Figure 2, a plot of hourly barometric measurements made at a station in Indonesia over 3 consecutive days. The pressure tends to peak around 9:00 or 10:00, both morning and night. There is also a considerably smaller 24-hour component apparent in the time series (note that the pressure minimum in the afternoon is deeper than the one in the morning). The actual pattern is not perfectly repeated each day, of course, because the surface pressure is also affected by other weather events.

The solar semidiurnal tide in the atmosphere is roughly 15 times stronger than the lunar semidiurnal tide, even though the gravitational tidal excitation by the Sun is less than half that by the Moon. The dominance of solar over

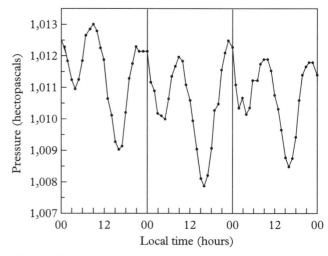

TIDES. Figure 2. *The surface atmospheric pressure measured at Jakarta, Indonesia (6° south latitude, 107° east longitude).* Results are for each hour during 3 consecutive days. The label 00 indicates local midnight, and 12 indicates local noon.

lunar oscillations is a reflection of the strong solar heating influence on the atmosphere, which has no lunar counterpart. The solar tides are almost entirely forced by the day-to-night contrast in heating rather than by gravitational effects. Therefore, the solar tides in the atmosphere are often referred to as *thermotidal oscillations.*

Although there is an obvious astronomical explanation for the twice-daily character of the gravitational tides, the dominance of the 12-hour period for the thermotidal oscillation in barometric pressure was not understood until the late 1960s. As one would expect, the atmospheric heating has a dominant 24-hour (or *solar diurnal*) period, but there are also higher-frequency components present. This is illustrated in Figure 3, which shows a typical curve of solar heating through the day, along with the mean, diurnal, and semidiurnal components of the total. Typically, the amplitude of the semidiurnal component of heating is less than one-fourth that of the diurnal heating. Thus, it is indeed surprising that the solar semidiurnal barometric oscillation is about 2 to 3 times as strong as the diurnal oscillation (at least at low latitudes). The explanation for this lies in the details of the dynamical response of the atmosphere to periodic heating. Just as in the oceanic case, one could imagine a kind of equilibrium response to the heating, with the atmosphere simply becoming warmer on the sunlit side and cooler on the night side. As in the ocean, however, the

actual response is really a dynamic, propagating global wave which involves changes in pressure, temperature, and wind. In the atmosphere, this wave can propagate vertically as well as horizontally. The vertical scale of solar atmospheric heating is well suited to excite the semidiurnal oscillation, and much less efficient in exciting the diurnal oscillation. The Coriolis effect also contributes to the reduced response seen at the surface to the diurnal atmospheric thermal forcing. It turns out that vertical propagation is inhibited strongly when the wave frequency is less than the Coriolis parameter ($2\Omega\cos\theta$, where Ω is the frequency of the Earth's rotation and θ is the latitude). For the diurnal tide this occurs over half the globe (everywhere poleward of 30° latitude), while for the semidiurnal tide this condition is nowhere met. [*See* Coriolis Force.]

Both the 24-hour and 12-hour thermotidal oscillations in barometric pressure are also present outside the tropics, but the amplitude drops with latitude. At a typical location at 45° north or south latitude, the peak-to-peak amplitude of the semidiurnal tidal oscillation is about 80 pascals, while that for the diurnal oscillation is about 50 pascals. By contrast, the strength of pressure fluctuations associated with other weather disturbances rises rapidly with latitude (at 45° latitude, to more than 1,000 pascals). Thus, outside the tropics, the atmospheric tides cannot usually be observed by simple inspection of barometric time series such as that in Figure 2, but they can be detected by statistical analysis of long records of observations—for example, by averaging the pressures observed at each hour of the day for several months.

Although the solar tides are a significant feature of the surface pressure signal in the tropics, they are less important for the wind and other aspects of weather in the lower atmosphere. In fact, tidal winds near the surface are generally less than 1 meter per second. This very small wind signal results from the global scale of the tidal pressure signal, because the forcing of wind motions depends on the horizontal gradient of the pressure. The daily oscillation in winds forced by very small horizontal-scale differences in heating—the source of, for example, land-sea or mountain-valley winds—generally dominates the global tidal oscillation in surface winds.

The situation is very different in the upper atmosphere. The atmospheric tides can propagate vertically, and as the tidal signal reaches higher altitudes, the magnitude of the associated wind fluctuations increases. This can be regarded as a consequence of the conservation of energy: as the atmospheric density drops with height, the strength of the tidal wind oscillations must increase to compen-

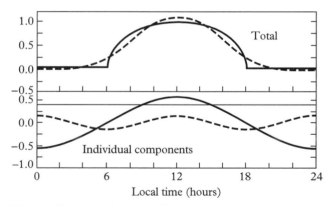

TIDES. Figure 3. *Solar heating.* The solid curve in the top panel is a schematic of the solar heating of the atmosphere through the day. The bottom panel shows the daily mean (light line), the diurnal component (heavy solid curve), and the semidiurnal component (dashed curve). The sum of these three components is shown as the dashed curve in the top panel. The good agreement between the actual total heating and the sum of components demonstrates that the heating is reasonably well represented by these three contributions. The agreement can be made even better by including additional higher-frequency components in the sum.

sate. The peak-to-peak amplitude of the diurnal and semidiurnal wind fluctuations rises to of the order of 10 meters per second at 50 kilometers height and can exceed 50 meters per second at 100 kilometers. The tides are therefore a very prominent aspect of the meteorology of the upper atmosphere.

[*See also* Moon.]

BIBLIOGRAPHY

Chapman, S., and R. S. Lindzen. *Atmospheric Tides*. New York, 1970.

Forbes, J. M. "Middle Atmosphere Tides." *Journal of Atmospheric and Terrestrial Physics* 46 (1984): 1049–1067.

Gross, M. G. *Oceanography: A View of the Earth*. 3d ed. Englewood Cliffs, N.J., 1982.

LeBlond, P. H., and L. A. Mysak. *Waves in the Ocean*. New York, 1978.

KEVIN HAMILTON

TILT AXIS. *See* Orbital Parameters and Equations.

TORNADOES. The word *tornado* is of Spanish origin and is related to English *turn*. Rapidly rotating vertical columns of air called tornadoes are pendant from cumulonimbus clouds and make contact with the ground (Figure 1). Tornadoes that occur over water, called *waterspouts*, are more frequent than those over land. Wind speeds near the ground in tornadoes may rarely reach 120 to 135 meters per second (265 to 300 miles per hour). Most tornadoes, however, are weak. The intensity of tornadoes is often determined on the basis of the damage inflicted using the Fujita scale, which ranges from F0 (18 to 32 meters per second or 40 to 72 miles per hour), characterized by light damage, to F5 (117 to 142 meters per second or 261 to 318 miles per hour), in which strong frame houses are blown away from their foundations. Tornadoes have killed many people and have completely destroyed portions of towns; the death toll was 271 in the Palm Sunday outbreak of 11 April 1965 in the midwestern United States. The largest tornado outbreak occurred on 3–4 April 1974, when more than 100 tornadoes struck parts of the central United States. The direction of rotation in tornadoes is usually cyclonic—the same direction as the rotation of the Earth about its axis (counterclockwise in the Northern Hemisphere); a few tornadoes, however, exhibit anticyclonic rotation.

Although they have been observed all over the world, and in all 50 states of the United States, at all hours of

TORNADOES. Figure 1. *Tornado near Hodges, Texas, during the early evening hours of 13 May 1989.* (Photo copyright Howard B. Bluestein.)

day and night and in all types of terrain, tornadoes are most frequent in the central section of the United States in the late afternoon during the spring. The damage path of tornadoes may be as wide as 3 kilometers and as long as 300 kilometers; most damage paths are only about 100 meters wide or less and less than 15 kilometers.

Tornadoes are visible as a rotating column of debris near the ground and by a condensation funnel aloft. However, a tornado may be inflicting damage on the ground even if a funnel cloud is not visible. Sometimes tornadoes are hidden by precipitation. Many first become visible as a rotating column of debris at the ground; a condensation funnel then forms, and appears to extend down to the ground. During the dissipation of a tornado, the condensation funnel may become narrow, contorted, and elongated, like a rope, and lean over so much that it appears to be nearly horizontal.

Most tornadoes consist of only one vertical column of rotating air, but some have two or more smaller vortices rotating around a common center. These smaller satellite vortices are called *suction vortices*. Multiple-vortex tornadoes often produce cycloidal swaths of damage, which move along with and rotate around the vortex.

There are several different types of tornadoes. The largest and most damaging are spawned in long-lived (longer than an hour), rotating thunderstorms called *supercells*. Because of the storm's duration the path associated with such a tornado may be very long. Tornadoes in supercell storms often go through a life cycle of 20 to 30 minutes, reappearing again and again along the path of the parent storm. During major tornado outbreaks, a line (or family) of tornadic supercells may produce parallel damage paths. Tornadoes in supercells usually form underneath a cyclonically rotating *wall cloud*, the lowered, precipitation-free cloud base that marks the ingestion of relatively humid, evaporatively cooled air into the main updraft of the thunderstorm; wall clouds are often found on the rear edge of the moving storm. Anticyclonically rotating tornadoes have been documented along a storm's *gust front*, the boundary between the evaporatively cooled downdraft of the storm and the warm moist air feeding the storm's updraft; cyclonic tornadoes may occur simultaneously under the same storm's wall cloud.

Tornadoes can also occur in ordinary, nonsupercell thunderstorms. Those that occur along the storm's gust front are not usually associated with condensation funnels. Sometimes called *gustnadoes*, they are short-lived and do not usually produce major damage.

Nonsupercell thunderstorms also can produce tornadoes during the developing stage of a convective cell; these tornadoes may be associated with relatively weak thunderstorms and do not last longer than 30 or 40 minutes. Because many nonsupercell tornadoes form along lines of growing cumulonimbus clouds and look like the waterspouts off the coast of Florida and along the Gulf coast, they are sometimes called *landspouts*. On rare occasion tornadoes have been reported in frontal rainbands. Short-lived tornadoes and funnel clouds are also seen in nonsupercell storms in the vicinity of relatively cold, upper-level troughs of low pressure; these vortices have been referred to as *cold-air funnels*.

When a hurricane makes landfall, tornadoes often occur in the hurricane's right-front quadrant. The parent storms sometimes have supercell characteristics.

Observations and Theory. Tornadoes are very difficult to study because they are small, short-lived, and difficult to predict. In the past, much of what we knew about tornadoes came from relatively few serendipitous observations by untrained people. In the early 1970s, however, scientists began to "chase" tornadoes, substantially increasing the number of visual observations. It was found that in supercell thunderstorms, tornadoes appear in specific locations, and that tornadoes do occur under other circumstances. This information was very useful for training severe-weather spotters.

Wind speeds may be estimated by photogrammetrically analyzing debris in movies taken of tornadoes; items of debris are tracked from frame to frame and are assumed to move along with the wind. Attempts have been made to obtain direct measurements of wind, pressure, and temperature by placing a device called TOTO (Totable Tornado Observatory) directly in the path of tornadoes. This proved to be difficult because tornadoes usually do not last long, and it is difficult to maneuver into a position directly in their path.

The wind associated with tornadoes and their parent storms has been measured more safely, and more successfully, using a remote Doppler radar. A Doppler radar, which uses precipitation and debris to backscatter the radar beam, measures the component of wind speed along the line of sight to the tornado. Although it is usually difficult to detect the tornado's circulation itself, because it is so small, the Doppler radar can detect the tornado's parent circulation in supercell storms, called the *mesocyclone*, which is approximately 10 to 15 kilometers across. The range of wind speeds in tornadoes has been measured remotely by Doppler radars. Although the first Doppler-radar measurements were made in 1958, it was not until the 1970s that more measurements became available in the southern plains of the United States through the efforts of the National Severe Storms Laboratory in Norman, Oklahoma. A portable Doppler radar from the Los Alamos National Laboratory in New Mexico has been used at the University of Oklahoma recently to measure the wind speeds in tornadoes from within visual range. The highest wind speeds measured to date by this technique were 120 to 125 meters per second. A high-frequency Doppler radar from the University of Massachusetts at Amherst is currently being used in a University of Oklahoma van to map out the wind field in tornadoes. In the future a Doppler laser radar, called a Doppler *lidar*, also may be used to determine the fine-scale structure of tornadoes. [*See* Radar.]

Because observational data on tornadoes have been so difficult to obtain, scientists have made laboratory models of tornadoes in rotating tanks and vortex chambers. The vortex chambers force air to swirl around, and the vortex is made visible by injecting dry ice or smoke into the chamber. There have also been computer simulations of tornado-like vortices using the equations of motion based on Newton's second law of motion applied to a fluid. Both the computer (or "numerical") simulations and the laboratory models have been used to reproduce under controlled conditions some of the phenomena observed in real tornadoes. For example, laboratory models and numerical simulations have defined the conditions under which a single vortex can degenerate into multiple vortices.

Data from real and simulated tornadoes suggest that the tornado vortex rotates like a solid body; that is, the azimuthal velocity is proportional to the distance from the center of the tornado. Outside the tornado, the azimuthal velocity drops off very rapidly. At the ground, air converges rapidly into the tornado. Under certain conditions, the air may actually descend at the center of the tornado and ascend away from the center. Although the pressure falls rapidly inside a tornado, it has been found that the damage inflicted by a tornado is not due to this rapid pressure change but rather to the wind stress, which is proportional to the square of the wind speed. Since tornadoes are so small and the wind speeds are so great, the outward-directed centrifugal force is much greater than the inward-directed pressure-gradient force; consequently, debris picked up by a tornado is often thrown out of the vortex. The vertical wind speeds may be as high as 80 meters per second (175 miles per hour). Pressure falls are at least 20 millibars and may be as high as 100 millibars, equal to 10 per cent of the total ambient pressure at the ground. Little is known about the temperature in tornadoes.

The maximum possible wind speed in tornadoes has long been a source of controversy. At one time it was thought that wind speeds as high as the speed of sound were possible. If the pressure drop in a tornado is due to the release of the latent heat of condensation in the thun-

TORNADOES. Figure 2. *Tornado in north central Oklahoma, north of Enid, late in the afternoon of 12 April 1991.* Portable Doppler radar recorded wind speeds as high as 80 meters per second (179 miles per hour). (Photo copyright Howard B. Bluestein.)

derstorm overhead, then the maximum wind speed possible depends significantly on whether the tornado circulation has descending air at the center and rising air farther outward, or ascending air at the center. It has been suggested that the interaction between the tornado circulation and the ground may lead to even higher wind speeds than those possible if the pressure drop associated with the tornado were due to heating alone. Although some think heating from lightning discharges may be important, tornadoes are often observed with no lightning in or adjacent to them.

The exact source of rotation in tornadoes is not known. Even though most tornadoes rotate cyclonically, the source of this rotation is not the Earth's rotation about its axis. Some possible sources of tornado rotation or *vorticity* (circulation per unit area) include vorticity about a horizontal axis induced along the edge of evaporatively cooled surface boundaries such as gust fronts; vorticity about a horizontal axis associated with vertical shear; and vorticity about a vertical axis associated with horizontal shear (lateral changes in wind speed). Vorticity about a horizontal axis can be tilted onto the vertical axis when there are narrow updrafts or adjacent updrafts and downdrafts. The convergence of air acts to increase the ambient vertical vorticity.

Forecasting and Detection. Systematic attempts to forecast tornadoes began in Oklahoma in the late 1940s after a tornado hit an air force base. Since then, weather forecasters have learned to recognize the large-scale conditions that precede tornadoes. Supercells, which spawn tornadoes, form in the presence of strong vertical wind shear and of a rapid increase in wind speed and/or change in wind direction in the lowest 6 kilometers of the atmosphere. Vertical wind shear may be strong in the vicinity of fronts, and the jet stream. Moderate to large amounts of buoyant energy, which are present when the temperature decreases relatively rapidly with height, are also required. Low-level moisture is needed to fuel the parent thunderstorm, and daytime heating from the sun or lift along boundaries—such as fronts, the dryline, or outflow boundaries—is necessary to initiate the storm. The *dryline* is the boundary between relatively moist maritime air and dry, warm continental air, found in the plains of the United States during the spring. The *outflow boundary* is the location where evaporatively cooled air in thunderstorms meets warm, moist air. [See Thunderstorms.]

Tornadoes in supercell thunderstorms are sometimes indicated by a hook-like configuration in the radar echo of the storm; the hook is usually associated with a mesocyclone and with the cyclonic wrapping of precipitation particles around the circulation. A Doppler radar can detect the mesocyclone of a supercell from a range of 160 kilometers or more. Owing to their rotation, supercells in the Northern Hemisphere usually follow a path that is to the right to the left in the Southern Hemisphere of the mean wind in the troposphere, while ordinary thunderstorms tend to move more closely in the same direction as the mean wind. Since tornadoes often occur after a mesocyclone has been detected, warnings can be issued as much as 20 or 30 minutes before a tornado occurs. Tornadoes in nonsupercell thunderstorms cannot be detected by radar unless they are very close to the radar, so these probably require a spotter network for reliable warnings.

The National Severe Storms Forecast Center in Kansas City, Missouri, issues convective outlooks for the United States several times a day. These outlooks give forecasts of the possibilities of severe thunderstorms over broad areas, covering many states, as much as 24 hours in advance. Later, *tornado watches* may be issued for approximately 6- to 8-hour periods if there is the possibility of tornadoes occurring within certain portions of states. If a tornado watch is issued for an area, residents should be alert and monitor their local television station, radar, or NOAA weather radio. If a conventional radar indicates a hook echo, if a Doppler radar shows a persistent mesocyclone "signature" throughout a substantial depth of the troposphere, or if a spotter or any other person reports a tornado, then the local National Weather Service office issues a tomado warning for counties in the projected path for approximately the next 30 to 45 minutes. When a tornado warning is issued, affected residents should seek shelter in basements or in small interior rooms, protecting their heads from flying debris.

With the advent of a new national network of Doppler radars in the mid-1990s in the United States, warnings of supercell tornadoes should be more accurate and timely. Short-term forecasting should improve because the radars can operate in clear air, and they can give meteorologists information on the vertical wind shear with much better time resolution than is possible with the present national network of twice-daily reports, called *rawinsonde data*. Doppler-radar wind profilers can yield similar information, which will be helpful to forecasters.

[See also Religion and Weather.]

BIBLIOGRAPHY

Bluestein, H. B. "Precipitation Systems in the Midlatitudes." In *Synoptic Dynamic Meteorology in Midlatitudes, vol. 2, Observa-*

tions and Theory of Weather Systems. New York, 1992, pp. 426–577.

Church, C., D. Burgess, C. Doswell, and R. Davies-Jones, eds. "The Tornado: Its Structure, Dynamics, Prediction, and Hazards." American Geophysical Union, Geophysical Monograph 79. Washington, D.C., 1993.

Cotton, W. R., and R. A. Anthes. "Cumulonimbus Clouds and Severe Convective Storms." In Storm and Cloud Dynamics. San Diego, Calif, 1989.

Davies-Jones, R. P., "Tornado Dynamics." In Thunderstorm Morphology and Dynamics, edited by E. Kessler. Norman, Okla., 1986, pp. 197–236.

Flora, S. D. Tornadoes of the United States Norman, Okla., 1954.

HOWARD B. BLUESTEIN

TORRICELLI, EVANGELISTA (1608–1647), Italian mathematician and scientist. Torricelli's discovery of the barometer (1643) gave seventeenth-century scientists a standardized instrument by which they could measure and monitor the weather accurately.

As a young student of physics and mathematics, Torricelli was intrigued by Galileo's writings. Galileo was in turn impressed by Torricelli's treatise on mechanics and invited him to Florence so they could work together. When Galileo died, Torricelli was appointed to take his place as mathematician and philosopher to the Medicis, the ruling family of Florence and Tuscany. The Grand Duke of Tuscany, Ferdinand II, was an influential patron of science who commissioned his master glassblower to manufacture dozens of thermometers in an effort to register changes in temperature identically.

Torricelli began conducting experiments on temperature and atmospheric density. He reasoned that air had weight, and he wrote at the time that, "We live submerged at the bottom of an ocean of elementary air." He developed new devices for further study, one of which was the mercury barometer—a mercury-filled glass tube, sealed at the top and inverted into a dish of mercury. He observed that the mercury level in the tube changed from day to day, which indicated to him that the air pressing on the cistern of mercury was heavier at certain times and lighter at others. He believed that the weight of the air pressed down on the mercury in the cistern, forcing it up into the tube. [See Instrumentation.]

Scholars of the time knew that the mercury usually stood at its greatest height in good weather, and at its lowest on rainy days. But they did not understand the reasons for the difference, concluding wrongly that air was heaviest on rainy days. Torricelli, however, experimented with progressively heavier liquids, working his way through sea water and honey to mercury. He used a tube that was 48 inches long and closed at one end. When it was filled with mercury and inverted in a dish of mercury, the mercury in the tube leaked out to only a height of 29 inches, leaving a vacuum in the upper end of the tube. Torricelli rightly concluded that the force maintaining the mercury was not an internal suctional "force of the vacuum" in the tube but the weight of the external air pressing on the mercury in the bowl. The device was called the *Torricelli tube* until the British scientist Robert Boyle named it *barometer* in 1667.

Torricelli wrote several comedies but published only one scientific work, *Opera Geometrica* (1644), on diffused geometry. He also studied the parabolic motion of projectiles, which Galileo had pioneered, and eventually converted Galileo's primitive air thermoscope (1592) to a liquid thermometer. He is also known for his machining techniques in producing telescope lenses.

BIBLIOGRAPHY

Gillispie, C. C., ed. Dictionary of Scientific Biography. Vol. 13. New York, 1976.

Schneider, S. H., and R. Londer. The Coevolution of Climate and Life. San Francisco, 1984.

DIANE MANUEL

TOURISM. *See* Recreation and Climate.

TRACE GASES. The bulk of the Earth's dry atmosphere—more than 99.9 percent—is composed of molecular nitrogen (N_2, 78.09 percent), molecular oxygen (O_2, 20.95 percent), and argon (Ar, 0.93 percent). All other constituent gases comprise only 0.03 percent. Despite their low concentrations, many of the other gases in the atmosphere, called the *trace gases,* play extremely important roles in the climate and habitability of our planet. Trace gases determine many of the physical and chemical characteristics of the Earth's atmosphere. Table 1 gives a list of many of the more important trace gases.

A number of trace gases are important to the radiative balance of the planet. The trace gases known as *greenhouse gases,* through their absorption and emission of the Earth's infrared radiation, contribute greatly to determining the average temperature of the planet. The primary greenhouse gases in the atmosphere are water vapor (H_2O), carbon dioxide (CO_2), ozone (O_3), methane

TRACE GASES. Table 1. *Some important trace gases that influence the Earth's atmosphere*

Trace Gas	Common Name	Importance to Atmosphere
CO_2	Carbon dioxide	Absorbs infrared (IR) radiation; affects stratospheric temperatures and ozone (O_3)
CH_4	Methane	Absorbs infrared radiation; affects tropospheric O_3 and OH; affects stratospheric O_3 and H_2O; produces CO_2
CO	Carbon monoxide	Affects tropospheric O_3 and OH cycles; produces CO_2
N_2O	Nitrous oxide	Absorbs infrared radiation; source of stratospheric nitrogen oxides; affects stratospheric O_3
NO_x $(=NO+NO_2)$	Nitrogen oxides	Affects O_3 and OH cycles; precursor of acidic nitrates (e.g., HNO_3)
$CFCl_3$	CFC-11	Absorbs infrared radiation; source of stratospheric chlorine; affects stratospheric O_3
CF_2Cl_2	CFC-12	Absorbs infrared radiation; source of stratospheric chlorine; affects stratospheric O_3
$C_2F_3Cl_3$	CFC-113	Absorbs infrared radiation; source of stratospheric chlorine; affects stratospheric O_3
C_2F_5Cl	CFC-115	Absorbs infrared radiation; source of stratospheric chlorine; affect stratospheric O_3
CHF_2Cl	HCFC-22	Absorbs infrared radiation; source of stratospheric reactive chlorine; affects stratospheric O_3
CCl_4	Carbon tetrachloride	Absorbs infrared radiation; source of stratospheric reactive chlorine; affects stratospheric O_3
CH_3CCl_3	Methylchloroform	Absorbs infrared radiation; source of stratospheric reactive chlorine; affects stratospheric O_3
CF_2ClBr	Ha-1211	Absorbs infrared radiation; source of stratospheric reactive bromine; affects stratospheric O_3
CF_3Br	Ha-1301	Absorbs infrared radiation; source of stratospheric reactive bromine; affects stratospheric O_3
SO_2	Sulfur dioxide	Forms aerosols, which scatter solar radiation and affect cloud properties
COS	Carbonyl sulfide	Forms aerosol in stratosphere which alters albedo
$(CH_3)_2S$	Dimethyl sulfide	Produces cloud condensation nuclei, affecting cloudiness and albedo
C_2H_2, etc.	Non-methane Hydrocarbons	Absorb infrared radiation; affect tropospheric O_3 and OH
O_3	Ozone	Absorbs ultraviolet (UV), visible and infrared radiation
OH	Hydroxyl	Important oxidizer; scavenger for many atmospheric pollutants, including CH_4, CO, CH_3CCl_3, and CHF_2Cl; affect stropospheric and stratospheric O_3
H_2O	Water vapor	Absorbs near-infrared and infrared radiation; important to production of OH

(CH_4), nitrous oxide (N_2O), and various halocarbons (particularly the long-lived chlorine- or bromine-containing compounds such as the chlorofluorocarbons, the CFCs). The atmospheric concentrations of many of these gases are presently changing, a process strongly influenced by human activities. Some greenhouse gases in the atmosphere, such as the chlorofluorocarbons, arise almost entirely from human industry. [*See* Greenhouse Gases; Radiation.]

Other trace gases, such as sulfur dioxide (SO_2) and hydrogen sulfide (H_2S), can affect climate through their roles as precursors to the production of aerosols (liquid droplets or dust particles) in the atmosphere. The resulting aerosols scatter solar radiation, preventing some of it from reaching the Earth's surface. They also serve as condensation nuclei for the production of water droplets in clouds.

Ozone is one of the most important trace gases. There has been much concern in recent years about effects on stratospheric ozone concentrations resulting from human-related emissions of several other trace gases. The primary reason for this concern has centered on the importance of ozone as an absorber of solar ultraviolet (UV) radiation. Because of this absorption by ozone, most of the biologically harmful UV radiation emitted by the Sun never reaches the Earth's surface. Absorption of solar radiation by ozone also explains the increase in temperature, with altitude in the stratosphere. Finally, ozone is an important greenhouse gas and can influence climate. [*See* Ozone.]

By far the largest amount of ozone (about 90 percent) is in the stratosphere. The primary source of stratospheric ozone is the dissociation of molecular oxygen (O_2), which results in atomic oxygen (O). The oxygen atoms then react with O_2 to from ozone. Destruction of ozone in the stratosphere occurs primarily by catalytic chemical reactions involving trace gases such as nitrogen oxides, chlorine oxides, and hydrogen oxides. In the troposphere, ozone can be both produced and destroyed by reactions involving trace gases.

Of particular interest in regard to stratospheric ozone destruction are the effects resulting from chlorine- and bromine-containing trace gases (such as chlorine monoxide, ClO, and bromine monoxide, BrO), which are produced in the stratosphere by the dissociation of CFC-11 ($CFCl_3$), CFC-12 (CF_2Cl_2), and other halocarbons. Industrial halocarbons are responsible for most of the reactive chlorine and bromine in the stratosphere. Note that although there are very large natural sources of atmospheric chlorine, only the ocean-produced methyl chloride (CH_3Cl) has a sufficiently long atmospheric residence time to contribute significantly to reactive chlorine in the stratosphere. Very explosive volcanic eruptions with significant amounts of hydrochloric acid (HCl) can also contribute to stratospheric chlorine for a few years following the eruption. [*See* Volcanoes]

Several trace gases in the troposphere, in particular hydroxyl (OH), ozone, and hydrogen peroxide (H_2O_2), are important in destroying many important trace gases emitted into the atmosphere from the Earth's surface. This self-cleansing feature of the atmosphere is called the *oxidizing capacity*. The most important oxidant is

hydroxyl; the OH radical has been referred to as the "tropospheric vacuum cleaner" because it reacts with hundreds of gases in the atmosphere, including many pollutants which are transformed to compounds more readily removed from the atmosphere. Hydroxyl is the primary scavenger for such important trace gases as atmospheric methane, most of the higher hydrocarbons (referred to as *nonmethane hydrocarbons*), carbon monoxide (CO), the hydrofluorocarbons (such as CHF_2Cl), methylchloroform (CH_3CCl_3), methyl bromide (CH_3Br), sulfur dioxide (SO_2), and hydrogen sulfide (H_2S). Therefore, the atmospheric concentrations of hydroxyl determine the atmospheric lifetimes and hence the abundances of these trace gases. Hydroxyl is generated primarily by the interactions of water vapor, ozone, and solar ultraviolet radiation. The amount of tropospheric hydroxyl may be changing in the current atmosphere as a result of changing levels of ozone and UV radiation, and increasing emissions of the gases it scavenges. However, the tropospheric concentration of the hydroxyl radical has not been well characterized because of difficulties inherent in measuring this trace constituent.

A number of other trace gases affect the atmosphere in important ways. For example, methane is a greenhouse gas, but it also affects the chemistry of the global atmosphere as an important reactant in determining the amount of hydroxyl and ozone in both the troposphere and the stratosphere. The oxidation of methane in the stratosphere, a significant source of water vapor, explains the increase in water vapor with altitude in the stratosphere. Methane also affects the reactivity of chlorine gases in the stratosphere.

The atmospheric concentrations of many trace gases are changing. The concentration of carbon dioxide is increasing 0.4 percent per year; it has increased by more than 25 percent since the beginning of the Industrial Revolution in the eighteenth century. Methane concentrations have more than doubled over that time period, with current concentrations increasing at a little less than 1 percent per year. Nitrous oxide (N_2O) concentrations have been increasing at about 0.3 percent per year for at least several decades. Although industrial production of chlorofluorocarbons is planned to cease within the next few years, their atmospheric concentrations have increased dramatically over the past few decades. Human-related emissions of sulfur dioxide appear to have increased substantially over the past century. Ozone concentrations in the stratosphere are decreasing, while tropospheric concentrations appear to be increasing;

however, the latter is less supported by measurements than is the former, particularly the stratospheric ozone decreases at high latitudes in the late winter and early spring.

[*See also* Ozone Hole; Pollution.]

BIBLIOGRAPHY

Graedel, T. E. *Chemical Compounds in the Atmosphere.* New York, 1978.

Hewitt, C., and W. T. Sturges. *Global Atmospheric Chemical Change.* Essex, 1992.

Kaye, J., and M. McFarland. "Atmospheric Ozone." In *Encyclopedia of Earth System Science,* edited by W. A. Nierenberg. New York, 1992, vol. 1, pp. 237–260.

Khalil, M. A. K. "Atmospheric Trace Gases, Anthropogenic Influences and Global Change." In *Encyclopedia of Earth System Science,* edited by W. A. Nierenberg. New York, 1992, vol. 1, pp. 285–305.

Moore, B., ed. *Trace Gases and the Biosphere.* Tucson, Ariz., 1989.

Ramanathan, V., et al. "Chemical-Climate Interactions and Effects of Changing Atmospheric Trace Gases." *Reviews of Geophysics* 25 (1987): 1441–1482.

Wuebbles, D. J., and J. Edmonds. *Primer on Greenhouse Gases.* Chelsea, Mich., 1991.

Wuebbles, D. J., and J. S. Tamaresis. "The Role of Methane in the Global Environment." In *Atmospheric Methane,* edited by M. A. K. Khalil. New York, 1992, pp. 469–513.

DONALD J. WUEBBLES

TRADE WINDS. *See* General Circulation; Teleconnections; *and* Zonal Circulation.

TRANSPORTATION. Climate and weather have affected almost every aspect of human evolution; they influence life's daily, seasonal, annual, and lifetime rhythms. Climate and weather determined the geographical regions in which the human species developed, found food, and satisfied other neccessities. Human responses to climatic extremes and temperature underlie the use of clothing, shelter, and fire. Weather and atmospheric conditions have a strong influence on commerce, the economy, recreation—and particularly, modes of transportation.

Climate has guided the evolution and selection of animals used in transportation: elephants and oxen in tropical countries, camels and llamas in arid climates, reindeer and certain breeds of dogs in Arctic climates. The use of horses for combat and transportation in the temperate Mediterranean region had a major impact on the spread of Greek and Roman civilizations and the later European conquest of the Americas.

Many effects of weather on transportation are matters of common knowledge, reported in the daily news. Both highways and automobiles are designed to function safely in extreme weather conditions. Presently, road weather information systems (RWIS) are being developed, including meteorological sensors to gather weather information in the highway environment, sensors in the roadway to report pavement conditions, and data analysis systems leading to forecasts helpful in initiating safety measures, such as snow control.

In air transportation, weather and forecasts influence the design of airports, schedules, routing, and ice control. Delays and extra costs are associated with bad weather. Fatal crashes are almost three times more likely if a flight does not have a weather briefing.

In the days of sailing vessels, global wind patterns were the major factors determining the selection of routes. With today's power-driven vessels, weather forecasts are used to avoid storm systems, to time invasions by sea, to plan recreational cruises, and to make other sea transportation decisions.

The controlled use of fire has made it possible to overcome many limitations imposed on transportation by weather and climate. The availability of fire led to the poleward march of societies, greatly extending the geographical range inhabitable by humans. Cooking with fire expanded the range of foods that could sustain humans, in turn influencing the development of trade. Processing clays and ores with fire led to the development of ceramics and metal implements—milestones in the development of civilization in general and of transportation in particular. The wheel, the chariot, and the wagon all evolved with the help of the controlled use of fire. Conversion of the heat of combustion to perform mechanical work led to the Industrial Revolution and modern modes of transportation. These mobile shelters, functional in all but the greatest extremes of weather, include automobiles, trucks, trains, ships, and airplanes, and even space vehicles. Most of these are now fueled by hydrocarbons, air-conditioned using chlorofluorocarbons (CFCs), and protected from unwanted fires by halon fire extinguishers, products of modern chemical technology.

Anthropogenic Emissions. Unfortunately, the combustion exhausts and other emissions from modern modes of transportation have significant adverse impacts on the atmosphere. Figure 1 displays problems caused by anthropogenic (arising from human activities) emissions. Most emissions associated with transportation are

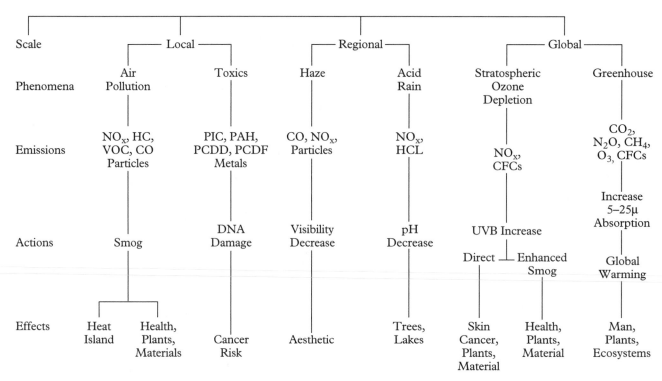

TRANSPORTATION. Figure 1. *Impacts of emissions.* This chart depicts the local, regional, and global impacts of emissions from the transportation sector, showing the phenomena, the main emissions, the action of these emissions, and their effects. (Adapted from Green, 1990.)

products of incomplete combustion (PIC) or unburned polyaromatic hydrocarbons (PAH), but the CFCs escaping from air conditioners and the halons from fire suppressants also have global environmental impacts.

Alternative fuels are now under serious consideration to minimize many of these detrimental effects. By international agreement, CFCs and halons are already being phased out, although the optimal replacements for presently used refrigerants, solvents, and fire suppressants have not yet been identified. The major alternative energy sources and fuel blending agents now being considered include reformulated gasoline, methanol, ethanol, some esters (MTBE, ETBE, and TAME), liquid petroleum gas (LPG), compressed natural gas (CNG), hydrogen, and electricity from storage batteries. Switching to compressed natural gas and liquid petroleum gas should lead to significant reduction in urban ozone and toxics. Using methanol should lower toxic hydrocarbons, but it will increase emissions of the carcinogen formaldehyde. While CNG vehicles may emit more methane, they produce 30 percent less carbon dioxide, giving a net lowering of about 8 percent in emission of greenhouse gases. The use of methanol produced from natural gas rather than from coal should also lower greenhouse emissions.

Figure 2 presents an overview of options for the conversion of various solid fuels to transportation fuels. Regulations intended to minimize harmful atmospheric emissions are becoming the primary motivating force in choosing alternative fuels, although there are still uncertainties in the assessments of their environmental impact. The biochemical or thermochemical reactors that convert coal, biomass, or municipal waste to more hydrogenated gaseous or liquid fuels have undergone extensive development. At this time the costs of feedstocks (source materials), fuel conversion, and environmental externalities (social costs not included in the fuel price) are major but uncertain factors in future choices of alternative transportation fuels.

Industrial nations are now intensively contemplating regulation to foster alternatives to hydrocarbon fuels, CFCs, and halons from the standpoint of their environmental, energy, and economic impacts. The solutions have not yet been identified, and many challenging problems remain for the coming generation of applied scientists and engineers.

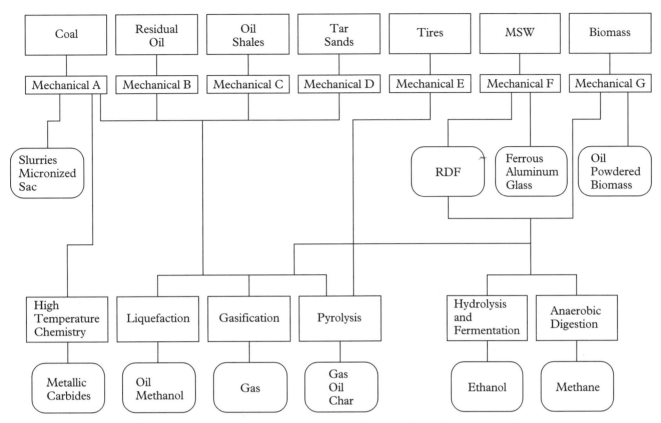

TRANSPORTATION. Figure 2. *Overview of solid fuel conversion options.* Boxes in row 1 are feedstocks; row 2, mechanical processes; row 3, mechanical products or byproducts; row 4, thermochemical or biochemical processes; row 5, products. (Adapted from Green, 1991.)

BIBLIOGRAPHY

Baselly, S. E., et al. *Road Weather Information Systems.* National Research Council, Strategic Highway Research Program, No. 350. Washington, D.C., 1988.

Golaszewski, R. *Weather Briefing Use and Fatal Weather Accidents.* National Research Council, Transportation Research Board No. 1158. Washington, D.C., 1988.

Green, A. E. S., ed. *Greenhouse Mitigation.* American Society of Mechanical Engineers, FACT vol. 7. New York, 1989.

Green, A. E. S., ed. *Solid Fuel Conversions for the Transportion Sector.* American Society of Mechanical Engineers, FACT vol. 12. New York, 1991.

Green, A. E. S., and W. E. Lear, Jr., eds. *Advances in Solid Fuel Technologies.* American Society of Mechanical Engineers, FACT vol. 9. New York, 1990.

Greene, D. L., and P. J. Santini, eds. *Transportation and Global Climate Change.* Washington, D.C., and Berkeley, Calif., 1993.

Probstein, R. F., and R. E. Hicks. *Synthetic Fuels.* New York, 1982.

Tabler, R. D. *Design Guidelines for the Control of Blowing and Drifting Snow.* National Research Council, Strategic Highway Research Program No. H-384. Washington, D.C., 1994.

ALEX E. S. GREEN

TREE-RING ANALYSIS. Tree rings are the concentric layers formed each year in the wood of trees. The rings can be seen in cross sections of tree trunks. The chemical composition, size, and structure of an annual tree ring constitutes a record of conditions in the tree's environment at the time the ring was being formed. The processes of cell division, enlargement, and maturation that determine the size and internal structure of the annual ring are subject to environmental control. When a factor (for example, moisture supply) limits one or more of these processes to a similar extent in most of the individual trees at one location, or over a larger region, and that factor varies from year to year, a common interannual pattern of tree-ring properties may be observed. For example, interannual variations in precipitation can result in characteristic patterns of broad and narrow rings in trees as much as 1,000 kilometers apart in southwestern North America. Such patterns are also found in regions with short, cool summers, such as in high mountains and at high latitudes. Here variations in summer temperatures produce fluctuations in ring width or wood density. If the common pattern of tree-ring properties is strong enough, and a sufficiently long series of rings is available in each case, it is possible to match the ring pattern of wood of unknown age against dated wood and thus attribute each growth ring in the former sample to a calendar year. This technique, called *cross-dating*, was developed early in the twentieth century by Andrew E. Douglass. Its application to fields including paleoclimatology, ecology, hydrology and archaeology is known as *dendrochronology*, a name derived from the Greek words for "tree" and "knowing the time."

Geography and Chronology. Trees that form the distinct, reliably annual rings (whose properties record interannual climate fluctuations) are found throughout the forested temperate and boreal regions. Trees close to the limits of their distribution tend to yield particularly strong records of past interannual climate variability because in such places they are responding to conditions that are extreme for them. A few cases of trees from tropical regions showing datable annual rings provide a basis for current efforts to extend dendrochronology into these regions. Trees at middle and high latitudes may live from several hundred to—in a few species—several thousand years. Their wood may be preserved in peat or other sediments, by cold dry conditions, or by incorporation in buildings or artifacts. This permits the establishment of cross-dated and replicated tree-ring measurement series, called *chronologies*, that are longer than the maximum life span of an individual tree.

The longest continuous examples of such chronologies extend over most of the Holocene period (roughly the past 10,000 years), notably for bristlecone pine (*Pinus longaeva*) in the White Mountains of California and for oaks (*Quercus* spp.) in western Europe. Networks of tree-ring chronologies based on ring width, and recently on wood density, exist for North America, Europe, Morocco, New Zealand, Tasmania, the western Himalayas, and temperate South America, and more are being prepared for Siberia and temperate Asia (see Figure 1). Such a network typically contains 10 to 250 climatically sensitive tree-ring chronologies, each containing series from two radii each from 10 to 40 trees of one species. Before these are combined into a site chronology, nonclimatic variation such as biological growth trend is removed from the measurements. Most of these networks provide useful climatic information for 250 to 500 years. The small number of climate-sensitive tree-ring chronologies with lengths of 1,000 years or more is increasing rapidly, with recent additions from Tasmania, southern South America, the Polar Urals, Scandinavia, the southeastern United States, the Sierra Nevada (California), New Mexico, Morocco, and the eastern Mediterranean region.

Climatic Knowledge. The multimillennial chronologies of bristlecone pine in North America and oaks in Europe made possible the calibration of the radiocarbon time scale, an achievement of great importance to many aspects of research on the climate of the Holocene. This permitted the discovery of hitherto unknown variations in atmospheric carbon-14 and opened up a whole new realm of geophysical investigation which some scientists believe will improve our understanding of the interaction between solar variability and climate. In addition to the radionuclide carbon-14, the stable isotopes of carbon, hydrogen, and oxygen in tree rings have been used to infer past climatic conditions.

Most tree-ring study of past climate has used ring width and, recently, maximum latewood density (see Figure 1). Long bristlecone pine chronologies were used to study century-scale climate fluctuation over the past 5,000 years (LaMarche, 1974). This was based on the inference that groups of decades with smaller rings in bristlecone pine near its upper elevational limit were associated with cooler periods, while small annual rings in the same species at its lower limit indicated a severe moisture deficit. The confidence that could be placed in tree rings as records of climate was enhanced by establishing the nature of the physiologic mechanisms controlling tree-ring formation and variability (Fritts, 1976).

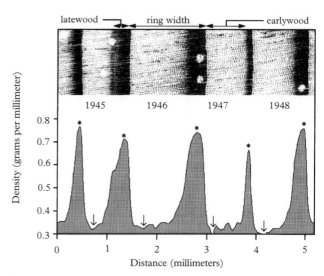

TREE-RING ANALYSIS. Figure 1. *Wood surface (top) of a pine from northern China showing the rings of 1945 through 1948; density trace (bottom) for the same rings as determined by X-ray densitometry.* Maximum latewood densities are indicated by asterisks (*); arrows (↓) indicate minimum earlywood densities. (From Hughes et al., 1994.)

In 1956 Edmund Schulman reported large regional patterns of ring-width variability in the semiarid regions of North America. His survey also revealed the great longevity of bristlecone pines.

Further sampling and more sophisticated analyses by others established the existence of synoptic-scale interannual variation in tree-ring widths. This has been exploited using techniques from multivariate statistics (Fritts, 1976, 1991; Cook and Kairiukstis, 1990) to reconstruct synoptic-scale interannual variation in temperature, precipitation, and sea-level pressure anomalies for recent centuries, notably in western North America (Fritts, 1991; Briffa et al., 1992) and in Europe (Briffa et al., 1988). This method has great power because reconstructed annual or seasonal maps for the past century or so can be tested directly against the instrumental record or other data for the same years. Other such reconstructions, usually of regional means or single-station records, have been made in many regions (Hughes et al., 1982; Bradley and Jones, 1992; Diaz and Markgraf, 1992). These and other reconstructions have cast light on the potential climatic role of explosive volcanic eruptions, the Southern Oscillation, and solar variability. They also provide a basis for comparing recent variability with that of preindustrial times. Similar networks of tree-ring records have been used to reconstruct the flow of rivers and the extent and frequency of droughts (Meko et al., 1991).

[See also Paleoclimates.]

BIBLIOGRAPHY

Bradley, R. S., and P. D. Jones, eds. Climate since A.D. 1500. London and New York, 1992.
Briffa, K. R., P. D. Jones, and F. H. Schweingruber. "Summer Temperature Patterns over Europe: A Reconstruction from 1750 A.D. Based on Maximum Latewood Density Indices of Conifers." Quaternary Research 30 (1988): 36–52.
Briffa, K. R., P. D. Jones, and F. H. Schweingruber. "Tree-ring Density Reconstructions of Summer Temperature Patterns across Western North America since 1600." Journal of Climate 5 (1992): 735–754.
Cook, E. R., and L. A. Kairiukstis, eds. Methods of Dendrochronology: Application in the Environmental Sciences. Dordrecht, 1989.
Diaz, H. F., and V. Markgraf, eds. El Niño: Historical and Paleoclimatic Aspects of the Southern Oscillation. Cambridge, 1992.
Fritts, H. C. Tree Rings and Climate. London, 1976.
Fritts, H. C. Reconstructing Large-Scale Climatic Patterns from Tree-Ring Data: A Diagnostic Analysis. Tucson, Ariz., 1991.
Hughes, M. K., P. M. Kelly, J. R. Pilcher, and V. C. LaMarche, Jr., eds. Climate from Tree Rings. Cambridge, 1982.
Hughes, M. K., X. Wu, X. Shao, and G. M. Garfin, "A Preliminary Reconstruction of Rainfall in North-Central China since A.D. 1600 from Tree-Ring Density and Width." Quaternary Research 41 (1994): 88–99.
LaMarche, V. C., Jr. "Paleoclimatic Inferences from Long Tree-ring Records." Science 183 (1974): 1043–1048.
Meko, D. M., M. K. Hughes, and C. W. Stockton. "Climate Change and Climate Variability: The Paleo Record." In Managing Water Resources in the West under Conditions of Climate Uncertainty. Washington, D.C., 1991, pp. 71–100.

MALCOLM K. HUGHES

TRIGGERED LIGHTNING. See Lightning.

TROPICAL CIRCULATIONS. The large-scale motion of the atmosphere in the tropics, the equatorial part of the Earth between 23°27' north and south latitudes, is characterized by two major components. The first, called the divergent component because it results in the movement of air away from the tropics, is driven by non-adiabatic heating processes, including the difference in solar radiative heating from equator to poles, the latent heat of condensation, sensible heating of the Earth's surface, and heat flux convergence resulting from atmospheric turbulence. The second component, the rotational wind, arises because the Earth's rotation generates vorticity (a measure of the rate of rotation of a fluid) and because the divergent circulation itself produces rotation. The distribution of the heat sources and sinks that drive the tropical circulations is governed by the placement of land masses and sea surface in the region. Among these heating processes, latent heating from condensation in the tropical atmosphere is the most important driving force of the circulations. [See General Circulation; Latent Heat; Vorticity.]

In the divergent component of the circulation a series of large-scale atmospheric cells exist permanently in the equatorial zone and move from east to west; these are called the Walker cells, collectively known as the Walker circulation. A second series, the Hadley cells or Hadley circulation, move air meridionally, from the equator toward the poles. Together, these circulations constitute the large-scale divergent component of the tropical circulation. Both are thermally direct; that is, they consist of rising warm air and sinking cold air, driven by latent heating. They are confined to the tropical zone by rotational forces. A full description of the tropical circulation would include not only the Walker and Hadley circulations but also the rotational wind component; however, this article concentrates on the former.

The center of the Walker circulation, on an annual mean, is represented by a broad region of strongly rising air over the warm surface of the tropical western Pacific

Ocean and the eastern Indian Ocean, and by widespread sinking (or *subsidence*) over the cold waters of the eastern Pacific and western Indian Ocean (Figure 1). Additional components of the Walker circulation are the ascending motions over northeastern Brazil and eastern Africa, which are linked to a descending motion over the eastern Atlantic.

Both the Walker and Hadley circulations are closely linked to the large-scale monsoon circulation, which varies seasonally with variations in insolation (solar heating), sea-surface temperature, and land-surface temperature. During summer in the Northern Hemisphere, the center of the Asian monsoon circulation lies over the northern part of the Bay of Bengal, with strong subsidence over the subtropical eastern Pacific and the Arabian Peninsula. Because of the Earth's rotation, the off-equatorial monsoon circulation is characterized by a strong cyclonic flow (counterclockwise in the Northern Hemisphere and clockwise in the Southern) in the lower troposphere (atmospheric layer closest to the planet's surface), and an anticyclonic flow (the reverse direction) at upper levels. During the Northern Hemisphere's winter, the center of this circulation shifts to the monsoon region of Indonesia and northern Australia, with strong subsidence over the equatorial eastern Pacific Ocean. In the latter season,

the circulation is more symmetrical with respect to the equator because the plane of the Walker circulation then lies closer to the equator. [*See* Monsoons.]

Large interannual variations in the Walker circulation are observed to be associated with the phenomenon of El Niño and the Southern Oscillation (ENSO), a recurring climatic event that begins in the tropics and influences most of the globe. In El Niño, abnormally warm sea-surface temperatures occur over the equatorial eastern Pacific at intervals of 2 to 7 years; the Southern Oscillation is a fluctuation in surface pressure in the tropical atmosphere, moving from east to west at similar intervals. This pressure fluctuation reflects the anomalous exchange of atmospheric mass between the Eastern and Western hemispheres as a result of changes in the Walker circulation. Although these two phenomena had been known independently since the 1920s, it was not until the late 1960s that scientists demonstrated a connection between them, pointing out that they could be considered two elements of a single, global-scale oscillation in the ocean–atmosphere system of the tropics. [*See* El Niño, La Niña, and the Southern Oscillation.]

During an ENSO event, the center of the Walker circulation shifts to the central and eastern Pacific, with anomalous rising motion over the eastern Pacific and

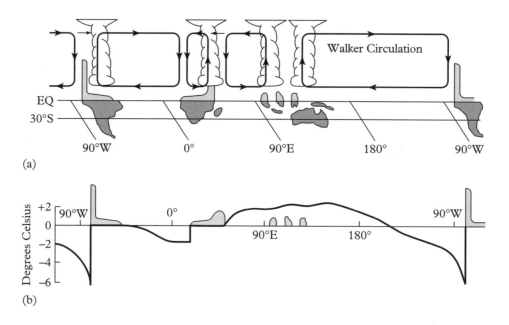

TROPICAL CIRCULATIONS. Figure 1. *Schematic diagram of (a) the Walker circulation and (b) sea-surface temperature along the equator.* Temperature variation is given relative to the zonal mean of 27°C. The strongest rising motion occurs over the warm pool of the western Pacific, and the strongest sinking motion over the cold eastern Pacific. (Adapted from Wyrthi, 1982.)

sinking motion over the western part. This shift causes enhanced surface westerly winds in the central Pacific and increased easterly winds in the western Pacific and Indian Ocean (Figure 2a). At the same time, changes in the Pacific Ocean's thermocline (temperature profile) generate increasing surface currents flowing eastward.

The event complementary to El Niño is termed *La Niña*, when large-scale atmospheric and oceanic anomalies follow a reverse pattern. During La Niña, the ascending branch of the Walker circulation over the western Pacific strengthens, and the descending branch intensifies and expands to cover the eastern and central Pacific. The pattern of the thermocline is reversed as well. The warm "pool" of surface water is confined to the western Pacific and Indian Ocean, while cold tongues over the equatorial eastern Pacific are well developed, extending farther west.

During an ENSO event, changes in both the Walker and Hadley circulations are responsible for increased rainfall over the eastern Pacific and western coastal South America, accompanied by drought conditions over Indonesia and Australia. In addition, the Indian monsoon weakens, and rainfall over Brazil decreases. In addition, changes in tropical convection and circulations have far-reaching consequences on climate and weather patterns outside the tropics. These anomalies include extremely cold weather in northeastern North America, typhoons over the Hawaiian Islands, and floods in California. Many of these events are due to alteration of the normal storm track forced by anomalous tropical heating. Understanding these long-distance connections has made it possible to incorporate data on tropical circulation and sea-surface temperatures into models for forecasting seasonal variation in North America.

In addition to its large interannual variability, the Walker circulation also experiences large fluctuations over shorter time scales. One of the most important intraseasonal fluctuations is the *30–60 day* or *Madden-Julian Oscillation* (MJO). This is characterized by the eastward shift of anomalous convection and wind along the equator from the Indian Ocean to the Central Pacific. This movement resembles the shift of the center of rising motion during El Nino, but it occurs on a time scale of 30 to 60 days. The rising branch of the MJO is associated with the organization of so-called *supercloud clusters,* or aggregations of a large number of smaller, westward-propagating cloud clusters and cloud elements, which are held together by some mechanism that is not yet fully understood. This phenomenon occurs throughout the year, although its amplitude and frequency is modified by the monsoon circulation. Theoretical and modeling studies have shown that the origin and modulation of the MJO are related to interaction among large-scale circulation, latent heating, and surface evaporation, modified by the Earth's rotation; however, the dynamical foundations of the MJO and its complex interactions with supercloud clusters and the monsoon are still not clear.

There is evidence that the MJO and related fluctuations in the tropical circulation are coupled to the extratropical circulation through waves propagating from certain tropical regions. The MJO may therefore influence the evolution of extratropical weather on time scales of months to seasons; hence, observations of the MJO may contribute to the predictability of weather elsewhere.

We now recognize that the MJO represents a fundamental oscillation mode of the Walker circulation, much as ENSO is a fundamental oscillation mode of the tropical ocean-atmosphere system. The spatial and temporal scales of the MJO are strongly affected by ENSO because of the differences in sea-surface temperatures and atmospheric mean states. It is possible that the timing and

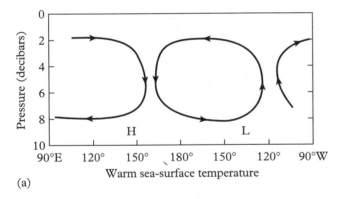

(a)

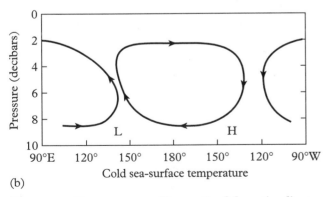

(b)

TROPICAL CIRCULATIONS. Figure 2. *Schematic diagram showing the departure from normal of the Walker circulation during (a) El Niño and (b) La Niña.* (Adapted from Julian and Chervin, 1978.)

detailed evolution of an ENSO event may, in turn, be influenced by variations in the MJO.

Daily fluctuations in the Walker circulation have been linked to the influence of cold surges from eastern Asia on convection in the "maritime continent" (Borneo and Indonesia). These fluctuations are good examples of short-range interactions between tropical and midlatitude circulations.

[*See also* Cyclones; *and* Tropical Climate.]

BIBLIOGRAPHY

Bjerknes, J. "Atmospheric Teleconnection from the Equatorial Pacific." *Monthly Weather Review* 97 (1969): 163–172.

Julian, P. R., and R. M. Chervin. "A Study of the Southern Oscillation and Walker Circulation Phenomenon." *Monthly Weather Review* 106 (1978): 1433–1451.

Lau, K. M., and P. H. Chan. "The 40–50 Day Oscillation and the El Niño/Southern Oscillation: A New Perspective." *Bulletin of the American Meteorological Society* 67 (1986): 533–534.

Madden, R., and P. R. Julian. "Detection of a 40–50 Day Oscillation in the Zonal Wind in the Tropical Pacific." *Journal of Atmospheric Science* 28 (1971): 702–708.

Walker, G. T. "Correlation in Seasonal Variation of Weather. IX: A Further Study of World Weather." *Memoirs of the Indian Meteorological Department* 24 (1924): 275–332.

Wyrthi, K. "The Southern Oscillation, Ocean-Atmosphere Interaction and El Niño." *Marine Technology Society Journal* 16 (1982): 3–10.

WILLIAM K.-M. LAU

TROPICAL CLIMATE. The zone of tropical climates, lying geographically between the tropics of Cancer and Capricorn, is defined dynamically as that part of the troposphere about the girdle of the Earth occupied by deep easterly flow. The deep easterlies of the tropics are bounded at Earth's surface by the centers of the large semipermanent subtropical anticyclones, and aloft by the westerlies of mid latitudes (Figure 1). Monsoonal climates, which fall within the bounds of the tropics, are treated as an important departure from this model of global tropical circulations. [*See* General Circulation; Monsoons.]

The distribution of ocean and land masses, together with the three-dimensional circulations of the oceans and atmosphere, leads to the subdivisions in tropical climate shown in Figure 2. *Wet* or *humid tropics* lie over the western ends of the tropical oceans and in the confluent flow of northeast and southeast *trades*. The equatorial confluence, called the *Intertropical Convergence Zone* or ITCZ, is a region of high rainfall and diminishing winds. Extension of the ITCZ over the continents of Africa and South

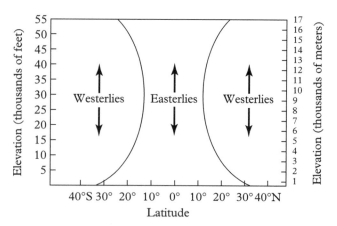

TROPICAL CLIMATE. Figure 1. *Boundaries between the planetary easterlies and westerlies; the region of deep easterlies defines the limits of the tropics.* The figure represents a vertical cross section through the atmosphere (troposphere) from the surface to 17,000 meters (55,000 feet), representing the tropics along an arbitrary line of longitude from about 40° south latitude to 40° north latitude.

America and the large islands of New Guinea, Borneo, and Sumatra, leads to extremes in continental humid equatorial climate. Land masses with elevated topography increase annual rainfall from lowland values of 200 centimeters per year (termed *wet*) to over 450 centimeters per year (*very wet*) on elevated slopes.

Dry tropics, lying at the eastern and poleward ends of the tropical oceans, are a product of the large-scale sinking of the Hadley-Walker circulations amplified by cold upwelling ocean waters. Rainfall decreases progressively from 100 centimeters per year in the west to low extremes over the eastern margins. This dry region includes the great subtropical deserts of the Sahara, Kalahari, and Namib in Africa and the Atacama in South America. Over the oceans, the dry tropics are reflected on such islands as the Cape Verdes in the Atlantic and the Line Islands in the Pacific. [*See* Meridional Circulations.]

Continental effects, together with elevated topography, modify the extremes of the wet and dry tropics to produce a more benign environment in high tropical *savannas*. An example is the Serengeti Plains in Tanzania.

Six of the 12 months of the year in the wet or humid tropics over the continents typically have rainfalls of 25 centimeters per month or more. Only 2 months of the year are likely to receive as little as 5 centimeters (2 inches). The closed rainforests of the humid tropics contain many plant species that cannot tolerate either marked changes in temperature or mean monthly temperatures below 17.8°C (64°F).

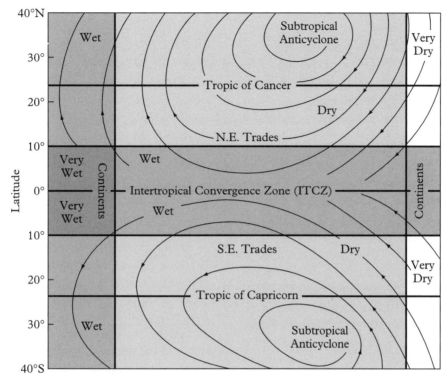

TROPICAL CLIMATE. Figure 2. *Surface winds around the two large subtropical anti-cyclone, with the northeast and southeast trades flowing into the intertropical convergence zone (ITCZ) for a model tropical ocean with continents at the eastern and western extrem-ities.* The wet tropics (≥ 200 centimeters per year) and very wet (> 400 centimeters per year) and the dry (50 to 100 centimeters per year) and very dry air are shown.

With the exception of the eastern ends of the tropical oceans, high rainfall (over 200 centimeters per year) extends across the oceans and up to 10° latitude on either side of the ITCZ. Rainfall decreases rapidly over the high tropical savannas, dropping from 50 to 100 centimeters per year to very low values over the subtropical deserts, where rain can be absent for years at a time.

The phenomenon known as *El Niño*, reflecting an abnormal increase in surface water temperatures in the eastern Pacific, can reverse the dry conditions typically prevailing over the cold waters off Peru and Ecuador, resulting in torrential rains over the Pacific coastal mountains of these countries and storms along the California coast. Responses to El Niño (called *teleconnections*) are felt in the Amazon Basin (with up to 20 percent reduction in rain) and in the tropical and subtropical regions of Argentina, Africa, and Australia in the form of drought—an explicit demonstration of the connection between the tropics and the global atmosphere. Other periodic or near-periodic oscillations reflected in rainfall include a near-18-year cycle in the Southern Hemisphere subtropics and the near-50-day periodicity of the Mad-den-Julian circulation in the Pacific. [*See* El Niño, La Niña, and the Southern Oscillation; Teleconnections.]

Tropical climates are the product of the general circulation of the planet's atmosphere. Air in contact with the tropical ocean, circulating around the subtropical anticyclone, acquires moderately high temperatures (25° to 26°C) and humidities (70 to 80 percent) with no appreciable changes over long horizontal distances. Thus the contrasting gradients in temperature and moisture characteristic of mid-latitude frontal cyclones are absent in the tropics. More radiation is received in tropical regions from the Sun than is reradiated by the Earth. This excess radiant energy, initially absorbed by the tropical oceans and land surfaces, must be returned to the atmosphere across the interface of air and surface in the form of latent and sensible heat, elevated to high altitudes by convective clouds and transported poleward by the large-scale circulations. [*See* General Circulation.]

Tropical climates are characterized by temperature and humidity fields that change little over space and time. The temperature changes that identify seasonal changes at high latitudes are absent, replaced by changes in rainfall—wet and dry seasons replace summer and winter. High sun angles combine with the cloud and wind fields to produce dual maxima and minima in rainfall in locations within about 10° latitude of the equator, and single maxima and minima toward the margins of the tropics. Thus primary and secondary wet and dry seasons occur each year at low latitudes, and only single wet and dry seasons away from the equator.

Latent heat supplied to the atmosphere from the ocean and from vegetated surfaces is the primary source of energy driving atmospheric systems in the tropics. The latent energy is removed from the surface by turbulent motions which feed convective clouds organized in squall systems or synoptic-scale storms. For much of the time, especially over the oceans, the tropical atmosphere is characterized by fair weather cumulus clouds and moderate temperatures (27° to 29°C daytime maxima; 21° to 24°C nighttime minima). Organized storm systems are confined to the wet months and occupy only a small fraction of the total time, so that more than half of the rainfall occurs in less than 10 percent of the time. Tropical climates are characterized by fair days with little cloudiness. Diurnal effects such as sea and land breezes and afternoon thunderstorms are pronounced over some tropical islands and land masses. [See Cyclones, article on Tropical Cyclones.]

The links and responses in the tropical system are highly nonlinear, so that the Earth's most severe storms, the *hurricanes* or tropical cyclones, occur intermittently in this benign environment. [See Hurricanes.] All hurricanes must develop from preexisting organized convection over warm (27°C or greater) seas that yield the latent heat necessary to drive the storm. Only about 10 percent of tropical storms capable of intensifying into hurricanes do so; thus the climate of the tropics is disrupted only infrequently in the wet season, and then mostly in the western ends of the tropical oceans by these storms, called *hurricanes* in the Atlantic and eastern Pacific and *typhoons* in the central and western Pacific.

MICHAEL GARSTANG

TROPICAL FORESTS. See Deforestation; and Rainforests.

TROPICAL PRECIPITATION SYSTEMS. The precipitation systems characteristic of the tropics, between the equator and 27° latitude, bear little resemblance to extratropical systems. Temperature variations are not significant within tropical air masses, nor are frontal mechanisms predominant.

Much tropical precipitation is produced by convective clouds. Usually there are wide gaps between the clouds, leading to spatially discontinuous distribution of rainfall. Tropical rainfall often appears to occur randomly, but it is nevertheless often caused by some kind of organized disturbance. As the tropical synoptic disturbances form within a uniform air mass, temperature and density differences do not supply much of the energy; instead, latent heat from the condensation of water vapor is the main source of energy.

Tropical rain-forming systems disturbances can be broadly classified into three groups, according to their spatial scales:

- Planetary-scale systems, equatorial trough, such as the trade winds, or the monsoon trough.
- Synoptic-scale or large-wave scale disturbances, such as easterly waves, tropical cyclones, depressions, or waves in the upper troposphere.
- Mesoscale convective disturbances, such as convective elements or cumulus cloud clusters.

Equatorial Trough. The northeasterly and southeasterly trade winds meet in the equatorial trough, a zone of relatively low pressure situated near the equator. The meeting of the trades, with its intrinsic convergence and uplift of air masses, is usually violent, with huge, towering cumulus clouds marking the area of convergence. The energy for this ascent originates from latent heat released during cumulus cloud development. This energy accumulates from evaporation in the subtropical maritime areas and increases within the deepening layer of air below the trade-wind inversion as the air moves equatorward and westward. Heating of air over land and orographic influences provide the necessary triggering force to release the accumulated energy. The development of convection over the oceans, however, is controlled by the difference between the temperatures of the sea surface and the air above. The equatorial trough is not characterized uniformly by continuous ascent, clouds, and rain; the term *Intertropical Convergence Zone* (ITCZ) is generally applied to the areas of greater activity in the equatorial trough, where high rainfall of long duration is experienced. The rainfall maxima in the equatorial trough zone are found over the continental areas of South America and Africa and the region of Indonesia. [See Intertropical Convergence.]

Tropical Monsoons. Thermal gradients set up by the different rates at which land and ocean respond to seasonal solar heating lead to a system of winds which blow from the sea to the land in summer and in the opposite direction in winter—the summer and winter monsoons. The regional features of the monsoons are known to be linked with the large-scale overturning of the atmosphere known as the Hadley and Walker circulations, but the coupling mechanism is not yet well understood. Short-period rainfall variations within the monsoon season are caused by synoptic disturbances whose dimensions are relatively smaller. [*See* Dynamics; Monsoons.]

In many parts of southern and eastern Asia, most of the annual rainfall is received in the summer months; the Indian summer monsoon is a well-known example (see Figure 1). The mighty Himalayan mountain range, acting as a barrier to the low-level circulation and as a high-level heat source for the upper-air circulation, plays a key role in drawing the monsoon current toward the Indian subcontinent. Observed records spanning more than a century in the past have shown that the Indian summer monsoon rainfall has been a very stable (see Figure 2) and highly dependable source of water for the region. However, there is significant year-to-year variability in the monsoon rainfall, resulting in occasional floods and droughts (Parthasarathy et al., 1994). Recent studies have shown that this variability is strongly modulated by the influences of El Niño and the Southern Oscillation on the monsoon circulation.

The two main monsoonal regions in the tropics are west of 100° east longitude (the Indian subcontinent and East Africa), where the main source of moisture is the Indian Ocean, and east of 100° east longitude (Southeast Asia, Australasia, and northern Australia), where the monsoon air originates in the subtropical high pressure cell over Australia. The typical rain-forming mechanisms involved in the monsoon systems include orographic lifting by elevated regions across the monsoon current (for example, the Western Ghat mountains along the western coast of India), depressions, lows, and troughs. A characteristic of the monsoon is pulsations in the strength of the main current—a few days of heavy rainfall followed by interruptions with weak monsoon activity. The dry periods within the monsoon season are sometimes prolonged and are known as *break-monsoon* periods in India, where they may lead to drought.

The magnitude of the monsoon circulation over Africa is much smaller than that of its Asian counterpart in terms of both the area covered and the thickness of the air layers involved. The monsoon over the sub-Saharan zone of Africa is directly related to the seasonal behavior of the quasi-permanent circulation features associated with the surface pressure trough, whose northward shift in summer brings the monsoon airstream into the continent. The monsoons over East Africa are closely associated with the seasonal shift of the ITCZ.

The monsoonal circulation over northern Australia constitutes an extension of the Asian system, but with the seasonal characteristics reversed. The Asian winter monsoon becomes the northwest monsoon of north Australia. The tropical parts of northern Australia receive a substantial part of their annual rainfall from November through April.

Tropical Low-Pressure Disturbances. The transport of water vapor and convective activity are organized in the tropics by low-pressure areas superimposed on the average flow patterns. These areas are called *disturbances, depressions,* or *cyclones,* depending on their intensity. The preferred areas for the formation and movement of tropical cyclones are often linked to warm ocean currents. Tropical storms do not occur within 5° latitude on either side of the equator. These storms decay rapidly after moving onto land. Rainfall in tropical cyclones is usually concentrated in a rather narrow zone around the core, but the actual amount received varies greatly, depending on the size, intensity, and movement of the individual disturbance; rainfall from a single storm can be anywhere between 25 and 100 centimeters. These storms are more frequent in the warmer months, usually in late summer and early fall.

Mesoscale Convective Systems. Much like the mesoscale convective complexes of the mid latitudes,

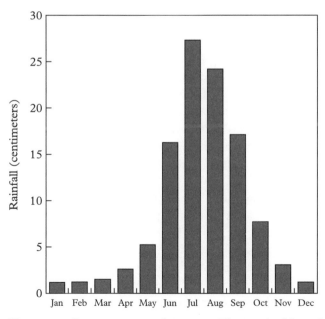

TROPICAL PRECIPITATION SYSTEMS. Figure 1. *Normal monthly area-weighted average rainfall over India, based on 124 years of data (1871–1994).*

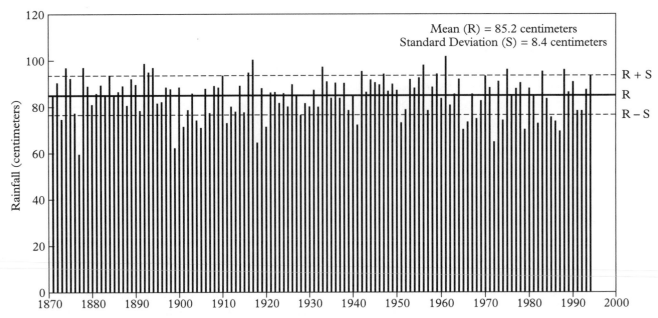

TROPICAL PRECIPITATION SYSTEMS. Figure 2. *Variation of all-India area-weighted average summer monsoon (June through September) rainfall from 1871 to 1994.*

tropical cumulonimbus convection frequently becomes organized into mesoscale systems. These systems, also called *cloud clusters,* contain both deep convective and stratiform clouds, producing considerable rainfall. A mesoscale convective system is typically identified in satellite imagery by a large cirrus shield 200 to 1,000 kilometers across. In its mature phase, the cluster consists partly of convective towers—which contain buoyant updrafts, negatively buoyant downdrafts, and heavy showers of rain—and partly of deep stratiform clouds with light precipitation spread over a horizontal distance of 100 to 200 kilometers. The mesoscale updrafts and downdrafts are distinctly more widespread and weaker than the localized convective drafts. Tropical mesoscale convective systems are occasionally associated with disturbances that develop into cyclonic storms. They are found over a major part of the equatorial trough but are more common over West Africa, the tropical Atlantic Ocean, and the South China Sea.

Squall Lines. Linear systems consisting of numerous thunderstorms organized in lines or bands develop and move as more-or-less organized systems known as *squall lines.* These systems can be hundreds of kilometers long and 10 to 30 kilometers wide, consisting of cumulonimbus clouds and zones of strong and weak thunderstorm activity. Squall lines are distinguished from the mesoscale convective systems by their rapid propagation, explosive growth, and lack of a cirrus shield. The lifetime of a squall line is considerably longer than that of its component cumulonimbus elements. The squall line consists of a row of cumulonimbus clouds preceding a large area of cold-air downdraft. This cold air is produced under a precipitating, trailing anvil cloud which emanates from the cumulonimbus clouds. The cold air spreading below the warm boundary-layer air ahead of the system initiates the convective clouds. The arrival of a squall line is marked by a sudden squall and a temperature drop which precede the rain shower from the cumulonimbus cloud. Most of these systems move faster than the surrounding air at all levels. Squall lines account for a large part of the annual rainfall over West Africa. They have also been observed, though less frequently, over the tropical North Atlantic, the Caribbean Sea, and northern South America. They also produce considerable rainfall at the times of onset and withdrawal of the monsoon circulation over India and southeast Asia. [*See* Squall Lines.]

Thunderstorms. Much tropical rainfall is produced by thunderstorms, which in many areas occur on more than 50 days per year. When the air is warm and humid, convection usually leads to widely scattered thunderstorms, developing over locations where surface heating is most intense or where convergence is most concentrated. The precipitation associated with the thunderstorms is intense, localized, and of short duration. Over the humid tropics, cumulonimbus clouds typically rise to a height of around 18 kilometers, with their tops spreading out to form anvil clouds. [*See* Thunderstorms.]

Easterly Waves. Another tropical system that produces heavy rainfall, particularly over the Caribbean area, is the easterly wave. Easterly wave disturbances develop in the dry trade air as it moves across the western Pacific, Indian, and central Atlantic oceans. These disturbances are weak, elongated, migrating troughs of low pressure moving from east to west in the trade winds, aligned approximately perpendicularly to the direction of the trades and usually proceeding more slowly than the trades themselves. Ahead of these disturbances clear weather prevails, but as the trough line approaches, concentrated convective activity with multiple rows of high cumulonimbus clouds develops, bringing showery spells lasting several hours.

[*See also* Tropical Circulations.]

BIBLIOGRAPHY

Ayoade, J. D. *Introduction to Climatology for the Tropics.* New York, 1983.
Boucher, K. *Global Climate.* London, 1975.
Hastenrath, S. *Climate Dynamics of the Tropics.* Dordrecht, 1991.
Houze, R. A., Jr., and P. V. Hobbs. "Organization and Structure of Precipitating Cloud Systems." In *Advances in Geophysics.* Vol. 24. New York, 1982, pp. 225–315.
McIlveen, R. *Fundamentals of Weather and Climate.* London, 1992.
Nieuwolt, S. *Tropical Climatology.* New York, 1977.
Parthasarathy, B., A. A. Munot, and D. R. Kothawale. "All-India Monthly and Seasonal Rainfall Series: 1871–1993." *Theoretical and Applied Climatology* 49 (1994): 217–224.
Riehl, H. *Introduction to the Atmosphere.* New York, 1978.
Riehl, H. *Climate and Weather in the Tropics.* London, 1979.
Tarakanov, G. G. *Tropical Meteorology.* Moscow, 1982.

B. PARTHASARATHY and K. RUPA KUMAR

TROPOSPHERE. *See* Atmospheric Structure.

TUNDRA CLIMATE. The word *tundra* is broadly used to refer to the landscapes that are found above the altitudinal or latitudinal tree line. In the introduction to their monumental book on arctic and alpine environments, Ives and Barry (1974) introduce the notion of tundra by citing R. E. Beschel's remark: "As we do not usually lump . . . non-polar, non-alpine, and non-tree covered vegetation of the remainder of the globe under one heading, one wonders why a collective term should be applied to the vegetation of arctic and alpine regions in the word *tundra,* unless we understand its use

as an involuntary expression of ignorance" (Beschel, 1970:86).

Tundra, indeed a many-faceted term, denotes extensive areas of the Arctic in North America and northern Asia, characteristically low-lying, treeless plains which range from wet meadow to dry heath (dwarf shrubland) and rocky fellfield. Alpine tundra has many similar characteristics, although the vegetation forms a more complex small-scale mosaic owing to the nature of the terrain, and the extensive wet tundras of the Arctic are largely lacking. This article will focus primarily on arctic tundra. [*See* Arctic.]

Characteristics. Tundra covers about 5.5 percent of the land surface of the Earth (Rodin et al., 1975). Figure 1 shows the extent of arctic tundra. The boundary between tundra and the boreal forests to the south of it, called the *tundra-forest ecotone,* is not well defined. The most common diagnostic feature of the tundra is treelessness; trees may be defined in this context as woody plants with a single central stem at least 2 meters tall. Other definitions of tundra regions include climatic parameters and their consequences, such as the presence of permafrost, although the correlation between treeline and permafrost extent is poor, as Figure 1 shows.

Figure 1 also shows that the Arctic Circle, which forms the southern boundary of the Arctic in an astronomical and cartographic sense, is a poor indicator of climatic or vegetational boundaries. Tundra extends well south of the Arctic Circle in North America, and the boreal forest boundary is well north of the Arctic Circle in Asia, North America, and Europe. Permafrost is very extensive in Asia, where it extends south into northern China, and also in North America. A little more than one-third of the land area north of the Arctic Circle, or about 3.5 million square kilometers, is covered by tundra. Boreal forests and ice (glaciers and ice sheets) cover the other two-thirds, in roughly equal proportions.

The vegetation of the tundra consists of low herbaceous plants, dwarf shrubs, and lichens, but wide variations occur within different physiographic provinces. Mountainous areas, foothills, and coastal plains differ in topography, geology, and climate, and consequently in fauna and flora. The diversity of vascular plant species is greatest in the foothills and least on the coastal plain of Alaska's Arctic Slope (Brown et al., 1974), for example. The arctic tundra is well known for its wildlife, which includes caribou, moose, grizzly and polar bear, wolverine, musk ox, and arctic fox, among others. Microtine rodents, such as lemmings, voles, and mice, are a substantial part of the mammalian fauna. During the brief

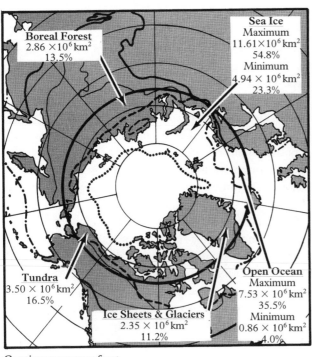

Continuous permafrost
Discontinuous permafrost
Treeline
Minimum pack ice extent
Maximum pack ice extent

TUNDRA CLIMATE. Figure 1. *Location and a real extent of the major terrain types north of the Arctic Circle.* (From Weller and Wendler, 1990. Courtesy of the International Glaciological Society.)

summer, large numbers of migrating waterfowl and shorebirds arrive to nest on the tundra, feeding on its huge insect populations.

Of the world's major ecosystems, tundra has the lowest temperatures and the shortest growing season. The limits of biological accommodation and adaptation to low temperature were extensively studied in the U.S. Tundra Biome program in the early 1970s (Brown et al., 1980). The tundra plants are well adapted to the cold winters and chill summers of arctic and alpine regions and seldom suffer irreparable, damage from weather, either physically or evolutionarily (Billings, 1974).

Climate. Traditionally, areas where the average temperature of the warmest month is below 10°C have been identified as having a tundra climate (Köppen, 1936), but the presence of arctic air masses may be a better indicator. Bryson (1966) showed that arctic tundra areas in North America experience arctic air on at least half the days in July. The median location of the arctic front in

July corresponds approximately with the northern limit of the boreal forest (Barry, 1967). It is commonly argued that short, cool summers are the basic cause of the treeless character of both alpine and arctic tundra environments. However, while these environments have similarities in temperature and length of snow cover, there are also major differences. The Arctic is dark for as much as 6 months each year, while lower-latitude alpine sites experience regular diurnal and seasonal regimes of solar radiation. The alpine tundra is also generally much windier and receives more snowfall than the arctic tundra, and alpine sites have extreme night cooling which affects growth.

The weather and climate of the arctic tundra are the result of net radiative losses, advection of heat into the Arctic through atmospheric and oceanic circulation, and the nature and conditions of the surface, including the presence of snow and ice. Stable surface inversions caused by radiative cooling form over most of the Arctic. Intense cold periods result from the combined effect of radiative cooling at the surface and anticyclonic conditions with clear skies. During summer, the temperature of the pack ice on the Arctic Ocean remains at 0°C, and there are large stretches of open water. Fog and low stratus clouds form over the ice and open water, persisting throughout the summer in spite of changes in atmospheric circulation. Inland from the coast and away from the chilling influence of the coastal waters, temperatures are much warmer—one factor that permits a much more diverse tundra flora in the foothills.

Winter in the arctic tundra is 6 to 9 months long, characterized by a relatively shallow (30 to 40 centimeters) snowcover, darkness, and January temperatures averaging about −30°C in North America, somewhat lower in Asia (Siberia), and much higher (−10°C) in northern Europe. Spring and fall are short transition periods, with rapid melting in spring. Late winter and spring are generally characterized by clear skies, permitting receipt of a high proportion of the incident solar radiation. Temperatures begin to rise, lagging 4 to 6 weeks behind the increase in solar radiation. The mean daily temperatures are above freezing during the short summer, but cloudiness and fog over the Arctic Ocean and the coastal areas prevent temperatures there from rising much above 5°C, even in July. Temperatures increase inland from the coast until one reaches the forest-tundra ecotone, with July mean temperatures of 10°C or higher.

Precipitation in the Arctic is difficult to measure, because it is generally light and associated with high winds; in addition, for the greater part of the year it falls as dry snow, which is redistributed by winds according to exposure and local topography. With an annual precipitation of 20 to 30 centimeters, and frequently less than 10 centimeters, arctic tundra is comparable to arid zones elsewhere. Only the permafrost barrier makes its dense vegetative cover possible.

Energy Balance and Microclimates. The environmental conditions within a few meters above and below the ground surface, where most biological activities take place, constitute the microclimate of a region. The microclimate is characterized by the radiation, temperature, and moisture regimes of these near-surface layers. It is determined largely by the regional climates, as modified by local topography and vegetation. In the tundra environment with its low, exposed vegetation, large variations in microclimate may exist within a small area. Even a few centimeters' difference in microtopography may produce marked differences in soil temperature, depth of

snow, wind effects, snow drifting, and resultant protection of plants.

Quantitatively, the microclimate is described by the energy and water balances at the surface. These balances are related in a complex way. Solar radiation provides the energy utilized by surface physical and biological processes. This radiation comprises shortwave visible and longwave thermal radiation, with both incoming and outgoing components at the surface. The energy balance expresses how the available net radiative energy is converted into sensible and latent heat to warm the air and ground, and to melt snow or evaporate water from the surface. The water balance is determined by the energy balance and in turn affects the energy balance through latent heat exchanges and through the effects of snow on surface conditions.

The energy balance of arctic wet tundra at Barrow, Alaska is shown in Figure 2, from Weller and Holmgren (1974). Six distinct seasonal phases of the tundra cli-

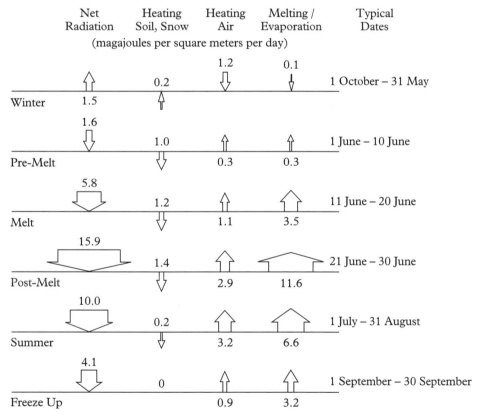

TUNDRA CLIMATE. Figure 2. *Energy balance of the arctic tundra at Barrow, Alaska, for six seasonal phases of the tundra climate.* The width and direction of the arrows and numbers indicate energy flux directions and rates in millions of joules per square meters per day. (From Weller and Holmgren, 1974. Courtesy of the American Meteorological Society.)

mate, each characterized by relatively large changes in the energy and water balance, were identified. The 8-month winter phase has an energy balance with net radiative losses, compensated primarily by sensible heat fluxes toward the surface, in the presence of steep temperature inversions. This is followed by three brief but distinct phases related to snow-melting. The most dramatic changes in the energy and water balances take place during this period, when the amount of solar radiation absorbed by the surface increases by an order of magnitude owing to the melting and disappearance of the snow cover. In the last of these phases, the tundra is completely wet and evaporation rates reach 6 millimeters per day, with mean values of 4.5 millimeters per day (Weller and Holmgren, 1974). Through the melting phase, within a span of 2 to 3 weeks, the net radiation increases by a factor of 10 and the latent heat flux increases by a factor of 40.

Evaporation continues throughout summer, but by the time freezing of the tundra occurs in early September, net radiation has been reduced appreciably and evaporation rates are down to about 1 millimeter per day. Soon snow covers the tundra again, and the radiation balance becomes negative.

Global Change and the Environment. Despite the severe climate and the remoteness of arctic tundra areas, increasing demands have been placed on them to provide energy and minerals, as well as recreation. Conflicting demands for wilderness, industrial development, recreation, and retention of the traditional lifestyles of the indigenous people have emerged. Large oil and gas fields are in production in Alaska, Canada, and Siberia, and environmental impact on the tundra, particularly in Siberia where nuclear waste disposal has added to the problem, are often severe. Controversy rages regarding the opening of new exploration areas and pipeline corridors for petroleum development.

Other problems are posed by global climatic change. The Arctic is a good indicator of global change, and feedback processes there also influence the global climate. The large carbon reservoirs of the arctic and subarctic tundra, peatlands, and boreal forests are potential sources of the greenhouse gases carbon dioxide and methane, released when the climate warms and permafrost melts, providing a positive feedback on climate (i.e., more warming). Likewise, if the surface reflectivity or albedo of these regions decreases owing to a warming climate, the added radiation absorbed provides a positive feedback. The geographical extent, composition, and diversity of the narrowly adapted flora and fauna of the arctic tundra is likely to change with the climate.

[See also Climatic Zones.]

BIBLIOGRAPHY

Barry, R. G. "Seasonal Location of the Arctic Front in North America." *Geographical Bulletin* 9 (1967): 79–95.

Beschel, R. E. "The Diversity of Tundra Vegetation." In *Productivity and Conservation in Northern Circumpolar Lands*, edited by W. A. Fuller and P. G. Kevan, IUCN Publ. 16. Morges, Switzerland, 1970.

Billings, W. D. "Arctic and Alpine Vegetation: Plant Adaptations to Cold Summer Climates." In *Arctic and Alpine Environments*, edited by J. D. Ives and R. G. Barry. London, 1974, pp. 403–443.

Brown, J., P. C. Miller, L. L. Tieszen, and F. L. Bunnell. "An Arctic Ecosystem: The Coastal Tundra at Barrow, Alaska." Stroudsburg, Pa., 1980.

Bryson, R. A. "Air Masses, Streamlines and the Boreal Forest." *Geographical Bulletin* 8 (1966): 228–269.

Ives, J. D. and R. G. Barry, eds. *Arctic and Alpine Environments*. London, 1974.

Köppen, W. "Das geographische System der Klimate." In *Handbuch der Klimatologie*, edited by W. Köppen and R. Geiger. Berlin, 1936.

Rodin, L. E., N. I. Bazilevich, and N. N. Rozov. "Productivity of the World's Main Ecosystems." In *Productivity of World Ecosystems*. National Academy of Sciences. Washington, D.C., 1975, pp. 13–26.

Weller, G., and B. Holmgren. "The Microclimates of the Arctic Tundra." *Journal of Applied Meteorology* 13 (1974): 854–862.

Weller, G., and G. Wendler. "Energy Budgets over Various Types of Terrain in Polar Regions." *Annals of Glaciology* 14 (1990): 311–314.

GUNTER WELLER

TURBULENCE. *[This article treats a technical aspect of climate and weather studies; some of it is intended for readers at an advanced level.]*

An inherent feature in most atmospheric phenomena, turbulence manifests itself through many fascinating observable phenomena. The random shapes of clouds, the erratic motion of particulates in the wind, the patterns of smoke or steam billowing from a stack, and the unpredictable nature of weather patterns are all evidence of the turbulent motions that affect the flow of fluids in the atmosphere (water and air are both "fluids" in this sense). Most fluid flows, both natural and technological, are experienced in a turbulent environment; therefore, the importance of understanding and predicting turbulent flow behavior is clear.

From a qualitative observational point of view, turbulent flow (or turbulence) can be described as the random or chaotic motions of fluids; however, this is a limited description and cannot be considered a definition. In fact, despite the importance of understanding the effects of turbulence in engineering and geophysical flows, there is no generally accepted definition of turbulence. Rather than debating a rigorous definition, it has proven more constructive to begin with a discussion of the features that are shared among flows considered turbulent (Tennekes and Lumley, 1972).

Characteristics of Turbulent Flows. Discussed below are various characteristics of turbulent flows, including their order and randomness, length and time scales, three-dimensional nature, unpredictability, and diffusive properties.

Order and randomness. Turbulent flows are often described in terms of their apparently random and unpredictable motions. In many predictive methods, velocity components, temperature, and other relevant properties in a turbulent flow are treated as random variables described by various statistical approaches. Over the past few decades, the idea that turbulence consists of completely random motions has been modified to account for the *coherent motions* that are now recognized as important features of many turbulent flows. For example, turbulent boundary layers and homogeneous turbulent shear flows exhibit horseshoe or hairpin vortices that affect fluid transport near solid surfaces. Free shear flows, such as the mixing layer, clearly exhibit coherent vortex structures, even for very high turbulence intensities (Brown and Roshko, 1974). The realization that coherent structures exist in many turbulent flows has led to some important developments in the numerical simulation of these flows, including direct numerical simulation and large eddy simulation, discussed below.

Although the concern with coherent motions in analytical descriptions and in modeling studies is fairly recent, their appearance in turbulence has long been noted. One of the first written descriptions of a turbulent flow by an engineer was that of Leonardo da Vinci, who made detailed studies of fluid motion. A careful observer of nature and a strong advocate of the of experimental approach, Leonardo studied and recorded flow patterns in many different configurations and formulated some of the earliest descriptions of fluid motion, including the following:

Observe the motion of the surface of the water, which resembles that of hair, which has two motions, of which one is caused by the weight of the hair, the other by the direction of the curls; thus water has eddying motions, one part of which is due to the principle current, the other to the random and reverse motion.

This is an interesting observation in that it identifies two types of eddying motions, one ordered and one random. This organized structure plays an important role in turbulent mixing and has been the subject of much recent research.

Turbulent length and time scales. In the study and prediction of turbulent flows, much reference is made to *length scales* and *time scales.* A length scale describes the length that characterizes specific features. Another way of thinking about length scales in a turbulent flow is to relate them to the size of eddies. In turbulent flows one encounters a wide range of eddy sizes. The size characterizing the largest eddies in the flow is generally constrained by geometry. In a cloud, the largest motions are on the order of the size of the cloud itself. If one is considering flows in the Earth's boundary layer (the region of fluid motion affected by the Earth's surface), the largest eddy at any location in the boundary layer will be on the order of the distance from the surface of the Earth at that location—a geometric constraint. We will refer to this largest eddy size as the *integral scale* and denote it by L. The largest eddies in the flow account for most of the transport of momentum and energy. Below the integral scale, turbulent motions exist over a wide range of sizes. The size of the smallest eddies of the flow is determined by viscosity, which acts to damp out perturbations in the flow and is most effective in the regions of highest gradients, which are at the smallest scales.

To obtain a better understanding of this important feature of turbulence, it is instructive to consider the forces acting on a parcel of fluid. If the flow speed is high or the length scale (physical domain) over which the flow is developing is large, inertial forces dominate. These forces tend to break the fluid up into smaller and smaller eddy sizes (note that other forces, such as buoyancy, can have the same effect). When a small enough size is reached, the viscous forces begin to play a more dominant role. As the viscous forces take over, the turbulent motions are damped out. The length scale where this damping is most prominent can be very small; in atmospheric flows, it is generally on the order of millimeters. This scale is called the *Kolmogorov length scale,* named after the scientist who developed the scaling laws relevant to these ideas—A. N. Kolmogorov (1903–1987), perhaps the most important contributor to the mathematical theory of turbulence.

The difference between the integral and Kolmogorov scales can be many orders of magnitude. This results in severe limitations in simulating the behavior of turbulent flows.

At this point, it is necessary to introduce the concept of the *Reynolds number*. The Reynolds number is a dimensionless number that has several interpretations. One of the most important interpretations (though not unique) is that it can be viewed to represent the ratio of inertial to viscous forces acting on an element of fluid. Mathematically, the Reynolds number, *Re,* is defined as $Re = LV/\nu$, where L is the integral scale, V is a characteristic velocity of the flow, and ν is the kinematic viscosity of the fluid. For very high Reynolds number flows, the viscous forces become increasingly small in comparison to the inertial forces. Smaller-scale motions are then necessarily generated until the effects of viscosity become important and energy is dissipated. Scaling analyses indicate that the ratio of the integral scale to the Kolmogorov scale is proportional to $Re^{3/4}$. Since atmospheric flows can have Reynolds numbers greater than 10^6, the ratio is clearly quite large.

This description is useful in introducing the idea of an *energy cascade.* In a turbulent flow, most of the kinetic energy is contained in the largest scales of the flow. Inertial forces tend to transport this energy to smaller and smaller scales, until the kinetic energy is converted to heat by viscous forces. These ideas are simply expressed by Richardson's (1922) poem:

> And little whorls have lesser whorls,
> Big whorls have little whorls,
>   That feed on their velocity.
> And little whorls have lesser whorls,
>   And so on to viscosity.
>   (In the molecular sense)

The other type of scale, a *time scale,* is the characteristic time period that defines a particular fluid motion. The time scale of an eddy, for example, can be defined as the time over which that eddy retains its coherence. Large eddies have much longer time scales than small eddies: the ratio of time scales between the largest and smallest eddies in a flow is proportional to $Re^{1/2}$. This also has implications in the solution of turbulent flows.

Three-dimensional nature. Turbulence is an inherently three-dimensional phenomenon. Some of the most important mechanisms related to the evolution of turbulent flows are identically zero in strictly two-dimensional flows. To illustrate, consider a swirling motion in a two-dimensional plane. Distortion of this motion—for example, by either a stretching or compression in the third dimension (normal to the plane of the motion under consideration)—results in a change in the length scale characterizing the initial motion. This is a qualitative description of the phenomenon of *vortex stretching,* which is a key mechanism in the transfer of energy to eddies of different sizes. This process cannot occur in a two-dimensional flow. As a result, a rigorous analytical or numerical representation of turbulence is quite difficult to achieve in practice because the full three-dimensional structure of the flow must be accounted for.

Unpredictability. No two turbulent flows are the same. Furthermore, the evolution of a turbulent flow is very sensitive to both its initial conditions and any applied perturbations. This is a result of the nonlinear dynamical nature of fluid flows. Over time, the instantaneous flow fields of two turbulent flows with nearly identical initial conditions differ considerably. This makes the long-term accurate prediction of transient turbulent flows extremely difficult. This dynamical nature of flows in the atmosphere contributes to the difficulty of obtaining accurate predictions of storm paths, the internal structure of clouds, the development of storm fronts, pollutant concentrations over distances far from their source, or the time and location of wind gusts, to mention only a few examples.

Diffusive properties. One of the most significant features of a turbulent flow is its effect on the transport of mass, momentum, and energy. Turbulent flows exhibit much higher transport rates of mass, momentum, and energy than do nonturbulent flows. In other words, turbulence is very effective at promoting mixing. In the atmosphere, this enhanced transport results from the fluctuations in the velocity and scalar fields in directions transverse to the mean wind direction. We often speak of the diffusivity of turbulence, or the *turbulent diffusivity,* in describing the enhanced mixing effects of turbulent flows.

Modeling and Simulation of Turbulent Flows. Because of the tremendous impact turbulence has on flows of technological, environmental, and economic importance, it is crucial that effects of turbulence be understood and realistically treated in models of the atmosphere. However, the features of turbulence discussed above show that there is significant complexity in the structure of turbulent flows. This complexity makes predicting their behavior an extremely difficult task. It also opens the door to one of the most exciting and active areas of engineering and physical science research—the scientific study and modeling of turbulence. A brief overview of the prediction methods used for treating turbulence is presented below.

A rigorous prediction of motions in the atmosphere would include a complete description of the evolution over time of the velocity field, temperature, pressure, and other thermodynamic properties (e.g., moisture content) at every point in space. In theory, this is possible because fluid and energy transport (including turbulent behavior) is described by a known set of equations. These include conservation equations for energy transport; a set of equations known as the *Navier-Stokes equations,* which describe the evolution of the fluid momentum (velocity); equations for the transport of individual chemicals and particulates; and a number of thermodynamic relationships which give information on the pressure, density, and other thermodynamic properties.

Let $\phi(x, y, z, t)$ represent a typical variable in a turbulent flow at any point in space (x, y, z) at any time t. ϕ could represent, for example, temperature or any of the velocity components. A typical evolution (transport) equation for ϕ can be expressed in words as follows:

The time rate of change of $\phi(x,t)$
    = convection of ϕ in space by velocity
    + diffusion by molecular motion   (1)
    + source terms.

The mathematics involved in describing these physical process results in a complex set of coupled nonlinear differential equations that exhibit random, time-dependent solutions. Analytically, exact solutions are only available for very simple configurations. The conditions under which analytic solutions are available are generally not directly applicable to predict flow behavior in configurations of practical importance. However, rapid developments in computer technology have permitted us to obtain, numerical or computer-generated solutions to these equations. The accuracy of these solutions depends on many factors, including the ability to specify correctly the initial and boundary conditions, the details of the numerical algorithm used to integrate the equations, the range of length and time scales important to a particular problem, and the sophistication of the models used to describe the physical processes involved.

There are many different approaches to achieving computational solutions of turbulent flow behavior. Several of them are discussed below. First, let us present some background on what type of information is obtained from a numerical solution of the equations that describe atmospheric fluid flow.

The exact equations of fluid motion describe the evolution of velocity, mass, and energy at every point in a defined space and at every instant in a defined time period. A properly posed problem defines a region of space and a period of time over which a solution is desired. Depending on the type of simulation employed, various levels of approximation must be made before a numerical solution can be obtained. The first approximation involves describing the *real* solution, which is continuous over space and time, by a finite representation at discrete *grid points* in the computational representation of the flow field. The flow field solution is obtained at each of these grid points. To initiate the solution, the initial conditions describing the flow at time zero are needed in this finite domain. (If a *steady state* solution exists—one in which the features of interest do not vary with time—then only the boundary conditions are needed.) In addition, the conditions that determine the solution of the flow on the boundary (called *boundary conditions*) are needed. The correct specification of the initial and boundary conditions is crucial for obtaining a realistic solution, but it is often difficult to achieve. Thus, the boundary and initial conditions reflect what is sought.

Consider a cloud model. It may be the objective of the investigator to predict the size range and number frequency of liquid droplets. This information can be important for precipitation predictions, or it may be needed to determine the radiation energy flux to the Earth (droplets may reflect or absorb incoming radiation). The growth of individual droplets is determined by the local cloud conditions, which are dependent on the manner in which air outside the cloud is entrained into the cloud, as well as on the behavior of the air within the cloud. An accurate description of these cloud processes requires knowledge of the flow behavior, which is generally turbulent. A solution may thus involve an initial condition that describes the conditions at the cloud base, with boundary conditions described by the environmental air profile. The numerical solution would then solve the governing equations for the evolution of the cloud properties with time. Coupled to this solution would be models describing microphysical processes such as droplet growth.

The numerical solutions can be conducted using a number of different approaches. For the study of turbulence in the engineering and physical sciences, three approaches are most commonly used: *direct numerical simulation* (DNS), *large-eddy simulation* (LES), and *moment methods* (also referred to as *Reynolds averaged Navier-Stokes* or RANS, although other averaging techniques are used). Another approach, used primarily in combustion applications, the *probability density function* (PDF) techniques. Each of these techniques is based on solutions of equations that are derived directly from the exact governing equations.

In addition to these approaches, many empirical models not based on the exact equations are regularly applied to describe the effects of turbulence. An example is the *Gaussian plume model*. The objective of this model is to describe the evolution of a constituent concentration in a plume—for example, the concentration of a chemical species or particulate matter in a spreading plume upon its injection into the atmosphere, as from a smokestack, as it is transported away from its injection point. As the name implies, a Gaussian form for the average concentration profile across the plume is assumed. The vertical and horizontal spreads of the plume are generally taken from empirical formulae for their dependence on downstream distance, stability of the atmosphere, surface roughness, and other factors (see Pasquill and Smith, 1983). The discussion below is limited to approaches based on the Navier-Stokes equations and will not include these more empirically based models, because solution approaches based on a representation of the exact equations provide a useful way to illustrate many of the features of turbulence.

Direct numerical simulation. The first approach, DNS, includes numerical approaches that attempt to solve explicitly the complete set of exact equations governing the motion of a turbulent flow. In a DNS, all details of flow are assumed to be accurately represented, from the largest geometrical constraints to the gradients at the smallest scales. This requires that the discretized domain be fine enough (contain enough grid points) to resolve accurately all scales of motion.

As an example, consider a DNS that attempts to solve for the unsteady flow in a 100-meter region of the boundary layer. For a mean velocity of 2 meters per second, a viscosity of 1.5×10^{-5} square meters per second, and an integral scale L of 100 meters, the Reynolds number works out to be $Re = 1.3 \times 10^7$. Applying the length scaling introduced earlier gives a ratio of integral scale to Kolmogorov scale of approximately 2×10^5. This is an indication of the number of grid nodes that would be needed to describe all scales of motion explicitly in one dimension. Because turbulent flows are three-dimensional, this means that at least 10^{15} grid points would be needed for a complete three-dimensional numerical simulation! This far exceeds the capacity of existing com-

puters. The difficulty becomes even more apparent when we consider the unsteady nature of the flow. For these reasons, DNS of realistic flows is not a foreseeable possibility.

DNS has, however, seen important applications in turbulence research. It is used to study fundamental features of turbulence in very idealized flows that exhibit canonical features of turbulent flows. Under these conditions, DNS has proven to be a valuable tool. Important examples include DNS of homogeneous shear flows and of turbulent mixing layers.

Turbulence modeling. The preceding example forcefully indicates that because turbulent flows exhibit a wide range of length scales, it is necessary to approach the solution differently. Instead of attempting to solve the exact equations of motion, we do not compute the detailed turbulent motions exactly, but rather account for them by *modeling* their effects on the flow. This is accomplished by modifying the governing equations and introducing modeling assumptions about the effects of turbulence. The goal of the modeler is to incorporate as much of the physics of turbulent transport as possible, yet to make them simple enough to be treated by modern computational techniques. Rather than providing an exact description of the time evolution at each point in space, these approaches usually provide only a statistical description of the flow. That is, the solution obtained at some point in space and at some time cannot be expected to be the exact solution, but rather an approximation that may involve an average of the real solution over some quantity of time and space. In these methods, explicit information about small-scale features (in both time and space) of the flow may be lost, but information about larger-scale features is retained. This is entirely acceptable, as the details of the small-scale turbulent motion are generally not of direct concern. What *is* of concern is adequately representing the effects of the turbulence on the observable features of the flow.

The most common mathematical means of achieving this type of description involves first assuming that the dependent variables (velocity components, temperature, and so on) can be treated as random variables. These random variables are considered to be composed of a mean, or average component, and a fluctuation about the mean. The objective then is to solve a set of evolution equations for the lower-order statistical moments (for example, the mean and standard deviation). Techniques that follow this approach are called *moment methods*. For a typical dependent variable—say temperature, T, or velocity, *V*—this decomposition into a mean and fluctuating component can be expressed as:

$$T = \bar{T} + T'$$
$$V = \bar{V} + V' \tag{2}$$

where the overbar denotes an appropriately defined average quantity, and the primed variables indicate the fluctuating component about the mean. (In general, both the mean quantities and fluctuations are functions of space and time.) The objective of this approach is to use the averaging process to remove the details of turbulence from the problem. Upon inserting the definitions of Equation 2 in the exact governing equations, and carrying out and averaging operation on the equations, we obtain a new set of equations describing the evolution of the averaged values. (For details, see Rodi, 1984; Launder and Spalding, 1972; Launder et al., 1984; a thorough discussion of applications in meteorology is Stull, 1988.) This relaxes a major constraint on the resolution requirements of the numerical solution, as the small-scale fluctuations are no longer explicitly computed.

A difficulty that arises through this averaging process in that additional terms appear in the governing equations. Although explicit treatment of the turbulence has been avoided, the additional terms that appear as a result of the averaging represent the effects of the turbulence on the evolution of the mean properties. Analogous to Equation 1 above, the equations that must be solved now can be expressed as:

The rate of change of $\bar{\phi}(\mathbf{x},t)$
   = convection of $\bar{\phi}(\mathbf{x},\mathbf{t})$ in space by the mean
      velocity
      + diffusion of $\phi(\mathbf{x},t)$ by molecular motion   (3)
      + source terms
      + turbulent transport of $\bar{\phi}(\mathbf{x},\mathbf{t})$.

When applied to the transport equations for the velocity components (momentum), these resulting equations are called the *Reynolds average Navier-Stokes equations,* or RANS. The turbulent transport term, which contains all the effects of the turbulence, represents an additional unknown variable. The job of the turbulence modeler is to provide a description of these effects in terms of the known mean variables to obtain a closed set of equations. All the effects of turbulence, with its associated complicating effects on the flow, must be accounted for in this model. This is the major challenge of the turbulence research community.

Eddy diffusivity models. The most popular methods by which the effects of turbulence on the flow are modeled are based on the key observation of the enhanced mixing and diffusive behavior of turbulence. This is accomplished though eddy motions and velocity fluctuations.

Some of the earliest ideas for modeling the effects of turbulence on the average flow quantities still play an important role in most turbulence models in use today. In an analogy to the role viscosity plays in smoothing out gradients and promoting mixing, it was suggested that the effects of turbulence could be treated by an empirical viscosity coefficient, the *eddy viscosity*. Molecular diffusion results from gradients in fluid properties. The rate at which these gradients are smoothed out depends on the magnitude of the gradients and on the specific diffusion coefficient (viscosity) of the property under consideration. Using similar reasoning, the effects of turbulence on the mean flow quantities (for example, a particular velocity component) can thus be postulated to be described mathematically as proportional to the average gradients of the velocity component under consideration, with the constant of proportionality being the eddy viscosity. This is called a *gradient diffusion* approach. In this approach, information about the details of the eddy motions is lost (at least at length and time scales below those over which the averaging process mentioned above is performed). However, if information about averaged properties is desired, the approach is very useful. As a result, much of the research in turbulence modeling over the past century has been directed at finding the correct expression for the eddy viscosity. One difficulty results from the fact that the diffusivity of turbulence is not an inherent property of the flow but depends on details of the flow at any instant and location. Another difficulty stems from the fact that using a gradient diffusion approach implies a particular physical behavior of the turbulence, and this physical description is not always correct. Nonetheless, for many applications, the approach is valuable. Below we discuss some classical ideas used in modeling the eddy viscosity.

Mathematical expressions for the eddy viscosity were originally developed by assuming that the eddies in a turbulent flow behave analogously to molecules in a fluid. A fluid exhibits viscous effects as a result of molecular collisions that result in exchanges of momentum. The analogy is to treat turbulent eddies as individual entities that also collide with each other and exchange momentum. In a gas, the molecular viscosity is proportional to the speed of the molecules (which is, in general, a function of temperature) and to the mean free path between molecular collisions. Relating these concepts to the eddy viscosity (denoted below by ν_t), it can be postulated that the eddy viscosity will be proportional to a characteristic velocity of the turbulence, and some length scale, termed the *mixing length;* that is,

$$\nu_t = Clv \tag{4}$$

where l is the characteristic length scale and v is the velocity scale of the turbulence. C is the constant of proportionality, which must be determined empirically. Therefore, the effects of turbulence stated in Equation 3 are absorbed into the molecular transport term, but with a diffusion coefficient that is based on the properties of the flow and not on the properties of the fluid. The details used to determine l and v distinguish the different modeling approaches, which are beyond the scope of this discussion.

The modeling philosophy described above has been successful in many applications, and variants of it are most common in predictive methods used in practical applications. However, there are some significant limitations to the approach. First, the analogy between molecular entities and turbulent motions is not really accurate. Molecules are generally spaced very far apart, and interactions among them do not change their internal structure. A turbulent eddy, on the other hand, is a conceptualization used to describe the swirling motion of turbulent flows. Eddies are not individual entities, and the interactions among these motions do not leave them unchanged. Furthermore, this modeling approach assumes that the action of all modeled motions has the same physical effect on the mean flow, which is not true. These observations indicate that results of turbulence modeling studies must be treated cautiously. For example, if one is using turbulence models to assist in description or prediction in engineering or atmospheric problems, the results should be viewed as a suggestion of expected behavior and not considered to be exact behavior. This simply reflects the tremendous complexity of turbulent flows. With an appreciation of this complexity, the informed user of turbulence models can use them to provide valuable information on the effects of turbulence in real situations.

Higher-level modeling: Large eddy simulation. The approach just discussed suffers from several limitations. Important among these are the domain over which the spatial or temporal averaging is applied, and

the physical correctness of the gradient diffusion assumption. Another simulation technique that has been developed to study turbulent flows is *large-eddy simulation,* or LES. This approach is now used in all areas of turbulence research, but it was pioneered in the early 1970s for application in meteorology. The idea is straightforward, but the application has its subtleties. Instead of using a temporal or spatial average as in moment methods, this technique applies a spatial *filter* to the independent variables (such as velocity and temperature). This spatial filter is generally some type of low-pass filter, by which the small-scale fluctuations of the turbulence are mathematically filtered out of the solution. The variables (say $v(x, y, z, t)$) are then decomposed into a filtered component, $\langle v(x, y, z, t) \rangle$, and a component referred to as the *subgrid value*, $v''(x, y, z, t)$. This new representation of the dependent variables is inserted into the exact governing equations, which are themselves then subjected to the same filtering process. The result is a set of equations for the filtered variables, which contain additional terms describing the interactions of the subgrid components on the large-scale or filtered variables. If we filter the equations of motion, the time development of the large-scale eddy features of the flow can be computed explicitly. Only the small-scale motions, and the effects they have on the large scales, must be modeled. As a result, the modeling of the physical processes is limited to small scales. Such models are called *subgrid-scale models* because the modeling is directed at motions not resolved numerically (motions at scales finer than the grid resolution). The advantage is that behavior of the small subgrid scales is generally considered to more universal than that of the large-scale features, which are functions of boundary and initial conditions; the models developed can thus be applied to a wider range of applications.

The original modeling philosophy behind the development of subgrid-scale models was based on the assertion that the primary physical effect of small-scale on large-scale features was as a mechanisms to dissipate kinetic energy. As such, the idea of using an eddy diffusivity approach for the small scales seems justified. Other parameters appearing in the subgrid model formulation include a length scale representative of the domain over which filtering is performed. It should be pointed out that many different types of spatial filters can be applied, and the modeling is certainly not unique. The collection of reviews edited by Galperin and Orszag (1994) and Schumann and Friedrich (1985) provide many details on this LES methodology and the subgrid-scale modeling approach.

Recently it has been recognized that a simple eddy viscosity approach is insufficient, because energy transport between the large and small scales is not a simple one-way process. More advanced models account for this two-way exchange of energy among eddies; Leith (1993) describes such a model.

Despite the attractiveness of LES and the advances that have been made since its introduction, LES is still used primarily in research application (Zang, 1993). Although the complexity of problems being solved is increasing as computational resources improve, LES is not applied on a regular basis to predict the behavior and motion of major meteorological phenomena.

Conclusion. The study of turbulence involves a wide variety of tools and the expertise of scientists and engineers in many disciplines. The tools employed involve experimentation, analysis and modeling, and numerical simulation. Experimental studies provide information on the physical mechanisms that lead to certain observed behaviors. The data obtained are also valuable in the assessment of turbulence models. DNS plays a similar role, but is restricted to simple canonical representations of turbulence. Most other numerical modeling approaches are primarily diagnostic tools, incapable of revealing much about the fundamental nature of turbulence. Analytical studies have provided much insight into statistical features of turbulence; they have been instrumental in the formulation of basic scaling laws and order-of-magnitude approximations and useful in suggesting some different avenues of experimental and numerical research.

Although our predictive capabilities and understanding of basic turbulent flow features continue to improve, turbulence remains among the major unsolved problems in the physical sciences. The importance of this feature in so many facets of the physical and biological sciences will keep the study of turbulence and development of turbulence modeling approaches among the major research areas for decades.

[*See also* Dynamics.]

BIBLIOGRAPHY

Brown, G. L., and A. Roshko. "On Density Effects and Large Structures in Turbulent Mixing Layers." *Journal of Fluid Mechanics* 64 (1974): 775–816.

Deardorff, J. W. "Numerical Investigation of Neutral and Unstable Planetary Boundary Layers." *Journal of Atmospheric Science* 29 (1972): 91–115.

Galperin, B. and S. Orszag, eds. *Large Eddy Simulation of Complex Geophysical and Engineering Flows.* Cambridge, 1993.

Launder, B. E., and B. D. Spalding. *Mathematical Modeling of Turbulence.* London, 1972.

Launder, B. E., W. C. Reynolds, and W. Rodi. *Turbulence Models and Their Applications.* Paris, 1984.

Leith, C. E. "Stochastic Backscatter Formulation for Three-Dimensional Compressible Flow." in *Large Eddy Simulation of Complex Geophysical and Engineering Flows,* edited by B. Galperin and S. Orszag. et al. Cambridge, 1993, pp. 105–118.

Pasquill, F., and F. B. Smith. *Atmospheric Diffusion.* 3d ed. Chichester, 1983.

Richardson, L. F. *Weather Prediction by Numerical Process.* Cambridge, 1922.

Rodi, W. *Turbulence Models and Their Application in Hydraulics.* Delft, Netherlands, 1984.

Schumann, U., and R. Friedrich, eds. *Direct and Large Eddy Simulation of Turbulence.* Braunschweig, 1985.

Stull, R. *An Introduction to Boundary Layer Meteorology.* Boston, 1988.

Tennekes, H., and J. Lumley. *A First Course in Turbulence.* Cambridge, Mass., 1972.

Zang, T. A. "Large Eddy Simulation as a Tool in Engineering and Geophysics: Panel Discussion." In *Large Eddy Simulation of Complex Geophysical and Engineering Flows,* edited by B. Galperin and S. Orszag. Cambridge, 1993, pp. 559–568.

PATRICK A. McMURTRY

TYPHOONS. *See* Asia; Cyclones, *article on* Tropical Cyclones; Hurricanes; *and* Religion and Weather.

U

UNIFORMITARIANISM. *See* Scientific Methods.

UPPER ATMOSPHERE. The upper atmosphere is that region of the Earth's atmosphere above the *tropopause*—the boundary between the *troposphere* (the lower atmosphere where most of us spend our time) and upper regions such as the stratosphere, mesosphere, thermosphere and exosphere. The upper atmosphere is accessible directly only by very high-flying aircraft, balloons, rockets, and orbiting satellites, or remotely via various sounding techniques such as ground-based lidar and radar (see Figure 1). The primary parameter defining these regions is temperature. In Figure 2, note that each boundary is distinguished by a temperature inversion or a reversal: the tropopause defines a minimum in temperature, the stratopause a maximum, and the mesopause another minimum. These regions of the atmosphere are further distinguished by their composition consisting of both neutral and charged molecules; the latter actually defines an overlapping region called the *ionosphere*. Although the ionosphere represents a tiny fraction of the total atmospheric concentration, it strongly affects radio propagation, including various forms of telecommunication, and is therefore an important component of the atmosphere. The discussion here concentrates on the *stratosphere* and *mesosphere*; meteorologists usually refer to these regions as the *upper atmosphere,* but aeron-

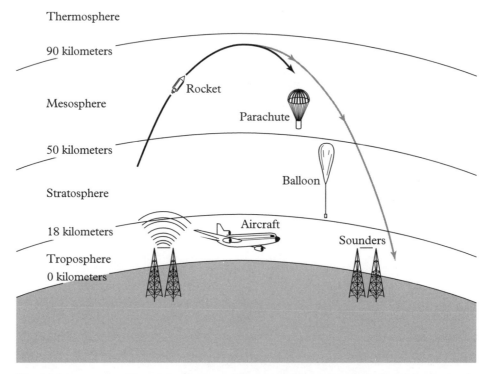

UPPER ATMOSPHERE. Figure 1. *Regions of the atmosphere and principal vehicles for in situ measurements.*

omers and dynamicists more commonly term them the *middle atmosphere*, as is done in the rest of this article. [*See* Atmospheric Structure; Ionosphere.]

For many years, the middle atmosphere was thought to be orderly and well defined. Standard models predicted the stratosphere to be horizontally layered (as the name implies), so that it was virtually impossible for particulates and other minor constituents to pass through it vertically, either upward or downward. The stratosphere exhibits a temperature increase with altitude because of the presence of ozone, an important minor constituent which absorbs solar ultraviolet (UV) radiation to cause

heating of the region. The absorption of solar UV in the stratosphere also acts as a shield against UV radiation, which would otherwise damage plant and animal life on the ground. Earlier, the mesosphere was thought to be quiet and passive, with atmospheric constituents having nearly the same relative concentrations as found near the ground owing to the process of atmospheric mixing. Above this height, within the thermosphere, the collision rates between molecules become very low because of the low density of the atmosphere, permitting diffusive separation to occur. In this process gravity causes the heavier molecules to separate out more rapidly than light mole-

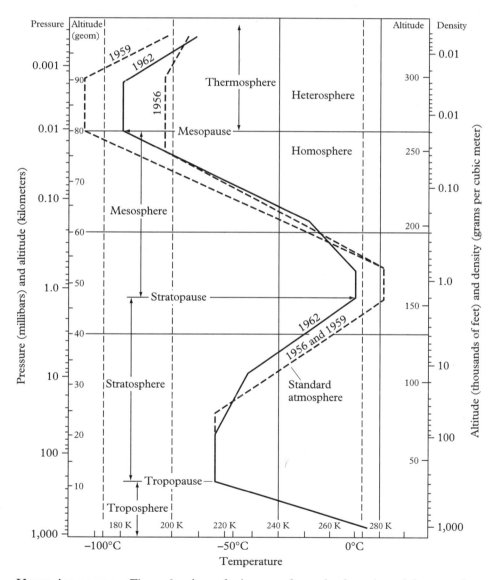

UPPER ATMOSPHERE. Figure 2. *Atmospheric nomenclature for the regions of the atmosphere up to 100 kilometers.* The temperature profiles for the U.S. Standard Atmospheres (1956, 1959, and 1962) are shown.

cules and atoms, so that the mean density of the atmosphere begins to decrease with altitude. The lower region is often called the *mixing region* or *homosphere,* whereas the region of diffusive separation is termed the *heterosphere.* [*See* Stratosphere.]

These simple concepts lasted for many years because of the difficulty of making measurements in the middle atmosphere. The mesosphere is too high for sampling by aircraft and balloons and too low for sampling by orbiting satellites. In situ measurements within the mesosphere can be made only by rockets. Within the lower stratosphere, we can use high-altitude balloons and very high-flying aircraft up to an altitude of about 30 kilometers. In addition, some parameters in both regions can be measured remotely from ground stations or satellites. Nonetheless, measurement difficulties have caused the middle atmosphere—particularly the upper stratosphere and mesosphere—to suffer from lack of attention, and this region has sometimes been dubbed "ignorosphere."

New results are now changing our understanding of these regions in a dramatic way. The middle atmosphere is not passive but highly dynamic. Furthermore, the various regions are not independently driven, as originally thought, but appear to interact both dynamically and electrodynamically. They also contain a multitude of newly discovered minor constituents on both a macroscopic and microscopic scale, which can strongly affect the physics and chemistry there. The remainder of this article concentrates on these concepts to demonstrate the emerging notion of the middle atmosphere, which alters the earlier classical picture.

Chemical Composition. The middle atmosphere (stratosphere and mesosphere) up to the homopause is dominated by nitrogen (N_2) and oxygen (O_2); it also contains many of the dominant minor constituents— such as argon (Ar) at about 1 percent—with the same abundances as in the troposphere. It is the more minor constituents, often representing less than one millionth of the atmosphere by volume or by weight, that cause many of the unique characteristics, which distinguish this region from the troposphere. Within the stratosphere, ozone (O_3) is a very important and well-known minor constituent. It normally peaks near 25 kilometers altitude and reaches mixing ratios of about 10 parts per million. It is responsible for the absorption of solar UV radiation below 200 nanometers wavelength, forming a global shield to protect us from this harmful radiation. The absorption of this solar UV radiation within the stratosphere is responsible for the temperature inversion seen at the tropopause and the heating of the middle atmo-

sphere up to the stratopause near 50 kilometers, as shown in Figure 2. For these reasons, an immense amount of scientific research has been dedicated to the study of ozone in the stratosphere, as well as to the other minor constituents involved in its production and destruction. Various chemical models involving hundreds of reactions have been constructed and applied in order to predict the future trend for this important constituent. Many minor species, such as odd nitrogen (NO_x), odd hydrogen (HO_x), and various compounds of bromine, chlorine, fluorine, and sulfur all play a role in ozone chemistry. Some of the compounds responsible for ozone loss, such as hydrofluorocarbons, are of anthropogenic (human-generated) origin, with consequent political implications regarding their manufacture and use. At present there is much concern regarding the growing ozone hole above Antarctica, as well as about the formation of similar holes in other locations, such as the Arctic. [*See* Ozone; Ozone Hole.]

The mesosphere also contains an abundance of minor species, less well understood owing to a lack of measurements, and also having less impact on our local environment. Many of the minor constituents found lower in the stratosphere are also present here, but in reduced quantity. Ultraviolet absorption by ozone is still responsible for local heating, but as can be seen in Figure 2, it is insufficient to maintain the temperature growth with altitude that is seen in the stratosphere. The mesosphere is the region of the upper atmosphere where higher-energy UV (called *solar Lyman alpha*) from the Sun, solar X-rays, and other sources of ionizing radiation begin to form the ionosphere. The lowest region of the ionosphere, termed the *D region,* overlaps with the mesosphere. The number of ions here is tiny compared to the number of neutral molecules (less than 1 part in 10^{12}), yet their presence can influence many of the processes occurring in the mesosphere. Mass spectrometric measurements via rocket have shown that O_2^+ and NO^+ are often the dominant ionic species affecting the ion-neutral chemistry of this region. However, as we penetrate the D region to lower altitudes, heavy water cluster ions of the form $XY^+(H_2O)_n$ occur, where XY could be O_2, N_2, NO, etc., and $n = 1, 2, 3, \ldots, m$; m is a large but unknown number. Because of the charge neutrality requirement, there must be one negative charge for each positive charge; in the daytime the negative charge usually exists in the form of free electrons, unattached to any molecules because of solar photodetachment. At night or at lower altitudes during the daytime, where solar UV radiation cannot reach to photodetach electrons from molecules, the electrons remain attached to form negative ions. There have been a few measurements of negative ions with various rocket techniques, but at present our knowledge of these species is poorly understood. It appears that many of the negative ions are relatively large, clustering with water in a manner similar to that of the positive ions.

Above 85 kilometers there is a constant injection of meteoric debris originating from various extraterrestrial bodies such as comets. This debris produces regions of enhanced metallic species, both neutral and charged. Lidar observatories have recorded layers of neutral sodium, iron, calcium, and other elements. In situ mass spectrometers on rockets have observed these species in ionic form, along with magnesium, sulfur, aluminum, and other metals, as well as related compounds such as their oxides, and attachment to water clusters.

Discussion of middle atmospheric composition would not be complete without mentioning the presence of aerosols and particulates, which probably permeate the entire region. Particulates may significantly affect the chemical structure of the middle atmosphere through surface interactions called *heterogeneous chemistry,* although our present knowledge of such atmospheric processes is rudimentary. At the lower altitudes within the stratosphere, the influx of particulates and aerosols from volcanoes, thunderstorms, and other strong tropospheric disturbances can penetrate the tropopause and lead to the deposition of layers of macroscopic or submacroscopic materials within the stratosphere. The long chemical lifetimes of these materials, coupled with the mainly horizontal transport within this region, cause them to remain for periods on the order of years. Layers at between 20 and 30 kilometers have been observed by balloons and have been found to show annual cycles in abundance. At higher altitudes within the mesosphere, particulates can also arise from ablation of meteorites and other forms of extraterrestrial bombardment, as well as from the formation of ice particulates. Nacreous clouds in the stratosphere and the noctilucent clouds seen in the summer polar mesosphere are visible manifestations of this type of layering by macroscopic particulates. [*See* Aerosols.]

Dynamical and Electrodynamical Considerations.
The middle atmosphere is subject to a wide range of forces affecting its circulatory patterns, varying in magnitude from global down to 1 meter or less. Planetary waves and solar and lunar tidal motions influence the large-scale structure, while mesoscale tropospheric disturbances such as hurricanes and thunderstorms produce

gravity waves that affect more localized regions. Small-scale turbulence is also produced in the mesosphere by the breaking of gravity waves in this region. Internal gravity (buoyancy) waves are produced by airflow over orographic features such as mountains; they are responsible for lee waves and clear air turbulence (CAT) and may also produce turbulence at higher altitudes.

Within the stratosphere, tropospheric weather disturbances decay rapidly with altitude, so that the flow becomes completely dominated by zonal mean circulation and planetary waves. This stability produces a flow which is predominently horizontal; vertical flow is usually negligible. Nonetheless, several unusual phenomena exist within this region. The quasi-biennial oscillation (QBO) in the mean zonal winds of the equatorial stratosphere shows a cycle with a period of 24 to 30 months, in which the winds are first westerly and then become easterly. The effect begins at higher altitudes and exhibits an increasing lag with depth down to the tropopause. A complex mix of vertically propagating waves is believed to drive the QBO. Another phenomenon is the sudden stratospheric warmings that arise every few years in the winter polar region. Within a period of a few days, the polar vortex breaks down, with an accompanying heating of the polar stratosphere. Since the warming is mainly a Northern Hemisphere phenomenon, it is concluded that orographically forced waves play an important role in its occurrence; the Southern Hemisphere has much smaller mid-latitude land masses, incapable of producing the required stationary planetary waves. [*See* Quasi-Biennial Oscillation.]

The mesosphere shows evidence for large-scale transport and wave structure, but it is also a region of small-scale turbulence, mainly between 70 and 90 kilometers altitude. These small-scale variations have time constants on the order of seconds; this, coupled with our inability to linger for in situ studies, makes this a very difficult region to study. Figure 3 shows the complex evolution of turbulence developing from a breaking gravity wave in this region, as described by a three-dimensional computer model developed by D. Fritts. The ionospheric plasma (electrons and ions) plays an important role in acting as a tracer for neutral transport and turbulence in this region. The turbulence was first discovered using releases of barium and other gases from rockets in the 1960s. More comprehensive surveys with MST (Mesosphere, Stratosphere, Troposphere) radars, usually near 50 megahertz, have shown the turbulence to be present at all latitudes, but most intense and frequent during the polar summer. The polar summer mesospheric echoes

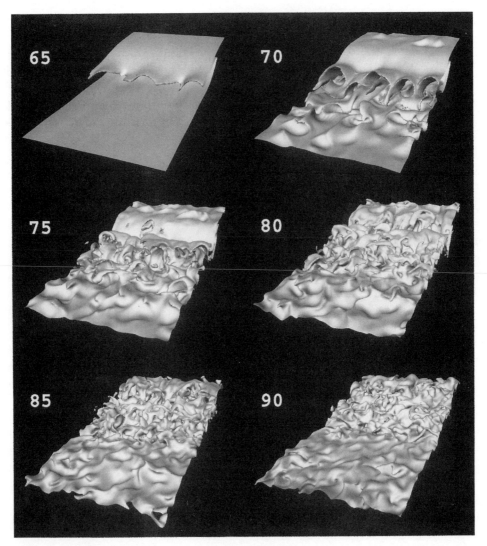

UPPER ATMOSPHERE. Figure 3. *Six snapshots showing the progression of a mesospheric gravity wave breaking from a three-dimensional model simulation.*

(PSMEs) have also been observed by higher-frequency radar observatories, sometimes at frequencies in the gigahertz range. The echoes are caused when the radio waves are scattered by free electrons, which are thought to map the neutral turbulence. Simple scattering theory precludes the possibility of electron scattering at these frequencies, necessitating new ideas on how to adjust the electron scattering cross-section to account for the observed features. These new theories invoke the need for heavy molecular ions, such as the water cluster ions discussed earlier, or the presence of aerosols and particulates, which might be found in the vicinity of noctilucent clouds. Our understanding of these phenomena is further clouded by recent findings that seem to imply that dif-

ferent scattering layers may originate from independent processes which do not all relate to turbulence. [*See* Scattering.]

Noctilucent clouds (recently renamed *polar mesospheric clouds*) are an unusual mesospheric phenomenon. These are the highest clouds in our atmosphere, occurring between 82 and 84 kilometers altitude during the polar summer, when the mesopause can reach temperatures as low as 110 K. These clouds were first discovered more than a century ago and have been extensively studied—remotely from ground and satellite, and in situ by rocket—yet today we still are unsure of their origin and composition. We suspect that they are formed by hydrated particulates (ice particles) that grow to visible size. The clouds, when observed from the ground, show all the features of mesospheric wave patterns and transport, exhibiting a structure that can best be described as resembling the surface of the ocean. Unfortunately, ground viewing of noctilucent clouds requires the Sun to be more than 5° below the horizon, and also the absence of lower, tropospheric cloud cover. Satellites can sense these clouds remotely from above; by this means they are found to blanket the polar cap during the polar summer, but to become less extensive and less frequent as lower latitudes are approached. Noctilucent clouds are evidence for the presence of particulates and may have a relationship to the origin of the PMSE echoes, but this is still uncertain. [*See* Mesosphere.]

Turbulence and sudden transport movements do not stop at the mesopause but continue into the lower thermosphere. For example, neutral sodium of meteoric origin has been observed by lidar to form sudden and very narrow layers (about 1 kilometer thick) near 95 kilometers, which appear and disappear in seconds after a lifetime on the order of minutes. There is no known explanation for these phenomena. In the same region, sporadic E layers (E_s) occur; during their lifetimes, the electron density is enhanced by several orders of magnitude in a narrow layer. The ions responsible for this enhancement are metallic species, and it is believed that a combination of wind shear forces and electric field anomalies drives this effect.

In addition to tracking neutral dynamics, the charged particle environment can also exhibit unique properties in an electrodynamical sense. Electric fields created in the magnetosphere propagate downward and provide information useful for reconstructing electrodynamic drifts and related processes occurring at much higher altitudes. Furthermore, the electric field of the middle atmosphere is perturbed by the presence of aerosols and particulates, by orographic effects, and by thunderstorms and other meteorological disturbances, and that at higher latitudes is perturbed by galactic cosmic radiation. The middle atmosphere is also a pass-through region for both direct and alternating electrical currents, so that it acts to couple electrically the upper and lower atmospheres. There is also evidence that particulates may be responsible for the formation of very large electric fields (on the order of volts per meter) within the upper stratosphere and lower mesosphere. These fields are 2 to 3 orders of magnitude larger than expected for this region. These events are infrequent and difficult to predict; furthermore, the rocket techniques used to measure them have been ques-

tioned. Nonetheless, if real, they could affect the general structure of the global electric circuit by adding significant current to the system. These fields appear to exist at heights where external radiation from energetic electrons and protons, such as might occur during solar active events at high latitude, could modulate the local electrical conductivity and thereby influence the large electric fields and their effects. If solar activity could regulate the current system of the global electric circuit in this manner, it could influence the tropospheric electrical environment, affecting aerosol growth, cloud formation, lightning frequency, and other processes; this would link solar activity to tropospheric weather. Assessing the impact of these electric fields, however, must await their validation and better understanding of their frequency and extent.

[*See also* Atmosphere; *and* Global Electric Circuit.]

BIBLIOGRAPHY

Banks, P. M., and G. Kockarts. *Aeronomy.* New York, 1973.
Danilov, A. D. *Chemistry of the Ionosphere.* New York, 1970.
Herman, J. R., and R. A. Goldberg. *Sun, Weather, and Climate.* New York, 1985.
Holton, J. R. *An Introduction to Dynamic Meteorology.* 2d ed. New York, 1979.
Rees, M. H. *Physics and Chemistry of the Upper Atmosphere.* Cambridge, 1989.

RICHARD A. GOLDBERG and KENNETH H. SCHATTEN

URBAN CIRCULATION. *See* Small-Scale or Local Winds.

URBANIZATION. Every aspect of weather and climate is affected by urban development; as a result, the atmosphere of cities provides some of the best examples of human-induced climate modification. The cause is the drastic disruption of surface soils, vegetation, stream networks, and the atmosphere that accompanies the inexorable process of site clearance, construction, and escalating human activity in cities—a process known in general as *urbanization*. These alterations lead to the creation of an environment that is a better store of heat, a poorer store of water, a greater drag on airflow, and a net source of heat and pollutants. As soon as weather observation networks were established in the nineteenth century, differences were noted between urban and rural levels of visibility, sunshine, wind, temperature, humidity, cloud, fog and precipitation. Such effects are now known to occur in all cities.

A city presents a relatively rough, warm surface that typically creates more turbulence than the surface of its rural surroundings. This leads to better mixing and a deeper boundary layer of air. When winds are strong, the extra drag of the city reduces wind speeds above roof-level, which in turn causes the flow to change direction as it enters the city and to resume its original direction after it exits downstream. In near-calm conditions, the pressure gradient created by the warmer city produces its own low-level convergent *urban circulation,* similar to the sea breezes flowing into an island from all sides. Both effects produce convergence and consequent uplift over the city. The frictional effect can also delay the passage of synoptic and mesoscale fronts across a large city. [*See* Small-Scale or Local Winds.]

The surface and atmospheric changes of urbanization alter the transfer of radiation, heat storage, and the availability of surface moisture for evaporation. This changes the cycling of heat and water and creates an urban climate with distinct surface temperatures and levels of humidity. The result is an atmosphere that is generally warmer (called an *urban heat island*), especially at night, and less moist, especially by day. The city may be more humid at night (and in the winter, in cold climates), partly owing to water vapor released by fuel combustion. Many of these urban effects increase as the city grows over time. [*See* Heat Island Effect.]

The increased warmth, as well as greater humidity and aerosol count, can alter the amount of nocturnal fog recorded so that it is not easy to generalize; however, many large cities are characterized by an increase in haze below cloud level in the urban boundary layer. It seems likely that the combination of greater uplift, a deeper boundary layer, and increased aerosol concentrations could produce more cloud and perhaps more precipitation over and downwind of a city. Several studies confirm this, but it is difficult to prove that extraneous influences such as topography are not involved and to pinpoint the reasons for specific phenomena.

The urban climate can have an impact on human health, economies, and societies. People living in cities usually experience added heat stress and exposure to pollutants. The altered climate must be considered in planning for the safety of built structures, building lighting, space heating and cooling, garden irrigation, urban water supply, flooding, road ice, snow removal, and the weathering of building materials. The transformation of pollutants through oxidation and reduction has become a major concern. Finally, forecasting urban weather events, crucial for millions of people, requires an understanding of the city's special climatic characteristics.

BIBLIOGRAPHY

Landsberg, H. E. *The Urban Climate.* New York, 1981.
Oke, T. R. *Boundary Layer Climates.* London, 1987.

T. R. OKE

V

VEGETATION AND CLIMATE. Vegetation patterns are strongly related to climate at several different spatial scales. As climate has changed over time, vegetation has changed with it. The different plant communities that appear as one travels from country to country (or even from one part of a mountain to another) are also strongly associated with variation in climate. The Ionian philosopher Theophrastus (ca. 370–285 B.C.E.) is sometimes called "the father of ecology" because of his observations relating climate and vegetation. In his *Enquiry into Plants* and minor works on *Odors* and *Weather Signs,* Theophrastus noted that when plants were transferred to regions with inappropriate climates, they did not grow or bear fruit, and that there was a relationship between "individual properties" (adaptations) of particular plants and climate.

Figure 1 shows the different vegetation communities at low, middle, and high elevations in an Australian mountain range. The climatic variations associated with rising terrain—primarily differences in temperature and precipitation—are among the factors causing this kind of zonation.

There is a surprising similarity in the patterns of vegetation in different parts of the world that have similar climates. This similarity can be observed at different scales. For example, the landscape-level patterns of spruce forests in Siberia and Canada look similar. Places that have recently been burned are colonized by deciduous trees such as aspen or birch; south-facing, warmer slopes have more productive forests. If one has learned to interpret the pattern of vegetation in Canada with respect to local climatic and topographic variation, then one can extend these interpretations to Russia. Of course, many species of the forests in Eurasia and those in North America are rather closely related. The same convergence, however, occurs in plant communities that are far apart in evolutionary terms. For example, the evergreen shrubland landscapes associated with the Mediterranean climates of Chile, California, Italy, and Australia look superficially similar, even though a different group of plants dominates each of them. [*See* Orographic Effects.]

Not only does vegetation display an overall similarity in similar climates, but the plants that make up these assemblages may also look alike even at the organ level. Thus different species of plants from different Mediterranean-climate shrublands may have thick, toughened *sclerophyllous* leaves with similar shapes, or they may produce *serotinous* seedpods that open only after a wildfire has passed through.

Plant Adaptations to Climate. Some of the patterns of similarity among plants in the same climates can be explained in terms of straightforward adaptations to climatic variables. Sometimes, however, apparent adaptations to one factor may actually be responses to entirely different factors. For example, producing tough seeds may appear to help the plants resist drying and wildfires in subtropical environments, but this may be most important as a barrier to the destruction of the seeds by insects.

One climatic factor that has a profound effect on the physiognomy (structural appearance) of the vegetation in different parts of the world is minimum temperature and their relation to various physiological adaptations to chilling. Table 1 lists cardinal minimum temperatures and the plant forms expected to be associated with these temperatures. (The term *cardinal* in this context refers to the absolute need for certain adaptations within certain temperature ranges.) These minimum temperatures are related to different mechanisms for adapting to low temperatures. For example, the freezing point of pure water (0°C) and the point at which supercooled water freezes (−40°C) are major divisions in the list of cardinal minimum temperatures. At temperatures below 0°C, water in plants is likely to freeze, with damage to plant tissues. At temperatures of −40°C, supercooled liquid water (water below the freezing point but still in liquid form) forms ice spontaneously. Different adaptations—such as using supercooling of water within tissues to prevent freezing

VEGETATION AND CLIMATE. Figure 1. *An elevational zonation of Eucalyptus forests in the Brindabella Mountains of Australia.* Three different zones can be seen: Next to the lake in the foreground is a "peppermint" forest (so named because the leaves of the dominant trees, *E. viminalis, E. robertsonii,* and *E. radiata,* smell like peppermint oil); in the middle is an alpine ash *(E. delegatensis)* forest; and in the background, at the top of the mountain, is snow gum *(E. pauciflora)* forest.

VEGETATION AND CLIMATE. Table 1. *Cardinal minimum temperatures and associated expected dominant vegetation*

Temperature Range (°C)	Response	Expected Physiognomy
>15	Temperature not limiting	Broad-leaved evergreen when rainfall adequate
−1 to 15	Chilling	Broad-leaved evergreen when rainfall adequate
−15 to 0	Freezing and supercooling	Broad-leaved evergreen
−40 to −15	Freezing and supercooling	Broad-leaved deciduous
<−40	Freezing and supercooling	Evergreen and deciduous needle-leaved

Source: Adapted from Woodward, 1986.

at low temperatures—allow plants to tolerate the lower temperatures.

Individual plants can also adjust their ability to tolerate stressful environmental conditions. In the process known as *hardening,* a plant becomes able to tolerate lower temperatures after it is exposed to progressively lower temperatures over a period of time—for example, during the gradually cooling fall period. If two plants of the same species, one protected in a greenhouse and the other exposed to outdoor conditions for increasing intervals, are exposed to the same low temperature, the greenhouse plant will be more damaged. Hardening thus involves an acclimation to low temperature at the tissue level.

Scale and Adaptations to Climate. The overall responses of plants to climatic conditions occur on several time scales. In the case of hardening, a plant's history in the preceding few weeks may determine whether cold stress is lethal to it. Furthermore, adaptations at one scale may fail owing to a lack of adaptations at other scales.

At the biochemical level, different plants employ three different biochemical pathways for photosynthesis: the C_3 or Calvin Benson cycle; the C_4 pathway; or the Crassulacean Acid Metabolism (CAM) pathway. Plants using the C_4 pathway (such as tropical grasses) have an advantage in that they can carry on photosynthesis in conditions of strong light, high temperature, and lower water availability than are optimal for C_3 plants (such as trees). Plants with the CAM pathway are not as efficient at photosynthesis as the other two kinds of plants, but they can survive in harsh, arid environments. The often bizarre succulents of Southern African deserts are a notable example. The proportion of plants in a community that employ each biochemical pathway is directly related to climate: there are more C_4 plants than other types in the humid tropics, and no C_4 plants in the tundra vegetation of high latitudes.

Leaves often vary in size and shape in response to climatic variables. In capturing light to provide energy for photosynthesis, the leaves of plants can experience significant heat loading, which is controlled by reradiation of heat from the surface, by convection of heat from the surface, and by evaporative cooling. The leaves maintain a complex balance by opening stomata (small pores, usually on the underside of the leaf) to admit the carbon dioxide needed for photosynthesis and to evaporate water vapor to cool the leaf. If too much water is lost too fast, though, the leaf wilts and may not recover. The size of the leaf as well as its surface properties (for example, shininess to reflect heat, or hairiness to alter the convective flow of heat and water vapor) can be important. The general advantages of these features are associated with the similarity of leaves in equivalent climatic regimes.

At longer time scales, plants at different life stages may respond differently to climatic conditions. Some plants can grow in environments in which they are unable to reproduce. Their flowers may be sensitive to frost, or a particular growing season may be too short for the plant to produce seed. Plants that often experience such rigors, for example in the Arctic, tend to be long-lived as individuals so that they have some optimal years. Conditions needed for seeds to germinate (begin growing) are often related to minimal, maximal, or cumulative temperatures, or to moisture. Seedlings of some species seem unable to survive conditions that can be tolerated by the adult plants. The smaller size of seedlings can make them more responsive and thus more sensitive to rapid variation in weather conditions than are the more massive mature plants.

At even longer time scales, if variations in the climate are sufficiently pronounced, the adaptations of plants in a given area may be unsuited to the new conditions. The development of a new plant community may be limited by the time required for the appropriately adapted flora to migrate to the location. The species most responsive to climatic variation on a time scale of millennia may be those best adapted to long-distance dispersal—an adaptation not often thought of as a response to climate. We have gained insight into the longer-term responses of vegetation to climate by studying past changes, particularly in the past 20,000 years.

Paleovegetation. The history of the vegetation at a given location is evident in fossils remaining at the site. In the reconstruction of past vegetation, the fossil remains of plants are often separated into *macrofossils* (remains of leaves, twigs, wood, and other large tissues) and *microfossils* (often plant pollen). The macrofossil evidence often allows one to determine the species of plants that were growing in a given locale; microfossils can usually be identified only to the level of genus. The different types of fossils are also clues to the spatial distribution of the vegetation. For example, a leaf or twig found in the sediments of a lake may have only blown a few meters to arrive in the lake; a large pollen grain may have traveled 100 meters; and a small, light pollen grain may have been transported 100 kilometers or more. The amount of pollen produced by various kinds of plants differs, so the presence of a few pollen grains may indicate either a distant source or the rare occurrence in the nearby landscape of an abundant pollen producer. A similar amount of pollen from a scant pollen producer would be interpreted quite differently. Some species are pollinated by insects rather than by wind, so their pollens are rarely found in sediments. These factors conspire to make the interpretation of pollen difficult. [*See* Paleoclimates.]

Nevertheless, paleoecologists have united to compare their data sets in attempts to understand the patterns seen at a given site. The use of dating methods based on radioactive traces (particularly the carbon-14 dating techniques) has proved invaluable for cross-comparing the vegetation at different sites. Important results of this work are the following:

• Species differ at the rate that they can disperse in their regional or continental patterns of distribution.

- Ecosystems do not appear to respond to climate change as a unit. Rather, the species of plants comprising the system appear to respond at different rates in rearranging their distributions.
- Within the past 10,000 years and certainly over longer time periods, vegetation associations can change greatly. Species that are almost always associated in the same ecosystems now were not always so. Combinations of species that formed past ecosystems are not found now.
- Climate variations in the tropics, especially moister fluctuations, seem to have displaced the vegetation substantially; this has also occurred in the higher northern latitudes in response to continental glaciation.
- There appears to be sufficient resolution in our ability to represent the patterns of vegetation change for continental areas at regular intervals over the past 18,000 years such that we can test models of climate over similar intervals.
- While individual plants can survive daily or seasonal variations, there can be significant ecosystem responses to small sustained climate change.

Current Challenges. One particularly challenging issue in the realm of vegetation–climate relations is prediction of the response of vegetation to climatic change. This interest is motivated both by concern about how vegetation may respond to a greenhouse warming, and by paleoecological work in reconstructing past vegetation. Predictive capability is being developed by using computers to solve mathematical models of ecological responses to climate variation or change at several different scales. Major problems center around the difficulty of testing models that make long-term predictions of change in response to previously undocumented conditions. The data being collected by paleoecologists is of major value as an analogue for the model conditions. Similarly, the vegetation models can be tested for their ability to predict changing patterns of vegetation with elevation or latitude. These patterns, which led Theophrastus to discourse on the relation between climate and vegetation more than 2,000 years ago, are today involved in testing computer models.

BIBLIOGRAPHY

Box, E. O. *Macroclimate and Plant Forms: An Introduction to Predictive Modeling in Phytogeography.* The Hague, 1981.
Budyko, M. I. *Climate and Life.* New York, 1974.
Muller, M. J. *Selected Climatic Data for a Global Set of Standard Stations for Vegetation Science.* The Hague, 1982.
Smith, R. L. *Ecology and Field Biology.* 4th ed. New York, 1990.
Stephenson, N. L. "Climatic Control of Vegetation Distribution: The Role of the Water Balance." *American Naturalist* 135 (1990): 649–670.
Walter, H. *Vegetation of the Earth and Ecological Systems of the Geo-Biosphere.* 2d ed. New York, 1979.
Woodward, F. I. *Climate and Plant Distribution.* Cambridge, 1986.

HERMAN H. SHUGART

VIRGA. *See* Precipitation.

VIRTUAL TEMPERATURE. *See* Conditional and Convective Instability; *and* Temperature.

VISIBILITY. Atmospheric visibility is the transparency of the atmosphere to visible light. Meteorologists have traditionally characterized visibility in terms of visual range, the greatest distance at which a normal observer can distinguish a dark object from the surrounding sky. Visual range is an important factor in the safety of aircraft operations, so it is among the meteorological variables routinely monitored at airports. Other more subtle aspects of image transmission, such as discoloration and loss of detail, are now receiving attention as the result of clean air laws aimed at mitigating the aesthetic impacts of air pollution.

Figure 1 illustrates the way the atmosphere modifies the appearance of a distant object. Some of the light reflected from a visual target toward an observer is absorbed along the way (a) or scattered in other directions (b), leaving only a fraction of the initial signal to reach the observer (c). At the same time, extraneous light from other directions is scattered into the observer's line of sight by the intervening atmosphere (d). This added *airlight* acts as visual noise which masks the diminished signal from the target. The visible sky is pure airlight

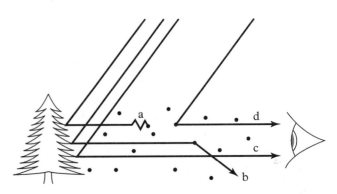

VISIBILITY. Figure 1. *Elements of daytime visibility.*

which we see when we look at the void of space. It is the great increase in airlight rather than any change in atmospheric transmittance that makes the stars disappear at dawn. [*See* Radiation; Scattering.]

The scattering and absorption sketched in Figure 1 arise from the interaction of light with individual air molecules, droplets, and suspended particles. Just as a macroscopic object's optical impact can be measured by its shadow, microscopic objects' impacts can be quantified in terms of their effective cross sections for scattering and absorption. The atmospheric concentration of scattering and absorbing cross sections is called the *extinction coefficient*, b_{ext}. With dimensions of cross section per air volume, or 1/length, b_{ext} represents the fraction of light intercepted in a layer of unit depth.

In 1925, Harald Koschmieder showed the visual range r_v in a uniform and uniformly illuminated atmosphere to be inversely proportional to the extinction coefficient b_{ext}:

$$r_v = \log_e \epsilon / b_{ext}.$$

The quantity ϵ in this expression is the minimal contrast between target and sky that the eye can detect. A typical observer can just barely spot a large object when it is about 5 percent darker than the sky, and this yields a value of about 3 for the dimensionless coefficient $\log_e \epsilon$.

Visibility is very sensitive to the atmospheric content of droplets and particles. A given concentration of material scatters much more light in the liquid or solid phase than it does as a gas; this is why clouds seem to appear out of nowhere when water vapor condenses. The most effective scatterers are droplets and particles with diameters comparable to the wavelength of visible light, around 0.5 microns. A droplet or particle concentration of only 10 microns per cubic meter can yield extinction of 3 to 5×10^{-5} per meter; for comparison, pure air at a concentration of 1 kilogram per cubic meter produces extinction of only 1×10^{-5} per meter. The droplets or particles in this example carry only 10 billionths of the total gas, liquid, and solid mass, but Koschmieder's formula shows that they reduce the visual range from about 300 kilometers to between 50 and 75 kilometers.

The most dramatic reductions in visibility are associated with weather events such as rain, snow, and fog, events involving high concentrations of liquid or frozen water. Other natural events that can sharply reduce visibility are wildfires and dust storms, which inject large quantities of particulates into the atmosphere.

More persistent and pervasive visual impairment arises from human activity. In urban-industrial regions such as eastern North America, Europe, and parts of Asia, combustion of coal and oil is an important source of visibility-reducing particles. Such combustion produces particles directly, in the form of soot and ash, and indirectly, through gaseous sulfur dioxide emissions that are converted in the atmosphere to sulfate particulate. Sulfate and other hygroscopic particles promote the condensation of ambient water vapor at humidities below saturation, enhancing the optical impact of these particles. In many rural areas, open burning of vegetation in a variety of agricultural and forestry operations is an important source of optically effective particles.

BIBLIOGRAPHY

Middleton, W. E. K. *Vision through the Atmosphere.* Toronto, 1952.
National Acid Precipitation Assessment Program. *Visibility: Existing and Historical Conditions.* State of Science and Technology Report 24. Washington, D.C., 1991.
National Research Council. *Haze in National Parks and Wilderness Areas.* Washington, D.C., 1992.

WARREN H. WHITE

VISIBLE ELECTROMAGNETIC RADIATION. *See* Shortwave Radiation.

VOLCANOES.
The emission of gases from the Earth, manifested as volcanic activity, helped create the atmosphere and continues to modify it. The gaseous composition of the atmosphere has evolved and has been responsible for climate change over long (geologic) periods of time; clouds and other atmospheric particulates or *aerosols* produce shorter-term but important effects on climate and weather. Volcanic eruption clouds are the most important of several intermittent and short-lived natural atmospheric perturbations that can influence the radiation budget, surface temperatures, and circulation patterns.

Long-term Trends in Atmospheric Composition. The Earth's atmosphere has developed over the past 4.5 billion years through volcanic emissions of carbon dioxide and water vapor, the most common volcanic gases, together with nitrogen (which occurs in low abundance in present-day volcanic gases) and perhaps methane. Both vigorous degassing during eruptions and quiescent degassing from active but non-erupting volcanoes has been and still is important. Volcanic volatile emissions also include sulfur, chlorine, and fluorine. Many other gases, such as argon, as well as trace metal species, are produced in minor amounts. Sulfur is indirectly of prime importance in modulating climate in the short term, as explained below; chlorine has been important in the development of the composition of the oceans.

The Earth's early atmosphere was dominated by carbon dioxide until the advent of life about 3.5 billion years ago, when photosynthesis and burial of organic carbon led to the accumulation of oxygen in the environment. To reach a significant level of oxygen took until about 2 billion years ago. Since that time the atmosphere has evolved to its present composition: nitrogen, approximately 78 percent by volume; oxygen, approximately 21 percent; and all other species in trace amounts totalling 1 percent. The nitrogen-dominated composition of our atmosphere is due to the fact that various biogeochemical cycles have developed which process oxygen and carbon and bind them in compounds, whereas nitrogen is relatively inert and has become concentrated in Earth's atmosphere over time. Despite continued outgassing of volcanic carbon dioxide, the CO_2 content of the atmosphere (at present 350 parts per million) seems to have been generally decreasing over geologic time. Periods of higher outgassing that correspond to elevated levels of volcanism—for example at subduction zones and at the mid-ocean ridges during periods of higher lithospheric plate production and spreading—may have been related to higher atmospheric CO_2 concentrations and periods of warmer climate during Earth's history. [See Atmosphere, article on Formation and Composition.]

Volcanic Aerosols. Sulfur is released into the atmosphere during eruptions mainly as sulfur dioxide gas, which is converted by photochemical oxidation to sulfuric acid via hydroxyl radicals. Sulfuric acid and water are the dominant components (70 percent and 30 percent respectively). Aerosols form liquid spheres (about 0.3 micron modal radius), with solid silicate (ash) cores dominant in the first month. Atmospheric residence time ranges from days, weeks, months, or years for regional, zonal, hemispheric, and global distribution, respectively.

Shortwave radiation scattering from aerosols leads to surface and lower tropospheric cooling, while longwave absorption warms the stratosphere. Injection of other volatile compounds, such as hydrochloric acid and water, alter air composition; composition is further altered by heterogeneous reactions on sulfate aerosol surfaces. Volcanic aerosols affect stratospheric stability and tropospheric dynamics and may provide nuclei for upper tropospheric cirrus clouds.

Latitude and time of year of eruptions influence dispersion of volcanic aerosol clouds. No anthropogenic influences to stratospheric aerosols have been noted, but future chlorofluorocarbon buildup could have an impact on ozone and hence on aerosol-derived atmospheric reactions. Significant eruptions, and therefore aerosol clouds, occur every 10 to 20 years on average, with a major eruption perhaps every 100 years.

Shorter-term Effects of Volcanic Aerosols. The impact of volcanic eruptions on Earth's climate system has been documented and debated for two centuries. Considerable evidence now exists for a connection between explosive volcanism, the generation of volcanogenic aerosols, and climatic cooling. Other atmospheric perturbations, such as ozone depletion, are also associated with periods of enhanced atmospheric aerosols. In addition to gases, volcanic eruptions emit volcanic ash and dust, tiny particles of silicate glass derived from fragmentation of the magma that rises from within the Earth and causes the eruptions.

Scientists first recognized and observed volcanic atmospheric dust veils covering vast areas in 1884, the year after Krakatau had erupted in Indonesia, but it was then thought that fine-grained silicate dust was the cause of the atmospheric perturbations. However, since the discovery of the stratospheric sulfate aerosol layer by Christian Junge and his associates in the 1960s, it has become evident that this layer in the atmosphere, which plays a large role in modulating the net incoming radiation, is strongly influenced by volcanogenic sulfuric acid droplets derived from the gaseous sulfur dioxide—and to a lesser extent hydrogen sulfide—released from the magma.

Silicate dust from eruptions seems to remain in the atmosphere for a shorter time (a few months) and to cause smaller effects than sulfuric acid aerosols, but the potential impact of this dust in great abundance is poorly known. Other volcanic volatiles such as water vapor and halogens may be important in determining ozone levels, particularly as concentrations of chlorofluorocarbons continue to increase in the atmosphere. Moreover, it has recently been realized that depletion of stratospheric ozone in temperate latitudes occurs after volcanic aerosol injections; processes involving sulfate aerosols as sites of heterogeneous chemical reactions are implicated, similar to those involving chlorofluorocarbons that are thought to cause the ozone hole. [See Aerosols; Ozone Hole.]

Historic Volcanism. Through study of historic eruptions and their atmospheric impacts, we have learned what types of eruption can influence the Earth's climate. Climatic effects appear to ensue if the eruption cloud is powerful enough to inject material into the stratosphere, or if the eruption is of sufficient duration that a higher-than-normal level of tropospheric aerosols is maintained. The impact on climate also depends on the sulfur content of the magma; some volcanoes erupt sulfur-rich magmas, some magma with average amounts, and some sulfur-poor magmas. The source of volcanic sulfur is yet to be determined; some may be primordial, and some may be recycled in the Earth's lithosphere. If an eruption is large enough, even with a low magmatic sulfur content, sufficient magma is erupted to cause a significant release of sulfur gas, leading to the generation of aerosols.

Benjamin Franklin is usually credited with first making the association between a volcanic eruption and climate or weather. While he was U.S. Ambassador in Paris in 1783, he theorized (published in an article in 1784) that the awful haze or dry fog, presumably sulfuric acid aerosols, that hung over the city in the latter part of 1783 was caused by an eruption in Iceland. In fact, one of the world's greatest historic lava-producing eruptions was happening at Laki, Iceland, from June 1783 to February 1784. In this event, most of the aerosols generated may have been limited to the troposphere; this is a good example of an eruption of long duration maintaining tropospheric aerosols. Under this haze, as with other aerosol veils, the Sun's rays were dimmed, and anomalously hot or cold weather occurred, including bitter cold in August and the coldest winter on record in New England. Crop failures were commonplace the next year and were also experienced after the next large eruption, which occurred in 1815.

The year 1816, "the Year without a Summer," is probably the best-known example of a volcanically induced climate cooling event. It was largely caused by the eruption of Tambora in Indonesia, in April 1815, which erupted 50 cubic kilometers of magma and yielded 100 megatons (10^{14} grams) of sulfuric acid aerosols, in one of the largest known eruptions of the past few millennia. The aerosol veil caused cooling of up to 1°C regionally and probably more locally, with effects lasting until the end of 1816 and extending to both hemispheres. It snowed in New England in June, and abnormally cool temperatures in Europe led to widespread famine and misery, although the connection to volcanic aerosols was not realized at the time.

The aftermath of Krakatau's climactic eruption in August 1883 was probably the first time that scientists widely took notice of a volcanically induced atmospheric event, because the developed world rapidly learned of the eruption by telegraphic communication. Observers could relate to the eruption the colorful sunsets and other optical phenomena that spread over the Northern Hemi-

sphere in the months following. The Royal Society of London published a volume in 1888 that documented the eruption and the ensuing dust veil. After Krakatau, scientists measured and reported decreases in incoming radiation following volcanic eruptions in 1902 and 1912.

The next eruption to produce a significant stratospheric aerosol veil was Gunung Agung on Bali, Indonesia, in 1963. This small but violent eruption with a large sulfur release occurred at a time of year when the circulation patterns caused most of the aerosol cloud to move into the Southern Hemisphere. Immediately following this event, which occurred soon after the existence of the stratospheric sulfate aerosol layer had been discovered, the first airborne samples of aerosol droplets were collected between Australia and Bali. Temperature changes after the Agung eruption were also measured widely for the first time; it was shown that the tropical troposphere cooled by as much as 0.5°C, while the stratosphere in the same region warmed by several degrees.

Some of the most significant data bearing on the relationship of volcanism to climate come from proxy sources, such as tree rings and ice cores. The widths of annual tree growth rings vary according to climatic conditions, becoming thinner in anomalously cold or dry years and thicker in unusually wet years. Several dendrochronologic studies have shown that tree rings from various areas around the world record anomalous growth conditions in years following major eruptions. Perhaps even more useful, though sometimes difficult to interpret, are annual layers in ice cores drilled in major polar ice caps such as Greenland and Antarctica. Because considerable amounts of atmospheric aerosols fall to the Earth's surface over the poles, in a process called *sedimentation,* acid fallout contaminates the surface snow. Eventually the snow becomes buried and compacted to glacial ice, preserving an annual record of acid fallout that can be detected by measuring the electrical conductivity of an ice layer. Almost every historic major eruption that has produced significant aerosols shows up in one of the ice cores; the biggest, such as Tambora, appear in both the northern and southern polar ice caps. The Greenland ice cores show a higher-than-average number of smaller acidity peaks in a period corresponding to the Little Ice Age. This suggests that increased frequency of sulfur-producing volcanism for several centuries might have a moderating effect on climate over such time periods. Some of the newest ice cores provide a record back more than 100,000 years, with errors in the age estimate of each layer increasing with time; these will eventually give us valuable information on ancient eruptive activity. [*See* Cores; Little Ice Ages; Tree-Ring Analysis.]

Several recent eruptions have proven invaluable in improving our understanding of the connection among volcanism, aerosols, and climate change. Two eruptions of very sulfur-rich magma—at El Chichón, Mexico, in 1982 and Mount Pinatubo, in Luzon, Philippines, in 1991—have been thoroughly documented in terms of the stratospheric aerosols produced, their spread and atmospheric residence time, and their climatic effects. El Chichón was a small-volume eruption of unusually sulfur-enriched magma which produced the first large aerosol veil to be tracked by instruments on satellites. Its impact on surface temperatures has, however, been difficult to assess. Pinatubo was the second-largest eruption in the twentieth century (the largest was in Alaska in 1912), and its aerosol veil hung over the Earth for more than 18 months, causing spectacular sunsets and sunrises worldwide. There was more material in the aerosol layer than at any time since Krakatau erupted. General circulation models of temperature changes expected under this aerosol cloud predicted Northern Hemisphere surface cooling of as much as 0.5°C, and these appear to have been verified by the meteorological data.

Extent of Climatic Effects. Although the largest historical eruptions have taught us a great deal about the connections between volcanism and climate, in historic times there has been no truly major volcanic eruption—one that compares with, for example, the Toba eruption in Sumatra some 74,000 years ago. This event produced about 2,800 cubic kilometers of magma, compared with 0.5 cubic kilometers for Mount St. Helens (in the Cascade Range, Washington) in 1980, or 5 cubic kilometers for Pinatubo. It occurred at a time when the world's climate was getting much colder at the onset of a glacial period. The connection between these two events, the great eruption and the abrupt cooling, may not be coincidental, but the available data show that the cooling started before the eruption occurred. Study of this and other great eruptions led to the concept of *volcanic winter,* a natural parallel to nuclear winter. [*See* Nuclear Winter.]

Recent studies have established a strong correlation between certain types of eruption and climate change, and some suggest that there may be a simple relationship between the sulfur yield of a volcanic eruption and decrease in surface temperature. It is known that enhanced levels of aerosols after volcanic eruptions coincide with periods of lowered incident solar radiation, lower surface and tropospheric temperatures, and stratospheric warming. In reality, however, the magnitude of the signal in the surface temperature record is difficult to detect against background variation, even for the biggest historic eruptions. Recently it has been suggested that for the El Chichón and Agung cases, subtracting the warming effects of El Niño from the temperature record makes the volcanic signal clearer. Simple models suggest that eruptions on the scale of large prehistoric events have the potential to alter climate to a much greater degree than has been observed following historic events. A mounting body of proxy data suggests that the climatic and environmental changes produced by eruptions in the early Middle Ages may have been dramatic. However, there may be self-limiting mechanisms on the loading of the stratosphere by aerosols.

How large an impact on climate can volcanic eruptions have in terms of affecting surface temperature, precipitation, and weather patterns—either short-term effects from individual events, or longer-term changes controlled by periods of enhanced volcanism? Eruptions have a frequency that varies according to type and magnitude of the event. Although the greater the eruption, the lower the frequency of occurrence. Large eruptions are a certainty in the future. We have an imperfect record of past eruptions and of changes in temperature and weather; future events will provide natural experiments in which we can study the impact of volcanic aerosols, an effort already begun.

Perhaps the most important challenge to the research effort on global change, for the purpose of making policy decisions, will be the early detection of the effects of global warming caused by human activities. It will be essential to recognize and remove effects resulting from natural phenomena, among which volcanic aerosols are significant. Nevertheless, we are still a long way from understanding fully the effects of volcanic eruptions on atmospheric composition, climate, and weather.

BIBLIOGRAPHY

Harington, C. R., ed. *The Year without a Summer? World Climate in 1816.* Ottawa, 1992.

Lamb, H. H. "Volcanic Dust in the Atmosphere; With Its Chronology and Assessment of Its Meteorological Significance." *Philosophical Transactions of the Royal Society of London,* Series A, 266 (1970): 425–533.

McCormick, M. P., L. W. Thomason, and C. R. Trepte. "Atmospheric Effects of the Mt. Pinatubo Eruption." *Nature* 373 (2 February 1995): 399–404.

Palais, J., and H. Sigurdsson. "Petrologic Evidence of Volatile Emissions from Major Historic and Pre-historic Volcanic Eruptions." In *Understanding Climate Change,* edited by A. Berger, R. E. Dickinson, and J. W. Kidson. AGU Geophysical Monograph 52. Washington, D.C., 1989, pp. 31–53.

Rampino, M. R., and S. Self. "The Atmospheric Impact of El Chichón." *Scientific American* 250 (January 1984): 48–57.

Rampino, M. R., S. Self, and R. B. Stothers. "Volcanic Winters." *Annual Review of Earth and Planetary Science* 16 (1988): 73–99.

Stommel, H., and E. Stommel. "The Year without a Summer." *Scientific American* 240 (June 1979): 73–99.

Stothers, R. B. "The Great Eruption of Tambora and Its Aftermath." *Science* 224 (1984): 1191–1198.

Toon, O. B., and J. B. Pollack. "Atmospheric Aerosols and Climate." *American Scientist* 68 (1980): 268–278.

STEPHEN SELF

VORTICITY.

[This article treats a technical aspect of climate and weather studies; it is intended for readers at an advanced level.]

Vorticity may be considered to be the local rotation of a fluid. Even in unidirectional flow in which speed varies in the cross-stream direction, vorticity and a tendency to turn or rotate exist. The definition of vorticity $\boldsymbol{\omega}$ equates it to the curl of the velocity:

$$\boldsymbol{\omega} = \nabla \times \boldsymbol{U}.$$

The distribution of vorticity in a fluid is a useful quantity with which to describe flows at high Reynolds number. (The Reynolds number R is defined by $R = UL/V$, where U is a characteristic velocity, L a characteristic length, and V the kinematic viscosity of the fluid; R is a measure of the ratio of the inertial force and the viscous force. The reason is that in a homogeneous fluid—that is, one of constant density—vorticity cannot be created or destroyed within the interior of the fluid. It is produced only at the boundaries, whence it is diffused by viscosity and convected into the fluid. The rules for the ways in which vorticity changes as a parcel of fluid moves are well known.

Consider a fluid where motion in an inertial reference frame is governed by the Navier-Stokes equation:

$$\rho \frac{Du_i}{Dt} = \rho \frac{\partial \Psi}{\partial x_i} - \frac{\partial p}{\partial x_i} + \frac{\partial}{\partial x_i} \left\{ 2\mu \left[e_{ij} - \frac{1}{3} \Delta \delta_{ij} \right] \right\} \quad (1)$$

where $\rho(\partial \Psi / \partial x_i)$ represents a conservative body force, μ is the viscosity coefficient, and

$$e_{ij} = \frac{1}{2} \left(\frac{\partial u_i}{\partial x_j} + \frac{\partial u_j}{\partial x_i} \right),$$

$$\Delta = \frac{\partial u_j}{\partial x_j},$$

$$\delta_{ij} = \text{Kroenecker delta}.$$

The fluid's motion must also satisfy conservation of mass:

$$\frac{1}{\rho} \frac{D\rho}{Dt} + \frac{\partial u_j}{\partial x_j} = 0. \quad (2)$$

Along with these is an energy equation which is not immediately needed.

The vorticity tendency equation is obtained by taking the curl of Equation 1. This results in the following:

$$\omega_a = \nabla \times \boldsymbol{u}$$

$$\frac{D\omega_a}{Dt} = \omega_a \cdot \nabla \boldsymbol{u} - \omega_a \nabla \cdot \boldsymbol{u} \quad (3)$$

$$+ \frac{\nabla\rho \times \nabla p}{\rho^2} + \nabla \times \left(\frac{\boldsymbol{F}}{\rho} \right).$$

The third item on the right-hand side is the baroclinic production term; the fourth term is the frictional production term, where $\boldsymbol{F}$ has components F_i given by:

$$F_i = \frac{\partial}{\partial x_j} \left\{ 2\mu \left[e_{ij} - \frac{1}{3} \Delta \partial_{ij} \right] \right\}$$

as in Equation 1.

We consider first a homogenous ($\nabla\rho = 0$) incompressible ($\nabla \cdot \boldsymbol{u} = 0$) fluid, for which Equation 3 reduces to

$$\frac{d\omega_a}{dt} = \omega_a \cdot \nabla \boldsymbol{u} + \nu\nabla^2\omega_a \quad (3a)$$

where $\nu = \mu/\rho$. The term $\omega_a \cdot \nabla \boldsymbol{u}$ in Equation 3a can be interpreted as changes in vorticity produced by the flow, with the vorticity moving with the material parcels. For example, if $\omega_a \cdot \nabla \boldsymbol{u}$ has only z-component, namely $\zeta_a(\partial w/\partial z)$, where ζ_a is the z-component of ω_a, w the z-component of $\boldsymbol{u}$, and if $\partial w/\partial z$ is positive, the fluid parcel must be elongating in the direction of ω_a, and contracting in the perpendicular direction (from Equation 2). The resulting increase in ω_a is said to be caused by vortex stretching.

The term $\omega_a \cdot \nabla \boldsymbol{u}$ in Equation 3a is obviously zero in the case of two-dimensional flow, and if, further, ν is negligible, then ω_a would be constant following the motion.

The circulation Γ is defined by

$$\Gamma = \oint_C \boldsymbol{u} \cdot d\mathbf{l} \quad (4)$$

where $d\boldsymbol{l}$ is an element of arc around the closed curve C. If C is a reducible curve, or any curve in a simply connected domain, then by Stokes' theorem,

$$\Gamma = \int_A \omega_a \cdot dA$$

where C is the perimeter of the area A. Then, by using Equation 3a, it is easily shown that

$$\frac{d\Gamma}{dt} = -\nu \oint_C (\nabla \times \boldsymbol{\omega}) \cdot d\boldsymbol{l}. \qquad (5)$$

For an inviscid fluid, Equation 5 reduces to Kelvin's theorem, which states that for an inviscid, incompressible, homogeneous fluid, the circulation Γ is constant following the motion.

For a fluid in solid-body rotation at angular velocity Ω, its vorticity is 2Ω. In general, the absolute vorticity ω_a can be written as

$$\omega_a = \omega_r + 2\Omega$$

where ω_r is relative vorticity.

Application. In meteorological and oceanographic applications, it is often the vertical component of vorticity, ζ, which is most important. Here, the absolute vorticity ζ_a is composed of the relative vorticity ζ_r and the Earth's vorticity f, that is:

$$\zeta_a = \zeta_r + f \qquad (6)$$

where $f = 2\Omega \sin \phi$; Ω is the angular velocity of the Earth's rotation, ϕ is the latitude, and f is the local Coriolis parameter. ζ_a denotes the local vertical component of total vorticity. In natural coordinates we can write

$$\zeta_a = -\frac{\partial V}{\partial n} + \frac{V}{R_s} + f$$

where s is along the flow and n is normal and to the left of the flow. Shear vorticity is denoted by $-\partial V/\partial n$, and V/R_s denotes the curvature vorticity, where R_s is the radius of curvature of the local flow. For barotropic nondivergent flows, the sum of shear, curvature, and Earth's vorticity is an invariant. These three components keep exchanging the vertical component of vorticity among themselves as the barotropic flow evolves. Thus, shear vorticity can end up as curvature vorticity, resulting in a local storm in barotropic nondivergent flows. Shear flow instabilities can result in the growth of curvature vorticities, and other mechanisms—such as cumulus convection—can result in the enhanced growth of disturbances, especially in tropical latitudes. Thus, the vorticity exchange from one species to another is an important process in fluid dynamics.

Figure 1 illustrates an example of the flow field and of the absolute vorticity over the entire globe on a particular day at 200 millibars (at a level about 12 kilometers above sea level). Here we note that over the Northern Hemisphere, cyclonic flows are generally characterized by positive vorticity, with the converse being the case over the Southern Hemisphere. Over the Northern Hemisphere, maxima of cyclonic vorticity are found over what are called *troughs* of the circulation, and minima are noted over *ridges*. The converse relation holds for the Southern Hemisphere. These systems in extratropical latitudes show, in general, an eastward propagation with the active weather systems. Along the equator, the Coriolis parameter f is zero; here the relative vorticity ζ_r is not defined as either cyclonic or anticyclonic. The line separating the largely positive vorticity of the Northern Hemisphere from the largely negative vorticity of the Southern Hemisphere does not lie along the equator; rather, it meanders north and south along the equatorial belt. This meandering is indicative of equatorial wave activity.

Twisting and vertical advection effects, also called the *tipping term*, convert the vertical shear of the horizontal wind to a horizontal shear (and hence to the vertical component of vorticity). This occurs via differential vertical motions; that is, upward motion in one region brings

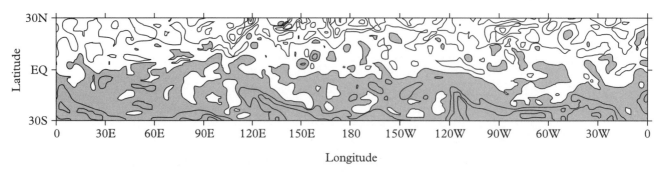

VORTICITY. Figure 1. *The distribution of absolute vorticity at 200 millibars on a typical day (26 July 1992).* The separation between shading and no shading denotes a zero isopleth. The unshaded regions are positive and the shaded regions are negative. Interval of isopleths is 4×10^{-5} per second.

up (advects) slower horizontal winds, and downward motion over an adjacent region brings stronger horizontal winds downward, such that strong horizontal shear develops via this twisting (or tipping) effect between the regions of ascent and descent. In large-scale flow problems, the twisting effect is usually counteracted by the vertical advection of vorticity. This is easy to see in a normal frontal synoptic setting involving cold air to the west and warm air to the east. The warmer air slopes over the cold air westward, and the frontal cyclonic vorticity is advected upward by the rising warm air. The effect of cold air sinking and warm air rising, via the twisting effect, brings in stronger southwesterly flow (momentum) in the cold air, and the rising warm air tends to bring up weaker southwesterly flow. The region between these regions of descent and ascent eventually acquires an anticyclonic shear, thus counteracting the effects of the twisting.

Potential Vorticity. Although changes in vorticity must obey Equation 3, there is a powerful constraint on these changes given by Ertel's theorem on the conservation of potential vorticity.

Suppose that some scalar property λ of the fluid is a conserved quantity:

$$\frac{D\lambda}{Dt} = 0. \qquad (7)$$

For example, for adiabatic motions λ may be chosen as the potential temperature of the air, or the density of a liquid. If $\nabla \cdot \boldsymbol{u}$ is eliminated from Equation 3 by using Equation 2, one can form an equation governing the rate of change of ω_a/ρ. Then, a scalar product of $\nabla\lambda$ with $\frac{D}{Dt}\left(\frac{\omega_a}{\rho}\right)$ gives the following:

$$\frac{D}{Dt}\left[\frac{\omega_a}{\rho} \cdot \nabla\lambda\right] = \nabla\lambda \cdot \left[\frac{\nabla\rho + \nabla p}{\rho^3}\right] + \frac{\nabla\lambda}{\rho} \cdot \left[\nabla \times \frac{\boldsymbol{F}}{\rho}\right]. \qquad (8)$$

If the fluid is barotropic ($\nabla\rho \times \nabla p = 0$) or if λ is a function of p and ρ only, and if frictional effects are negligible, then

$$\frac{D}{Dt}\left[\frac{\omega_a}{\rho} \cdot \nabla\lambda\right] = 0. \qquad (9)$$

The conserved quantity is called the *potential vorticity* Π:

$$\Pi = \frac{\omega_a}{\rho} \cdot \nabla\lambda$$
$$= \left(\frac{\omega_r + 2\Omega}{\rho}\right) \cdot \nabla\lambda. \qquad (10)$$

In shallow water theory, for example, the rank or relative position of a fluid parcel in a column is a conserved quantity. It is then easily shown that

$$\Pi = \frac{\zeta_r + f}{H}$$

where H is the total depth, is a conserved quantity.

This is a most useful constraint. If H is constant, then $\zeta_r + f$ must be constant; motion to another latitude ϕ must be accompanied by a corresponding change in relative vorticity ζ_r. On the other hand, if motion is confined to small changes in latitude so f is nearly constant, analogous changes in ζ_r must accompany depth variations. In fact, laboratory modeling of f-variations is made possible through use of depth variation.

In meteorological applications, the formation of a lee trough where a westerly flow impinges on a mountain is a classical example. Here, the descending air in the lee of the mountain experiences an increase of H; this calls for an increase of $\zeta_r + f$ to provide a conservation of $\zeta_r + f/H$, thus making the flow more cyclonic, and producing the lee trough.

[*See also* Dynamics; General Circulation; Gravity Waves; *and* Potential Vorticity.]

RUBY KRISHNAMURTI and T. N. KRISHNAMURTI

W

WALKER CIRCULATION. *See* Drought; El Niño, La Niña, and the Southern Oscillation; General Circulation; *and* Tropical Circulations.

WARFARE AND WEATHER. The courses of individual battles and entire wars have often been affected, sometimes decisively, by climate or weather. This article discusses meteorological effects on three of the great battles of history.

Battle of the Spanish Armada against the Winds. In the later sixteenth century, relations between Spain and England became strained. At the time, Spain ruled Belgium and the Netherlands whose population strove to throw off the foreign occupation. England supported the rebels, concerned that a rival great power controlled the coast of mainland Europe opposite her own coast, angering Philip II (1527–1598), king of Spain since 1556. An additional irritant was the successful British privateering campaign against Spanish shipping.

About 1580, Philip began building up a fleet, or an armada, to transport 30,000 Spanish troops stationed in the Low Countries to invade England. By May 1588 a fleet of about 130 ships was assembled at the port of Lisbon. On 30 May, the "Invincible Armada," as the Spanish called it, weighed anchor; on 30 July it sailed into the English Channel.

The timing of the naval attack on England could have hardly been more unfortunate for Spain. The year 1588 fell in a cold and stormy phase of Europe's climatic history. Figure 1 is a short segment of a chart published by F. H. Schweingruber (1983), based mainly on Swiss studies of glaciers and tree-ring density (or width) published in 1982. Figure 1 shows that the period between about 1570 and 1600 was a cold period in the Alps (the shaded section of the curve around the top line of the diagram indicates a cold period when the tree-ring densities were below the average of a long period), and

undoubtedly in western Europe as well. The last 17 years of the century, including 1588, were particularly cold overall. Such cold periods are usually stormy and rainy. Lamb (1988: 386) makes the interesting suggestion that Shakespeare's lines, "The spring, the summer, the chiding autumn, angry winter, change their wonted liveries, and mazed the world . . . now knows not which is which" (*A Midsummer Night's Dream,* II.i) refer to the unusually cold and stormy 1590s, when the play was composed.

The summer of 1588 was particularly cold and stormy in the general area of the North Sea. Lamb (1988: 389) remarks that tentative weather maps for most dates between 18 June and 3 July 1588 have "much of the appearance of winter bad weather maps." In his records of the sailing, the Duke of Medina Sidonia, commander of the Armada, writes that on July 27, "It blew a full gale with very heavy rain squalls and the sea was so heavy that all the sailors agreed that they had never seen its equal in July. Not only did the waves mount to the skies but some sea broke clean over the ships" (Lamb et al., 1987: 31). Although there were sea battles between the Armada and the English navy near Plymouth and Gravelines, the Armada had to fight the winds rather than the English. In fact, the English navy managed to sink only one Spanish man-of-war to capture and three or four more.

Most of the storm winds were from the south or southwest, making it almost impossible for the Spanish ships to reach the English coast. Nor was it possible for them to take abroad troops from the Low Countries. Ships of the Armada were driven northward and, after they rounded Scotland, the gales in the Irish Sea took the heaviest toll on them. Especially violent was the gale of 21 September in the Irish Sea, which alone accounted for the loss of some 17 ships. Only about one-half to two-thirds of the Armada was able to return to Spain.

Much invaluable information about the wind and weather encountered by the Armada has been preserved in the records kept and letters sent to Philip by the Duke of Medina Sidonia and other officers. These sources were

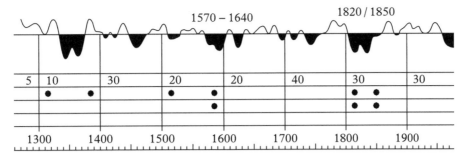

WARFARE AND WEATHER. Figure 1. *Fluctuations in temperature over the years 1300 to 1900, estimated from analyses of tree-ring densities and glacier positions in the Swiss and Austrian Alps.* The oscillating curve in the upper part of the figure shows variations in tree-ring density about an 8,000-year average; the solid portions below the baseline indicate colder-than-average periods, and the portions above, warmer periods. The solid circles denote times of glacial advance. (Adapted from a figure by F. H. Schweingruber.)

evaluated by H. H. Lamb and a team of scientists of the University of East Anglia, England. The results of their investigations are summarized by Lamb et al. (1987). This and a subsequent paper by Lamb (1988) offer tentative pressure maps, including one for 21 September, the day of the great gale.

Queen Elizabeth I ordered the striking of a commemorative medal indicating the decisive role played by the winds. The inscription of the medal, drawn from the Latin Bible, is translated: "Yahweh blew up a wind and they are [were] dispersed."

Battle of Waterloo. The Waterloo campaign was fought on 15 to 18 June 1815 between the armies of Napoleon I and the allies England and Prussia. The scene of the campaign was an area extending over 14 to 40 kilometers south-southeast of Brussels. On 16 June, a thunderstorm prevented Napoleon from dealing a decisive blow to the Prussian force. The most important weather effects, however, developed on 17 June and the night of 17/18 June, with grave consequences for the French army on 18 June, the day of the Battle of Waterloo.

The afternoon thunderstorm of the 17th and the subsequent night softened the ground, turning it into a quagmire. Early on the 18th, Napoleon and his artillery experts surveyed the ground conditions and concluded that it would not be possible to begin the battle with an artillery barrage until the soil dried sufficiently. In addition to the difficulty of moving artillery across soaked ground, the shot would have been buried in the soft ground, rather than ricocheting and hitting the enemy.

Instead of 9 A.M., as planned by the Emperor, the artillery barrage began at 11:30 A.M. At first the battle went in favor of the French, but there was not enough time

to defeat Wellington before the arrival of remnants of the Prussian force at about 4 P.M. The Prussians helped to turn the battle into a decisive victory for the allies. In consequence of the defeat, Napoleon finally lost his throne and was exiled; the French lost control of the territories conquered in the Napoleonic wars, and the anti-Napoleon coalition redrew the political map of Europe at the Congress of Vienna (1815). Several prominent historians and military experts have stated that had Napoleon been able to join battle earlier in the morning, Wellington's army would have been defeated before the coming of the Prussians.

Battle for Moscow. The German World War II attack on the Soviet Union began in June 1941. After a victorious advance in July and August 1941, the German generals pressed Hitler to order an offensive aimed at capturing Moscow. The generals believed that the fall of Moscow would deal a grave blow to the Russians' morale; they forgot that in 1812, the capture and burning of Moscow by Napoleon only steeled the Russians' determination to fight back. The generals succeeded in convincing Hitler, and "Operation Typhoon," the campaign for the capture of Moscow, started on 2 October. Soon, however, two major meteorological factors frustrated the German effort.

The first weather factor was the so-called *rasputiza,* the rainy season that nearly every fall turned the fields and roads of Russia into a quagmire, rendering vehicular or even foot traffic nearly impossible. In 1941 the rasputiza set in about 10 October and lasted for a month, totally blocking the German advance. The German high command pinned its hopes on the coming of the first frost, when the ground would become firm again, but the advent of the cold in autumn 1941 was not that of a typical Russian winter. Winter 1941–1942 set in rather early, and it turned out to be the coldest winter (December through February) known in the continuous instrumental meteorological record begun at St. Petersburg in 1752. On 5 December there was a severe cold outbreak: air temperatures in the open, outside the "heat island" effect of cities, dropped to between −30° and −38°C. The German troops were not prepared for such cold in terms of clothing, food, and fuel supply. Moreover, some items of the German armament failed to function at low temperatures. No less disastrous was the fact that rail services tended to break down at temperatures of −15°C and lower. At such temperatures, about half of the rail engines ceased operating; after snowstorms, rail service stopped, so that the German army fighting in the Battle for Moscow could not be resupplied. At the Warsaw, Poland, railroad station, some 10,000 tons of supplies accumulated without the possibility of shipping on.

The rasputiza, and especially the cold, caused a loss of momentum for the German offensive. On 6 December, the Battle for Moscow was abandoned by the Germans. This enabled the Soviet army to launch its first major counteroffensive of the war. In the continued fighting, the number of casualties caused by cold increased to such an extent that in the last days of December, the number of German troops lost through frostbite and other consequences of cold exceeded the number lost through enemy action.

BIBLIOGRAPHY

Lamb, H. H. "The Weather of 1588 and the Spanish Armada." *Weather* 43 (1988): 386–395.

Lamb, H. H., et al. "Flavit et Dissipati Sunt. It Blew and They Were Scattered. The Spanish Armada Storms: A Weather Perspective of July to October, 1588." *Chinook* 9 (1987): 28–43.

Neumann, J. "Great Historical Events that Were Significantly Affected by the Weather. 11: Meteorological Aspects of the Battle of Waterloo." *Bulletin of the American Meteorological Society* 74 (1993): 413–420.

Neumann, J., and H. Flohn. "Great Historical Events that Were Significantly Affected by the Weather. 8.I: Long-Range Weather Forecasts for 1941–42 and Climatological Studies." *Bulletin of the American Meteorological Society* 68 (1987): 620–630.

JEHUDA NEUMANN

WARM FRONTS. A front represents the boundary or, more precisely, the zone of rapid transition between air masses with differing characteristics of temperature and moisture. Warm fronts are distinguished from other fronts by having a direction of movement such that warm air following the front replaces colder air. The term *warm front* denotes the side of the zone of temperature change with warmer air, and the zone of temperature change itself is referred to as the *frontal zone*. In addition to the changes in surface air temperature and humidity that define the warm front, the surface wind direction also changes across a warm front. This wind shift from cold to warm air in a clockwise direction can be used to locate the warm front accompanying the zone of temperature change. A warm front is symbolized on a surface analysis chart by a heavy line with semicircular symbols that point toward the cold air, which is the direction of frontal movement. Warm fronts were first identified in the early twentieth century by the Norwegian meteorologists as the boundary between warm, moist air within the warm sector of a developing cyclone and the cooler air to the north and northeast of the cyclone (Bjerknes, 1919). Although most of characteristics defining of warm fronts were first identified in relation to mid-latitude cyclones, warm fronts also occur in situations not directly associated with these, for instance along coastlines and mountain ranges.

Structure. Although warm fronts are defined principally by their temperature and wind characteristics at the Earth's surface, they do possess significant three-dimensional structure (see Figure 1). The warm front slopes toward adjacent colder air as the altitude increases. The slope of the front is a result of the physical requirement that the horizontal temperature gradient must approximately balance the vertical gradient of the horizontal winds (termed *thermal wind balance*). The slope of the warm front defines a shallow wedge of cold air below the warm air mass. The temperature gradient associated with this sloping frontal boundary decreases with height as a result of a decreased wind shift across the front at higher altitudes. The wind shift observable at the surface decreases with height as the winds gradually turn to become parallel to the frontal zone in the upper troposphere. This turning of the winds with height is due to the thermal wind balance in the frontal zone.

The warm, moist air on the warm side of a warm front is forced upward by certain structural and dynamic factors associated with the sloping warm frontal zone. The wind shift across the frontal zone produces convergence of air at the front at all levels. This flow of air toward the front is largest at the surface, so surface air is forced upward. Because air converges toward the front at all levels, the air continues to be forced upward at all levels up to a considerable height, at which point the winds

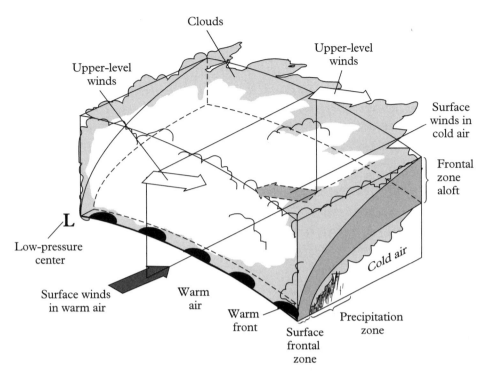

WARM FRONTS. Figure 1. *Three-dimensional structure of a warm front.*

become more nearly parallel across the front. This lifting forced by the wind shift across the front makes the front behave somewhat like a material boundary across which air on one side cannot be transported to the other side. Although warm fronts are not really material boundaries, the frontal uplift and the tendency of air to conserve its potential temperature strongly limit mixing of air across the warm front. Consequently, warm air appears to "ride up" the warm front: surface air moves toward the surface warm front, is lifted, and turns clockwise to flow parallel to the front at upper levels. This process is referred to as *overrunning* because the warm air is lifted and overruns the wedge of cold air below it. [*See* Low Pressure; Potential Temperature.]

Precipitation and Clouds. As a result of frontal lifting, the warm, moist air that originated in the warm sector is cooled as it rides up the warm front. The cooling results in condensation, producing extensive clouds and precipitation along the warm front as it extends vertically upward. Because the warm front slopes and is moving toward the cold air, the clouds and precipitation occur ahead of the surface warm front. As a warm front approaches, a surface observer first sees upper-level clouds such as cirrus and cirrostratus, followed by progressively lower clouds (altostratus or altocumulus, stra-

tus or stratocumulus, and nimbostratus or cumulonimbus). The distance from the surface warm front to the first upper-level clouds associated with the warm front may be from several hundred to more than a thousand miles. The dominant cloud types associated with warm fronts tend to be layered, stratiform clouds as a result of the high atmospheric stability induced when warm air overlies cold air. However, convective clouds (altocumulus, stratocumulus, and cumulonimbus) may occur when the warm air being lifted becomes destabilized, either through the release of inherent potential instability as the air is lifted, or through modification of the stability in the frontal zone by cold air flowing over the front at upper levels.

The precipitation associated with warm fronts occurs over a broad area ahead of the surface warm front. With higher stratiform clouds, precipitation is light and usually does not reach the ground. As the surface front approaches, the precipitation becomes increasingly heavier, reaching a maximum just prior to the passage of the front. The increasing precipitation is caused by a progressive increase in the depth of the cloud layer from which the precipitation is falling. Recent studies have found that the intensity of the precipitation varies substantially on small scales within the broad area of precipitation ahead of the warm front. This smaller-scale variability has been found to occur in bands, referred to as *rainbands* or *banded precipitation*. The cause of this phenomenon is not completely understood, but instabilities along the warm front seem to be the primary factor.

The type of precipitation associated with warm fronts ranges from rain to forms of frozen or partially-frozen precipitation, depending on vertical temperature variation in the precipitating air. The precipitation usually begins as snow in the upper atmosphere and may reach the ground as rain, freezing rain, ice pellets, or snow. If the lower atmosphere is at or below freezing and the warm air mass at the frontal boundary is relatively cool, then precipitation that starts as snow will fall through cool air and reach the ground as snow. If, however, the warm air mass at the frontal boundary is not sufficiently cold, then the snow may melt before falling into the colder air near the ground and refreezing. This results in either freezing rain or ice pellets, depending on the depth and temperature of the lower cold air mass. If the air in the cold air mass is well above freezing, then rain is usually observed. The exact type of precipitation and its location relative to the warm front are complex functions of the vertical temperature structure and present a very challenging problem to forecasters. Typically, snow is observed well ahead of a warm front where the underlying cold air mass is relatively deep; rain is observed near the warm front where the cold air is very shallow; and a band of freezing rain and ice pellets separates the regions of snow and rain.

Formation. Warm fronts form when the zone of temperature change between two air masses intensifies and begins to move toward colder air. Two key aspects of the formation process are the development of the initial temperature gradient and the intensification of this gradient over time to become a warm front. Warm fronts most commonly form when cold and stationary fronts associated with a previous storm persist and become the warm front associated with a new storm. In this situation, a boundary between air masses already exists, although its temperature gradient may be very weak. The warm front forms when the winds in the warm air intensify the flow of air toward the air-mass boundary. The increased winds in the warm air coupled with weak winds in the cold air produce confluence along the temperature gradient or frontal zone. This confluence increases the temperature gradient to generate the front and initiate its movement. The confluent winds are typically established by the falling pressure associated with a developing low to the west or southwest along the frontal zone.

Warm fronts also develop when air mass boundaries are established along coastlines or mountain ranges. In the case of a coastline, warm ocean temperatures modify a cold air mass over the ocean to generate a temperature gradient along the coast. When the winds shift so that the flow of air is onshore (toward the land), shallow warm fronts form. These shallow warm fronts, termed *coastal fronts* are frequently observed along the eastern coast of North America. Because these fronts tend to be shallow, their clouds and precipitation patterns differ from those of the warm fronts associated with cyclones. Downslope flow in the lee of mountains can also develop a new air mass boundary within a cold air mass. Air flowing downslope is warmed by compression and establishes a temperature gradient in the lee of the mountains. This temperature gradient can become a warm front when a low-pressure system also develops in the lee of the mountains and the warm air starts to move away from the mountains. This situation is common in the lee of the Rocky Mountains during the winter.

[*See also* Cold Fronts; *and* Fronts.]

BIBLIOGRAPHY

Bjerknes, J. "On the Structure of Moving Cyclones." *Geofysiske Publikasjonen* 1 (1919): 1–8.
Newton, C. W., and E. O. Holopainen, eds. *Extratropical Cyclones: The Eric Palmén Memorial Volume.* Boston, 1990.
Wallace, J. M., and P. V. Hobbs. *Atmospheric Science: An Introductory Survey.* New York, 1977.

WENDELL A. NUSS

WARM SECTOR. The warm sector of an extratropical cyclone (a low-pressure system in the mid latitudes) is the portion containing the warm air; thus, it is the area extending from the rear of the warm front to the leading edge of the cold front. The warm sector is the portion of the tropical air mass that participates in the circulation of the mid-latitude storm. This region is important as the major source of the moisture that enters the storm circulation, ascends, and then precipitates. Wave cyclones develop in the presence of fronts, so it is in this class of extratropical cyclone that a warm sector is usually identified. [*See* Cyclones.]

The movement of the cyclone is usually in the direction of the air flow in the warm sector. In the Northern Hemisphere, the air flow in this region is typically from the southwest, because the low center is to the northwest and the subtropical high is to the southeast. In the Southern Hemisphere, the air flow in this region is typically from the northwest, because the low center is to the southwest and the subtropical high is to the northeast.

The warm sector is clearly defined at the surface during the development stage of the cyclone. Aloft, it extends poleward above the warm and cold fronts. As the cyclone matures, this warm moist inflow from lower latitudes stops, or is *occluded*. The cyclone is then said to be in its *occluded stage*.

Because the warm sector is a part of the storm circulation, warm air farther from the storm, or well above the storm circulation, is not part of the warm sector. The poleward edge of the warm sector is marked by the warm and cold fronts, but its low-latitude boundary is not distinct. The concept of air masses is rarely applied to the upper troposphere (above 5 kilometers); for this reason, the term *warm sector,* above 5 kilometers, usually is limited to cloudy air drawn poleward ahead of the storm and ascending above the warm front.

The weather within the warm sector at the surface is typically warm and windy and usually moist, except in arid continental areas. It is warm but not windy to the southeast of the warm sector. Showers, thunderstorms, or lines of thunderstorms (called *squall lines*) can occur in the warm sector, either in advance of the cold front or above the warm front.

Convective storms—showers and thunderstorms—often develop within a deep layer of warm air that is moist at low levels and dry aloft. When this deep layer is lifted in the extratropical cyclone, condensational heating in the lower, moist region destabilizes the layer so that the moist air boils upward in convective clouds.

When the low advances and deepens and the high ahead of it moves away only slowly, the pressure difference increases, so the air flow then increases in a *low-level jet* from lower latitudes that very rapidly pumps moisture into the storm, allowing it to develop faster.

In oceanic storms with unlimited surface moisture and little surface friction, the warm sector winds can be strong and can extend for a considerable distance with little change in direction along the flow. Therefore, ocean waves become high. When such storms make landfall, wind damage can be severe.

Another feature of oceanic storms is that the wind shift across the warm front and the cold front can be very sharp. In the Northern Hemisphere, strong southwest flow in the warm sector can be extensive, with an abrupt change to flow from the east poleward of the warm front, and change to flow from the north poleward of the cold front. In the Southern Hemisphere, strong northwest flow in the warm sector can be extensive with a sharp

change to flow from the east poleward of the warm front, and change to flow from the south poleward of the cold front.

Although warm sectors are usually identified in extratropical cyclones that form in the presence of fronts, cyclones that arise within a cold air mass can develop regions with straight flow on the warmer side of the low-pressure center, from which warmer air ascends ahead of the cyclone. The term *warm sector* can also be used in reference to these storms, as well as to classical wave cyclones.

The concept of the *occluded front* was formulated along with the concept of the occluded cyclone, but these two concepts can be treated as distinct. Because fronts exist on scales of tens of kilometers while the warm sector has a typical width of hundreds of kilometers, the warm sector can be more easily observed and understood. It is easy to determine when this warm air flow into the storm is cut off or occluded, and the cyclone is in its occluded stage. The nature of the occluded front remains a subject of considerable uncertainty, partly because it is narrow and difficult to identify with conventional observations and in conventional data and in conventional numerical prediction models.

[*See also* Occluded Fronts; *and* Occlusion.]

CARL W. KREITZBERG

WATER. *See* Hydrological Cycle.

WATER RESOURCES. Fresh water is an essential natural resource, necessary for all environmental and societal processes. It is renewable, made continuously available by the constant flow of energy from the Sun to the Earth, which evaporates water from the oceans and land and redistributes it around the globe. More water evaporates from the oceans than falls on them as precipitation; thus there is a continuous transfer of fresh water from the oceans to the continents. This water runs off in the rivers and streams that sustain natural ecosystems and human societies. Human beings require water to grow food, provide energy, and run industries. In order to mobilize water for human needs, we build huge reservoirs to store water for dry periods and to hold back floods; we build aqueducts to transport water from water-rich to water-poor regions; we desalinate salt water in arid regions; and we dream of towing icebergs from the polar regions and of reversing the flow of massive rivers.

As we approach the twenty-first century, we must acknowledge that many of our efforts to harness water have been inadequate or misdirected. We remain ignorant of the functioning of basic hydrologic processes. Rivers, lakes, and groundwater aquifers are increasingly contaminated with biological and chemical wastes. Vast numbers of people lack clean drinking water and rudimentary sanitation. Millions of people die every year from water-related diseases such as malaria, typhoid fever, and cholera. Massive water developments have destroyed many of the world's most productive wetlands and other aquatic habitats. The economic and hydrologic resources for major new irrigation projects cannot be found. Finally, predicted changes in global climatic conditions, particularly those associated with the greenhouse effect, will alter future water supply, demand, and quality.

The Earth has been called the "water planet"; two-thirds of its surface is covered with water. Yet the vast bulk of the water on Earth is too salty for growing crops or drinking. What freshwater is available depends largely on the behavior of the atmosphere, which transfers energy and water around the globe in the form of clouds, storms, and other atmospheric features. How the atmosphere moves water from one place to another determines why the Sahara Desert is dry, why the tropics are wet, why droughts and floods occur, and ultimately how human societies develop.

There are about 1,400 million cubic kilometers of water on Earth, but 97 percent of it is in the oceans. The remaining 3 percent is the freshwater found in rivers, lakes, underground aquifers, and the massive icecaps of Greenland and Antarctica, and as water vapor in the atmosphere. Table 1 lists the world's water resources. Less than 100,000 cubic kilometers of water—just 0.3 percent of Earth's total freshwater reserves is found in the rivers and lakes that constitute the bulk of our usable supply.

Global freshwater resources are very unevenly distributed. Some places receive enormous quantities; others receive almost none. The Atacama Desert in South America receives no rain in many years. Pictures from

WATER RESOURCES. Table 1. *Water reserves on the Earth*

	Distribution			Percentage of Global Reserves	
	AREA (10³ square kilometers)	VOLUME (10³ cubic kilometers)	LAYER (meters)	OF TOTAL WATER	OF FRESH WATER
World ocean	361,300	1,338,000	3,700	96.5	—
Ground water	134,800	23,400	174	1.7	—
Fresh water		10,530	78	0.76	30.1
Soil moisture		16.5	0.2	0.001	0.05
Glaciers and permanent snow cover	16,227	24,064	1,463	1.74	68.7
Antarctic	13,980	21,600	1,546	1.56	61.7
Greenland	1,802	2,340	1,298	0.17	6.68
Arctic islands	226	83.5	369	0.006	0.24
Mountainous regions	224	40.6	181	0.003	0.12
Ground ice/permafrost	21,000	300	14	0.022	0.86
Water reserves in lakes	2,058.7	176.4	85.7	0.013	—
Fresh	1,236.4	91	73.6	0.007	0.26
Saline	822.3	85.4	103.8	0.006	—
Swamp water	2,682.6	11.47	4.28	0.0008	0.03
River flows	148,800	2.12	0.014	0.0002	0.006
Biological water	510,000	1.12	0.002	0.0001	0.003
Atmospheric water	510,000	12.9	0.025	0.001	0.04
Total water reserves	510,000	1,385,984	2,781	100	—
Total fresh water reserves	148,800	35,029	235	2.53	100

Source: I. A. Shiklomanov, "World Fresh Water Resources," in Gleick, 1993.

space show the cloudless skies over the great Sahara of Africa. At the other extreme, parts of the island of Kauai, Hawaii, have recorded more than 11 meters of rainfall in a single year—some 435 inches.

Other examples of the uneven distribution of water abound. Twenty percent of global average runoff comes from a single river system, the Amazon, while runoff from the entire continent of Australia constitutes just 1 percent of global runoff. Thirty percent of the total runoff from Africa comes from a single river basin, the Congo/Zaire. Moreover, many regions get nearly all their annual precipitation during a brief, intense rainy season. In Cheerapunji, India, over 10.5 meters of rain per year may fall during the short period of the monsoon. Much of California gets no rainfall at all from May through September, precisely the period of greatest water demand.

This tremendous variability is a natural feature of our climate and helps define many of the problems facing us today. Solving our water problems requires that we expand our study of these natural characteristics, improve our ability to assess both static and dynamic hydrologic processes, and incorporate the effects of human economic activities and growing populations into our understanding of the global water cycle.

Global Freshwater Problems. The Earth's increasing population is rapidly outgrowing its finite supply of water, leading to declining water quality and human health, limits on food production, increasing species extinction, constraints on energy production, and questions about the viability of unlimited economic development. Today, well over a billion people throughout the world are served by sewage-treatment systems not up to the standards of ancient Rome, or by no systems at all. Similar numbers lack access to clean drinking water. While we have continued to improve our understanding of the links among water quality, water-related diseases, and the hydrologic cycle, we have failed to reduce the suffering these problems cause. Among the problems are microbiologic contamination, responsible for many of the world's most persistent and widespread diseases, and chemical contamination, which poses risks for both human beings and aquatic ecosystems.

New epidemics of waterborne disease continue to appear in many parts of the world because of inadequate sanitation, lack of access to safe drinking water, and ignorance about the links between water and disease. Worldwide, more than 250 million new cases of water-related diseases are now reported each year, resulting in approximately 10 million deaths annually. The vast majority of

these deaths occur in the tropical countries of the developing world. In Latin America, an unprecedented outbreak of cholera beginning in 1991 and continuing through the mid-1990s shows how vulnerable poorer regions are to these diseases.

Other water-quality problems abound: nitrate contamination from the runoff of agricultural fertilizers and industrial wastes; heavy-metal contamination from mine drainage, industrial processing of ores, and leaching from solid wastes; and the contamination of water by organic materials from many manufacturing processes and the production of pesticides. There is an urgent need to provide low-cost technologies already proven effective in preventing disease and improving public health; and new and better data must be collected on even the most common water-quality problems, such as fecal contamination, microbiologic disease, and the types and extent of chemical contamination.

Without dependable water supplies, growing food for the Earth's population is at best a marginal affair. The earliest civilizations developed in the now arid lands of the Middle East, Asia, and the Americas along major rivers and streams such as the Tigris, Euphrates, Nile, Indus, and Colorado, which provided reliable irrigation water. In Egypt, irrigation was being practiced along the Nile 5,400 years ago. The civilizations that arose in the foothills of the Himalayas used the waters of the Indus basin for irrigation as long as 5,000 years ago, constructing networks of canals for this purpose. In the New World, the ancestors of the Incas were using irrigation 3,000 years ago, and there is archeological evidence that irrigation canals were built and maintained by the ancient Hohokam in the Salt River valley of the lower Colorado River basin some 2,000 years ago.

The Green Revolution of the twentieth century was accomplished in large part by expanding the use of irrigation technology and water resources. Between 1950 and the end of the 1980s, the Earth's global irrigated area more than doubled, from 94 million hectares to over 230 million hectares. One-third of the total global harvest of food comes from the 17 percent of the world's cropland that is irrigated. Figure 1 shows the percent of cropland irrigated on each continent. Today, two-thirds of the fresh water consumed worldwide goes to produce food, and more will be required in the future to meet the growing needs of the Earth's swelling population.

There are signs that society is having difficulty meeting the water needs of a growing population. Irrigated area per capita increased through the twentieth century until the late 1970s, when it peaked at 48 hectares per 1,000

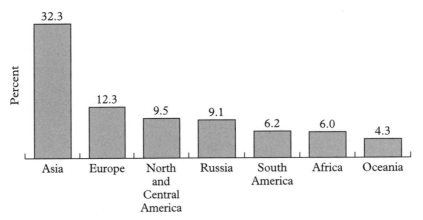

WATER RESOURCES. Figure 1. *Percentage of cropland that is irrigated, by region.*

people. Since then, the amount of irrigated area per person has fallen as population growth has outstripped additions to irrigated area. Significant increases in the amount of irrigated area in the future are unlikely, in part because of the rising economic and environmental costs of irrigation. Although the efficiency of existing irrigation systems may be improved so that more food can be grown with less water, dramatic improvements will be needed to meet the present rapid population growth.

In addition to using water to grow food, we use it to help produce energy, and we use energy to help us clean, pump, and transport water. In recent years we have begun to encounter physical and environmental constraints in our use of both energy and water. Limitations on the availability of fresh water in some regions of the world may restrict the type and extent of energy development, and limitations on the availability of energy will constrain our ability to provide adequate clean water and sanitation services to the billions of people who lack those basic services.

Water is needed for producing almost every form of energy, from cooling water for nuclear, fossil-fuel, and geothermal plants, to water for building and operating solar-thermal electric and other renewable-energy facilities. As water available for energy production decreases in the future, owing to growing demands from other sectors, we are likely to begin to face regional constraints on our ability to produce and use energy. At the same time, growing energy demands will place additional pressures on water supplies in regions where water is supplied by pumping ground water or by aqueduct systems that rely on energy to move water from one region to another.

There are some bright spots. Many renewable energy sources—particularly photovoltaics, wind generation,

and some solar-thermal systems—require far less water per unit of energy produced than do conventional systems. For example, the severe drought in California during the late 1980s and early 1990s sharply curtailed hydroelectric production, but it had no effect on the operation of the world's largest wind electric and solar-thermal electric plants. In water-short regions, sources of energy with low water requirements may increasingly be the systems of choice.

In addition to using water to produce energy, we use energy to produce, move, and clean water. Energy permits us to make use of water that would otherwise be undrinkable or unobtainable. We now remove salt and other contaminants using desalination and wastewater treatment techniques, and we pump water from deep underground aquifers or distant sources. The availability and price of energy and technology sets limits on the extent to which these unusual sources of water can be tapped. As a result, understanding the links between water supply and quality and energy will help us evaluate future constraints on meeting future water needs.

Climate Changes. Because the availability of fresh water is so closely tied to the behavior of the atmosphere, managing our water resources in the future will be greatly complicated by global climatic changes—the so-called *greenhouse effect*. Higher temperatures will increase evaporation, change snowfall and snowmelt patterns, and lead to alterations in water demand. Precipitation will increase in some areas and decrease in others, in ways we understand only poorly. These changes will affect water availability in rivers and lakes, hydroelectricity generation, and agricultural productivity. A rising sea level will contaminate coastal freshwater aquifers with salt water. [*See* Greenhouse Effect.]

Many uncertainties remain about climate changes and their impacts on water resources. This results in part from limitations in the ability of large-scale climate models (general circulation models, or GCMs) to incorporate and reproduce important aspects of the hydrologic cycle. Many hydrologic processes, such as the dynamics of clouds and rain-generating storms, occur on spatial scales far smaller than can yet be accurately modeled. At the same time, the hydrologic processes included in the models are far simpler than those in the real world. We thus know far less than we would like about how the global water cycle is likely to change. Despite these limitations, research in the past few years has revealed some important things about how hydrology and water supplies may be affected by climatic changes. [*See* General Circulation Models.]

Temperature and precipitation are key factors in determining the distribution of natural ecosystems, what crops we grow in what places, the characteristics of human habitations and energy use, and how much water we use for different aspects of human development. The best current estimate is that doubling the concentration of greenhouse gases in the atmosphere, which is expected to occur within the next several decades given current trends, would cause the global average temperature of the Earth to increase by about 3°C. This increase will occur at a rate far faster than any comparable change seen over the past 10,000 years, and it will raise the Earth's temperature higher than any experienced over the past 2 to 3 million years. As temperatures rise, the evaporation of water from land and water surfaces will increase, as will global average precipitation. Most recent estimates are that global evaporation and precipitation will increase by between 3 and 15 percent, for an equivalent doubling of atmospheric carbon dioxide concentration. The greater the warming, the larger these increases will be.

Regional changes will differ substantially from these global ones, though they are substantially harder to predict. Climate models do, however, seem to indicate some consistent changes. For example, air temperatures over land are likely to increase faster than those over water; warming in the winter is predicted to be greater in the northern latitudes than toward the equator; and temperature increases in southern Europe and central North America may be higher than the global average. Precipitation is expected to fall more consistently and intensely throughout the year at high latitudes and in the tropics, and in mid-latitudes in winter. In certain zones, such as between 35° and 55° north latitude, precipitation may increase by as much as 10 to 20 percent, though summer rainfall may decrease over much of the northern mid-latitude continents. There are few consistent and large-scale changes in precipitation predicted for subtropical arid regions, but even small changes in these arid zones can lead to large changes in ecological and human systems.

Among the other important impacts of climatic change will be changes in soil moisture, in storm frequency and intensity, and in the behavior of snowfall and snowmelt. Soil moisture determines which plants can grow in different regions and how much additional water needs to be applied with irrigation to grow a crop. While precipitation overall may increase, this does not necessarily mean that land surfaces and soils will get wetter, because increases in evaporation may exceed the increases in precipitation. Indeed, one important recent research finding was that the incidence of droughts in the United States, measured by an index that looks at soil wetness conditions, may increase dramatically as temperatures go up—despite an accompanying increase in precipitation—because of the increased losses from evaporation.

Most climate models also suggest large-scale drying of the Earth's surface over northern mid-latitude continents in the northern summer, owing to higher temperatures and either insufficient precipitation increases or actual reductions in rainfall. Drying in these regions would affect both agricultural production and other water demands.

In basins with substantial snowfall and snowmelt at some time during the year, snowpack acts as a large storage reservoir that redistributes runoff over time. Climate impact assessments suggest that temperature increases in these basins will have three effects: they will increase the ratio of rain to snow in cold months; they will decrease the overall length of the snow season; and they will increase the rate and intensity of snowmelt. As a result, average winter runoff and average peak runoff will increase, peak runoff will occur earlier in the year; and there will be a faster and more intense drying out of soil moisture. One consequence may be a substantial increase in the risk of floods. In many mid- and high-latitude river basins, the worst floods occur during snowmelt runoff periods or when rain falls on snow. One of the greatest worries about higher temperatures is therefore the increased probability and intensity of flood flows. When these are combined with possible precipitation increases in many regions, flooding becomes a critical concern.

Another vitally important question concerns effects on the variability of the climate—the frequency and intensity of extremes. Extreme weather events, including typhoons, hurricanes, and monsoons, bring both death and life to many regions. Single typhoons in southern Asia have killed hundreds of thousands of people in Bangladesh through flooding of low-lying lands. Floods along the Huang River in China have killed more people than any other kind of natural disaster: in 1887, between 900,000 and 2 million Chinese died from flooding or subsequent starvation; in 1931, between 1 million and 3.7 million people died. At the same time, intense monsoons bring vital rain to India during the growing seasons, and without this rain, the reliability of crop production would decrease. Although not enough research has been done on this complex issue, there are some indications that the variability of the hydrologic cycle increases when mean precipitation increases, and vice versa. In one model study, the total global precipitation increased, but the area over which it fell decreased, implying more intense local storms and hence more runoff. [See Monsoons.]

Other changes in variability are also likely, though we cannot yet predict what they will be. There is some indication of possible reductions in day-to-day and interannual variability of storms in the mid-latitudes. At the same time, there is evidence from both model simulations and empirical studies that the frequency, intensity, and area of tropical disturbances may increase. Far more modeling and analytical efforts are needed in this area.

Perhaps the most important effect of climatic change on water resources will be a great increase in the overall uncertainty associated with the management and supply of freshwater resources. Rainfall, runoff, and storms are all natural events with a substantial random component: in the language of hydrologists, they are *stochastic*. In many ways, therefore, the science of hydrology is the science of estimating the probabilities of certain types of events. But these estimates almost always assume that the patterns of climate seen in the recent past will continue into the future. Indeed, hydrologists have few analytical tools with which they can incorporate future changes of uncertain magnitude.

Future climatic changes effectively make obsolete all our old assumptions about the behavior of the water supply. Perhaps the greatest certainty about future climatic changes is that the future will not look like the past. We may not know precisely what it will look like, but changes are coming.

Conclusions. Since the beginning of civilizations, humans have harnessed portions of the global water cycle for food production, transportation, and energy. In arid and semiarid regions, the earliest irrigation systems supported civilizations that were stable for centuries. The ready availability of energy from falling water helped to power the Industrial Revolution. The Green Revolution, which permitted us to feed our rapidly growing population in the 1960s through the 1980s, relied in large part on the availability of freshwater for the dramatic expansion of irrigated agriculture.

As we now look to the twenty-first century, several challenges face us. Foremost among them is how to satisfy the food, drinking water, sanitation, and health needs of 10 to 15 billion people, when we have failed to do so in a world of 5 billion. More water will be required for new industrial developments, new energy projects, and expanded agricultural efforts. Yet the total amount of irrigated land per person has already begun to fall, and it is poised to fall even farther and faster as population growth continues to outpace efforts to bring new land under irrigation. Meeting these human needs without conflicts over resources will take considerable ingenuity and commitment.

Adding to the difficulty is a problem of uncertain but potentially enormous magnitude—the alteration of the Earth's atmosphere, including both the destruction of the ozone layer and global climatic change. Human beings have become such a large force on the Earth that we are now able to manipulate, albeit unintentionally, the very atmosphere that permits life and shapes our lives, our civilizations, and our environment.

We now know that we are on the verge of changing our climate in an uncontrolled manner through the emission of trace gases from burning fossil fuels as we cut down our tropical forests. Unless we begin quickly to reduce our emissions of these gases, significant changes are likely. Among the expected changes are higher global and regional temperatures; increases in global average precipitation; changes in the regional patterns of rainfall, snowfall, and snowmelt; changes in the intensity, severity, and timing of major storms; and a wide range of other geophysical effects. These changes will have many secondary impacts on freshwater resources, altering both the demand for and the supply of water, and changing its quality.

Many uncertainties remain about the timing, direction, and extent of these climatic changes, as well as about their societal implications. These uncertainties greatly complicate rational water-resource planning for the future, and they have contributed to intense political debate over how to respond to the problem of climate change. But the water problems facing us can be solved if we can muster the wisdom to take the long view, the political will to take actions today to reduce the risks of future catastrophes, and the willingness to invest in our future.

[*See also* Ground Water; Hydrological Cycle; *and* Population Growth and the Environment.]

BIBLIOGRAPHY

Gleick, P. H. "Climate Change, Hydrology, and Water Resources." *Review of Geophysics* 27 (1989): 329–344.

Gleick, P. H., ed. *Water in Crisis: A Guide to the World's Fresh Water Resources.* New York, 1993.

Mintzer, I. M., ed. *Confronting Climate Change: Risks, Implications and Responses.* Cambridge, 1992.

Postel, S. *Last Oasis: Facing Water Scarcity* New York, 1992.

Waggoner, P. E., ed. *Climate Change and U.S. Water Resources.* New York, 1990.

PETER H. GLEICK

WATERSPOUTS. *See* Tornadoes.

WATER VAPOR. The atmospheric component water vapor is the gaseous form of dihydrogen oxide (H_2O). It is also referred to as *moisture* and, in older literature, as *aqueous vapor.* Although water exists in the atmosphere as vapor, liquid, and solid, the vapor phase dominates. Water vapor is the atmospheric constituent with the greatest effect on the dynamics, thermodynamics, and radiation balance of the atmosphere. It strongly affects the dynamics of the atmosphere because very large amounts of energy are involved in phase changes between water vapor and either liquid water or ice. Its effect on the radiation balance is important because water vapor is a very good absorber of infrared radiation. Despite the importance of water vapor, we still do a poor job of measuring it and predicting the processes that involve it, such as rain formation.

The distribution, transports, and phase changes of water vapor are parts of the hydrological cycle. Water enters the atmosphere mostly through evaporation and transpiration at the Earth's surface (the combination of the two processes is referred to as *evapotranspiration*). Water leaves the atmosphere mostly through condensation followed by precipitation. The atmospheric concentration of water vapor varies enormously even within the troposphere—from several percent by mass in the tropical boundary layer to only a few parts per million near the tropopause. Concentrations of water vapor also vary greatly on small spatial and temporal scales.

The water vapor content of a parcel or volume of air can be expressed using many different measures, a practice which can cause confusion. These conventional measures include the following:

- *Absolute humidity* or *vapor density* is the mass of vapor in a given volume.
- *Mixing ratio by mass* means the mass of vapor divided by the mass of dry air in a given volume.
- *Mixing ratio by volume* means the number of water molecules divided by the number of molecules from dry air in a given volume.
- *Specific humidity* is the mass of water vapor in a moist air parcel divided by the total mass of the parcel.
- *Vapor pressure* is the partial pressure attributed to the water vapor.
- *Relative humidity* is the actual vapor partial pressure divided by the saturation vapor pressure.
- *Dew point* is the temperature at which the saturation vapor pressure over water of a parcel equals the actual vapor pressure of the parcel.
- *Frost point* is the temperature at which the saturation vapor pressure over ice of a parcel equals the actual vapor pressure of the parcel.

The choice of measure is based on convenience or habit. *Integrated water vapor* (or the misleading term *total precipitable water*) is the amount of water vapor in a given area integrated over a column that extends the full height of the atmosphere. *Total column water* includes water in all phases but is usually dominated by water vapor. (For more detailed definitions and usages of these terms see Huschke 1959; Iribarne and Godson 1981). [*See* Humidity.]

Distribution. The composition of the atmosphere is quite uniform up to very high altitudes with respect to all constituents *except* water vapor. The relative concentration of water vapor varies enormously. In the tropics near the ocean surface, water vapor makes up almost 4 percent of the atmosphere by volume (molecular fraction); about 15 kilometers above the surface, near the tropopause, the fraction drops to about 3 parts per million by volume. Because the pressure at the tropopause is about one-tenth of the surface value, and the temperature is much colder, the absolute humidity drops by more than 4 orders of magnitude (a factor of about 10^{-4}) just within the troposphere. In the stratosphere, the mixing ratio is usually in the range 3 to 8 parts per million. Figure 1 shows average conditions for the water vapor concentration of the atmosphere as a function of altitude and latitude.

Water vapor concentration closely follows temperature; changes in relative humidity are therefore not nearly as great as changes in mixing ratio. The highly smoothed

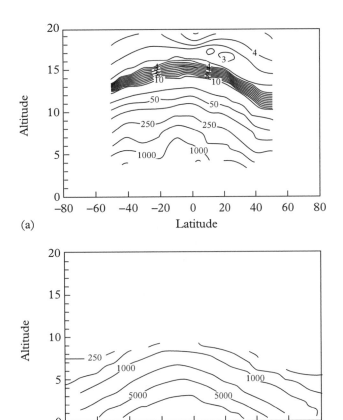

(a)

(b)

WATER VAPOR. Figure 1. *Zonally averaged, monthly mean mixing ratios of water vapor as a function of latitude and altitude.* (a) Measurements made by a satellite-based instrument for the upper troposphere and lower stratosphere for January 1987. (b) Measurements made by radiosondes for the lower troposphere for January (1963–1973); contour lines are 250, 500, 1,000, 2,500, 5,000, and 10,000 parts per million by volume. (From Larsen et al., 1993. Courtesy of the American Geophysical Union.)

distribution in Figure 1 hides enormous variations on small spatial and temporal scales. A satellite picture of radiation in a band near 6.7 microns, which provides an approximate measure of total column water, is shown in Figure 2. Very strong horizontal gradients of water vapor are clearly evident. Strong gradients of water vapor also occur in the vertical. Figure 3 provides a sample vertical sounding of the troposphere, showing temperature and dewpoint as a function of pressure. From this information, all other measures of water vapor can be calculated; some of these are plotted alongside the sounding. Above a thin surface layer, the mixing ratio of water vapor is nearly constant in the so-called *mixed layer*. Vertical

soundings can vary greatly, depending on the time and place at which they are taken.

Measurement. Water vapor is difficult to measure accurately over the full range of atmospheric conditions. Errors in regular water vapor measurements are probably on the order of 10 percent near the surface, increasing to about a factor of 2 near the tropopause. Currently, our best sources of information on water vapor are instruments on radiosondes, (weather balloons) launched once or twice daily from fewer than 1,000 sites around the world. Such instruments must be disposable and therefore inexpensive, and hence they are usually inaccurate. They do a very poor job, or even fail, at measuring the low humidities at high altitudes. The slow progress in predicting precipitation is in part the result of the poor quality, infrequency, and coarseness of water vapor measurements. In the future, water vapor measurements may be made by a combination of more sophisticated and accurate aircraft-based and remote-sensing instruments, providing a more accurate and detailed description of atmospheric water.

Devices for measuring water vapor are called *hygrometers*. Many different physical processes are used as the bases of water vapor measurements. *Psychrometers* are devices with two thermometers: one measures ambient temperature, and one has a wet, muslin-covered bulb to measure the *wet-bulb temperature*. The wet-bulb temperature is lower than the air temperature because of cooling by evaporation from the muslin cover; for a given temperature, the lower the wet-bulb temperature, the lower the humidity. Psychrometers are cumbersome and used only for surface measurements. Dewpoint or frost-point hygrometers depend on the condensation of moisture, usually on a chilled mirror. [*See* Instrumentation.]

Other devices rely on materials that undergo a change in physical dimension, chemical properties, or electrical properties when they absorb moisture. These devices use materials such as hair, goldbeater's skin (a prepared animal membrane), carbon films, and lithium chloride. On U.S.-launched radiosondes, carbon-film *hygristors* (making use of changes in electrical resistance) are currently used. Another popular sensor, used on the radiosondes of many other countries, measures humidity-induced changes in the electrical capacitance of a thin film made of an organic polymer.

Spectral techniques for measuring water vapor are becoming increasingly sophisticated, employing emission as well as absorption spectra. These techniques are particularly useful for remote sensing from satellites or

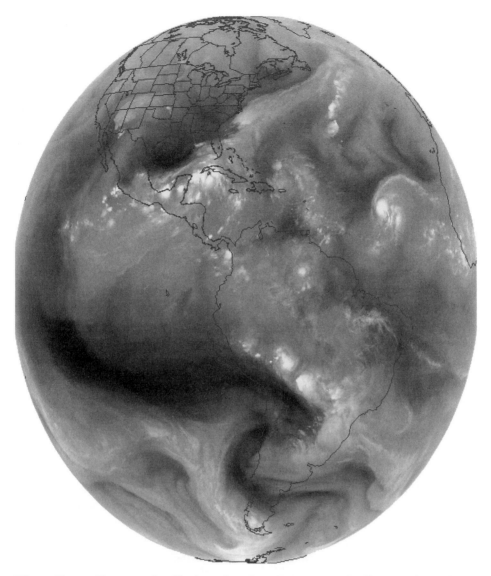

WATER VAPOR. Figure 2. *Satellite image based on measurements of radiation with wavelengths (a band around 6.7 microns) that closely follow total column water vapor.* Lighter areas indicate more water vapor. The large variations on small spatial scales are evident. The image is from GOES-8 for 28 September 1995 at 1145 coordinated universal time. (Courtesy of the National Oceanic and Atmospheric Administration.)

from the ground. Microwave radiometers are the instruments used in many satellited-based measurements of water vapor, while new ground-based systems make use of lasers (*lidar* instruments). A new class of techniques infers concentrations of water vapor from variations in the atmosphere's index of refraction and the associated variation in the speed of radio waves.

Thermodynamics and Dynamics. Like dry air, water vapor in the atmosphere behaves, to a good approximation, as an ideal gas. Water molecules are less massive than most other molecules in air; water has an average molecular weight, M_w, of 18.015, while the average molecular weight of dry air is 28.964. Therefore, a moist parcel of air is less dense than a dry parcel at the same temperature and pressure. The *virtual temperature* provides a convenient way of accounting for water vapor in a parcel when one is most concerned with comparing parcel densities. A moist parcel has a virtual temperature slightly greater than its actual temperature. [*See* Temperature.]

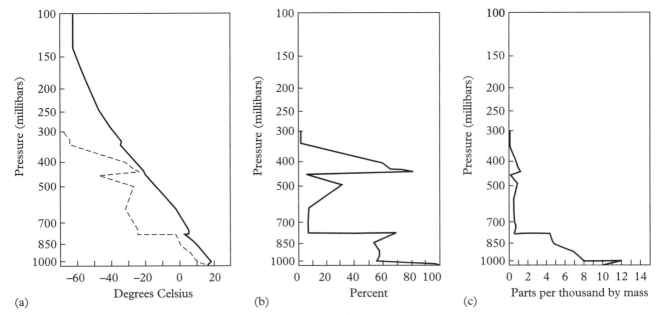

WATER VAPOR. Figure 3. *Vertical profile of water vapor.* (a) An example of a vertical sounding from a radiosonde showing temperature and dew point as a function of pressure; from these, the concentration of water vapor can be calculated. (b) Relative humidity, in percent. (c) Mixing ratio by mass in parts per thousand. Reliable estimates cannot be made for the upper troposphere using radiosondes. Note the large variations on very small spatial scales and the very rapid decrease in the mixing ratio of water vapor with increasing height. Data are for a station near Wallops Island, Virginia, at nearly the same time as Figure 2. (Courtesy of Michael C. Morgan.)

Water exists in the atmosphere in all three of its common phases—vapor, liquid, and solid (there are actually several different phases of ice, but those details are beyond this discussion). It is constantly changing between them. In the atmosphere, there are four phase changes involving water vapor: *evaporation* or *vaporization* is the change from liquid to vapor; *condensation* is the change from vapor to liquid; *sublimation* is the change from solid to vapor; and *deposition* is the change from vapor to solid (deposition is often subsumed into condensation). The amounts of energy, or latent heat, released by or required for these phase changes are quite large. At 0°C, the latent heat of vaporization (or condensation), $L_v = 2.50 \times 10^6$ joules per kilogram. The latent heat of sublimation (or deposition), $L_s = 2.83 \times 10^6$ joules per kilogram. (The difference between L_v and L_s is equal to the latent heat of fusion, L_f, which is the energy needed for the phase change between liquid water and ice.) By comparison, the amount of heat needed to change the temperature of a parcel is small; the specific heat of water vapor, with constant pressure, $c_p = 1,850$ joules per kilogram per kelvin, even though the specific

heat of water is greater than that of most other substances (almost double the value of dry air).

The maximum vapor pressure of water (at thermodynamic equilibrium) is limited by the temperature of the vapor. However, because the rest of the dry air makes up the bulk of the mass of the atmosphere, the temperature of the water vapor is determined largely by the temperature of the air with which it coexists. This sometimes leads to the not-quite-correct statement that air of a given temperature can "hold" a certain amount of water. The Clausius-Clapeyron equation relates changes in the saturation vapor pressure of water, e_s, to the (absolute) vapor temperature, T:

$$\frac{de_s}{dT} = \frac{e_s L M_w}{RT^2}, \tag{1}$$

where $R \approx 8.3 \times 10^3$ joules per kelvin per mole is the ideal-gas constant. This equation applies for water vapor in equilibrium with either liquid water or ice, using L_v or L_s, respectively, for L. If one assumes that the latent heat of vaporization is constant (L_v and L_s vary by less than

5 percent over temperatures of meteorological interest, decreasing slowly with increasing temperature), Equation 1 can be integrated to yield

$$e_s = e_s(T_0) \exp\left[\frac{LM_w}{R}\left(\frac{1}{T_0} - \frac{1}{T}\right)\right]. \qquad (2)$$

The constants of integration can be determined by the observation that for $T_0 = 273$ K $= 0°C$, $e_s = 6.11$ millibars. For temperatures below the freezing point, the satuation vapor pressure over ice is less than the saturation vapor pressure over water; this indicates that ice is formed preferentially over liquid water at freezing temperatures. The most important result from Equation 2 is that saturation vapor pressure increases very rapidly (roughly exponentially) with increasing temperature.

Although the integrated Clausius-Clapeyron equation is a good approximation of the maximum vapor pressure near the surface of water bodies, it cannot easily be applied to the equilibrium of water droplets in clouds and the surrounding water vapor. If the air were very clean (that is, with very few particles of any sort), supersaturation of several hundred percent would be possible. On the other hand, condensation can occur on hygroscopic nuclei at relative humidities much lower than 100 percent—for example, about 75 percent on sodium chloride (common salt). In practice, the available supply of particles—called *condensation nuclei*—allows condensation to begin very close to the saturation vapor pressure at thermodynamic equilibrium. The liquid and solid products of condensation and sublimation are called *hydrometeors,* including droplets, snowflakes, graupel, and hail. [*See* Precipitation.]

When a parcel of dry or subsaturated air is lifted adiabatically (with no heat moving into or out of the parcel) to levels of lower pressure, it expands and cools. The rate of cooling, called the *dry adiabatic lapse rate,* is about 1 K for every 100 meters increase in altitude. The presence of water vapor complicates the situation. Once the parcel has been cooled to the point of saturation, water vapor condenses as the parcel is cooled further, releasing latent heat. The level at which condensation begins is called the *lifting condensation level.* If a parcel is lifted or rises farther, the release of latent heat partially compensates for the cooling from expansion; as a result, the *moist adiabatic lapse rate* (which is nearly identical to the *pseudoadiabatic lapse rate*) is only about half the rate of the dry one. This has enormous consequences for the dynamics of the atmosphere. [*See* Adiabatic Processes; Convection.]

Unless the vertical profile of the ambient atmosphere cools at less than the moist adiabatic lapse rate, a parcel that has water condensing and releasing latent heat as it rises becomes warmer than its environment and thus positively buoyant. The parcel then accelerates upward in the process called *moist convection.* This process can continue until the parcel reaches a part of the atmosphere that is very stably stratified, such as at the tropopause, resulting in cumulus towers more than 20 kilometers high. If there are parcels that will accelerate upward when the water vapor in them condenses, but not otherwise, the air column is said to exhibit *conditional instability.* A measure that helps to calculate the buoyancy of a parcel and the height to which it can rise is the *equivalent potential temperature.* Dynamics involving condensation of water vapor and the resulting formation of deep convective clouds are largely responsible for the development of mesoscale systems with spatial scales of 10 to 100 kilometers. (For a more detailed discussion on the role of water vapor in atmospheric convection, see Rogers 1979.) [*See* Potential Temperature.]

Processes involving evaporation are also important and occasionally dramatic. Very intense downdrafts, called *microbursts,* are driven by the cooling from the evaporation of raindrops. Microbursts have been implicated in several aircraft crashes. Evaporation is most important at the surface, where it cools the land and ocean and transfers energy and water vapor to the atmosphere. [*See* Evaporation; Small-Scale or Local Winds.]

Chemistry. Water vapor is intimately involved in many atmospheric chemical reactions. For atmospheric chemistry, however, water is more important in clouds as either a liquid or a solid than as a vapor. Water vapor is the primary source of the hydroxyl radical (OH), which serves as an oxidant in reactions with many trace gases.

In the upper troposphere and above, chemical reactions are a significant source of water vapor. The most important reactions for water production are the oxidation of methane (CH_4) and the combustion of jet fuel. In the stratosphere, OH, HO_2, and other radicals derived from water vapor participate in reactions that destroy ozone. [*See* Atmospheric Chemistry and Composition.]

Radiation. Water has a greater effect than any other substance on the radiation balance of the atmosphere. It is the dominant *greenhouse gas*—the dominant absorber of infrared radiation (heat) emitted from the Earth's surface and from within the atmosphere. Some incoming solar radiation is absorbed by water vapor. Solar radiation is also scattered by water droplets in clouds and by hygroscopic aerosols, but this process does not directly involve water in vapor phase. [*See* Greenhouse Gases; Radiation.]

Water vapor is a very good absorber of infrared radiation (see Figure 4). Its absorption spectrum has strong bands near 1.4, 1.8, 2.7, and 6.3 microns, caused by vibration-rotation modes. A series of rotational absorption bands begins at 11 microns, and the bands increase in strength with increasing wavelength past the wavelength at which there is significant infrared radiation present in the atmosphere (around 30 microns).

Although most of the water vapor resides in the lower levels of the atmosphere (about 90 percent is below 4,000 meters), the small amounts that exist at higher elevations have a disproportionately large effect on the radiation balance. For example, a 5 percent increase in the relative humidity at 10 kilometers from the surface has about the same effect on the radiation balance of the atmosphere as a 5 percent increase in the relative humidity at 1 kilometer, even though the latter change involves much more water vapor.

Climate and Climate Change. The climate of a particular location is in large part determined by the amount of available water—the humidity, precipitation, and evapotranspiration—and by how that availability normally varies over the course of the year. Water vapor has a profound effect on our planet's climate through the normal *greenhouse effect*. John Tyndall, in the 1860s, thought that differences in water vapor above different

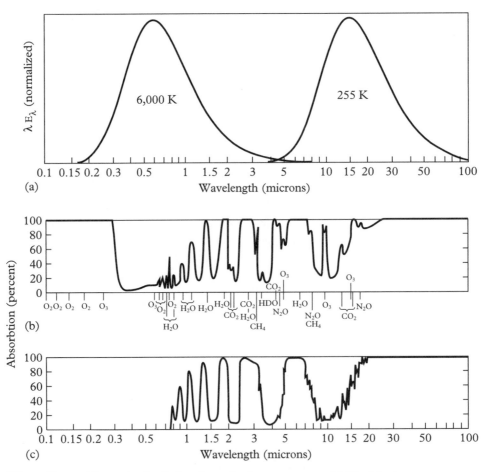

WATER VAPOR. Figure 4. *Radiative absorption by water vapor.* (a) Black body curves approximating the spectra of the incoming solar radiation (6,000 K) and the outgoing terrestrial radiation (255 K). (b) Absorption spectrum of water vapor for the entire vertical extent of the atmosphere (all gases). (c) Part of the absorption spectrum of water vapor, integrated over the depth of the atmosphere. (Adapted from Peixoto and Oort, 1992. Courtesy of the American Institute of Physics.)

locations were largely responsible for all the diverse climates on Earth.

Changes in the water vapor content of the atmosphere are closely linked to global climate change. Increases in atmospheric water vapor will increase the greenhouse effect. Numerical models clearly demonstrate a positive feedback between surface temperature and water vapor, especially for vapor in the upper troposphere: increases in temperature lead to increases in water vapor, which in turn lead to further increases in temperature. (Negative feedbacks also exist, so a "runaway greenhouse" evaporating all water is not likely on Earth.) There is already some evidence that atmospheric water vapor has increased recently; however, it is uncertain how the distribution of atmospheric water vapor will change in response to temperature changes resulting from increases in other greenhouse gases such as carbon dioxide and methane. These unanswered questions lie at the center of most scientific disagreements about human-caused climate change.

[See also Condensation; and Greenhouse Effect.]

BIBLIOGRAPHY

Deepak, A., T. D. Wilkerson, and L. H. Ruthke, eds. *Atmospheric Water Vapor.* New York, 1980.
Huschke, Ralphe E., ed. *Glossary of Meteorology.* Boston, 1959.
Iribarne, J. V., and W. L. Godson. *Atmospheric Thermodynamics.* 2d ed. Dordrecht, 1981.
Larsen, J. C. et al. "A Comparison of the Stratospheric Aerosol and Gas Experiments." *Journal of Geophysical Research* 98, D3 (1993): 4897–4917.
Peixoto, J., and A. H. Oort. *Physics of Climate.* Woodbury, N.Y., 1992.
Rogers, R. R. *A Short Course in Cloud Physics.* 2d ed. Oxford, 1979.

MARK DAVID HANDEL

WAVES. *See* General Circulation; Gravity Waves; *and* Turbulence.

WEATHER. *See* Climate and Weather, *overview article.*

WEATHER FORECASTING. *[This entry consists of two articles:*

Models and Methods
History of Study
The first article provides discussions of models and methods involved in forecasting processes and in subjective, statistical, numerical, and specialized weather forecasts. The second

article treats early efforts of data collection and exchange, early forecasting methods, numerical weather prediction, and current directions in weather forecasting. For a discussion of marine weather forecasting, see Marine Weather Observations and Predictions.*]*

Models and Methods

Weather forecasting is the process of projecting the state of the atmosphere forward in time. The final products of a forecast may include numerical quantities such as temperature or air pressure, as well as more subjective descriptions such as "partly cloudy" or "sunny."

Attempts to predict weather almost certainly go back as far as human history. People involved in weather-sensitive activities, especially mariners and farmers, have traditionally become skilled at recognizing and interpreting weather signs.

Over the past century, the practice of weather forecasting has changed from an art to a science. Today, much as physicians apply the science of medicine to diagnose and treat patients, the weather forecaster applies the science of meteorology to diagnose and predict the weather.

This change was made possible by modern technological advances, beginning with the telegraph and later radio, both means of exchanging weather information rapidly over long distances. Beginning in the 1920s, the radiosonde provided the first systematic observations of temperature, pressure, and humidity above the Earth's surface. More recent advances include weather radar, weather satellites, and the electronic computer. Weather forecasters today use computers for data collection, data analysis, weather prediction and data display. Weather satellites provide global views of weather and have introduced new ways of observing and predicting storms. [See Instrumentation; Satellite Observation; *and the article on* History of Study *in this entry.*]

Process of Forecasting. Any weather forecast process begins with observations of the weather. These include ground-level weather observations made by human observers and automated systems, along with radiosonde measurements, radar reports, and observations from aircraft and weather satellites.

The principal processes used in weather forecasting may be categorized as *subjective, numerical,* and *statistical.* In subjective processes, forecasters work directly from observations of the nature of present weather, interpreting these based on their experience and knowledge of principles. The type of process used depends upon the time period of the forecast. Very short-range predictions (forecasts for a few hours) utilize mainly subjective techniques. Short-range forecasts (from a few hours to a few days) use a combination of numerical, statistical, and subjective techniques. Medium-range (3-day to 10-day) forecasts are based almost entirely on numerical techniques. Long-range predictions of average conditions for a month or a season have until recently been largely statistical and subjective; however, they now use numerical techniques as well.

Subjective Weather Forecasts. These forecasts involve the interpretation of current weather observations by trained weather forecasters. Forecasters, who usually have a bachelor's or master's degree in meteorology, use their training and experience to predict the weather. Until the 1950s, almost all forecasts were subjective; now, numerical and statistical techniques predominate in the short- and medium-ranges.

Subjective techniques are most important in the preparation of very short-range forecasts and weather warnings. For forecasts of a few minutes to a few hours, human abilities for pattern recognition are far more important than the ability to process quickly vast amounts of numerical data. The most important tools for very short-range prediction—radar and satellite observations of clouds and precipitation—are more susceptible to human interpretation than to the quantitative requirements of numerical and statistical methods.

The role of the skilled weather forecaster is also critically important for the prediction of such phenomena as severe thunderstorms, tornadoes, hurricanes, and tropical storms. In the United States, the National Hurricane Center and the National Severe Storms Forecast Center are staffed by forecasters with special expertise in these areas.

Statistical Forecasts. Forecasts can also be made by employing equations that express statistical relationships between the quantity to be forecast and other variables (or the same variable) at some earlier time. For example, one approach to forecasting tomorrow's temperature would be to examine many years of weather data to identify variables that correlate regularly with temperature changes from day to day. Various statistical methods can be used to identify these variables and to relate them numerically to the quantity to be predicted. [*See* Statistical Methods.]

The most important use for statistical methods in weather forecasting couples the statistical method with the output from numerical weather prediction. This approach, called *Model Output Statistics* (MOS), predicts variables not from previous weather data, but from variables forecast by models. A forecast of tomorrow's temperature, for example, might be made from a statistical equation that includes numerical forecasts of temperature, humidity, wind direction, and clouds. The MOS technique improves the accuracy of variables directly forecast by models. It also allows the objective prediction of quantities—such as probability of precipitation or surface visibility—that models do not forecast directly.

Statistical techniques are important for long-range prediction as well. For monthly and seasonal predictions, forecasters rely strongly on "lagged" statistical correlations. These are, for example, correlations between the average temperature in the month or season of the forecast and averages of the temperature or other parameters in previous months or seasons. Prediction of the phase of the El Niño-Southern Oscillation has become an important component of seasonal weather prediction because of correlations between ocean surface temperatures in the tropical Pacific and seasonally averaged weather in many parts of the world. [*See* El Niño, La Niña, and the Southern Oscillation.]

Numerical Weather Forecasts. Numerical weather prediction (NWP) produces forecasts by solving systems of equations that describe the behavior of the atmosphere. These equations are expressions of the laws of physics for fluids. They include Newton's second law of motion, the first law of thermodynamics, the law of conservation of mass, and the equation of state for an ideal gas. Forecasting the weather in this way was first envisioned in the early 1900s and attempted in the 1920s, but the very number of calculations required was so great that the process became feasible only with the development of the electronic computer. [*See* Primitive Equations.]

The first step in NWP is to collect the data. Weather observations taken all over the world at the same time each day are entered into a global telecommunications network, maintained by individual countries under international agreements for the exchange of weather data. Within a few hours of recording, weather observations pass through the network and into the computers of major weather forecast centers around the world.

Once sufficient data have been collected, the next step in the process is to analyze the data. The most important function of analysis is to use the information gathered at irregularly spaced observation points to estimate the values at the points of a regularly spaced array for use in forecast calculations. The analysis defines the state of the atmosphere at the starting time for the forecast. [*See* Numerical Weather Prediction.]

The next step is the forecast itself. Equations are solved to produce a new value of each variable at each grid point. The process marches forward in time to the desired length of the forecast—usually 48 or 72 hours for short-range forecasts and 5 to 10 days for forecasts of medium range. The forecasts can be displayed at any interval needed, but the intervals are usually 6 or 12 hours.

The equations, together with the mathematical methods and the grid points and levels, are referred to as a *weather forecast model.* A sample of the output from such a model is shown in Figure 1.

For mathematical reasons, the time steps in the forecast must be very short—usually only a few minutes. Thus, a weather forecast for 10 days may require the calculation of 1,000 or more individual forecasts. Forecasts must be produced in a reasonable amount of time, so the complexity of the model used is limited by the power of the computer. Modern weather forecast models may require more than 1 trillion calculations to produce a forecast of a few days.

Accuracy. There are many ways to measure the accuracy of weather forecasts. The accuracy of a series of forecasts—for example, forecasts of temperature—is often expressed as an average error (Figure 2). Forecast accuracy may also be expressed as the percentage of a series of forecasts that proved correct. Assertions that weather forecasts are, for example, 85 percent correct are

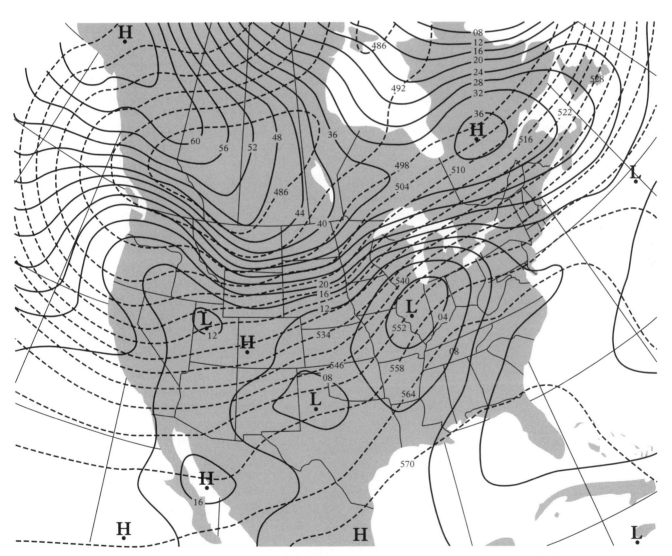

WEATHER FORECASTING: Models. Figure 1. *A sample output of numerical weather prediction.* This is a 36-hour forecast of sea-level pressure from a numerical model. The dashed lines are predictions of the average temperature in the lower parts of the troposphere.

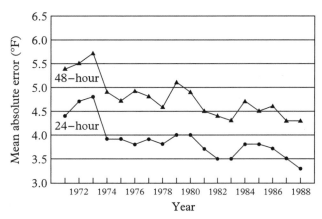

WEATHER FORECASTING: Models. Figure 2. *An example of a long-term record of forecast accuracy.* The graph shows the average value of the absolute error of 24- and 48-hour temperature forecasts at 90 U.S. cities. Accuracy figures are often stratified by season and these data are for the winter months (October through March) from 1970–1971 through 1987–1988.

impossible to interpret without a definition of what *correct* means in this context.

It is important to distinguish between forecast *accuracy* and forecast *skill*. A forecast of "no rain" in Los Angeles on each summer day is likely to be correct on about 97 percent of the days; however, the only skill required to make such a forecast is experience in living in Los Angeles or access to climatological records. Another highly accurate but "zero-skill" forecast might be that the weather will not change over the next few minutes or hours. Sometimes such a forecast is dramatically wrong, but most of the time, it is correct. The first type of forecast is called a *climatological forecast;* the second is a *persistence forecast.* The skill of a weather forecast method is usually expressed by comparing the method's accuracy with that of a climatological or persistence forecast. The longer the range of the weather forecast, the less accurate it is likely to be. The accuracy of current day-to-day weather forecasts typically decreases to about the level of a climatological forecast (zero skill) in 5 to 7 days.

The atmosphere is a *chaotic* system, which means that after some period of time, it "forgets" its initial conditions and becomes inherently unpredictable. (Something is called *unpredictable* if the best available prediction is no better than chance.) The most commonly accepted limit of atmospheric predictability is approximately 2 weeks. This limit applies to day-to-day weather forecasts, but not to predictions of average conditions or climate.

Specialized Weather Forecasts. Specialized forecasts are issued for commercial or recreational activities

that are especially sensitive to weather. For example, frost warnings are issued to warn farmers to protect their crops. Forecasts of winds and waves are issued for mariners. Flying is a very weather-sensitive activity, and special forecasts of turbulence, icing, and terminal weather are issued for aviation. These are only a few of the specialized kinds of weather forecasts available.

In the broadest sense, all specialized forecasts are byproducts of the forecast process required to produce general forecasts for the public. The difference between general and specialized forecasting is primarily in the interpretation of the forecast in terms of the weather sensitivities of a particular industry or activity.

River and Flood Forecasts. Flooding can be caused by heavy rains, spring snowmelt, or ice jams on rivers and streams. Flooding of major rivers, such as the Mississippi River in the United States, is experienced as a gradual rise in water level over a period of days to weeks. Small stream flooding can take place in minutes or hours after a heavy rain. In extreme cases, a wall of water can form and move rapidly downstream, causing great devastation and loss of life. An example of such a "flash flood" is the wall of water that raced down the Big Thompson River in Colorado in 1976, causing 139 deaths and property damage of $35 million. [*See* Floods.]

River and flood prediction is a two-stage process. The first stage involves observation and prediction of precipitation—or, in the case of snowmelt or ice breakup on rivers, unusually warm temperatures. The second stage involves predicting what will happen to the water. Some of it will be absorbed into the ground, some will evaporate, and some will enter streams, then rivers, and then the ocean. Flood-prediction models try to simulate these processes and predict the water levels at locations on streams and rivers where measurements are made.

Warnings for small-stream or urban flooding require fast response times. The forecast process in this case is usually condensed to a comparison of observed and forecast precipitation amounts with prior estimates of the amount of rainfall required to saturate the ground.

The most important observations for river and flood forecasting are precipitation, river and stream levels, and ground wetness. Precipitation observation and short-range prediction is greatly aided by radar observations and cloud observations from weather satellites.

BIBLIOGRAPHY

Atkinson, B. W., and A. Gadd. *Weather.* New York, 1988.
Bosart, L. "Weather Forecasting." In *Encyclopedia Americana.* Danbury, Conn. 1990.

Neiburger, M., J. Edinger, and W. Bonner. *Understanding Our Atmospheric Environment.* 2d ed. San Francisco, 1982.

Tribbia, J., and R. Anthes. "Scientific Basis of Modern Weather Prediction." *Science* 237 (July, 1987): 493–499.

Williams, J. *The Weather Book.* New York, 1992.

WILLIAM D. BONNER

History of Study

Weather forecasting developed in response to the needs of societies to protect themselves from storms, severe heat and cold, and economic losses caused by adverse weather. Organized efforts to prepare weather forecasts achieved very little success at first because the problem was too difficult; its physical and mathematical nature was not at all understood. The history of weather forecasting presented here will relate early and subsequent organized attempts to achieve conceptual models of the atmosphere and means of employing them in making weather forecasts.

Society's requirements for forecasts did not permit weather forecasting to be postponed until the necessary understanding, observations, and scientific methods were available, so many ad hoc methods were employed. At first, simple extrapolation tempered by experience achieved some success. The subsequent history of forecasting led to the concept of air masses and fronts. Theoretical developments in understanding atmospheric motions, together with the availability of the electronic computer by the mid-twentieth century, made possible a scientific and more successful approach to weather forecasting.

Early Efforts: Data Collection and Exchange. By the middle of the nineteenth century, the invention of the telegraph and the establishment of national telegraph networks linking cities made possible the exchange of weather data quickly and over considerable distances. It was already known that storms were mobile and that the same large storm could appear in several different cities on successive days. Benjamin Franklin, in 1743, had noted the progress of storms by comparing data from several colonial cities. James P. Espy prepared synoptic weather charts in 1841 from data received by mail from several American cities; subsequent telegraphic exchange of data allowed a look at weather systems on a current basis, as they formed and moved.

In 1849 a synoptic weather chart—a chart displaying weather conditions as they exist simultaneously over a large area—was prepared by A. J. Henry for the current weather. In 1861, the British Admiral Fitzroy prepared a weather chart using data obtained by telegraph. France issued regular daily telegraphic weather charts beginning in 1863. In 1870 a measure passed the U.S. Congress and was signed by President Ulysses S. Grant establishing a national weather service to be conducted by the Army Signal Service. The first task of the service was to issue storm warnings for the coasts and the Great Lakes. Forecasts were issued three times a day, beginning in 1871. These landmark events are described by Mitchell and Wexler (1941) and by Whitnah (1961). In 1873 the International Meteorological Committee was established to standardize and synchronize international exchanges of weather data.

From these beginnings, national and international communication networks were established to exchange weather data among the nations of the world. These arrangements are now coordinated by the World Meteorological Organization in Geneva, Switzerland, the successor to the International Meteorological Committee, and are implemented by the individual nations. In almost any country of the world it is now possible to have access to a vast array of meteorological data from fixed weather stations, ships at sea, meteorological satellites, balloon-borne observing instruments, and vast numbers of commercial aircraft in flight. These data are the raw material for the preparation of weather forecasts.

Early Forecasting Methods. Little has been written on early methods of weather forecasting for a simple reason; no one could describe exactly how it could be done. Experience in forecasting made the forecaster. The surface weather chart displayed what the early forecaster wanted to know. High or low (sea-level) pressure systems moved across the map in somewhat irregular ways, and these pressure systems were often characterized by certain types of weather; for example, in the forward part of a moving high pressure system there was often cooling and clearing weather. The forecaster noting an approaching high would examine its past motion, try to remember similar events, and then forecast "typical" weather. This approach might succeed well enough for a day or so in advance, but sudden change in motion, shape, or intensity of highs or lows in the forecast area could result in a missed forecast.

The use of weather types was a formal analogue method. Its best-known proponent, Franz Bauer of Germany, published long-range forecasts based on a selection of analogues of the current weather map. To find analogues he used an extensive file of daily weather maps covering a period of about 100 years. He called his method the *method of the similar case.* In the United States,

I. P. Krick of the California Institute of Technology taught his students forecasting by reference to analogues. Certain initial *types* of weather patterns were defined, which were represented as destined to develop along certain typical paths. Unfortunately, some applications of Krick's types involved day-by-day projections out to 5 days or even longer. This long-range forecasting by types found a home in at least one region of the U.S. Air Force's Air Weather Service during World War II, and its application could be seen on a major network television station in Washington, D.C. into the 1970s. It proved however, to be without merit.

In the first two decades of the twentieth century, the Norwegian Vilhelm Bjerknes and his students recognized that masses of air of different origins and thermal properties could be brought into proximity while still retaining rather sharp transition zones between them. These zones, called *fronts*, were often noted to be associated with precipitation as the warmer air was lifted up over the air on the colder side of the front. Perturbations, called *waves*, sometimes traveled along the front, developing into cyclones and bringing bad weather with them. While still not free from controversy about details, this Norwegian view of the atmosphere had sufficient merit that by the end of the twentieth century, fronts and cyclone waves are still standard and accepted concepts in weather analysis. Forecasters learned to recognize the frontal waves on their daily weather maps and to forecast their development in accordance with the Norwegian models, which were widely accepted during World War II by meteorologists worldwide. Today television audiences in most countries are shown weather maps in which the extratropical regions are well populated with high and low pressure areas, air masses of differing origins, fronts between different types of air, and frontal waves occluding and developing into cyclones. [*See* Fronts.]

World War II and its demand for meteorologists strongly influenced the development of forecasting throughout the world. Many technically educated young people were given thorough university training in order to enter their armed forces as weather forecasters, leading to a substantial increase in the number of university departments offering such courses. University departments in the United States were strengthened by adding emigré European scientists to their staffs. The result was the addition of several thousand well-trained meteorologists to the military services and eventually to the available force of meteorologists after the war. These people staffed old and new university meteorology departments,

did weather forecasting for airlines and the government services, and performed research financed by governments.

World War II had many military operations in the tropics. U.S. military forecasters with practically no training in tropical forecasting were required to do the best work they could with little understanding of the meteorological problems they faced. In 1943 the Office of Naval Research financed a joint effort by the University of Chicago and the University of Puerto Rico at San Juan. Its objective was to learn more about tropical circulations and forecasting. This effort produced a number of reports clarifying the nature of tropical circulations and perturbations, as well as new and appropriate tropical weather analysis procedures. The clarification of the nature of waves in the easterlies and the preparation of an analysis workbook were of considerable value to tropical analysis and forecasting. The problem of forecasting hurricane development and movement is a complicated and difficult one. [*See* Tropical Circulations; Tropical Precipitation Systems.]

Numerical Weather Prediction. The predominant transformation of weather forecasting in the twentieth century was the development of numerical weather prediction. By the beginning of the century, the equations for atmospheric state and motions were known. In 1903–1904, Vilhelm Bjerknes proposed integration of the hydrodynamical equations from an initial atmospheric state for the calculation of a forecast. In 1911 Lewis Fry Richardson of England considered the integration of the finite difference equations to solve the forecast problem; in 1916–1918 he made a calculation of the initial derivatives of the atmospheric variables with time. The completely unrealistic nature of his result is commonly attributed to the inadequacies of his initial data. The volume of the required calculations was so formidable that this line was not pursued further for some years.

Richardson's effort showed that since the equations of state and motion were well enough known, the calculation of a forecast by dynamic methods was possible in principle; however, the volume of the calculations required would not permit the calculation of a forecast, especially in real time. He estimated that 64,000 computers (meaning, in his day, people doing computing by hand) would be required to "race the weather for the whole globe."

The work of Jerome Namias, heavily influenced by Carl-Gustaf Rossby, was a first effort to use dynamical principles in forecasting. Leading the U.S. Weather Bureau's 5-day forecast activity, Namias introduced the use of constant vorticity trajectories and Rossby's equation for the propogation of long atmospheric waves as guidance for forecasts of 5-day mean flow patterns. In 1952, Ragnar Fjoesrtoft published a method for the integration of simple dynamic atmospheric models by graphical methods. The graphical integration methods were slow and difficult to use in practice, but they were used to some extent in Europe and in several weather centers of the U.S. Air Force in the early 1950s. [*See* Vorticity.]

The development of electronic computers offered exciting possibilities of reviving the old idea of using dynamic prediction methods, which has come to be called *numerical weather prediction*. A small group was formed for this purpose at the Institute for Advanced Study at Princeton, New Jersey, under the leadership of John von Neuman and Jule Charney. Their first numerical integration of the barotropic vorticity equation, in 1950, was followed by integration of a three-level quasigeostrophic model, which successfully showed the development of a severe East Coast storm that had baffled forecasters when it occurred unexpectedly. This accomplishment created a sensation and soon led to the formulation of plans by the weather bureau, air force, and navy to procure a computer and jointly enter into operational numerical weather forecasting, with eventual success. [*See* Dynamics; General Circulation Models; Numerical Weather Prediction.]

Before numerical prediction was proven feasible, a substantial effort at forecasting by statistical means was started, almost in competition with the former. Subsequent to the success of the dynamic approach, statistical efforts were focused on those weather elements not explicitly contained in the dynamic approach, using the dynamic predictions as input parameters. This approach, called *model output statistics*, has been very fruitful. [*See* Statistical Methods.]

Current Directions. In the 1990s, a large-scale effort is being made to improve short-range forecasts and warnings in the United States by the installation of Doppler radars, vertical profilers of wind, automatic surface weather stations, and other technological means, with emphasis on providing short-range warnings of severe weather. This effort will undoubtedly find a prominent place in the future of weather forecasting.

BIBLIOGRAPHY

Ashford, O. M. *Prophet or Professor? The Life and Work of Lewis Fry Richardson.* Bristol and Boston, 1985.
Friedman, R. M. *Appropriating the Weather: Vilhelm Bjerknes and the Construction of a Modern Meteorology.* Ithaca, N.Y., and London, 1989.
Mitchell, C. L., and H. Wexler. "How the Daily Forecast Is Made." In *Climate and Man,* Yearbook of Agriculture. Washington, D.C., 1941, pp. 579–598.
Whitnah, D. R. *A History of the United States Weather Bureau.* Urbana, Ill., 1961.

GEORGE P. CRESSMAN

WEATHER HAZARDS. The Earth's atmosphere is both a resource and a hazard. Many attributes of the atmospheric system are beneficial, and others are dangerous. The hot and humid climate of the tropics makes possible the growth of the tropical rainforests with their enormously rich diversity of species; at the same time, this climatic regime spawns severe tropical cyclones (hurricanes in the Atlantic and Caribbean, typhoons in the Pacific) that can cause great loss of life and property damage. It is the hazardous behavior of the atmosphere that attracts public attention, and much scientific research. The fundamental value of the atmosphere as a resource and as part of the global life-support system is frequently taken for granted: it is as essential to human life as is the sea to the fish that swim in it. No choice is involved, and no decisions have to be made.

For many practical purposes, however, the atmospheric system's weather and climate offers choices. In making these choices we determine the extent to which the atmospheric system will be a beneficial resource or a dangerous hazard.

Weather-related hazards may be associated with temperature, in the form of heat waves, cold snaps, and unseasonal frosts. Their effects may come through precipitation—as snow, heavy rain, hail, or freezing rain—and resultant floods and avalanches. Thunderstorms may also present lightning danger. High winds create such hazardous conditions as blizzards (blowing snow), dust or sand storms, clear air turbulence, tornadoes, and the tropical cyclones known as hurricanes and typhoons; wind also increases the effect of cold through wind chill.

There are two different and complementary ways of viewing and defining these phenomena. From atmospheric science they are more or less uncommon or extreme events and may be defined in terms of their physical magnitude and frequency. For human activities, they are defined in terms of the threat they pose to life, health, or economic value. Thus it is useful to distinguish extreme or abnormal events and hazards. In meteorological terms, a hurricane formally comes into existence when its mean wind speed reaches 64 knots (120 kilometers per hour). Hydrologists commonly define a flood as discharge above a certain level, or flood stage, usually the level at which the river channel is full from bank to bank; thus the actual discharge that corresponds to flood stage varies according to the shape and slope of the local topography. Flood stage can also indicate the point at which damage begins, further reducing the value of the definition for the purposes of comparative hydrometeorological studies. For this reason, hydrometeorologists prefer to describe and analyze floods in terms of frequency. The definition of drought offers even greater difficulties. It is essentially a relative term, and any definition in terms of rainfall deficit or water balance must also refer to normal rainfall or moisture-related activity. A shortage of rainfall (or soil moisture) in the growing season may be described as an agricultural drought. Its impact depends on the particular crop, and drought hazard may be alleviated by the choice of drought-resistant varieties. Attempts have been made to develop precise meteorological definitions, such as "absolute" and "partial" drought, but these have now been discontinued. [See Drought; Floods.]

Common to these scientific definitions is the concept of a threshold or critical value at which the hazard event may be said to begin. If the magnitude and frequency of extreme events is to be compared from place to place or time to time for the purposes of atmospheric research, it is essential to have a common definition. For most atmospheric hazards there is no common definition, even in purely physical terms. One exception is fog, for which there is an internationally agreed definition as visibility of less than 1 kilometer. Even here local usage varies: in Britain, for example, the use of the term "fog" in forecasts for the general public relates to visibility of less than 180 meters.

Common definitions for extreme or abnormal atmospheric events viewed as hazards are even more impractical. The onset of a flood hazard depends not only on the magnitude and frequency of discharge, but also on the choices that have been made with respect to the use of flood-prone land. Where floodplains have been left undeveloped, or where vulnerable property has been restricted to a certain height above flood stage, a hydrological flood may constitute no hazard. The point at which the hazard begins (called *flood-damage stage*) is thus different from the extreme event defined in hydrological terms.

The same can be said of all the hazards listed above: their definition is a matter of human choice. Build your ski chalet in the path of a potential avalanche and you have created a hazard. Build your resort hotel out on the edge of the beach, just above mean high tide on the east coast of the United States or on a Caribbean island, and you are probably vulnerable to hurricane storm damage and coastal flooding.

In many such cases it is possible to reduce or even eliminate the hazard by judicious choice of location. More often, however, hazard is an inevitable accompaniment of using resources or seeking a livelihood in a particular place. There is no way to grow citrus fruit in Florida without encountering frost hazard. Sailing across the North Atlantic in winter inevitably involves exposure to the hazard of icebergs. Tornadoes occur most frequently in the highly productive agricultural region of the Midwest, and most Canadian cities experience severe blizzards during the course of each winter.

Even in these cases there are alternative choices that can be made to reduce hazard impact or lower the risk. For example, Florida citrus growers can choose sites carefully to avoid cool air drainage. Alternative methods include direct heating of the air near the ground, the production of a smoke screen over the crop (smudging), and the flooding or sprinkling of crops to add the thermal content of the water and increase the effective specific heat of the soil.

The actual choice of adjustments to atmospheric hazards is a complex social process that may involve decisions at all levels from the individual and local, to corporate and governmental, to international. The type of decision may range from ad hoc (as in restricting building permits) to resource management policies (as in planning floodplain land use) to broad national strategy (as in funding comprehensive flood damage reduction). Research on the choice of these adjustments, and the actual mix adopted in different jurisdictions has led to some useful insights.

A common observation is that natural hazards in developing countries tend to involve a much higher loss of life and a relatively lower level of property damage. Factors contributing to this contrast include the effective deployment of technology for hazard control, hazard warnings, and evacuation in the more highly developed countries, which has helped reduce the number of deaths. In the absence of the same effective deployment of technology in developing countries, loss of life has continued to grow, as seen in such natural disasters as the droughts in the Sahel region of Africa, tropical cyclones in Bangladesh, and earthquakes in China and elsewhere. In some cases greater population pressure on the land, patterns of unsustainable resource use, poverty, and uncontrolled urbanization expose large populations to extraordinary danger.

In some developed regions the technological control of natural hazards has given rise to a greater sense of security, which has in turn encouraged the invasion of hazard-prone lands and the exposure of more property to risk than would otherwise have occurred. Thus increased safety and reduction in loss of life have been accompanied by rising economic damage.

In more traditional societies, and in the more traditional sectors of modernizing economies, vulnerability to natural hazards may be less. In rapidly changing societies, traditional ways of coping with hazard events may be lost before the advantages of modern technology are effectively realized. Economic development is usually sought by the introduction of directly productive activities, while the necessary safeguards are brought in later in a react-and-cure approach. Thus human-hazard interactions are moving from a pattern of low damages and low deaths in traditional societies, through high deaths in developing societies, to high damages in developed industrial societies. The challenge for hazard research and hazard management now is to find ways to move to more comprehensive policies and practices worldwide that will reduce damages in developed industrial countries and reduce deaths in developing countries without a corresponding increase in damages. The react-and-cure approach is being replaced by an anticipate-and-prevent approach. The development of such resilient responses is broadly in keeping with the goals and methods of sustainable development. The extent to which numbers of fatalities and damages from natural hazards are reduced in the future, therefore, may be taken as a rough indicator of progress toward sustainable development. The 1990s have been designated the International Decade of Natural Disaster Reduction. While it is still too early to expect results of Decade activities to be reflected in statistics, the path of progress in adjusting to atmospheric hazards is clear.

[*See also* Avalanches; Hurricanes; Lightning; Snowstorms; *and* Tornadoes.]

BIBLIOGRAPHY

Burton, Ian, R. W. Kates, and G. F. White. *The Environment as Hazard.* New York and London, 1993.
el-Sabh, M. I., and T. S. Murty, eds. *Natural and Man-Made Hazards.* Dordrecht and Boston, 1988.
Kirby, A. *Nothing to Fear.* Tucson, Ariz., 1990.
Meteorological Office, *Meteorological Glossary.* 6th ed. London, 1991.
O'Riordan, T. "Coping with Environmental Hazards." In *Geography, Resources and Environment,* vol. 2, edited by Robert W. Kates and Ian Burton. Chicago, 1986, pp. 272–309.
White, G. F., ed. *Natural Hazards: Local, National, and Global.* New York, 1974.

IAN BURTON

WEATHERING. *See* Erosion and Weathering.

WEATHER LORE. Lore has been described as knowledge gained through study or experience. In no field has the subject received as much attention as in the study of weather. Since ancient times people have observed the local weather and tried to formulate proverbs and sayings that would supply some key to its behavior and serve as a guide to the morrow's conditions. [*See* Proverbs.]

Two Greek philosopher-poets, Theophrastus of Eresus on Lesbos and Aratus of Macedonia, composed collections of weather sayings. These were preserved by the Romans and eventually passed on to Western culture. Theophrastus (371/70–288/87 B.C.E.), a leader of the Peripatetic school of philosophy, compiled two collections of brief weather indications, *On Weather Signs* and *On Winds.* He was a student of Aristotle (384–322 B.C.E.), whose *Meteorologica* was a summary of the Greeks' knowledge of the atmosphere. Theophrastus's collection gives around 80 indications of rain, 45 of wind, 50 of storm, 24 of fair weather, and 7 of the weather for longer periods of a year or less. Many of them duplicate each other and some are contradictory, yet they have persisted in the culture of many nations for 2,000 years.

Aratus (ca. 315–ca.245 B.C.E.) is best remembered for his poem on astronomy, *Phaenomena,* the second part of which is devoted to a treatment of weather signs. He adopted the study of weather signs as a legitimate object of concern for the Greeks, who were still loyal to the idea of religious control over the weather. [*See* Religion and Weather.]

Both collections were translated into Latin and influenced future writers on nature. By far the most notable presentation of the subject in Latin was that of Virgil (70–19 B.C.E.), whose lengthy poem *Georgica,* glorifying peasant life and duties, contains many quotable lines about agricultural pursuits, the signs of the seasons, and the recommended times for plowing, sowing, and harvesting. Other Roman writers on nature, such as Pliny

(23–79 C.E.) and Seneca (4 B.C.E.–65 C.E.), include many references to weather types.

During the late Middle Ages, the pseudoscience of astrology became popular throughout Europe. Though condemned by religious authorities, it took hold among the general populace and was employed for weather prediction. After the invention of the printing press, a means was at hand to spread weather lore through printed almanacs. These became best-sellers, second only to the Bible.

In general, the almanacs included three types of weather lore. That related to winds and clouds had some scientific basis. A second type connected with saints' days possessed doubtful validity but was quite popular nonetheless. A third type treated the behavior of birds and animals, which has been found to be controlled more by past and present weather rather than to be a true indication of the future.

In the British Isles the most popular book of weather lore in the late seventeenth century was *The Shepherd's Legacy,* which presented "The Shepherd of Banbury's Rules" to guide agriculturists through the seasons. Later on, in the eighteenth and nineteenth centuries, the most popular of the yearly publications was *Old Moore's Almanac,* which carried the usual astronomical and meteorological lore.

In America during colonial days, the most widely distributed weather-related publication was *An Almanack for the Year* 17*xx,* published by Nathaniel Ames, father and son, from 1726 to 1775. It had a distribution of 80,000. After the Revolution its place was filled by the *Farmer's Almanac* (later *The Old Farmer's Almanac*), edited by Robert B. Thomas. Its two-hundredth anniversary was celebrated with an enlarged edition in 1992. More memorable, and still quoted today, was Benjamin Franklin's *Poor Richard's Almanac* (1732–1757). His droll presentation of weather proverbs won an appreciative audience over the years. [*See the biography of Franklin.*]

The foremost collector of classic weather lore in England in the modern age was Richard Inwards (1840–1937), a mining engineer and editor with wide scientific interests. His book entitled *Weather Lore* (1869) originally contained 91 pages but had expanded to 251 pages in the fourth posthumous edition published in 1950. Inwards bequeathed ownership of *Weather Lore* to the Royal Meteorological Society, which published the expanded volume in 1950 to celebrate its centennial.

In the United States, the National Weather Service in Washington, D.C., took the lead in collecting local weather wisdom. This resulted in two government publications: *Weather Proverbs* (1883) by Lieutenant Henry Harrison Dunwoody and *Weather Folklore and Local Weather Signs* (1903) by Professor Edward B. Garriott. These presented the results of surveys conducted by local weather officials throughout the country.

In the middle years of the present century, Eric Sloane, a Connecticut artist, took to painting cloud and weather scenes. His authentic murals of the atmosphere now adorn the Museum of Natural History in New York City and the National Air and Space Museum in Washington, D.C. He also produced a number of books on Americana and on the weather. One of his most most popular was *Folklore of American Weather,* in which he attempted to assess the validity of popular proverbs as true, false, or possible. An analysis of Sloane's judgment of his selected list of proverbs showed 44 true, 35 false, and 45 possible.

A remarkable collection of American weather lore is *Weather Wisdom* (1976; rev. ed., 1990) by Albert Lee. His original subtitle is explanatory: *Being an Illustrated Practical Volume Wherein Is Contained [a] Unique Compilation and Analysis of Facts and Folklore of Natural Weather Prediction.* Against a background of modern meteorological theory and practice, the author attempts to show the correctness of many time-honored words of weather wisdom.

The division of weather lore still holding the most interest of the public has a connection with the seasonal calendar, such as the annual saints' days, the passage of the months, and weather predictions made from animal behavior. Some frequently mentioned items are the following:

• *Groundhog Day.* The groundhog (woodchuck) is believed to emerge from hibernation on 2 February; if he sees his shadow, he goes back into his burrow for another 6 weeks, indicating 6 more weeks of winter. This belief was transferred from Germany to Pennsylvania, and through annual media attention has since spread throughout the United States. In church calendars, Groundhog Day coincides with Candlemas Day, formerly called the Feast of the Purification of the Virgin and now called the Presentation of Our Lord. The date 2 February also marks the halfway point between the winter solstice and the spring equinox.
• *Saint Swithin's Day.* In England it is said that if it rains on Saint Swithin's Day (15 July), it will rain for a total of 40 consecutive days. This belief arose from a long delay in the canonization ceremony of Saint Swithin,

which had been scheduled for 15 July 964. Swithin, bishop of Winchester, had died in 862 and had expressly requested to be buried in the churchyard so that the "sweet rain of heaven might fall upon his grave." When in 964 the monks attempted to remove his body into the cathedral choir for the canonization ceremony, it rained for 40 days, delaying the proceedings. The tradition of a rainy saint is found in varied form in many European countries and was brought to North America by English settlers. British climatologists have tested it for London and have found no occasion of 40 consecutive days of rain in the summer.

• *Dog Days.* The Dog Days, a period in the Northern Hemisphere from about 3 July to 10 August, get their name from the Latin *caniculares dies,* which is what the ancient Romans called the hottest weeks of summer. Their belief was that the "Dog Star," Sirius, the brightest star in the heavens, rising in conjunction with the Sun, added its heat to the solar strength. In fact, the increase is infinitesimal.

• *Wooly Bear Caterpillar.* The caterpillar of the tiger moth is called a wooly bear because of its dense coat of fine hairs. Measurements of the width of the brown band around its middle is supposed to give an indication of the severity of the coming winter. This is pure fantasy. Although much is made of it in the American press each autumn, the prediction is never verified in the following spring.

BIBLIOGRAPHY

Dolan, E. F. *The Old Farmer's Almanac Book of Weather Lore.* New York, 1988

Dunwoody, H. H. C. *Weather Proverbs.* Signal Service Notes IX. Washington, D.C., 1883.

Garriott, E. B. *Weather Folklore and Local Weather Signs.* Bulletin 33, Weather Bureau no. 294. Washington, D.C., 1903.

Inwards, R. *Weather Lore: A Collection of Proverbs, Sayings, and Rules Covering the Weather.* 4th ed., edited by E. L. Hawke for the Royal Meteorological Society. London, 1950.

Leach, M., and J. Fried, eds. *Funk and Wagnalls Standard Dictionary of Folklore, Mythology, and Legend* (1949). 1-vol. ed. San Francisco, 1984.

Lee, A. *Weather Wisdom: Facts and Folklore of Weather Forecasting.* Rev. ed. Chicago, 1990.

Opie, I., and M. Tatem, eds. *A Dictionary of Superstitions.* Oxford and New York, 1989.

Smith, E. L. *Weather Lore and Legend.* Lebanon, Pa., 1972.

Swainson, C. *A Handbook of Weather Folk-Lore.* Edinburgh, 1873; Detroit, 1974.

Wurtele, M. G. "Some Thoughts on Weather Lore." *Folklore* 82 (1971): 292–303.

DAVID M. LUDLUM

WEATHER MAPS. *See* Charts, Maps, and Symbols.

WEATHER MODIFICATION. In contemporary usage, the term *weather modification* typically refers to the modification of clouds and precipitation. The most direct and simplest form of weather modification is the treatment of cold fogs to improve visibility. A more complex activity is the treatment of orographic (mountain) clouds, generally to increase snowfall or rain. Still more complex treatments attempt to modify convective (cumulus) cloud systems to increase precipitation or decrease hail. The most elaborate efforts are directed at storm systems such as hurricanes or extratropical storms.

The modification of the atmosphere can result in short-term atmospheric changes in weather or long-term changes in climate, either intentionally carried out or the inadvertent results of other human activities. Here we deal primarily with intentional activities, focusing on the modification of clouds and precipitation. Inadvertent weather modification, however has been shown to result from local heating caused by urban influences, pollutants released into the atmosphere, changes in the Earth's surface, irrigation, and other factors. [*See* Heat Island Effect; Pollution.]

Principles. Clouds present a visual tracer of atmospheric air currents. When air moves upward, it expands and cools. Because the mass of water vapor in an air parcel (given volume) is fixed, and because the amount of water that air can hold in vapor form is less at colder temperatures, the relative saturation of an air parcel increases as the parcel rises, expands, and cools. When the temperature is reached at which the parcel is vapor-saturated, liquid (or ice) water must condense out as further cooling takes place, and a cloud is formed. The cloud acts as a visual tracer of the upward rising air current. The edge of the downwind (lee) side of mountain clouds, for example, shows where descending air has heated and acquired an increased capacity to hold water vapor, so that the cloud's liquid, visible water evaporates.

The water condensed out of an air parcel is generally in the form of very small water droplets. Each droplet forms on a small particle called a *condensation nucleus* and grows very rapidly for a short time (a few tens of seconds). There are always many condensation nuclei in the atmosphere (a few hundred per cubic centimeter near the oceans and many more over continents), and so few of the droplets can acquire enough water vapor to grow beyond a critical small size (about 10^{-3} centimeter). The goal of most weather-modification efforts is to provide a more efficient mechanism for the mass of many small cloud droplets to combine into hydrometeors (or particles) large enough to fall out as precipitation. [See Precipitation.]

In some cases, the formation of precipitation-sized particles can be facilitated artificially by producing or providing some larger droplets. These can then collide and coalesce with the smaller droplets. This technique has sometimes been successful in atmospheric experiments, but it has not been generally practical owing to the difficulty and cost of delivering large quantities of water, or large nuclei to form larger droplets. In addition, the time required for the collision process to produce large drops is frequently longer than the lifetime of the cloud.

The main mechanism for the natural or artificial growth of large, precipitation-sized hydrometeors in atmospheric clouds results from the coexistence of ice particles and supercooled liquid droplets at temperatures colder than 0°C. The ice crystals grow at the expense of the supercooled water droplets because the vapor pressure over water is greater than that over ice. Small liquid droplets do not freeze except at very cold temperatures (around −40°C) unless ice nuclei are present. While condensation nuclei are always present in the atmosphere, ice nuclei are typically not present in large numbers; for 1 million condensation nuclei, there may be only 1 ice nucleus. Natural concentrations of ice nuclei are thus usually too low to use all of the available supercooled liquid water, particularly at cloud temperatures warmer than about −20°C.

We can modify clouds by supplying a smoke of artificially prepared nuclei that are both more efficient and more abundant than natural ice nuclei. Smokes of silver iodide compounds can be used to produce very large numbers (around 10^{15} per gram) of ice nuclei that are effective at −20°C. A variety of silver iodide complexes can be produced with different nucleation characteristics. This permits tailoring of the nuclei for the type of cloud treatment desired. Ice crystals can also be produced in a cloud by injecting very cold substances such as dry ice (frozen carbon dioxide) or liquid nitrogen. Ice particles then form spontaneously owing to the extreme cooling, without the need for ice nuclei, in a process called *homogenous nucleation*. [See Cloud Seeding.]

Results. The effectiveness of artificial treatment can be demonstrated in the laboratory and readily seen in simple atmospheric clouds. There is no real controversy about this, but there has been considerable disagreement over the years as to whether the amount of precipitation thus formed is significantly more than what would ultimately have formed if the cloud had been left untreated. An unequivocal answer has been difficult to obtain because of the complexity and variability of cloud systems. Precipitation changes induced by present technology appear to be of the order of 10 to 15 percent, while natural variability in precipitation in most regions, even on an annual basis, is of the order of 400 percent. Despite this problem, physical observations and studies and actual experiments provide us with important information on the effects of intentional weather modification.

Improvement in atmospheric visibility from treating some supercooled fogs has been clearly demonstrated, and this method is presently used at some airports. Clearing warm fogs at temperatures above 0°C, using heating or mixing techniques, has proven difficult and expensive.

A number of experiments have shown artificial treatment to increase mountain snowfalls from 10 to 15 percent. Many researchers and water-users have concluded that this is the type of weather modification most ready for practical, beneficial application.

The results of experiments from seeding convective clouds to augment precipitation or decrease hail have been inconsistent. Some data have indicated precipitation increases, while others suggest decreases. It seems that some clouds have the potential for precipitation increase, while others do not. The characteristics of treatable clouds are more difficult to specify for the more complex convective cloud types than for mountain clouds. Hail suppression is very complex; most of the evidence supporting reduction of hail from weather modification have come from crop-loss insurance studies. There have been only a few experiments involving artificial treatments of large-scale weather systems such as tropical and extratropical storms, and the results of these are inconclusive.

The scientific tools for studying the effects of cloud treatments and for developing the technology have greatly improved recently. These advances include better basic understanding of cloud microphysical processes, increasingly useful cloud conceptual and numerical models, and vastly improved in situ and remote sensing instrumentation. Funding support for weather modification research, however, was restricted during the 1980s owing to a generally wet period, with reduced pressure to augment water supplies, and to tight national budgets. The greatest present need for research and technology development relates to systems and procedures for delivering seeding materials in proper concentrations, at the right time, and in the proper cloud locations.

Social Issues. Economic, legal, and environmental impacts from weather modification are important issues. A basic problem is that the effects of cloud seeding cannot be restricted to a specific site. Seeding materials move and disperse, and cloud systems develop and move. Even in areas specified for seeding treatments, some people want precipitation modification and others do not. For example, in mountainous areas local residents may not want more precipitation, but cloud seeding is performed there to provide more water for downstream users. Another question is that of who owns the additional water. Local, state, national, and international interests have all laid claim, in specific cases, to water resulting from precipitation augmentation.

Changes in precipitation might also affect the flora and fauna of an area. These effects probably are not great, however, because natural variation in precipitation in most areas is 10 to 100 times greater than the changes expected from deliberate cloud seeding.

A number of U.S. states passed laws regulating weather modification. Most of these laws specify that the program directors must have a permit that verifies their professional qualifications and a license to treat a specific area. Although most of the state laws do not address social issues, they provide a forum for considering these issues during the licensing process.

In summary, the physical basis for modifying clouds, and in some cases precipitation, is well established, but the magnitude of the results is often difficult to quantity. Consistent indications of beneficial results using present technology have most frequently been associated with the treatment of mountain cloud systems.

BIBLIOGRAPHY

Davis, R. J. "Four Decades of American Weather Modification Law." *Journal of Weather Modification* 19 (1987): 102–106.

Hess, W. N., ed. *Weather and Climate Modification.* New York, 1974.

James, J. "Teaching Weather Modification in Lower Division/College Classes." *Journal of Weather Modification* 23 (1991): 83–85.

Lambright, H. W. "Weather Modification, Drought, and Public Policy: A Case History." *Journal of Weather Modification* 18 (1986): 119–126.

Steinhoff, H. W., and J. D. Ives, eds. *Ecological Impacts of Snowpack Augmentation in the San Juan Mountains of Colorado.* Division of Atmospheric Water Resources Management, Bureau of Reclamation, U.S. Department of Interior. Denver, 1976.

Super, A. B., ed. *Physics of Winter Orographic Precipitation and Its Modification.* Bureau of Reclamation, U.S. Department of Interior. Denver, 1986.

Weather Modification Advisory Board. *The Management of Weather Resources.* 2 vols. U.S. Department of Commerce. Washington, D.C., 1978.

World Meteorological Organization. *WMO Statement of the Status of Weather Modification.* Geneva, 1991.

LEWIS O. GRANT

WEATHER SERVICES. Nearly all the nations of the world have established government weather services for the purposes of observing, recording, and forecasting the weather. Additional nongovernmental weather services operate in some countries to provide weather data, forecasts, and specialized weather services or studies to industries or other private activities. In general, the private weather services are not competitive with the government services but supplement their information with specialized information, including custom forecasts and studies for paying clients.

In the United States there exist both a well-developed government weather service and a private weather industry, working in cooperation. A longstanding example is the aviation industry, which has supplemented government weather services with its own since about 1930. Many countries also have separate weather services for the military. It is customary for public and military services to minimize duplication and to cooperate closely.

Government Weather Services. The basic functions of a government weather service are to observe the weather, to maintain weather records, to prepare and issue weather forecasts and warnings, and to supply specialized weather information for the use of agriculture, aviation, air pollution control, and navigation in coastal waters and at sea. It must also maintain comprehensive weather records for studies of the climate. Because weather systems move without regard to national boundaries, governments must exchange weather data, and to some extent forecasts, on a current basis. The planning and decisions for such exchanges are coordinated by the World Meteorological Organization in Geneva, Switzerland.

The weather-observing function of government weather services includes the full range of weather observation and the collection, communication, and archiving of data. These data include observations of temperature, dewpoint, wind, pressure, and other features made at the Earth's surface, from ships at sea, by weather radar, and by upper-air soundings using balloon-borne instruments. A few governments or groups of governments operate weather satellites to obtain regional and global information on cloud structure and movements as well as indirect vertical temperature soundings.

New methods of observing the atmosphere are beginning to come into use, involving sensing of winds and temperatures in the free atmosphere by indirect methods. These data too will be shared with private and other governmental weather services.

A vast amount of high-quality wind and temperature data in the upper atmospheric levels is obtained by direct measurements made by commercial air traffic through aircraft navigational systems. These data are collected through air traffic communications systems and made available for use by all. They are used by government weather services for forecasting, further distributed to private and foreign weather services, and retained in permanent archives. There is a direct return of benefits to aviation through more accurate forecasts of temperature and winds at flight levels.

Any commercial firm in the United States can have access to current weather data from any part of the globe, either directly from a foreign weather service or from domestic meteorological circuits. Although the primary task of weather observation and data collection is carried out by governments, some private weather services arrange and operate local observing networks for specific clients. The nature of these depends on the requirements of the clients.

Each national weather service is responsible for issuing weather forecasts and warnings to meet national requirements. These needs include general weather forecasts for the use of the public, forecasts and warnings to air, sea, and highway traffic, forecasts and climatological data for agriculture and forestry, data for air pollution control, and warnings of severe weather to protect life and property.

Some meteorological services are responsible for issuing river forecasts and flood warnings, including flash floods. This obliges them to maintain hydrological organizations within the meteorological service. The advantage of placing responsiblity for river and flood warnings in the meteorological service is that most floods follow heavy rains, and an integrated organization can cope rapidly with the full complexity of the problem. The main contribution of the meteorological service to river and flood forecasting is the provision of quantitative precipitation forecasts, nationally organized river forecasting methods, and highly-developed warning dissemination services. Flash floods, which can cause a heavy death toll, have a tendency to occur at night, and they can catch a forecast and warning service unaware if no offices are open during those hours.

The World Meteorological Organization designated Washington, Moscow, and Melbourne as World Meteorological Centers, responsible for making global weather forecast guidance generally available. In addition, The European Center for Medium Range Forecasting was established by the European Council of Directors (of meteorological services). The Washington National Meteorological Center and the U.K. Meteorological Office have been designated World Aviation Area Forecast Centers. All these centers have advanced numerical prediction capability. They furnish forecast guidance to other nations as well as to their own services. In addition, nearly every national weather service operates a national weather prediction center. In many of these centers, numerical weather prediction, accomplished locally or cooperatively with neighboring countries, is the basis for their forecasting.

Aviation of all types needs forecasts of wind and temperature at flight levels, as well as for severe weather such as thunderstorms or icing conditions both during flight and at terminals. These are prepared mainly by government weather services. Occasionally wind shear just above the ground during landing or takeoff is a serious hazard to flight safety; new instrumentation around airports is expected to help detect these dangerous local winds before aircraft fly into them.

Weather forecasts for aviation first assumed importance during the 1920s and 1930s, when a series of aviation disasters caused by weather shocked the public and led to some improvements in the weather-reporting capabilities of the U.S. Weather Bureau. Even now, public funds tends to be appropriated for improvements in weather services in greater amounts following weather-related disasters.

Weather services to the general public consist of forecasts of local and regional weather for a few hours to several days in advance. The most important are forecasts and warnings of weather that threatens life or property—for example, hurricanes, tornadoes, flash floods, slowly-rising floods, or cold waves. It is thus necessary for meteorological services to have means of rapid and effective dissemination of information. Commercial radio and television stations are effective partners in this activity. In the United States many of these stations employ their own meteorologists, who explain the weather to the public and give vital assistance in disseminating forecasts and warnings of severe weather or floods.

For many years the National Weather Service has maintained the National Hurricane Center at Miami as a focal point for forecasts of hurricane development and movement on the Atlantic and Gulf coasts and adjacent regions. These storms have some unique characteristics and are so threatening to life and property that specialized attention is required.

Daily weather information is needed for many agricultural operations, particularly at times of the year when crops or livestock are critically sensitive to the weather. In order for weather data and forecasts to be effective, the provider of data and forecasts needs to know the changing sensitivity of the agricultural operations in the local region. This can best be accomplished by having a specified liaison person in the meteorological service with this responsibility; generalized forecasts prepared without such information may prove to be of little value to agricultural operations. Examples of events critical to agriculture include heavy frost in fruit-growing regions, a frost or freeze in Florida or southern Texas, or a late winter storm on the Great Plains during calving time. Agricultural operations tend not to employ private meteorology and generally expect to be supported by government weather services.

In the United States forest fires cause much destruction of valuable timber and some loss of life. The federal, state, and local governments share responsibility for fighting these fires, depending on the local situation. To do this they need close-in meteorological advice to warn of wind shifts or general weather changes; otherwise, the strategy of firefighting or even the lives of the firefighters can be endangered. The U.S. National Weather Service employs a number of experienced forecasters who can be detailed to the site of forest fires to provide this advice.

Military Weather Services. In the United States and many other countries, the military services have their own special weather services, with facilities for data collection and processing and their own observing and forecasting staffs. These services have access to the data used by the civil government as well as to special data from military sources, which are not always available to the civilian weather service. In some countries the military weather services have their own data processing and numerical forecasting operations. At first glance this may seem to be an unnecessary duplication of services, but the military services need the flexibility to move their resources quickly on short notice. A military service must deal with some classified matters. In the United States, civil and military meteorological services coordinate their activities thoroughly in order to minimize total cost. Meteorologists who leave the military services often find employment in the civilian service.

Private Weather Services. There are many private weather services in the United States. The National

Weather Service does not undertake specialized studies for individuals or corporations, or prepare forecasts for their special needs, so it encourages the development of the private sector in meteorology. Probably the first legitimate private weather services were formed by airlines, which were having severe difficulties with weather hazards in the 1920s and 1930s. Now private firms furnish weather services of many kinds, with emphasis on special meteorological work for firms with unique weather problems.

Recent legislative actions to regulate air pollution have had a considerable impact on private meteorology, as manufacturers and other firms have turned to private consultants for advice on design and monitoring. In late 1992, 26 meteorological consulting firms were advertising specialties in air quality monitoring and analysis in the *Bulletin of the American Meteorological Society*. Another 12 firms advertised capabilities in modeling the dispersion of pollutants. Twenty-three firms advertised specialties in forensic meteorology, some of which certainly related to legal disputes over air pollution. Other private firms do specialized consulting on meteorological problems, including forecasting unique to a particular company's activities, as well as special consulting and forecasting for power firms on dispersion of pollutants, for construction firms, or in the production of specialized graphics for use in television weather programs. In late 1992, 75 private meteorological consulting firms advertised their services, giving some idea of the size of this industry, which is likely to grow. In most other nations, private meteorology is either nonexistent or developing more slowly.

[*See also* Professional Societies.]

BIBLIOGRAPHY

Ahrens, C. D. *Meteorology Today: An Introduction to Weather, Climate and the Environment.* 4th ed. St. Paul, Minn., 1991.
File, D. *Weather Watch.* North Pomfret, Vt., 1991.
Burroughs, W. J. *Watching the World's Weather.* Cambridge, 1991.

GEORGE P. Cressman

WEATHER VANE. *See* Instrumentation.

WET BULB POTENTIAL TEMPERATURE. *See* Potential Temperature; *and* Temperature.

WIND CHILL. A number of climatological indices are used to measure the response of humans to severe weather. The term *wind chill* was originally coined by

WIND CHILL. Table 1. *Wind chill equivalent temperatures (°F), from Siple's formula*

Temperature		Wind Speed (miles per hour)							
°F	CALM	5	10	15	20	25	30	35	40
32	66	32	22	16	11	7	5	3	2
30	64	30	20	13	8	5	3	1	−1
28	63	28	18	11	6	3	0	−2	−4
26	63	26	16	9	3	0	−3	−5	−7
24	61	24	13	6	0	−3	−6	−8	−10
22	61	22	11	3	−2	−6	−9	−11	−13
20	60	20	8	0	−5	−9	−12	−14	−16
18	59	18	6	−2	−8	−12	−15	−17	−19
16	58	16	4	−5	−10	−15	−18	−20	−22
14	57	14	1	−8	−13	−18	−21	−23	−25
12	56	12	−1	−10	−16	−21	−24	−26	−28
10	56	10	−4	−12	−19	−24	−27	−29	−31
8	54	8	−6	−15	−21	−26	−30	−32	−34
6	53	6	−8	−18	−24	−29	−33	−35	−37
4	52	4	−11	−20	−27	−32	−35	−38	−40
2	51	2	−13	−23	−30	−35	−38	−41	−43
0	50	0	−15	−26	−32	−37	−41	−44	−46
−2	49	−2	−17	−28	−34	−40	−44	−47	−49
−4	48	−4	−20	−30	−40	−43	−47	−50	−51
−6	47	−6	−23	−32	−43	−45	−50	−53	−54
−8	46	−8	−25	−35	−46	−48	−53	−56	−57
−10	45	−10	−27	−38	−48	−51	−56	−59	−61
−12	45	−12	−29	−40	−51	−54	−59	−62	−64
−14	44	−14	−32	−42	−54	−56	−62	−65	−67
−16	43	−16	−35	−45	−56	−59	−64		
−18	42	−18	−37	−48	−59	−62	−67		
−20	41	−20	−39	−52	−62	−65	−70		
−22	40	−22	−42	−53	−65				
−24	39	−24	−44	−56					
−26	38	−26	−46	−58					
−28	37	−28	−48	−61					
−30	37	−30	−50	−64					
−32	36	−32	−52	−67					
−34	35	−34	−54						
−36	34	−36	−57						
−38	33	−38	−59						
−40	32	−40	−62						

Paul A. Siple to describe variations in human comfort caused by wind at cold temperatures. He specifically defined wind chill as the measure of the quantity of heat that the atmosphere is capable of absorbing within an hour from an exposed surface one meter square. Siple computed wind chill through experiments in Antarctica during the early 1940s, in which he measured the freezing rate of water in a small plastic cylinder at various wind speeds and temperatures. The resulting formula determines loss of heat from the cylinder in units of kilocalories per square meter per hour.

A wind chill index, developed from these initial exper-

WIND CHILL. Table 2. *Wind chill equivalent temperatures (°F), from Steadman's formula*

Temperature °F	CALM	\multicolumn{8}{c}{Wind Speed (miles per hour)}							
		5	10	15	20	25	30	35	40
32	34	32	27	24	21	17	14	12	10
30	32	30	25	21	18	15	12	10	7
28	30	28	23	19	15	12	9	6	4
26	28	26	21	17	13	9	6	3	1
24	26	24	19	14	10	7	3	0	−3
22	25	22	16	12	8	4	1	−3	−6
20	23	20	14	9	5	1	−2	−6	−9
18	21	18	12	7	2	−2	−5	−9	−13
16	19	16	10	5	0	−4	−8	−12	−17
14	17	14	8	2	−3	−7	−12	−16	−20
12	15	12	6	0	−5	−10	−15	−19	−24
10	13	10	4	−2	−8	−13	−18	−23	−28
8	11	8	1	−5	−11	−16	−21	−26	−32
6	9	6	−1	−7	−13	−19	−24	−30	−36
4	7	4	−3	−10	−16	−22	−28	−34	−40
2	5	2	−5	−12	−19	−25	−31	−38	−44
0	3	0	−7	−15	−22	−28	−35	−42	−49
−2	1	−2	−10	−17	−25	−31	−39	−46	−54
−4	−1	−4	−12	−20	−28	−35	−42	−50	−58
−6	−3	−6	−14	−22	−30	−38	−46	−54	−63
−8	−4	−8	−16	−25	−33	−41	−50	−59	−67
−10	−6	−10	−19	−28	−36	−45	−54	−63	
−12	−8	−12	−21	−30	−39	−48	−58	−68	
−14	−10	−14	−23	−33	−42	−51	−62		
−16	−12	−16	−26	−36	−45	−55	−66		
−18	−14	−18	−28	−38	−49	−59			
−20	−16	−20	−30	−41	−52	−63			
−22	−18	−22	−32	−44	−55	−66			
−24	−20	−24	−35	−47	−58				
−26	−22	−26	−37	−49	−62				
−28	−24	−28	−39	−52	−65				
−30	−26	−30	−42	−55	−68				
−32	−27	−32	−44	−58					
−34	−29	−34	−47	−61					
−36	−31	−36	−49	−64					
−38	−33	−38	−51	−67					
−40	−35	−40	−54	−69					

iments, estimates a "wind chill equivalent temperature" (when winds are blowing at 5 miles [8 kilometers] per hour) that is equal in cooling power to a given combination of actual temperature and windspeed Table 1 shows that similar wind chill temperatures can be obtained from different combinations of air temperature and windspeed; for example a 0° (32°F) temperature with a wind speed of 15 miles per hour yields the same wind chill temperature as a −9°C (16°F) temperature with a wind speed of 5 miles per hour. A person would be equally uncomfortable in either combination of air temperature and wind speed.

There are several problems with this concept. The wind chill temperature assumes that heat loss occurs from a bare skin surface; it does not consider heat loss via human respiration; and it is considered inaccurate at wind speeds greater than 40 miles (64 kilometers) per hour. In addition, when winds are calm, Siple's calculations yield an unrealistically high wind chill equivalent temperature. In the 1970s, R. G. Steadman improved the wind chill index by considering the influence of air temperature and wind speed combinations on appropriately clothed individuals. Steadman's wind chill equivalent temperatures (Table 2) are higher than Siple's for a given combination of air temperature and wind speed and are generally considered more realistic.

Many more weather indices have been developed to evaluate human comfort in both summer and winter. These indices are useful in determining human response to weather, because the sensation of temperature that the human body feels is often quite different from the actual temperature of the air. The human body is constantly releasing energy, and anything that influences the rate of heat loss from the body also influences comfort. Factors other than temperature that affect human comfort include humidity, wind speed, solar radiation, and even pollution levels. Some of these indices have been given imaginative names such as *humiture*, *humidex*, *humisery*, *apparent temperature*, the *discomfort index*, and the *temperature-humidity index* (THI). [*See* Humidity.]

Virtually all the human comfort indices developed, to date, suffer from a serious shortcoming in that they assume that a given index value produces identical levels of discomfort at all locations. However, a temperature of 12°C (10°F) with a wind speed of 15 miles (24 kilometers) per hour would be perceived as very uncomfortable in Atlanta, Georgia, but rather uneventful in Minneapolis, Minnesota. For this reason, new relative indices have been developed, based on the premise that human beings are most affected by weather that is unusual in their own locale. The best-known is the *weather stress index* (WSI), which compares a particular weather event to average conditions at a location and evaluates how unusual that event is. The WSI is based on the calculation of an apparent temperature, which is the perceived air temperature based on combined air temperature, relative humidity, and wind speed. In summer, it then defines the proportion of days with apparent temperatures lower than the day under review; however, in winter the WSI defines the proportion of days with apparent temperatures higher than the day under review. Thus a day with a WSI of 99

WIND CHILL. Table 3. *Apparent temperatures corresponding to 99 percent weather stress index (WSI) values at 3 P.M. in selected U.S. cities in summer*

	°C	°F
Albuquerque, N.Mex.	39	102.2
Atlanta, Ga.	41	105.8
Bismarck, N.Dak.	41	105.8
Denver, Colo.	38	100.4
Key West, Fla.	40	104
Little Rock, Ark.	46	114.8
Mobile, Ala.	43	109.4
Oklahoma City, Okla.	46	114.8
Philadelphia, Pa.	42	107.6
Phoenix, Ariz.	45	113
San Francisco, Calif.	25	77
Seattle, Wash.	34	93.2
Topeka, Kans.	46	114.8
Tulsa, Okla.	47	116.6

percent in summer would prove to be quite stressful to people, as 99 percent of the days during that time of year would possess a lower apparent temperature. In winter, 99 percent of the days would have a higher temperature if the WSI were 99 percent for a given day. Of course, the apparent temperatures corresponding to a 99 percent WSI would vary according to time of year and locale, as shown by Table 3.

Further research is necessary to determine weather's impact on human beings. For example, what is the influence of several consecutive days of stressful weather? Is there a threshold condition above which everyone is uncomfortable, regardless of his or her ability to acclimatize? Future indices of human comfort must begin to address these and other uncertainties.

[*See also* Health.]

BIBLIOGRAPHY

Court, Arnold. "Windchill Origins and Validity." In *Proceedings of Polar Meteorology Symposium, IAMAP, IUGG*. Hamburg, 1983.

Jendritzky, G. "Selected Questions of Topical Interest in Human Bioclimatology." *International Journal of Biometeorology* 35 (1991): 139–150.

Kalkstein, L. S., and K. M. Valimont. "An Evaluation of Summer Discomfort in the United States Using a Relative Climatological Index." *Bulletin of the American Meteorological Society* 67 (1986): 842–848.

Kalkstein, L. S., and K. M. Valimont. "An Evaluation of Winter Weather Severity in the United States Using the Weather Stress

Index." *Bulletin of the American Meteorological Society* 68 (1987): 1535–1540.

Lowry, W. P. *Atmospheric Ecology for Designers and Planners.* Eugene, Ore., 1988.

Lutgens, F. K., and E. J. Tarbuck. *The Atmosphere: An Introduction to Meteorology.* 4th ed. Englewood Cliffs, N.J., 1989.

Mather, J. R. *Climatology: Fundamentals and Applications.* New York, 1974.

LAURENCE S. KALKSTEIN

WINDSTORMS. *See* Religion and Weather; *and* Small-Scale or Local Winds.

WINTER. *See* Seasons.

WINTER SOLSTICE. *See* Orbital Parameters and Equations.

Y–Z

YOUNGER DRYAS. The rapid and dramatic cooling known as the *Younger Dryas* was the most recent of several climatic reversals that occurred after the significant global warming following the last ice age of the Pleistocene. Evidence of this reversal was first discovered in Scandinavia in the form of fossil fragments of *Dryas,* an herb of the rose family, which is an indicator of an arctic-alpine climate. In the 1930s, tundra pollen as well as macrofossils were identified, and the Younger Dryas became an established European pollen zone spanning approximately a millennium. Geomorphological, lithological, faunal, and isotopic data provide additional evidence that this climatic fluctuation occurred at numerous sites throughout Europe from approximately 11,000 to 10,000 years ago.

In the 1980s, the discovery of changes in faunal assemblages in marine cores concurrent with the changes already noted on land led to the conclusion that the North Atlantic polar front readvanced to its ice age position during this period (see Figure 1). Geochemical analyses showed that the production of North Atlantic Deep Water (NADW), which today is responsible for bringing heat to northern Europe, was greatly reduced during this time, just as it was during the ice age. There is still dispute regarding the cause for the cold, fresh meltwater lid (glacial meltwater which overlies saltier water) that resulted in the NADW reduction and the dramatic return to colder conditions. Recent evidence points to meltwater from the retreating Laurentide ice sheet, entering the Atlantic through the St. Lawrence River, as a likely source because the timing of this eastward flow corresponds to the millennium in question. Other possibilities

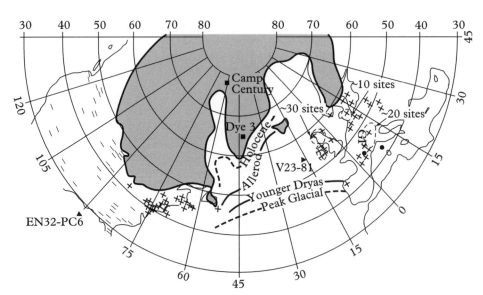

YOUNGER DRYAS. Figure 1. *Sites at which sediments spanning the time interval 13,000 to 9,000 years before present have been studied.* At sites designated +, an oscillation in climate corresponding to the Younger Dryas is found; at sites designated −, this oscillation has not been reported. The shaded area corresponds to the area covered by ice just before the onset of the Allerod warm period. Positions of the polar front in the northern Atlantic, during peak glacial, Allerod, Younger Dryas, and Holocene times, as reconstructed by Ruddiman and McIntyre (1981), and the locations of two long ice borings in Greenland and two key ocean cores are shown. (From Peteet, 1992.)

848

for the cooling include catastrophic melting of marine-based icebergs, solar variations, and volcanic eruptions.

The geographic distribution of the Younger Dryas event is currently a topic under scrutiny. Greenland ice cores show that a major Younger Dryas shift in oxygen isotope ratios took place, with atmospheric temperatures cooling possibly as much as 7°C. In northwestern Europe, tree species that had been migrating northward after the glaciers' retreat were killed, and tundra conditions prevailed. Scottish glaciers actually formed anew. Deciduous trees such as oak and ash in northeastern North America were replaced by the boreal conifers larch, spruce, and fir. It is not known to what degree large animal extinctions were influenced by climate during this interval. In regions far from the North Atlantic, there are hints of the Younger Dryas in data from isolated ocean, land, and ice sites. Nonetheless, conclusive evidence for an extra-Atlantic *regional* signal remains to be established.

Recent research has addressed the rapidity of the temperature changes and the actual calendar-year timing of the Younger Dryas event. Greenland ice cores indicate that the return to warm conditions at the close of the Younger Dryas took place in about 50 years. Sediment changes on land also point to a very rapid warming period of between 50 and 100 years. The latest Greenland ice-core estimate of the onset of this event is 12,850 ± 100 years ago, with its close at 11,650 ± 70 years ago. Changes in carbon dioxide and methane concentrations suggest that greenhouse-gas feedbacks were involved in the shift from warm to cold conditions.

[*See also* Ice Ages; *and* Paleoclimates.]

BIBLIOGRAPHY

Broecker, W., D. Peteet, and D. Rind. "Does the Ocean-Atmosphere System Have More Than One Stable Mode of Operation?" *Nature* 315 (1985): 21–26.

Dansgaard, W., J. W. C. White, and S. J. Johnsen. "The Abrupt Termination of the Younger Dryas Climate Event." *Nature* 339 (1989): 532–534.

Lehman, S. J., and L. D. Keigwin. "Sudden Changes in North Atlantic Circulation during the Last Deglaciation." *Nature* 356 (1992): 757–762.

Mott, R. J. "Late-Glacial Climate Change in the Maritime Provinces." *Syllogeus* 55 (1985): 281–300.

Peteet, D. M. "The Polynological Expression and Timing of the Younger Dryas Event: Europe versus Eastern North America." In *The Last Deglaciation: Absolute and Radiocarbon Chronologies*, edited by E. Bard and W. S. Broecker. New York, 1992, pp. 327–344.

Peteet, D. M. "Younger Dryas Climatic Reversal in Northeastern USA? AMS Ages for an Old Problem." *Quaternary Research* 33 (1990): 219–230.

Peteet, D. M., guest ed. "Global Younger Dryas?" *Quaternary Science Reviews* 12 (1993): 277–356.

Rind, D., et al. "Impact of Cold North Atlantic Sea Surface Temperature on Climate: Implications for a Younger Dryas Cooling (11-10k)." *Climate Dynamics* 1 (1986): 3–33.

Ruddiman, W. F., and A. McIntyre. "The North Atlantic Ocean during the Last Glaciation." *Palaeogeography, Palaeoclimatology, Palaeoecology* 35 (1981): 145–214.

Taylor, K. C., et al. "The 'Flickering Switch' of the late Pleistocene Climate Change." *Nature* 361 (1993): 432–436.

Watts, W. A. "Regional Variation in the Response of Vegetation of Lateglacial Climatic Events in Europe." In *Studies in the Late Glacial of North-West Europe*, edited by J. J. Lowe et al. Oxford, 1980, pp. 1–22.

DOROTHY M. PETEET

ZONAL CIRCULATION. The zonal circulation is the general west-to-east motion of air in the troposphere, approximately parallel to latitude circles or zones of latitude. At a given time, tropospheric flow may be parallel to latitude circles over a particular region; this is known in weather forecasting as *zonal flow*. The term *zonal circulation*, however, refers to the westerly (west-to-east) component of tropospheric winds, averaged over a period of time and/or with respect to longitude.

A wind observation may be partitioned into zonal (west-to-east) and *meridional* (equator-to-pole) components. The average of the zonal component of the wind observations over several years and around latitude circles—the *zonal average*—reveals that the strongest tropospheric zonal winds, about 40 meters per second, are found in the upper troposphere over subtropical latitudes of the winter hemisphere (December, January, and February in the Northern Hemisphere, and June, July, and August in the Southern). This zonal circulation shifts poleward and weakens in each hemisphere during the transition from winter to summer seasons. (See Figure 1).

The energy source for the zonal circulation is the equator-to-pole temperature gradient in each hemisphere. Because of the net radiative heat gain in the tropics and heat loss at the poles, the temperature generally decreases from equator to pole in each hemisphere. The heating of the tropical atmosphere induces convection; the rising air associated with this convection diverges aloft and flows poleward. The Coriolis effect of the Earth's rotation deflects this poleward flow to the east in each hemisphere, yielding the westerly, zonal circulation. This rising tropical air, coupled with poleward-flowing air aloft, averaged over time and longitude, is a type of mean meridional circulation. Specifically, the circulation over the tropical

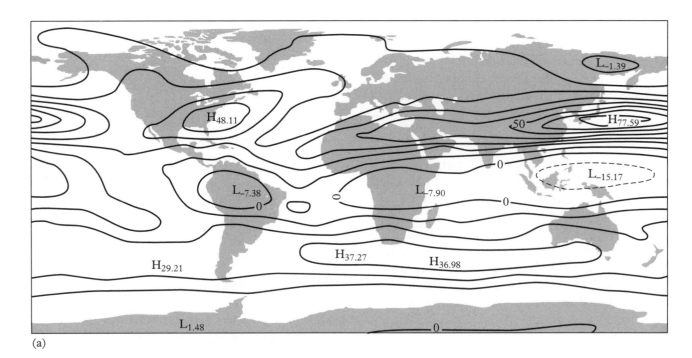

(a)

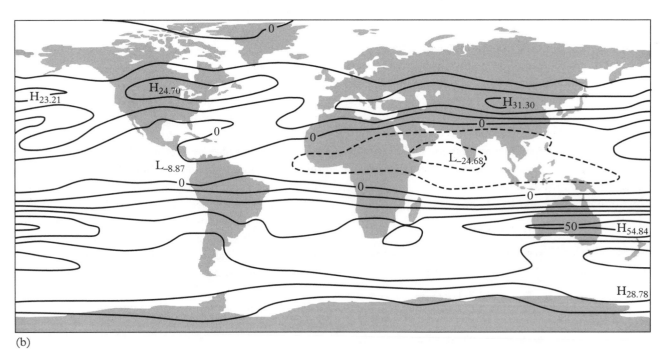

(b)

ZONAL CIRCULATION. Figure 1. *Ten-year (1979–1988) averaged zonal wind speeds, in meters per second, at the 200-millibar pressure level for (a) January and (b) July.* Solid contour lines at 10-meters-per-second intervals denote westerly (eastward) winds; dashed contour lines correspond to easterly (westward) winds. (The data were provided by the European Centre for Medium-Range Weather Forecasts.)

and subtropical latitudes is called the *Hadley cell*. This circulation cell is completed by sinking air in the subtropics and returning equatorward flow at the surface, deflected by the Coriolis effect into the tropical easterlies or trade winds. [*See* Coriolis Force; Meridional Circulations].

Embedded in the zonal circulation are eddies (troughs and ridges) with local meridional winds; the eastward motion of these eddies through the zonal circulation is responsible for day-to-day weather changes at the Earth's surface. Averaged around latitude circles, these eddies impart westerly momentum to the zonal circulation in a process known as *eddy momentum flux convergence*. The relative importance of the Hadley cell and eddy momentum flux convergence in maintaining the zonal circulation has been debated for many years. Both processes contribute importantly to the maintenance of the zonally averaged westerly circulation, but their relative contributions can differ substantially depending on longitude.

The eddies also transport heat poleward, acting to balance the net radiative heat loss at the poles. The poleward eddy momentum flux induces rising air in the troposphere over middle-to-high latitudes, in a zonally averaged sense. The divergence of this air aloft causes a zonally averaged equatorward flow (averaged around latitude circles) in the upper troposphere from middle and high latitudes to the subtropics. This mean meridional circulation is known as the *Ferrel cell*. The Coriolis effect of this equatorward flow imparts easterly momentum to the zonal circulation, opposing the westerly accelerations due to the Hadley cell and to the momentum flux convergence of the eddies. The net effect of these processes is to keep the zonal circulation in thermal wind balance, which is the theoretically expected proportionality between the decrease in equator-to-pole temperature and the increase in the westerly wind speed from the Earth's surface to the tropopause.

Averaged over time but not over longitude, the picture of the westerly or zonal component of the wind in the upper troposphere is more complex than that discussed above. That is, the zonal circulation is not uniform around latitude circles. The strongest westerly winds in the upper troposphere are found in subtropical to middle latitudes over the western Atlantic and Pacific oceans in the Northern Hemisphere and over the Atlantic and Indian oceans in the Southern Hemisphere. Maximum upper tropospheric westerly wind speeds exceed 70 meters per second over the western North Pacific during the Northern Hemisphere winter.

Likewise, the long-term averaged meridional circula-

tions are not uniform around latitude circles but vary with longitude. The strongest upper tropospheric flows from equator to subtropics are found just west of the westerly wind maxima of the wintertime Northern Hemisphere, suggesting that these meridional circulations, and the action of the Coriolis effect upon them, are primarily responsible for the maintenance of these zonal wind maxima. The upper tropospheric convergence of momentum owing to eddies is more uniformly distributed around subtropical to middle latitudes of the Northern Hemisphere, with maxima just east of the zonal wind maxima. This indicates that the eddies play a lesser role than the mean meridional winds in maintaining the zonal circulation.

On a day-to-day basis, there is considerable variability in the strength and location of zonal wind maxima. Typically, periods of zonal tropospheric flow over a mid-latitude location alternate irregularly with periods of highly amplified tropospheric waves with attendant meridional flow. The westerly current may be deflected north or south of the mid-latitudes, or it may even be split into several branches by persistent stationary circulations in the troposphere. The zonal flow can strengthen or weaken over a region during the course of a few days. It is thought that, at least over the North Pacific Ocean, zonal flows can be more persistent than other types of flow regimes. Nonetheless, the zonal flow regime can change abruptly, in a matter of days, to a wavy regime with accompanying meridional flow. These regime changes are very difficult to predict more than a few days in advance. Thus, understanding the nature and variability of the zonal tropospheric wind is of critical importance for studies of the atmospheric circulation and for extratropical weather prediction.

[*See also* General Circulation.]

BIBLIOGRAPHY

Holton, J. R. *An Introduction to Dynamic Meteorology*. 3d ed. San Diego, 1992.
Horel, J. D. "Persistence of Wintertime 500 mb Height Anomalies over the Central Pacific." *Monthly Weather Review* 113 (1985): 2043–2048.
Lau, N. C. "The Three-dimensional Structure of Observed Transient Eddy Statistics of the Northern Hemispheric Wintertime Circulation." *Journal of the Atmospheric Sciences* 35 (1978): 1900–1923.

STEPHEN J. COLUCCI

ZONES. *See* Climatic Zones.

GLOSSARY

※

Edited by GERALD R. NORTH

CONTRIBUTORS

Gordon D. Carrie, Matthew S. Gilmore, Chris Lucas, Donna W. Smith, and Mark J. Stevens

ablation (ice) The combined processes of melting and sublimation, which reduce the mass of the ice or snow in a glacier or snow field.

absolute humidity A measure of humidity; the amount of water vapor in the air, computed as mass of water vapor per unit mass of air and usually measured in grams per kilogram.

absorptivity The frequency-dependent ratio of the irradiance absorbed to that incident upon a particular substance.

accumulation rate The rate at which snow is accumulated by deposition and sublimation. Can also refer to the rate at which sediments are accumulated on the ocean floor. The term is usually used in connection with glaciers or snow fields.

acid precipitation Because pure precipitation is slightly acidic (due to the reaction between water droplets and carbon dioxide, creating carbonic acid) with a potential hydrogen (pH) of 5.6, acid precipitation refers to precipitation with a pH less than 5.6. Acid precipitation includes rain, fog, snow, and dry deposition. Anthropogenic pollutants (carbon dioxide, carbon monoxide, ozone, nitrogen and sulfur oxides, and hydrocarbons) react with water vapor to produce acid precipitation. These pollutants come primarily from burning coal and other fossil fuels. Sulfur dioxide, which reacts readily with water vapor and droplets (i.e., has a short residence time in the atmosphere as a gas), has been linked to the weathering (eating away) of marble structures and the acidification of freshwater lakes (consequently killing the fish). Natural interactions within the biosphere can also lead to acid precipitation.

adiabatic lapse rate The variation of temperature with height of an air parcel lifted adiabatically from the surface.

adiabatic process A process not involving exchange of heat with the surroundings. A substance undergoing an adiabatic process can only exchange energy with its environment through mechanical work.

advection The movement of some field which is variable in space (e.g., temperature) by the wind. The amount of advection is related to the gradient of the field in question and the velocity of the wind. Usually refers to transports in the horizontal direction.

advection fog This type of fog develops when warm humid air flows horizontally inland. The water vapor cools to the dew-point temperature because the ground is colder than the air. It is common in the wintertime in the central United States.

aeronomy The science treating the physics and chemistry of a planet's upper atmosphere.

aggregation (ice particles) A method by which ice crystals grow, involving collisions between crystals of different sizes and the sticking together of the crystals after the collision. Analogous to coalescence for liquid particles.

air mass A large body of air in which the temperature, humidity, and other properties are nearly constant. Air masses are usually characterized by the region of their origin. With time and movement, air masses can be modified by changes in latitude, insolation, and characteristics of the region. Air-mass boundaries are generally associated with frontal zones.

air-mass thunderstorm An air-mass thunderstorm develops in a thermodynamically unstable and weak vertical wind shear environment, where there are no nearby fronts or storm systems. An air mass is characterized by similar moisture and temperature characteristics over an area of several hundred thousand square miles. One typical thermodynamically unstable environment is the summertime Maritime Tropical (mT) air mass found from the Midwest to Gulf Coast states. A storm forming in this environment has a lifetime of about 30 minutes, going through the stages of developing, mature, and dissipating. It is also referred to as a *garden variety thunderstorm*.

air parcel A mass of air in the atmosphere assumed to contain the same air during some process, with uniform thermodynamic properties. Many discussions of atmospheric processes are presented as descriptions of the behavior of air parcels.

albedo The fraction of incident radiation upon a surface which is reflected back to space. Can refer to the local surface or to the planet as a whole.

altocumulus (Ac) White and/or gray patches, sheets, or layers of clouds generally with shading composed of rounded masses or rolls, which are sometimes partially fibrous or diffuse and which may or may not be merged. Altocumulus clouds exist in the mid level of the sky between 2 to 6 kilometers above the ground (in meteorology, the prefix *alto* is used with other cloud types to describe other mid-level clouds as well). They are one of four genera that are hybrids of cumulus clouds. [*See also* **cumulus**.]

altostratus (As) High, sheet-like (layered) clouds which are found at an average height of about 3,000 meters (10,000 feet) or 1,700 to 6,000 meters (6,000 to 20,000 feet). They appear fibrous, striated, or lightly striped. The clouds are composed of water droplets and/or ice crystals (unlike the higher cirrostratus which are primarily made of ice crystals). [*See also* **stratus**.]

ampere (A) The accepted International System of Units (SI) measure for an electric current's strength, which equals the rate of flow of charge of 1 coulomb per second in a conductor or conducting medium.

anaerobic Refers to processes which do not use oxygen directly from the atmosphere or water (such as bacterial growth). Can also refer to environmental conditions in the ocean or lakes where all the dissolved oxygen has been removed.

anafront A cold front that has upward motion over the position of the surface front. [*Compare* **katafront**.]

anemometer An instrument which measures the speed and direction of the wind. Rotating anemometers, such as the cup or propeller variety, are typically used to measure wind speed. The cup anemometer is the standard for operational surface wind measurements, but the greater rotation rate associated with a propeller anemometer makes it better for measuring light winds. The rotation rate of the propeller or cups is linearly related to the wind speed for typically observed surface winds. Vane anemometers have a wind vane mounted to point the propeller into the wind so that wind direction is also measured.

angular momentum Measure of the rotation of a particle about some axis. Larger for a particle that is more massive, rotating faster, or farther from the center of rotation. The term is conserved for a closed system.

annual cycle The yearly cycle of changes in the climate system owing to the variation of solar radiation incident upon the Earth as it orbits the Sun.

anomaly Difference between a particular measurement and its mean, taken over time and/or space.

Antarctic circumpolar current A large wind-driven ocean current which flows completely around the Antarctic continent from west to east.

anthropogenic Attributed to human activities. Anthropogenic climate change refers to all climatic change attributed to human activity, such as the climatic effects of the increase of certain greenhouse gases in the atmosphere.

anticyclonic Rotation around a center of curvature that is opposite the rotation of the Earth. In the Northern (Southern) Hemisphere anticyclonic circulation is clockwise (counterclockwise). In the atmosphere, anticyclonic circulations are associated with regions of high pressure and fair weather.

anvil cloud The top portion of a cumulonimbus cloud which spreads horizontally outward from the block-like updraft. It spreads out, faster if there are strong jet stream winds, near the equilibrium level where the updraft air is no longer positively buoyant. The cloud, which can occur upwards of 18,000 to 22,000 meters (60,000 to 75,000 feet), is primarily made of water droplets near the updraft; those that move farther away turn to ice crystals. The cloud got its name because the shape is similar to a blacksmith's anvil.

aphelion The point in the orbit of a planet which is farthest from the Sun.

apparent temperature *See* **heat index**.

approaching-severe thunderstorm A thunderstorm with hail diameter greater than 12 millimeters (0.5 inches) but less than 19 millimeters (0.75 inches) and surface winds between 40 and 57 miles per hour (64 and 92 kilometers per hour).

aquifer A subsurface permeable geologic formation which contains water.

Arctic front The boundary between the very cold dense arctic air mass associated with the cold arctic anticyclone and the less cold polar air mass. The Arctic front is a passive front in which movement is a result of gravity and the weight of the air mass.

arcus cloud A thick accessory cloud with uneven boundaries sometimes located on the lower front portion of cumulus and cumulonimbus clouds.

astronomical unit (AU) A unit of measure based on the mean distance of the Earth from the Sun (about 149.6 million kilometers or about 93 million miles).

atmosphere The gaseous mass surrounding stars, planets, and satellites. Also, a unit of pressure that is equal to 101,325 newtons per square meter. In meterology, pressure is usually expressed in terms of millibars (mb) where 1 atmosphere is equal to 1,013.25 millibars. In the International System of Units (SI), 1 atmosphere equals 1,013.25 hectopascals (hPa); 1 atmosphere can also be expressed as 760 millimeters (29.2 inches) of mercury.

atmospheric lapse rate Rate of change of temperature with height, usually measured in kelvins per kilometer.

available buoyant energy Energy available for heating or doing work due to air parcels or masses of different densities being adjacent. Ultimately derived from potential energy.

azimuth Compass bearing to a point on the ground under some object in the sky, usually the Sun, a distant star, or a satellite.

backdoor cold front The term for a cold front along the eastern coast of the United States that moves from the northeast rather than the west or northwest, as do ordinary cold fronts in that region.

bar A unit of pressure that is equal to 10,000 newtons per square meter. In meteorology, pressure is usually expressed in terms of millibars (mb) where 1,000 millibars is equal to 1 bar. In the International System of Units (SI), 1 bar 1,000 hectopascals (hPa).

baroclinic Description of an atmospheric condition in which constant pressure surfaces intersect surfaces of constant temperature. In weather analysis, a baroclinic atmosphere exists when isotherms exist in an isobaric surface analysis. In a baroclinic atmosphere, there is a horizontal temperature gradient and variation with height of the geostrophic wind (thermal wind).

baroclinic instability An amplification of a perturbation owing to the vertical shear and high wind speeds found in the jet stream. Only waves with wavelengths on the synoptic scale (1,000 kilometers) tend to amplify with this type of instability. Baroclinic instability converts potential energy to kinetic energy and is an important mechanism in the global energy balance.

baroclinic wave A wave that develops in a highly baroclinic atmosphere as a result of baroclinic instability. Baroclinic waves develop in the region of strong temperature gradients. Horizontal temperature advection is required for baroclinic waves to develop. Baroclinic waves are also referred to as *baroclinic cyclones* or *baroclinic disturbances*. During baroclinic wave development, waves tilt westward with height.

barometer An instrument that measures atmospheric pressure (the weight of the atmosphere per unit area). One type is the liquid barometer (mercury is the standard liquid) which balances the weight of the atmosphere against the liquid column (the height of which rises with increasing atmospheric pressure). The barometric pressure is read from a calibrated scale. Another type is the aneroid (meaning without liquid) which is made of a partially evacuated container with an internal spring. It compresses (expands) with increased (decreased) atmospheric pressure, causing a calibrated pointer to indicate the pressure.

biogeophysical feedback A feedback mechanism which connects the biological and geophysical domains of the climate system.

biosphere The regions of the continents and oceans containing life forms that directly interact with the climate system.

black body Idealized system that perfectly absorbs all radiation incident upon it and that radiates at the theoretical maximum rate at all wavelengths. Used as a prototype in discussions of radiative transfer.

blizzard Falling snow over a sustained duration with wind speeds greater than 15 meters per second (35 miles per hour) and visibility less than 400 meters (0.25 miles) over an extended period.

blocking Interruption of predominately wavy flow in the westerlies. Distortion of westerly wind flow results in a meridional flow and the development of warm anticyclonic vortices poleward, with cool cyclonic vortices equatorward of the original flow. Blocking patterns tend to persist for days or even weeks.

Boltzmann constant 1.38×10^{-23} joules per kelvin.

boreal forest The forested region south of the arctic tree line and adjacent to the tundra.

bottom water (ocean) The sea water just above the ocean bottom with physical properties which distinguish it from the overlying deep water. These properties include the potential temperature, salinity, and dissolved oxygen content.

boundary layer The layer of air near Earth's surface where frictional forces are important. Typical heights range from 500 meters over the ocean to 1,500 meters over the land.

British thermal unit (Btu) The amount of heat or energy needed to increase the temperature of 1 pound of water 1°F (about 252 calories).

bulk aerodynamics A method of parameterizing the amount of flux based on a specified transfer or drag coefficient and a representative value of the wind speed.

buoyancy Vertical force on a system owing to a density difference between the system and its surroundings.

calorie The quantity of heat or energy required to raise the temperature of 1 gram of water 1°C.

calving The process of forming icebergs where a glacier or ice shelf meets the ocean. As the glacier or ice shelf is pushed out over the water it becomes unstable and pieces of ice break off (are calved) and become icebergs.

canopy An overarching cover consisting of the tree or shrub tops of a vegetated area. The canopy is of an approximately uniform thickness, separating the moving air of the atmospheric boundary layer from a less active layer among vegetation below.

carbon cycle The cycle of storage, transformation, and release of carbon as it moves through the climate system. Carbon appears throughout the climate system in many forms, from free molecular carbon dioxide and methane in the atmosphere, to calcium carbonate in the shells of living and dead marine organisms, to the carbon in decaying plants. The carbon cycle refers to the continuous process of converting carbon from one molecular form to another as it cycles through the biogeochemical system.

carbon-14 (^{14}C) A radioactive isotope of carbon with an atomic weight of 14 that is formed in the upper atmosphere by cosmic ray bombardment. The half-life of carbon-14 is 5,700 years. Radioactive carbon isotopes are incorporated into all organic matter during the lifetimes of plants and animals. Radiocarbon dating of fossil material can therefore be done by measuring the amount of carbon-14 left in samples of, for example, shells or fossil pollens.

catchment area This is another name for a drainage area. It is a structure, like a basin, which collects and drains water. It is typically used to indicate a basin collecting precipitation. The water collected is absorbed by the soil or runs off to a common river or reservoir.

celestial mechanics The development of mathematical theory used to predict the motions of the planets, based on the gravitational interactions of the planets and the Sun.

Celsius The temperature scale developed in 1742 by Swedish astronomer, Anders Celsius (1701–1744). The scale defines the boiling temperature of water as 100°C and the freezing temperature of water as 0°C at standard atmospheric pressure. It is the standard temperature scale for science and industry. Multiply the Celsius temperature by 9/5 and add 32 to the result to obtain the equivlent Fahrenheit temperature.

Cenozoic era The current geologic era, which began 65 million years ago, following the Mesozoic. The Cenozoic is subdivided into the Tertiary and Quaternary periods. The Quaternary period began about 1.8 million years ago. The Quaternary is subdivided into the Pleistocene and Holocene epochs.

chaos A process that is fundamentally unpredictable and does not exhibit any regular behavior. Despite the limitations on predictability, chaotic systems often have near-periodic behaviors that can be described precisely.

chemical oceanography The study of the chemical processes that occur in the ocean.

chemical weathering The effects on the Earth's surface and on humankind's structures by contact with chemicals in the atmosphere. These chemicals can be naturally occurring, such as sulfate aerosols caused by volcanic eruptions, or anthropogenic, such as acid rain caused by coal-burning power plants.

chinook wind The warm dry wind which descends downslope on the lee side of a mountain range. It is warm from the compressional heating of the air as it sinks, and it is dry because the air retains the low moisture properties of the higher altitude source region. The Rocky Mountains' chinook wind, Los Angeles basin's Santa Ana wind, and the Swiss Alps' foehn wind are all different names for a wind with the same characteristics. The Native Americans living on the eastern side of the Rockies referred to the wind as the "snow eater." The chinook wind has been known to increase the air temperature by 34°F in 7 minutes and melt 5 feet of snow in as little as 12 hours.

chlorofluorocarbons (CFCs) Chemicals consisting of chlorine, fluorine, and carbon atoms in various combinations, manufactured for use as coolants and solvents. CFCs are known to break down and destroy stratospheric ozone.

circulation A measure of the tendency of a fluid to move around a closed curve. The circulation gives a sense of the rotation speed and direction of a fluid.

cirrocumulus (Cc) Thin white patches, sheets, or layers of cloud with no shading seen at high altitudes (greater than 6 kilometers above the ground) and composed of small elements in the form of grains or ripples, merged or separated, more or less regularly arranged. Cirrocumulus clouds are one of four genera that are hybrids of cumulus clouds. [*See also* **cumulus**.]

cirrostratus (Cs) A layer of high clouds composed of ice crystals. Cirrostratus are usually very thin and appear as a white fibrous cloud layer. The layer may be so thin that it may only be detected by the presence of a halo around the Sun. [*See also* **stratus**.]

cirrus (Ci) Deriving their name from the Latin for "a lock of curly hair," cirrus clouds are white fibrous clouds composed of ice crystals. These ice crystals may be large and may fall out of the cloud before evaporating. This fall out of ice crystals extends the cloud in the vertical. A cirrus cloud, one of the three major cloud types that comprise the ten main groups or *genera*, is not a cloud layer but is a single cloud element.

Clausius-Clapeyron relation Equation that relates saturation vapor pressure, temperature, and latent heat of vaporization.

climate The weather conditions of a region averaged over an extended time period.

climate model A mathematical model of the climate based on physical principles. Such a model can range in complexity from a single equation to many thousands of equations. At each level of complexity more physical processes are incorporated into the model.

climate modification The modification of the climate can occur by either natural processes or by human activities. Natural processes include volcanic eruptions, El Niño-Southern Oscillation (ENSO) events, solar variability, or random variations. Human-caused climate modifications include the effects of pollution and deforestation.

cloud base The lowest level in the atmosphere in which cloud particles are visible to humans, for any particular cloud or layer of clouds.

cloud–climate feedback Clouds have a large effect on the climate by increasing the albedo of the Earth and also by decreasing the outgoing longwave radiation to space. The two effects nearly cancel. If the distribution of cloud amount and type is sensitive to the current climate state, then a change in the climate state will affect the clouds, which will in turn affect the climate state through a change in the net global energy balance. The mechanisms of such a cloud–climate feedback are uncertain.

cloud climatology The study of the spatial and temporal distribution of cloud types.

cloud condensation nuclei (CCN) Aerosols upon which water vapor condenses to form cloud drops. The ability for an aerosol particle to be a CCN is related to its size, water solubility, and the extent of saturation in the surrounding atmosphere. Particles that serve as CCN include sea salt, byproducts of combustion, and particles produced by plants.

cloud droplet Liquid water particles which are a result of the condensation of water vapor onto a cloud condensation nucleus. Cloud droplets range in size from a few to a few hundred microns in diameter. Cloud drops differ from raindrops only in size, with cloud drops generally being smaller.

cloud seeding The process in which stratus clouds or fog made of supercooled liquid water are seeded with small particles of dry ice or with smoke containing silver iodide. The dry ice particles (which are −78.5°C) act to freeze the water droplets into ice crystals by direct contact. Silver iodide provides an artificial ice nucleus owing to its ice-like crystal structure, which encourages freezing at higher supercooled temperatures. Either substance causes a large number of ice crystals to be produced. These ice crystals then grow via vapor deposition while the surrounding cloud droplets evaporate to maintain saturation. The ice crystals grow because the saturation vapor pressure over ice is less than that over water (the Bergeron process).

cold front A boundary between two air masses in which the colder air is displacing the warmer air. Represented on weather maps with a blue line studded with triangles pointing in the direction of motion. The temperature of air in back of a cold front is cooler and is usually more arid than the air it displaces.

cold-rain process The formation of rain from the melting of precipitation-sized ice particles. Also called the *Bergeron process*.

collision and coalescence Because vapor deposition is less effective for larger drops, collision and coalescence is the main process for forming warm raindrops greater than 12 to 20 microns. Large rain droplets (the collector drops) overtake and collide with smaller droplets (the collected drops) owing to their greater terminal fall velocity. If the droplets are charged or an electric field exists (which commonly does in natural clouds), the droplets will nearly always coalesce (stick together to form a single larger drop) if they collide.

collision efficiency The ratio between the number of drop collisions actually occurring with a larger collector drop to the total number of smaller droplets within the volume swept out by the larger drop. It can also be thought of as the probability of collision between any smaller drop within the volume swept out by the larger drop. In the case where the drops are charged or there is an electric field present, any drops which collide will nearly always coalesce so that the collision and collection efficiencies are nearly equal.

collision rate The collision rate is the average rate at which water droplets collide in a cloud. It depends on the number of smaller droplets present and the size of the larger droplets relative to the smaller drops. The fall speed depends on the size of the drop. In order for collisions to occur there must be a dispersion in the drop sizes. Tiny drops will not collide with very large ones because they will be carried around them by the air flow. Collision between very large drops may lead to fragmentation. The growth rate of raindrops in a cloud depends on the collision rate.

computational fluid dynamics Study of fluid motions by creating a simplified model of a fluid system in a computer, and then simulating its evolution from one moment to the next. Often used when the motion cannot be obtained by solving mathematical equations analytically, or if the system is too large or too complex to reproduce in a laboratory.

computer climate model Computer simulation of the evolution of the state variables of the atmosphere–ocean system in a computer. The system is described on a set of grid points, at each of which is defined a temperature, pressure, wind velocity, etc. The state at some instant is used to calculate the state a short time later, and the process is repeated for the duration of the numerical "experiment." Climate models are designed for the study of long-term behavior, whereas forecast models seek to describe the atmosphere over short intervals into the future.

condensation Phase transition from gas to liquid, accompanied by release of heat. In meteorology, condensation (the opposite of evaporation) usually refers to the method by which water vapor becomes a liquid.

condensation nuclei Particle of soil, meteoritic or volcanic dust, salt, or combustion products can serve as condensation nuclei, upon which the condensation of water vapor occurs. Although, hygroscopic particles (those nuclei which dissolve into solution) are effective because the vapor pressure is lowered over the droplet surface, solid nuclei are preferred owing to the high radius of curvature over their surface to which the water vapor molecules can adhere. Larger condensation nuclei are responsible for producing the larger cloud condensation, and so they are referred to as *cloud condensation nuclei* (CCN). Some condensation nuclei allow condensation at a relative humidity as low as 65 percent (which is why polluted air near cities contains haze). The abundance of CCN cause most clouds never to exceed saturations of 101 percent. If there were no condensation nuclei, clouds and rain would not occur, because laboratory experiments show that saturations of at least 200 percent are required for homogeneous nucleation.

conditional instability Stability in the atmosphere which is lost when the air is heated by water vapor condensing.

conservation law An assertion that some quantity describing

an isolated system does not change in time. Useful for predicting the later state of a system.

conservative quantity Quantity that describes some aspect of an isolated system that does not change in time. It can be a fairly complicated, empirical mathematical expression involving temperature, pressure, speed, location, and so on.

constant stress layer The lowest layer in a neutrally buoyant atmosphere over a smooth surface typically has the property of exhibiting constant stress (force of horizontal resistance per unit area across planes parallel to the surface) as a function of altitude above the surface. This condition allows for a simple logarithmic solution for the wind profile as a function of altitude, and this solution agrees well with observations.

continental air mass An air mass that develops over a continent or a large land mass. With little surface water accessibility, these air masses tend to have low moisture content. They are quickly modified when they move over water that is warmer than land.

continental climate A climate that is usually found in the interior of continents. This climate is characterized by low humidities, low rainfall amounts and large diurnal temperature variations. A desert experiences the extremes in a continental climate.

continental drift The movements of the continental plates across the Earth's crust, as explained by the theory of plate tectonics.

continental glaciation The formation of continental-sized ice sheets during glacial periods.

contrail A narrow white trail of ice-crystal clouds produced by a flying aircraft, which discharge water vapor or droplets that condense in the atmosphere at temperatures colder than freezing.

convection The transport and mixing of heat (and other air characteristics) by way of vertical air motions. Convection usually refers to the vertical air motions accelerated by a local density (and therefore buoyancy) difference between some parcel of air and the environment. Both dry and moist convection occur in the Earth's atmosphere. Dry convection is common near the Earth's surface where the temperature lapse rates are superadiabatic (less than $-10°C$ per kilogram), therefore providing positive buoyancy to an unsaturated parcel rising and cooling at the dry adiabatic rate of $-10°C$ per kilogram. Greater buoyancy differences are achieved for moist (saturated) convection where latent heat (released during cloud condensation) warms the updraft air to increase the positive buoyancy and (where absorbed during the evaporation of rain/cloud) cools the downdraft air to increase the negative buoyancy.

convective available potential energy (CAPE) CAPE provides a measure of the total increase in kinetic energy (per unit mass) that an updraft could realize through positive buoyant acceleration. In addition to the conversion of potential energy associated with interchanging less dense air below with more dense air, a positive buoyancy is caused by a warming of the updraft parcel of air owing to the latent heat released when water changes phase (condenses) from a vapor to a liquid, thus making the parcel warmer than the environment. Since the buoyant acceleration acting on the parcel is not constant with height, a line integral must be used to compute the work. CAPE is proportional to a maximum theoretical updraft velocity at the level near cloud top where the cloud is in equilibrium (zero buoyancy) with the environment.

convective cloud Cloud formation makes the convection process visible. An unstable atmospheric column will lead to updrafts. If the air below is moist, the updraft will lead to condensation into droplets. The condensation process accelerates the updraft owing to the release of latent heat. Examples are the various types of cumulus clouds (e.g., cumulus, cumulus congestus, cumulonimbus). [*See also* **cumulus.**]

convective instability Also known as *potential instability*. If the wet-bulb or equivalent potential temperature decreases with height, then the atmosphere is convectively (potentially) unstable. It is an indication that the atmosphere has potential for convection if lifting of the atmosphere as a whole occurs. This lifting, typical ahead of shortwaves or synoptic-scale wave cyclones, can change an absolutely stable layer in a sounding (for example, one associated with a temperature inversion) to one that is conditionally unstable.

convergence Negative divergence or the rate at which a material field contracts. In a horizontal field, convergence occurs when there is more flow into a defined region than there is out of the region. Owing to the requirement of conservation of mass, horizontal convergence must result in vertical motion.

Coriolis force Tendency of a free body or parcel to turn when it is in a rotating frame of reference. The Coriolis force caused by the Earth's rotation makes bodies moving long distances turn to the right in the Northern Hemisphere and the left in the Southern Hemisphere. Only occurs when the body is viewed in a rotating frame of reference.

coulomb (C) Meter-kilogram-second unit of electric charge equal to the charge of 6.28×10^{18} electrons; 1 coulomb equals 1 ampere-second.

cryosphere The glaciers, icecaps, and sea ice on Earth.

cumulonimbus (Cb) A cumulus cloud of significant vertical extent which is producing heavy precipitation. The cloud often has an anvil-shaped appearance on top, and it is bubbly (cauliflower) shaped on the sides. The top of the cloud is sufficiently dense to filter out sunlight so that the cloud base looks dark. Cumulonimbus clouds are one of four genera that are hybrids of cumulus clouds. [*See also* **anvil cloud.**]

cumulus (Cu) Widely known as "fair weather clouds," cumulus clouds are one of the three major cloud types that comprise the ten main groups or *genera*. Cumulus (a heap, pile, or mound) or *convective* clouds are defined as many individual clouds detached and separated by clear skies. The clouds are dense, developing vertically into domes or towers that resemble cauliflower. The sunlit upper portion is brilfjliant white, and the bases are dark and gray. The outlines of the clouds are sharp when they consist of water droplets and fuzzier or diffuse when they consist of ice crystals. Besides the genera cumulus clouds, there are four other genera that are hybrids of cumulus clouds and differ from the most generic kind of cumulus cloud: cirrocumulus, altocumulus, stratocumulus, and cumulonimbus. In addition to the more general cumulus formation, when cumulus clouds have only slight vertical ascent and are usually confined to the lowest 2 kilometers of the atmosphere they are referred to as *cumulus humilis*; when cumulus clouds reach moderate vertical extent (generally between 2 to 6 kilometers above the ground) and exhibit small protuberances they are termed *cumulus mediocris*; finally, when cumulus clouds extend high up into the atmosphere (generally 6 kilometers or more above the ground) and their bulging parts resemble cauliflower they are called *cumulus congestus*. [*See also* **cirrocumulus**; **altocumulus**; **stratocumulus**; *and* **cumulonimbus.**]

cyclogenesis The process by which a localized low pressure forms and/or strengthens in cyclonic circulation on a stationary front or surface trough.

cyclone A vast area with winds circulating inward to a low atmospheric pressure center. The winds circulate counterclockwise in the Northern Hemisphere and clockwise in the Southern Hemisphere.

cyclonic circulation Rotation around a center of curvature that is in the same direction as the rotation of the Earth. In the Northern (Southern) Hemisphere cyclonic circulation is counterclockwise (clockwise). In the atmosphere, cyclonic circulations are associated with regions of low pressure and stormy weather.

dart leader (lightning) A dart leader is a dart of light which is seen to propagate down the same channel as the first flash (return stroke) channel in a lightning strike. Occurring 40 to 60 milliseconds after the first flash, it propagates at about one-hundredth the speed of light and is associated with a 1,000-ampere current. Once it reaches near the ground, the increase in the electric field causes a return stroke along the path of the dart leader (but in the opposite direction). It is distinguished from a stepped leader in that there is no branching off the original channel path.

deep convection When moist air is lifted in a potentially unstable atmosphere, there is energy available (CAPE) for the air to rise to its level of neutral buoyancy, usually somewhere in the upper troposphere. This lifting and the unstable atmosphere result in clouds, precipitation, and deep towering convective cells.

dendrochronology The compilation of a time chronology based on trees' annual formation of growth rings. The width of each ring can depends on the temperature and precipitation during the year in which the ring was formed. In this dating technique, tree patterns are matched with those of known age. Using overlapping patterns, dating can be carried back about 3,000 years.

deposition *See* **sublimation.**

depression An area of low pressure. In the tropics, a minimally organized tropical disturbance with wind speeds up to 34 knots.

desert A region which has very little precipitation and cannot support appreciable vegetation.

desertification The conversion of nondeserts into deserts owing to natural climatic change or human activity. Deforestation and overuse of farmland are examples of human activities which can lead to changes in the local climatic conditions resulting in desertification.

deterministic Predictable; a system is deterministic if its state at any later time can be determined to any desired accuracy from its present motion and properties.

detrainment The transfer of air from a cloud updraft or other organized air current to the surrounding environment by way of mixing. It is the opposite of entrainment.

dew Water that condenses onto surfaces that are cooler than the dew-point temperature of the surrounding air.

dew point The thermodynamic condition at which humid air would reach saturation if that air cooled without changing its pressure.

dew-point temperature The temperature at which a parcel of air with a given moisture content would saturate.

diabatic heating (cooling) A temperature change brought about by the direct transfer of heat energy. Examples of diabatic processes include solar or terrestrial radiation and the release of latent heat.

diffuse radiation Radiation traveling in all directions, as opposed to traveling in a beam.

dimethyl sulfide (DMS) A molecular gas which is produced in the surface waters of the ocean by phytoplankton. This gas escapes to the atmosphere where it can form a sulfate-based aerosol.

direct circulation A circulation in which the primary driver is warm air rising and cold air sinking. An example of a direct circulation is the land/sea breeze.

discretization error Calculation error owing to description of a system at discrete times and locations, whereas the system properties exist over a continuous range of times and locations.

diurnal cycle This is another name for daily cycle. For a given location on the Earth, daily cycles in meteorological variables such as temperature, relative humidity, precipitation, wind speed, and direction are all due to the Earth's rotation (resulting in insolation differences in space and time). These daily cycles follow the insolation cycle. They are most pronounced at mid latitudes and in the tropics and least pronounced (or missing) near the poles at the time of the solstices.

divergence The rate at which a material field expands. In a horizontal field, divergence occurs when there is more flow out of a defined region than there is into the region. Due to the requirement of conservation of mass, horizontal divergence must result in vertical motion.

Dobson unit (DU) A measure of atmospheric ozone; 1 Dobson unit corresponds to 0.01 millimeters of ozone, assuming all of the ozone in the atmosphere at a location has been confined to one layer at 1 atmosphere of pressure and 0°C.

Doppler broadening Tendency of atoms or molecules to emit a slightly wider range of frequencies because motion changes the apparent wavelength of the emitted radiation.

Doppler radar A Doppler radar not only records the echo intensity and the time between the sent and received microwave pulses but also their frequency difference. Conventional radar just records the echo intensity and time between pulses. The greater the frequency shift, the higher the component speed of the precipitation either directly toward or away from the radar transmitter. Most Doppler radars are stationary, but some are mobile, mounted in airplanes or

trucks which can then take them to observe interesting storms.

drag coefficient Number used to predict the force exerted by a fluid (for example, air) on a body moving through it.

drainage wind Downslope winds that occur at night when air is cooled as it makes contact with mountainous surfaces. As air cools its density increases, and it slides down the side of the mountains into the valleys below.

drizzle Liquid precipitation consisting of numerous water droplets with diameter less than 0.5 millimeters (0.02 inches) falling or sometimes appearing to float closely spaced and eventually reaching the ground. Drizzle is sometimes described as a mist.

dropwinsonde An instrument similar to a rawinsonde except that it is dropped with a parachute from an aircraft to take measurements of the atmosphere's vertical profile as it falls. It is typically used over the oceans during hurricane reconnaissance when the surface launch of a rawinsonde would otherwise be impractical and dangerous. The tracking of the transmitter by 3 or more ground-based stations allows the wind velocity and direction to be calculated.

drought A protracted, abnormal period of dry weather for a certain region.

dry adiabatic lapse rate Rate of change of temperature with height for a parcel that undergoes an isentropic compression or expansion with no water substance involved. The numerical value is 9.8°C per kilometer.

dry line A boundary analogous to a front separating moist air from dry air. Typically, little temperature contrast is observed across the dry line. It is primarily observed during the spring over the Great Plains of the central United States.

dust devil A vigorous rotating column of air made visible by dust, sand, and debris. They have been observed as tall as 500 meters (1,645 feet) and are typically between 3 and 30 meters (10 and 100 feet) wide. A column forms during dry convection when the near-ground temperature lapse rate is superadiabatic; typical in dry desert regions during the day. As the air flows horizontally toward a local hot spot, it will start spiraling either clockwise or counterclockwise.

dust storm These storms are typical in the Great Plains of the United States behind a surface cold front or dry line when the winds are strong and can blow topsoil into the air. The topsoil is most easily blown when it is loose from recent plowing and dry from a lack of rainfall. The National Weather Service (NWS) criterion for a dust storm is winds greater than or equal to 48 kilometers per hour (30 miles per hour) causing blowing dust to reduce the visibility to 0.8 kilometers (0.5 miles) or less. When visibility is reduced to 400 meters (0.25 miles) or less, the NWS will issue an advisory to warn motorists that driving conditions are hazardous.

dust veil (volcanic) Volcanic eruptions can inject vast quantities of ash, dust, and sulfur dioxide into the stratosphere. Only the smallest particles will remain in the atmosphere for longer than a few weeks. These fine particles can serve as the condensation nuclei for the formation of sulfate aerosols.

dynamical system A system with well-defined properties that vary in time.

dynamic meteorology The study of air motions in the atmosphere.

dyne The force in the centimeter-gram-second system required to give a mass of 1 gram an acceleration of 1 centimeter per second per second.

easterlies Name given to winds that blow from east to west. Regions of easterlies include equatorial regions (trade wind region) and polar regions (polar easterlies).

eccentricity A measure of how much a planetary orbit departs from a perfect circle, which by definition has an eccentricity of zero.

eddy A small volume of air that rotates differently than the primary flow of the layer of air in which it exists. A tornado is an extreme example of an eddy, having a circulation different from the large-scale flow of air surrounding the thunderstorm in which the tornado forms.

eddy viscosity Force on a body owing to momentum exchange with small eddies. Behaves like molecular viscosity but on a larger scale.

Ekman layer Layer in the atmosphere or ocean where surface friction is an important determinant of the motion.

electromagnetic radiation See **radiation**.

El Niño A complex state of the equatorial ocean–atmosphere system associated with anomalously warm ocean temperatures in the central and eastern Pacific Ocean, anomalously cold ocean temperatures in the western Pacific Ocean, and a weakening of the southeasterly trade winds. Changes in the ocean temperatures result in increased convection over the warmer waters of the central and eastern Pacific and decreased convection over the western Pacific. Weather patterns are also affected in other regions of the world. [See **teleconnection**.] El Niño occurs every three or four years.

emissivity How well a system radiates at a given wavelength.

energy flux Measure of energy flow; given as the amount of energy traveling through a unit area in a unit of time.

ENSO El Niño-Southern Oscillation.

entrainment The mixing of relatively cool, dry air from outside a cumulus cloud into its interior. Entrainment helps limit the growth of cumulus clouds.

environmental pressure The actual weight of the air per unit area. No correction to sea level pressure is applied.

equation of state Relation between state variables in a thermodynamic system. For example, in a gas the equation of state relates the temperature, pressure, and density.

equilibrium line (ice sheet) For given climatic conditions, the latitude where a balance is maintained between the ablation of the ice sheet and the accumulation of new ice.

equilibrium state Thermodynamic state in which the properties of a system are constant, or assumed constant, and there are no net exchanges of mass or energy with the surroundings over time.

equinox The two points in the sky where the projection of the Earth's equatorial plane intersects the apparent annual path of the Sun. In the Northern Hemisphere, the Sun appears at the vernal equinox on about March 21 and at the autumnal equinox on about September 22. On these two days the Sun appears directly overhead at noon on the equator.

erg In the centimeter-gram-second system, the unit of work or energy that 1 dyne imparts when acting through a distance of 1 centimeter.

evaporation Transformation from a liquid to a gas that is usually dependent on the amount of saturation of the gas. The process of evaporation results in a heat loss or cooling of the liquid. The reverse process of evaporation is condensation.

evapotranspiration The sum of evaporation and transpiration.

exosphere The outermost portion of a planet's atmosphere, consisting of a hot layer of light atoms often moving at escape velocity.

extinction coefficient Rate at which energy traveling through a medium is lost due to scattering and absorption.

extratropics Regions of the Earth that are poleward of the tropics or the tropical easterlies. The dynamics of extratropical weather systems are generally governed by large pressure and temperature gradients.

Fahrenheit The temperature scale developed by German physicist Gabriel Daniel Fahrenheit (1686–1736) from 1708 to 1724. The scale was first derived using the temperature of an ice and ammonia salt solution as 0°F and the normal temperature of human blood as 100°F under standard atmospheric pressure. It was revised several times, and today water's freezing and boiling temperatures are 32°F and 212°F, respectively, and human blood is 98.6°F under standard atmospheric pressure. This scale is the standard for everyday use in the United States. Subtract 32 from the Fahrenheit temperature and multiply the result by 5/9 to obtain the equivalent Celsius temperature.

fair-weather electric current The fair-weather electric current is equivalent to a leaking current occurring between two spherical plates. That is, the space charge moves downward through the atmosphere (dielectric) between the ionosphere and Earth (plates). This net downward transport of positive charge occurs in a current between 1,350 to 1,800 amperes over the entire Earth's surface. Those studying atmospheric electricity hypothesize that lightning causes this fair-weather electric current by bringing negative charge to ground (keeping the spherical capacitor charged). If lightning did not occur continuously, the Earth–ionosphere capacitor would lose its current in less than 10 minutes.

fair-weather electric field The fair-weather electric field is measured to be an average of 120 volts per meter directed downward to the ground. It decreases exponentially with height with the air density (because the dielectric strength decreases causing the atmosphere's conductivity to increase).

feedback A process which changes the relationship between a forcing of the climate and the response of the climate to that forcing. This type of process is often referred to as a *feedback mechanism*. Climate feedbacks can amplify the response of the system to an externally imposed perturbation.

Ferrel Cell Indirect meridional cell that is poleward of the tropical Hadley circulation in zonally averaged observations. The equatorward sinking branch is representative of the air motions in the subtropical highs. The poleward and ascending branch corresponds to the stormy region of the polar jet stream.

finite difference methods Description of a continuously varying system in terms of its values on a system of discrete grid points. Finite difference methods are used in the approximate numerical solution of ordinary and partial differential equations.

flow instability Tendency of a fluid to evolve rapidly from one flow state (usually steady) to a very different state with little energy input from the environment. For example, strong shear flows tend to be unstable.

flux Transport of a material or property through an interface. Mathematically, the flux is defined as the surface or volume integral of the dot product of the function and the normal component of the surface or volume. Meteorological examples of fluxes of interest include heat, momentum and moisture fluxes.

fog A cloud having its base in contact with the ground which reduces visibility to distances less than 1 kilometer. It can form either due to cooling of the air to its dew-point temperature (either by orographic lifting or radiative cooling) or by evaporating water into the air so that the temperature falls and the dew point rises to the wet-bulb temperature.

foraminifera An order of marine protozoans that live near the surface (planktonic) or on the bottom (benthic) of the ocean. The foraminifera are microorganisms with hard shells of calcium carbonate and other calcium-containing compounds. The remains of these and other organisms form the sediments of the ocean floors.

forecast skill A number which measures the "goodness" or accuracy of a set of forecasts or a forecaster. Skill scores usually range from 0 to 1, with 1 being perfect.

fractional cloud cover The expression of the amount of cloud cover at an observing site as a fractional number; usually reported in amounts from 0 to 1 in increments of 0.1.

free convection A vertical displacement of a parcel of air which occurs as a result of an unstable lapse rate of virtual temperature.

freezing rain Rain that falls but then freezes on contact with a surface which is cooler than the rain's freezing temperature.

friction drag Force exerted by a fluid on a body due to eddy formation and surface energy exchanges.

front The transition zone between two air masses, which differ in temperature or moisture content. Cold fronts and warm fronts are two kinds of front, but other more specific terms used are Arctic front, backdoor cold front, gust front, occluded front, polar front, stationary front, and storm front.

frontal lifting The lifting occurring when a cold front plows under a warmer air mass or when warm air flows upslope over the warm frontal surface where colder air is below. In both cases, the warmer and more moist (less dense) air mass is lifted.

frontal system A frontal system is the synthesis of several fronts which are all acting together in the cyclone. There may initially be only a stationary front comprising the system. After cyclogenesis, a warm front, cold front, and stationary fronts will make up the frontal system. When the cyclone begins to occlude, an occluded front will be added to the frontal system. These fronts will have varying strengths and orientations relative to the extratropical cyclone during the cyclone's life cycle.

frontal zone A transition zone 10 to 200 kilometers wide between air masses

frontogenesis The process by which horizontal temperature gradient and vertical wind shear are increased such that a front is formed where there was none (or only weakly) before.

frost A deposit of ice crystals onto vegetation and other surface objects by reverse sublimation (also called *deposition*) when the dew-point temperature is below freezing (0°C/32°F) and the actual temperature of the object falls to the dew-point temperature owing to longwave radiative cooling.

Fujita scale A scale created by Theodore Fujita to measure wind damage; the scale ranges from F0 (light damage) with a wind force up to 32 meters per second (72 miles per hour) to F5 (incredible damage) with a wind force above 117 meters per second (261 miles per hour).

funnel cloud A rotating funnel-shaped smooth cloud extension suspended from a cumulus or cumulonimbus cloud with the circulation not in contact with the ground. A funnel cloud does not imply that a tornado will form; however, funnels are often seen before, during, and after tornado formation. During a tornado, a funnel cloud may extend to the ground, but at this point, the term *tornado* supersedes the term *funnel cloud*.

gamma radiation Electromagnetic radiation produced by a nuclear or particle process. In theory, it can be of any energy and of any wavelength. The radiation is emitted as a photon, the elementary particle that is the quantum of electromagnetic radiation, and this emission is called a *gamma ray*. The highest energy/shortest wavelengths of electromagnetic radiation are the hard X-rays produced only by nuclear processes and emitted as gamma radiation. Gamma radiation is generated by radioactivity, the decay of radioactive (unstable) atoms both in Earth's crust and in the atmosphere, and in the reactions caused by cosmic ray collisions with atmospheric constituents, as well as by the decay of elementary particles produced by these collisions.

gauss In the centimeter-gram-second system, a unit of measure for magnetic induction or magnetic flux density equal to 1 line of magnetic flux per square centimeter.

general circulation model (GCM) Computer program that simulates the state of the entire atmosphere and/or ocean at discrete points from one moment in time to a later time, usually with a large number of small time steps.

geophysical fluid dynamics Study of fluid motions on a rotating planet, specifically in the atmosphere, ocean, and molten subsurface material.

geopotential The work required to raise a unit mass from mean sea level to a specified height. Geopotential often replaces the geometric height in the hydrostatic equation. Geopotential variation with respect to height is only dependent on temperature.

geostrophic wind The horizontal velocity field that approximates the wind field with the assumption that the Coriolis force exactly opposes the pressure gradient force. The geostrophic wind vector is parallel to isobaric or height contours with low heights or pressure to the left. The geostrophic wind is a good approximation of the wind only in large-scale systems in the absence of friction and is still only an approximation of the actual wind.

glacial period A period of time in Earth's history when there were many glaciers and ice sheets covering parts of the Earth. The global climate was then much cooler than it is at present, during this past-Pleistocene interglacial called the *Holocene*.

glaciation The process by which cloud particles change from water droplets to ice crystals. The upper area of cumulonimbus clouds are glaciated.

glacier A slowly moving mass of ice which is formed in mountainous regions by the accumulation of snow over thousands of years.

glaze *See* **rime ice**.

global warming The increase in global annual mean surface temperature which has occurred over approximately the last century. The magnitude of this increase has been estimated to be about 0.6 K.

GOES Geostationary Operational Environmental Satellite; a U.S. weather satellite in a synchronized orbit with the Earth that holds the satellite over the same location on the equator.

Gondwanaland A large continental mass that at one time was part of a supercontinent formed during the late Precambrian era more than 500 million years ago. It consisted of Africa, South America, India, Antarctica, and Australia. Gondwanaland then became the major part of the supercontinent Pangea, which formed during the Paleozoic more than 200 million years ago.

gradient A vector normal to isolines (also called *level surfaces*) pointing toward higher values. Denoted mathematically as ΔF, where F is the field for which the local gradient vector is being calculated. For example, pressure, geopotential height, and temperature gradients are important in meteorology.

gradient wind balance In curved flow, the resulting wind can be approximated by the balance between the centripetal acceleration, the Coriolis acceleration and the horizontal pressure force. Because the geostrophic approximation is not valid for curved flow, gradient wind is generally more useful, but still only an approximation of the wind field.

graupel The nucleus of graupel is an ice crystal which falls and collects supercooled liquid cloud droplets. Those droplets freeze onto the ice crystal's surface such that the original crystal shape is hidden.

gravity wave A wave which results from the buoyancy force acting to restore parcels displaced from hydrostatic equilibrium by some external force. They are often formed at the tropopause interface at the top of thunderstorms and travel horizontally outward in all directions; transverse to the restoring force. This type of wave prefers to travel in stable layers (where the lapse rate is less than moist adiabatic).

greenhouse effect The additional warming of the Earth's surface caused by the absorption of outgoing longwave radiation by certain molecular species (gases) in the atmosphere, which then emit longwave radiation back to the surface. At the same time these molecular species are transparent to the incoming shortwave radiation, which can then heat the surface.

greenhouse gas Any atmospheric gas which is an effective infrared absorber. The most important of these are water vapor, carbon dioxide, methane, and the chlorofluorocarbons (CFCs).

Greenwich mean time (GMT) The mean solar time at 0° longitude which passes through Greenwich, England. This time is the basis for standard time throughout the world. Local standard time can be determined by adding 1 hour for each 15° of longitude that is west of and between 0° longitude and the local longitude.

grey body A body which has an absorptivity independent of the frequency of the incident radiation.

ground fog Any fog which hides less than 60 percent of the sky.

ground water Water which has accumulated below the ground surface.

Gulf Stream The ocean current which flows northward from Florida to Newfoundland along the eastern coast of North America. The Gulf Stream is part of the western boundary current of the North Atlantic ocean gyre.

gust front Cooler air flowing out from the leading base of a cumulonimbus cloud.

Hadley circulation Direct meridional circulation with an upward branch located in the region of the Intertropical Convergence Zone and a subsiding branch corresponding to the region of the Subtropical Highs. The Hadley circulation is important in the global energy balance by transporting heat and momentum poleward and moisture equatorward.

hail Balls or chunks of ice (starting at 0.635 centimeters/0.25 inches in diameter) that are produced in and fall from cumulonimbus clouds with strong updrafts. The hail begins as an ice pellet (sleet) or graupel which is suspended by the storm's updraft. Supercooled liquid water within the updraft freezes onto the growing hailstone (accretion) in alternating sheets of rime and glaze. The hailstone eventually falls when the updraft can no longer suspend its weight.

halo One of the rings or arcs that we seen from Earth around the Sun or Moon; they are generated by the refraction of light through high-altitude ice crystal clouds (e.g., cirrostratus clouds) in Earth's atmosphere.

halocarbon This is a halogenated hydrocarbon, the molecule that results from the replacement of one, some, or all of the hydrogen atoms in a hydrocarbon with atoms of the halogen group (fluorine, chlorine, bromine). The chlorofluorocarbons (CFCs) that have become pollutants in the atmosphere are halocarbons.

haze Droplets of water, much smaller than cloud droplets, which have condensed onto cloud condensation nuclei (CCN). Some of these nuclei become wetted at a relative humidity as low as 65 percent. Pollution (smoke, dust, exhaust emissions) is the major source of many CCN that form the solute for haze droplets. Haze results in a reduction of visibility because it scatters visible light.

heat conduction Heat flow through a rigid material body.

heat convection Heat flow owing to a material receiving heat and then moving with the heat somewhere else.

heat engine Device that receives heat and uses some of the energy to perform work; or one that has work done on it and uses the energy to induce heat flow.

heat index An *apparent temperature* (a measure of how a person would feel) that results from certain combinations of high temperature and relative humidity. When the relative humidity is high, the body is not as effective at converting sensible heat from the skin into latent heat by the evaporation of perspiration. Since it feels warmer to the person, the heat index temperature is higher than the actual temperature. When the relative humidity is very low, the body is able to cool by evaporation of perspiration quite readily. This results in a heat index temperature lower than the actual temperature. For example, a day with a temperature of 95°F (35°C) and a relative humidity of 70 percent would call for an "extremely hot" rating with an apparent temperature near 130°F (54°C). For many people, physical activity could result in heatstroke and should be avoided on such a day.

heat island A localized high temperature over a city, higher than the surrounding region. The temperature perturbation compared to the surrounding countryside is typically +1°C for small cities to +4°C for larger cities. This higher temperature is due to the replacement of natural vegetation in the city with asphalt, concrete, buildings, and other human-made structures. These human-made structures typically have low albedo, a high heat capacity, and no evapotranspiration, so that most of the heating is sensible. This has been studied in cities, such as St. Louis, Missouri, where it was found to cause large increases in the amount of hail, thunderstorms, and rainfall downwind of such cities. Columbia, Maryland, however, has a heat island of only +0.5°C, because the city was developed without eliminating much of the natural vegetation.

hertz (Hz) The unit of frequency equal to 1 cycle per second.

Holocene The current geologic epoch, which began after the Pleistocene, about 10,000 years ago, when the last ice age ended.

homogeneous nucleation In this process, a water droplet or ice crystal forms without a condensation or freezing nucleus, respectively. For water droplet formation, a sufficient number of water vapor molecules spontaneously come together and condense to form a liquid droplet embryo of critical size such that the water droplet continues to grow in a saturated environment. For ice crystal formation, a sufficient number of liquid water molecules spontaneously aggregate inside a water droplet to form an ice embryo of critical size such that the entire droplet turns to ice. Homogeneous nucleation of water droplets does not occur in nature because a relative humidity of 200 percent is required. Homogeneous nucleation of ice crystals does, however, occur in nature and is more likely for lower air temperatures and smaller droplet volumes.

horizontal advection Transport by the horizontal wind field of a material or property. On weather maps, advection can be found in regions where height or pressure contours intersect isopleths of the property or material of interest.

hurricane An Atlantic or eastern Pacific tropical cyclone with surface wind speeds in excess of 65 knots (32.5 meters per second/74 miles per hour). Hurricanes typically appear as closed, circular cloud bands on satellite imagery. Often an eye (a clear region near the center of the cloud bands) is also visible. Hurricanes are called *typhoons* in the western Pacific and by the generic term *tropical cyclones* in Australia and the Indian Ocean.

hydrologic cycle Transport of water substance within and between the Earth, atmosphere, and ocean. The hydrologic cycle is a closed cycle and includes all states of water.

hydrology The study of the waters on the Earth. Focus of hydrology is on evaporation and precipitation and the effects that these processes have on surface water, ground water, soil moisture, and the ocean. Hydrology is useful in water management practices, agriculture, and construction.

hydrosphere The water on Earth, which includes, glaciers, lakes, oceans, rivers, and so on. Clouds and water vapor can be regarded as either parts of the hydrosphere or atmosphere.

hydrostatic equation Equation representing the balance between gravity and the vertical pressure gradient force. If these forces are equal, there is no vertical motion (in the absence of other atmospheric motions). The atmosphere is approximately in hydrostatic balance.

ice age A period of approximately 100,000 years during which large ice sheets cover much of North America and parts of Europe. The global mean temperature may be as much as 10 K cooler than at present. The current cycle of ice ages began about 3 million years ago and is studied by researchers of the Cenozoic era. The last ice age ended about 10,000 years ago.

icecap Ice which covers the polar regions. In the Arctic this is sea ice, while in Antarctica it is a continental ice sheet.

ice fog A type of cloud near ground which is made of tiny particles of ice instead of water droplets. The dew-point temperature in this case must be well below freezing (less than −40°C) or else supercooled liquid droplets are likely to form instead. Like fog, it forms in clear and calm weather conditions. Ice fog is most common at high latitudes where those extremely cold temperatures and dew-point temperatures are most common.

ice sheet The covering of a land surface by a continuous sheet of ice which exists throughout the year. Ice sheets are formed by the accumulation of snow over thousands of years. Antarctica and Greenland are almost entirely covered by ice sheets.

ice shelf A thick layer of ice which has been pushed out onto the ocean by a glacier. The Ross ice shelf in Antarctica is about 200 feet thick.

incompressible (air) Having constant density even at different pressures. Often a useful and sufficiently accurate assumption in theoretical descriptions of air flow.

inertial flow Motion resulting from the balance between the Coriolis force and the centrifugal force. This flow is called *inertial* because both forces are apparent forces, due only to the motion relative to the coordinate system.

infrared (IR) Electromagnetic radiation with wavelengths between visible red light and microwaves; it ranges from 750 to 1 million nanometers (1 millimeter). Infrared is subdivided into near (NIR, 750 to 2,500 nanometers), middle (MIR, 2,500 to 30,000 nanometers), and far (FIR, 30,000 to 1 million nanometers). [*See also* **radiation**.]

infrared absorber A molecule which absorbs electromagnetic radiation at infrared wavelengths. Water vapor and carbon dioxide are strong infrared absorbers.

initialization (model) The process of specifying the initial data or initial conditions for numerical forecast models. Proper initialization is vital to insuring optimal model forecasts. Initialization is difficult because of the lack of data available on proper time and space scales and the effect and compounding importance of small initial errors.

in situ Usually referring to comparable observations. In situ observations are observations of different types but taken at comparable times and locations. Geologists and archeologists use it in the original Latin sense, meaning "in position" or "in its original place."

insolation The daily incoming solar radiation.

interannual Refers to processes which occur on a time scale of more than 1 year but usually less than 10 years.

interglacial The time period between the end of a previous ice age and the onset of the next ice age. The periods have lasted for approximately 10,000 to 20,000 years. The Earth is currently in an interglacial period which is known as the *Holocene*.

internal energy Energy in a body due to random microscopic molecular motions and chemical bonds. In equilibrium thermodynamics, it is a function only of the state of the system.

Intertropical Convergence Zone (ITCZ) The equatorial region where trade winds converge, supplying moisture and heat, to form an irregular line of clouds, convection, and precipitation. The ITCZ represents the upward branch of the Hadley circulation. The ITCZ is important to the global energy balance.

intraannual Refers to processes which occur on a time scale of less than 1 year but more than 1 month.

inversion *See* **temperature inversion.**

inviscid Frictionless motion of a fluid. Often assumed in descriptions of large-scale atmospheric processes.

ionosphere The portion of Earth's atmosphere defined compositionally not thermally; characterized by the presence of ions. It extends from about 60 or 70 to 1,000 kilometers, and beyond, where free electrons and ions exist for long time periods and affect the propagation of radio waves. There are several layers of the ionosphere, which differ in their electron and ion density: D region, E region, and F region.

irreversible change A change from one thermodynamic state to another along a path which cannot be constructed by a series of infinitesimal steps, each of which is reversible. A typical irreversible change is a spontaneous one in nature; for example, rapid expansion of a gas.

isobar A line of constant pressure on a weather map. Isobars are found on a surface of constant height. Pressure gradients and isobar patterns are used to infer wind fields.

isobaric saturation point *See* **dew point.**

isotherm A line of constant temperature on a weather map.

jet stream Any of the narrow, fast-moving (25 meters per second/57 miles per hour) belts of air that weave through the mid latitudes of both hemispheres. They are essential to upper- atmospheric circulation.

joule (J) A unit of energy equal to the work done by a force of 1 newton through a distance of 1 meter or 10^7 ergs or 0.2389 calories. An accepted unit of measure in the International System of Units (SI).

katabatic wind Dense air which drains down the slopes of elevated plateaus into the valleys below. The air mass grows cold and dense from radiative cooling or contact with snow fields. The katabatic wind which flows from the Austrian Alps toward the Adriatic Sea is known as a *bora*, while that which flows down the Rhone Valley toward the Mediterranean coast of southern France is known as a *mistral*.

katafront A cold front that has downward motion over the position of the surface front. [*Compare* **anafront.**]

Kelvin An absolute temperature scale named after Lord Kelvin (William Thomson, 1824–1907), who devised it in 1848.

It has the same degree size as the Celsius scale, but was extended down to absolute zero (0 K is −273.15°C), which refers to the temperature at which all molecular motion would theoretically cease (i.e., the zero pressure for a gas). It is a resetting of the Celsius scale to reflect thermodynamic principles and eliminate the need for negative numbers. In this scale the freezing point of water is 273 K and the boiling point is 373 K. Temperatures very close to absolute zero have been produced in the laboratory.

Kelvin wave A wave in the ocean or atmosphere created by an imbalance between pressure gradients and Coriolis forces near a boundary. An important form of Kelvin wave is along the equator band. These gravitationally induced equatorial Kelvin waves travel from west to east.

knot A unit of velocity of 1 nautical mile (6,076.12 feet) per hour or 1.15 statute miles per hour or 0.51 meters per second.

land breeze A circulation induced by differential nighttime cooling between a land surface and a water surface, causing a wind flow from land to water. It is common on calm nights in the summertime. It forms near the coast or on islands because the land is cool relative to the sea. A typical horizontal length scale is 100 kilometers.

landspout A violently rotating column of air extending downward from a cumulus or cumulonimbus cloud and having circulation in contact with the ground. Landspouts are distinguished from tornadoes because they form from nonsupercell thunderstorms in a fashion similar to the waterspout. Although the majority of landspouts are weaker than the Fujita wind damage intensity rating F2—wind speeds of 50 to 70 meters per second (113 to 157 miles per hour)—and move slowly, they should still be taken seriously. [*See* **Fujita scale.**]

La Niña The cold phase of El Niño, or when convection in the western Pacific and southeasterly trades are anomalously high. If you consider that El Niño is one extreme of the Southern Oscillation with increased convection over the eastern Pacific and decreased convection over the western Pacific, then La Niña is the opposite extreme with intensified convection over the western Pacific so convection over the eastern Pacific is minimized.

lapse rate The rate of decrease of temperature with height. Knowledge of the atmospheric lapse rate is used in determining the stability of the atmosphere.

latent heat Heat associated with the isothermal change in the state of water. Latent heat is absorbed when ice melts and water evaporates. Latent heat is released in the freezing and condensation processes.

Laurentide ice sheet The ice sheet that covered much of eastern North America during the last ice age.

lee wave A stationary gravity wave created by the forcing of air over high mountains of width 1 to 20 kilometers (0.6 to 12 miles). This lee wave appears downwind of the mountain and occurs due to the oscillation of the air in a stable atmosphere; the air flow streamlines, tracing out the wave. Lee waves are seen near the Rocky Mountains, Sierras, Cascades, and Appalachians. The properties of the wave depend on the atmospheric stability, the width of the mountain range, and the wind shear. These waves are most favored when the angle between the winds and mountain ridge is perpendicular and when the winds increase and the stability decreases with height. Lenticular or rotor clouds can form in the crests of the waves.

lifting condensation level The level to which a parcel of air must be lifted before decompressional cooling (at the dry adiabatic lapse rate) allows that parcel (with constant mixing ratio) to reach saturation with respect to a plane surface of pure water.

light (visible) *See* **visible light.**

lightning A natural electric discharge which occurs most commonly in the atmosphere owing to the charge separation within cumulonimbus clouds. The lightning discharge is a luminous short circuit between two locations; cloud-to-ground, intracloud, intercloud, or cloud-to- air. It is visible to humans as a thin, branching channel of light (literally a long spark) and may appear to flicker if there is more than one stroke of current. A cloud-to-ground discharge is of most interest because of its power; it can kill people, start forest fires, cause property damage, and knock out electrical power. Temperatures in a lightning discharge can reach some 28,000°C (50,000°F), the highest naturally occurring temperature on Earth. Lightning produces electromagnetic radiation—from ultraviolet to visible to infrared to radio waves.

light waves Electromagnetic radiation in the middle third of the electromagnetic spectrum, between the short wavelength third (X-rays) and the long wavelength third (radio waves). They range from 4 to 1 million nanometers (1 millimeter) and include ultraviolet light (4 to 400 nanometers), visible light (400 to 750 nanometers), and infrared (750 to 1 million nanometers).

linewidth (absorption spectrum) Width of a dark absorption line (interval) in the spectrum of light from a gas, caused by gas molecules absorbing at that particular wavelength. Finite linewidth can be caused by the finite lifetime of the molecular substance, Doppler broadening, and collision broadening (shortening the lifetime of molecular states by collisions).

little ice age That period from approximately 1400 to 1850 C.E. during which much of Europe had a climate cooler than it does at present. During that time many of the glaciers in the Alps, Scandinavia, and Greenland were more extensive than they are today.

longwave radiation The electromagnetic radiation emitted by the Earth is in the infrared band of the electromagnetic spectrum and is referred to as *outgoing longwave radiation*.

low-level jet A strong current of air with limited horizontal and vertical extent found in the lower troposphere. Often associated with the occurrence of severe weather.

low pressure Often used to refer to a surface cyclone and its associated fronts. Generally, any region where the air pressure is a local minimum.

macroburst A "large" downburst of wind with outflow greater than 4 kilometers (2.5 miles) in horizontal diameter, lasting from 5 to 30 minutes. They can be either the wet variety (with rain) or the dry variety (associated with virga). Damaging curved or straight-line winds from a microburst

can be as high as 60 meters per second (134 miles per hour). Their damage was often blamed on tornadoes before they were well understood. The strongest macrobursts can cause damage similar to that of an F2 tornado. [*See* **Fujita scale.**]

magnetosphere The level of the atmosphere (above about 500 kilometers) where particles are sparse and density is so low that collisions between particles rarely happen; the motion of the particles is also constrained by the magnetic field of the Earth.

mammatocumulus (mamma) cloud A pouch-like cloud that protrudes several thousand feet beneath the anvil of a cumulonimbus cloud. The pouches are formed by descending air.

mare's-tail cirrus (Ci) Long, white, hair-like wisps of cirrus clouds, thicker at one end then tapering to the other. They are named after their close resemblance to actual tails of female horses. The clouds themselves are made up of primarily ice crystals. The comma- shape to the clouds is due to the fallout of larger ice crystals in an environment with vertical wind shear.

maritime Description of an air mass that develops over a large body of water. Maritime air is characterized by its high moisture content. Maritime climates have moderate diurnal and annual temperature ranges, such as Hawaii.

mass conservation Air conserves its mass in the atmosphere even though the volume may change (by way of compression, expansion due to sensible heating, etc.). This principle is invoked in the continuity equation. The Lagrangian form of the continuity equation expresses a balance between the divergence of the horizontal winds and the rate of change of density following an air parcel. The Eulerian form of the continuity equation expresses a balance between the mass divergence and the local rate of change of density. The Eulerian equation states basically that if there is continual convergence of air at the surface (such as in a low-pressure system), that air must rise such that new air can continue to replace the old.

meridional circulation A circulation that sets up in the plane defined by the vertical and north–south (longitudinal or meridional) coordinate. Motions are confined to this plane.

mesocyclone *See* **supercell.**

mesopause The transition layer between the mesosphere and the thermosphere. The mesospause is located at the top of the mesosphere at approximately 70 to 80 kilometers and is the level of minimum temperature in the atmosphere.

mesoscale meteorology The study of weather features possessing the horizontal dimensions from 2 to 2,000 kilometers (1.2 to 1,250 miles) and the time scale from 1 hour to several days. A dense array of weather stations, such as the Oklahoma Mesonet, can resolve the larger weather features of this scale whereas the conventional National Weather Service stations are meant to resolve synoptic-scale features. Examples of mesonet-observable features (20 to 2,000 kilometers) within the mesoscale range are secondary frontal circulations, meso- lows, hurricane rain bands, lee waves, squall lines, low-level jet, and inertial waves. Examples of smaller mesoscale features (2 to 20 kilometers), requiring a mobile mesonet or radar observation, are urban heat islands, supercell mesocyclones, mesoscale convective vortices, and orographic disturbances.

mesosphere The atmospheric region located above the stratopause at an altitude of about 55 to 80 kilometers, characterized by a decrease in temperature with increasing altitude.

microburst A "small" downburst of wind with outflow less than 4 kilometers (2.5 miles) in horizontal diameter. They can be either the wet variety (with rain) or the dry variety (associated with virga). Damaging curved or straight-line winds from a microburst can be as high as 75 meters per second (168 miles per hour). Their damage was often blamed on tornadoes before they were well understood. The strongest microbursts can cause damage similar to that of an F3 tornado, which has wind speeds ranging from 71 to 92 meters per second (158 to 206 miles per hour). [*See* **Fujita scale**.]

microclimate The local climate which distinguishes the physical characteristics of the local region from other regions. The vertical extent of such a climate ends where it can no longer be distinguished from other climates. This is often related to the extent to which the vegetation and so forth affects the air just above it—which is basically the Planetary Boundary Layer (PBL). Trees and plants produce microclimatic variations due to their effect on the air's moisture supply and wind speed. An absence of trees and plants in a city may create a microclimate where it is locally warmer than the surroundings. [*See* **heat island**.] Lakes may also vary the microclimate—cooling the surface air in the summertime and warming it in the winter.

micrometeorology The study of weather features possessing the horizontal dimensions from 2 to 2,000 meters (6.56 to 6,560 feet) and the time scale from 1 to 3,600 seconds. It is considered the study of turbulence, diffusion, and heat transfer within the atmosphere. It includes weather phenomena such as tornadoes, short gravity waves, dust devils, thermals, wakes, and plumes.

microwaves Electromagnetic radiation with wavelengths longer than infrared. They are the shortest radio waves, with wavelengths ranging from 1 millimeter to 1 meter.

Mie scattering Scattering of electromagnetic radiation by particles roughly the same size as the wavelength of the radiation. Amount and direction of scattering is very sensitive to the relative sizes, making predictions of color and intensity difficult.

Milankovitch theory Theory which explains the occurrence of ice ages over the last million years as the result of long-term variations in the Earth's orbital parameters. These parameters are the eccentricity of the Earth's orbit, the obliquity of global rotation, and the precession of the equinoxes. The insolation at a particular latitude and time of year are determined by these parameters. The theory predicts that a glacial cycle can begin only when there are successive cold summers at high latitudes.

mixing length Distance a typical parcel is assumed to travel before being absorbed or indistinguishable from the surrounding medium.

mode An oscillatory or decay pattern associated with a single pure frequency. Complex motions and behavior are often represented as a combination of modes.

moist adiabatic lapse rate The rate of decrease of temperature with height of a parcel undergoing ascent during a moist process. As a parcel is lifted, saturation occurs and latent heat is released. The increase in the moist adiabatic lapse rate over the adiabatic lapse rate is due to the latent heating.

moist convection Buoyant ascent of warm, moist air. When the air is lifted to a level where it is cooled enough to saturate, clouds and possibly precipitation result. During ascent, moist parcels cool at the same rate as seen in the moist adiabatic lapse rate. The depth of the convection is determined by the lapse rate of the ambient air.

momentum The product of the mass and velocity for an object in motion. Momentum can only be changed by the application of an external force, which is expressed by Newton's second law of motion.

monsoon Seasonal winds. The Intertropical Convergence Zone (ITCZ) is the region where northeasterly and southeasterly trade winds converge. In regions where the ITCZ follows the path of the Sun, winds blow from the northeast or southeast according to the location of the ITCZ. These shifts in prevailing winds are known as *monsoons*. Monsoons often are associated with regions that experience wet and dry seasons.

near infrared (NIR) Electromagnetic radiation with a wavelength from 750 to 2,500 nanometers, usually created by molecular vibrations.

net radiative heating The net heating effect of the radiative forcing of the climate. This can be positive or negative.

newton (N) A unit of force. A force of 1 newton applied to a mass of 1 kilogram causes an acceleration of 1 meter per second per second.

NEXRAD NEXt generation weather RADar, now known as the 1988 Weather Surveillance Doppler Radar (WSR-88D), is part of the National Weather Service's modernization and restructuring. A network of 150 WSR-88Ds uniformly covers most of the United States. The WSR-88D is used primarily for detecting short-lived, life-threatening, hazardous weather events such as severe thunderstorms, tornadoes, and flash floods. The WSR-88D is an improvement over the previous Weather Service radars in that it is more sensitive to weak echo returns, provides finer detail of storms and sees them at greater distances, and is automated in its scanning strategy. Additionally, it has several special products like wind velocity, spectrum width, storm-total precipitation accumulation estimates, and vertical wind profiles.

nimbostratus (Ns) Gray, dark clouds with no clearly defined base which are associated with continuous precipitation that may or may not reach the ground. The clouds are several thousand feet thick and will completely block the Sun. Their precipitation may consist of snow, sleet, rain, or drizzle, but not hail (since there is no convection). [*See also* **stratus**.]

noctilucent cloud A cloud that forms in one of the ionized layers of the ionosphere at about 80 kilometers above the Earth's surface, recently renamed *polar mesospheric clouds*. Noctilucent clouds look similar to cirrus clouds. They can only be seen at high latitudes, usually in the summer when the Sun is below the horizon.

nuclear winter The name given to the possible climatic effect resulting from a nuclear war. In the 1980s it was hypothesized that vast quantities of dust and soot would be injected into the atmosphere by a nuclear war and its aftermath. These particles would block out the Sun for an extended period of time, creating a "winter" climate over the whole Earth.

nucleus A particle (of dust, salt) on which water molecules or ice accumulate forming raindrops, snow, hail, and so on. [*See also* **cloud condensation nuclei**.]

numerical weather prediction (NWP) A method of forecasting the weather through the use of mathematical models and computers which predict the state of the atmosphere at some future time.

obliquity The angle between the plane of the Earth's orbit (the ecliptic) and the plane of the Earth's equator. This angle varies from about 22° to 24.5° over a period of 40,000 years. Currently the obliquity is about 23.5°. This "tilt" of the Earth's axis is responsible for the seasons.

occluded front A type of front which forms in the late stages of cyclogenesis as a surface low moves away from the junction of its warm and cold fronts. It appears on a weather map as a line with alternating warm- and cold-front symbols.

ohm (Ω) In the meter-kilogram-second system, the unit used to measure electrical resistance, which equals the resistance of a circuit with an electromotive force of 1 volt that maintains a current of 1 ampere.

optical depth Distance through which electromagnetic radiation traveling through a medium is diminished by a given factor, usually about a half to a third.

orbital forcing The forcing of the climate by the variation in the solar radiation incident upon the Earth as determined by the Earth's orbital parameters.

orographic effect The effect that the varying topography has on atmospheric circulations. Deflection of air flow by mountains lifts air, resulting in cloud formation. Mountains also block or disturb atmospheric circulations and are regions of Earth–atmosphere momentum transfer. Topography also has an orographic effect on climate by influencing precipitation and temperature.

ozone Molecule made of three oxygen atoms. Most ozone in Earth's atmosphere is in the ozone layer of the stratosphere, where it is generated by shortwave solar radiation after the dissociation of molecular oxygen. Ozone in the troposphere is also generated by humans in photochemical reactions associated with urban air pollution. This tropospheric ozone is an irritant to the eyes and throat and is detrimental to plant surfaces.

ozone layer Layer of the stratosphere about 20 kilometers up where most of the stratospheric ozone exists. The ozone absorbs much of the harmful ultraviolet light from the Sun, making life on Earth's surface possible.

pack ice Floating sea ice that covers at least half of the visible ocean surface.

paleoclimate A climate of the geologic past, as opposed to a climate of the historical past for which instrumental records are available.

parcel A theoretical volume of air used in describing the vertical ascent and decent of air according to the stability of the atmosphere. A parcel consists of a uniform volume of air that ascends according to the adiabatic and moist adiabatic lapse rates. It is assumed for study purposes that the environment does not affect or modify the parcel and that the parcel does not affect the motions of the atmosphere.

pascal (Pa) A unit of pressure in the International System of Units (SI); 1 pascal is equal to 1 newton per square meter; 100 pascals 1 millibars 1 hectopascal.

perihelion The point closest to the Sun in the orbit of a planet.

pH The potential of hydrogen ions—a measure of acidity or alkalinity. It is the log of the reciprocal of the hydrogen ion concentration. The pH scale runs from 0 to 14, 0 being very acidic and 14 very alkaline.

photochemical smog The natural dissociation of nitrogen dioxide (by light below 385 nanometers) into atomic oxygen and nitric oxide sets the stage for ozone to be produced. Normally, that ozone would react again with nitric oxide to give back oxygen and nitrogen dioxide (no net reaction). However, the burning of fossil fuels increases the nitric oxide in the atmosphere so that abnormally high levels of ozone are produced. In turn, when the atmosphere is polluted by substances easily oxidized by ozone (such as hydrocarbons), complex reactions take place that result in irritants to eyes and skin as well as show phytotoxicity (plant poisons). The main smog irritants, formaldehyde, acrolein, and peroxyacetylnitrate (PAN), can also act to decrease visibility.

photochemistry Chemical reactions in the atmosphere which involve either absorption or emission of the Sun's radiation. One typical example is the formation of ozone in the stratosphere, which requires the absorption of the Sun's ultraviolet light.

photosynthesis The conversion of incident sunlight to chemical energy by plants, blue-green algae, and certain types of bacteria. The photosynthesis reaction converts water and carbon dioxide into molecular oxygen and carbohydrates. It is believed that most of the oxygen in the Earth's atmosphere is produced in this way.

physical oceanography The study of the world ocean by applying the principles of physics and mathematics. The major areas of physical oceanography are the measurement and characterization of the ocean's water masses and their motions.

Planck's function A mathematical function used to calculate the intensity of electromagnetic radiation emitted by a black body at a particular frequency and temperature. Named after its discoverer, the German physicist Max Planck (1858–1947).

planetary boundary layer (PBL) The atmospheric layer closest to a planet's surface. On Earth, for average mid-latitude conditions, the PBL is about 1 kilometer deep. The layer is characterized by mechanical eddies which mix momentum down from aloft and thermal eddies which mix moisture and heat up from the surface. Therefore, highly stable environments may have only a 30 meters deep PBL while highly convective environments may have a 3 kilometers deep PBL. The PBL can be further subdivided into a bottom surface layer and the top Ekman layer.

planetary waves Large-scale waves generally associated with the jet stream. There are generally four or five planetary waves spanning the circumference of the Earth at one time. Planetary waves are forced by large north–south oriented mountain chains as well as continent–ocean temperature differences. Planetary waves also propagate vertically affecting circulations in the stratosphere.

Pleistocene *See* **Cenozoic era**; *and* **ice age**.

polar front Discontinuity or boundary between the cool polar air mass and the warmer tropical air mass. The polar front is associated with the polar jet stream and often spans the circumference of the Earth. Cyclonic disturbances are often associated with the polar front owing to the strong temperature gradients in the region on the polar front.

polarization The plane of vibration for a wave, usually electromagnetic. Can be a factor determining scattering and absorption properties.

positive feedback When the feedback mechanism increases the magnitude of the climate response to the climate forcing.

positive flash (lightning) A positive flash is a lightning discharge which brings positive charge to ground in the stepped or dart leader. Comprising only 4 percent of all lightning flashes detected annually, the median return-stroke peak current is 45 kiloamperes (15 kiloamperes larger than for a negative flash). A positive flash usually has only one return stroke but that stroke is responsible for 100 coulombs or more of charge transfer. It is known to be one cause for forest fires.

potential energy A property attributable to a particle of mass in a conservative force field such as gravity. In such a field the sum of potential and kinetic energy of the particle are conserved. In the atmosphere, it is the energy required to lift a parcel of air to a given height above mean sea level. In a hydrostatic atmosphere, this energy is proportional to the internal energy of the parcel. Only a small amount of the potential energy reservoir is available for conversion to kinetic energy.

potential evapotranspiration The evapotranspiration which would occur if the soil conditions were wet. In this case the evapotranspiration is at its maximum value for the given atmospheric conditions.

potential temperature The temperature a parcel of air would have if the parcel were brought to 1,000 millibars by an adiabatic process according to the dry adiabatic lapse rate.

precipitation The liquid or solid form of water which falls and reaches ground underneath clouds. Examples of precipitation are rain (light, moderate, heavy), drizzle, snow, hail, ice pellets, and graupel.

pressure Defined as force per unit area, pressure is the meteorological variable which is measured with a barometer. Atmospheric pressure is the force per unit area exerted by the air due to gravity. Standard units for expressing sea level pressure are 14.7 pounds per square inch (English), 101.3 hectopascals (metric), 1,013 millibars or 29.91 inches of mercury. The last three units are most commonly used by meteorologists. Since pressure decreases monotonically with height, it is often used as a vertical coordinate on thermodynamic soundings or vertical cross sections of the atmosphere.

pressure gradient The change in pressure over some distance (either horizontally or vertically) with respect to a point in space. This is to be distinguished from the pressure gradient force which acts to accelerate the wind from high to low pressure (i.e., a portion of which is expressed as minus the pressure gradient).

primitive equations Mathematical description of atmospheric motion and energy propagation with minimal approximations.

propagate To transport energy or information to another place.

psychrometer An instrument used to measure the moisture content of the air. This instrument includes two thermometers, one of which measures the dry bulb temperature and the other measures the wet-bulb temperature. The difference between the wet and dry bulb temperature is used to interpret the water vapor content.

quad A unit of energy equal to 10^{15} (1 quadrillion) British thermal units.

Quaternary period *See* **Cenozoic era**.

radar RAdio Detection And Ranging began as a tool to detect enemy aircraft during World War II. Microwave energy (typically between 3 to 10 centimeters wavelength) is transmitted from a radar dish in beam pulses and reflected off precipitation particles. Since the beam travels at the speed of light, the storm-to-radar distance is known by counting the time between when each pulse was sent and received. The amount of energy reflected is proportional to the type of precipitation and precipitation size; this gives an indication of where the more intense precipitation is occurring. The radar is typically rotated 360° so that a plan-view composite of the surrounding storms can be seen.

radar reflectivity Is a measure of the amount of microwave energy reflected by precipitation or other particulates in the atmosphere. It depends primarily on the type of precipitation and its size. Radar reflectivity is simply a measure of the radiation returned from a given surface to that sent out incident upon the surface. This energy is converted into a logarithmic (base 10) unit so that a wide range of reflectivity can be expressed with a short number scale.

radiation The term includes electromagnetic, acoustic, and particle radiation, and all forms of ionizing radiation. Climate and weather studies focus on electromagnetic radiation, which is energy propagated through space by oscillating electric and magnetic fields. Radiation is characterized by wavelength and frequency. Electromagnetic radiation wavelengths range from the shortest (10^{-14} meters) to the longest (10^8 meters): these are X-rays (hard—includes those emitted as gamma radiation—and soft), light waves (ultraviolet, visible, infrared), and radio waves (including microwaves). [*See also* **infrared**; **microwaves**; **ultraviolet**; *and* **visible light**.]

radiation balance The Earth maintains a constant global annual temperature by a balance between the incident solar radiation which is absorbed by the climate system and the outgoing longwave radiation.

radiation fog A cloud which forms along the ground when the air is moving less than 8 kilometers per hour (5 miles per hour) and the longwave radiation emitted from the ground is greater than the insolation absorbed, so that net radiative cooling occurs. Night has the greatest radiative cooling, and radiation fog will form then. During the night air cools to the dew-point temperature so that condensation (fog) forms.

radiative cooling The process by which the surface of the Earth (and the air in contact with it) cools by the emission of longwave radiation. More radiative cooling can occur when atmospheric greenhouse gasses (such as water vapor or droplets and carbon dioxide) exist in smaller concentrations. The air is limited to cool only to its dew-point temperature. This happens at night when the longwave emission cannot be offset by insolation.

radiative forcing The forcing of the climate by radiative processes. These can include the effects of aerosols, clouds, and greenhouse gases through the absorption and emission of both shortwave and longwave radiation.

radiosondes Instrument packages containing a thermistor, aneroid barometer, and an electric resistivity hygrometer that measure the atmospheric vertical profile of temperature, pressure, and relative humidity, respectively. The battery-powered measurements are simultaneously transmitted (using a radio frequency) to a ground-based receiving station as a helium-filled balloon carries it upward. Some of the larger models come with parachutes to lower it safely to ground after the balloon bursts.

rainbow A 42°-radius arc or circle with several bands of visible light (from red to violet) caused by refraction and reflection of sunlight by water droplets; seen in the sky when falling rain is illuminated by direct sunlight. A muted, secondary rainbow with a 51° radius is sometimes visible as well. [*See* **visible light**.]

rain gauge A measuring instrument which measures precipitation. One standard automated type is called the *tipping bucket*, which releases the collected precipitation at designated increments. A heated variety can also be used in the wintertime to melt snow into its rain equivalent amount. The amount of rain is then digitally recorded.

Rankine An absolute temperature scale named after William Rankine (1820–1872) that has the same degree size as the Fahrenheit scale but was extended down to absolute zero (0°R is −459.6°F), which refers to the temperature at which all molecular motion would theoretically cease (i.e., the zero pressure for a gas). In this scale the freezing point of water is 492°R and the boiling point is 672°R. It is a resetting of the Fahrenheit scale to reflect thermodynamic principles and eliminate the need for negative numbers. Temperatures very close to absolute zero have been produced in the laboratory.

rawinsonde The same as radiosonde data, except that wind speed is also measured. If 3 or more ground-based stations can track the radiosonde transmitter, then the velocity and direction of the winds (with which the radiosonde is moving) can be calculated (hence the name rawinsonde).

Rayleigh scattering Scattering of long wavelength electromagnetic waves by small particles. For example, scattering of sunlight by air molecules falls in the Rayleigh scattering regime. Short (blue) wavelengths scatter more, giving the sky its blue color, and also red sunset, because the blue light has been scattered out in the longer path to the setting Sun.

reflectivity Fraction of electromagnetic radiation reflected (not absorbed) by a surface.

refraction The bending of light passing from one medium to another. Light refracted by ice crystals in cirrostratus clouds can be seen from Earth as a halo encircling the Sun or Moon. [*See also* **halo**.]

refractive index Ratio of the vacuum speed of light to its speed in a medium. The larger the refractive index, the slower the wave. Also can be used to determine how much a ray bends when it enters the medium obliquely.

relative humidity (RH) Ratio that compares the amount of water vapor in the air to the saturation amount. The ratio of the mixing ratio of the ambient air and the saturation mixing ratio of the same air with the same pressure and temperature. Relative humidity is usually expressed as a percentage. When using relative humidity as an indication of absolute moisture quantity, caution must be used, because colder air has a smaller saturation mixing ratio than warmer air.

remote sensing The application of distant devices to infer local properties, usually through electromagnetic radiation. For example, instruments on board Earth-orbiting satellites determine atmospheric properties such as temperature, cloudiness, and moisture content.

resolution Measure of the smallest distance between two objects or points that a sensor can perceive or a model can describe.

reversible change Process that can return to its original state with negligible change in its surroundings.

Richardson number Ratio of some measure of the atmosphere's stability to a measure of wind shear. Small Richardson number implies turbulence is likely.

rime ice The opaque and granular ice covering which forms on aircraft as it flies through clouds of supercooled liquid water. The rime forms due to a rapid freezing of small supercooled water drops as they touch the wings or other exposed surfaces. The drops freeze with pockets of air in between. If the supercooling is not as great and the supercooled drops are larger, the water will take longer to freeze and will form a transparent ice called *glaze*.

Rossby waves Large-scale transverse waves in a flow in a rotating frame whose primary restoring force is due to the gradient of the Coriolis force in the poleward direction.

Saffir-Simpson scale A scale from 1 (minimal) to 5 (catastrophic) developed by Herbert Saffir and Robert Simpson to gauge the severity (damage potential) of hurricanes.

salinity The number of grams of inorganic salts dissolved in a kilogram of sea water. Salinity is usually expressed in units of parts per thousand (ppt). The average salinity of the world ocean is about 34.7 parts per thousand.

satellite altimetry The determination of surface elevation by use of a radar altimeter placed on a satellite. Over the ocean, the topography can be used to infer ocean currents.

saturation Temperature and pressure at which water can exist in equilibrium as gas and liquid, or as gas and solid. The amount of water vapor in a volume of space when in equilibrium with the liquid phase at a given temperature. When the relative humidity is 100 percent, the air has reached saturation. The vapor pressure in the air is in equilibrium with the vapor pressure over a plane surface of pure water so that the rate of evaporation and condensation are equal.

saturation vapor pressure The vapor pressure that the water vapor in a parcel of air would have if the water vapor were in equilibrium with a pure liquid surface. The saturation vapor pressure is temperature dependent, which differs for a flat or a curved liquid surface (such as a spherical droplet).

scattering Changes in an electromagnetic wave velocity, or distribution of velocities, owing to interaction with particles or interfaces. Scattering can also refer to the deflection of a particle's path due to interaction with some object.

sea breeze A circulation induced by differential daytime heating between a land surface and a water surface, causing a wind flow from water to land. It is common on calm days in the summertime.

sensible heat flux The transfer of sensible heat through a surface of unit area during a unit of time by turbulent fluid motions.

severe thunderstorm A thunderstorm which produces a tornado, hail at least 19 millimeters (0.75 inches) in diameter, and/or winds at least 93 kilometers per hour (58 miles per hour). Even though lightning and heavy rainfall may occur from such storms, they are not directly considered as criteria for National Weather Service severe thunderstorm warnings. A separate flash flood warning may be issued if heavy rain occurs.

shear-driven turbulence Continuous creation of irregular eddy motions in a flow due to shear instability. In plane parallel flow the motion will become unstable if the shear is sufficiently large. If the flow is stably stratified (dense air below sparse), the amount of shear necessary for shear-driven turbulence will be larger.

shear flow A fluid motion in which there is a gradient in the direction perpendicular to the flow direction, either horizontally or vertically in the velocity field. Large shears can lead to flow instabilities.

shear stress The stress exerted on a parcel of fluid either by a stationary object or adjacent flowing parcels that are moving either slower or faster. Shear stress is a function of the viscosity of the fluid.

simulation Mathematical reproduction of some aspect of atmospheric or oceanic behavior in a computer.

skill (statistics) Usually refers to the degree of accuracy or improvement that a personal weather forecast has compared to that of forecasts based on chance, persistence, or climatology alone.

sleet Another name for ice pellets. It consists of frozen raindrops or refrozen melted snowflakes which are 7.6 millimeters (0.3 inches) or less in diameter. The pellets may be spherical or irregular in shape and are either translucent or transparent.

smog A combination of smoke and photochemically created pollution with or without fog. It is often associated with urban areas where automobile hydrocarbon and nitric oxide emissions and coal-burning power plant sulfur dioxide emissions can be large. Smog is made of phytotoxic chemicals, skin irritants, and lung irritants which may aggravate persons with breathing problems. It can also reduce visibility.

solar evolution The Sun is a star formed approximately 4.5 billion years ago from a gravitationally contracting cloud of interstellar dust and gas. Theories of solar evolution predict that since its formation the Sun's luminosity has increased by about 30 percent.

solar wind A stream of charged particles (mostly protons and electrons, with the ions of many elements) which emanates from the Sun at speeds varying from 300 to 1,000 kilometers per second. Interactions of the solar wind and Earth's upper atmosphere produce auroras and geomagnetic storms.

solar zenith angle Angle between a line pointing to the Sun and the vertical.

solstice Time twice a year when the point directly under the Sun is farthest south or north. Conventionally marks the start of summer and winter.

Southern Oscillation An oscillation in surface pressure differences between Tahiti and Darwin, Australia. This oscillation has a period of a few years. The Southern Oscillation is linked to El Niño and the Walker circulation. When the sea level pressure is anomalously low at Tahiti and anomalously high at Darwin, El Niño conditions occur.

squall line The unbroken line of thunderstorms or squalls extending over several hundred miles in length which often forms over or ahead of fast-moving cold fronts at mid latitudes. These squall lines often produce strong straight-line winds and hail near their leading convective edge. A trailing stratiform precipitation region is found behind the leading edge. They are a type of organized mesoscale convective system (MCS), because the latent heat release associated with the convection develop the pressure gradients that induce quasi-steady circulation of rear-to-front and front-to-rear air motions. The rear-to- front flow then reinforces lifting and convection strength along the leading edge.

stability In mathematics, a tendency of the results of a numerical simulation to remain well behaved (usually bounded) even after many time steps. Unstable models are definitely inaccurate. Stable models may or may not be accurate.

stable isotope Atom whose nucleus is not radioactive and therefore does not alter (decay) with time. Such a nucleus has a stable configuration of protons and neutrons.

Standard Atmosphere Hypothetical vertical profile of the atmosphere that was developed to represent the time and horizontally averaged distribution of temperature, pressure, and density. In the standard atmosphere, pressure, density and temperature are a function of height only. The standard atmosphere is used internationally to standardize calibrations and measurements.

standing wave A wave in which there is no movement of the nodes with respect to a fixed reference point. In a standing wave in a fluid, parcel motion may flow through the wave without movement of the wave structure itself. Atmospheric planetary waves are generally standing waves because the orography forcing the wave is stationary.

steering A specific observation of wind speed and direction that is indicative of the motion of a larger storm or circulation in which it is embedded.

Stefan-Boltzmann law Relation between total radiation emitted by an ideal black body and its temperature.

stepped leader (lightning) A stepped leader is a faint luminous light which is seen to propagate in discrete 50 meter steps with a short pause between each step; it branches downward from the cloud base to the ground before a lightning flash. Propagating at 1/3,000 the speed of light, it deposits about 5 coulombs of charge (usually negative) along its path and is associated with a 100- to 200-ampere current. Once one of the stepped leader's branches nears the ground, the increase in the electric field causes a return stroke along the path of that branch (but in the opposite direction).

stochastic Describes a random process whose overall behavior can be described statistically.

stratigraphy (paleoclimate) Branch of geology that studies the nature, distribution, and relations of rock strata, especially of the Earth's crust.

stratocumulus (Sc) Gray and/or whitish patches, sheets, or layers of clouds composed of rounded masses or rolls, which are nonfibrous and which may or may not be merged. Stratocumulus clouds can result when layers of clouds are heated from below and convection follows. Stratocumulus clouds, one of four genera that are hybrids of cumulus clouds, are made up of water drops and possibly hail and snow. They are very common over portions of the world's oceans and are thought to be very important in determining the Earth's radiation budget. [*See also* **cumulus**.]

stratopause The top of the stratosphere, where the transition occurs between the stratosphere and the mesosphere above it. The stratopause is located at a height of about 50 kilometers.

stratosphere The atmospheric region that lies above the tropopause. Temperature generally increases with height in the stratosphere. The upper stratosphere is warm due to the absorption by ozone of ultraviolet rays. There is very little vertical mixing in the stratosphere and residence times of the atmospheric particles are very long.

stratospheric model A representation of the atmosphere with mathematical equations or computer programs, optimized to display phenomena in the stratosphere (about 10 to 50 kilometers above the surface).

stratospheric warming Abnormally high temperatures in the polar stratosphere during winter. Usually occurs in the north polar region.

stratus (St) One of the three major cloud types that comprise the ten main groups or *genera*, stratus clouds organize themselves into sheets or layers. They often form when stable air possessing water vapor is forced to rise (i.e., the cloud is not convective) such as ahead of a synoptic-scale cyclone system over the warm frontal zone. They can cover areas as big as several hundred thousand square kilometers. Stratus (the act of spreading or strewing) clouds form below 2,440 meters (8,000 feet) and are water droplets, unless the freezing level extends lower, as in the wintertime and at higher latitudes. Altostratus is a uniform layer of clouds which form between about 2,440 and 6,100 meters (8,000 and 20,000 feet) and is usually a mixture of water droplets and ice crystals. [*See* **altostratus**.] Cirrostratus is a uniform layer of ice-crystal clouds which can be a mile thick and forms above about 6,100 meter (20,000 feet). [*See* **cirrostratus**.]

strong thunderstorm Thunderstorms which are defined as having small hail and outflow winds less than or equal to 40 miles per hour (64 kilometers per hour).

sublimation The process by which a substance changes phase from vapor to solid or vice-versa. An equivalent term used is *deposition*. In meteorology, we usually speak of the changing of ice to water vapor or water vapor to ice. While sublimation is always occurring, vapor to solid sublimation dominates under saturated (or supersaturated) conditions, whereas solid to vapor sublimation dominates under subsaturated conditions.

subtropics The region of transition poleward of the tropics and equatorward of the extratropics. The subtropical highs are located in the subtropics.

superadiabatic A level in the atmosphere occurring right above the surface, where the decrease in temperature with height is greater than the dry adiabatic lapse rate. This condition is absolutely unstable for both dry and moist convection. The atmosphere will immediately overturn to reduce this instability and return the lapse rate to dry adiabatic, but continued insolation and surface heating can act to preserve this superadiabatic lapse rate.

supercell A thunderstorm characterized by persistent, mesoscale rotating updrafts (referred to as *mesocyclones*) that are deep relative to the cumulonimbus cloud top (usually 8 to 15 kilometers). Lasting as long as 6 hours, supercells may produce every type of hazardous weather: large damaging hail, frequent cloud-to-ground lightning, flash-flooding, damaging outflow winds, and tornadoes. Most strong or violent tornadoes are spawned by supercells. The supercell updraft remains horizontally displaced from the downdraft so that the storm is typically longer lived than the air-mass thunderstorm. Supercells form in environments with large vertical wind shear and high instability.

supercooled droplet Water droplets typically found in the lower layers of clouds which are cooler than the freezing temperature. Typically the temperature range is between 32° and 0°F (0° and −17.8°C), but if there is an absence of nuclei, supercooled droplets can be as cold as −40°F (−40°C).

supersaturation When the air over a cloud droplet is supersaturated (relative humidity greater than 100 percent), the vapor pressure of the air is greater than the saturation vapor pressure on the droplet's surface. In such an environment, water droplets grow by vapor deposition; more water vapor molecules are condensing to the droplet than are evaporating from it. Greater supersaturation occurs in stronger storm updrafts because the process of vapor deposition cannot keep up with the decompressional cooling with ascent. Ice crystals in the presence of abundant cloud droplets are said to be *supersaturated* with respect to ice even though relative humidity may only be 100 percent. This is because the saturation vapor pressure over ice is less than that over water, allowing vapor deposition onto the ice.

surface layer The lowest few meters of the atmosphere where the velocity profile is approximated as having a constant horizontal frictional stress (using the surface value). In this layer, the wind profile from the top of the roughness length (typically 1 to 4 centimeters above the land surface) to the top of the surface layer is logarithmic and varies only with the roughness of the surface. Rougher surfaces (larger roughness lengths) have wind profiles which asymptotically approach a smaller wind speed at the top of the surface layer.

surface stress The surface stress in the atmosphere has been measured to be about 0.1 newton per square meter. The surface stress is a force per unit area exerted on the surface layer winds by the surface itself. It can be thought of as a retarding force (frictional drag) that results in a logarithmic reduction in the wind speed the closer one gets to the surface.

surging glacier Abrupt local increase in a glacier's speed owing to a wave propagating through the ice.

synoptic meteorology The branch of meteorology concerned with the collection and analysis of simultaneous observations over a broad area. This area, called the *synoptic scale*, typically ranges from 500 to 10,000 kilometers.

teleconnection The phenomenon in which weather patterns in one region influence the weather patterns in a distant location. One of the most noted teleconnection patterns is the Pacific North American (PNA) pattern, where tropical circulations associated with El Niño are correlated with weather pattern anomalies across North America.

temperature inversion Layers in the atmosphere where the temperature increases with height. There is a negative lapse rate in an inversion. These layers are stable and they inhibit vertical mixing.

terminal fall velocity The maximum velocity a mass will reach according to the properties of the fluid that the mass is falling through. The mass reaches its terminal fall velocity when the frictional and buoyant forces of the mass are in equilibrium with the force of gravity.

Tertiary period *See* **Cenozoic era**.

thermal A rising bubble of warm air which forms at the surface where heating is large and dry convection is occurring under superadiabatic conditions. Clouds start out as thermals before they reach their convective (lifting) condensation level but not all thermals become clouds. This is because the surface characteristics (supplying the thermal) are either too dry or too cold for the thermal to reach saturation aloft.

thermal equilibrium State in which the thermal properties of a system are constant in time, even with small disturbances from the environment.

thermal wind balance The thermal wind balance occurs between the horizontal temperature gradient and the vertical shear of the geostrophic wind (i.e., thermal wind). During frontogenesis or frontolysis, there is not a thermal wind balance because ageostrophic motions are occurring to cause the wind parallel to the isobars to depart from geostrophic. During frontogenesis, for example, the horizontal pressure gradient increases with height due to the increase in the low-level horizontal temperature gradient. This causes the total wind at a particular level to accelerate in an across-isobar direction toward lower pressure; the acceleration being more dramatic with height. The Coriolis force responds to this new wind speed and eventually adjusts its direction back parallel to the isobars. The result is a larger thermal wind (or geostrophic wind shear) in order to balance the larger low-level temperature gradient.

thermocline Ocean depth at which temperature, density, and salinity abruptly change with increasing depth, then become relatively uniform again.

thermodynamics Branch of science concerned with the nature of heat and its conversion into other states of energy.

thermohaline circulation Large-scale vertical motions in the ocean, driven by density differences due to variations in salinity and temperature.

thermometer An instrument with a graded scale used to measure temperature. Conventional thermometers are filled with a liquid (alcohol, mercury, or water) that expands or contracts with a change in the temperature. Some devices make use of the temperature's dependence on electrical resistance to measure changes.

thermopause The transitional layer between the thermosphere, where temperature increases with height, and the exosphere. The height of the thermopause varies between about 200 and 500 kilometers depending on solar activity.

thermosphere The region of the atmosphere above the mesopause where temperatures increase with height. The temperature of the thermosphere can reach 2,000 K in active Sun periods. The thermosphere extends above about 80 kilometers.

thunderstorm A storm involving one or more cumulonimbus clouds which are accompanied by lightning and thunder. The amount of rain, precipitation, hail, lightning, and outflow wind speed can vary according to the microphysical characteristics (cloud condensation nuclei), the thermodynamic instability, and wind shear characteristics of the environment.

time constant Time required for a decaying signal or process to decrease by about two thirds.

topography The features of the Earth's surface including mountains, water surfaces, forests, and deserts. The topography affects atmospheric circulations by providing surface drag and variability in surface heating.

tornado A violently rotating column of air extending downward from a cumulonimbus cloud and having circulation in contact with the ground. Dust and debris usually make the tornado visible near the ground and these may be the only indication of a tornado in the case when a funnel cloud (funnel, column, or rope-shaped) is not present. Supercell thunderstorms are believed to produce the strongest tornadoes, whereas strong nonsupercell convection can produce weaker tornadoes (like landspouts or waterspouts). Tornadoes are categorized on the Fujita wind damage intensity rating scale from F1 to F5, where F5 is associated with the most damage and highest estimated winds. [*See* **Fujita scale**.]

transpiration Emission of water vapor by plants.

tree ring Growth ring in the cross section of a tree trunk, indicating growth during 1 year. The series of variations in tree-ring thickness give information about weather during the life of a tree. [*See also* **dendrochronology**.]

tropical cyclone A generic term for an intense storm which forms over the warm ocean in the tropics. Tropical cyclones are characterized by relatively warm temperatures at their centers, strong winds located in the lower troposphere, and copious amounts of rainfall.

tropical depression A disturbance in the tropics with a closed circulation and surface wind speeds below 35 knots (39 miles per hour; 17.5 meters per second). The weakest type of tropical cyclone.

tropical storm A tropical cyclone with surface wind speeds between 35 and 65 knots (39 to 74 miles per hour; 17.5 to 32.5 meters per second). Tropical storms are named alphabetically, starting from "A" each year. Tropical storms typically appear as curved cloud bands on satellite imagery.

tropical wave An area of disturbed weather in the tropics with a horizontal scale of 500 to 1,000 kilometers without a closed circulation.

tropics Although there is no universally accepted definition of the tropics, they can generally be considered to be the region of the Earth between 30° north latitude and 30° south latitude, the approximate latitudes of the subtropical ridges (in pressure). Other criteria used to define the tropics include, but are not limited to, insolation, temperature ranges, and characteristics of the ocean.

tropopause The atmospheric boundary layer separating the troposphere and the stratosphere. The height of the tropopause varies from 8 kilometers in polar regions to up to 20 kilometers in the tropics.

troposphere The lowest region of the atmosphere, where the temperature generally decreases with height and where atmospheric moisture and mixing is prevalent. Approximately 80 percent of the Earth's atmosphere is found in the troposphere.

truncation error Error in a repetitive calculation owing to finite difference approximations in the equations.

tundra Arid region in high latitudes.

turbulence Flow which is highly irregular, exhibiting time-dependent fluctuations and intermittency. The tendency for turbulence to occur in a flow depends on the (lack of) viscosity, shearing rates, (lack of stable) stratification, and differential heating. Fundamental to turbulence is the nonlinear transfer of energy across a spectrum of spatial scales.

typhoon See **hurricane**.

ultraviolet (UV) Electromagnetic radiation with wavelengths between that of visible violet light and X-rays; it ranges from 400 to 4 nanometers. Ultraviolet is subdivided into near (NUV) 400 to 300 nanometers, far (FUV) 300 to 200 nanometers, extreme (XUV) 200 to 100 nanometers, and vacuum (VUV) 100 to 4 nanometers. Most of the potentially hazardous UV radiation emitted by the Sun is absorbed by the ozone layer. [*See also* **radiation**.]

updraft A rising air current.

upslope fog This type of fog is caused by the humid air flowing over hills and mountains (orographic lifting). The decompressional cooling of the air as it rises causes it to eventually reach its dew-point temperature (or condensation point) and fog forms.

upwelling Rising of water from the deep ocean. Often brings vital minerals to organisms near the surface.

vapor pressure A portion of the atmospheric pressure or the partial pressure attributed to the presence of water vapor in the atmosphere. The vapor pressure is a function of the density of water vapor in air and the temperature of the air.

vector A mathematical construction for describing quantities that require both a magnitude and a direction for their complete specification. A vector may be specified by giving its components along coordinate directions.

virga Streak or wisp-looking sheets of precipitation that trail downward from clouds; these are rain or ice particles falling and evaporating before reaching the ground.

virtual temperature The virtual temperature of a moist air sample is the temperature of dry air which has the same density as the moist air for a given pressure. The virtual temperature and the actual temperature are equivalent if no water vapor is in the air. Otherwise, the virtual temperature is always higher than the actual temperature because air is less dense for warmer air or for air with more water vapor in it.

viscosity Internal forces across planes in a moving fluid that oppose motion and deform parcels.

visible light Electromagnetic radiation with wavelengths between infrared and ultraviolet. This is also called *visible radiation*, the portion of the electromagnetic spectrum processed by the human eye. It is subdivided into red (750–610 nanometers), orange (610–590 nanometers), yellow (590–570 nanometers), green (570–500 nanometers), blue (500–450 nanometers), indigo (450–425 nanometers), and violet (425–400 nanometers). [*See also* **radiation**.]

visibility Can be expressed as a maximum distance at which objects can be discerned with the unaided eye. In the daytime, one evaluates a dark object against the horizon whereas at night, one evaluates a bright object which is emitting or reflecting light. Haze and aerosols cause light coming toward the eye to be scattered. This results in an object losing contrast in the daytime (due to increased air brightness from sunlight scattering off the air in front of the object) or becoming dimmer at night (because light emitted or reflected from the object is being scattered on its way to the eyes).

volcanic ash Solid material emitted during volcanic eruptions. Possible contributor to long-term weather and climate change by blocking sunlight.

volt (V) In the meter-kilogram-second system, the unit used to measure electromotive force or difference in potential between two points in an electric field needing 1 joule of work to shift a positive charge of 1 coulomb from the point of lower potential to the point of higher potential.

vortex A circulation that has closed streamlines around a center or any flow that has vorticity.

vorticity A measure of the rotation of a fluid. Mathematically, the vorticity is the curl of the velocity field. The vorticity gives a measure of the rotation about a single point whereas the circulation gives a measure of the rotation of a fluid over a finite area.

Walker circulation A circulation in the equatorial vertical plane. The upward branches of the circulation are found in the convectively active region of the western Pacific Ocean, the equatorial African region, and the equatorial South American region. The descending branches are found over the Atlantic Ocean, Indian Ocean, and eastern Pacific Ocean. The strongest cell is the Pacific cell. The Walker circulation enhances the easterlies over the tropical Pacific Ocean and changes location with El Niño.

warm front A boundary between air masses as classified by movement. If the front is moving such that the cold air is being displaced by the warm air, then the front is classified as a warm front.

warm occlusion A type of occluded front in which the cold air overrides the warm air. The cold front is only apparent at upper levels. Static stability decreases after passage of a warm occlusion.

warm sector The area behind the warm front and ahead of the cold front. Typically to the southeast of the surface cyclone. Conditions here are relatively warm and moist, with southerly winds.

waterspout A violently rotating column of air extending downward from a cumulonimbus cloud and having circulation in contact with a body of water. Waterspouts are distinguished from tornadoes because they form from nonsupercell thunderstorms over water bodies (high instability and low vertical wind shear environments). Note however that supercell tornadoes which form over land may also move over lakes or other bodies of water. Although the majority of waterspouts are weaker than the Fujita wind damage intensity rating F2—wind speeds of 50 to 70 meters per second (113 to 157 miles per hour)—and move slowly, they should still be taken seriously by mariners and coastal residents. [See **Fujita scale**.]

watt (W) A unit of power in the International System of Units (SI); 1 watt is equal to 1 joule per second.

wavelength The distance along the direction of propagation between corresponding points on a wave (for example, from crest to crest).

weathering Disintegration of surfaces from friction of wind and moving water pollutants such as acid rain, and by expansion of freezing water.

westerlies Name given to winds that blow from west to east.

Regions where predominant wind patterns have a westerly component. Regions of westerlies are found in the mid latitudes.

western boundary current Strong concentrated current on the west side of an ocean basin, caused by wind stress and Earth's rotation.

wet-bulb temperature The temperature a parcel of air would have if it were brought to saturation at constant pressure by evaporating water into it. The wet-bulb temperature is used as a measure of water vapor content. The drier the air is, the more depressed will be the wet from the dry bulb temperature. Wet-bulb temperature is usually measured with a psychrometer.

wind shear An abrupt change in wind speed or direction. Wind shear occurs at all altitudes.

zonal winds Winds that generally blow from the east or west along latitude circles without a significant meridional (north–south) component. Zonal winds are usually associated with time-averaged circulations.

DIRECTORY OF CONTRIBUTORS

Thomas P. Ackerman

Professor of Meteorology, Pennsylvania State University, University Park

Shortwave Radiation

Marsha Ackermann

Writer, Ann Arbor, Michigan

Drama, Dance, and Weather; Music and Weather; Painting, Sculpture, and Weather

Syun-Ichi Akasofu

Director and Professor of Geophysics, Geophysical Institute, University of Alaska, Fairbanks

Aurora Borealis

Richard B. Alley

Associate Professor of Geosciences and Associate of the Earth System Science Center, Pennsylvania State University, University Park

Greenland Ice Sheet

Ian Allison

Principal Research Scientist, Glaciology, Antarctic Cooperative Research Centre and Australian Antarctic Division, Hobort, Tasmania, Australia

Icebergs

Richard L. Armstrong

Senior Research Associate, Climatology, University of Colorado, Boulder

Avalanches

Roni Avissar

Associate Professor of Meteorology and Physical Oceanography, Rutgers University, New Brunswick, New Jersey

Scales

Robert C. Balling, Jr.

Director of the Office of Climatology, Arizona State University, Tempe

Seasons

Roger G. Barry

Director of World Data Center—A for Glaciology, and Professor of Geography, University of Colorado, Boulder

Arctic

Werner A. Baum

Dean Emeritus of the College of Arts and Sciences, and Professor Emeritus of Meteorology, Florida State University, Tallahassee

Professional Societies

P. R. Bell

Staff Member, Retired, Oak Ridge National Laboratory, Tennessee

Moon

André Berger

Ordinary Professor of Meteorology, Université Catholique de Louvain, Belgium

Orbital Parameters and Equations; Orbital Variations

W. H. Berger

Professor of Oceanography, Scripps Institution of Oceanography, La Jolla, California

Isotopes

Jody Berland

Associate Professor of Humanities, York University, Ontario, Canada

News Media

Mark S. Binkley

Associate Professor of Geography, and Director of the Broadcast Meteorology Program, Mississippi State University, Mississippi State

Oceans

Warren Blier

Assistant Professor of Atmospheric Sciences, University of California, Los Angeles

Charts, Maps, and Symbols; Temperature

Howard B. Bluestein

Professor of Meteorology, University of Oklahoma, Norman

Hailstorms; Tornadoes

Gordon B. Bonan

Research Scientist, National Center for Atmospheric Research, Boulder, Colorado

Taiga Climate

William D. Bonner

Senior Research Associate, National Center for Atmospheric Research, Boulder, Colorado; and Former Director of the National Meteorological Center, Camp Springs, Maryland

Weather Forecasting, *article on* Models and Methods

Ian K. Bradbury

Lecturer in Biogeography, University of Liverpool, United Kingdom

Biosphere

Roscoe R. Braham, Jr.
Scholar in Residence, North Carolina State University, Raleigh
Lake Effect Storms

Anthony J. Brazel
Professor of Geography, Arizona State University, Tempe
Microclimate

Ian Burton
Director of the Environmental Adaptation Research Group, Atmospheric Environment Services, Environment Canada, Downsview, Ontario
Weather Hazards

Steven Businger
Professor of Meteorology, University of Hawaii, Honolulu
Diurnal Cycles

Eduardo Cadava
Professor of English, Princeton University, New Jersey
Literature and Weather

Brian Cairns
Associate Research Scientist, Department of Applied Physics, Columbia University, New York
Statistical Methods

Wallace H. Campbell
Research Scientist, U.S. Geological Survey, Denver, Colorado
Ionosphere

Randall S. Cerveny
Associate Professor of Geography, Arizona State University, Tempe
Small-Scale or Local Winds

Mark A. Chandler
Associate Research Scientist, Center for Climate Systems Research, Columbia University, New York
Global Warming

James A. Coakley, Jr.
Professor of Atmospheric Science, Oregon State University, Corvallis
Cloud Climatology

Judah Cohen
Research Associate, National Research Council, New York
Cumulus Clouds; Cyclones, *article on* Explosive Cyclones; Emissivity; Enthalpy; Gravitational Instability; Inertial Instability; Jet Streaks; Occlusion; Potential Temperature; Sleet; Snowstorms; Static Stability

Stewart J. Cohen
Impacts Climatologist, Environmental Adaptation Research Group, Atmospheric Environment Services, Environment Canada, Downsview, Ontario, Canada
Climate Impact Assessment

Samuel C. Colbeck
Senior Research Scientist, U.S. Army Cold Regions Research and Engineering Laboratories, Hanover, New Hampshire
Snow; Snow Cover

Stephen J. Colucci
Associate Professor of Atmospheric Science, Cornell University, Ithaca, New York
Anticyclones; Cyclogenesis; Cyclones, *overview article*; Zonal Circulation

Curt Covey
Physicist, Lawrence Livermore National Laboratory, Livermore, California
General Circulation Models

George P. Cressman
Director of the National Weather Service, Retired, Washington, D.C.
Weather Forecasting, *article on* History of Study; Weather Services

Walter F. Dabberdt
Associate Director of the National Center for Atmospheric Research, Boulder, Colorado
Instrumentation

Gretchen C. Daily
Bing Interdisciplinary Research Scientist, Center for Conservation Biology, Stanford University, California
Population Growth and the Environment

J. M. Anthony Danby
Professor of Mathematics, North Carolina State University, Raleigh
Gravity

Christopher A. Davis
Scientist, National Center for Atmospheric Research, Boulder, Colorado
Potential Vorticity

Henry F. Diaz
Research Climatologist, Environmental Research Laboratories, National Oceanic and Atmospheric Administration, Boulder, Colorado
Low Pressure; Temperate Climate

Russell R. Dickerson
Professor of Meteorology, University of Maryland, College Park
Atmospheric Chemistry and Composition

Rodolfo Dirzo
Professor of Ecology, Centro de Ecología, Universidad Nacional Autónoma de México, Mexico City
Rainforests

Nolan J. Doesken
Assistant State Climatologist, Colorado State University, Fort Collins
Recreation and Climate

Randall M. Dole
Acting Director of the Climate Diagnostic Center, Environmental Research Laboratories, National Oceanic and Atmospheric Administration, Boulder, Colorado
Blocking

Leo Donner

Physical Scientist, Geophysical Fluid Dynamics Laboratory, National Oceanic and Atmospheric Administration, and Associate Professor, Princeton University, New Jersey

Conditional and Convective Instability

Charles A. Doswell III

Head of the Mesoscale Applications Group, National Severe Storms Laboratory, and Adjunct Associate Professor of Meteorology, University of Oklahoma, Norman

Storms; Thunderstorms

Philip G. Drazin

Professor of Applied Mathematics, University of Bristol, United Kingdom

Shear Instability

David J. Drewry

Director of Science and Technology, Natural Environment Research Council, Swindon, United Kingdom

Antarctic Ice Sheet

Leonard M. Druyan

Senior Research Scientist, Center for Climate Systems Research, Columbia University, New York

Arid Climates; Asia; Coriolis Force; Drought; Geography and Climate

Tim Eastman

Associate Research Scientist, Institute for Physical Science and Technology, University of Maryland, College Park

Magnetosphere

Timothy P. Eichler

Graduate Student in Atmospheric Science, Columbia University, New York

Cyclones, *article on* Subtropical Cyclones; Hurricanes

William P. Elliott

Research Meteorologist, Air Resources Laboratory, National Oceanic and Atmospheric Administration, Silver Springs, Maryland

Atmospheric Structure

Kerry A. Emanuel

Professor of Meteorology, Massachusetts Institute of Technology, Cambridge

Cyclones, *article on* Tropical Cyclones

John Firor

Director of the Advanced Study Program, National Center for Atmospheric Research, Boulder, Colorado

Climate Modification

Inez Y. Fung

Professor of Earth and Ocean Sciences, University of Victoria, British Columbia, Canada

Charney, Jule Gregory

Dian J. Gaffen

Research Meteorologist, Air Resources Laboratory, National Oceanic and Atmospheric Administration, Silver Spring, Maryland

Humidity

Rolando R. Garcia

Senior Scientist, Atmospheric Chemistry Division, National Center for Atmospheric Research, Boulder, Colorado

Stratosphere

Michael Garstang

Professor of Meteorology, University of Virginia, Charlottesville

Tropical Climate

Stanley David Gedzelman

Professor of Atmospheric Sciences, City College of New York, New York

Fog

Marvin A. Geller

Professor and Director of the Institute for Terrestrial and Planetary Atmospheres, State University of New York, Stony Brook

Dynamics

Steven J. Ghan

Senior Research Scientist, Battelle Pacific Northwest Laboratory, Richland, Washington

Cloud-Climate Interactions; Clouds; Cloud Seeding, *article on* Processes

Peter H. Gleick

Director of the Global Environment Program, Pacific Institute for Studies in Development, Environment, and Security, Oakland, California

Water Resources

Richard A. Goldberg

Senior Staff Scientist, Laboratory for Extraterrestrial Physics, NASA Goddard Space Flight Center, Greenbelt, Maryland

Sun; Upper Atmosphere

Vivien Gornitz

Associate Research Scientist, Columbia University, and NASA Goddard Institute for Space Studies, New York

Ice Age Sea Level

Russell W. Graham

Curator and Head of Geology, Illinois State Museum, Springfield

Animals and Climate

Lewis O. Grant

Professor Emeritus of Atmospheric Science, Colorado State University, Fort Collins

Orographic Effects; Weather Modification

Alex E. S. Green

Graduate Research Professor of Mechanical Engineering, University of Florida, Gainesville

Transportation

Lee Michael Grenci

Meteorologist and Instructor of Meteorology, Pennsylvania State University, University Park

Meteorology

Jean M. Grove
Life Fellow, Girton College, Cambridge, United Kingdom
Little Ice Ages

Nathaniel B. Guttman
Research Climatologist, National Climatic Data Center, Asheville, North Carolina
North America

Timothy M. Hall
Research Fellow, Cooperative Research Centre for Southern Hemisphere Meteorology, Monash University, Notting Hill, Victoria, Australia
Ozone; Ozone Hole

Kevin Hamilton
Research Meteorologist, Geophysical Fluid Dynamics Laboratory, National Oceanic and Atmospheric Administration, Princeton, New Jersey
Tides

Mark David Handel
Senior Program Officer, National Research Council, Washington, D.C.
Sensible Heat; Water Vapor

Kenneth Hartt
Professor of Physics, University of Rhode Island, Kingston
Chaos

David Hawkins
Distinguished Professor Emeritus of Philosophy, University of Colorado, Boulder
Scientific Methods

James D. Hays
Professor of Geological Sciences, Columbia University, Lamont Doherty Earth Observatory, Palisades, New York
Milankovitch, Melutin

R. K. Headland
Archivist and Curator, Scott Polar Research Institute, University of Cambridge, United Kingdom
North Pole

Merete Heggelund
Director of Economics, Saskatchewan Energy Conservation and Development Authority, Saskatoon, Canada
Environmental Economics

James A. Henry
Associate Professor of Geography, Tennessee State University, Nashville
Frost

Andrew J. Heymsfield
Research Scientist, National Center for Atmospheric Research, Boulder, Colorado
Cirrus Clouds

Katherine K. Hirschboeck
Associate Professor of Climatology, Laboratory of Tree-Ring Research, University of Arizona, Tucson
Floods; Runoff

Mitchell K. Hobish
Consulting Synthesist, and Independent Scholar, Baltimore, Maryland
Satellite Instrumentation and Imagery

Jennifer Hopwood
Lecturer in Mathematics, University of Western Australia, Nedlands
Dust; Dust Storms; Erosion and Weathering

Robert M. Hordon
Associate Professor of Geography, Rutgers University, New Brunswick, New Jersey
Icelandic Low

David D. Houghton
Professor of Atmospheric and Oceanic Sciences, University of Wisconsin, Madison
Teleconnections

Malcolm K. Hughes
Professor of Dendrochronology, and Director of the Laboratory of Tree-Ring Research, University of Arizona, Tucson
Tree-Ring Analysis

C. Ian Jackson
Director of Chreod Ltd., Ottawa, Ontario, Canada
Hare, F. Kenneth

Laurence S. Kalkstein
Professor of Climatology, University of Delaware, Newark
Health; Wind Chill

Sally M. Kane
Senior Economist, Office of Policy and Strategic Planning, National Oceanic and Atmospheric Administration, U.S. Department of Commerce, Silver Springs, Maryland
Economics of Climate Change

Akira Kasahara
Senior Scientist, National Center for Atmospheric Research, Boulder, Colorado
Primitive Equations

James F. Kasting
Professor of Geosciences, Pennsylvania State University, University Park
Atmosphere, *article on* Formation and Composition

David W. Keith
Postdoctoral Fellow, Atmospheric Research Project, Harvard University, Cambridge, Massachusetts
Energetics

William W. Kellogg
Senior Research Associate, Climate and Global Dynamics Division, National Center for Atmospheric Research, Boulder, Colorado
Albedo; Carbon Dioxide; Greenhouse Effect

J. T. Kiehl

Senior Scientist, National Center for Atmospheric Research, Boulder, Colorado

Black Body and Grey Body Radiation

Michael D. King

Senior Project Scientist, Earth Observing System, NASA Goddard Space Flight Center, Greenbelt, Maryland

Satellite Instrumentation and Imagery

Rudolf Klige

Professor of Hydrology, and Corresponding Member of the Russian Academy of Sciences, Moscow State University

Hydrological Changes

Margaret Kneller

Post Doctoral Associate, Lamont Doherty Earth Observatory, Palisades, New York

Cores; Ice Ages; Lakes

Nancy C. Knight

Associate Scientist, National Center for Atmospheric Research, Boulder, Colorado

Hail

Paul G. Knight

Instructor of Meteorology, Pennsylvania State University, University Park

Ice Storms

Carl W. Kreitzberg

Professor of Physics and Atmospheric Science, Drexel University, Philadelphia

Warm Sector

Ruby Krishnamurti

Professor of Oceanography, Florida State University, Tallahassee

Vorticity

T. N. Krishnamurti

Robert O. Lawton Professor of Meteorology, Florida State University, Tallahassee

Monsoons; Vorticity

K. Labitzke

Professor of Meteorology, Free University of Berlin, Germany

Quasi-Biennial Oscillation

William K.-M. Lau

Head of the Climate and Radiation Branch, Laboratory for Atmospheres, NASA Goddard Space Flight Center, Greenbelt, Maryland

General Circulation; Meridional Circulations; Subsidence; Tropical Circulations

David R. Legates

Associate Professor of Geography, University of Oklahoma, Norman

Condensation; Latent Heat; Precipitation

Donald H. Lenschow

Senior Scientist, Mesoscale and Microscale Meteorology Division, National Center for Atmospheric Research, Boulder, Colorado

Boundary Fluxes; Planetary Boundary Layer

John M. Lewis

Senior Scientist, National Severe Storms Laboratory, Norman, Oklahoma

Conservation Laws

J. G. Lockwood

Senior Lecturer in Geography, Retired, University of Leeds, United Kingdom

Europe

William P. Lowry

Professor Emeritus of Ecology, University of Illinois, Urbana-Champaign

Biometeorology

David M. Ludlum

Founding Editor, Weatherwise Magazine, Retired

Weather Lore

Walter A. Lyons

President and Senior Scientist, Forensic Meteorology Associates, Inc., Fort Collins, Colorado

Coastal Climate; Forensic Meteorology

K. B. MacGregor

Senior Scientist, High Altitude Observatory, National Center for Atmospheric Research, Boulder, Colorado

Solar Wind

Diane Manuel

Independent Scholar, San Jose, California

Aristotle; Bjerknes, Vilhelm F. K. and Jacob A. B.; Climate and Weather, *article on* History of Study; Fahrenheit, Gabriel Daniel; Franklin, Benjamin; Jefferson, Thomas; Köppen, Wladimir; Lorenz, Edward; Richardson, Lewis Fry; Rossby, Carl-Gustaf; Torricelli, Evangelista

Gregg Marland

Senior Staff Scientist, Environmental Sciences Division, Oak Ridge National Laboratory, Tennessee

Geoengineering

C. Marquardt

Research Scientist, Free University of Berlin, Germany

Quasi-Biennial Oscillation

Douglas G. Martinson

Senior Research Scientist, Lamont Doherty Earth Observatory, Palisades, New York; and Adjunct Professor of Geological Sciences, Columbia University, New York

Paleoclimates

John R. Mather

Professor of Geography, University of Delaware, Newark

Thornthwaite, Charles Warren

John McCarthy

Special Assistant for Program Development, National Center for Atmospheric Research, Boulder, Colorado

Aviation

Patrick A. McMurtry

Associate Professor of Mechanical Engineering, University of Utah, Salt Lake City

Turbulence

Kevin McSweeney

Professor of Soil Science, University of Wisconsin, Madison

Soils

Mary McVicker

Staff Member, North Carolina State University, Raleigh

Diurnal Cycles

Mark F. Meier

Professor of Geological Sciences, and Fellow of the Institute of Arctic and Alpine Research, University of Colorado, Boulder

Cryosphere; Glaciers

Paulette Middleton

Principal and Division Director of the Environmental Quality and Integrated Assessment Division, Science and Policy Associates, Boulder, Colorado

Acid Rain; Pollution

Wolfgang Mieder

Professor of German and Folklore, University of Vermont, Burlington

Proverbs

Ron L. Miller

Associate Research Scientist, Department of Applied Physics, Columbia University, New York

Intertropical Convergence; Inversion

Paul H. Moser *(deceased)*

Chief of Ground Water Section, Hydrogeology Division, Geological Survey of Alabama, Tuscaloosa

Ground Water; Hydrological Cycle

Stephen E. Mudrick

Associate Professor of Atmospheric Science, University of Missouri, Columbia

Fronts

Norman Myers

Visiting Fellow, Green College, Oxford, United Kingdom

Deforestation

B. Naujokat

Senior Faculty Member in Meteorology, Free University of Berlin, Germany

Quasi-Biennial Oscillation

Jehuda Neumann *(deceased)*

Professor Emeritus of Atmospheric Sciences, Hebrew University, Jerusalem, Israel; and Visiting Scholar, Department of Meteorology, University of Helsinki, Finland

History, Climate, and Weather; Warfare and Weather

Melville E. Nicholls

Research Scientist, Department of Atmospheric Science, Colorado State University, Fort Collins

Gravity Waves

Neville Nicholls

Senior Principal Research Scientist, Bureau of Meteorology Research Centre, Melbourne, Victoria, Australia

Australia; Subtropical Climate

Sharon E. Nicholson

Professor of Meteorology, Florida State University, Tallahassee

Africa; Desertification; Savanna Climate

John W. Nielsen-Gammon

Assistant Professor of Meteorology, Texas A&M University, College Station

Cold Fronts; Divergence; Gust Front

John M. Norman

Professor of Soil Science, University of Wisconsin, Madison

Soils

Gerald R. North

Distinguished Professor of Meteorology, and Head of the Department of Meteorology, Texas A&M University, College Station

Models and Modeling; Thermodynamics

Wendell A. Nuss

Associate Professor of Meteorology, Naval Postgraduate School, Monterey, California

Warm Fronts

Johannes Oerlemans

Professor of Meteorology, Institute for Marine and Atmospheric Research Utrecht, University of Utrecht, Netherlands

Sea Level

Astrid E. J. Ogilvie

Associate Professor, Institute of Arctic and Alpine Research, University of Colorado, Boulder

Lamb, Hubert Horace

T. R. Oke

Professor of Geography, and Head of the Department of Geography, University of British Columbia, Vancouver, Canada

Heat Island Effect; Urbanization

John E. Oliver

Professor of Physical Geography, and Director of the Climate Laboratory, Indiana State University, Terre Haute

Climatic Zones; Climatology; Maritime Climate

Harold D. Orville

Distinguished Professor of Meteorology, South Dakota School of Mines and Technology, Rapid City

Cloud Physics

Richard E. Orville

Director of the Cooperative Institute for Applied Meteorological Studies, Texas A&M University, College Station

Lightning

Lionel Pandolfo

Assistant Professor of Atmospheric Sciences, University of British Columbia, Vancouver, Canada

Rossby Waves

Claire L. Parkinson

Research Climatologist, NASA Goddard Space Flight Center, Oceans and Ice Branch, Greenbelt, Maryland

Agassiz, Louis; Croll, James; Sea Ice

B. Parthasarathy

Deputy Director of the Climatology and Hydrometeorology Division, Indian Institute of Tropical Meteorology, Pune

Humid Climate; Tropical Precipitation Systems

Dorothy M. Peteet

Research Scientist, NASA Goddard Institute for Space Studies, New York; and Adjunct Research Scientist, Lamont Doherty Earth Observatory, Palisades, New York

Younger Dryas

S. George Philander

Professor and Director of the Atmospheric and Oceanic Sciences Program, Princeton University, New Jersey

El Niño, La Niña, and the Southern Oscillation

C. Martin R. Platt

Senior Principal Research Scientist in Atmospheric Radiation, Commonwealth Scientific and Industrial Research Organization, Mordialloc, Victoria, Australia

Haze; Radiation; Scattering

Colin G. Price

Assistant Professor of Geophysics, Tel Aviv University, Israel

Atmospheric Electricity; Convection

Lynn K. Price

Energy Analyst, Energy Analysis Program, Lawrence Berkeley Laboratory, Berkeley, California

Climate Control

V. Ramaswamy

Research Scientist, Geophysical Fluid Dynamics Laboratory, National Oceanic and Atmospheric Administration, Princeton University, New Jersey

Longwave Radiation

Erik A. Rasmussen

Associate Professor of Meteorology, Niels Bohr Institute of Astronomy, Physics and Geophysics, Copenhagen, Denmark

Polar Lows

Peter Sawin Ray

Professor and Chair of Meteorology, Florida State University, Tallahassee

Radar; Squall Lines

Elmar R. Reiter

Professor Emeritus of Atmospheric Science and Civil Engineering, Colorado State University, Fort Collins

Jet Stream

William E. Riebsame

Associate Professor of Geography, University of Colorado, Boulder

Human Life and Climate

Raymond G. Roble

Senior Scientist, National Center for Atmospheric Research, Boulder, Colorado

Global Electric Circuit; Thermosphere

Alan Robock

Associate Professor of Meteorology, University of Maryland, College Park

Nuclear Winter

Juan G. Roederer

Professor Emeritus of Physics, Geophysical Institute, University of Alaska, Fairbanks

Geophysics

Daved C. Rogers

Research Scientist, Atmospheric Science Department, Colorado State University, Fort Collins

Orographic Effects

Gerald J. Romick

Senior Professional Staff, Applied Physics Laboratory, John Hopkins University, Laurel, Maryland

Aeronomy; Mesosphere

Arthur H. Rosenfeld

Professor of Physics, University of California, Berkeley; and Director of the Center for Building Science, Lawrence Berkeley Laboratory, California

Climate Control

Cynthia Rosenzweig

Research Agronomist, NASA Goddard Institute for Space Studies, New York

Agriculture and Climate

Marc Ross

Professor of Physics, University of Michigan, Ann Arbor

Energy Resources

Jonathan D. Roughgarden

Professor of Biological Sciences and of Geophysics, and Director of the Earth Systems Program, Stanford University, California

Ecology

K. Rupa Kumar

Assistant Director of the Climatology and Hydrometeorology Division, Indian Institute of Tropical Meteorology, Pune

Humid Climate; Tropical Precipitation Systems

William S. Russell

Associate Research Scientist, Columbia University, and NASA Goddard Institute for Space Studies, New York

Numerical Methods

Vincent J. Schaefer *(deceased)*

Professor Emeritus of Atmospheric Sciences, State University of New York, Albany

Cloud Seeding, *article on* Historical Perspective

Kenneth H. Schatten

Astrophysicist, Laboratory for Terrestrial Physics, NASA Goddard Space Flight Center, Greenbelt, Maryland

Sun; Upper Atmosphere

Stephen H. Schneider

Professor of Biological Sciences, Stanford University, California

Coevolution of Climate and Life; Gaia Hypothesis

David M. Schultz

Research Assistant, Department of Atmospheric Science, University at Albany, New York

Cyclones, *article on* Mid-latitude Cyclones; Occluded Fronts

Stephen Self

Professor of Volcanology, University of Hawaii at Manoa, Honolulu

Volcanoes

William D. Sellers

Professor of Atmospheric Sciences, University of Arizona, Tucson

Freezing Rain

Kathryn P. Shah

Associate Research Scientist, Center for Climate Systems Research, Columbia University, New York

Planetary Atmospheres

Glenn E. Shaw

Professor of Physics, Geophysical Institute, University of Alaska, Fairbanks

Aerosols

Herman H. Shugart

W. W. Corcoran Professor of Environmental Science, University of Virginia, Charlottesville

Vegetation and Climate

Lisa C. Sloan

Assistant Professor of Earth Sciences, University of California, Santa Cruz

Geology and Weather

John T. Snow

Professor of Meteorology, and Dean of the College of Geosciences, University of Oklahoma, Norman

Geopotential; Rain

Richard C. J. Somerville

Professor of Meteorology, and Director of the Climate Research Division, Scripps Institution of Oceanography, University of California, San Diego

Adiabatic Processes; Climate and Weather, *overview article*; Diabatic Processes

Graeme L. Stephens

Professor of Atmospheric Science, Colorado State University, Fort Collins

Entropy

Peter H. Stone

Professor of Meteorology, Massachusetts Institute of Technology, Cambridge

Baroclinic Instability

D. V. Subba Rao

Research Scientist, Bedford Institute of Oceanography, Dartmouth, Nova Scotia, Canada

South America

Edward J. Szoke

Associate Scientist, National Center for Atmospheric Research, Boulder, Colorado

Severe Weather

Peter K. Taylor

Head of Meteorology, James Rennell Division, Southhampton Oceanography Centre, United Kingdom

Marine Weather Observations and Predictions

Owen E. Thompson

Professor of Meteorology, University of Maryland, College Park

Barometric Pressure; Dew; Dew Point; Heat Low; High Pressure

Philip D. Thompson *(deceased)*

Senior Scientist, National Center for Atmospheric Research, Boulder, Colorado

Numerical Weather Prediction

Gregory J. Tripoli

Associate Professor of Atmospheric and Oceanic Sciences, University of Wisconsin, Madison

Cumulonimbus Clouds

George Tselioudis

Associate Research Scientist, Columbia University, NASA Goddard Institute for Space Studies, New York

Clouds and Radiation; Stratocumulus Clouds; Stratus Clouds

Anastasios A. Tsonis

Professor of Atmospheric Sciences, University of Wisconsin, Milwaukee

Climate

Thomas H. Vonder Haar

University Distinguished Professor of Atmospheric Science, Colorado State University, Fort Collins

Satellite Observation

James C. G. Walker

Arthur F. Thurnau Professor of Atmospheric Science, University of Michigan, Ann Arbor

Atmosphere, *overview article*

Stephen G. Warren

Professor of Geophysics and Atmospheric Sciences, University of Washington, Seattle

Antarctica

Gunter Weller

Professor of Geophysics, Geophysical Institute, University of Alaska, Fairbanks

Tundra Climate

Douglas A. Wesley

Science and Operations Officer, National Weather Service, Cheyenne, Wyoming

Cold Air Damming

Warren H. White

Senior Research Associate in Chemistry, Washington University, St. Louis

Visibility

Donald A. Wilhite

Professor of Agricultural Climatology, and Director of the National Drought Mitigation Center, University of Nebraska, Lincoln

Palmer Drought Severity Index

Cort J. Willmott

Professor and Chair of Geography, University of Delaware, Newark

Evaporation; Evapotranspiration

Donald J. Wuebbles

Professor of Meteorology, and Head of the Department of Atmospheric Sciences, University of Illinois, Urbana-Champaign

Greenhouse Gases; Trace Gases

John A. Young

Professor of Atmospheric and Oceanic Sciences, University of Wisconsin, Madison

Friction

Serinity Young

Visiting Scholar, Southern Asian Institute, Columbia University, New York

Religion and Weather

Minghua Zhang

Assistant Professor of Atmospheric Sciences, State University of New York, Stony Brook

Conditional Instability of the Second Kind

INDEX

N.B.: Page numbers printed in boldface indicate a major discussion. Italicized page numbers refer to a figure or table.

Eddies (*continued*)
 viscosity of, 267, 282, 317–318, 319, 791–792, 793, 860
 zonal circulation and, 851
Edwards, Jonathan, 473
Effective temperature, 585
Egypt, 25, 129–130, 133, 254, 374, 394, 396, 639–640, 643, 819
Ehrlich, Paul, 177, 269
Eichler, Timothy P., *as contributor,* 231, 407–411
Einstein, Albert, 362, 363, 661
Ekman, Vagn Walfrid, 191, 193, 591
Ekman balance, 320
Ekman-CISK, 191, 267
Ekman layer theory, 223, 267, 320–322
Ekman pumping circulation, 321
Ekman spiral, 320, 321, 548, 591
Elamites, 640
Elastic barometers, 436
El Chichón eruption (1982), 808
Electrical absorption hygrometers, 440
Electrical charge
 conservation of. *See* Conservation laws
 lightning, 465, 466–467
Electricity. *See* Atmospheric electricity; Global electric circuit
Electric power, 284, *285,* 286, 287, 414, 772, 820
Electric Power Research Institute, 470
Electromagnetic (EM) spectrum, 629, 630, 631, 632, 652
Electromagnetic fields. *See* Magnetosphere
Electromagnetic waves, longwave radiation and, 478–481
Electrons, 9, 10, 450, 451, 485, 487, 498, 709, 710, 754–755, 797, 798, 799
Electrosphere. *See* Ionosphere
Eliassen, Arnt, 110, 191
Elijah (prophet), 641
Eliot, T. S., 474
Elizabeth I (queen of England), 813
Ellenton, G. E., 223
Ellesmere Island, 351, 426, 522
Elliott, William P., *as contributor,* 67–70
El Niño, La Niña, and the Southern Oscillation (ENSO), **273–277,** 860. *See also* Teleconnections
 arid climates and, 49
 in Australia, 72–74, 128, 258, 275, 744, 777, 779
 buoy forecasting and, 489
 chaos theory and, 109
 climatology and, 124, 125, 128, 129, 148
 clouds and, 245
 core analysis related to, 203, 577
 crop growth and, 23
 cyclones and, 236
 droughts and, 23, 73–74, 75, 128, 258, 275, 276, 777, 779
 as flooding factor, 310, 777
 as forecasting factor, 719, 743–744, 830
 as hurricane frequency factor, 409
 intertropical convergence and, 447, 448
 inversions and, 449
 monsoons and, 514, 777, 781
 in North America, 277, 744, 777
 precipitation and, 611, 777, 779
 quasi-biennial oscillation and, 625

 as Rossby wave factor, 649
 satellite observation of, 656
 in South America, 714, 777, 779
 in subtropical climates, 734
 teleconnections and, 742, 743–744, 779
 tree-ring analysis and, 775
 tropical circulations and, 776–778
 in tropical climates, 779
 upwelling and, 322, 549
El Norte (wind), 529
Elsner, J. B., 109
Emanuel, Kerry A.
 as CISK theory challenger, 192
 as contributor, 231–236
 explosive cyclone theory, 224
Emerson, Ralph Waldo, 472, 473, 474
Emiliani, Cesare, 452
Emissivity, **277–278,** 860
 Antarctic levels of, 35
 clouds and, 35, 153–154, 156, 157, 158, 159, 161, 170, 172
 of longwave radiation, 478, 480, 481
 microclimatological study of, *505,* 506
 monitoring and recording of, 653, 656
 radiation and, 92–93, 247, 277–278, 633–634, 750
Emotions, climatic metaphors for, 472
EM spectrum. *See* Electromagnetic spectrum
Energetics, **278–283.** *See also* Dynamics; Kinetic energy
 baroclinic instability and, 80, 266, 283
 barotropic instability and, 82, 266
 conditional and convective instabilities and, 188, 189, 190, 266, 267, 590
 cyclones and, 235
 enthalpy and, 289
 entropy and, 290–293
 evaporation and, 305
 gravitational instability and, 361
 inertial instability and, 433–434
 latent heat and, 465
 modeling of, 653
Energy, conservation of. *See* Conservation laws
Energy cascade, 788
Energy resources, **284–289.** *See also* Atmospheric electricity; *specific types*
 Arctic as source of, 786
 for climate control, 131–135
 for climate modification, 140, 141, 399
 and environmental economics, 294, 600, 601
 and explosive cyclones, 221
 and human adaptation, 398
 and motor vehicles, 771, 772, *773*
 and water, 820
 and weather hazards, 836–837
England. *See* Great Britain
Enlightenment, 470, 473
Enlil (god), 640
Enquiry into Plants (Theophrastus), 801
ENSO. *See* El Niño, La Niña, and the Southern Oscillation
Enthalpy, 190, 278, 279, **289–290,** 682
Entisols, 705, *706*
Entrainment, 102, 190, 213, 391, 449, 591, 724, 860
Entropy, **290–293,** 752
 chaos theory and, 105, 108–109
 Charney's studies, 541
 energetics and, 278, 279, 282
Enuma Elish (creation story), 640

Environment. *See* Animals and climate; Coevolution of climate and life; Ecology; Environmental economics; Pollution; Population growth and the environment; Rainforests
Environmental Defense Fund, 390
Environmental economics, **293–295**
 of acid deposition phenomena, 3
 antipollution strategies and, 597–598
 climate change issues, 271–273
 coevolutionary considerations, 178
 ecological issues, 268–270
 energy resource costs, 284, 285, 286, 287, 288, 837
 geoengineering costs, 339
 human adaptation considerations, 398, 399
 population growth and, 600–601, 836
 weather hazard issues, 836–837
 weather modification and, 841
Environmental pressure, 753, 860
Environment Canada, 520
Eolian landforms, 697
Eolian processes. *See* Arid climates; Small-scale or local winds
EOS. *See* Earth Observing System
Eötvös, Roland Von, 364
Epistemology, 472
Epstein, E., 542
Equations of continuity, 539, 542
Equation of state, 192–193, 334, 436, 539, 612, 613, 830, 860
Equatorial circulation. *See* Tropical circulations
Equatorial climate. *See* Humid climate
Equatorial trough, 780, 782
Equilibrium level, 188, 683, 757, 758, 761
Equilibrium tide, 762
Equinoxes, 860. *See also* Orbital parameters and equations
 diurnal cycles and, 247
 Earth's tilt and, 553
 historical study of, 130
 orbital changes and, 421, 559
 precession of, 421–422, 423, 507, 555, 556, 557, 558, 559–561, 562, 580
 radiation received and, 556–557
 seasonal changes and, 680–681
 temperature changes and, 700
Equipotential surfaces, 365
Equivalent potential temperature, 602, 827
Eratosthones, 145
ERBE. *See* Earth Radiative Budget Experiment
ERB satellites. *See* Earth Radiation (ERB) Budget satellites
Ergs (sand seas), 297
Erosion and weathering, **295–299**
 aerosol production and, 65, 384
 carbon dioxide and, 59, 103, 583
 climate and, 123–124, 343
 desertification and, 241
 dust storms and, 262
 energy resource reduction of, 287
 by floods, 296, 308
 geological, 295, 296, 297, 298, 342–345, 697
 human activity and, 50, 89
 during ice ages, 418, 419, 420, 580, 697
 isotope evidence of, 453
 planetary atmospheres and, 583
 rainforests and, 637

Helium (*continued*)
    in planetary atmospheres, 9, 585, 587, 588
    Sun and, 418, 708, 709, 710
Helmholtz, Hermann von, 192, 193, 645,
    686–687
Henry, A. J., 833
Henry, James A., *as contributor*, 326–328
Henze, Hans Werner, 255
Hepburn, Katharine, 255
Herbertson (thermal zone theorist), *142*, 143
"Here Comes the Rain Again" (song), 518
"Here Comes the Sun" (song), 518
Herodotus, 395, 473
Herschel, William, 500–501
Heterogeneous chemistry, 797
Heterogeneous reactions, 62
Heterosphere, 68, 796
Heymsfield, Andrew J., *as contributor*,
    116–122
Hibernation, 31–32
High-energy particles, 7
High pressure, **392–393**
    anticyclones and, 41–43, 218, 320, 324
    in Arctic, 46
    arid climates and, 48
    in Asia, 51
    cold air damming and, 178, 702
    dynamics and, 265, 392–393
    in Europe, 302
    historical long-term shift of, 394
    hurricane formation and, 408
    meridional circulations, 482
    monitoring and recording of, 114, 392, 393
    teleconnections and, 742
High-pressure areas, 392
High-resolution Infrared Radiation Sounder
    (HIRS), 654
Highstands, 424
High tide, 515, 516, 761, 762
Himalaya Mountains, 51, 217, 219, 447,
    513, 565, 612, 774, 781, 819
Hindcast skill, 720
Hinduism, 642
Hinkelmann, K., 541
Hippocrates, 130, 145, 499
Hiroshige, Ando, 575
Hiroshima, 534
HIRS. *See* High-resolution Infrared
    Radiation Sounder
Hirschboeck, Katherine K., *as contributor*,
    308–312, 649–651
Historical climatology studies, 124, 129–131,
    464, 473. *See also specific personal names
    and subjects*
Histories, The (Herodotus), 395
History, climate, and weather, **393–397**
    acid rain toxicity, 2
    in Africa, 17–18
    climate control, 130–132
    famous battles, 812–814
    famous snowstorms, 702–703
    famous volcanic eruptions, 535, 622, 806,
        807, 808
    hurricane destruction, 409–411
    Little Ice Ages, 475–476
Histosols, 705, *706*
Hitler, Adolf, 813
Hittite empire, 394
Hoar frost. *See* Frost

Hobish, Mitchell K., *as contributor*, 652–655
Hoggar (Africa), 13
Hokkaido (Japan), snowstorms, 462
Hokusai, Katsushika, 575
Hölderlin, Friedrich, 472
Holiday Inn (film), 256
Holmgren, B., 785
Holocene, 476, 862
Holton, J. R., 624, 625, 627
Homer, 642
Homer, Winslow, 575
Homogeneous nucleation, 11, 119, 840, 862
Homopause, 796
Homosphere, 68, 796
Honshu (Japan), snowstorms, 462
Hooke, Robert, 130
Hooke's law, 318
Hooke swinging plate anemometer, 490
Hopper, Edward, 575
Hopwood, Jennifer, *as contributor*, 260–262,
    262–263, 295–299
Hordon, Robert M., *as contributor*, 427–430
Horizontal convergence and divergence, 251,
    252
Horne, Lena, 256
Horn of Africa, 53
Horse latitude calms, 143–144
Horton, R. E., 650
Hortonian overland flow, 650
Hoskins, Brian, 606
Hot lightning, 468, 469
Hough, Sydney Samuel, 614, 645
Houghton, David D., *as contributor*, 741–744
Housing. *See* Architecture and climate;
    Human life and climate
Howard, Luke, 165, 210, 390
Howell, W. E., 12
Huang River, 822
Hudson Bay, 384, 522, 527, 671
Hudson River School (art), 574–575
Hudson's Bay Company, 44
Hughes, Malcolm K., *as contributor*, 773–775
Hugo, Victor, 471
Huhwuhli Nieh (Native American song), 517
Human life and climate, **397–400**, 854. *See
    also* Biometeorology; Biosphere;
    Clothing and climate; Coevolution of
    climate and life; Deforestation;
    Desertification; Gaia hypothesis;
    Geophysics; Heat island effect;
    Pollution; Population growth and the
    environment
    in Africa, 18, 659
    agriculture, 20–25
    albedo change effects, 27
    artificial avalanches, 76, 77
    atmospheric composition, 56, 805
    in Australia, 75
    climate control, 131–135
    climate impact assessment, 135–138, 354,
        374
    climate modification and, 139–141,
        399–400
    climatology and, 148
    clouds and, 156, 164, 170–171
    drought effects, 257, 258, 259
    dust storm activity and, 263, 697
    ecological considerations, 268–270,
        637–638
    economics of climatic change and,
        270–273

energy resource use, 284, 285
environmental economics and, 294
explosive cyclones and, 221–222
flooding adaptation, 311
fog and, 314
geoengineering and, 338–339
greenhouse gas production and, 63, 103,
    104, 356, 369–370, 371, 372, 373, 597,
    728, 769, 770, *772*
health risks, 386–390
historical events and trends, 393–397
hurricanes and, 483
hydrological changes, 414
lake effect storms and, 462
lightning and, 466
microclimatological study of, 505
ozone depletion activities, 62, 569, 572,
    573, 596, 597, 796
savannas and, 659
soils and, 707–708
temperature factor, 751
transportation and, 771–773
weather hazards and, 835–837
weather modification and, 839
wind impact on, 697
Humboldt, Alexander von, 142
Humboldt Current. *See* Peru Current
Humid climate, **400–404**. *See also* Tropical
    climate
    in Africa, 15, 16, 17–18, 401, 402, 778
    in Asia, 401, 778
    soils, 705, 707
    in South America, 401, 403, *712*, 778
    vegetation, 803
Humidex, 846
Humidity, **404–407**
    adiabatic processes and, 5–6, 186, 188,
        189, 196
    in arid climates, 48, 55
    cloud formation and, 119, 120, 162,
        163–164, 167, 169, 170, 175, 184, 185,
        186, 592
    condensation and, 404, 405–406, 608, 827
    cyclones and, 234, 235
    dew formation and, 242, 243, 244, 404
    diurnal cycles of, 246, 247, 249, 250
    drought and, 258
    as erosion and weathering factor, 343
    evaporation and, 305, 404, 405–406, 753
    evapotranspiration and, 306, 404
    fog formation and, 313
    fronts and, 183, 322, 323, 325
    geography and, 340, 341, 342
    global warming and, 412
    haze and, 385
    health and, 388
    heat island effect and, 391
    measurement of, 823
    monitoring and recording of, 130,
        438–440, 444, 445, 489, 490, 503, 652,
        824–825
    physical discomfort and, 150–151,
        406–407, 751, 846
    in planetary boundary layer, 591
    relative, 119, 186, 404, 405, *406*,
        438–439, 608, 718, 732, 734, 753, 823,
        827
    in subtropical climate, 733, 734
    in tropical climates, 497, 778, 779
Humisery, 846
Humiture, 846
Hungary, 394

Lake Ngami, 18
Lake Ontario snowbelt, *462*
Lake Rudolf, 13
Lakes, **462–463**. *See also* Great Lakes
  acid rain effects, 2, 3
  African, 13–14, 17, 18
  Antarctic, 33, 40
  coastal erosion and weathering, 296
  core analysis of, 198, 199–200, 577, 578
  evaporation rates, 304, 406, 416
  flooding and, 310–311
  hydrological changes, 413
  hydrological cycle and, 317, *818*, 820
  ice formation on, 208
  motor vehicle emissions, *772*
  pollution, 818
  radiation absorption by, 58
  size and extent, 463
  snowstorm formation and, 176, 461–462,
    463, 701, 702
Lake Superior snowbelt, *462*
Lake Tanganyika, 14
Lake Victoria, 13, 14, 380
LaMarche, V. C., 394, 396
Lamarck, Jean-Baptiste, 165
Lamb, Hubert Horace, 394, **463–464**, 812,
  813
Lambert, Johann Heinrich, 130
Lamb's Dust Veil Index, 464
Lamb weather types, 464
Lancaster, Burt, 255
Land Art (Earthworking), 575–576
Land-breeze fronts, 760
Land breezes, 176, 180, 249, 693, 780
Land ice. *See* Glaciers
Landsat satellites, 654, 669, *670*, 673–674
Landsat Thematic Mapper (TM), 654
Landspouts, 765
Language, weather as, 471
Language Mesh (Celan), 472
La Niña. *See* El Niño, La Niña, and the
  Southern Oscillation
Laos, 53
Laplace, P., 614, 645
Laplace equation, 537, 538, 645
Lapse rate, 4, 5, 69, 178, 196
Large eddy simulation (LES), 663, 664, 789,
  792–793
Large scale. *See* Scales
Large-scale meteorology. *See* Meteorology
Laser radar, 119, 122, 765, 794, 797, 799,
  825
Laskar, J., 564
Latent cooling, 178
Latent heat, **465**. *See also* Sensible heat
  albedo and, 359, 741
  boundary fluxes and, 100, *101*
  clouds and, 117, 152, 166, 167, 169, 501,
    565–566, 727
  condensation and, 5–6, 58, 152, 196, 197,
    275, 305, 318, 330, 341, 497, 602, 718,
    751, 752, 753, 760, 766, 775, 780, 826,
    827
  conditional instability of the second kind
    and, 191, 192
  convection and, 196, 331, 727
  cyclone formation and, 223, 224, 225,
    226, 231, 234–235, 606
  diabatic transfer of, 244, 281, 754

energetics and, 279, 280, 281, 283
entropy and, 292
evaporation and, 305, 753, 826
evapotranspiration and, 306
general circulation and, 330, 331, 779
heat island effect and, 391
historical study of, 501
hurricane development and, 407
intertropical convergence and, 448, 565,
  753–754
meridional circulations and, 497
microclimatological study of, 505, 506,
  785
monsoon release of, 513
oceanic transfer of, 45, 550, 551
precipitation formation and, 609–610
snowstorm development and, 702
temperature modeling role, 509, 510
thunderstorm activity and, 760
tropical circulations and, 775, 777, 780
water and transfer of, 58, 129
Latitude. *See* Geography and climate
Latitude of declination, 680
Lau, William K.-M., *as contributor,* 330–334,
  496–498, 732–733, 775–778
Laurentian uplands, 527
Laurentide Ice Sheet, 47, 353, 373, 848
Lavoisier, Antoine, 61
Law of Atmospheres, 668
Lawrence Livermore National Laboratory,
  535
Lax's equivalence theorem, 538
LCL. *See* Lifting condensation level
Lead pollution, 3, 12, 41, 90
Leads (energy transfer), 45–46
Leaf Area Index, 506
Lebanon, 394
Lebedeff, S., 676
Le Corsaire (Mazilier), 254
Lecture on the Weather (Cage), 518
Lee, Albert, 838
Lee-side lows, 215, 217, 219, 221, 229–230,
  564–565, 816
Lee troughs, 221, 229–230, 564, 566, 811
Lee waves, 367
Legates, David R., *as contributor,* 184–186,
  465, 608–612
Leighton, Robert, 735
Leith, C. E., 542, 793
Lemonnier, P.-C., 66
Length scales, 787–788, 789, 790, 792
Lenschow, Donald H., *as contributor,*
  99–102, 590–593
Lent, 643
Lenticular clouds, 566
Leonardo da Vinci, 130, 787
Leovy, C. B., 624
Lepanto, battle of (1571), 640
Le Sacre du Printemps (The Rite of Spring)
  (ballet), 255
LES models. *See* Large eddy simulation
Les Patineurs (The Skaters) (ballet), 255
Les Quatres Saisons (The Four Seasons)
  (ballet), 255
"Let It Rain" (song), 518
Le Treut, H., 562
Let's Get Married (film), 256
Letter on the Blind (Diderot), 472
Letters from an American Farmer
  (Crèvecoeur), 474
Leucippus, 668
Leverrier, Urbain, 131

Lewis, John M., *as contributor,* 192–195
Lewis and Clark expedition, 454
Lewy, Hans, 537, 615
LI. *See* Lifted Index
Lichenometry, 475–476
Lichens, 740
Lidar. *See* Laser radar
"Life Cycle of Cyclones and the Polar Front
  Theory of Atmospheric Circulation"
  (J. Bjerknes and Solberg), 544
Lifted Index (LI), *684*
Lifting condensation level (LCL), 188, 189,
  196, 212–213, 683, 684, 718, 753, 757,
  827
Light. *See* Radiation; Photosynthesis
Light, speed of, 631
Lightfoot, Gordon, 517
Lighting, and climate control, 133–134, 800
Lightning, **465–470**
  atmospheric electricity and, 65, 66–67,
    355, 799
  as aviation hazard, 77, 468
  clouds and, 157, 158, 164, 170, 209, 212
  convection and, 198, 757
  cyclones and, 723
  as fire source, 90–91
  flash, 466–468, 470, 868
  historical study of, 130, 315
  as literary device, 472, 473
  nitrogen production by, 64
  religious beliefs about, 640, 642
  during snowstorms, 702
  thunderstorms and, 465, 466, 467, 468,
    470, 757, 758, 759, 761
  tornado wind speed and, 767
  types, 468–469
  as weather hazard, 835
Lightning Field, The (sculpture), 575–576
Limb sounders, 653
Limestone, 103, 376, 586
Limpopo River, 14
Lindzen, R. S., 624, 626
Line Islands, 778
Linss, W., 65
Lion in Winter, The (Goldman), 255–256
Liquid petroleum gas (LPG), 772
Liquid water path, 691
Literature and weather, **470–475**
  plays, 254, 255–256, 812
  proverbs and lore, 129, 617–621, 838
Lithification, 343
Little Ice Ages, **475–478**
  Arctic ice during, 47
  climatology and, 149
  crop growth effects of, 23
  glacier expansion during, 353, 373
  historical events affected by, 812–813
  in South America, 714
  temperature change during, 735, 736–737
  volcanic emission during, 807
Livy, 396
LLJs. *See* Low-level jets
Local climates, 148
Local winds. *See* Small-scale or local winds;
  specific wind types
Lockwood, J. G., *as contributor,* 299–303
Locusts, 23
Loder, J. W., 515
Loess, 261, 262, 297, 298, 420, 422, 697,
  707
Loess Plateau, 261, 263, 297
Log normal distribution, 10

Specific humidity, 404, 405, 823
Specimen Days (Whitman), 473
Spectral irradiance, 736
Speed of escape. *See* Escape velocity
Spenser, Edmund, 471, 474
Spin-down circulation, 321
Spin-up circulation, 321
Spodosols, 705, *706*, 707
Sports. *See* Recreation and climate
Spring. *See* Seasons
Springs (water resource), 376, *377*, 416, 417
Spring tide, 761
Spruce trees, 801, 849
Sputtering, 589, 590
Squall lines, **716–718**, 782
 anticyclones and, 41
 coastal climates and, 176
 cyclonic activity and, 218, 230
 dust storms and, 263
 erosion, weathering, and, 298
 fronts and formation of, 325, 378, 716
 gravity waves and, 368
 mesoscale modeling of, 511
 monitoring and recording of, 114, 716, 717
 severe weather and, 685, 716
 tropical climates and, 484, 780
 warm sectors and, 817
Squalls, 16, 445, 701
Sri Lanka, 401
SST. *See* Sea-surface temperature
Stability indices, 684
Stable stratification, 591, 718
Standard atmosphere, 482
Stars. *See also* Sun; *specific stars*
 gravity and, 363–364
Star Wars (Strategic Defense Initiative), 536
Static instability, 361, 602, 718
Static neutrality, 602, 718
Static potential energy, 188
Static stability, **718**. *See also* Baroclinic instability
 blocking activity and, 98
 convection and, 198, 602, 718
 gravitational, 318
 inversions and, 449, 450
 potential temperature and, 602, 718
 thunderstorms and, 757
Stationary eddies, 332
Stationary fronts, 324, 502, 816
Stationary waves, 332, 333
Station models, 113, 115
Statistical mechanics, 752–753
Statistical methods, **718–720**, 743, 830, 835
Steadman, R. G., 846
Steam devils, 217, 461
Steam fog, 313–314
Stefan, Josef, 92
Stefan-Boltzmann law, 92, 93, 277–278, 478, 542, 631
Steinbeck, John, 472
Stellar Evolution and Its Relation to Geological Time (Croll), 207
Stephens, Graeme L.
 as contributor, 290–293
 entropy findings, 292
Steppe climates, 48, 49, 398, 657
Stepped leader, (lightning), 466–467
Stevens, Wallace, 472, 474

Stewart, Balfour, 451
Stochastic system, 481
Stone, Peter H., *as contributor,* 81–83
Storm-Cloud of the Nineteenth Century, The (Ruskin), 470
Storm in the Mountains (painting), 574
"Storm of the Century" (1993), 702
Storms, **720–723**. *See also* Cyclones; Dust storms; Explosive cyclones; Hailstorms; Hurricanes; Ice storms; Sandstorms; Severe weather; Snowstorms; Thunderstorms; Typhoons
 in Africa, 16
 animals and, 27
 in Antarctica, 38
 in Asia, 54
 classification of, 218, 232
 climate modification and, 139, 140, 821–822
 clouds and, 157, 160, 164
 convergence, divergence, and, 253–254
 flooding and, 308, 309, 310, 311
 Franklin's research on, 833
 fronts and, 324–325
 general circulation as factor, 332
 historical events affected by, 812–813
 historical study of, 131, 501, 502, 503–504
 inversions and, 450
 lake effect, 461–462, 463
 latitude as factor, 340
 lightning, 465–470
 as literary devices, 472, 474
 low pressure, 483–484
 in North America, 527, 528
 radar tracking of, 627, 628–629
 religious beliefs about, 640, 641, 642
 weather modification and, 839
Storm surges, 177, 232, 236, 296, 310, 409, 411, 488, 491, 675, 721
Storm tides. *See* Storm surges
Stormy Weather (film), 256
Stowe, Harriet Beecher, 474
Strahler, Arthur, 48
Straits of Magellan, 712
Stranger, The (Camus), 472
Strategic Defense Initiative (Star Wars), 536
Stratiform clouds. *See* Altostratus clouds; Cirrostratus clouds; Nimbostratus clouds; Stratocumulus clouds; Stratus clouds
Stratigraphic superposition, 198
Stratigraphy, 578
Stratocumulus clouds, **723–725**
 over Arctic, 46
 characteristics of, 165, 167, 169, 211, 723
 cyclones and, 233
 formation of, 214, 461, 592, 723–724, 732
 global warming and, 358–359
 inversions and, 450, 723–724
 lake effect storms and, 461
 radiation and, 170, 172, 211, 691, 723, 724–725
 warm fronts and, 815
Stratopause, 6, 57, 69, 498, 584, 725, 794, 796
Stratosphere, 6, 7, 8, 36, 57, 62, 63, **725–730**, 794
 aerosols in, 65, 384, 728, 797, 806, 808
 blocking activity in, 95
 Charney's research on, 110
 climate modification and, 139, 141

general circulation patterns, 332–333, 726–728
 greenhouse gases and, 63, 64, 726
 inversions in, 448
 jet stream and, 457, 458
 meridional circulation patterns, 498, 727, 728–729
 oxidants and, 770, 827
 ozone and, 480, 571, 572, 573, 596–597, 725–726, 727, 728–730, 770–771, 772, 795, 796, 806, 827
 planetary wave dynamics in, 649
 pollution, 827
 quasi-biennial oscillation, 622–626, 797
 structure and composition of, 69–70, 568–569, 584, 690, 725–726, 728–730, 795–797, 823
 temperatures in, 68, 69–70, 248, 332, 333, 498, 499, 725, 726, 728
 thunderstorm activity and, 761
Stratospheric and Mesospheric Sounder (SAMS), 146
Stratospheric clouds, 38
Stratospheric warmings, 8, 499, 573, 624, 649, 727
Stratus clouds, 120, 121, 154, 155, 636, 691, **730–732**
 over Arctic, 46, 531–532, 784
 characteristics of, 157, 165, 167, 170, 210, 730
 in coastal climates, 176–177
 fog and, 314, 732
 formation of, 43, 179, 212, 592, 730
 global warming and, 358–359, 732
 seeding of, 174
 temperature changes and, 325, 730
 warm fronts and, 815
Stravinsky, Igor, 255
Streak lightning, 468
Streams. *See* Rivers
Streets (cloud formation), 214
Strontium, 59, 453, 757
Sturgis, William, 12
Subadiabatic lapse rate, 5
Subgrid-scale models, 792–793
Sublimation, 162, 196, 327, 465, 586, 588, 589, 590, 671, 826, 827
Submarines, Arctic exploration, 533
Subpolar glaciers, 348
Subpolar regions, 482
Subsidence, **732–733**, 734, 750, 776
"Substances that Deplete the Ozone Layer" (Montreal Protocol), 62
Subsurface stormflow, 650
Subtropical climate, **733–734**
 in Africa, 15, 16, 403
 agricultural products, 21
 anticyclonic circulations, 42, 43, 733, 734
 in Asia, 51, 53, 403, 733
 in Australia, 73, 403
 clouds in, 212
 deserts in, 48–49, 734, 778, 779
 drought in, 258–259
 ENSO effects in, 734
 erosion and weathering in, 297–298
 evaporation rates, 305, 307, 734
 fog in, 314
 friction's impact on, 320
 humid-type, 403, 733, 734
 Mediterranean, 299, 300, 302, 340, 733, 801
 in North America, 403, 733, 734

Temperature (*continued*)
in Australia, 73, 74
and barometric pressure, 83–84
blocking activity and, 95, 97
boundary fluxes and, 99, 100, *101*
climate control and, 131–135
climate modification and, 138–139,
    140–141, 237, 359–361, 820, 821
climatic zones and, 142–143, 145, 148,
    400
clothing considerations, 150–151, 751
clouds and, 154, 155, 156, 169–170,
    171–173, 634, 730
condensation and, 184, 186, 608
convection and, 195, 196, 197, 584
core analysis of, 204, 577, 580
and crop growth, 21–22, 23, 24
and cyclones, 215–216, 218, 220–231,
    234–235, 605, 606
dew formation and, 242, 243
diabatic processes and, 3, 223–224,
    244–246, 279, 280, 281, 283
diurnal cycle of, 176, 246, 247–249, 250,
    314, 745
drought and, 259
dynamics and, 266, 727
effective, 585
emissivity and, 277–278
as erosion and weathering factor, 296,
    343
in Europe, 299–300, 302, 303, *813*
evaporation and, 304–305, 820
exospheric, 8
fluctuation in Europe (1300–1900), *813*
fluctuation of global mean temperatures,
    476, *477*
fog and, 313–314
freezing rain formation and, 316, 430–432
friction and, 320, 322
fronts and, 180–184, 322, 323, 324, 325,
    326, 377, 378, 546, 816
as frost determinant, 327
as general circulation factor, 330, 331, 332
general circulation models of, 334, 337
geography and, 339, 340, 341, 342
geopotential and, 348
greenhouse gases' effects on, 480
and hailstorm formation, 383
haze effects on, 385, 690–691
health relationship, 386, 387–388,
    389–390, 800
as heat low factor, 391–392
historical events affected by, 394, 395,
    396–397, 813–814
human adaptation to, 397, 740, 751, 771
in humid climates, 400, 401–402, 403
and humidity, 244, 404, 405, 406–407,
    751
instabilities and, 186–187, 188, 189, 191
intertropical convergence and, 447, 448
lake effect storms and land-water
    contrasts, 461, 463
latent heat effects on, 465
light-aerosol interactions and, 11, 13
of lightning, 465–466
lowest recorded, 34
microclimatological study of, 505, 506
modeling of, 508–510

monitoring and recording of, 111, 113,
    115, 130, 146, 437–438, 444, 445, 489,
    490, 500, 503, 652, 653, 654, 656, 841,
    842, 843
monsoon climates and, 51
in North America, 522, *523–524*, 526,
    527–528, 529, 530
ocean's effect on, 359–361, 491, 493, 494,
    550–551, 733, 734
of planetary atmospheres, 9, 583,
    584–585, 586, 587, 588, 589
primitive equations and, 613–614
quasi-biennial oscillation and, 622, 624
radiation and, 634
in savanna climates, 658
sea-ice effect on, 342, 359, 670–671
as snow formation factor, 700–701, 702
in South America, 712–713
in subtropical climates, 733–734
in taiga climates, 738, 740
in temperate climates, 745–746
tidal activity and, 516–517
in tropical climates, 779–780
at tropopause, 498, 823
in tundra climates, 784
in upper atmosphere, 6, 7–8
vegetation adaptation to, 801–802, 803
virtual, 100, *101*, 187, 188, 441, 825
as weather hazard, 835
wind chill factor, 150–151, 391, 530, 835,
    843–847
Temperature-humidity index (THI), 846
Tempest, The (Shakespeare), 254, 470, 474
Temporales, 484
Ten-to-twelve-year oscillation (TTO), 625
Terminal Doppler Weather Radar, 78
Terpenes, 101
Terre altos (winds), 696
Terrestrial infrared radiation. *See* Longwave
    radiation
Tetley, Glen, 254
Teton Mountains, 344
Thailand, 25, 53, 600
Thales of Miletus, 130, 416, 499
Thatched roofing, 18
Theatrical performance. *See* Drama, dance,
    and weather
Theodolite measurements, 444
Theophrastus, 145, 619, 801, 804, 837
Theory of Moral Sentiments (Smith), 295
"Theory of Weather, The" (Goethe), 474
Thermal balance, 25, 368
Thermal cells. *See* Meridional circulations
Thermal conductivity, 491
Thermal infrared radiation. *See* Longwave
    radiation
Thermal low-pressure system. *See* Heat low
Thermally direct/indirect circulation, 733
Thermal radiation. *See* Longwave radiation
Thermals, 197, 212–214, 248, 250, 367,
    500, 684
Thermal wind balance, 331, 332, 814, 851
Thermistors, 437–438, 444, 445
Thermocline, 276, 320, 322, 550–551, 777
Thermocouple devices, 437
Thermodynamic disequilibrium, 234
Thermodynamics, **751–754**. *See also*
    Entropy; Latent heat; Sensible heat
adiabatic processes and, 3–6, 341,
    601–602
atmospheric, 278, 280–282, 392
climatology and, 147

cloud formations and, 119
cloud physics and, 162
conditional and convective instability and,
    186, 190, 191
cyclone formation and, 234–235
dew point and, 243
diabatic processes and, 244–246
energy production and storage, 88, 89
energy resource use and, 288
enthalpy, 289
general circulation models and, 334
gravity waves and, 366
historical study of, 131
numerical weather prediction and, 539,
    830
ozone and, 62
primitive equations and, 612, 613, 614,
    615
snow and, 697
temperature modeling role, 509
turbulence and, 789
water vapor and, 823, 825–827
Thermohaline circulation, 359–360, 361
Thermometers, 124, 130, 146, 308, 434,
    435, 437, 454, 490, 500, 748, 768
Thermopause, 70, 584
Thermoscope, 768
Thermosphere, 6–7, 57, 68, 70, 498, 584,
    586, 754–756, 794, 799
Thermotidal oscillation, 762
THI. *See* Temperature-humidity index
Thin-film hygrometers, 439, *440*
30–60 day Oscillation. *See* Madden-Julian
    Oscillation
Thomas, Robert B., 838
Thompson, C. W., 143
Thompson, J. J., 10
Thompson, Owen. E., *as contributor,* 83–84,
    242, 243–244, 391–392, 392–393
Thompson, Philip D., 541, 542
    as contributor, 539–543
Thomson, James, 470, 474, 517
Thomson, William. *See* Kelvin, Lord
Thor (god), 642
Thoreau, Henry David, 474, 518
Thorium, 66, 198, 735
Thornthwaite, Charles Warren, 48, 384,
    400, **756–757**
Thornthwaite climate classification system,
    299, 400, 756, 757
Thorntree-tall grass savannas, 657
Thorogood, George, 518
Threshold response, instrumentation, 435
Threshold speed, of wind, 261
Throughflow, 650
Thunder
convection and, 198, 757. *See also*
    Thunderstorms
historical study of, 130
lightning and, 465, 466, 467, 468, 470,
    757, 758, 759, 761
religious beliefs about, 639, 640, 642
during snowstorms, 702
"Thunder" (song), 518
Thunderheads. *See* Cumulonimbus clouds
"Thunder Rolls, The" (song), 518
Thunder Shower (painting), 575
Thunder snows, 702
"Thunderstorm in Big Sur" (song), 519
Thunderstorm in the Rocky Mountains
    (painting), 574
Thunderstorm Project (1949), 683, 758

Visibility (*continued*)
  monitoring and recording of, 111, 113, 490
  snowstorms and, 700
Visible radiation, 629, 632, 633, 634, 689, 690–691, 750
"Visit to the Clerk of the Weather, A" (Hawthorne), 474
Vitruvius, 132
Vivaldi, Antonio, 517
Volatile organic carbons (VOCs), 1
Volcanoes, **805–809**
  aerosols and, 9, 64, 65, 70, 184–185, 260, 261, 343–344, 385, 480, 596, 610, 624, 728, 770, 797, 805–808
  ash as aviation hazard, 80
  carbon dioxide and, 103, 126, 358, 805, 806
  climate modification by, 735, 736, 806–808
  core analysis and, 199, 453, 807–808
  dust ejection by, 261–262, 464
  energetics and, 278
  as fire source, 91
  flooding and, 310, 311
  Franklin's research on, 315, 807
  in geologic cycle, 343–344
  global warming and, 808
  greenhouse gases and, 480
  ground water resources and, 376
  haze formation and, 385, 690–691
  impact on instrumentation data, 490
  lightning and, 465
  Little Ice Ages ending and, 477
  as nuclear winter analog, 535, 808
  outgassing by, 59, 583, 589, 805–806
  satellite monitoring of, 653
  scattering and, 667
  soil formation and, 704, 705
  sulfur levels and, 90, 261, 610, 805–806, 807, 808
  temperature modification by, 807, 808
  tree-ring analysis of, 775, 807
Volterra, Vita, 268
Voluntary Observing Ships (VOSs), 488–489, 490
Vonder Haar, Thomas H., *as contributor,* 655–657
Von Koch snowflake, 107–108
Vonnegut, Bernard, 173
von Neumann, John, 110, 537, 540, 541, 834
Vortex chambers, 765–766
Vortex stretching, 788
Vorticity, **809–811**. *See also* General circulation; Potential vorticity
  absolute, 433, 456, 602, 603, 810
  as aviation hazard, 80
  blocking activity and, 95, 98
  clouds and, 167
  conservation laws and, 193–195, 266
  convergence and divergence impact on, 253
  cyclone development and, 222, 224, 225, 227, 236, 810–811
  historical study of, 131, 645
  jet stream and, 456
  Rossby waves and, 645–649
  as severe weather factor, 685–686
  shear instability and, 687, 810–811

of tornadoes, 767
weather forecasting and, 835
VOSs. *See* Voluntary Observing Ships
Voyager (artificial satellite), 587, 588, 589

W

Wagner, Richard, 254
Walden (Thoreau), 474
Wales, 703
Walker, James C. G., *as contributor,* 55–58
Walker circulation, 8, 257, 275, 332, 449, 733, 775–778, 781
Wall clouds, 209, 765
Ward Hunt Ice Shelf, 426
Warfare and weather, 473, **812–814**
  chaff, 11, 13
  Coriolis force effect on weaponry, 206
  forecasting development, 481, 834
  historical events, 393, 395, 396–397, 502, 812–814
  jet stream discovery, 455, 502
  marine observations and predictions, 488
  nuclear winter, 534–536, 808
  radar development, 626
Warm air advection, 43, 178, 221, 222, 227, 228
Warm anticyclogenesis, 43
Warm-based glaciers, 348, 350
Warm conveyor belt, 224, 228
Warm-core anticyclones, 95, 98
Warm fronts, **814–816**
  classification of, 323–324
  clouds and, 815
  convergence, divergence, and, 254
  cyclones and, 91–92, 220, 228, 229, 230, 324, 544–545, 546, 547, 814
  freezing rain and, 316, 431, 816
  historical study of, 502
  precipitation and, 325, 815–816
  warm sectors and, 817
Warm seclusion, 229
Warm sector, **816–817**
Warm-type occlusion, 545, 547
Warren, Stephen G., *as contributor,* 32–39
Wasatch (wind), 693
Washington, George, 454
Was (Not Was) (musical group), 518
Waste Land, The (Eliot), 474
Water, drinking, 3, 818, 819, 822
Waterfalls, African, 14
Waterloo, battle of (1815), 393, 813
Water resources, 91, **817–823**. *See also* Hydrological cycle
  agriculture and, 22, 23, 24–25, 374, 819–820, 821, 822
  albedo of, 26, 27, 491, 634, 671
  animal life and, 28
  Antarctic Ice Sheet and, 39, 414, *818*
  climate modification and, 399, 411–415, 818, 820–822
  drought effects on, 257, 258
  Earth's reserves, *818*
  erosion and weathering role, 296–297, 298, 343, 344
  evaporation rates, 304–305, 411–412
  ground water, 374–377, 412, 413, 417, 651, 818
  human adaptation and, 398
  icebergs as, 427, 817
  lake, 462–463
  microclimatological study of, 505, 785–786

nitrate levels in, 90, 819
satellite monitoring of, 653
Thornthwaite's research on, 756–757
in urban areas, 800
Waterspouts, 217, 501, 685, 764, 765
Water stress, 22, 28, 48
Water table, 374, 376, 417
Water vapor, **823–829**. *See also* Condensation; Dew; Dew point; Evaporation; Evapotranspiration; Frost; Humidity; Hydrological cycle
  adiabatic processes and, 5–6, 187–189, 191, 196, 244, 753–754, 827
  aerosols and, 11
  Antarctic levels of, 36
  as atmospheric component, 56, 57–58, *61*, 68–69, 404, 596, 726, 728, 753, 754, 805, 823–824
  boundary fluxes and, 100
  climate dynamics and, 124, 129, 177
  climate modification and, 357, 358, 359, 828–829
  cloud formation and, 156, 159, 162, 164, 165, 166, 168–169, 170, 171, 209, 212, 275, 635- 636
  cloud seeding and, 175
  deforestation impact on, 237, 238
  fronts and, 322, 323
  frost formation and, 327
  general circulation models of, 334
  as greenhouse gas, 7, 247, 249, 271, 359, 368, 370, 371, 372, 404, 418, 498, 501, 653, 656, 768, 827–828
  heat transfer and, 751
  in humid climates, 401
  latent heat and, 465
  as modeling factor, 512
  oxidants and, 770, 827
  photosynthesis and, 87, 305, 803
  planetary atmospheres and, 583, 585, 586, 587
  radiation absorption by, 36, 57, 58, 69, 139, 177, 247, 359, 368, 372, 479, 480, 629, 633, 634, 652, 689, 691, 692, 823, 827–828
  sea ice impact on, 671
  as snow development factor, 701
  sources of, 404
  temperature and, 826–827
Wave-CISK, 192, 267
Wave clouds, 367
Wave cyclones, 816, 817
Wavenumber, longwave radiation, 478
Wave packets, 632, 647, 666
Waves. *See specific types*
WBF mechanism. *See* Wegener-Bergeron-Findeisen (WBF) mechanism
WCIP. *See* World Climate Impacts Program
Weapons, nuclear. *See* Nuclear winter
Weather. *See* Climate and weather; *specific conditions*
Weather and Forecasting (publication), 521
Weather Bureau, U.S., 454, 502, 834–835, 842
Weather Channel (cable television), 521
Weather charts, 111
Weather Folklore and Local Weather Signs (Garriott), 838
Weather forecasting, **829–835**. *See also* Charts, maps, and symbols; Marine weather observations and predictions; Numerical weather prediction; Statistical methods